# Microbial Forensics

Visit the Microbial Forensics, Second Edition companion Web site at:

http://www.elsevierdirect.com/companion.jsp?ISBN=9780123820068

The companion Web site contains supplementary materials as well
as all the images (in color, where available) from the volume.

*Microbial Forensics 2nd Edition* is fully revised and updated and serves as a complete
reference of the discipline. It describes the advances, as well as the challenges and
opportunities ahead, and will be integral in applying science to help solve future biocrimes.

# Microbial Forensics

## Second Edition

### EDITORS

**Bruce Budowle**
Institute of Investigative Genetics
Department of Forensic
and Investigative Genetics
University of North Texas Health Science Center
Fort Worth, Texas

**Steven E. Schutzer**
Department of Medicine
University of Medicine and Dentistry of New Jersey
New Jersey Medical School
Newark, New Jersey

**Roger G. Breeze**
Centaur Science Group
Washington, DC

**Paul S. Keim**
Microbial Genetics and Genomics Center
Northern Arizona University
Flagstaff, Arizona
and
Pathogen Genomics Division
The Translational Genomics Research Institute
Flagstaff, Arizona

**Stephen A. Morse**
National Center for Emerging and Zoonotic Infectious Diseases
Centers for Disease Control and Prevention
Atlanta, Georgia

Amsterdam • Boston • Heidelberg • London • New York • Oxford
Paris • San Diego • San Francisco • Singapore • Sydney • Tokyo
Academic Press is an imprint of Elsevier

Academic Press is an imprint of Elsevier
30 Corporate Drive, Suite 400, Burlington, MA 01803, USA
525 B Street, Suite 1800, San Diego, California 92101-4495, USA
84 Theobald's Road, London WC1X 8RR, UK

Cover: Image of coronavirus cells on the cover courtesy of Dr. Frederick Murphy, CDC.

**Library of Congress Cataloging-in-Publication Data**
Application submitted.

**British Library Cataloguing-in-Publication Data**
A catalogue record for this book is available from the British Library

ISBN 978-0-12-382006-8

For information on all Academic Press publications
visit our website at www.elsevierdirect.com

Printed in the United States of America

Transferred to Digital Printing, 2013

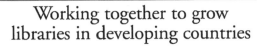

Working together to grow
libraries in developing countries

www.elsevier.com | www.bookaid.org | www.sabre.org

ELSEVIER    BOOK AID International    Sabre Foundation

# Contents

# Contributors

**Jonathan Allen (493)**   Lawrence Livermore National Laboratory, Livermore, California

**Elizabeth L. Bahr (649)**   Office of General Counsel, Department of the Navy, Alexandria, Virginia

**Hazel Bailey (393)**   U.S. Government Accountability Office

**S. Arunmozhi Balajee (297)**   Molecular Epidemiology, Mycotic Diseases Branch, Centers for Disease Control and Prevention, Atlanta, Georgia

**Neel G. Barnaby (89)**   FBI Laboratory, Quantico, Virginia

**John R. Barr (405)**   Centers for Disease Control and Prevention, National Center for Environmental Health, Division of Laboratory Sciences, Emergency Response and Air Toxicants Branch, Atlanta, Georgia

**Roger G. Breeze (693)**   Centaur Science Group, Washington, DC

**Luke N. Brewer (421)**   Computational Materials Science and Engineering, Sandia National Laboratories, Albuquerque, New Mexico

**Bruce Budowle (15, 545, 561, 667, 693)**   Institute of Investigative Genetics, Department of Forensic and Investigative Genetics, University of North Texas Health Science Center, Fort Worth, Texas

**Robert L. Bull (315)**   FBI Laboratory, Quantico, Virginia

**James P. Burans (89, 619)**   National Bioforensics and Analysis Center, Ft. Detrick, Maryland

**Robin M. Bush (109)**   Department of Ecology and Evolutionary Biology, University of California, Irvine, California

**Consuelo Carrillo (75)**   Plum Island Animal Disease Center, Animal and Plant Health Inspection Service, U.S. Department of Agriculture, Greenport, New York

**Thomas A. Cebula (29, 479)**   Johns Hopkins University, Baltimore, Maryland and Institute for Genome Sciences, University of Maryland School of Medicine, Baltimore, Maryland

**Ranajit Chakraborty (561)**   Institute of Investigative Genetics, Department of Forensic and Investigative Genetics, University of North Texas Health Science Center, Fort Worth, Texas

**Brian H. Clowers (449)**   Pacific Northwest National Laboratory, Richland, Washington

**Gregory A. Dasch (277)**  Rickettsial Zoonoses Branch, Division of Viral and Rickettsial Diseases, Centers for Disease Control and Prevention, Atlanta, Georgia

**David D. Duncan (155)**  Ibis Biosciences Inc., a subsidiary of Abbott Molecular Inc., Carlsbad, California

**David J. Ecker (155)**  Ibis Biosciences Inc., a subsidiary of Abbott Molecular Inc., Carlsbad, California

**David M. Engelthaler (297)**  The Translational Genomics Research Institute, Flagstaff, Arizona

**Marina E. Eremeeva (277)**  Rickettsial Zoonoses Branch, Division of Viral and Rickettsial Diseases, Centers for Disease Control and Prevention, Atlanta, Georgia

**Mark W. Eshoo (155)**  Ibis Biosciences Inc., a subsidiary of Abbott Molecular Inc., Carlsbad, California

**Jacqueline Fletcher (89)**  Department of Entomology and Plant Pathology, National Institute for Microbial Forensics and Food and Agricultural Biosecurity, Oklahoma State University, Stillwater, Oklahoma

**Jeffrey T. Foster (259, 315)**  Center for Microbial Genetics and Genomics, Northern Arizona University, Flagstaff, Arizona

**W. Florian Fricke (29, 479)**  Institute for Genome Sciences, University of Maryland School of Medicine, Baltimore, Maryland

**Shea Gardner (493)**  Lawrence Livermore National Laboratory, Livermore, California

**Rockne P. Harmon (681)**  Forensic/Cold Case Consultant, Alameda, California

**Karen H. Hill (259)**  Los Alamos National Laboratory, Biosciences Division, Los Alamos, New Mexico

**Crystal Jaing (493)**  Lawrence Livermore National Laboratory, Livermore, California

**Rudolph C. Johnson (405)**  Centers for Disease Control and Prevention, National Center for Environmental Health, Division of Laboratory Sciences, Emergency Response and Air Toxicants Branch, Atlanta, Georgia

**Suzanne R. Kalb (405)**  Centers for Disease Control and Prevention, National Center for Environmental Health, Division of Laboratory Sciences, Emergency Response and Air Toxicants Branch, Atlanta, Georgia

**Arnold F. Kaufmann (5)**  United States Public Health Service (retired), Stone Mountain, Georgia

**Paul S. Keim (5, 15, 259, 315, 545, 667, 693)**  Center for Microbial Genetics and Genomics, Northern Arizona University and Pathogen Genomics Division, The Translational Genomics Research Institute, Flagstaff, Arizona

Ali S. Khan (239)   National Center for Zoonotic, Vector-Borne, and Enteric Diseases, Centers for Disease Control and Prevention, Atlanta, Georgia

Paul G. Kotula (421)   Materials Characterization Department, Sandia National Laboratories, Albuquerque, New Mexico

Melissa Kramer (461)   Cold Spring Harbor Laboratory, Cold Spring Harbor, New York

Terrance Leighton (581)   Children's Hospital Research Institute, Oakland, California

W. Ian Lipkin (173)   Center for Infection and Immunity, Mailman School of Public Health and College of Physicians and Surgeons of Columbia University, New York

Ronan T. Loftus (59)   IdentiGEN, Dublin, Ireland

James D. Marks (327)   Department of Anesthesia and Pharmaceutical Chemistry, University of California, San Francisco, and San Francisco General Hospital, San Francisco, California

Robert F. Massung (277)   Rickettsial Zoonoses Branch, Division of Viral and Rickettsial Diseases, Centers for Disease Control and Prevention, Atlanta, Georgia

W. Richard McCombie (461)   Cold Spring Harbor Laboratory, Cold Spring Harbor, New York

Ciaran Meghen (59)   IdentiGEN, Dublin, Ireland

Ulrich Melcher (89)   Department of Biochemistry and Molecular Biology, Oklahoma State University, Stillwater, Oklahoma

Jack Melling (393)   U.S. Government Accountability Office

Joseph R. Michael (421)   Materials Characterization Department, Sandia National Laboratories, Albuquerque, New Mexico

Stephen A. Morse (199, 693)   National Center for Emerging and Zoonotic Infectious Diseases, Centers for Disease Control and Prevention, Atlanta, Georgia

Randall S. Murch (581, 649)   Center for Technology, Security and Policy, Virginia Polytechnic Institute and State University, Alexandria, Virginia

Catherine L. Murray (137)   Laboratory of Virology and Infectious Disease, Center for the Study of Hepatitis C, The Rockefeller University, New York

Forrest W. Nutter Jr., (89)   Department of Plant Pathology, Iowa State University, Ames, Iowa

Francisco M. Ochoa Corona (89)   Department of Entomology and Plant Pathology, National Institute for Microbial Forensics and Food and Agricultural Biosecurity, Oklahoma State University, Stillwater, Oklahoma

Thomas S. Oh (137)   Laboratory of Virology and Infectious Disease, Center for the Study of Hepatitis C, The Rockefeller University, New York

Richard T. Okinaka (259)   Center for Microbial Genetics and Genomics, Northern Arizona University, Flagstaff, Arizona

Jennifer Parla (461)  Cold Spring Harbor Laboratory, Cold Spring Harbor, New York

Talima Pearson (259, 545)  Center for Microbial Genetics and Genomics, Northern Arizona University, Flagstaff, Arizona

Peter T. Pesenti (605)  Chemical and Biological Division, Science and Technology Directorate, Department of Homeland Security, Washington, DC

Nicki Pesik (239)  National Center for Zoonotic, Vector-Borne, and Enteric Diseases, Centers for Disease Control and Prevention, Atlanta, Georgia

John Picuri (155)  Ibis Biosciences Inc., a subsidiary of Abbott Molecular Inc., Carlsbad, California

Jacques Ravel (15, 29, 479)  Institute for Genome Sciences, University of Maryland School of Medicine, Baltimore, Maryland

Charles M. Rice (137)  Laboratory of Virology and Infectious Disease, Center for the Study of Hepatitis C, The Rockefeller University, New York

James Robertson (627)  Forensic and Data Centres, Australian Federal Police, Australian Capital Territory, and the Faculty of Applied Science, University of Canberra, Australian Capital Territory, Australia

Paul E. Roffey (627)  Forensic and Data Centres, Australian Federal Police, Australian Capital Territory, and the Faculty of Applied Science, University of Canberra, Australian Capital Territory, Australia

Gregory B. Saathoff (221)  Critical Incident Analysis Group, University of Virginia School of Medicine, Charlottesville, Virginia

Ronald Schouten (221)  Law and Psychiatry Service, Massachusetts General Hospital, Harvard Medical School, Boston, Massachusetts

Steven E. Schutzer (357, 667, 693)  Department of Medicine, University of Medicine and Dentistry of New Jersey, New Jersey Medical School, Newark, New Jersey

Sushil K. Sharma (393)  U.S. Government Accountability Office

Evan Skowronski (173)  U.S. Army Edgewood Chemical Biological Center, Aberdeen Proving Ground, Maryland

Tom Slezak (493)  Lawrence Livermore National Laboratory, Livermore, California

Jenifer A.L. Smith (189, 379)  Bioforensics Consulting, LLC, Edgewater, Maryland

Noel H. Smith (43)  Veterinary Laboratories Agency, Surrey, UK and Centre for the Study of Evolution, University of Sussex, Sussex, UK

Carla Thomas (89)  Department of Plant Pathology, University of California, Davis, California

Clinton Torres (493)  Lawrence Livermore National Laboratory, Livermore, California

Marisa Torres (493)  Lawrence Livermore National Laboratory, Livermore, California

**Apichai Tuanyok** (259)  Center for Microbial Genetics and Genomics, Northern Arizona University, Flagstaff, Arizona

**Stephan P. Velsko** (509, 527)  Lawrence Livermore National Laboratory, Livermore, California

**Elizabeth Vitalis** (493)  Lawrence Livermore National Laboratory, Livermore, California

**Amy J. Vogler** (259)  Center for Microbial Genetics and Genomics, Northern Arizona University, Flagstaff, Arizona

**David M. Wagner** (259, 545)  Center for Microbial Genetics and Genomics, Northern Arizona University, Flagstaff, Arizona

**Karen L. Wahl** (449)  Pacific Northwest National Laboratory, Richland, Washington

**Elizabeth Weirich** (199)  National Center for Emerging and Zoonotic Infectious Diseases, Centers for Disease Control and Prevention, Atlanta, Georgia

**Mark Wilson** (545)  Forensic Science Program, Western Carolina University, Cullowhee, North Carolina

**David S. Wunschel** (449)  Pacific Northwest National Laboratory, Richland, Washington

# Foreword

In an increasingly complex world, the fields of science, engineering, and technology advance at an ever more rapid pace. At the same time, the means by which crimes may be committed have become more sophisticated, with greater capacity to inflict harm. Scientists, as professionals in their respective fields, are dedicated to discovery, with intellectual curiosity driving their research, as well as motivation to improve the quality of life for all of humanity. Unfortunately, just as in every walk of life there will be a malicious few who will work in an opposite direction, namely to do harm to their fellow human beings for whatever reasons, a maladjusted mind will justify. An example of a perverse usurpation of science occurred at the beginning of the 21st century in the form of mailings of the anthrax pathogen, *Bacillus anthracis*, in an unprecedented biological attack.

From the anthrax mailings, with their disastrous impact on human health, society, and the economy, has developed the entirely new discipline of microbial forensics. Few are aware of this development, especially of its significant value behind the scenes in tracking down perpetrators of what might be termed "high tech crime," namely those who employ a biological weapon.

My experience with the anthrax attacks was personal in that I served as Director of the National Science Foundation at the time the event occurred. Because my training had been in microbiology and molecular biology, I was able to bring to bear the needed science and scientists who had the capability to assist in analyses of evidence from this tragic event. With colleagues from all of the relevant agencies, including the National Institutes of Health, the Department of Homeland Security, the Department of Defense, the Department of Justice, and the Central Intelligence Agency (CIA), we were able to provide the Federal Bureau of Investigation (FBI) with advice and assistance, as appropriate, in their task of tracking the source of the biological material used in the attacks. We were, in effect, serving as midwives to this new discipline.

Interestingly, the origin of microbial forensics, as a disciplinary activity, traces to the foresight and pioneering efforts of colleagues at both the FBI and the CIA. The need for microbial forensics certainly heightened with the anthrax mail attacks, and development of the field of microbial forensics since has been impressive, especially so through the cooperation of government agencies in circumscribing the vision and the mission of microbial forensics. Scientific tools and practices were employed to determine the perpetrator of the bioterrorist act and, at the same time, work was done to ensure that innocent individuals would be protected from unwarranted suspicion. Many contributing to the first and second editions of this book were among those who joined in the scientific collaboration and contributed to the maturation of microbial forensics. During the years of the investigation, scientific meetings were organized by various organizations, including the American Academy of Microbiology (1) and the Cold Spring Harbor Laboratory–Banbury Center (2–4). A meeting of the Scientific Working Group on Microbial Genetics and Forensics also was hosted by the FBI (5,6).

Clearly, there is now a shared vision from which the field has evolved and continues to mature, illustrated by the recent National Research and Development Strategy for Microbial Forensics prepared by the National Science and Technology Council and Executive Office of the President (7).

The extraordinary growth of microbial forensics is articulated in this updated volume. I am pleased to have played a part in the emergence and maturation of this very important field of science and to have had the honor of working closely with my dedicated colleagues. I share the sentiments of my friends and colleagues that the science of microbial forensics will serve humankind in ways both beneficial and humanitarian now and in the future.

Rita R. Colwell, Ph.D., D.Sc.
Distinguished University Professor, University of Maryland,
College Park and Johns Hopkins University,
Bloomberg School of Public Health

## REFERENCES

[1] P. Keim, Microbial Forensics: A Scientific Assessment, American Academy of Microbiology, Washington DC, 2003.

[2] B. Budowle, S.E. Schutzer, M.S. Ascher, R.M. Atlas, J.P. Burans, R. Chakraborty, et al., Toward a system of microbial forensics: from sample collection to interpretation of evidence, Appl. Environ. Microbiol. 71 (2005) 2209–2213.

[3] B. Budowle, S.E. Schutzer, J.P. Burans, D.J. Beecher, T.A. Cebula, R. Chakraborty, et al., Quality sample collection, handling, and preservation for an effective microbial forensics program, Appl. Environ. Microbiol. 72 (2006) 6431–6438.

[4] B. Budowle, S.E. Schutzer, S.A. Morse, K.F. Martinez, R. Chakraborty, B.L. Marrone, et al., Criteria for validation of methods in microbial forensics, Appl. Environ. Microbiol. 74 (2008) 5599–5607.

[5] B. Budowle, S.E. Schutzer, A. Einseln, L.C. Kelley, A.C. Walsh, J.A. Smith, et al., Public health, Building microbial forensics as a response to bioterrorism. Science 301 (2003) 1852–1853.

[6] B. Budowle, S.E. Schutzer, A. Einseln, L.C. Kelley, A.C. Walsh, J.A. Smith, et al., Scientific Working Group on Microbial Genetics and Forensics, Quality assurance guidelines for laboratories performing microbial forensic work, Science 301 (2003) SOM 1–SOM 17.

[7] National Science and Technology Council–Executive Office of the President. National Research and Development Strategy for Microbial Forensics, 2010. Available from: http://www.whitehouse.gov/files/documents/ostp/NSTC%20Reports/National%20MicroForensics%20R&DStrategy%202009%20UNLIMITED%20DISTRIBUTION.pdf.

# Acknowledgments

This book is dedicated to all who have lost their lives or suffered as a consequence of terrorism. It is hoped that this book contributes to efforts to deter and combat terrorism and to better prepare all of us to respond to such nefarious acts. To make this book a reality, we have relied upon the experience and input from our many colleagues who have contributed so much to the field of microbial forensics. We thank all of them immensely. We also express our deep gratitude to Elizabeth Brown, Kristi Anderson, Renske van Dijk, and other staff members of Elsevier and Academic Press for their dedication, input, and organization without which this book would never have come to fruition.

# Introduction – The Rapidly Evolving Discipline of Microbial Forensics

**Steven E. Schutzer,[a] Bruce Budowle,[b] Roger G. Breeze,[c] Paul S. Keim,[d] and Stephen A. Morse[e]**

[a]*Department of Medicine, University of Medicine and Dentistry of New Jersey, New Jersey Medical School, Newark, NJ*
[b]*Institute of Investigative Genetics, Department of Forensic and Investigative Genetics, University of North Texas Health Science Center, Fort Worth, TX*
[c]*Centaur Science Group, Washington, DC*
[d]*Center for Microbial Genetics and Genomics, Northern Arizona University and Pathogen Genomics, The Translational Genomics Research Institute, Flagstaff, AZ*
[e]*National Center for Emerging and Zoonotic Infectious Diseases, Centers for Disease Control and Prevention, Atlanta, GA*

In the five years since the publication of the first edition of this book, the field of microbial forensics has evolved substantially in its ability to attribute the source of microorganisms and toxins that have been used in cases of bioterrorism and in biocrimes. The 2001 anthrax mail attack (often referred to as "Amerithrax") fortunately remains the worst and best known example of successful bioterrorism on U.S. soil (Chapter 2). At the time, the nation was ill prepared to investigate the unique types of microbiological evidence for attribution that were generated by that investigation, although traditional types of forensic evidence could readily be collected, analyzed, and interpreted within the framework of a statistically sound scientific foundation. The first edition of this book, which was published in 2005, had only limited reference to the anthrax mail attack because much of the scientific examination was still in progress and methodologies were still under development. Essentially, our capabilities were initially limited to detection and identification and did not include detailed characterization and comparative analyses. The first edition was developed as a foundational text to stimulate scientists, legal experts, and decision makers responsible for analyzing and interpreting evidence from a bioterrorism act, biocrime, or inadvertent microorganism/toxin release for attribution purposes and to describe the discipline and some of the opportunities and challenges ahead. Several chapters in this new edition bring us up to date on anthrax. The anthrax mail investigation, one of the most intense and expansive investigations to date by the Federal Bureau of Investigation and U.S. Postal Service (http://www.fbi.gov/anthrax/ameri-thraxlinks.htm), is now considered closed yet provides lessons learned on what should be done in microbial forensics to support an investigation and equally important what should be avoided.

Many of the elements discussed in the first edition and other articles related to microbial forensics are echoed in the 2009 U.S. National Research and Development Strategy for Microbial Forensics, (1) which identifies threat awareness, prevention and protection, surveillance and detection, and response and recovery as the essential pillars of U.S. national biodefense policy. "Attribution"—the investigative process by which the U.S. government links the identity of a perpetrator or perpetrators of illicit activity and the pathway leading to criminal activity—is part of the surveillance and detection pillar. Making a determination of attribution for a covertly planned or actual biological attack would be the culmination of a complex investigative process drawing on many different sources of information, including technical forensic analysis of material evidence collected during the course of an investigation of a planned attack or material evidence resulting from an attack. One of the key sources of attribution information in a biological attack is microbial forensics.

Future investigations of bioterrorism events or biocrimes are likely to be as multifaceted as the anthrax attack investigation, if not more so, and will demand more integration and better communication among government agencies. At the same time, microbial forensics is still a nascent field facing broad and complex scientific challenges. For these and other reasons, the U.S. government is making investments to provide a robust capability to detect, identify, and characterize biological agents. Advances in this area have been sufficiently exciting and significant to make a new edition of Microbial Forensics both timely and necessary. We also see growing interest and application of microbial forensics principles in food safety and environmental attribution and thus included relevant chapters in the second edition to cover a broader audience interest than national security. We recognize that traditional detective investigation, coupled with scientific analysis of evidence and frequently additional ongoing and novel experimentation, can generate investigative leads (2). Such lead data could support decision making or a legal proceeding.

We developed the first edition of this book for individuals entering the field of microbial forensics who were looking for a single source for initial guidance and information. We intend this new edition to continue to support that need with a combination of basic texts and chapters on more sophisticated technologies, such nonbiological analytical tools, high-resolution DNA sequencing, "Next" or "Now Generation" methods, and necessary companion bioinformatics methods to address the monumental amounts of data that now can be generated (3). A combination of diverse disciplines must be exploited to analyze evidence, including biology, microbiology, medicine, chemistry, physics, statistics, population genetics, and computer science. New techniques must be employed to extract the most information from forensic evidence used in terrorist and criminal events, especially when more traditional forms of evidence are either not available or very limited in content. These concepts are also explained in this book. Yet microbial forensics—the maturing discipline—also depends very heavily on some traditional and frequently overlooked scientific values: willingness to share often priceless samples with others across international boundaries, rigorous curation of microbial repository samples over decades, and organization and execution of international collaborative studies with recognition of all involved. We cite examples of all these too.

Over the past century, science has played an increasingly greater role in criminal investigation (4,5). Microbial forensic science will continue this tradition. But scientific analysis is only part of the process. Forensic science results must be integrated with other information and attention must be paid to steps that will assure admissibility of results in a court of law (6). We stress quality control and quality assurance as the means to ensure integrity of the evidence. Practices such as adherence to chain of custody procedures, documentation of activities, and the use of validated reagents, calibrated equipment, negative and known positive control samples, validated procedures, standard operating procedures, and so on are the essence of reliability and confidence. These in turn ensure admissibility in court.

The scientific foundations of Microbial Forensics will be strengthened, built upon, and likely remodeled by our present and future colleagues. Their accomplishments over the last five years have led to this new edition. We look forward to their future input, interaction, and insight.

# REFERENCES

[1] National Science and Technology Council – Executive Office of the President. National Research and Development Strategy for Microbial Forensics, 2010 http://www.whitehouse. gov/files/documents/ostp/NSTC%20Reports/National%20MicroForensics%20R&DStrategy% 202009%20UNLIMITED%20DISTRIBUTION.pdf.

[2] S.E. Schutzer, P. Keim, K. Czerwinski, B. Budowle, Use of forensic methods under exigent circumstances without full validation, Sci. Transl. Med. 1 (2009) 8cm7.

[3] E.R. Mardis, Next-generation DNA sequencing methods, Annu. Rev. Genomics Hum. Genet. 9 (2008) 387–402.

[4] B. Budowle, M.C. Bottrell, S.G. Bunch, R. Fram, D. Harrison, S. Meagher, et al., A perspective on errors, bias, and interpretation in the forensic sciences and direction for continuing advancement, J. Forensic Sci. 54 (2009) 798–809.

[5] B. Budowle, S.E. Schutzer, S.A. Morse, K.F. Martinez, R. Chakraborty, B.L. Marrone, et al., Criteria for validation of methods in microbial forensics, Appl. Environ. Microbiol. 74 (2008) 5599–5607.

[6] R. Harmon, Admissibility standards for scientific evidence, in: R.G. Breeze, B. Budowle, S.E. Schutzer (Eds.), Microbial Forensics, Academic Press, San Diego, 2005, pp. 381–922.

# Investigative Genetics

# A

SECTION

## Criminal

# CHAPTER 1

# The Kameido Anthrax Incident: A Microbial Forensics Case Study

**Arnold F. Kaufmann[a] and Paul S. Keim[b]**
[a] United States Public Health Service (Retired), Stone Mountain, Georgia
[b] Center for Microbial Genetics and Genomics, Northern Arizona University and Pathogen Genomics Division, The Translational Genomics Research Institute, Flagstaff, Arizona

## INTRODUCTION

The Aum Shinrikyo, an apocalyptic religious sect based in Japan, first came to worldwide attention in 1995 as the result of the sect's deadly sarin gas attack on the Tokyo subway system (1–3). Subsequent investigations revealed that the Aum Shinrikyo had launched earlier attacks with both chemical and biological agents. The biological attacks utilizing *Bacillus anthracis* spores and botulinum neurotoxin were notably unsuccessful, with failure to produce any casualties in at least seven alleged attempts over several years beginning in 1990. This chapter discusses an attack in 1993 that was launched from the Aum Shinrikyo headquarters building, then located in Kameido, a Tokyo suburb, with emphasis on laboratory and epidemiological studies (4,5).

## THE AUM SHINRIKYO: A BRIEF HISTORY

To put the Kameido incident into context, a brief history of the Aum Shinrikyo and its founder, adapted primarily from reviews by Hudson, Smithson, and Tu, is useful (1–3). The Aum Shinrikyo was founded by Shoko Asahara, whose birth name was Chizuo Matsumoto. Born into poverty in 1955 and suffering severe visual impairment due to infantile glaucoma, Matsumoto was sent at a young age to a government-subsidized boarding school for the blind. He purportedly felt abandoned by his family, which may have later led to an Aum Shinrikyo rule that followers were to cut off relationships with their parents to attain the supreme truth.

Having limited vision in one eye, Matsumoto developed influence over the other blind students, who paid him for various services. During his student years, he developed a reputation as a bully and con artist. After high school

5

Microbial Forensics. DOI: 10.1016/B978-0-12-382006-8.00001-3

graduation in 1975, Matsumoto established a successful acupuncture clinic, but he had to move to Tokyo in 1977 due to his involvement in a fight that resulted in injury to several persons. About this time, his stated ambitions included becoming the leader of a robot kingdom or the Prime Minister of Japan.

In Tokyo, he found work as an acupuncturist and enrolled in a preparatory school for the Japanese college entrance examination, with a goal of qualifying for matriculation in Tokyo University. Failing the entrance examination, Matsumoto married and established an acupuncture clinic and a natural foods shop. As a sideline, he concocted an alcohol extract of tangerine skins that he marketed as a miracle drug for weight loss and a variety of other conditions. His success in sales of this product attracted the attention of the police and subsequent arrest and imprisonment for violating the Japanese Cosmetics and Medical Instruments Act. This experience may have contributed to his animosity toward established authority.

After his move to Tokyo, Matsumoto became interested in religion and, in 1981, joined Agon Shu, a new religion based on Buddhism and yoga. In 1984, Matsumoto quit the Agon Shu and established Aum Shinsen, a yoga club that grew rapidly from 15 to more than 1000 members. He also changed his name to Shoko Asahara or Bright Light in Japanese. Following a trip to India in 1986–1987, Asahara changed the name of the yoga club to Aum Shinrikyo. Aum is a Hindu syllable representing the spoken essence of the universe, and Shinrikyo is derived from the Japanese words for "supreme truth." In 1989, Aum Shinrikyo was officially recognized as a religious sect in Japan, giving the sect tax advantages as well as the ability to claim the members' work in the sect's various enterprises as voluntary. The sect's growth continued with spread to other countries, including the United States, Germany, and Russia.

Sect members lived a spartan life and were expected to cut off all associations from their past life, to take a chastity vow, and to turn over all their assets to the Aum Shinrikyo. They were subjected to a heavy diet of their master's "wisdom," often simultaneously undergoing food and sleep deprivation. Members were expected to labor voluntarily in the sect's various commercial enterprises, such as sales of herbal teas and natural medications, operation of noodle shops, health clubs and babysitting services, and computer-related services. Those who balked were driven ever harder, drugged, and confined. In some extreme cases, defectors were murdered.

To carry out its activities, the Aum Shinrikyo developed a complex organizational structure consisting of 22 ministries plus the Offices of Religious Members. The latter was charged with recruitment of persons having needed skills, such as members of the Japanese Self Defense Forces and scientists. In effect, the Aum Shinrikyo assumed the form of a shadow government, which could supplant the existent Japanese government if Asahara's ambitions were fulfilled.

By 1990, with membership in the tens of thousands spread over six countries and an estimated 300 million to a billion dollars available, the Aum Shinrikyo was well positioned to further Asahara's ambitions and delusions. The Aum Shinrikyo initially attempted a conventional approach to political power by fielding a slate of 25 candidates for the national elections that year. None of the candidates, including Asahara, was even modestly successful. Asahara believed that the Japanese government had cheated him rather than that the electorate was put off by the doomsday overtones of the candidate's speeches. His belief was reinforced by the fact that the number of votes received by all the candidates was far fewer than the number of Aum Shinrikyo members in Japan.

Based on a pastiche of apocalyptic scenarios drawn from various religions, Asahara preached that Japan was destined to suffer a number of overwhelming catastrophes, including a poison gas attack by the United States. Asahara and his followers would survive the looming Armageddon and evolve into a super-race dominating the world. He became more vocal in expressing this belief after the humiliating electoral defeat in 1990. Not content to allow the catastrophes to occur in their own time, Asahara initiated development of chemical and biological weapons to speed up the process.

Only the chemical weapons program had some success. After overcoming initial production problems, the Aum launched an attack with sarin gas in Matsumoto City in June 1994. The attack targeted judges presiding over a land use dispute between the Aum Shinrikyo and a local real estate agent. Suspecting that the judges would make a decision unfavorable to the cult, Asahara ordered their assassinations. This gave an opportunity to test the effectiveness of their sarin on humans. The sarin release utilized a spray device and resulted in 311 known casualties, with 58 hospitalized, including seven deaths. The judges were unharmed.

Investigation of the Matsumoto City attack proceeded slowly, without definitive evidence linking the crime to a specific individual or group. The Aum Shinrikyo was suspected, and a sarin degradation product was detected in soil near a building on an Aum Shinrikyo compound, the Seventh Satayan, in Kamakiuishiki. The police did not seek a warrant to search the facility because of a conservative interpretation of pertinent laws. In an unrelated kidnapping case, however, the police found fingerprint evidence that an Aum Shinrikyo member was involved. This gave justification for obtaining a warrant to investigate the facility. In March 1995, Asahara learned of the plans for a police raid from Aum members within the Japanese Self Defense Forces. In a ploy to distract the police and buy time, Asahara decided to release sarin in the Tokyo subway system. Two days later, the attack was carried out, resulting in several thousand casualties, including 12 fatalities.

The Aum Shinrikyo cult had been thought to be an odd group and even a nuisance on occasion. Investigations after the Tokyo sarin attack revealed a more sinister aspect of the sect and its leader. In addition to the chemical weapons development program, the Aum Shinrikyo was found to have been actively pursuing biological weapons, albeit without success due to incompetence. In particular, the ineffective release of *B. anthracis* spores in Kameido was discovered, leading to the investigations discussed in this chapter.

## THE KAMEIDO ANTHRAX INCIDENT

On June 29–30, 1993, complaints about foul odors were registered with local environmental health authorities in Kameido in the Tokyo metropolitan area (4). The odors originated from the eight-story headquarters building of the Aum Shinrikyo. Some of the exposed persons reported appetite loss, nausea, and vomiting. Birds and pets were also reportedly ill, but the nature of these illnesses was not defined. The environmental health officials requested permission to inspect the headquarters building, but Aum Shinrikyo members at the scene refused. The officials checked the building's exterior, collected air samples, and began surveillance of activities at the building. Other than the nuisance posed by the odor, definitive human health risks could not be identified.

On the morning of July 1, neighbors began to complain about loud noises and an intermittent mist originating from one of two structures on the roof which were thought to be cooling towers (Figure 1.1). As the day progressed, 118 complaints about foul odors were received from nearby residents, primarily to the south of the building. Winds (2–4 m/sec) that day blew from a northeasterly direction (4). Light rain (7 mm total) fell in the early morning, with cloudy conditions thereafter. The temperature ranged from 16.9 to 19.9°C throughout the day.

A "gelatin-like, oily, gray-to-black" fluid from the mist coming off the "cooling towers" collected on the side of the building (4). Samples of the fluid were collected by the environmental health officials and stored in a refrigerator (4°C) for later testing.

**FIGURE 1.1**

Mist being dispersed from a spray device on the roof of the Aum Shinrikyo headquarters building, Kameido, Japan, July 1, 2006 (4).

The next day, July 2, Shoko Asahara agreed to stop using the rooftop devices and to clean and vacate the building (4). An environmental inspection found no equipment, including the rooftop devices, remaining in the building, and black stains on the walls were the only notable observation.

The problem, apparently being resolved, was largely forgotten until 1996. Police investigations of the sarin attack on the Tokyo subway system revealed that the Aum Shinrikyo was also involved in bioterrorism. Following the conservative Japanese policy of not revealing criminal evidence until the time of trial in court, the true nature of the Kameido incident was first disclosed to the public in May 1996 when Asahara was arraigned (2,4). Aum Shinrikyo members had confessed that the odors resulted from efforts to aerosolize a liquid suspension of B. anthracis spores. The motive was to trigger an inhalational anthrax epidemic and a subsequent world war. The war would culminate in the Aum Shinrikyo members becoming a super-race that would rule the world in accord with Asahara's preaching (4).

Many questions about the incident remained unanswered. For example, did the attack actually occur? If so, were B. anthracis spores utilized? Could the specific B. anthracis strain and its origin be identified? Was the strain virulent? Why did the attack apparently fail? Had illnesses occurred but gone undetected? Investigations were initiated to better characterize the alleged attack and its consequences.

## MICROBIAL FORENSICS INVESTIGATION

Fluid that had been collected from the Aum Shinrikyo headquarters building in July 1993 and subsequently stored at $4°C$ was examined in January 2000 for bacterial content (5). Polymerase chain reaction (PCR) screening of the fluid was positive for B. anthracis. Microscopic examination of the fluid with malachite green and safranin staining revealed spores, nonspecific debris, and bacterial cells other than large bacilli. The fluid was cultured by spreading on sheep blood agar plates and incubating at $37°C$ under ambient $CO_2$ concentration (Figure 1.2). Based on the number of bacterial colonies observed on the plates after incubation, the fluid contained approximately $4 \times 10^4$ bacterial colony forming units (CFU) per milliliter. Most colonies grew only weakly and were morphologically inconsistent with normal B. anthracis characteristics when grown under these conditions. The poorly growing bacteria were not further characterized.

**FIGURE 1.2**
Blood agar plate of sampling from the Kameido site (4). (See Color Insert)

About 10% of the colonies on the plates were typical of B. anthracis, being large and having a nonhemolytic, "gray ground glass"

**FIGURE 1.3**

Multiple-locus, variable-number tandem repeat analysis of a Kameido isolate and the Sterne strain of *Bacillus anthracis* (4). (See Color Insert)

appearance (Figure 1.2). The number of observed colonies consistent with *B. anthracis* was consistent with a concentration of $4 \times 10^3$ CFU/ml of the fluid. Forty-eight of these colonies were purified by single-colony streaking and subjected to the eight-locus multiple-locus variable-number tandem repeat analysis (MLVA) (6). All 48 colonies were *B. anthracis* and had an identical MLVA genotype, although the VNTR marker on the pX02 plasmid failed to amplify. The genotype of all 48 strains was vrrA, 313 bp; $vrrB_1$, 229 bp; $vrrB_2$, 162 bp; $vrrC_1$, 583 bp; $vrrC_2$, 532 bp; CG3, 158 bp; pX01-aat, 129 bp; pX02-at, no amplification. The lack of PCR amplification at the pX02 markers is consistent with strains that are missing the pX02 plasmid entirely. Amplification of these loci can occur in closely related *Bacillus cereus* strains, but actual amplicon sizes had been observed previously only in *B. anthracis*. The MLVA genotype observed was consistent with results obtained with the Sterne anthrax vaccine strain (Figure 1.3). The Sterne strain is a member of the A3.b diversity cluster, and in a study of 419 isolates, only four naturally occurring *B. anthracis* strains in the electronic database of worldwide isolates had the same seven-marker genotype, although these strains were pX02 positive (6).

The Sterne $34F_2$ anthrax vaccine strain is available commercially in Japan for veterinary use. It had been reported previously that the Aum Shinrikyo had obtained a veterinary vaccine strain of *B. anthracis*, which may have been used by them for bioterrorist attacks (1). Results are consistent with this previously unsubstantiated report. The Sterne strain has low virulence due to lack of the pX02 plasmid, which is the location of genes coding for the ability to produce capsule, a major virulence factor of *B. anthracis*.

## EPIDEMIOLOGICAL INVESTIGATION

Culture-confirmed anthrax is a nationally notifiable disease in Japan, with physicians being required to report all cases (4). Only four human anthrax cases were reported during the 1990s, with a single case being reported in Tokyo. The Tokyo case was diagnosed in a man who was in his eighties and resided in Sumida-ward, which is adjacent to Kameido-ward. The case occurred in August 1994 and had no obvious association with the 1993 Kameido incident.

Could additional anthrax cases from the 1993 attack have gone unrecognized or unreported? A retrospective case-detection survey was conducted in 1999 to address this question (4). Using the official registry of "foul odor" complaints, the residences of the 118 complainants were mapped to identify the area of presumed highest risk. The 0.33-km$^2$ high-risk area determined by this approach contained approximately 3400 households and 7000 residents. A telephone survey was conducted of physicians at 39 medical facilities (15 internal medicine, 7 dermatology, and 15 other specialties) serving the area. None of these physicians had treated cases of anthrax, unexplained serious respiratory illnesses, or hemorrhagic meningitis, a common complication of systemic anthrax in residents of the high-risk area (7).

## DISCUSSION

A number of lessons can be learned from the Kameido incident. The investigation suffered from a failure to detect the incident at the time. The Aum Shinrikyo did not attract much official attention until the sarin attack in Matsumoto City in 1994. The Japanese culture is very tolerant of varying religious beliefs, an attitude reflecting Japanese constitutional guarantees of religious freedom. The Aum Shinrikyo was but one of more than 180,000 minor religions active in Japan (1). The police policy of conservative interpretation of pertinent laws was another factor. Also, the policy of not revealing details about a criminal investigation until the time court procedures are initiated slowed communications between pertinent agencies, delaying investigation even after the Aum Shinrikyo's attempt to utilize biological weapons first became known during the investigation of the 1995 sarin attack on the Tokyo subway (3). Awareness of potential threats, a low threshold of suspicion, and active sharing of information between governmental agencies having pertinent expertise and/ or authority are some key components in early detection of terrorism incidents.

Early characterization of the Aum Shinrikyo biological weapons program was based primarily on statements by the perpetrators and expert opinions (1–3). Physical evidence and independent corroboration of the claims were notably absent. Fortunately, health officials had collected a specimen of fluid from the building at the time of the Kameido incident, although it had not been analyzed at the time. The fluid was kept as potential evidence in the event disease could be associated with the incident. A policy of keeping evidence for a period of years allowed its examination in 2000, more than 6 years after the incident. From a forensic perspective, however, the lack of formal chain-of-custody documentation might be an issue in some jurisdictions.

The MLVA analysis of the *B. anthracis* strain isolated from the fluid from the building revealed a genotype consistent with that of the Sterne $34F_2$ anthrax vaccine strain. A direct comparison, however, was not made with the Sterne $34F_2$ strain used to produce anthrax vaccine in Japan. In addition, the genotypic match was not unique to Sterne, as the published database contained additional samples with identical (excluding pX02) genotypes. No samples from Japan had ever been characterized using MLVA and, as such, the natural background *B. anthracis* was unknown. Newer DNA or other analytic techniques might have found strain similarities or variances, but their importance would still need to be established through a comparison to Japanese vaccine stocks and natural-occurring *B. anthracis*. Also, no effort was made to isolate *B. anthracis* from the building's environment or adjoining areas of Kameido to provide evidence that the strain isolated from the fluid had been dispersed and the extent of the dispersal. The ability of *B. anthracis* to survive many years in contaminated environment makes such an effort potentially productive (8).

Why did no disease result from the attack? One possibility is a small inhaled dose of spores for exposed persons. Although presumably strain dependent, the minimum inhaled dosage capable of causing human disease is unclear. However, unknowns such as aerosol dispersal device efficiency in generating a fine particle aerosol, concentration of spores in the fluid being dispersed, and aerosol cloud movement preclude making defensible estimates of inhaled dosage for persons exposed during the attack.

Another possible factor in lack of detected disease is the low virulence of the Sterne strain for humans. Despite frequent accidental inoculations of the vaccine during immunization of domestic animals, no documented cases of associated disease have been reported. The Sterne strain lacks a capsule, and other strains of *B. anthracis* lacking a capsule rarely produce human illness. Only three reports of illness associated with nonencapsulated strains have been made, with one case being in an immunocompromised person and the cause–effect relationship of the isolate to the illness observed in the other two cases being uncertain.

Serologic studies were not done at the time of the event but could have been potentially useful. For example, antibody and cell-mediated immune responses in persons who were exposed in the 2001 bioterrorism attack at the United States Capitol were used to demonstrate infection without resultant disease (9). If a similar study had been done in Kameido at the time of the attack, the question of whether infection had occurred in the absence of disease could have been addressed. The long lapse of time between the Kameido event and the epidemiological study would compromise the validity of such studies at this time.

The isolation of *B. anthracis* and the results of the MLVA testing might have proven useful in prosecuting some Aum Shinrikyo members, but this will have to remain conjectural. Criminal charges related to the biologic weapons development and attacks were not made, and, hence, the microbiological evidence was never tested in court. However, 13 Aum Shinrikyo members, including Asahara, have been convicted for perpetrating the sarin attack on the Tokyo subway and condemned to death. Despite the convictions of many leaders, the Aum Shinrikyo continues to exist, having been renamed Aleph, the first letter of the Phoenician alphabet. A splinter group, Hikari no Wa or Ring of Light, broke off Aleph following discord within the Aleph leadership. Surveillance of the two groups for potential terrorist activities has been maintained.

## CHALLENGES

The Kameido incident underscored a number of essential concepts in microbial forensics. Implementing these concepts must be a high priority for law enforcement jurisdictions. Early communication and information sharing between appropriate governmental agencies, particularly law enforcement and public health, is essential to an optimal outcome of microbial forensic studies. Public health agencies have a major role in the investigation of suspected and documented bioterrorism incidents, but the similarities and differences in public health and law enforcement investigations must be understood and coordinated so that both can be most effective (10). Investigations must, of necessity, be multidisciplinary and draw upon the best available expertise whether or not located in a governmental agency. Specimen testing must follow established chain-of-custody procedures, and all involved groups must be trained in these procedures. Strain subtyping is a powerful investigative tool for tracing the origins of microbial agents, but the procedures must be validated and have yet to be accepted by the courts. Epidemiological studies are essential to putting laboratory findings into the context of a bioterrorist event.

# REFERENCES

[1] R.A. Hudson, The sociology and psychology of terrorism: Who becomes a terrorist and why? Library of Congress, Federal Research Division report. Washington, DC, Library of Congress, 1999.

[2] A.E. Smithson, Rethinking the lessons of Tokyo Stimson Center report No. 35, in: A.E. Smithson, L.E. Levy (Eds.), Ataxia: The Chemical and Biological Terrorism Threat and the US Response, Simpson Center, Washington, DC, 2000, pp. 71–111.

[3] A.T. Tu, Chemical Terrorism: Horrors in Tokyo Subway and Matsumoto City, Alaken Inc., Fort Collins, CO, 2002.

[4] H. Takahashi, P. Keim, A.F. Kaufmann, C. Keys, K.L. Smith, K. Taniguchi, et al., *Bacillus anthracis* incident, Kameido, Tokyo, 1993, Emerg. Infect. Dis. 10 (1) (2004) 117–120.

[5] P. Keim, K.L. Smith, C. Keys, H. Takahashi, T. Kurata, A. Kaufmann, Molecular investigation of the Aum Shinrikyo anthrax release in Kameido, Japan. J. Clin. Microbiol. 39 (12) (2001) 4566–4567.

[6] P. Keim, L.B. Price, K.L. Klevytska Smith, J.M. Schupp, R. Okinaka, P. Jackson, et al., Multiple-locus VNTR analysis (MLVA) reveals genetic relationships within *Bacillus anthracis*, J. Bacteriol. 182 (10) (2000) 2928–2936.

[7] J.E.C. Holty, D.M. Bravata, H. Liu, R.A. Olshen, K.M. McDonald, DK. Owens, Systemic review: A century of inhalational anthrax cases from 1900 to 2005, Ann. Intern. Med. 144 (4) (2006) 270–280.

[8] P. Turnbull (Ed.), Anthrax in Humans and Animals, 4th Ed., World Health Organization, Geneva, 2008.

[9] D.L. Doolan, D.A. Freilich, G.T. Brice, T.H. Burgess, M.P. Berzins, R.L. Bull, et al., The US capitol bioterrorism anthrax exposures: Clinical epidemiological and immunological characteristics, J. Infect. Dis. 195 (2) (2007) 174–184.

[10] J.C. Butler, M.L. Cohen, C.R. Friedman, R.M. Scripp, C.G. Watz, Collaboration between public health and law enforcement: New paradigms and partnerships for bioterrorism planning and response, Emerg. Infect. Dis. 8 (10) (2002) 1152–1156.

# Microbial Forensic Investigation of the Anthrax-Letter Attacks

**Paul S. Keim,[a] Bruce Budowle,[b] and Jacques Ravel[c]**

[a]Center for Microbial Genetics and Genomics, Northern Arizona University and Pathogen Genomics, The Translational Genomics Research Institute, Flagstaff, Arizona
[b]Institute of Investigative Genetics, Department of Forensic and Investigative Genetics, University of North Texas Health Science Center, Fort Worth, Texas
[c]Institute for Genome Sciences, University of Maryland School of Medicine, Baltimore, Maryland

## THE ANTHRAX-LETTER EVENT

The United States was reeling in the immediate aftermath of the terrorist attacks on the World Trade Center and the Pentagon in 2001. Public fear and disruption were rampant, yet the government and the public coped and responded to the new challenge threatening national security. However, the country's vulnerability was exploited when, within less than 1 month after the worst terrorist attack perpetrated on U.S. soil, a bioterrorist attack was carried out that resulted in 22 infections and five deaths. The bioterrorism attack first became evident on October 2, 2001, when Robert Stevens, a previously healthy 63-year-old employee of American Media, Inc. (AMI) in Boca Raton, Florida, awoke from sleep with fever, emesis, and confusion. At the emergency department, a lumbar puncture was performed for presumed bacterial meningitis (1,2). Microscopic examination of the Gram stain of the cerebrospinal fluid revealed chains of large gram-positive bacilli. *Bacillus anthracis* was subsequently cultured from both his cerebral spinal fluid and blood (Figure 2.1) and confirmed by a laboratory response network laboratory within the Florida Department of Health on the following day (October 4). That day, the Florida Department of Health informed the public that a case of inhalational anthrax had been confirmed. This case immediately raised concern epidemiologically because the last reported case of inhalational anthrax in the United States occurred in 1976 (3). Moreover, this case was too close to the recent terrorist attacks to not suspect it to be another attack. Initially, the attack was covert but was eventually confirmed by the discovery of *B. anthracis* spores on Robert Stevens's computer keyboard in his place of work, which was an exceedingly uncommon place to find spores of *B. anthracis*. [A couple of points that should be noted: (i) Mr. Stevens

**15**

Microbial Forensics. DOI: 10.1016/B978-0-12-382006-8.00002-5

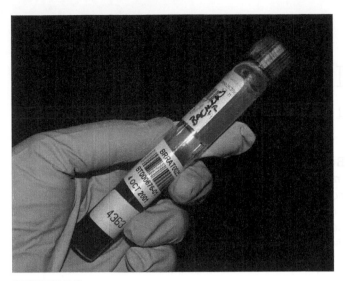

**FIGURE 2.1**

A cerebrospinal fluid culture from the index case contained *B. anthracis.*

was the index case, that is, first recognized case. There were actually several undiagnosed cases prior to the recognition of Mr. Stevens (e.g., Ernesto Blanco and a few in New York City). (ii) The epi-investigation and hospital surveillance identified Ernesto Blanco who was hospitalized with a severe respiratory infection. Mr. Blanco worked in the mailroom at AMI and further sampling revealed spores in that mailroom.] Following this incident, letters containing spores (Figure 2.2) were identified that had been sent to media outlets (NBC to Tom Brokaw and the *New York Post*) on September 18, 2001, and two U.S. congressional offices on October 9, 2001 (Senators Tom Daschle and Patrick Leahy; the letter to Senator Leahy was intercepted before delivery, after mail delivery was stopped in Washington, DC). The subsequent discovery of these *B. anthracis* spore-laden letters changed the status from a covert attack to an overt attack. Furthermore, it confirmed that a bioterrorist attack had occurred and might still be under way, which resulted in the first major bioterrorism investigation. Additionally, the fact that dissemination of a bioweapon did not require sophisticated technology changed our view regarding our nation's security system; the U.S. mail provided a simple mechanism to expose people to a deadly pathogen.

The government and the public were shocked and surprised at the attack both from our vulnerability and from our lack of preparedness (from a forensic perspective) to investigate this bioterrorist attack. The Federal Bureau of Investigation (FBI) had the responsibility to investigate this crime. For the greater part of the first decade of the 21st century, the FBI major case 184 investigation (also known as the "Amerithrax" investigation) was carried out by a minimum of 17 special agents and 10 U.S. postal inspectors, entailed more than 9100 interviews, more than 70 searches, and involved the cooperation of foreign governments. The intelligence from these investigations would be combined with forensic science evidence to help identify the perpetrator(s) of the anthrax attack. However, the FBI did not have in its forensic toolbox any validated analytical tools, let alone research assays, to forensically characterize the evidence for clues in order to build viable investigative leads to identify the perpetrator(s) of such a heinous act. The government would have to rely, and rightly so, on the country's assets and scientific prowess to pry out the forensic clues necessary to characterize microbial evidence for attribution (4).

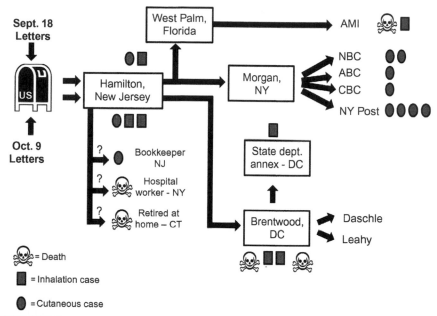

**FIGURE 2.2**

Tracking the anthrax letters. Anthrax spore–laden letters were mailed on September 18 and October 9 from Princeton, New Jersey, through the Hamilton, New Jersey, postal facility (19). The September 18 letters subsequently passed through the West Palm, Florida, and the Morgan, New York, postal facilities before arriving at their ultimate destinations. The October 9 letters also passed through the Hamilton, New Jersey, and the Brentwood facility before arriving at the Hart Senate office building. At least 22 individuals were infected and demonstrated symptoms of anthrax. Eleven of these were cases of inhalational anthrax, while the remaining 11 were cutaneous infections. Five of the individuals with inhalational anthrax died. Anthrax spores identified as the Ames strain were found at multiple locations along these routes. It is speculated that the deaths in New Jersey, New York, and Connecticut resulted from secondary spore contamination of letters in the Hamilton, New Jersey, facility that later came into contact with the victims; likewise for the State Department case.

The scientific team comprised representatives from the FBI, the U.S. Postal Service, the Department of Defense, the Centers for Disease Control and Prevention (CDC), the National Institutes of Health, the National Science Foundation, the Department of Homeland Security, the National Laboratories, and other government agencies, academia, and industry (academia and industry would contribute substantially to the genetic characterization of the evidence). Their talents were pooled in an attempt to characterize the forensic evidence. Microbial forensics employs detailed characterization assays to identify clues to the origin of a pathogen or toxin and/or its preparation for use in a criminal act (i.e., attribution). Analyses can entail microbiological analyses (e.g., culture), physical analyses, chemical analyses, and molecular biological analyses of microbial evidence, as well as analysis of traditional forensic evidence (e.g., fibers, fingerprints, handwriting, and human DNA). The FBI's forensic investigation pursued several different analytical avenues, ranging from physical and chemical to molecular approaches.

The microbial forensics of the Amerithrax investigation relied heavily on genetics and comparative genomics to provide invaluable investigative leads, which suggested that (i) the strain of *B. anthracis* used in the attack was more likely obtained from a laboratory source than from the environment and (ii) that a *B. anthracis* spore preparation known as RMR1029 at the U.S. Army Medical Research Institute for Infectious Diseases (USAMRIID), Fort Detrick, Maryland, was a potential source (5) or at least a direct lineage source from which the spores in the letters originated.

The DNA-based assays for microbial identification, characterization, and attribution in themselves were not novel and had been applied widely in many scientific endeavors, including forensic analyses. However, identification of specific evolutionary and individualizing genetic markers was not trivial and very demanding, particularly back in 2001. The technologies and methodologies were still nascent, laborious, time-consuming, and costly. There was little expectation that the level of individualization enjoyed for human DNA analyses would be possible when applied to microbial genetic evidence (4). The vast numbers of microorganisms, their complex biological and ecological diversities, and their capacity to mutate and evolve rapidly complicates analyses and interpretation of evidence in ways that do not impact human DNA forensics. There is substantial uncertainty about the microbial world. However, any reduction in possible sources of the spores in the letters was deemed helpful in eliminating unlikely leads and any signature markers could provide possible leads. Microbial forensics works hand-and-hand with traditional police investigative work.

## IDENTIFYING THE ATTACK STRAIN

The identity of the *B. anthracis* strain involved in the attacks was a critical piece of evidence that was available very early in the criminal investigation (6,7). The occurrence of an inhalational anthrax case was very suspicious in the wake of 9/11, but there was little direct evidence of a crime in the first few days of October 2001. While suspicious and unusual, it was possible that this first case (i.e., index case) was a naturally occurring infection, of public health concern but not a crime. However, the *B. anthracis* cultured from the cerebral spinal fluid (Figure 2.1) was identified quickly (October 5, 2001) as the Ames strain concurrently at both Northern Arizona University (NAU) and the CDC in Atlanta (6,7). The Ames strain was common in many research laboratories where it was used as a vaccine challenge strain due to its high virulence. Involvement of a laboratory strain in the index case reinforced suspicions that this was a nefarious event and not a case of naturally acquired anthrax.

However, the strength of this conclusion was difficult to gauge because of sparse knowledge about the Ames strain and a lack of knowledge about natural *B. anthracis* populations and about the discriminatory power of the genetic markers used to identify and differentiate *B. anthracis* strains.

*Bacillus anthracis* is a global pathogen but it is highly genetically homogeneous and methods to discriminate among isolates had little success until molecular genetic markers were applied in greater numbers (8). The high resolution of multiple locus variable number tandem repeat (VNTR) analysis (MLVA) was first developed for *B. anthracis* (9,10) and *Mycobacterium tuberculosis* (11) (driven by the lack of success of other approaches on highly homogeneous pathogens). MLVA was modeled after similar approaches long established in eukaryotic genetics, including the short tandem repeats used by forensic crime laboratories for human identification. By October 2001, an eight-locus system (MLVA8) had been developed (9,10), published along with a large database (10) and was being used actively at both the CDC and NAU. The allelic profiles between the isolate from the index case and the Ames strain were consistent at all eight loci; however, it was not a unique profile as there was a Texas goat isolate from 1997 with the same MLVA8 genotype. For forensic analysis and interpretation, a higher resolution approach was necessary for unambiguous identification of the strain used in the attack. At the time, whole genome sequencing was the only method able to provide such a high level of resolution.

## WHOLE GENOME SEQUENCING OF THE "FLORIDA" AMES STRAIN

In 2001, whole genome sequencing was a slow and expensive process, but the importance of the strain identification was high. Thus, a project was initiated quickly at the Institute for Genomic Research (TIGR) in Rockville, Maryland. Ironically, the genome sequence of the isolate from the index case, also known as the Florida Ames strain, which was completed in early 2002 (7), provided little investigative value as there was no genome database for comparison. At the time, the only other *B. anthracis* genome available was that of an attenuated strain, the Porton Down Ames isolate, which was partially completed and was found to be highly similar to that of the Florida Ames strain. This genomic comparison identified a few single-nucleotide polymorphisms (SNPs) and indels, but they were of little value because the Porton Down strain had been cured of its plasmids by mutagenic treatments and the differences could have been the results of this treatment. Subsequently, the whole genome sequence of *B. anthracis* Ames Ancestor (12) demonstrated that these differences provided no investigative value when compared with the Porton Down sample; they were proven to be unique to the Porton Down strain and, thus, provided no insights into the origin of the Florida Ames strain, and the Florida Ames strain showed no differences from the ancestral Ames isolate.

## DEVELOPMENT OF AMES-SPECIFIC ASSAYS

While early genome sequences provided no investigative leads, they were used to develop highly robust assays for SNP-based markers for the Ames strain (13–15). As mentioned before, the MLVA8 subtyping system was not 100% specific and even the addition of more VNTR markers (MLVA15) (16) improved resolution only slightly. In addition, the MLVA was a multiplex polymerase chain reaction (PCR) system that involved several manipulations and acrylamide gel or capillary electrophoresis that was tedious to perform. The resulting fragment sizes required careful scoring by an experienced team to assign alleles. Subsequently , the Ames strain genome was compared to sequences of other strains (e.g., Sterne) to identify 32 SNP marker alleles that might be unique to the Ames strain (14). These potential private alleles (i.e., autapomorphic characters) separating the genomes of these two closely related strains were converted into real-time PCR assays and then validated against a larger strain panel to verify their ability to identify the Ames strain. Many of the 32 SNPs proved to be less than 100% specific, but four in particular had alleles unique to all Ames strains within the data set tested (14). A globally representative panel of 88 strains was genotyped (14) to validate the "Ames specificity." Even when a panel of isolates from Texas, the origin of the Ames strain, was examined (15), these four SNP markers were able to differentiate the laboratory Ames strain from any other strain.

Real-time PCR SNP assays for the Ames strain were less expensive, faster, and had greater specificity than the MLVA15 system. This result was expected, as SNPs are far more stable markers than VNTRs. Along with their allele specificity, these assays were also subjected to multiple other challenges to test their robustness, sensitivity, and fidelity. Several of these assays proved capable of correctly genotyping from single DNA molecules (14). When inhibitors such as melanin and humic acid were added to the reactions, inhibition did occur and the sensitivity was affected, but they did not affect specificity (17). Likewise, an increased or decreased $Mg^{2+}$ concentration did not affect genotyping fidelity (17). Thousands of single-molecule reactions were performed to test for stochastic misgenotyping events, but none was observed (14,17). Altering the allele-specific probes was the only factor that would generate a false genotyping result, necessitating standard controls for Ames and non-Ames alleles. Extensive validation and ease of use were great assets to the investigation due to the large number of samples collected.

## THE FBI REPOSITORY

A reference population is critical to any match or nonmatch between samples. Soon after the event, the FBI began collecting information from anthrax research laboratories concerning their collections of strains. This was done under subpoenas requesting detailed inventories of strain collections. In 2001, the Select Agent regulation (42 C.F.R. §§72.6, 72.7) regulated possession of these agents but did not require individual laboratories to declare strain inventories to the CDC. Because there was no central database of *B. anthracis* strain inventories, the only way to identify entities with the Ames strain was through traditional investigative approaches followed by subpoenas for detailed lists. In the end, about 16 laboratories in the United States were identified as working with and/or housing the Ames strain. These entities were subsequently served with subpoenas to provide live cultures. The subpoena instructions were to sample each Ames strain isolate in their collection in a specific manner. The instructions specified generous portions of materials to allow for repeated and different analyses to be performed. This was done in duplicate for every sample and shipped to the FBI-directed laboratory effort at USAMRIID. One replicate from each submission was subsequently sent to NAU to confirm the identity of the Ames strain. In all, the 1077 samples collected and analyzed in such a fashion may have represented every single culture of the Ames strain in the United States. Although it cannot be verified that this repository was 100% complete, it certainly was the most comprehensive collection possible and reasonable assumptions could be drawn from analysis of these samples.

## MORPHOLOGICAL VARIANTS

Morphological variation was observed among colonies grown from spores found in the anthrax letters early in 2002. Handling and culturing of these spores were carried out at USAMRIID by experienced microbiologists. It was fortuitous that these staff members were very experienced at spotting and characterizing morphological variants of *B. anthracis* colonies. As the spores were being cultured, they observed morphological variants similar to what had been characterized previously at USAMRIID (18). The color of the colonies differed from the gray/white appearance of the wild-type Ames strain toward yellow and yellow/gray. Some variant colonies also had a more spreading and flat morphology than the wild-type Ames, with concentric rings of growth. In another variant, the colonies were more opaque and shiny, with a very compact shape. Perhaps most importantly was that these variants exhibited altered sporulation phenotypes. All of the anthrax-letter variants studied were poorly sporogenic when compared to the wild-type Ames ancestor. This characteristic is referred to as an *oligosporogenic phenotype*, which has secondary effects upon other phenotypes, including colony morphology (18). In all, four morphological variants were purified, studied extensively, and used investigatively to eliminate potential sources of the letter spores (Figure 2.3).

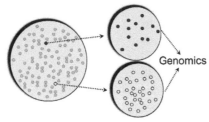

Genomics

**FIGURE 2.3**

Colony morphology. When spores were plated from the attack letters, morphological variants were observed at low frequencies. A subculture of these variants demonstrated stability of the morphological variants and allowed for isolation of DNA specific to the variants for whole genome sequencing.

The morphological variants were purified by USAMRIID staff and transferred to NAU for DNA extraction and then the genome was sequenced at TIGR. The whole genome sequences were compared to a very high-quality genomic sequence of the Ames ancestor strain (12). The Ames ancestor was the earliest known archived culture (May 1981) of the Ames strain and is believed to represent the original stock from which all other laboratory stocks were derived. Importantly, this sequence was identical to the Florida Ames isolate. Indeed, when sequences from the letter isolates were generated, they were all identical to the Ames ancestor. In contrast, the genome sequences from the morphological variants each contained one minor difference, likely to be the basis of the phenotypic variations. This variation included SNPs, indels, and large duplications. In three of the four cases, this variation was in or near genes involved in sporulation: phosphorylation of spoOF/spoOA, dephosphorylation of spoOF, and near the *spoOF* gene itself. Sequence differences were not limited to the chromosome as in one case the mutation was found on the plasmid pXO1. Critically, these genomic variations were amenable to the development of PCR-based assays.

The FBI contracted with both commercial and nonprofit laboratories to develop highly sensitive, specific, and quantitative PCR assays to detect the variants. These laboratories included TIGR, Midwest Research Institute, and Commonwealth Biotechnologies Inc. After extensive validation to establish protocols applicable to forensic analyses, development of reference standards, and demonstrated staff proficiency, assays were performed on DNA from the 1077 FBI repository samples. All analyses were done blindly and independently by the different contractors. Data interpretation and conclusions were then made by FBI scientists, independent of the performance laboratories. All DNAs extracted from spores recovered from the attack letters contained signatures for all four variants. Critically, 8 of the repository samples contained signatures for all four of the morphological variants. These 8 samples were all derived from the spore stock at USAMRIID known as RMR1029 and included RMR1029 itself.

RMR1029 was an unusual type of spore stock for many laboratories but not for USAMRIID where vaccine challenge trials required large quantities of stabilized spores (5). RMR1029 was a pooled and concentrated liquid spore suspension derived from twelve 10-liter fermentation batches and an additional twenty smaller batch cultures, a total of 164 liters of culture concentrated by centrifugation and resuspended to a 1-liter volume. The use of multiple cultures, which was intended to produce a large quantity of spores, apparently gave ample opportunity for variants to arise and become a stable portion of the total spore population. The spores were stored with phenol as a stabilizing agent, which would kill any vegetative cells or nonspore-forming bacteria. This stock was used as an inoculum for additional amplification culturing. Physical characterization of the letter spores supported the hypothesis that the actual spores found in the letters resulted from a subculture of RMR1029 and not from this stock directly.

## CONCLUSION AND REMAINING ISSUES

The variant compositional match to RMR1029 narrowed the Amerithrax investigation to a small number of suspect samples. However, the microbial forensic evidences did not restrict the possibilities to a single person or group of persons. It only showed that the only samples displaying all four variant signatures in the entire repository were derived from RMR1029 or RMR1029 itself. Thus, all other sources of Ames isolates were considered unlikely sources of the letter spores. Much forensic analysis is based on neutral models that allow for exclusionary probabilities to be calculated. It is possible that these morphological variants had a selective advantage under the growth conditions in the fermenter or in batch culture. If so, calculating match statistics is problematic, as random assumptions would be violated. However, it is common for observational evidence to be admitted into court that does not have probabilistic characterization. The extensive strain repository built as a part of the case would certainly have been supportive of an association between RMR1029 and the spores found in the letters.

The Amerithrax case investigation and the microbial forensic analysis mirrored technological developments occurring in genomics during this time period. Many more questions could be answered today in a relatively rapid and economical fashion because of advances in technology, such as next-generation sequencing. The morphological variants were present at low concentration in the samples, and standard DNA sequencing in 2001 would have only covered each genome to 8–12× coverage, not nearly enough to identify variants comprising 1% or less of the sample. Now because of much higher coverage, decreased costs and >150 gigabase sequencing throughput, it is conceivable that next-generation sequencing would have been a valuable tool to directly detect low-level components of mixtures. Indeed, it might have avoided any need for assay development, yielded higher quality data, and reduced potential error (predominantly false-negative data). Even though the capabilities were limited in 2001, the Amerithrax investigation was groundbreaking and pioneered new approaches to the investigation of microbial-based crimes. Future microbial investigations will doubtless capitalize upon advance genomics earlier, although the development of comprehensive strain archives and databases will still be a laborious but essential evidentiary resource.

# REFERENCES

[1] L.M. Bush, B.H. Abrams, A. Beall, C.C. Johnson, Index case of fatal inhalational anthrax due to bioterrorism in the United States, N. Engl. J. Med. 345 (2001) 1607–1610.

[2] M.S. Traeger, S.T. Wiersma, N.E. Rosenstein, J.M. Malecki, C.W. Shepard, P.L. Raghunathan, et al., Florida Investigation Team. First case of bioterrorism-related inhalational anthrax in the United States, Palm Beach County, Florida, 2001, Emerg. Infect. Dis. 8 (2002) 1029–1034.

[3] S.C. Suffin, W.H. Carnes, A.F. Kaufmann, Inhalation anthrax in a home craftsman, Hum. Pathol. 9 (1978) 594–597.

[4] P. Keim, Microbial Forensics: A Scientific Assessment, American Society for Microbiology, Washington, DC, 2003.

[5] FBI. FBI Amerithrax website. http://www.fbi.gov/anthrax/amerithraxlinks.htm.

[6] A.R. Hoffmaster, C.C. Fitzgerald, E. Ribot, L.W. Mayer, T. Popovic, Molecular subtyping of Bacillus anthracis and the 2001 bioterrorism-associated anthrax outbreak, United States, Emerg. Infect. Dis. 8 (2002) 1111–1116.

[7] T.D. Read, S.L. Salzberg, M. Pop, M. Shumway, L. Umayam, L.X. Jiang, et al., Comparative genome sequencing for discovery of novel polymorphisms in Bacillus anthracis, Science 296 (2002) 2028–2033.

[8] P. Keim, A. Kalif, J. Schupp, K. Hill, S.E. Travis, K. Richmond, et al., Molecular evolution and diversity in Bacillus anthracis as detected by amplified fragment length polymorphism markers, J. Bacteriol. 179 (1997) 818–824.

[9] P. Keim, A.M. Klevytska, L.B. Price, J.M. Schupp, G. Zinser, K.L. Smith, et al., Molecular diversity in Bacillus anthracis, J. Appl. Microbiol. 87 (1999) 215–217.

[10] P. Keim, L.B. Price, A.M. Klevytska, K.L. Smith, J.M. Schupp, R. Okinaka, et al., Multiple-locus variable-number tandem repeat analysis reveals genetic relationships within Bacillus anthracis, J. Bacteriol. 182 (2000) 2928–2936.

[11] R. Frothingham, W.A. Meeker-O'Connell, Genetic diversity in the Mycobacterium tuberculosis complex based on variable numbers of tandem DNA repeats, Microbiology 144 (1998) 1189–1196.

[12] J. Ravel, L. Jiang, S.T. Stanley, M.R. Wilson, R.S. Decker, T.D. Read, et al., The complete genome sequence of Bacillus anthracis Ames "Ancestor", J. Bacteriol. 191 (2009) 445–446.

[13] T. Pearson, J.D. Busch, J. Ravel, T.D. Read, S.D. Rhoton, J.M. U'Ren, et al., Phylogenetic discovery bias in Bacillus anthracis using single-nucleotide polymorphisms from whole-genome sequencing, Proc. Natl. Acad. Sci. USA 101 (2004) 13536–13541.

[14] M.N. Van Ert, W.R. Easterday, T.S. Simonson, J.M. U'Ren, T. Pearson, L.J. Kenefic, et al., Strain-specific single-nucleotide polymorphism assays for the Bacillus anthracis Ames strain, J. Clin. Microbiol. 45 (2007) 47–53.

[15] L.J. Kenefic, T. Pearson, R.T. Okinaka, W.K. Chung, T. Max, C.P. Trim, et al., Texas isolates closely related to Bacillus anthracis Ames, Emerg. Infect. Dis. 14 (2008) 1494–1496.

[16] M.N. Van Ert, W.R. Easterday, L.Y. Huynh, R.T. Okinaka, M.E. Hugh-Jones, J. Ravel, et al., Global genetic population structure of Bacillus anthracis, Plos ONE (2007) 2.

[17] J. Beaudry, P. Keim, Unpublished data, 2010.

[18] P.L. Worsham, M.R. Sowers, Isolation of an asporogenic (*spo*OA) protective antigen-producing strain of *Bacillus anthracis*, Can. J. Microbiol. 45 (1999) 1–8.

[19] J.A. Jernigan, D.S. Stephens, D.A. Ashford, C. Omenaca, M.S. Topiel, M. Galbraith, et al., Anthrax Bioterrorism Investigation Team. Bioterrorism-related inhalational anthrax: The first 10 cases reported in the United States, Emerg. Infect. Dis. 7 (2001) 933–944.

SECTION

B

# Civilian

# CHAPTER 3

# Food-Borne Outbreaks: What's New, What's Not, and Where Do We Go from Here?

**Thomas A. Cebula,**[a,b] **W. Florian Fricke,**[b] **and Jacques Ravel**[b]

[a]*Johns Hopkins University, Baltimore, Maryland*
[b]*Institute for Genome Sciences, University of Maryland School of Medicine, Baltimore, Maryland*

> Things are not always what they seem; the first appearance deceives many; the intelligence of a few perceives what has been carefully hidden....
>
> *Phaedrus* (1)

## PROLOGUE

Appearances sometimes cloud an incident, and circumstances surrounding the event can lead one to plausible, but erroneous, conclusions. Consider Claudius I, the fourth Emperor of Rome, who succumbed on October 13, 54 AD, after ingesting "poisoned" mushrooms (presumably) at his own dinner table. Historians relating the "facts" of this incident have concluded that dastardly deeds led to the Emperor's death, although a more recent account suggests that an innocent, but deadly, ingestion of a single *Amanita phalloides*, a most poisonous mushroom known as death cap, could have, just as likely, caused his death (2). Was Claudius' death a case of intentional, criminal food poisoning—a plot among Agrippina, Claudius' fourth wife, Halotus, his food taster, and perhaps Xenophon, his attending physician—to assassinate the emperor and secure the throne for Agrippina's son Nero or was it Claudius' voracious appetite for exotic foods, which led him to the death cap and his premature death (2)?

Now weigh the 751 cases of human salmonellosis in The Dalles, Oregon, in September and October of 1984, traced to bacterial contamination of several salad bars in the town (3). It was most unfortunate and untimely for the citizenship of The Dalles to have a pathogen such as *Salmonella enterica* subspecies I, serovar Typhimurium (*S. typhimurium*) taint their salads, especially at a time when an important local election was looming. Was this but another example of an accidental contamination caused by negligent food handling

Microbial Forensics. DOI: 10.1016/B978-0-12-382006-8.00003-7

or was this to be the first bioterrorism event to be recorded in modern U.S. history? Buoyed by a tardy but full confession more than a year after the incident, the public learned that the illnesses were undeniably due to willful and intentional contamination, perpetrated by the religious cult known as the Rajneeshees (4). Although prompted by different agents, the two incidents should remind the reader that what was true in 54 AD remains true today—when food is consumed "I, Claudius" and "I, John Q. Public" may be the ultimate "canaries in the coal mine."

Food-borne illness, one of humankind's oldest recognized maladies, can be caused by any of more than 250 microbiological, physical, and chemical agents (5). Illness can be the result of intoxication by a toxin contained naturally within a food or beverage, for example, the potent RNA polymerase II inhibitor $\alpha$-amanitin (6,7) in the incident that led to Claudius' death (2). As illness was limited only to Claudius, this would be regarded today as a sporadic case of food-borne illness. So too illness can occur by infection caused by ingestion of a microbial contaminant in food such as *S. typhimurium* in The Dalles episode (3,4). Here, a cluster of illnesses was traced to a common source, and thus this was deemed a food-borne outbreak, as the U.S. Centers for Disease Control and Prevention (CDC) defines an outbreak of food-borne illness as "a cluster of two or more infections caused by the same agent (pathogen or toxin), which upon investigation are linked to the same food" (8). The examples just cited reveal the many nuances and intricacies that may challenge public health and law enforcement professionals in discerning between an accidental and an intentional contamination of our food supply.

Herein, the discussion centers on food-borne illnesses arising because of contamination of a foodstuff with a bacterial pathogen. In particular, the dialogue focuses on *Salmonella enterica* and *Escherichia coli* O157:H7, zoonotic pathogens that were responsible for several recent, accidental, and large-scale food-borne outbreaks in the United States.

## FOOD-BORNE ILLNESS: AN OLD BUT PERSISTENT PROBLEM

Throughout history, humankind has had a fascination and desire to consume novel and exotic foods. With that predilection came the potential risk of ingesting substances that could cause illness or death, but over time—by proscription, prescription, and trial and error—foods were chosen for consumption, and processes were developed to ensure, with some certainty, their wholesomeness and safety. In more recent times, sound scientific and regulatory principles have combined so that today the United States enjoys one of the safest food supplies. Even so, food-borne illness still presents as a major food safety challenge in the United States where annually still more than 76 million cases occur (8–10).

Framing this statistic in a more personal way, if you reside in the United States today, you will probably experience at least one food-borne illness episode within the next 4 years. While you are likely to recover without any long-term effects, about 325,000 individuals will suffer acute effects severe enough to be hospitalized, some will be plagued with long-term complications (11,12), and about 5200 U.S. consumers will die each year because of food-borne illness (13). The morbidity and mortality tragedies also wield a significant economic burden—estimated to be in the range of $40 to $100 billion each year (14)—but the true societal costs are much higher when the rippling effects to the consumer, industry, and government are assessed properly. These factors, coupled with the countless opportunities for microbial contamination to occur as food is transported and transformed from "farm to fork," illustrate why food-borne pathogens are formidable public health concerns.

## THE CHANGING LANDSCAPE OF FOOD-BORNE OUTBREAKS

A half century ago, when food distribution was more limited, a typical food-borne outbreak tended to be local, a focus of cases linked to a picnic, party, and other family or social events. Within such intimate settings, illnesses were diagnosed and linked more readily, making for prompt tracing of cause. The root of these outbreaks usually was mishandling of a particular food at or near the consumption endpoint (10,15). Although such outbreaks still occur today, foods and food products have expanded well beyond a local environment, where food production, food processing, and food transportation are no longer cottage industries. Large concentrated production and processing areas have changed the complexion of the types of outbreaks now being reported.

A shifting U.S. diet over the past 50 years, one now filled with more fruits and vegetables, also has had an impact on the kinds of foods being implicated and the types of outbreaks now occurring. That is, this dietary shift has been accompanied by a dramatic increase in the number of food-borne outbreaks traced to raw produce consumption. Whereas, for example, contamination of fruits and vegetables accounted for less than 1% of the outbreaks in the 1970s, they were responsible for 13% of the total outbreaks by 2007 (16). The U.S. Food and Drug Administration (FDA), in issuing industry guidance, states that one-fourth of the 72 food-borne outbreaks associated with the consumption of fresh produce in the years 1996 to 2006 were attributed to fresh-cut leafy greens, a statistic reinforcing the fact that food vehicles requiring minimal processing before consumption are especially prone to contamination (17).

Moreover, a global food supply now sates our penchant for year-round availability of fresh fruits and vegetables; our cosmopolitan palate; and our desire to buy wholesome foods economically (18). Today, a travel time of about 48 hours or less is all one needs to sample new foods in even the most remote parts of the world, and those foods likewise can make a similar-timed trek to our borders to satisfy the U.S. appetite. This too complicates today's outbreak investigations. That is, in reporting recent outbreaks linked to *E. coli* O157:H7 contamination of ground beef and the ensuing recalls of millions of pounds of beef and beef products, the popular press has reminded the consumer that hamburgers purchased at their local supermarket might be an amalgam of beef products derived from different states or countries (19). Even a food product such as bread made and baked in the United States may contain key ingredients from perhaps as many as 17 countries (20). All of these factors thus contribute to the anatomy of today's outbreaks, ones that are diffuse, dispersed widely geographically, and linked to food vehicles and ingredients that carry a low pathogen load likely introduced at a much earlier step in the food production cycle (21). Unlike a localized cluster of cases, which might elicit an immediate response, a multistate diffuse outbreak of illnesses may take weeks even to be recognized as a food-borne incident (18,21).

For example, in addition to the beef incidents already mentioned, *E. coli* O157:H7 has been responsible for several widely publicized food-borne outbreaks and recalls of food commodities. Between August 19 and October 6 of 2006, *E. coli* O157:H7, associated with consumption of contaminated bagged baby spinach, sickened 205 individuals in 26 states. Approximately 50% of the ill were hospitalized, 31 individuals subsequently developed hemolytic uremic syndrome (HUS), and three people died (22–24). Then, between November 20 and December 6, 2006, 71 *E. coli* O157:H7 cases were associated with eating at Taco Bell restaurants in five northeastern states and resulted in about 75% of the afflicted being hospitalized and 9% of the cases advancing to HUS (25,26). Within about the same time frame, a separate outbreak was associated with eating at Taco John restaurants in two Midwest states and accounted for 81 additional illnesses and 26 hospitalizations (25,27). Both of these outbreaks were traced to contamination of fresh-cut California-grown iceberg lettuce, although the fresh-cut lettuce supplied to each of the restaurant chains came from distinct growers and suppliers (24,26,27).

Three recent outbreaks of salmonellosis also are worth noting. Over a five-and-a-half-month period, from August 1, 2006, to February 16, 2007, more than 420 illnesses in 44 states, resulting in about one in five of the ill individuals being hospitalized, were traced to two brands of peanut butter manufactured in a single facility that was contaminated with *Salmonella Tennessee* (28). Then, in April 2008, an outbreak was identified when illnesses in Texas and New Mexico were traced to the same strain of *S. Saintpaul* (FDA, 2008, June 3) (29). Although tomatoes were originally suspected, the outbreak continued through the summer, such that about 1440 reported illnesses in 43 states, the District of Columbia, and Canada were attributed to *S. Saintpaul* by the end of August 2008 (30,31). The eventual food vehicle for this outbreak was determined to be contaminated Mexican jalapeno and, perhaps, serrano peppers, and the source of *S. Saintpaul* was traced to contaminated water used in irrigation (31).

Illnesses that occurred in late 2008 and the early months of 2009 attributed to peanut paste and peanut butter contaminated with *S. typhimurium* (32,33) in one facility serve as further evidence of the changing landscape of food-borne outbreaks. Whereas this processing plant was estimated to supply less than 2% of the total peanuts and peanut products in the United States (34), foods containing the tainted peanut product were linked epidemiologically to 714 illnesses and nine deaths in 46 states (33). The ensuing class I recall by the FDA impacted over 3200 food products produced by more than 300 companies in the United States and at least 20 other countries (34,35). Revelations such as these have heightened concerns about accidental and deliberate contamination of the food supply, prompting a *caveat cenans* (36) anxiety among consumers.

Yet, "it must have been something I ate" is still the all too often repeated refrain in the overwhelming numbers of food-borne illness cases occurring in the United States each year. As infections with enteric pathogens such as *S. enterica* or *E. coli* O157:H7 usually are self-limiting, they simply are not reported and therefore can go largely unnoticed (9,10). It is estimated, for example, that only about 12% of the afflicted seek medical attention, and when they do only about one in five is asked to provide a stool sample for laboratory follow-up (21). Even assuming 100% compliance with physicians' requests, this implies that only about 2.5% of the estimated cases of food-related microbial infections reach the attention of public health laboratories (21), an obvious requisite for proper identification and reporting of food-borne illness cases. Thus, just as a high number of sporadic cases are not diagnosed (10,15), so too a number of food-borne outbreaks, especially those involving smaller numbers, stand a good chance of not being noticed. That is, with a 2.5% incidence of reporting and assuming a binomial distribution, the probability of linking two or more cases in a given outbreak involving 50 individuals is only 36%. If one were to wait to find a cluster of three or more, or five or more, cases from this outbreak, the probability diminishes to 13% and to about 1%, respectively.

A key first step in delimiting risk of a food-borne outbreak, be it caused by accident or design, is timely recognition that one indeed is occurring.

Obviously, in recognizing an outbreak, time is of the essence if large numbers of illnesses are to be averted. The need for a coordinated effort of active surveillance of food-borne illnesses across the United States is thus obvious, and these monitoring systems indeed are in place.

FoodNet (the Foodborne Diseases Active Surveillance Network), for example, is a shared enterprise of the CDC, the FDA's Centers for Food Safety and Applied Nutrition and Veterinary Medicine, particular state health departments, and the U.S. Department of Agriculture's Food Safety and Inspection Service that monitors the trends of particular food-borne pathogens and the types of foods that caused illness over time (37,38). PulseNet is a similarly comprised cross-agency, CDC-coordinated group of laboratories that performs standardized molecular subtyping of enteric pathogens derived from the follow-up of clinical cases (39,40). These two consortia working with CDC's OutbreakNet team, a coalition of public health officials and epidemiologists at the local, state, and federal level (8), provide a commanding surveillance and response network for averting large outbreaks of food-borne illnesses.

Although these systems have proved useful in identifying clusters of hemorrhagic *E. coli* and salmonellosis cases, it must be recognized that they are not real-time reporting systems. Even with these systems in place then, as the aforementioned cases exemplify, an active outbreak might be ongoing for some time, perhaps even have ended, before individual cases are linked to that outbreak (18). Consequently, ample opportunity exists for the contaminated food to be exhausted or discarded and therefore lost to public health or criminal trace-back investigations. This may well translate to the failure of adequate public health measures being put into place or, in a criminal investigation, to the loss of critical evidence of probative value. Indeed, in less than half of the food-borne outbreaks of human illnesses occurring in the United States has a pathogen been identified and a definitive food been implicated (40,41). Thus, augmenting these monitoring systems with real-time reporting resources such as BioSense, Internet-based infectious disease surveillance, and avant-garde groups such as Minnesota's "Team D" can only fortify our ability to recognize an outbreak sooner and thus enhance trace-back investigations into source and cause (31,42).

## ENTERIC PATHOGENS AS BIOTERRORISM AGENTS

Bioterrorism is defined as the use of a biological organism or its toxin as a weapon to induce fear, violence, or intimidation in order to achieve a desired end. A major concern in our post–September 11, 2001, world is that the food supply might become the target for a bioterrorist attack (4,43–46). Both *Salmonella enterica* and *E. coli* O157:H7 are considered Category B bioterrorism agents because they can be obtained without much difficulty, grown and manipulated readily, and disseminated easily (44,46). Such properties have made enteric pathogens agents of choice for biocrimes in the past (4,44). Moreover, because of the inherent diversity of these enteric pathogens (45,46), unless scrupulous attention is paid to sample collection, sample handling, and pathogen isolation (47), it becomes a daunting task to fingerprint and distinguish between individual isolates within an expanding "clonal" population and individuals from nearly identical, but different, "clonal" strains.

Food-borne outbreaks attributed to these pathogens have caused considerable anxiety among U.S. consumers, industry, and government alike. If an intentional attack on our food supply were to occur, it would help subvert trust between and among people and their government, engendering mistrust in our food supply and in the people who produce, provide, and safeguard it. Against a backdrop of naturally occurring food-borne illnesses, detection of intentional, covert events might well be more protracted and thus lead to an increased number of people afflicted (18). As witnessed by the action and attention paid to the unintentional outbreaks summarized earlier, one can only imagine what surreptitious seedings of these pathogens into the food supply might provoke. The tragic outcomes from infections with these enteric agents of course are the paramount concern of a furtive attack on the food supply. Such attacks also would be quite disruptive to day-to-day activities, with sizable direct and indirect economic costs.

Consider the financial outlays precipitated by some of the recent unintentional food-borne outbreaks. The *E. coli* O157:H7 2006 spinach outbreak alone, for example, was estimated to have cost the leafy green industry more than $350 million, and it was reported that sales of packaged spinach were still off by about 20% from preoutbreak figures 1 year later (48). The *S. Saintpaul* jalapeno pepper outbreak, attributed first to tomatoes, negatively impacted the U.S. tomato growing industry, which reported losses of more than $200 million near the end of the outbreak investigation (49). Also, the *S. typhimurium* peanut paste and peanut butter outbreak is expected to affect economic losses in excess of one billion dollars in peanut-producing states alone, with losses to other industries such as restaurants, grocery stores, candy, ice cream, and other attendant small businesses as yet not calculated (50).

Because the link of two or more food-borne illnesses is likely to be made by public health investigators, it is important to weigh how similar the molecular epidemiological and microbial forensic investigations are in establishing strain attribution. Both clearly share common goals of identifying and recognizing the patterns of a particular outbreak and determining the pathogen responsible as quickly as possible. Vital to both, likewise, are the aims of containing the outbreak and communicating outbreak specifics to the public health community at large. Notably, trace-backs of cause and origin are crucial to both epidemiological and microbial forensic investigations, yet the molecular armamentarium employed to assign ultimate attribution is likely not to be the same (44).

Pulsed-field gel electrophoresis (PFGE) remains the gold standard for the CDC for DNA fingerprinting of microbes (39,40), although, at times, the CDC has used multilocus variable number tandem repeat analysis (MLVA) to augment a particular investigation (51–53). In PFGE, a rare base-cutting restriction enzyme such as *XbaI* or *BlnI* is used to cut genomic DNA into a limited number of fragments. The restricted DNA is then resolved through an agarose gel with an alternating current into 10–30 fragments ranging in size from about 30 to 800 kb (1 kb = 1000 bp). The resulting pattern of DNA fragments becomes the identifier or fingerprint for that microbial isolate. MLVA, however, takes advantage of small repeated sequences occurring at discrete loci within the bacterial genome. These are a rich source of genetic polymorphisms as the repeat region can expand or contract because of slipped strand mispairing (54,55). Isolates can be distinguished one from another by measuring in each the relative sizes of the polymerase chain reaction amplicons spanning these repeat regions. It is important to point out that whereas PFGE surveys the landscape of an entire chromosome, MLVA samples but a tiny fraction of a chromosome. Although it is not the intent to debate the relative merits of these techniques, it is imperative to emphasize that PFGE and MLVA, like any method, possess inherent strengths but also suffer inherent weaknesses (56).

For example, in the 2008–2009 *Salmonella* outbreak, *S. typhimurium* isolates yielded at least three similar, but distinct, PFGE patterns (33). Yet, they were linked as one outbreak strain based on other epidemiological evidence and the fact that MLVA yielded only one pattern for these PFGE types (33). In the 2006 spinach outbreak investigation, however, particular *E. coli* O157:H7 isolates, believed not to be part of the outbreak, although isolated at about the same time frame, yielded PFGE profiles indistinguishable from those of outbreak isolates. In this case, MLVA patterns of these suspected "outlier" cases of *E. coli* O157:H7 infection were distinct from those obtained for outbreak isolates (52).

Although these analyses and requisite follow-up clearly were sufficient to limit the public health impact of these outbreaks, they do raise pragmatic legal and scientific questions such as "how similar might two strains be, yet yield distinct PFGE or MLVA types" or "how different might two strains be, yet yield indistinguishable PFGE or MLVA types?" Do the methods effectively sample the diversity that exists within the microbial population being studied? Also, as these methods were developed to address epidemiological concerns, do they provide sufficient information about the strain or isolate that would allow law enforcement officials to ascertain the most probable source and would the conclusions withstand microbial forensic evidentiary proceedings (43,44)?

The three 2006 outbreaks of *E. coli* O157:H7 linked to fresh-cut produce provided an opportunity at least to attempt to address these questions. As the outbreaks occurred within a relatively narrow time frame, concern was raised early on that the three episodes were linked and might be due to the same strain. PFGE analyses, however, revealed that isolates obtained from patients and foods in the spinach- and the two lettuce-associated incidents were distinct, thus indicating three independent clusters of infections (23,25,26), although the PFGE patterns from the spinach and Taco Bell isolates were remarkably similar (57).

Employing whole genome DNA arrays (58) and optical mapping (59) as a triaging strategy, we identified strains within the spinach-associated outbreak that contained distinct genomic differences, differences later confirmed by whole genome sequencing of these strains. Indeed, whole genome analyses based on single nucleotide polymorphisms (SNPs) of their core genomes indicated that these strains were very closely related, although individually distinctive, phylogenetically. These data indicate that more than one *E. coli* O157:H7 genotype was at the heart of the spinach-associated outbreak (60). Whole genome sequencing also revealed that only about 25 SNP differences distinguished the core genomes of spinach-associated and Taco Bell lettuce-associated strains. Moreover, comparisons of Taco John lettuce-associated strains showed interesting genome rearrangements, with the core genome differing from the Taco Bell- and spinach-associated counterparts by well over 500 SNPs (60).

## EPILOGUE

The whole genome sequencing data summarized here emphasize that different microbial genotypes can be isolated from patients, food, and environmental samples collected from the same food-borne outbreak. These data, therefore, call into question whether single-colony isolation, a key first step in a public health outbreak investigation, might thwart or bias a forensic investigation for criminal attribution by underestimating the population diversity that exists among extant microbial populations.

The genomic era has spawned a plethora of DNA fingerprinting methods to type and distinguish bacterial strains and isolates, but, as pointed out by van Belkum and colleagues (56), they must be standardized, validated, and applied appropriately if they are to be useful.

Next-generation, non-Sanger sequencing technologies have increased the capacity for rapid genomic analyses, allowing for each individual base pair of the genome to be interrogated efficiently and economically. Today's technologies allow us to delve into whether particular members of the population are culled by selection pressures within an individual host; selected by the microbiologist because of particular morphological characteristics; or enriched by conditions of storage, handling, and/or growth. Sequencing information that speaks to the diversity of the extant population in question, rather than a single reference genome, will significantly enhance microbial forensic investigations.

It is thus important to question what methods will be used to underpin proper attribution; how bacterial diversity will be assessed; and what will be the common microbial forensic lexicon to discriminate contextually words such as *strain, variant,* and *isolate* and descriptors such as *rare, indistinguishable, most likely,* and *clonal.* If indeed the outbreaks discussed in this chapter had been intentional, answers to such questions certainly would loom large should a Daubert challenge (61) be satisfied, and a jury of peers decide whether strain attribution had been ascribed correctly.

# REFERENCES

[1] Phaedrus, in: H.T. Riley, H.G. Bohn (Eds.), The Comedies of Terence and the Fables of Phaedrus. York Street, Covent Garden, London, 1853.

[2] V. Grimm-Samuel, On the mushroom that deified the emperor Claudius, Class. Q. New Ser. 41 (1991) 178–182.

[3] T.J. Török, R.V. Tauxe, R.P. Wise, J.R. Livengood, R. Sokolow, S. Mauvais, et al., A large community outbreak of salmonellosis caused by intentional contamination of restaurant salad bars, JAMA 278 (1997) 389–396.

[4] W.S. Carus, Bioterrorism and biocrimes: The illicit use of biological agents since 1900. Working paper: National Defense University, 2001; 1–219 (December 19, 2009).

[5] F.L. Bryan, Diseases Transmitted by Foods: A Classification and Summary, 2nd ed., U.S. Department of Health and Human Services, Centers for Disease Control, Atlanta, GA, 1982 HHS publication No. (CDC) 84-8237.

[6] D.A. Bushnell, P. Cramer, R.D. Kornberg, Structural basis of transcription: α-Amanitin–RNA polymerase II cocrystal at 2.8 Å resolution, Proc. Natl. Acad. Sci. USA 99 (2002) 1218–1222.

[7] X.Q. Gong, Y.A. Nedialkov, Z.F. Burton, α-Amanitin blocks translocation by human RNA polymerase II, J. Biol. Chem. 279 (2004) 27422–27427.

[8] CDC OutbreakNet Team (December 19, 2009) http://www.cdc.gov/foodborneoutbreaks/.

[9] P.S. Mead, L. Slutsker, V. Dietz, L.F. McCaig, J.S. Bresee, C. Shapiro, et al., Food-related illness and death in the United States, Emerg. Infect. Dis. 5 (5) (1999) 607–625.

[10] R.V. Tauxe, Emerging foodborne diseases: an evolving public health challenge, Emerg. Infect. Dis. 3 (1997) 425–434.

[11] J.L. Smith, S.A. Palumbo, I. Walls, Relationship between foodborne bacterial pathogens and the reactive arthritis, J. Food Safety 13 (1993) 209–236.

[12] P. Schiellerup, K.A. Krogfelt, H. Locht, A comparison of self-reported joint symptoms following infection with different enteric pathogens: effect of HLA-B27, J. Rheumatol. 35 (2008) 480–487.

[13] Summary and assessment, in: Institute of Medicine's Workshop Summary Forum on Microbial Threats. Addressing foodborne threats to health: Policies, practices, and global coordination, National Academy Press, 2006, pp. 1–2 (December 1, 2009).

[14] C.S. DeWaal, Testimony before the U.S. Senate Committee on Health, Education, Labor, and Pensions, Washington, DC October 22, 2009 (December 10, 2009). Available from: cspinet.org/new/pdf/senate_help_testimony-_oct22.pdf.

[15] R.V. Tauxe, Emerging foodborne pathogens, Int. J. Food Microbiol. 78 (2002) 31–41.

[16] I.B. Hanning, J.D. Nutt, S.C. Ricke, Salmonellosis outbreaks in the United States due to fresh produce: Sources and potential intervention measures, Foodborne Pathogens Dis. 6 (2009) 635–648.

[17] U.S. Food and Drug Administration. Guidance for industry: Guide to minimize microbial food safety hazards of fresh-cut fruits and vegetables (February 15, 2009). Available from: http://www.fda.gov/Food/GuidanceComplianceRegulatoryInformation/ GuidanceDocuments/ProduceandPlanProducts/ucm064458.htm#ch1.

[18] M.T. Osterholm, The food supply and biodefense: The next frontier of the food safety agenda, in: Institute of Medicine's Workshop Summary Forum on Microbial Threats. Addressing Foodborne Threats to Health: Policies, Practices, and Global Coordination. Washington, DC: National Academy Press, 2006, pp. 32–42.

[19] M. Moss, E. Coli path shows flaws in beef inspection, 2009 (October 6, 2009). Available from: www.nytimes.com/2009/10/04/health/04meat.html?_r=1&scp=1&sq= o157moss&st=cse.

[20] A. Schoenfeld, A multinational loaf. New York Times. June 15, 2007 (December 10, 2009). Available from: http://www.nytimes.com/imagepages/2007/06/15/business/20070616_FOOD_ GRAPHIC.html.

[21] The Council to Improve Foodborne Outbreak Response (CIFOR). Guidelines for Foodborne Disease Outbreak Response, 2009, pp. 1–200.

[22] J. Grant, A.M. Wendeboe, A. Wendel, B. Jepson, P. Torres, C. Smelser, et al., Spinach-associated Escherichia coli O157:H7 outbreak, Utah and New Mexico, 2006, Emerg. Infect. Dis. 14 (10) (2008) 1633–1636.

[23] A.M. Wendel, D.H. Johnson, U. Sharapov, J. Grant, J.R. Archer, T. Monson, et al., Multistate outbreak of Escherichia coli O157:H7 infection associated with consumption of packaged spinach, August-September 2006: The Wisconsin investigation, Clin. Infect. Dis. 48 (2009) 1079–1086.

[24] C. Smith DeWaal, X.A. Tian, D. Plunkett, Outbreak alert! Analyzing foodborne outbreaks 1998 to 2007. Closing the gaps in our food-safety system. Center for Science in the Public Interest, 11th edn., December 2009, pp. 1–24.

[25] Centers for Disease Control and Prevention CDC health update. Multi-state outbreak of E-coli O157 infections, November–December 2006 (CDCHAN-00255-2006-12-14-UPD-N). Available from: http://www2a.cdc.gov/HAN/ArchiveSys/ViewMsgV.asp?AlertNum=00255.

[26] California Food Emergency Response Team. Environmental investigation of Escherichia coli O157:H7 outbreak associated with Taco Bell restaurants in northeastern states. Final report redacted November 19, 2007, pp. 1–17. Available from: http://www.dhs.ca.gov/fdb/local/ PDF/Taco_Bell_final_report_redacted_11_19_2007.pdf.

[27] California Food Emergency Response Team. Investigation of the Taco John's Escherichia coli O157:H7 outbreak associated with iceberg lettuce. Final report redacted, 15 February, 2008, pp. 1–41. Available from: http://www.dhs.ca.gov/fdb/local/PDF/TJreport_FINAL_021508_ Redacted_Compressed.pdf.

[28] Centers for Disease Control and Prevention, Multistate outbreak of Salmonella serotype Tennessee infections associated with peanut butter—United States, 2006–2007, MMWR Morb. Mortal Wkly. Rep. 56 (2007) 521–524.

[29] U.S. Food and Drug Administration News Release. FDA warns consumers in New Mexico and Texas not to eat certain types of raw red tomatoes, 3 June 2008. Available from: http:// www.fda.gov/NewsEvents/Newsroom/PressAnnouncements/2008/ucm116904.htm.

[30] Centers for Disease Control and Prevention, Outbreak of *Salmonella* serotype Saintpaul infections associated with multiple raw produce items—United States, MMWR Morb. Mortal Wkly. Rep. 57 (2008) 929–934.

[31] R. Berg, *Salmonella* Saintpaul: What went wrong? J. Environ. Health 71 (2008) 50–52.

[32] Centers for Disease Control and Prevention, Multistate outbreak of *Salmonella* infections associated with peanut butter and peanut butter-containing products—United States, 2008–2009, MMWR Morb. Mortal Wkly. Rep. 58 (2009) 85–90.

[33] Centers for Disease Control and Prevention. Investigation update: Outbreak of *Salmonella typhimurium* infections, 2008–2009, 29 April 2009. Available from: http://www.cdc.gov/salmonella/typhimurium/update.html.

[34] K. Wittenberger, E. Dohlman, Peanut outlook: Impacts of the 2008–09 foodborne illness outbreak linked to *Salmonella* in peanuts. A report of the US Department of Agriculture's Economic Research Service (OCS-10a-01).

[35] L.R. Schiller, A germy world—foodborne infections in 2009, Nat. Rev. Gastroenterol. Hepatol. 6 (2009) 197–198.

[36] *Caveat cenans* is *Banqueter beware*.

[37] F. Angulo, A. Voetsch, D. Vugia, J. Hadler, M. Farley, C. Hedberg, et al., FoodNet Working group. Determining the burden of human illness from foodborne diseases: CDC's emerging infectious disease program foodborne disease active surveillance network (FoodNet), Vet. Clin. North Am. 14 (1998) 165–172.

[38] T.F. Jones, E. Scallan, F.J. Angulo, FoodNet: overview of a decade of achievement, Foodborne Pathog. Dis. 4 (2007) 60–66.

[39] B. Swaminathan, T.J. Barrett, S.B. Hunter, R.V. Tauxe, CDC PulseNet Task Force, PulseNet: The molecular subtyping network for foodborne bacterial disease surveillance, United States, Emerg. Infect. Dis. 7 (2001) 382–389.

[40] P. Gerner-Smidt, K. Hise, J. Kincaid, S. Hunter, S. Rolando, E. Hyytiä-Trees, et al., A five-year update, Foodborne Pathog. Dis. 3 (2006) 9–19.

[41] R.V. Tauxe, Real burden and potential risks from foodborne infections: The value of multi-jurisdictional collaborations, Trends Food Sci. Technol. 19 (2008) S18–S25.

[42] Center for Disease Control and Prevention, Syndromic surveillance: Reports from a national conference, 2003, MMWR Morb. Mortal Wkly. Rep. 53 (2004) S1–S264.

[43] B. Budowle, Genetics and attribution issues that confront the microbial forensics field, Forens. Sci. Int. 146 (2004) S185–S188.

[44] B. Budowle, R. Murch, R. Chakraborty, Microbial forensics: The next forensic challenge, Int. J. Legal Med. 119 (2005) 317–330.

[45] T.A. Cebula, S.A. Jackson, E.W. Brown, B. Goswami, J.E. LeClerc, Chips and SNPs, bugs and thugs: A molecular sleuthing perspective, J. Food Prot. 68 (2005) 1271–1284.

[46] T.A. Cebula, E.W. Brown, S.A. Jackson, M.K. Mammel, A. Mukherjee, J.E. LeClerc, Molecular applications for identifying microbial pathogens in the post-9/11 era, Expert Rev. Mol. Diagn. 5 (2005) 431–445.

[47] B. Budowle, S.E. Schutzer, J.P. Burans, D.J. Beecher, T.A. Cebula, R. Chakraborty, et al., Quality sample collection, handling, and preservation for an effective microbial forensics program, Appl. Environ. Microbiol. 72 (2006) 6431–6438.

[48] E. Weise, Food safety: No guarantees. Spinach recall: 5 faces. 5 agonizing deaths. 1 year later, USA Today, updated 24 September 2007. Available from: http://www.usatoday.com/money/industries/food/2007-09-20-spinach-main_N.htm.

[49] D.G. Maki, Coming to grips with foodborne infection—peanut butter, peppers, and nationwide *Salmonella* outbreaks, N. Engl. J. Med. 360 (2009) 949–953.

[50] E. Fredrix, Salmonella recall could cost peanut producers $1B. Associated Press, March 11, 2009. Available from: http://abcnews.go.com/Business/wirestory?id=7053494&page=2.

[51] E. Hyytia-Trees, S.C. Smole, P.A. Fields, B. Swaminathan, E.M. Ribot, Second generation subtyping: A proposed PulseNet protocol for multiple-locus variable-number tandem repeat analysis of Shiga toxin producing *Escherichia coli* O157 (STEC O157), Foodborne Pathog. Dis. 3 (2006) 118–131.

[52] E. Hyytia-Trees, Summary of MLVA results for the *E. coli* O157 spinach outbreak isolates. Conference Proceedings, 11th Annual PulseNet Update Meeting 2007. Available from: http://www.aphl.org/profdev/conferences/proceedings/Pages/11thAnnualPulseNetUpdateMeeting.aspx.

[53] E. Hyytia-Trees, P. Lafon, P. Vauterin, E.M. Ribot, Multilaboratory validation study of standardized multiple-locus variable-number tandem repeat analysis protocol for Shiga toxin-producing *Escherichia coli* O157: A novel approach to normalize fragment size data between capillary electrophoresis platforms, Foodborne Pathog. Dis. 7 (2010) 129–136.

[54] G. Streisinger, J. Owen, Mechanisms of spontaneous and induced frameshift mutation in bacteriophage T4, Genetics 109 (1985) 633–659.

[55] G. Levinson, G.A. Gutman, Slipped-strand mispairing: A major mechanism for DNA sequence evolution, Mol. Biol. Evol. 4 (1987) 203–221.

[56] A. van Belkum, P.T. Tassios, L. Dijkshoorn, S. Haeggman, B. Cookson, N.K. Fry, et al., Guidelines for the validation and application of typing methods for use in bacterial epidemiology, Clin. Microbiol. Infect. Dis. 13 (Suppl. 3) (2007) 1–46.

[57] G.A. Uhlich, J.R. Sinclair, N.G. Warren, W.A. Chmielecki, P. Fratamico, Characterization of Shiga toxin-producing *Escherichia coli* isolates associated with two multistate food-borne outbreaks that occurred in 2006, Appl. Environ Microbiol. 74 (2008) 1268–1272.

[58] S.A. Jackson, I.R. Patel, M.K. Mammel, T.J. Barnaba, J.E. LeClerc, T.A. Cebula, Use of a genotyping DNA microarray representing diverse pathotypes of *E. coli* and *Shigella* spp. for strain identification and discrimination between and within closely related species. Available from: ftp://ftp.jcvi.org/pub/data/.../ASM-2008%20ECSG%20Array5.28.08-4.pdf.

[59] M.L. Kotewicz, M.K. Mammel, J.E. LeClerc, T.A. Cebula, Optical mapping and 454 sequencing of *Escherichia coli* O157:H7 isolates linked to the US 2006 spinach-associated outbreak, Microbiology 154 (2008) 3518–3528.

[60] M. Eppinger, M. Mammel, J.E. LeClerc, J.Ravel, T.A. Cebula, manuscript in preparation.

[61] Daubert v. Merrell Dow Pharmaceuticals, 509 US 579, 1999.

References

# CHAPTER 4

# Genotype and Geography: The Global Distribution of Bovine Tuberculosis

**Noel H. Smith**

*Veterinary Laboratories Agency, Surrey,
and Centre for the Study of Evolution, University of Sussex, Sussex, United Kingdom*

Molecular epidemiology has "come of age" and is now well placed to explore the evolutionary history and global distribution of entire bacterial species (1–4). These new approaches build upon the results of over 20 years of bacterial population genetics showing that identifying bacterial strains by genotypic methods—that is, changes that occur directly in the chromosomal DNA—is far superior for the unambiguous identification of species or strains than phenotypic methods that assay such features as a sugar fermentation profile, phage resistance, or antigenic reaction of surface structures. For those bacterial species for which a globally standardized multilocus genotyping technique has been developed, any isolate can be assigned an unambiguous numerical "fingerprint." Standardized genotype results can be interpreted, repeated, transferred, or communicated among laboratories with ease and stored in globally available databases. A strain that is "fingerprinted" by a genotyping method in one laboratory can be compared with an online global database of genotypes within seconds (see, for example, MLST databases at http://www.mlst.net/).

Globally standardized molecular epidemiology using genotyping has also opened up investigations into the "phylogeography" of bacterial pathogens, the relationship between geographical location and phylogenetic history (4–9). Phylogeography is an increasingly common theme of bacterial diversity analysis (10) and has been driven by advances in high-throughput genomic genotyping methods; only recently has it become practical to analyze multiple molecular markers in population size samples of bacterial strains.

## WHAT CAN BACTERIAL GENOTYPING OFFER THE FORENSIC MICROBIOLOGIST?

With our current, and improving, understanding of the global distribution of diversity within bacterial species there are two important questions that phylogeography and molecular epidemiology of bacteria can address that are

43

Microbial Forensics. DOI: 10.1016/B978-0-12-382006-8.00004-9

of interest to microbial forensics (11). First, "Has this bacterial strain been recently imported, either intentionally or unintentionally, from another country?," with a follow-up question, "Can we identify the country of origin of an imported strain?"

This chapter outlines the conditions, analysis, and data collection necessary to answer these questions and calls upon our experience with the veterinary pathogen, *Mycobacterium bovis*—the cause of cattle tuberculosis (TB)—to show how the phylogeography and molecular epidemiology of a well-understood bacterial pathogen can assist in microbial forensics.

## HAS THIS STRAIN BEEN RECENTLY IMPORTED?

For any bacterial disease, this question cannot be answered unless there is both temporal stability and, more important, geographical localization of the molecular type; if all strains are everywhere or if the dominant molecular types change rapidly, then identification of an imported strain will be challenging. What is emerging from the analysis of the global distribution of many bacterial diseases is that the degree of geographical localization varies between species (4,5,12). For those species that frequently travel as harmless commensals with humans (such as the gut commensal *Escherichia coli* or bacterial meningitis serogroup B), there may be insufficient localization of genotype for anything but the broadest conclusions. However, remarkably, for many important bacterial diseases there seems to be significant and stable geographical localization of molecular type (4,5,12), although this characteristic may be quite subtle for some pathogens (13).

If geographical localization of a genotype is identified then the next important step is to understand the diversity, population structure, and stability (both temporal and molecular) of the genotypes. In any country with a limited number of unique genotypes, identifying imported strains may be relatively easy, but if there are many different genotypes present, then identifying an imported genotype will be more difficult. Furthermore, isolation of a previously unreported genotype in a country must be interpreted with caution. Completely new genotypes can be generated within a population by both mutation and bacterial sex (recombination); therefore, the rate at which both these processes generate new genotypes must be understood (14–16). To identify a new genotype as a "homegrown" mutation or recombinant is not an insurmountable problem, provided several, chromosomally dispersed, loci are used in genotyping. This is because, in general, evolution happens one step at a time and therefore novel genotypes generated by these mechanisms will be closely related to other genotypes in the population (usually the most common genotype). In bacterial population genetics, this has led to the concept of "clonal complexes" of bacteria—a group of related strains that differ by no more than a single or a few changes in the molecular markers used (17,18). Clonal complexes have the property of all being descended from a most recent common ancestral strain and bear a genotype related to that found in the most recent common ancestor—although some loci may have changed by mutation or recombination, the majority of loci define the strain as a member of the clonal complex. Finally, it is important to appreciate that any survey of bacterial diversity is merely a sampling of the diversity present in a country; the best one may be able to say is that a genotype is unusual or has never been seen before.

## WHERE DID THIS STRAIN COME FROM?

Analysis of the global distribution of bovine tuberculosis has shown that it is possible, broadly, to identify the country of origin of strains imported into Great Britain by either infected people or cattle. We are able to do this because of the acute geographical localization of genotype at the national level. We believe that a number of factors have contributed to the global geographical localization of genotype for this veterinary pathogen. For example, we suspect that bovine tuberculosis was distributed internationally only in historical times and therefore the population structure has not had time to be remixed by subsequent introductions. Furthermore, the movement of infected cattle between countries can be limited by quarantine, followed by testing and culling if necessary; these options are not as readily available for human diseases. Bovine tuberculosis, therefore, may not be a good paradigm for many human pathogens and it remains to be seen if a similar attribution to source is possible for human pathogens that can migrate frequently by jet airplane. Nevertheless, I shall first introduce the background to the disease and use our experience with bovine tuberculosis to describe the methods that were applied to take a first look at the global genotype distribution of the disease and then show how this understanding can have real practical benefits for microbial forensics.

## BOVINE TUBERCULOSIS

Tuberculosis in cattle is caused primarily by *Mycobacterium bovis* and is a severe wasting disease, leading, eventually, to death. More important, the disease can be transmitted to humans where it causes a disease indistinguishable from tuberculosis caused by *Mycobacterium tuberculosis* (19–21). In Australia, the United States, parts of South America, Cuba, and most of Europe, a "test-and-slaughter" protocol for cattle was implemented in the mid-20th century that has virtually eliminated the disease (22). In the British Isles, the same procedure reduced the disease to very low levels by the 1970s; however, since then, the incidence of disease has inexorably risen so that today almost 100,000 reactor cattle are culled each year (20). It has been suggested that maintenance of bovine tuberculosis in an alternative host, the Eurasian badger, the largest native carnivore in the British Isles, may be responsible for the inability of the test-and-slaughter policy to control bovine TB in British cattle (23). For the rest of the world, wherever there are cattle there is bovine TB; the disease has been reported from every continent, and for most countries neither surveillance nor control programs exist (22).

For most bacteria, the best method for genotyping is multilocus sequence typing (MLST), a technique based on nucleotide sequencing of eight or so chromosomally encoded housekeeping genes (24,25). However, for many important bacterial diseases, there is not enough diversity present for MLST to provide sufficient resolution. Many human pathogens are single, globally distributed clones that would all tend to be of the same MLST type (4); therefore, other techniques based on more variable loci than housekeeping genes are used. For bovine and human tuberculosis, two techniques, spoligotyping (a technique virtually unique to these pathogens) and variable number tandem repeat (VNTR) typing, a form of minisatellite typing, have been developed and are gaining global acceptance (20).

Spoligotyping is a polymerase chain reaction and hybridization technique that identifies a spoligotype pattern for each isolate, which is very similar to a "bar code" measuring the presence or absence of 43 unique spacer sequences found in the direct variable repeat region of the chromosome. To assist international communication, each spoligotype pattern of *M. bovis* is given a name, such as SB0120, by www.Mbovis.org. Strains with identical spoligotype patterns are considered related but this conclusion can be misleading if the spoligotype patterns have arisen independently in different lineages of *M. bovis* (homoplasy). There is accumulating evidence that spacers in spoligotype patterns are lost and never regained, which supports the results of phylogeny analysis based on single-nucleotide polymorphisms (SNPs) and deletions that show that *M. bovis* is totally clonal; there has never been a well-documented case of recombination (the transfer of genomic material between strains) in this group of organisms.

Variable number tandem repeat typing measures the number of highly variable minisatellite repeats present at various locations on the genome. Results are given as a simple string of numbers, which represent the number of repeats at each locus. There is no doubt that VNTR typing identifies more variation than spoligotyping alone, although for *M. bovis* it turns out that spoligotyping has a level of diversity that is more applicable to a global analysis than VNTR typing.

## GEOGRAPHICAL LOCALIZATION OF *M. BOVIS* GENOTYPES IN GREAT BRITAIN

The population structure of bovine tuberculosis in Great Britain (England, Scotland, and Wales) is probably better understood than any other bacterial disease in the world. The spoligotype database at the Veterinary Laboratories

Agency, Weybridge, United Kingdom, contains spoligotype data on over 56,000 strains of *M. bovis* isolated in Great Britain. Over 41,000 of these isolates, from 1987 to 2009, have also been genotyped by six locus VNTR (ETR-A to F). The genotype of an *M. bovis* strain is a combination of its spoligotype pattern and its VNTR pattern.

Genotyping of *M. bovis* in Great Britain shows that the population consists of a small number of related genotypes not found in the remaining mainland European hot spots of bovine tuberculosis: Belgium, France, Portugal, Spain, or Italy (14,26–30). If a strain from mainland Europe were imported to Great Britain, it would be readily identifiable by its spoligotype pattern; however, no strain originating from mainland Europe has ever been identified in British cattle. The same is not true for strains originating from Northern Ireland or the Republic of Ireland.

Genotypes of *M. bovis* within Great Britain are localized geographically in "home range" regions in which the genotype is dominant. The geographical localization of *M. bovis* genotypes is extreme in Great Britain, less pronounced in Northern Ireland, and much less evident in other European countries. Using genotype home ranges, we can, in general, predict where any strain of *M. bovis* came from to within 40 km or less. However, about 15% of strains are located outside of their relevant home range and, in conjunction with cattle movement data, we can identify cattle that were probably infected with the home range genotype prior to movement to a new area. This microbial forensics approach to "out-of-home range" breakdowns is now an important tool for estimating parameters of the bovine TB epidemic in Great Britain.

## BOVINE TUBERCULOSIS IN THE BRITISH ISLES

Although much of Europe has controlled bovine tuberculosis, our nearest neighbors and important trading partners, Northern Ireland and the Republic of Ireland, still have a severe problem with this disease (31,32). In collaboration with our colleagues at the Agri-Food and Biosciences Institute Belfast, we have surveyed the populations of bovine tuberculosis throughout the British Isles. The populations of *M. bovis* in the three regions of the British Isles (Great Britain, Northern Ireland, and the Republic of Ireland) are dominated by strains related to spoligotype pattern SB0140 (VLA type 9), and more advanced analysis of the population structure using phylogenetically informative SNPs and chromosomal deletions has shown that a single lineage of *M. bovis* is present throughout these islands. This lineage is marked by a deletion, called RDEu1, with the linked loss of spacer 11 in the spoligotype pattern, and therefore forms a clonal complex of strains. This clonal complex has been named Europe 1 (Eu1), it is rare in other parts of Europe (see later), and 99% of *M. bovis* isolates in the British Isles are members of this clonal complex.

When VNTR analysis was added to the spoligotyping of strains from these three regions of the British Isles, we were surprised to find that each region had a distinct population of *M. bovis*. For example, over 60% of strains from Great Britain had genotypes (spoligotype plus VNTR type) that were unique even though the spoligotypes were shared with Irish strains. The molecular types of *M. bovis* in Northern Ireland were so distinct compared to Great Britain that it is possible to estimate how frequently bovine tuberculosis is imported from the province to the rest of the United Kingdom. For example, a rare genotype in Great Britain, called 17:k, is very common in Northern Ireland, and if we assume that all cattle breakdowns (outbreak within one herd) in Great Britain of type 17:k are caused by infected cattle imported from Northern Ireland, then it is easy to show that the maximum number of detected imports from Northern Ireland of infected cattle of any genotype is unlikely to exceed 20 per year. This conclusion is supported by the analysis of several genotypes that are common in Northern Ireland yet do not have a home range in Great Britain.

In general, any import of bovine TB into Great Britain by cattle can be identified by its genotype; however, humans can also carry bovine tuberculosis, and to identify the origin of strains acquired in other countries by humans, we need to know the global distribution of *M. bovis* genotypes.

## GLOBAL DISTRIBUTION OF *M. BOVIS* GENOTYPES

To unravel the global distribution and phylogeography of *M. bovis* genotypes, we have started to identify epidemiologically important clonal complexes of *M. bovis*. These clonal complexes are epidemiologically important, rather than phylogenetically important, because they are present, or dominant, in several countries; we do not know yet the phylogenetic relationship between them (33). A clonal complex is a group of strains descended from a recent common ancestor; all members of a clonal complex can be identified by a molecular marker present in the recent common ancestor and is present in all members of the clonal complex by descent (33) and we assume that all members of a clonal complex will have important characteristics in common.

To assay the global distribution of *M. bovis* genotypes, we have developed a "ping-pong" approach of transferring control backward and forward between international collaborators and VLA, Weybridge, United Kingdom. Initially, we rely on our international colleagues to collect strains and carry out spoligotype surveys of the *M. bovis* population in their own country. Spoligotype patterns are analyzed at VLA, Weybridge, and used to identify possible clonal complexes; then representative strains are sent to VLA, Weybridge, by our collaborators for advanced molecular examination. At the VLA we identify a molecular marker specific for the clonal complex and develop a simple molecular assay. The assay protocol, suitable materials, and control strains are packaged into a kit, which is then distributed to collaborating scientists throughout the world; they are asked to survey their national collection of strains and return the data to us for collation and publication. In this way scientists in each country retain ownership of their own strains and experimental results and we, at VLA, can coordinate the data collection, quality control the results, and organize publication of data. There are a number of important points to assure success. The organizing laboratory must be of sufficient stature and expertise to organize the survey and must also be scrupulously fair in dealing with collaborators from other countries. The collaborating laboratories, for their part, must be able to store permanent stocks of the local strains so that conflicting results can be resolved. Most importantly, the assay protocol developed must be sufficiently robust so that it can be carried out in any laboratory in the world.

In our experience, deletion typing is the most suitable method for identifying clonal complexes of *M. bovis*. The assay protocol for a specific chromosomal deletion is very simple; furthermore, deletions with specific endpoints are unlikely to be generated independently in different lineages—a well-characterized deletion is likely to be identical by descent from a recent common ancestor. We have used this approach to identify three clonal complexes of *M. bovis*. The first, Europe 1, is dominant in the British Isles but is distributed globally, whereas the other two, African 1 and African 2, are confined to West and East Africa, respectively.

## EUROPE 1—A GLOBALLY IMPORTANT CLONAL COMPLEX OF *M. BOVIS*

Over 99% of *M. bovis* strains isolated in the British Isles are deleted for a chromosomal region named RDEu1 and are therefore members of a clonal complex named Europe1 (Eu1). Members of this clonal complex are defined by the specific deletion RDEu1 and all have the property of lacking spacer 11 in their spoligotype patterns. We assume that the most recent common ancestor of the Eu1 clonal complex was deleted for RDEu1 and also lacked spacer 11 in its spoligotype pattern. In contrast to the dominance of Eu1 in the British Isles, deletion analysis of population-sized surveys shows that strains of Eu1 are rare in France, Belgium, and Italy but present in about 6% of strains from Spain and Portugal (Figure 4.1). However, an analysis of spoligotype patterns from all over the world showed that a typical Eu1 spoligotype pattern (SB0140—VLA type 9) was common in many other countries. To determine the global distribution of the Eu1 clonal complex, representative collections of strains from many countries were surveyed for the Eu1-specific deletion—usually in the country of origin.

Results of deletion typing and spoligotyping over 800 globally representative strains are shown in Figure 4.2. Strains of the Eu1 clonal complex are

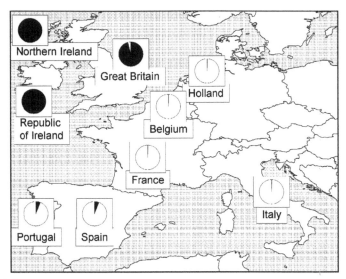

**FIGURE 4.1**

Distribution of the Eu1 clonal complex of *Mycobacterium bovis* throughout Europe. Pie charts show the proportion of strains that are members of the Eu1 clonal complex: black, Eu1; white, other clonal complexes. Strains of Eu1 are dominant in the British Isles, at about 6% in the Iberian Peninsula, and rare in the other countries surveyed. Most mainland European countries have controlled bovine tuberculosis and therefore population size samples are not available.

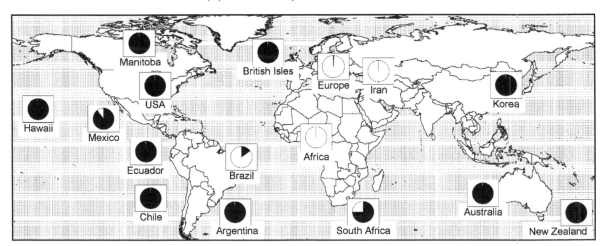

**FIGURE 4.2**

Global distribution of the Eu1 clonal complex. Pie charts show the proportion of strains that are members of the Eu1 clonal complex: black, Eu1; white, other clonal complexes. Eu1 dominates in the British Isles, former British colonies, and the New World (except Brazil).

common in former British colonies, such as South Africa, Australia, New Zealand, Canada, and the United States. However, Mexico, as well as several South American countries, is also dominated by strains of Eu1. In general, the British Isles, former British colonies, and the New World (with the exception of Brazil, which has colonial ties to Portugal) are dominated by the Eu1 clonal complex. In contrast, most of Europe and Africa, with the exception of South Africa, are dominated by other clonal complexes of *M. bovis* not related to Eu1 (see later). The Eu1 clonal complex of *M. bovis* therefore represents a globally important clonal complex of *M. bovis*.

The dominance of Eu1 in the British Isles and its presence in former British colonies suggest a simple phylogeographic explanation for the distribution of these strains throughout the world. The suggestion that the British Isles may have been the epicenter for the distribution of this clonal complex can be supported by the large number of modern cattle types that were bred in the British Isles (34). However, the presence of Eu1 strains in South Korea suggests that dispersal of this clonal complex may be more complicated than a simple bovine diaspora from the British Isles. South Korea has no real history of cattle trade with the British Isles but did repopulate its cattle herds with stock from Canada, Australia, and the United States after the Korean war (35). Therefore, the presence of the Eu1 clonal complex in South Korea may represent reintroduction of the disease from a secondary source rather than directly from the British Isles and illustrate one of the pitfalls of phylogeographic analysis of the origin of disease; the movement of people and domesticated animals in the past three centuries may obscure the origins of disease.

We have shown that Eu1 is an important globally distributed clonal complex of *M. bovis*, and its phylogeography suggests that it was distributed to many countries only in historical times. However, in Africa, other clonal complexes of *M. bovis* can be identified.

## AFRICAN 1—DOMINANT IN WEST-CENTRAL AFRICA

Bovine tuberculosis has been shown to be present in most countries in Africa, but in general, due to economic constraints, the true extent of the disease has not been evaluated and genotype surveys are often limited in scope (36). Surveys of *M. bovis* from Cameroon, Nigeria, Mali, and Chad showed that lack of spacer 30 was a common feature of *M. bovis* spoligotype patterns from these West-Central African countries (37–39). At VLA we identified an informative deletion of chromosomal DNA in strains from Mali that lacked spacer 30. This deletion was called RDAf1. When tested in population size surveys, an identical deletion was present in 96% of 338 strains sampled from these four West-Central African countries; in all RDAf1-deleted strains, spacer 30 was also deleted in the spoligotype pattern. We concluded that the majority of strains of *M. bovis* in Cameroon, Mali, Chad, and Nigeria formed a single clonal complex of strains descended from a common ancestor in which

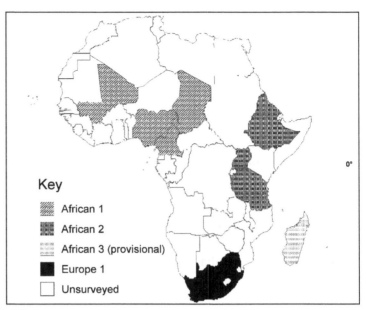

**FIGURE 4.3**

Clonal complexes of *Mycobacterium bovis* identified in Africa. Countries where each of the clonal complexes dominates are shown. The African 3 clonal complex in Madagascar has been identified by its spoligotype signature alone.

RDAf1 and spacer 30 were deleted (Figure 4.3). We called this clonal complex African 1 (Af1) (33).

To establish the geographical range of African 1 strains, we surveyed for the presence of the Af1 deletion in small collections of strains available from other countries in Africa (Algeria, Burundi, Ethiopia, Madagascar, Mozambique, South Africa, Tanzania, and Uganda). No strain of the Af1 clonal complex was identified by deletion typing in this survey; although strains with spacer 30 missing were identified, they did not carry the defining deletion of the RDAf1 region. Furthermore, large collections of *M. bovis* strains from Europe, the New World, and Iran were examined for the characteristic loss of spacer 30 in the spoligotype pattern and, where possible, tested for the African 1-specific deletion. This analysis concluded that strains of the Af1 clonal complex were not at high frequency in any region outside of West Central Africa.

Prior to analysis it had been assumed that strains of *M. bovis* would be dispersed throughout West Central Africa by the transhumance movement of cattle. However, when we VNTR typed the strains from these four West Central African countries, surprisingly, we found that each country had a unique population structure. That is, given the genotype of a strain one could, with reasonable accuracy, determine the country of origin (33). We have used this property to identify the country of origin of strains isolated from humans in Europe (see later).

## AFRICAN 2—DOMINANT IN EAST AFRICA

Spoligotype data collected for the African 1 study indicated that another clonal complex was present at high frequency in strains isolated from East African countries (Uganda, Burundi, Tanzania, and Ethiopia). In collaboration with Dr. S. Berg (VLA, Weybridge, UK), a specific deletion of chromosomal DNA, called RDAf2, was identified in strains from East Africa. Strains with the RDAf2 deletion were found in over 55% of *M. bovis* isolates from Uganda, Burundi, Tanzania, and Ethiopia; in all cases these strains had a specific spoligotype pattern signature. We named this clonal complex African 2 (Af2) (Figure 4.3). The RDAf2 region is intact in strains of Af1 and Eu1, suggesting that these clonal complexes are phylogenetically distinct and not related. Strains with the spoligotype signature of Af2 were not found in cattle samples from other African countries or in Europe and the New World. Like Af1, each West African country seems to have a unique population of this clonal complex, suggesting again that a single strain spread throughout these West African countries and subsequently developed country-specific population structures.

## OTHER CLONAL COMPLEXES OF *M. BOVIS* IN AFRICA

Af1 and Af2 are not the only clonal complexes of *M. bovis* present in Africa. Although virtually all strains are members of the Af1 clonal complex in Chad, Mali, and Nigeria, in Mali a second clonal complex, provisionally called Af5, can be recognized by its distinct spoligotype signature. The Af5 clonal complex makes up 40% of strains isolated in Mali, and other minor clonal complexes are found in the West African countries dominated by Af2. However, in Madagascar, all strains of *M. bovis* lack spacers 4, 5, 8, and 10, which would suggest the presence of another dominant clonal complex of *M. bovis* on this island; we have provisionally named the clonal complex Af3 but have been unable to investigate it further (Figure 4.3).

The distribution of different clonal complexes of *M. bovis* in Africa suggests a rich history of introduction, expansion, and population change, which is amenable to analysis by molecular epidemiology and phylogeography. Obviously country borders are political entities and we do not know the extent to which there is geographical localization of a genotype within each country. However, these preliminary results clearly demonstrate the potential value of further molecular epidemiological analysis of bovine TB within Africa.

## "OF WHAT USE IS A BABY?"

The identification of *M. bovis* clonal complexes is interesting but what use is it? The answer to this question is given above, however, there are a number of practical uses for these data. In the first instance, these clonal complexes may represent groups of strains with different selective advantages or behaviors. Comparing and contrasting phenotypic differences between these distinct divisions within *M. bovis* may elucidate the molecular mechanisms of these differences and illuminate the selective forces operating on both bovine tuberculosis and its cattle host. For example, *Bos taurus* (European cattle) are common in West Africa, where the African 1 clonal complex of *M. bovis* dominates, whereas *Bos indicus* (Zebu, Asian cattle) are common in East Africa where African 2 dominates (40,41). It remains to be seen if these two clonal complexes of *M. bovis* have specialized to different cattle hosts.

For the African clonal complexes, our analysis also shows that development of simple genotype schemes within these countries is worthwhile and will aid eradication schemes by identifying strains imported from neighboring countries (33). Furthermore, now that the dominant African 1 and the African 2 clonal complexes have been identified it is a simple matter to chromosome sequence representative isolates and gather a rich harvest of specific molecular polymorphisms to use in local epidemiological analysis. Genome sequencing will also resolve the phylogenetic status of these clonal complexes and may show that the majority of bovine tuberculosis found in Africa originated elsewhere and has been imported to the continent relatively recently. This, in turn, will develop our understanding of the historical and phylogeographical bases of bovine tuberculosis in Africa and feed back to our understanding of the disease in Europe.

For the Eu1 clonal complex, practical uses are even more compelling. It has been suggested that the Eu1 clonal complex has become dominant in the British Isles either by changing primary host from cattle to badgers or by avoidance of the test and slaughter protocol (20). These suggestions can now be tested by comparing the progress of Eu1 in other countries; many of these countries, for example, do not have badgers. Furthermore, instead of concentrating on the dominance of this clonal complex in the British Isles, it is now germane to ask why the Eu1 clonal complex became dominant on a global scale. This clonal complex may have a specific selective advantage (such as an ability to infect and maintain in alternative hosts or an ability to avoid disclosure by the test and slaughter) that increases its fitness compared to other clonal complexes of bovine TB or it may just have been the "lucky" clone that was widely distributed into regions where bovine TB was previously unknown. These suggestions are amenable to experimental analysis.

## CAN WE IDENTIFY THE ORIGIN OF IMPORTED STRAINS?

We can, in general, identify strains of *M. bovis* in cattle originating from outside of Great Britain because of the acute geographical localization of *M. bovis* genotypes and our understanding of the diversity and molecular evolution of the *M. bovis* population. Our analysis clearly shows that imports in cattle from mainland Europe, and farther afield, probably do not occur and that imports from Ireland are rare but can be identified. For this disease in cattle, therefore, we can answer the two forensic microbiology questions posed at the start of this chapter. However, humans can also carry bovine TB, and a more interesting application of this approach is to identify the country of origin of strains isolated from humans. We have several examples that suggest this is possible. For example, a strain isolated from a human in Birmingham in 2005 was identified as a member of the Af1 clonal complex and from the genotype we were able to correctly identify the country of birth of the patient. In a similar manner, strains of Af1 isolated from Chadian immigrants to France were shown to have the genotype common in Chad (33).

Analysis of the global distribution of *M. bovis* has shown that, in principle, microbial forensics can identify imported genotypes and may be able to identify the country of origin of the strains. We have argued that *M. bovis* may be unusual in its extreme geographical localization of genotype. However, it is the challenge of the next few decades to determine how useful this approach is when dealing with other human pathogens introduced into a country either unintentionally or maliciously.

## ACKNOWLEDGMENT

This work would not have been possible without the collaboration of many scientists from all over the world. I thank them all for their enthusiasm for this project.

## REFERENCES

[1] M. Monot, N. Honore, T. Garnier, R. Araoz, J.Y. Coppée, C. Lacroix, et al., On the origin of leprosy, Science 308 (5724) (2005) 1040–1042.

[2] R. Hershberg, M. Lipatov, P.M. Small, H. Sheffer, S. Niemann, S. Homolka, et al., High functional diversity in M. tuberculosis driven by genetic drift and human demography, PLoS Biol. 6 (2009) 12.

[3] Y. Moodley, B. Linz, Y. Yamaoka, H.M. Windsor, S. Breurec, J.Y. Wu, et al., The peopling of the Pacific from a bacterial perspective, Science 323 (5913) (2009) 527–530.

[4] M. Achtman, Evolution, population structure, and phylogeography of genetically mono-morphic bacterial pathogens, Annu. Rev. Microbiol. 62 (2008) 53–70.

[5] P.S. Keim, D.M. Wagner, Humans and evolutionary and ecological forces shaped the phylo-geography of recently emerged diseases, Nat. Rev. Microbiol. 7 (11) (2009) 813–821.

[6] A.G. Hoen, G. Margos, S.J. Bent, M.A. Diuk-Wasser, A. Barbour, K. Kurtenbach, et al., Phylogeography of Borrelia burgdorferi in the eastern United States reflects multiple independent Lyme disease emergence events, Proc. Natl. Acad. Sci. USA 106 (35) (2009) 15013–15018.

[7] A.J. Vogler, D. Birdsell, L.B. Price, J.R. Bowers, S.M. Beckstrom-Sternberg, R.K. Auerbach, et al., Phylogeography of Francisella tularensis: Global expansion of a highly fit clone, J. Bacteriol. 191 (8) (2009) 2474–2484.

[8] S. Gagneux, P.M. Small, Global phylogeography of Mycobacterium tuberculosis and impli-cations for tuberculosis product development, Lancet Infect. Dis. 7 (5) (2007) 328–337.

[9] V. Duchene, S. Ferdinand, I. Filliol, J.F. Guégan, N. Rastogi, C. Sola, Phylogenetic recon-struction of Mycobacterium tuberculosis within four settings of the Caribbean region: Tree comparative analyse and first appraisal on their phylogeography, Infect. Genet. Evol. 4 (1) (2004) 5–14.

[10] L.L. Knowles, The burgeoning field of statistical phylogeography, J. Evol. Biol. 17 (1) (2004) 1–10.

[11] B. Budowle, S.E. Schutzer, M.S. Ascher, R.M. Atlas, J.P. Burans, R. Chakraborty, et al., Toward a system of microbial forensics: From sample collection to interpretation of evidence, Appl. Environ. Microbiol. 71 (5) (2005) 2209–2213.

[12] M. Achtman, S. Suerbaum, Sequence variation in Helicobacter pylori, Trends Microbiol. 8 (2) (2000) 57–58.

[13] D. Falush, T. Wirth, B. Linz, J.K. Pritchard, M. Stephens, M. Kidd, et al., Traces of human migrations in Helicobacter pylori populations, Science 299 (5612) (2003) 1582–1585.

[14] N.H. Smith, J. Dale, J. Inwald, S. Palmer, S.V. Gordon, R.G. Hewinson, et al., The popula-tion structure of Mycobacterium bovis in Great Britain: Clonal expansion, Proc. Natl. Acad. Sci. USA 100 (25) (2003) 15271–15275.

[15] J. Maynard Smith, N.H. Smith, M. O'Rourke, B.G. Spratt, How clonal are bacteria? Proc. Natl. Acad. Sci. USA 90 (10) (1993) 4384–4388.

[16] G. Morelli, B. Malorny, K. Müller, A. Seiler, J.F. Wang, J. del Valle, et al., Clonal descent and microevolution of Neisseria meningitidis during 30 years of epidemic spread, Mol. Microbiol. 25 (6) (1997) 1047–1064.

[17] J.M. Smith, E.J. Feil, N.H. Smith, Population structure and evolutionary dynamics of patho-genic bacteria, Bioessays 22 (12) (2000) 1115–1122.

[18] E.J. Feil, Small change: Keeping pace with microevolution, Nat. Rev. Microbiol. 2 (6) (2004) 483–495.

[19] R. de la Rua-Domenech, Human Mycobacterium bovis infection in the United Kingdom: Incidence, risks, control measures and review of the zoonotic aspects of bovine tuberculo-sis, Tuberculosis (Edinb.) 86 (2) (2006) 77–109.

[20] N.H. Smith, S.V. Gordon, R. de la Rua-Domenech, R.S. Clifton-Hadley, R.G. Hewinson, Bottlenecks and broomsticks: The molecular evolution of Mycobacterium bovis, Nat. Rev. Microbiol. 4 (9) (2006) 670–681.

[21] N.H. Smith, K. Kremer, J. Inwald, J. Dale, J.R. Driscoll, S.V. Gordon, et al., Ecotypes of the Mycobacterium tuberculosis complex, J. Theor. Biol. 239 (2) (2006) 220–225.

[22] C.O. Thoen, J. Steele, M.J. Gilsdorf (Eds.), Mycobacterium bovis Infection in Animals and Humans, 2nd ed., Blackwell Publishing, Ames, IA, 2006.

[23] J. Gallagher, R.S. Clifton-Hadley, Tuberculosis in badgers; a review of the disease and its significance for other animals, Res. Vet. Sci. 69 (3) (2000) 203–217.

[24] J.E. Cooper, E.J. Feil, Multilocus sequence typing: What is resolved? Trends Microbiol. 12 (8) (2004) 373–377.

[25] M.C. Maiden, J.A. Bygraves, E. Feil, G. Morelli, J.E. Russell, R. Urwin, et al., Multilocus sequence typing: A portable approach to the identification of clones within populations of pathogenic microorganisms, Proc. Natl. Acad. Sci. USA 95 (6) (1998) 3140–3145.

[26] N. Haddad, A. Ostyn, C. Karoui, M. Masselot, M.F. Thorel, S.L. Hughes, et al., Spoligotype diversity of *Mycobacterium bovis* strains isolated in France from 1979 to 2000, J. Clin. Microbiol. 39 (10) (2001) 3623–3632.

[27] S. Rodriguez, B. Romero, J. Bezos, L. de Juan, J. Alvarez, E. Castellanos, et al., Spanish Network on Surveillance and Monitoring of Animal Tuberculosis. High spoligotype diversity within a *Mycobacterium bovis* population: Clues to understanding the demography of the pathogen in Europe, Vet. Microbiol. 141 (1–2) (2009) 89–95.

[28] M.B. Boniotti, M. Goria, D. Loda, A. Garrone, A. Benedetto, A. Mondo, et al., Molecular typing of *Mycobacterium bovis* strains isolated in Italy from 2000 to 2006 and evaluation of variable-number tandem repeats for geographically optimized genotyping, J. Clin. Microbiol. 47 (3) (2009) 636–644.

[29] C. Allix, K. Walravens, C. Saegerman, J. Godfroid, P. Supply, M. Fauville-Dufaux, Evaluation of the epidemiological relevance of variable-number tandem-repeat genotyping of *Mycobacterium bovis* and comparison of the method with IS6110 restriction fragment length polymorphism analysis and spoligotyping, J. Clin. Microbiol. 44 (6) (2006) 1951–1962.

[30] E.L. Duarte, M. Domingos, A. Amado, A. Botelho, Spoligotype diversity of *Mycobacterium bovis* and *Mycobacterium caprae* animal isolates, Vet. Microbiol. 130 (3–4) (2008) 415–421.

[31] D.A. Abernethy, G.O. Denny, F.D. Menzies, P. McGuckian, N. Honhold, AR. Roberts, The Northern Ireland programme for the control and eradication of *Mycobacterium bovis*, Vet. Microbiol. 112 (2-4) (2006) 231–237.

[32] S.I. Moore, M. Good, The tuberculosis eradication programme in Ireland: A review of scientific and policy advances since 1988, Vet. Microbiol. 112 (2006) 239–251.

[33] B. Muller, M. Hilty, S. Berg, M.C. Garcia-Pelayo, J. Dale, M.L. Boschiroli, et al., African 1, an epidemiologically important clonal complex of *Mycobacterium bovis* dominant in Mali, Nigeria, Cameroon, and Chad, J. Bacteriol. 191 (6) (2009) 1951–1960.

[34] J.E. Decker, J.C. Pires, G.C. Conant, S.D. McKay, M.P. Heaton, K. Chen, et al., Resolving the evolution of extant and extinct ruminants with high-throughput phylogenomics, Proc. Natl. Acad. Sci. USA 106 (44) (2009) 18644–18649.

[35] S.H. Wee, C.H. Kim, S.J. More, H.M. Nam, *Mycobacterium bovis* in Korea: An update, Vet. J. (2009).

[36] W.Y. Ayele, S.D. Neill, J. Zinsstag, M.G. Weiss, I. Pavlik, Bovine tuberculosis: An old disease but a new threat to Africa, Int. J. Tuberc. Lung Dis. 8 (8) (2004) 924–937.

[37] B. Muller, B. Steiner, B. Bonfoh, A. Fané, N.H. Smith, J. Zinsstag, Molecular characterisation of *Mycobacterium bovis* isolated from cattle slaughtered at the Bamako abattoir in Mali, BMC Vet. Res. 4 (2008) 26.

[38] S. Cadmus, S. Palmer, M. Okker, J. Dale, K. Gover, N. Smith, et al., Molecular analysis of human and bovine tubercle bacilli from a local setting in Nigeria, J. Clin. Microbiol. 44 (1) (2006) 29–34.

[39] M. Hilty, C. Diguimbaye, E. Schelling, F. Baggi, M. Tanner, J. Zinsstag, Evaluation of the discriminatory power of variable number tandem repeat (VNTR) typing of *Mycobacterium bovis* strains, Vet. Microbiol. 109 (3–4) (2005) 217–222.

[40] O. Hanotte, C.L. Tawah, D.G. Bradley, M. Okomo, Y. Verjee, J. Ochieng, et al., Geographic distribution and frequency of a taurine Bos taurus and an indicine Bos indicus Y specific allele amongst Sub-saharan African cattle breeds, Mol. Ecol. 9 (4) (2000) 387–396.

[41] O. Hanotte, D.G. Bradley, J.W. Ochieng, Y. Verjee, E.W. Hill, JE. Rege, African pastoralism: Genetic imprints of origins and migrations, Science 296 (5566) (2002) 336–339.

# CHAPTER 5

# Tracing Meat Products through the Production and Distribution Chain from Farm to Consumer

**Ronan T. Loftus and Ciaran Meghen**

*IdentiGEN, Dublin, Ireland*

## INTRODUCTION

A common theme during food safety incidents is the difficulty in following the source of contamination back through the processing and distribution chain. For instance, it took from April 2008 when the Centers for Disease Control and Prevention detected a *Salmonella saintpaul* outbreak that led to a wide-scale recall of tomatoes, to July 2008 when it was decided that Mexican chilies were a more likely source of the infection (1). In meat processing, similar difficulties arise when there is a food safety incident. In 2010, Cargill, one of the largest meat processors in North America, is being sued for $100 million for allegedly selling *Escherichia coli*-tainted hamburger in 2007, which left a 22-year-old dance instructor from Cold Spring, Minnesota, paralyzed. Cargill has been unable to identify the source of the contaminated raw material from among its suppliers, leaving it to take sole responsibility should the lawsuit be successful (2). Even in cases where the manufacturing source is known, such as during the 2009 *Salmonella* outbreak in peanuts, the collateral economic damage to unaffected peanut producers and manufacturers was significant (3).

These cases illustrate the challenge associated with conducting a timely, precise, and accurate recall, and indeed the scale of economic risk associated with outbreaks or a bioterrorist attack on the food industry. Technologies that can facilitate more precise identification of the contributing hazard have an important role in biosecurity, both as a deterrent and as an investigative tool. This chapter looks at how genetic identification can be used effectively in animal production systems to trace meat products themselves, and contaminants affecting those meat products.

## MEAT CHAIN STRUCTURE AND COMPLEXITY

Supply chain structures for major meat-producing species vary considerably at the producer level; cattle production is highly fragmented, whereas hog and

59

Microbial Forensics. DOI: 10.1016/B978-0-12-382006-8.00005-0

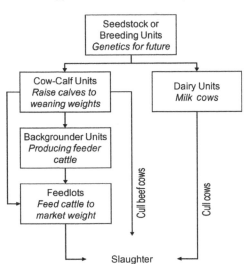

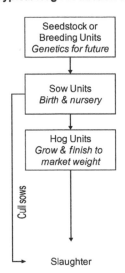

**FIGURE 5.1**

Comparison of cattle and hog production systems.

poultry production is more concentrated and vertically integrated. Figure 5.1 compares, in a generalized scheme, major differences between cattle and hog production in modern livestock economies.

While the differences in the schemes may not appear very large, it is the difference in absolute number of livestock producers that is especially significant. In the United States, for instance, it is estimated that there are more than 1 million cattle holdings yielding an annual cattle crop of some 33 million head. Sixty percent of these have been reared on some 986,000 holdings with fewer than 500 head. This compares to 65,000 hog producers and an annual crop of more than 100 million hogs, of which 50% are produced on just 110 holdings with greater than 50,000 head (4).

Tracing animals through these livestock production systems is aided by the identification of animals and the registration of premises. Identification may be individual or batch based and the technologies may be conventional numeric/alphanumeric tags, bar code tags, radio-frequency identification tags/transponders, tattoos, brands, biometrics, or genetic identification. However, at present there is no agreed animal identification system in the United States. Despite a concerted effort, the National Animal Identification System (NAIS), proposed by the U.S. Department of Agriculture (USDA) in 2004 in the immediate the aftermath of the Washington State bovine spongiform encephalopathy (BSE) case, has not been implemented effectively (5). The role that DNA identification technology can play in these circumstances

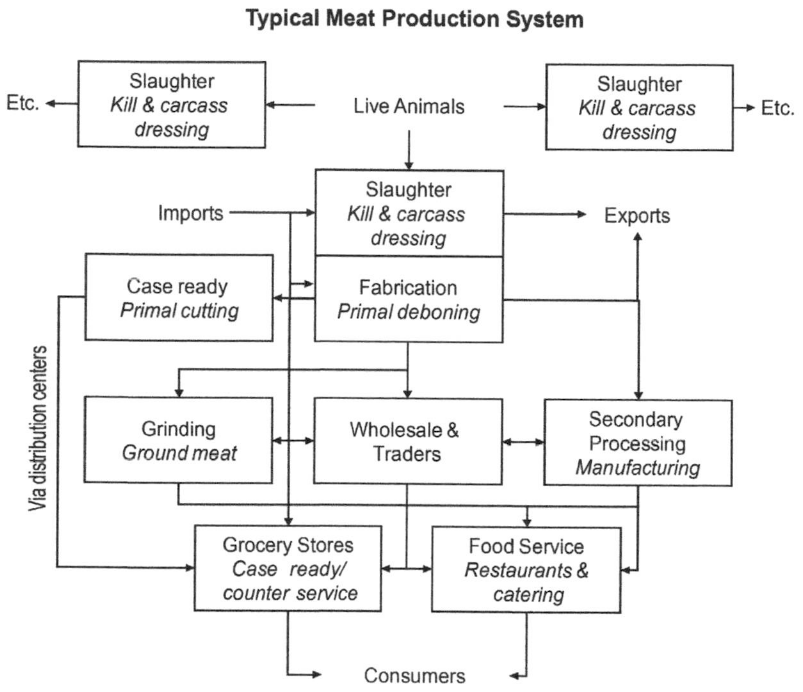

**Typical Meat Production System**

**FIGURE 5.2**
The meat processing chain, a generalized scheme.

has been described elsewhere. However, many developed livestock economies have implemented animal identification at some level (6–9) and so the U.S. position does not provide a general framework for thinking about the role of genetic identification at the producer level. Importantly, where animals are identified by conventional means, such as ear tags, the value of DNA methods is in verification rather than in identification. Where DNA technology has an especially useful role to play is from point of slaughter to final consumption. The meat processing side of the chain is more similar between species; animals are killed, fabricated, packed or further processed, and ultimately distributed through grocery or food service outlets. The flow of product information through these different transformation stages is generally operated at batch level and is often paper based and nonstandardized. Figure 5.2 illustrates some of the structural complexity that exists in meat processing chains, where individual animals are first batched and then comingled at each subsequent stage. Maintaining accurate physical identification, at the individual animal level, through this complexity is not possible in modern industrial-scale meat production facilities. This underscores the difficulty of tracking specific production information forward or backward through the chain for a given meat product.

Consider the question of how wide a (theoretical) recall might be required if the Washington State BSE cow had been fabricated into raw material for hamburger that had subsequently been supplied to, say, a large number of secondary processors that each supply the major food service multiples and grocery chains. The answer is uncertain but it is vastly more than the actual meat from the infected source animal. Indeed a record 143 million lbs of beef was recalled by Westland/Hallmark Meat Company in February 2008 (10) due to concerns over the slaughter of sick "downer" cattle. It was not possible to identify which products contained meat from the limited number of animals that were deemed unfit for human consumption.

Conversely, consider a steak bought at the meat counter of a grocery store where the consumer finds a veterinary needle or needle fragment in the final cooked product. The steak originated from a single source contributing animal on a specific source feedlot or ranch, but how many animals would be involved if a traceback was required? Again, while the answer is uncertain and depends on the particulars of the supply chain structure, it is likely that anything from 50,000 to 500,000 or more cattle could be involved and these could have been born and reared on any of the one million or more cattle ranches and any of 100 or more feedlots. It would not be possible to identify the specific contributing source by conventional means.

It is the lack of reliable information on product origins that requires large-scale recalls when a small volume of product is contaminated or affected. Despite the scares and incidents in the meat industry in recent years, it has been difficult to change the standard mode of operation to facilitate better information tracking for recall purposes. Meat processing has achieved high levels of efficiency in order to produce meat cheaply, but in order to do so it is conducted on an industrial scale and on narrow operating margins. A consequence is that the industry is poorly configured to invest in the information systems and technology required to allow more accurate recalls and tracebacks. Genetic identification technology offers distinct capabilities that may facilitate the investigating authorities in the event of a bioterrorist or other biological incident.

## GENETIC IDENTIFICATION OF THE HOST ANIMAL AND ITS DERIVED PRODUCTS

The clonal nature of many biological agents generally means that biological/microbial evidence is a class characteristic, as such evidence is circumstantial and does not readily allow source attribution. In contrast, autosomal DNA profiles provide individual-level identification, which can be compared to a source reference sample—the basis of current genetic fingerprinting. Genetic identification, as distinct from parentage inference, makes full use of genetic data and is among the most powerful evidence for source attribution. Matches between the evidence—the meat product—and the source individual animal can be declared with very high levels of statistical probability. This provides the foundations for an approach to forensic traceability in meat production (11). The level of statistical power is a function of the DNA marker system polymorphism, the number of DNA markers used, and several population genetic and other evidential parameters. For more information on the statistical issues involved in the use of DNA profiles for forensic inference, the reader is referred to Balding (12) and in particular formula 3.3.

In human forensics, short tandem repeat (STR) markers have been used widely for identification purposes since the mid-1990s (13). Other marker systems, developed more recently, such as single-nucleotide polymorphisms (SNP), provide advantages of cost and accuracy. However, a legacy, or technology lock-in, problem occurs because SNP and STR data are not interchangeable. When significant numbers of reference samples have been DNA typed previously using STRs and source samples are no longer available for retesting, there are challenges in migrating to SNPs. This is particularly relevant in human forensics and indeed animal parentage. In meat production, where there are no legacy issues with established STR profile databases, SNP markers are finding increased application for animal identification purposes (14–16).

The term "SNP," coined in the early 1990s, is simply another way of referring to a "point mutation" or a "base substitution." A single-nucleotide polymorphism is a DNA sequence variation occurring when a single nucleotide—A, T, C, or G—in the genome differs between paired chromosomes in an individual. For example, two DNA sequences at an autosomal bovine locus, AAGCCTA and AAGCTTA, contain a difference in a single nucleotide. In this case we say that there are two alleles—C and T—and that the individual is heterozygous CT at that locus. Another individual may be homozygous CC or TT. Almost all common SNPs have only two alleles (17). The utility of SNPs for identification purposes is based on the autosomal, codominant nature of the variation between individuals and the low mutation rate between generations (18). For a variation to be considered an SNP, it must have a minor allele frequency (MAF) of at least 0.01 in the population or, conversely, a major allele frequency of 0.99. It is estimated that complex mammalian genomes contain many millions of SNPs; however, those that are optimally useful for genetic identification, those with an MAF of between 0.4 and 0.5 across different populations in the same species, may be numbered in the thousands. Figure 5.3 illustrates the impact of minor/major allele frequency (MAF) on the individual locus power. This is calculated as the sum of the square of the genotype frequencies for a given allele frequency, assuming Hardy–Weinberg proportions (19).

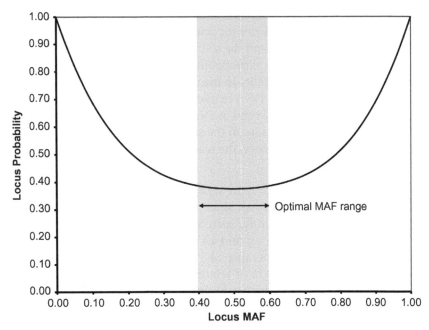

**FIGURE 5.3**
Single-nucleotide polymorphism locus probability based on varying allele frequency (MAF).

The cumulative profile probability for a typical series of SNPs optimized for identification purposes is given in Table 5.1, where the combined probability is the product of the individual locus probabilities (19). As can be seen, a moderate number of SNPs yield a high level of statistical power equivalent to that used in STR forensics.

A wide variety of analytical methods have become available that allow for the convenient and cost-effective assaying of large numbers of samples and the 30–50 SNP markers required for reliable identification. It is beyond our scope to characterize the available technologies; however, the reviews of Kim and Misra (20) and *Genetic Engineering and Biotechnology News* (21) give a general overview. At 2010 prices, full commercial service SNP genotyping is performed at about 20% the cost of STR genotyping and SNP genotyping may be expected to continue to fall in cost.

## TRACING MEAT PRODUCTS THROUGH THE SUPPLY CHAIN

Having established that an SNP profile can be generated for a suspect—evidence—meat sample, the ability to identify the contributing animal requires

**Table 5.1** Locus Power and Combined Power for a Generalized SNP Panel Optimized for Forensic Identification

| SNP Locus | MAF | Locus Probability | Cumulative Probability | Likelihood Ratio |
|---|---|---|---|---|
| 1 | 0.43 | 0.4192 | $4.19^{-01}$ | 2.39 |
| 2 | 0.48 | 0.3858 | $1.62^{-01}$ | 6.18 |
| 3 | 0.40 | 0.4432 | $7.17^{-02}$ | $1.40^{01}$ |
| 4 | 0.39 | 0.4518 | $3.24^{-02}$ | $3.09^{01}$ |
| 5 | 0.42 | 0.4269 | $1.38^{-02}$ | $7.23^{01}$ |
| 6 | 0.44 | 0.4118 | $5.69^{-03}$ | $1.76^{02}$ |
| 7 | 0.47 | 0.3917 | $2.23^{-03}$ | $4.48^{02}$ |
| 8 | 0.47 | 0.3917 | $8.74^{-04}$ | $1.14^{03}$ |
| 9 | 0.44 | 0.4118 | $3.60^{-04}$ | $2.78^{03}$ |
| 10 | 0.40 | 0.4432 | $1.59^{-04}$ | $6.27^{03}$ |
| 11 | 0.37 | 0.4700 | $7.49^{-05}$ | $1.33^{04}$ |
| 12 | 0.50 | 0.3750 | $2.81^{-05}$ | $3.56^{04}$ |
| 13 | 0.31 | 0.5311 | $1.49^{-05}$ | $6.70^{04}$ |
| 14 | 0.45 | 0.4048 | $6.04^{-06}$ | $1.66^{05}$ |
| 15 | 0.40 | 0.4432 | $2.68^{-06}$ | $3.74^{05}$ |
| 16 | 0.48 | 0.3858 | $1.03^{-06}$ | $9.68^{05}$ |
| 17 | 0.38 | 0.4608 | $4.76^{-07}$ | $2.10^{06}$ |
| 18 | 0.40 | 0.4432 | $2.11^{-07}$ | $4.74^{06}$ |
| 19 | 0.41 | 0.4349 | $9.17^{-08}$ | $1.09^{07}$ |
| 20 | 0.45 | 0.4048 | $3.71^{-08}$ | $2.69^{07}$ |
| 21 | 0.48 | 0.3858 | $1.43^{-08}$ | $6.98^{07}$ |
| 22 | 0.50 | 0.3750 | $5.37^{-09}$ | $1.86^{08}$ |
| 23 | 0.49 | 0.3802 | $2.04^{-09}$ | $4.90^{08}$ |
| 24 | 0.40 | 0.4432 | $9.05^{-10}$ | $1.11^{09}$ |
| 25 | 0.42 | 0.4269 | $3.86^{-10}$ | $2.59^{09}$ |
| 26 | 0.45 | 0.4048 | $1.56^{-10}$ | $6.40^{09}$ |
| 27 | 0.41 | 0.4349 | $6.80^{-11}$ | $1.47^{10}$ |
| 28 | 0.44 | 0.4118 | $2.80^{-11}$ | $3.57^{10}$ |
| 29 | 0.46 | 0.3981 | $1.11^{-11}$ | $8.97^{10}$ |
| 30 | 0.46 | 0.3981 | $4.44^{-12}$ | $2.25^{11}$ |

that the source animal be DNA sampled and profiled and that the profile be available in a database. Systematic sampling of live animals at the ranch or feedlot level can be logistically complex, expensive, and error prone. While breeding stock can be DNA profiled, in practice this provides limited scope for source attribution. In the event of a bioterror or food safety incident, it is of limited value to know the parents of the animal without also knowing the pathway by which that animal moved through the supply chain. As shown in Figures 5.1 and 5.2, there are complex pathways from source breeding animals to the meat product sold to the consumer and contamination may occur at any point across this full chain.

At point of slaughter, animals are presented sequentially as carcasses on a production line in a more controlled environment. Each carcass is identified by a unique kill number that is generated by the slaughterhouse information system. A number of sampling systems have been developed that integrate the collection of a tissue sample from the carcass with the correlation of the sample to the kill number. For payment purposes, the kill number correlates to the livestock supplier and the linkage is therefore of high integrity. In cases where the live animal is identified individually, as required by European Union legislation, for instance (6), the kill number can also be associated directly with ear tag data and movement history of the animal. Figure 5.4 illustrates the reduced complexity of a DNA-based system.

The authors' experience of operating genetic identification programs across the international and commercial meat processing arena allows us to make some observations on the potential for this approach to be adopted as a tool in microbial forensics. However, to be clear, the motivation for commercial users of this technology today is not bioterrorism but rather validating product quality attributes or commercial specifications, such as organic or grass fed; angus, hereford, or other breeds; and maturation period, among many others.

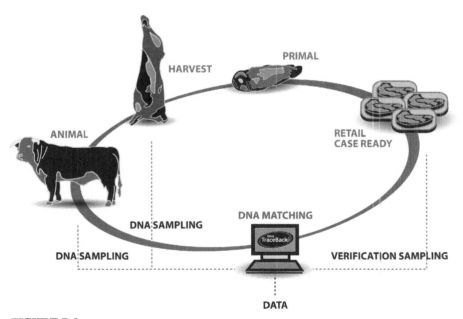

**FIGURE 5.4**
Illustration of DNA identification concept in meat tracking.

It has been our experience, from first implementing industrial-scale DNA traceability programs for beef in 1998 (11), that systematic sampling of carcasses and subsequent source identification of derived meat products, up to point of consumption, is straightforward. Moreover, using the principles of Hazard Analysis and Critical Control Point, we sample at risk points in the meat supply chain to monitor conformance to specification and to help pinpoint the source of contaminated product in the event of an incident. Routine validation of system conformance mitigates risks for commercial users.

## WASHINGTON STATE BSE-POSITIVE COW

The Washington State BSE-positive cow was *not* slaughtered for human consumption; if it had been, there would be little or no public health risk because the specified risk material should be removed postmortem. It is therefore unlikely that a recall would be invoked even if meat from this animal had entered the human food chain. However, it is helpful to consider this hypothetical scenario to better understand the capability of genetic identification to assist in tracking the affected meat.

At first we must assume that the cow was slaughtered in a slaughterhouse that has systematic DNA sampling on the production line. This is a straightforward process and is already operated in many slaughter facilities. Figure 5.5 shows a picture of the types of sampling devices used routinely, in large volumes, for this purpose.

DNA sampling may be performed by the commercial operator, by a third-party contractor, or by USDA Food Safety and Inspection Service inspectors (subject to necessary approval). Samples are typically aggregated on a daily basis and sent to the testing laboratory for DNA analysis.

Cull cows, as distinct from prime beef cattle, are often processed for ground beef for the food service, further manufacturing, and retail sectors and we assume that meat from the infected animal is destined for these channels. From slaughter the carcass is fabricated into primal cuts, along with all of the other carcasses slaughtered that day. This may be anything from 100 to 5000 animals. Typically, fabricated primals and the trim from these primals come off the manufacturing line into 2000-lb combo bins, and the combo bins are filled by primal type and trim, graded by fat content. This variation of combos allows ground beef of a desired final fat content to be blended or a defined cut, for example, sirloin ground beef, to be prepared. At this point, the combos may be joined by deliveries from other fabrication plants and subsequently ground on site or transported to a separate grinding facility. As the meat from the BSE-infected animal moves through the process, there are batch and blend records prepared to indicate the production source of the raw material, and so within the fabrication and grinding phase we are able to determine what batches are likely to contain meat from the BSE cow. In order to determine whether a particular product is contaminated, at, say, the postgrinding stage, we sample the ground beef against a statistical plan. If we wanted to be confident that the product contained less than 5% meat from the source cow, 60 subsamples would provide 95% confidence of detecting the BSE animal if present at ≥5%.

The ability to trace individual contributing animals from a complex mixture, such as ground beef, is entirely feasible and is conducted routinely by IdentiGEN Inc. in the performance of its DNA traceability services. Individual muscle fibers are removed from the meat sample and DNA is extracted. Figure 5.6 illustrates schematically the approach taken, and Table 5.2 provides sample data from an actual analysis of ground beef to illustrate the concept. Dates and sample codes have been changed to fit the hypothetical case and to make actual data anonymous.

As we know that the infected cow was slaughtered on December 23, 2003, and its slaughter number is 611-122303-1234 (the combination of slaughter number, date, and establishment number creates a unique carcass code), we see

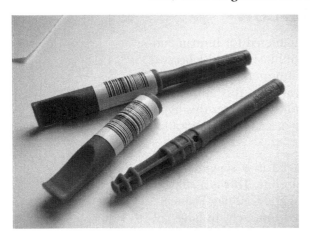

**FIGURE 5.5**
Single-use device for DNA sampling of carcasses.

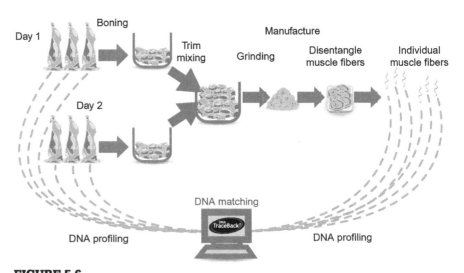

**FIGURE 5.6**
DNA identification of source animals for ground meat products.

from Table 5.2 that 611 is present in the batch of 80% lean chuck prepared on December 29.

Beyond simply identifying the specific animal of origin, routine DNA monitoring is used to track production parameters used for process performance optimization and quality control. For instance, in Table 5.2, carcass 425 is overrepresented, which may indicate an uneven blending of raw material. More specifically, it is possible to estimate the actual number of unique carcasses represented in 1 lb of ground beef or other unit volume. We have conducted experiments where extensive subsampling is performed and the count of recurrent DNA profiles is used to predict the total number of contributing individuals.

For instance, 54 unique profiles were detected among 87 subsamples taken from a 1-lb pack of ground beef. Based on a binomial statistical model, and given observed data, the expected total number of animals in the pack is predicted to be ≈90. This result can be confirmed by simulation. Figure 5.6 compares the pattern of "recurrence" of individual profiles to simulated data from three ground beef products, where $N$ is the number of known contributing animals. The correlation between observed data and $N = 90$ can be seen to be very strong.

This allows the investigator to consider sampling finished ground beef products in the market at large, whether food service, manufactured, or at the grocery

**Table 5.2** Individual Carcass Matches from 10 Subsamples of Ground Beef[a]

| Ground Beef Subsample ID | Matched Carcass Sample ID | Carcass Kill Date | Kill No. | Producer Code | Match Likelihood | Interval Day |
|---|---|---|---|---|---|---|
| 1 | 100115108 | 23 Dec 03 | 425 | 771 | $3.60^{10}$ | 5 |
| 2 | 100063218 | 23 Dec 03 | 598 | 725 | $7.02^{11}$ | 5 |
| 3 | 100017306 | 22 Dec 03 | 974 | 711 | $2.25^{14}$ | 6 |
| 4 | 100017212 | 23 Dec 03 | 425 | 771 | $8.63^{10}$ | 5 |
| 5 | 100115108 | 23 Dec 03 | 425 | 771 | $7.20^{10}$ | 5 |
| 6 | 100081108 | 23 Dec 03 | 1004 | 711 | $7.42^{13}$ | 5 |
| **7** | **100081373** | **23 Dec 03** | **611** | **725** | **$4.40^{13}$** | **5** |
| 8 | 100141475 | 22 Dec 03 | 648 | 702 | $1.36^{13}$ | 6 |
| 9 | 100081656 | 23 Dec 03 | 602 | 725 | $4.63^{13}$ | 5 |
| 10 | 100114910 | 23 Dec 03 | 123 | 788 | $3.56^{10}$ | 5 |

[a]Each subsample is an individual muscle fiber removed from the ground beef product. "Interval day" refers to time elapsed between the product pack date and kill date.

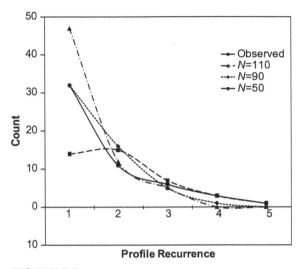

**FIGURE 5.7**

Recurrence of individual DNA profiles in a sample of ground beef, where $N$ is the actual number of contributing animals.

store, and positively including or excluding meat products from contamination by the affected animal. This provides a practical means by which a high level of precision can be achieved. Combined with available production records, DNA identification therefore offers unparalleled ability to assist in the detection of contamination. A more extensive analysis will help characterize the distribution pathway of the affected product and help narrow the scope of recall.

While it is likely that an animal such as the Washington State BSE-positive cow would be manufactured into hamburger, genetic identification technology is applied more easily to the whole muscle cuts and steaks. Whole muscle cuts can generally be traced directly to the reference carcass without recourse to probabilistic models.

## FOOD SAFETY

A further dimension that has particular relevance to microbial safety and quality control is the ability to identify the production age of raw material used in a meat-grinding operation. Figure 5.7 illustrates the distribution of days between slaughter and grinding and packing. It can be seen clearly that there is a small quantity of older material being introduced to the system. This may be "rework" that has persisted in the grinding operation or it could be frozen material introduced intentionally. The ability to detect and measure this type quality attribute allows meat processors to better control their production systems.

## PRACTICAL LIMITS

Naturally there are limits to what DNA identification can achieve. If the meat product has been consumed and there is no residual product for testing, it will not be possible to track the source, although blood residue from retail packs removed from trash has provided us with a usable source of DNA evidence. Further processed and cooked products such as hotdogs, sandwich meats, and ready meals require such a high level of forensic analysis as to be economically or technologically impractical.

There are also operational issues associated with large-scale sampling programs such as suggested here, particularly when reference sample acquisition is conducted in an industrial rather than forensic setting. Figure 5.8 shows unexpected DNA matches between carcasses. Data are taken from one slaughter

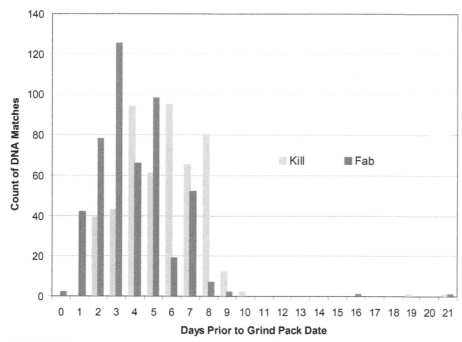

**FIGURE 5.8**
Time intervals between grind pack date (0), slaughter, and fabrication for a batch of ground beef samples.

plant, where, over a 3-week period, ≈30,000 cattle were slaughtered and DNA profiled. Following a pair-wise comparison of each carcass profile (Figure 5.9), a small number of identical profiles were detected. In examining data, it became clear that the matches are generally found between carcasses killed on the same day and within 1 or 2 carcasses of each other on the slaughter line. This strongly suggests that the operator has mistakenly resampled the same carcass twice and missed the intended carcass—an example of operator error. At the level of 47 "missed" carcasses out of ≈30,000, or 0.15%, the effect is negligible on system performance. For the small number of matches detected that do not occur close together on the same day, it can be shown by tracing the kill numbers to ear tag data that these animals were born on the same farm and are most likely to be identical twins. This genetic artifact would not confound any real-world investigation. Moreover, when matching criteria are relaxed in order to allow a certain level of mismatching between profiles, we have identified a cohort of carcass pairs that represent close relatives, usually siblings or half-siblings (data not shown).

Missing reference sample data, which may be accounted for by sampling error such as that outlined above or simply samples not taken, can be accommodated within a Bayesian statistical framework that allows inferences to be made when the evidence meat sample does not match to a reference carcass.

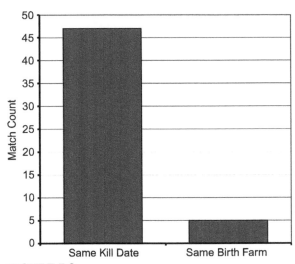

**FIGURE 5.9**

Carcass-to-carcass matches following pair-wise comparison among 29,987 source carcass samples.

## SUMMARY AND FUTURE DIRECTIONS

Genetic identification is being used routinely, albeit in clearly defined commercial contracts, to provide effective traceability of meat products, particularly pork and beef. The conceptual framework for these programs acts as a model for microbial forensics. In order to harness the individual-level identification afforded by genetic analysis, reference samples from the point of slaughter are required. It is not for the authors to suggest how the investigating authorities might access or mediate sampling of carcasses, but in the European case it would be relatively straightforward for officials currently sampling carcasses for BSE testing to be tasked with DNA sample collection.

To operate DNA traceability for meat at a national scale in the United States would be a significant undertaking; as stated previously, there are approximately 33 million cattle slaughtered each year and more than 100 million hogs. The largest human DNA database, that of the UK Home Office, contained 3.4 million individual profiles in 2005, gathered over a 10-year period, using STR technology (22). However, taking cattle alone, and at cost, it would take perhaps $60–100 million to DNA sample and profile each individual animal using present technology. Nonetheless, operating at that scale would drive rapid innovation in the automation of sample collection technology, DNA extraction, and SNP genotyping analysis and inevitably deliver substantial cost savings.

Indeed, molecular genetic detection systems that can be deployed at point of use are creating new opportunities for timely diagnosis in the clinical setting. Development of real-time, point-of-use genetic identification technology would allow DNA identification to form part of the process records generated in the ordinary course of meat processing operations. Combined with enterprise information systems, DNA could eventually be used in a manner similar to bar codes and radio-frequency label technology. That day remains some considerable time in the future.

## REFERENCES

[1]  M. Maxwell, FDA clears tomatoes of salmonella, now suspects chili peppers. Informify News, July 21, 2008. Available from: www.informify.com/top-stories/48-health/320-fda-clears-tomatoes-of-salmonella-now-suspects-chili-peppers.

[2] W.T. Minnesota woman sues Cargill for $100 million in E. coli case. Pioneer Press December 4, 2009. Available from: www.allbusiness.com/legal/legal-services-litigation/13528664-1.html.

[3] E. Fredrix, Salmonella recall could cost peanut producers $1B. Brattleboro Reformer, March 11, 2009. Available from: www.allbusiness.com/company-activities-management/company-structures-ownership/12223580-1.html.

[4] USDA. Livestock Operations 2003 Summary. National Agricultural Statistics Service 2004. Available from: http://usda.mannlib.cornell.edu/usda/current/LvStkOp/LvStkOp-04-30-2004.pdf.

[5] J. McGeary, Growing opposition to the National Animal Identification System. Wise Traditions in Food, Farming and the Healing Arts 2009. Available from: www.westonaprice.org/NAIS-Update-Fall-2009.html.

[6] Council Regulation No. 1760/2000 of the European Parliament and of the Council of 17 July establishing a system for the identification and registration of bovine animals and regarding the labelling of beef and beef products and repealing Council Regulation (EC) No. 820/97. Official Journal of the European Community 2000 L 268, 24–28.

[7] Can-Trace. Canadian Food Traceability Data Standard Version 2.0. Can-Trace Secretariat 2006. Available from: http://www.can-trace.org/portals/0/docs/CFTDS%20version%202.0%20FINAL.pdf.

[8] Meat and Livestock Australia. National Livestock Identification System 2010. Available from: http://www.mla.com.au/TopicHierarchy/IndustryPrograms/NationalLivestockIdentificationSystem/default.htm.

[9] T. Ozawa, N. Lopez-Villalobos, H.T. Blair, An update on beef traceability regulations in Japan, Proc. New Zealand Soc. Anim. Prod. 65 (2005) 80–84.

[10] A. Martin, Largest recall of ground beef is ordered, The New York Times (2008). Available from: www.nytimes.com/2008/02/18/business/18recall.html.

[11] C. Meghen, C.S. Scott, D.G. Bradley, D.E. MacHugh, R.T. Loftus, E.P. Cunningham, DNA based traceability techniques for the meat industry, Anim. Genet. 29 (1998) 48–59.

[12] D.J. Balding, Weight-of-evidence for forensic DNA profiles. 2005. Statistics in Practice.

[13] J.M. Butler, Fundamentals of Forensic DNA Typing, Elsevier Academic Press, 2009.

[14] G.A. Rohrer, B.A. Freking, D.J. Nonneman, Single nucleotide polymorphisms for pig identification and parentage exclusion, Anim. Genet. 38 (2007) 253–258.

[15] M.P. Heaton, G.P. Harhay, G.L. Bennett, R.T. Stone, W.M. Grosse, E. Casas, et al., Selection and use of SNP markers for animal identification and paternity analysis in U.S. beef cattle, Mamm. Genome 13 (2002) 272–281.

[16] F.A.O. Werner, G. Durstewitz, F.A. Habermann, G. Thaller, W. Kramer, S. Kollers, et al., Detection and characterization of SNPs useful for identity control and parentage testing in major European dairy breeds, Anim. Genet. 35 (2004) 44–49.

[17] SNPs: Variations on a theme. National Center for Biotechnology Information 2007. Available from: www.ncbi.nlm.nih.gov/About/primer/snps.html.

[18] A. Amorimab, L. Pereiraa, Pros and cons in the use of SNPs in forensic kinship investigation: A comparative analysis with STRs, Forensic Sci. Int. 150 (1) (2005) 17–21.

[19] Committee on DNA Forensic Science: An Update, National Research Council. The Evaluation of Forensic DNA Evidence, p. 19. National Academy Press, 1996.

[20] S. Kim, A. Misra, SNP genotyping: Technologies and biomedical applications, Annu. Rev. Biomed. Eng. 9 (2007) 289–320.

[21] SNP genotyping methods surge ahead. Genet. Eng. Biotechnol. News 26 (2006) 11.

[22] The National DNA Database. Home office. Available from: www.homeoffice.gov.uk/science-research/using-science/dna-database/index.html.

# CHAPTER 6

# Microbial Forensics of RNA Viruses: Foot-and-Mouth Disease Virus

**Consuelo Carrillo**

*Plum Island Animal Disease Center, Animal and Plant Health Inspection Service,
U.S. Department of Agriculture, Greenport, New York*

## CHALLENGES POSED BY RNA VIRUSES

Over the past decades, it has been well established that RNA viruses share high mutability and rapid evolutionary capabilities (1,2). Contrary to DNA systems, RNA viral pathogens exhibit a degree of genetic plasticity that may, in part, explain why there are frequently "emerging pathogens" in new hosts (3–7). The high mutation rates and rapid evolutionary capabilities of RNA viral genomes require completely different approaches from those taken for DNA forensics.

This chapter mostly focuses on the microbial forensic challenges of highly variable RNA viruses using the example of foot-and-mouth disease (FMD), an economically devastating and highly transmissible disease of livestock and other animals. The FMD virus (FMDV) was the first animal virus discovered and belongs to the Picornaviridae family (8–10). Like other important human pathogens, such as poliovirus, hepatitis A, and rhinovirus, FMDV contains a positive-sense, single-stranded RNA genome of approximately 8000 nucleotides that is translated directly into a single polyprotein upon entry into the host cell cytoplasm. FMDV exhibits high mutation rates that result in genetically heterogeneous populations in continuous competition and that allow selection of best-adapted variants for each particular environment in a quasi-deterministic fashion (11–13). Despite the large body of molecular knowledge concerning biochemical and genetic properties of FMDV, there is a lack of systematic and statistically supported genetic studies directed toward reaching a consensus about meaningful parameters to be included in a microbial forensic investigation. Generation of this knowledge has started recently, but our capabilities to define, analyze, and obtain conclusions from data obtained at the molecular level are still very limited. Acquiring and understanding the necessary data will depend on new technologies (14–16), new bioinformatics tools (17–19), and global accessibility of data (20). Newly developed tools,

Microbial Forensics. DOI: 10.1016/B978-0-12-382006-8.00006-2

such as pyrosequencing (21–23), ultradeep parallel sequencing (24), microarray technology (25,26), nanotechnology (biosensors) (27), and sequencing by oligonucleotide ligation and detection (28–30), will be a great help in molecular forensic studies.

## A CASE STUDY: FOOT-AND-MOUTH DISEASE IN THE UNITED KINGDOM (2001)

In 2001, after being free of FMD for more than 20 years, the United Kingdom suffered its worst FMD epidemic ever. In just 11 months, at least six million animals were destroyed and losses were calculated to exceed £12 billion (31–35). Losses involved not only the agricultural sector of the economy, but also sectors integrated with agricultural and tourism-related industries. The agricultural, financial, and social costs of controlling the outbreak and the long-term effects on farming and rural communities were incalculable. The exact origin of the infection and the transmission pathways were never determined. However, over the past 25 years, a strain of FMDV serotype Pan Asia O has spread from India throughout Southern Asia and the Middle East. During 2000, this strain caused disease outbreaks in the Republic of Korea, Japan, Russia, Mongolia, and South Africa (36,37). In February 2001, the Pan Asia O strain spread to Europe (38). Studies of sequences of the viral capsid protein VP1 (also known as 1D protein) coding region from approximately 30 Pan Asia O isolates demonstrated that the U.K. virus was closely related to all of them and nearly identical to the South African isolate O/SAR/1/2000 (39). Notably, nucleotide sequences of the VP1 coding regions of 30 Pan Asia O viruses isolated over an 11-year period differed by no more than 5%, making meaningful phylogenetic and forensic analyses extremely difficult (36,40). Complete genome sequence analysis of eight Asian, African, and European isolates of Pan Asia O strains confirmed the close relationship between the South Africa and the U.K. outbreaks, but failed to identify, or even imply, the mechanism of introduction or the source attribution for the latter outbreak. These results were consistent with either a common source for both the 2000 South Africa and the 2001 U.K. outbreaks or that O/SAR/1/2000 is the source of the strain that caused the U.K. outbreak (41). To provide a definitive answer required new methods that allow a statistical evaluation of the likelihood of a virus belonging to a specific lineage.

After the outbreak was controlled, a number of questions were asked: What was the source of the virus? How was it introduced? Was its introduction accidental or deliberate? Who was responsible if the introduction was deliberate? If it were a deliberate release, could natural virus or artificially manipulated isolates be distinguished? Could the isolate have enhanced virulence or adapted host range? None of these questions can yet be answered with confidence.

Lack of critical understanding of the 2001 outbreak can be contrasted with detailed knowledge of the 2007 FMDV outbreak. In that incident, complete genome sequence analysis of the isolates allowed full identification of the initial and intermediate sources of the outbreak, revealed a probable chain of transmission events, predicted undisclosed infected premises, and connected a second cluster of outbreaks in September to those in August (42).

## WHAT INFORMATION IS NEEDED

Highly polymorphic genomic areas are usually attractive for forensic analysis because of their high powers of discrimination. However, because these areas undergo frequent mutation and are subject to an unknown number of natural selective pressures, interpretation of their genetic profiles requires knowledge of genetic variability, which includes mutation frequency, mutation rates, mutation and recombination processes, tolerance and constraints of genes and proteins, and evolution and memory of the microbial population. This information is critical for choosing the most informative genomic areas, or a combination of them, for analysis.

The evolution of FMDV has been studied extensively in cell culture, indicating that features of RNA replication are as follows: high mutation rates, short replication times, and high progeny yields. Misincorporations of at least 1 nucleotide every 1000 to 100,000 nucleotides have been reported as normal values for FMDV. Mutation rates per nucleotide site in the range of $10^{-3}$ to $10^{-5}$ for a 10-kb genome ensure that each progeny genome includes an average of 0.1–10 mutations (11). There is an impressive amount of sequence information from the carboxy-terminal region of the FMDV capsid protein VP1 gene (43), driven in part because this region carries neutralizing antibody epitopes important for the selection of suitable vaccines. Rates of nucleotide substitution for VP1 ranging from 0.5 to $1.5 \times 10^{-2}$ nucleotide substitutions/per site/per year occur on this short fragment of about 633 nucleotides (data are almost exclusively from tissue culture-adapted isolates). This capsid protein carries the viral signal to interact with the cellular receptor and many linear and conformational epitopes (8). Therefore, VP1 is known to be under strong host selective pressure: combined with the high adaptability of the virus to genetic change makes this region of VP1 theoretically appropriate for analysis of phylogenetic relationships between isolates (44–46).

Currently, approximately 2400 partial or complete VP1 sequences of the seven FMDV serotypes are available in GenBank. In general, these sequences show that viral populations circulating and causing disease outbreaks around the world are genetically heterogeneous but related, being distributed by regions in genetically distinct virus populations known as "topotypes" that can be differentiated based on nucleotide sequence differences of up to 15% (46).

These sequences enable scientists to unambiguously identify the strain of virus responsible for an outbreak, but do not allow precise identification of strain origin. High selective pressure and the plasticity of the capsid protein gene accumulate sequence changes that prevent the use of VP1 phylogenetics as a forensic tool.

Although FMDV complete genome sequences have become more abundant in recent years, the amount of data available is still limited. In most cases, restricted genomic areas are characterized extensively and redundantly, whereas many others are underrepresented. Similarly, some geographic regions have submitted large numbers of closely related isolates (resulting in almost identical sequences), whereas viruses from other endemic areas are absent from the common databases. Complete genome sequences for some FMDV serotypes are still not available in significant number or are in access-restricted databases. Therefore, there is a pressing need to establish a global-shared FMDV database, with universal codes of information and rules for annotation.

Experiments comparing genomic positions, numbers of mutations, and kinds of substitutions accumulated in the consensus sequence of the same FMDV isolate during replication in pigs and in BHK-21 cells (47–49) demonstrate that FMDV sequences from tissue culture are affected greatly by host change and should be considered carefully in analyses of phylogenetics and evolution. We lack understanding of the principles that regulate these restrictions and the differences between *in vivo* and *in vitro* mutation tolerance. Although the evolution of RNA viruses is the product of the combination of an unknown number of factors, ultimately there must be a number that represents the probability that a given nucleotide and/or amino acid can be substituted and still allow the virus to succeed as the new master sequence of a given viral population. Complete genome sequence analysis of distantly related FMDV isolates will better define the probability of substitution, even though the bottlenecks and selective pressures affecting the viruses are unknown. A comparison of complete genomes of FMDV isolates from very distant and unrelated sources has provided a wider perspective of the means of genetic variability for FMDV and knowledge of genetic constraints and limits of variation within phylogenetic groups (50,51).

In the future, using standardized and detailed annotation rules for more complete FMDV genomes will allow collection of data with the capability to reveal functional domains and molecular markers shared by sets of individuals with specific characteristics as well as the shape of their phylogenetic relationships. Random mutations and transient substitutions of no biological value would be excluded easily during analysis of a new circulating strain and will allow knowledge of genetic imprints of selective value that confer selective advantage under specific circumstances. However, from a forensic point of view, there will be those "improbable" changes, which are unusual, deleterious for the virus, or out of the range of natural probability, that will be the ones with real informative value. Thus, knowledge of conserved genomic sequence regions, as well as the probability of a given mutation occurring in nature, would be of great value in determining whether a virus has been altered unnaturally.

A common observation in sequence alignments is the high frequency of certain specific nucleotide and amino acid substitutions. For example, during viral evolution, variable positions in FMDV capsid proteins are alternatively occupied by a small subset of all possible amino acids, and a true accumulation of amino acid substitutions is not observed. This feature of natural adaptation and/or evolution of FMDV could be used for attribution and origin interpretation of new isolates. Identification of less variable genomic areas with high informative content would be of great interest for current phylogenetic procedures. Consensus must be reached about the most informative areas of the genome for each investigatory purpose in order to elaborate a universal standard operating protocol for microbial forensic analysis. In some cases, an abundance of certain types of nucleotide and amino acid substitutions or an absence of many others may reflect constraints on allowable variation for the viral RNA genome, probably due to the complexity of viral functions required for cell–host interactions. However, when comparing nonstructural proteins, the paucity of available sequences for comparison may lead to wrong conclusions, suggesting a constraint where there is just underrepresentation (51,52). Clarification of the tolerance for change in different genomic regions is absolutely necessary for further progress in viral forensic analysis and interpretation.

The FMDV studies of González-Candelas and Moya (53) strongly suggest that phylogenetic reconstruction methods can infer erroneous phylogenies due to nucleotide convergences between isolates belonging to different experimental lineages that join by accident in time or space sampling. These authors also point out that diverse evolutionary mechanisms acting under different experimental dynamics can generate alterations and change the frequencies of genetic variants, which may lead to misinterpretation of the real evolutionary history. Procedures for statistical evaluation in classical molecular epidemiology may not be useful for forensic purposes. For example, when variability of the system is too high, as is the case for many RNA viruses, bootstrap support for identification of the relevant node defining the monophyletic clade becomes useless from a forensic point of view because this only shows the probability of the two alternative hypotheses. For some forensic analyses, maximum likelihood testing is the method of choice for evaluating competing phylogenetic hypotheses when linked to other types of evidence (e.g., other genetic and/or epidemiological information) and provides quantitative criteria for deciding between alternative possibilities. Similarly, statistical evaluation of alternative phylogenetic hypotheses may be used to single out the likelihood of an isolate sharing a parental virus with other isolates as compared to the existence of a monophyletic clade that includes all the isolates related to a common source. More data are needed to reach conclusions and establish genetic markers of viral growth in cell cultures. Similar studies need to be enlarged to host range *in vivo* and *in vitro* and also under partial immune protection.

## WHERE WE ARE NOW

A few important contributions to FMDV microbial forensics have been published since the previous edition of this book, but much more information is needed. Rapid sequencing of complete FMDV genomes has proven to be extremely useful, not only in demonstrating informative capability and value relevant to outbreak control (42), but also by raising a series of important questions about the epidemiology and biology of FMDV that need urgent attention. For example, discovery of FMDV SAT viruses with new genomic signatures that place them as a mixture of Euroasiatic and African lineages (50,51) exposes how necessary it is to obtain sequences from more isolates of all three SAT serotypes. This information would further understanding of the evolutionary dynamics of the sublineages circulating in Africa and how they relate to other sublineages. Another important observation was rapid accumulation of mutations during replication in natural hosts in the absence of specific immune pressure that lead to a gradual loss of virulence (but not of viral titer in BHK-21) and, finally, interruption of disease transmission (47). This unexpected result, reminiscent of transmission bottleneck and the effects of the Muller Ratchet *in vitro*, suggests a whole new interpretation of transmission of FMDV between natural hosts in a confined environment. Even more important from a forensic perspective is demonstration of totally different patterns of evolution when the virus replicates *in vivo* or *in vitro*. This absolute difference in evolution was predicted during the search and evaluation of monoclonal antibody-resistant mutants in viral swarms of FMDV serotype C, strain Santa Pau (C-S8c1) from each one of the cell culture and pig hosts (48,49). In those experiments, the frequency of mutants *in vivo* was more than 1 log lower (10 times more difficult to find) than the frequency of mutants *in vitro*. More interesting was the finding that the amino acid positions able to be substituted, the number of nonsynonymous substitutions, and the identity of the residues that were able to replace the parental sequence were notoriously different. These results have been confirmed in a second set of experiments taking advantage of new methodologies (47). Measurements of the effects of selective pressure in the total genome of FMDV (dN/dS and transition/transversion rate ratios between others) show an up to 10 times difference between replicates in pig or in BHK-21 cells, and the distribution of changes in the genome is different, indicating the possibility of using such differences to establish forensic markers with specific values.

Another important contribution was the spatial distribution and quantification of FMDV genetic diversity over the course of the U.K. 2001 epidemic, which provided new information on the nature of FMDV transmission during an outbreak without influence of cell culture adaptation of the virus. This work showed that having the right samples and full-length genome sequences can lead to inference of the trajectory of an epidemic even on premises where no actual samples were taken (41).

An additional problem, never resolved by partial genome analysis, is the high frequency of homologous and nonhomologous recombination of FMDV (and genome segment reassortment of other RNA viruses). Recombination occurs at a high frequency in field isolates and vaccines, probably acting as a major evolutionary force in the expansion of many viruses. Significantly, many emerging viruses belong to virus families that recombine actively (54–56). Naturally occurring FMDV recombinant viruses have long been described. However, genetic evidence from full-length sequences of natural FMDV isolates of all serotypes and different origins shows that recombination is much more common than previously thought and plays a significant role in the evolution of FMDV in nature (50,57). Thus, classification and identification of circulating FMDV isolates based solely on VP1 may lead to highly erroneous interpretations.

## WHAT NEEDS TO BE DONE

The scarce data collected from *in vivo* experiments, epidemiological data, functional analysis, and structural studies indicate that, in contrast to *in vitro* results, kinds of substitutions, number of changes in the genome, and positions affected by mutations at any specific time point in evolution are restricted: in other words, data suggest existence of thresholds for the expression of phenotypic traits, which may underlie the *in vivo* restriction of variability rates. Thus, while high mutation and recombination rates may lead to survival and adaptability of the species, variation is limited by functional and structural constraints to preserve continuity of the "core" information. However, despite such evidence, data to advance phylogenetic hypotheses for forensic purposes (with a confidence interval of 95% or greater) and to provide the necessary statistical support to convert data into evidence are lacking (53,58–60), including mutation range and characteristics in natural infection. Since the initial development of multiplex polymerase chain reaction tests and microarrays of the 1990s, scientific advances, combined with robotics and bioinformatics, have made it possible to improve the new molecular technologies to the point of being robust clinical diagnostic tools.

For forensic purposes, a globally representative set of informative signatures should be used as genetic profiles of viral populations. In that sense, limited FMDV sequence information from genomic regions other than VP1 has hindered the search for other genetic markers with informative value in other areas of the genome. Identification of such informative areas of the genome, other than VP1, will improve routine epidemiological analyses and meet the new demands of forensic microbiology. Discovery of functional signatures and genetic markers of positive selection will permit assessment of the probability that two or more mutations can occur within a given viral master sequence in a similar proportion and in two distinct viral populations, thus providing a tool to distinguish between two very closely related individuals. The challenge now is to identify the right combination of genetic markers that cover a wide range of possible phenotypic traits. While genes with slower rates of mutation fixation may be useful for analyzing natural evolution, genes with higher rates may be useful for analysis of recent events. By careful choice of the genomic region to study, based on evolutionary divergence levels for the issue under consideration, the necessary information will be provided to find the best genetic targets for analysis of alternative hypotheses, which is why the use of high-throughput DNA sequencing for routine analysis of complete genomic sequences of FMDV isolates is now imperative. As explained earlier, a public database of full-length FMDV genomes will allow (i) assessment of mutation frequency, mutation rates, mutation and recombination processes, tolerance and constraints of genes and proteins, and evolution and memory of FMD viruses in natural isolates; (ii) precise identification of strains circulating around the world for attribution; (iii) estimation of the probability of a given mutation occurring in nature for identification of unnaturally altered viruses and specific signatures; and (iv) statistical support for hypothesis development and evolutionary model selection.

Pyrosequencing has opened the possibility for rapid analysis of minority genetic variants or "memory genomes," which are defined as specific accompanying mutants of the major FMDV population in cell culture (ranging from about 0.1 to 20% of the total number of genomes in the viral population analyzed). These memory genomes arise as a consequence of population dynamics, when viral swarms are subjected to discontinuous selective pressures, and their analysis helps in understanding the biological history of the virus. These genomes remain undetectable by consensus sequencing and have been analyzed by cloning or other elaborate techniques (61). New multiparallel pyrosequencing techniques and biosensors will allow their application to forensic purposes (62).

Microarray-based techniques offer simultaneous detection of a wide variety of genome contents at levels of efficiency, redundancy, sensitivity, specificity, and reproducibility that are most promising for microbial forensics (63–65). Pseudo-pan viral arrays have been used for detection and identification of viruses (66), pathogen discovery (67), genome recombination (68), and strain typing (69) and will revolutionize our current concepts of laboratory testing for known and unknown pathogens, including host-specific responses to infection. Incorporation of host genome-specific markers to the array adds forensic informative power about the source of the sample. Resequencing arrays allow analysis of viral variants and sequences of the genetic signatures of previously characterized microorganisms (70,71) that indicate viral isolate origin.

New algorithms are needed for forensic investigations to examine the cladistic structure of trees as a function of the distance of each case from the first infection source and to evaluate the precision and robustness of the historical reconstruction. An estimate of the history of transmission events in a particular epidemic can be extracted from reconstructed "epidemic trees" in which the case-reproduction ratio for the spread of the infectious disease has statistical reliability (42). Selection of the appropriate evolutionary model is a critical premise for forensic attribution, keeping in mind that phylogenetic reconstructive methods involve mathematical simulations that grossly simplify the relationship between organisms. The unavoidable translation of the biological properties of a virus in nature to numerical units is not an absolute and exact process. Data used in these analyses do not contain the complete historical record of the virus and are far from ideal in representing all possible evolutionary pathways of the population, which can result in erroneous phylogenies. We must take into account the stochastic nature of transmission and lack of knowledge about its consequences in viral population genetics, as well as the biases and gaps involuntarily inflicted during the collection and processing of epidemiological data. For instance, FMDV infects up to 70 species of cloven-hoofed mammals, but the extent of lesions and clinical disease is host dependent: sheep present with much milder disease than cattle or swine. In the 2001 U.K. epidemic, sheep played a prominent role in the early stages of the outbreak, laying the groundwork for the subsequent epidemic as the infection passed unobserved from sheep to sheep. Passage of virus in sheep for an unknown number of generations has the potential to distort the shape of the evolutionary tree, possibly masking parenthood relationships with previous isolates. Thus, knowledge of genetic changes associated with replication in specific host species needs to be incorporated into the design of new methods for genetic analysis.

There are still unexplored fields in FMDV microbial forensics for which discovery by application and standardization of old and new generation technologies will be critical. Our goal is not just to distinguish species and strains, but even individuals within complex microbial populations.

# REFERENCES

[1] E.D. Domingo, Virus evolution, in: B.N. Fields, D. Knipe, P. Howley (Eds.) Field's Virology, Lipincott-Williams and Wilkins, Philadelphia, PA, 2007.

[2] E. Domingo, V. Martin, C. Perales, A. Grande-Pérez, J. García-Arriaza, A. Arias, Viruses as quasispecies: Biological implications, Curr. Top. Microbiol. Immunol. 299 (2006) 51–82.

[3] E. Domingo, Viruses at the edge of adaptation, Virology 270 (2000) 261–263.

[4] S. Cleaveland, D.T. Haydon, L. Taylor, Overviews of pathogen emergence: Which pathogens emerge, when and why? Curr. Top. Microbiol. Immunol. 315 (2007) 85–111.

[5] M.E. Woolhouse, D.T. Haydon, R. Antia, Emerging pathogens: The epidemiology and evolution of species jumps, Trends Ecol. Evol. 20 (5) (2005) 238–244.

[6] R. Antia, R.R. Regoes, J.C. Koella, C.T. Bergstrom, The role of evolution in the emergence of infectious diseases, Nature 426 (2003) 658–661.

[7] C.R. Parrish, E.C. Holmes, D.M. Morens, E.C. Park, D.S. Burke, C.H. Calisher, et al., Cross-species virus transmission and the emergence of new epidemic diseases, Microbiol. Mol. Biol. Rev. 72 (2008) 457–470.

[8] E. Domingo, E. Baranowski, C. Escarmís, F. Sobrino, Foot-and-mouth disease virus, Comp. Immunol. Microbiol. Infect. Dis. 25 (2002) 297–308.

[9] M. Gromeier, E. Wimmer, A.E. Gorbalenya, Genetics, pathogenesis and evolution of Picorna-viruses, in: Origin and Evolution of Picornaviruses, Academic Press, 1999, pp. 287–343.

[10] R.R. Rueckert, J.l. Melnick, R.B. Couch, F. Blaine-Hollinger, J.R. Ticehurst, Picornaviridae, in: B.N. Fields, D.M. Knipe, P.M. Howley (Eds.), Fields Virology, vol. 1, Lippincott-Raven Publishers, Philadelphia, PA, 1995, pp. 609–810.

[11] E. Domingo, J.J. Holland, Mutation rates and rapid evolution of RNA viruses, in: S.S. Morse (Ed.), The Evolutionary Biology of Viruses, Raven Press, New York, 1994, pp. 161–184.

[12] E. Domingo, J.J. Holland, P. Ahlquist, RNA Genetics, vols. 2 and 3. CRC Press, Boca Raton, FL, 1988.

[13] A. Moya, E.C. Holmes, F. González-Candelas, The population genetics and evolutionary epidemiology of RNA viruses, Nat. Rev. Microbiol. 2 (2004) 279–288.

[14] L. Bodrossy, A. Sessitsch, Oligonucleotide microarrays in microbial diagnostics, Curr. Opin. Microbiol. 7 (2004) 245–254.

[15] M. Schena (Ed.), Microarray Biochip Technology, Eaton Publishing, Sunnyvale, CA, 2000.

[16] M. Margulies, M. Egholm, W.E. Altman, S. Attiya, J.S. Bader, L.A. Bemben, et al., Genome sequencing in microfabricated high-density picolitre reactors, Nature 437 (2005) 376–380.

[17] C. Rödelsperger, C. Dieterich, CYNTENATOR: Progressive gene order alignment of 17 vertebrate genomes, PLoS One 5 (2010) e8861.

[18] J.A. Nylander, J.C. Wilgenbusch, D.L. Warren, D.L. Swofford, AWTY (are we there yet?): A system for graphical exploration of MCMC convergence in Bayesian phylogenetics, Bioinformatics 24 (2008) 581–583.

[19] J.P. Huelsenbeck, P. Joyce, C. Lakner, F. Ronquist, Bayesian analysis of amino acid substitution models, Philos. Trans. R. Soc. Lond. B Biol. Sci. 363 (2008) 3941–3953.

[20] F.M. McCarthy, T.J. Mahony, M.S. Parcells, S.C. Burgess, Understanding animal viruses using the gene ontology, Trends Microbiol. 17 (2009) 328–335.

[21] V.M. Deyde, T. Nguyen, R.A. Bright, A. Balish, B. Shu, S. Lindstrom, et al., Detection of molecular markers of antiviral resistance in influenza A (H5N1) viruses using a pyrosequencing method, Antimicrob. Agents Chemother. 53 (2009) 1039–1047.

[22] C. Hoffmann, N. Minkah, J. Leipzig, G. Wang, M.Q. Arens, P. Tebas, et al., DNA bar coding and pyrosequencing to identify rare HIV drug resistance mutations, Nucleic Acids Res. 35 (2007) e91.

[23] N. Eriksson, L. Pachter, Y. Mitsuya, S.Y. Rhee, C. Wang, B. Gharizadeh, et al., Viral population estimation using pyrosequencing, PLoS Comput. Biol. 4 (2008) e1000074.

[24] F.D. Bushman, C. Hoffmann, K. Ronen, N. Malani, N. Minkah, H.M. Rose, et al., Massively parallel pyrosequencing in HIV research, AIDS 22 (2008) 1411–1415.

[25] J.G. Hacia, W. Makalowski, K. Edgemon, M.R. Erdos, C.M. Robbins, S.P. Fodor, et al., Evolutionary sequence comparisons using high-density oligonucleotide arrays, Nat. Genet. 18 (1998) 155–158.

[26] S.Y. Wen, H. Wang, O.J. Sun, S.Q. Wang, Rapid detection of the known SNPs of CYP2C9 using oligonucleotide microarray, World J. Gastroenterol. 9 (2003) 1342–1346.

[27] R. Levicky, A. Horgan, Physicochemical perspectives on DNA microarray and biosensor technologies, Trends Biotechnol. 23 (2005) 143–149.

[28] L. Kaderali, A. Deshpande, J.P. Nolan, P.S. White, Primer design for multiplexed genotyping, Nucleic Acids Res. 31 (2003) 1796–1802.

[29] M. Mokry, H. Feitsma, I.J. Nijman, E. de Bruijn, P.J. van der Zaag, V. Guryev, et al., Accurate SNP and mutation detection by targeted custom microarray-based genomic enrichment of short-fragment sequencing libraries, Nucleic Acids Res. Feb (2010) 17.

[30] B. Lin, A.P. Malanoski, Z. Wang, K.M. Blaney, N.C. Long, C.E. Meador, et al., Universal detection and identification of avian influenza virus by use of resequencing microarrays, J. Clin. Microbiol. 47 (2009) 988–993.

[31] M.J. Martin, F. Gonzalez-Candelas, F. Sobrino, J. Dopazo, It has been a difficult year for the British Ministry of Agriculture, Nat. Immunol. 2 (2001) 565.

[32] The Royal Society of Edinburgh. Inquiry into foot and mouth disease in Scotland, 2002. Available from: www.ma.hw.ac.uc.

[33] The Royal Society. Infectious diseases of livestock. Summary and recommendations, 2002. Available from: www.royalsoc.ac.uc.

[34] Foot and Mouth Disease 2001: Lessons to be learned inquiry. House of Commons, 2002. Available from: http://archive.cabinet-office.gov.

[35] D. Thompson, P. Muriel, D. Russell, P. Osborne, A. Bromley, M. Rowland, et al., Economic costs of the foot and mouth disease outbreak in the United Kingdom in 2001, Rev. Sci. Tech. 21 (3) (2002) 675–687.

[36] P.W. Mason, J.M. Pacheco, Q.Z. Zhao, N.J. Knowles, Comparisons of the complete genomes of Asian, African and European isolates of a recent foot-and-mouth disease virus type O pandemic strain (PanAsia), J. Gen. Virol. 84 (2003) 1583–1593.

[37] T. Kanno, M. Yamakawa, K. Yoshida, K. Sakamoto, The complete nucleotide sequence of the PanAsia strain of foot-and-mouth disease virus isolated in Japan, Virus Genes 25 (2002) 119–125.

[38] I. Nobiron, M. Rémond, C. Kaiser, F. Lebreton, S. Zientara, B. Delmas, The nucleotide sequence of foot-and-mouth disease virus O/FRA/1/2001 and comparison with its British parental strain O/UKG/35/2001, Virus Res. 108 (2005) 225–229.

[39] Origin of the Foot and Mouth Disease epidemic in 2001, Department of Environment, Food and Rural Affairs, 2002 and 2003. Available from: www.defra.gov.uk.

[40] N.J. Knowles, A.R. Samuel, P.R. Davies, R.J. Midgley, J.F. Valarcher, Pandemic strain of foot-and-mouth disease virus serotype O, Emerg. Infect. Dis. 11 (2005) 1887–1893.

[41] E.M. Cottam, D.T. Haydon, D.J. Paton, J. Gloster, J.W. Wilesmith, N.P. Ferris, et al., Molecular epidemiology of the foot-and-mouth disease virus outbreak in the United Kingdom in 2001, J. Virol. 80 (2006) 11274–11282.

[42] E.M. Cottam, J. Wadsworth, A.E. Shaw, R.J. Rowlands, L. Goatley, S. Maan, et al., Transmission pathways of foot-and-mouth disease virus in the United Kingdom in 2007, PLoS Pathog. 4 (2008) e1000050.

[43] N.J. Knowles, A.R. Samuel, Molecular epidemiology of foot-and-mouth disease virus, Virus Res. 91 (2003) 65–80.

[44] M.A. Fares, A. Moya, C. Escarmís, E. Baranowski, E. Domingo, E. Barrio, Evidence for positive selection in the capsid protein-coding region of the foot-and-mouth disease virus (FMDV) subjected to experimental passage regimens, Mol. Biol. Evol. 18 (2001) 10–21.

[45] D.T. Haydon, A.D. Bastos, N.J. Knowles, A.R. Samuel, Evidence for positive selection in foot-and-mouth disease virus capsid genes from field isolates, Genetics 157 (2001) 7–15.

[46] A.R. Samuel, N.J. Knowles, Foot and mouth disease type O viruses exhibit genetically and geographically distinct evolutionary lineages (topotypes), J. Gen. Virol. 82 (2001) 609–621.

[47] C. Carrillo, Z. Lu, M.V. Borca, A. Vagnozzi, G.F. Kutish, D.L. Rock, Genetic and phenotypic variation of foot-and-mouth disease virus during serial passages in a natural host, J. Virol. 81 (2007) 11341–11351.

[48] C. Carrillo, J. Plana, R. Mascarella, J. Bergadá, F. Sobrino, Genetic and phenotypic variability during replication of foot-and-mouth disease virus in swine, Virology 179 (1990) 890–892.

[49] M.A. Martínez, C. Carrillo, F. González-Candelas, A. Moya, E. Domingo, F. Sobrino, Fitness alteration of foot-and-mouth disease virus mutants: Measurement of adaptability of viral quasispecies, J. Virol. 65 (1991) 3954–3957.

[50] C. Carrillo, E.R. Tulman, G. Delhon, Z. Lu, A. Carreno, A. Vagnozzi, et al., Comparative genomics of foot and mouth disease virus, J. Virol. 79 (2005) 6487–6504.

[51] C. Carrillo, E.R. Tulman, G. Delhon, Z. Lu, A. Carreno, A. Vagnozzi, et al., High throughput sequencing and comparative genomics of foot-and-mouth disease virus, Dev. Biol. (Basel) 126 (2006) 23–30.

[52] H. van Rensburg, D. Haydon, F. Joubert, A. Bastos, L. Heath, L. Nel, Genetic heterogeneity in the foot-and-mouth disease virus leader and 3C proteinases, Gene 289 (2002) 19–29.

[53] F. González-Candelas, A. Moya, Time and rate of evolution are the key to establish transmission cases, AIDS 19 (2005) 1552–1553.

[54] J.H. Strauss, Recombination in the evolution of RNA viruses, in: S.S. Morse (Ed.), Emerging Viruses, Oxford University Press, Oxford, 1993, pp. 241–251.

[55] L. Heath, E. van der Walt, A. Varsani, D.P. Martin, Recombination patterns in aphthoviruses mirror those found in other picornaviruses, J. Virol. 80 (2006) 11827–11832.

[56] J. García-Arriaza, S. Ojosnegros, M. Dávila, E. Domingo, C. Escarmís, Dynamics of mutation and recombination in a replicating population of complementing, defective viral genomes, J. Mol. Biol. 360 (2006) 558–572.

[57] A.L. Jackson, H. O'Neill, F. Maree, B. Blignaut, C. Carrillo, L. Rodriguez, et al., Mosaic structure of foot-and-mouth disease virus genomes, J. Gen. Virol. 88 (2007) 487–492.

[58] M.A. Bracho, M.J. Gosalbes, D. Blasco, A. Moya, F. González-Candelas, Molecular epidemiology of a hepatitis C virus outbreak in a hemodialysis unit, J. Clin. Microbiol. 43 (2005) 2750–2755.

[59] F. Lesourd, J. Izopet, C. Mervan, J.L. Payen, K. Sandres, X. Monrozies, et al., Transmissions of hepatitis C virus during the ancillary procedures for assisted conception, Hum. Reprod. 5 (2000) 1083–1085.

[60] J. Izopet, C. Pasquier, K. Sandres, J. Puel, L. Rostaing, Molecular evidence for nosocomial transmission of hepatitis C virus in a French hemodialysis unit, J. Med. Virol. 58 (2) (1999) 139–144.

[61] C.M. Ruiz-Jarabo, A. Arias, E. Baranowski, C. Escarmís, E. Domingo, Memory in viral quasispecies, J. Virol. 74 (2000) 3543–3547.

[62] C. Briones, E. Domingo, Minority report: Hidden memory genomes in HIV-1 quasispecies and possible clinical implications, AIDS Rev. 10 (2008) 93–109.

[63] P.A. Bryant, D. Venter, R. Robins-Browne, N. Curtis, Chips with everything: DNA micro-arrays in infectious diseases, Lancet Infect. Dis. 4 (2004) 100–111.

[64] J. Shendure, R.D. Mitra, C. Varma, G.M. Church, Advanced sequencing technologies: Methods and goals, Nat. Rev. Genet. 5 (2004) 335–344.

[65] M.B. Miller, Y.W. Tang, Basic concepts of microarrays and potential applications in clinical microbiology, Clin. Microbiol. Rev. 22 (2009) 611–633.

[66] T.J. Gentry, J. Zhou, Microarray-based microbial identification and characterization, in: Y.W. Tang, C.W. Stratton (Eds.), Advanced Techniques in Diagnostic Microbiology, Springer Science and Business Media, New York, 2006, pp. 276–290.

[67] R.W. Barrette, S.A. Metwally, J.M. Rowland, L. Xu, S.R. Zaki, S.T. Nichol, et al., Discovery of swine as a host for the Reston ebolavirus, Science 325 (2009) 204–206.

[68] E. Cherkasova, M. Laassri, V. Chizhikov, E. Korotkova, E. Dragunsky, V.I. Agol, et al., Microarray analysis of evolution of RNA viruses: Evidence of circulation of virulent highly divergent vaccine-derived polioviruses, Proc. Natl. Acad. Sci. USA 100 (16) (2003) 9398–9403.

[69] D. Metzgar, C.A. Myers, K.L. Russell, D. Faix, P.J. Blair, J. Brown, et al., Single assay for simultaneous detection and differential identification of human and avian influenza virus types, subtypes, and emergent variants, PLoS One 5 (2) (2010) e8995.

[70] Z. Wang, L.T. Daum, G.J. Vora, D. Metzgar, E.A. Walter, L.C. Canas, et al., Identifying influenza viruses with resequencing microarrays, Emerg. Infect. Dis. 12 (2006) 638–646.

[71] A.P. Malanoski, B. Lin, Z. Wang, J.M. Schnur, D.A. Stenger, Automated identification of multiple micro-organisms from re-sequencing DNA microarrays, Nucleic Acids Res. 34 (2006) 5300–5311.

# CHAPTER 7

# Forensic Plant Pathology

Jacqueline Fletcher,[a] Neel G. Barnaby,[b] James P. Burans,[c] Ulrich Melcher,[d] Forrest W. Nutter Jr.,[e] Carla Thomas,[f] and Francisco M. Ochoa Corona[a]

[a]Department of Entomology and Plant Pathology, National Institute for Microbial Forensics and Food and Agricultural Biosecurity, Oklahoma State University, Stillwater, Oklahoma
[b]FBI Laboratory, Quantico, Virginia
[c]National Bioforensics and Analysis Center, Ft. Detrick, Maryland
[d]Department of Biochemistry and Molecular Biology, Oklahoma State University, Stillwater, Oklahoma
[e]Department of Plant Pathology, Iowa State University, Ames, Iowa
[f]Department of Plant Pathology, University of California, Davis, California

## INTRODUCTION

Plant resources in the United States, which include crops, forests, range, nurseries, and orchards, as well as natural and landscaped spaces, are essential for human and animal life. In addition to providing food, feed, fiber, and recreational opportunities they harness sunlight energy, utilize carbon dioxide, and recycle oxygen. Plants are affected naturally by a host of microbial pathogens that colonize their surfaces, invade their interior spaces, compete with them or metabolize their tissues for nutrients, upset the balance of their growth hormones, and trigger or suppress their gene activity. The science and practice of plant pathology are targeted to the prevention, detection and diagnosis, response, and recovery from such naturally induced disease outbreaks.

Heightened biosecurity concerns in the early 2000s brought focus to the possibility that crops and other plant resources could be targeted directly by individuals or groups motivated to cause harm. Intentional targeting of plants by the release of significant pathogens could not only reduce crop yield and quality, but also could erode consumer confidence, affect economic health and the environment, and possibly impact human nutrition and international relations (1–3). Since that time a number of countries have implemented steps to enhance agricultural biosecurity. In the United States, new

89

Microbial Forensics. DOI: 10.1016/B978-0-12-382006-8.00007-4

programs in microbial forensics and criminal attribution have strengthened national security capabilities (4).

## NATURALLY CAUSED VERSUS INTENTIONAL INTRODUCTION?

Farmers, foresters, and other plant producers know that the vast majority of plant disease outbreaks are incited through sequences of natural events. In most cases, a familiar set of diseases for any given crop will appear repeatedly in a given location, depending upon weather and cropping conditions. However, even an unfamiliar set of symptoms is unlikely to cause alarm—a phenomenon that could be termed *suspicion inertia*.

What features of a plant disease outbreak might trigger concern, on the part of a first detector, that a crime had occurred? What would prompt a call to law enforcement, and when would that call be made? Certain indicators, alone or—more likely—in combination, are most likely to trigger a consideration that a disease should be examined more closely, and that a criminal investigation is appropriate (5,6). Factors such as a new geographical location (disease not seen in this area before), absence of an insect vector required for natural introduction, presence of a pathogen not seen before in this location, unusual pattern of disease in the field, weather history nonconducive to pathogen survival or disease development, disease occurring at an unusual time of year, disease present in one field but not in surrounding ones, physical evidence of inoculation (spray equipment, inoculum containers, gloves or masks, etc.) or of unauthorized human visitors (tire tracks, footprints, gates left open, etc.), or recognized motivation (recent argument, firing of an employee, money owed, etc.) are all potential indicators of human involvement in a pathogen release.

To assist law enforcement personnel in determining if an agricultural crime has occurred, a decision tool was developed (6) in which criteria were assigned weights and values to assess the probability of intent. An accompanying worksheet and fact sheet aid inexperienced users to apply the tool. Evaluations in both natural and intentional field settings in Oklahoma show promise for the utility of the tool in a field investigation to support decision making related to criminal activity (Figure 7.1).

## HISTORY OF AGRICULTURAL BIOWEAPONS

Motives for intentional plant pathogen introduction could include economic gain (within a farm community, between residents of different states, perhaps between nations) due to effects on marketing and trade, revenge (the disgruntled neighbor or employee), or publicity (making a statement about

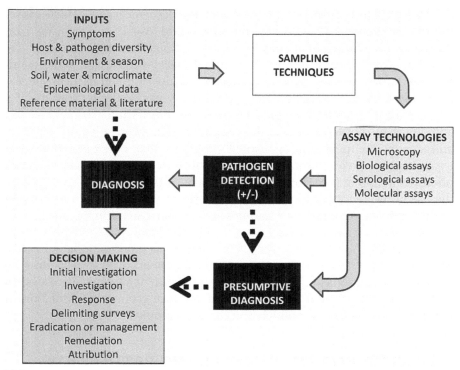

**FIGURE 7.1**

Flow of activity and information for decision making based on pathogen detection and disease diagnosis. Field data, including symptoms and epidemiological information, are compared with reference material and databases to determine appropriate sampling and analysis techniques. Samples are subjected to laboratory assays that detect and identify microbes present; further testing may be used to discriminate among strains or isolates of a pathogen. The compiled field information and test results inform a final diagnosis. A "presumptive diagnosis" is sometimes made when circumstances require a quick response (before or in the absence of conclusive diagnosis), allowing responders to act based on the evidence at hand until a definitive diagnosis is completed.

an ideological position such as genetic engineering, stem cell research, or animal rights). It is also possible to deploy plant pathogens criminally, yet unknowingly. Some introductions of the citrus canker bacterium into Florida were likely to have occurred due to the illegal importation of citrus planting stock from canker-affected countries; those responsible probably knew that bringing the plant stock into the United States was illegal, but it is unlikely that they knew that the plant pieces carried bacterial inoculum.

The history of state-sponsored programs to develop and weaponize biological agents for use against agricultural targets is well documented (3,7–9). The Germans are believed to have used biological weapons in World War I against the United States, inoculating horses with *Burkholderia mallei*, which causes the disease glanders (8). During and after World War II, research was conducted on the efficacy of various bioagents, optimal dissemination methods, and defensive countermeasures. The United States, Russia, and other countries are known to have generated weapons against numerous crop species, including corn, rice, wheat, potatoes, soybeans, sugar beets, and cotton (8,9). Because most antiplant biological weapons are not harmful to humans and animals, they are therefore safer than zoonotic pathogens to handle, develop, and deploy. In most state-sponsored programs that developed biological weapons, they were considered to serve more as deterrents than as actual offensive weapons (9). However, in the wake of the 2001 anthrax mailings, the use of biological weapons for nonstate-sponsored terrorism was brought to the forefront. An attack on a nation's agricultural systems in the furtherance of political or social objectives, known as agroterrorism, was suddenly considered a real possibility. Because the goals of a terrorist group or lone individual are often different from those of a nation, there is no need to reproduce an extensive bioweapons program. Simple introduction of a foreign disease agent to a nation's agricultural enterprise could produce economic destruction or panic in a population as confidence in the food supply is lost (2,7,10).

## THE NEED FOR FORENSIC PLANT PATHOLOGY

If plant pathogens or their products are used deliberately to cause social or economic damage or are introduced inadvertently by illegal actions, law enforcement officials are responsible for determining the source, method, and time of the introduction and for identifying those responsible by forensic investigation and analysis (11–17). Forensic science provides scientific analytical support for the ultimate goal of *attribution* of a criminal act (11–13,18,19). The significant legal ramifications resulting from criminal attribution and prosecution necessitate higher degrees of scientific validation and stringency than those normally used in disease diagnosis and plant pathogen identification (18,20).

The ideal bioforensic investigation will support the identification and characterization of a specific microbe, determinations of how the microbe was produced, and reconstruction of its method of introduction, thereby providing scientific data that will be useful to investigators to link it to the perpetrator(s). The bioforensic investigation should consist of a collection of defined and validated techniques that minimize the time between on-site sample collection and arrival at a forensics laboratory and the time required for controlled laboratory analysis. It may be easier to generate data that an investigator can use to establish *exclusion* (that a particular pathogen or person is *not* involved in the incident) than absolute attribution (evidence that uniquely associates a particular pathogen or person to the incident).

Although a subdiscipline of forensics targeted specifically toward microbial pathogens and toxins associated with bioterrorism and biocrimes involving humans and animals has been developing over the past several years, few specific methods or standard operating procedures (SOPs) have been developed and validated rigorously for application to plant pathogens. The emerging science of forensic plant pathology requires the adaptation and validation of protocols for crime scene sampling, evidence handling, laboratory testing, and analysis. As plant pathogen forensics takes shape, existing methods, SOPs, and protocols are being assessed, standardized, and validated so that their use will be defensible in a criminal investigation. Plant pathologists and forensic scientists (especially those in microbial forensics) are working together closely in both group environments (such as the American Phytopathological Society's Microbial Forensics Interest Group) and in small collaborative projects with the National Bioforensics Analysis Center (NBFAC).

## PATHOGEN DETECTION AND DIAGNOSTICS

*Detection* of a microbe in a plant sample by observation of disease symptoms or pathogen signs or by a molecular assay establishes that an organism is present but implies nothing about a causative role for that microbe in the disease. Detection technologies based on symptoms are relatively simple, but challenges arise when several pathogens induce similar symptoms or when multiple pathogens occur in the same plant. One pathogen can mask symptoms of another, infect several hosts, causing different symptoms in each, or act synergistically with another pathogen, producing a distinctive and sometimes more severe disease than either pathogen alone (21–23). Plant disease *diagnosis* is establishment of the cause of observed damage, generally accomplished by a combination of careful observations of plant and pathogen growth, signs and symptoms, soil, water and environmental conditions, seasonality, host and pathogen diversity, epidemiological data, and serological, DNA- or RNA-based assays. For new diseases, an additional requirement is the fulfillment of Koch's postulates (24). In the past three decades, serological and nucleic acid-based assays have allowed precise but inconclusive *presumptive diagnosis* of a plant disease (associating the presence of a pathogen with a disease but falling short of proof of cause, Figure 7.1) (21), a service frequently offered by plant diagnostic clinics and used at the farm level for making crop management decisions.

Presumptive diagnosis is insufficiently rigorous for applications in agricultural biosecurity and forensic plant pathology, in which sample handling follows a chain of custody and each sample has a legal identity. Diagnostic and detection procedures for agricultural biosecurity and forensics should include multiple methods: light and/or electron microscopy, biological assays (culturing, indexing, and mechanical transmission), and serological and molecular tests (25). The number of methods applied in a given case will depend on the pathogen type, the availability of validated methodologies, and the genomic stability of the pathogen (26).

Diseases reported most often are those occurring in plant populations that have monetary value to humans: crops, orchards and vineyards, nurseries, forests, rangelands, or ornamental landscapes. Also of concern are pathogens that slip across a nation's borders (ports of entry) during international trade of produce, bulbs, ornamentals, seeds, and wood or other biological products, or that are found at quarantine transitional facilities or mail centers. Cases of biocrime or agroterrorism also would require forensic analysis. In all cases, rapid detection is critical to effective response and timely mitigation (27,28).

Symptomatology alone is too variable for reliable diagnosis. Data from biological assays or indexing can be highly accurate, but also costly, time-consuming, and unsuitable for high throughput (27). ELISA, polymerase chain reaction (PCR) (and sequence validation), and microarrays allow rapid and sensitive detection and timely decision making (27,29–32). ELISA and PCR are economical, and ELISA allows high numbers of predetermined tests to be processed. Although not high throughput, PCR, real-time PCR, and their variants provide high sensitivity with limited capability for multiplex applications (27,28,30–32). The high sensitivity of PCR makes it the preferred method for samples collected out of season or carrying pathogens in low titers. Microarray sensitivity is comparable to that of ELISA and for the method can provide high throughput and high specificity (29).

Despite recent impressive advances in diagnostic technologies, accurate and timely plant disease diagnosis—in the end—is a human interpretation of a preponderance of evidence. No technology can replace the hands-on experience of a diagnostician, information available from databases and journals, and consultation and validation with external laboratories (22).

## EPIDEMIOLOGY IN FORENSIC INVESTIGATION

Plant disease epidemiology can provide objective, quantitative data, data analyses, and science-based data interpretation for the attribution of biocrimes involving plant pathogens (7,33–35). A critical early decision during a new plant disease incident is whether the pathogen was introduced deliberately (i.e., a biocrime). The integration of global positioning systems (GPS), geographic information systems, and satellite imagery can provide valuable data to make such decisions in near real time. A $10 \times 10$-km, high-resolution ($<1\,m^2$) satellite image, taken as soon as an outbreak is confirmed, can provide the following forensics-relevant information:

- A permanent "fixed" record of a suspect field that can be digitally stored, retrieved, and analyzed years after the event.

- Detection and geospatial analysis of pathogen-specific anomalies in fields of the same crop within the same $10 \times 10$-km scene (e.g., what other fields are likely to be affected and should be examined). These analyses can help investigators determine if the incident was a natural event or a deliberate attack.
- Within-field anomalies (such as primary and secondary disease foci) can be geospatially referenced and analyzed to determine if their spatial pattern is indicative of a natural or deliberate introduction. Ground crews could then be directed to sample at the exact GPS coordinates of primary disease foci to assess whether the pathogen genetic structure is typical or atypical of the population (i.e., whether the pathogen is a natural population or an artificial mixture of multiple biotypes).
- Analysis of spatial patterns of disease foci using spatial statistics applied at within-field and multifield scales. Precise GPS coordinates for the epicenters of primary disease foci could inform ground investigators where to look for physical evidence of a deliberate introduction. For example, image intensity contour maps generated using ArcGIS (ESRI, Redlands, CA) were used to locate the exact GPS epicenter of disease foci of Asian soybean rust (Figure 7.2). Such maps depict areas of lower image intensities (i.e., crop canopies showing severe soybean rust symptoms in the center of the disease focus) relative to areas with higher image intensities (healthier areas of the crop canopy). Using the contour map method, the location of nine focal epicenters was predicted within $1.5 \pm 0.92$ m of the actual locations where soybean rust point sources were introduced by researchers into soybean field plots (35).

## MUTATION, EVOLUTION, AND FORENSIC PLANT PATHOLOGY

Like human and other animal pathogens, plant pathogens undergo mutations that, when they are not repaired, become variations on which selection acts to produce evolution (7). Such mutations are at once a boon to and a problem for the microbial forensic investigator. On the one hand, evolution means that differences between strains of an organism are plentiful enough that many sources of a phytopathogen can be excluded from consideration simply on the basis of their genetic distance from the crime scene organism. On the other hand, evolution may be so rapid that genomes of the crime scene and suspect source organisms are not identical due to changes occurring since their derivation from their most recent common ancestor or to selection of different individuals from the pool that is the source strain. Such differences make reliable attribution

**FIGURE 7.2**

Contour map for a primary focus of Asian soybean rust based on 2-unit interval pixel intensity values extracted from an IKONOS satellite image obtained August 27, 2006, over Quincy, Florida (35). Image consists of 22 × 22 pixels, each providing 1-m² resolution. (See Color Insert.)

more difficult and suggest that methods beyond those based on DNA need also be pursued in an investigation.

The considerations given earlier apply when deciding on which method to use to compare a crime scene organism with a suspect phytopathogen. The effects of mutation and evolution on results are strongly method dependent. In important enough cases, the ultimate analysis, from which data on all the other kinds of DNA tests can be derived, and the one recommended for courtroom presentation, is nucleotide sequencing of the entire genome of crime scene, suspect, and control organisms. However, in initial investigations where exclusion is the principal objective, less sensitive, but less expensive, methods to survey the phytopathogen genomes are useful. These include allele-specific PCRs, single strand conformation polymorphism, multilocus variable number tandem repeat analysis, amplified fragment length polymorphism, and restriction fragment length polymorphism of PCR products.

Investigators must keep in mind that mutation and evolution do not stop after a crime is committed or discovered. They continue as organisms continue to live and replicate their genomes. Even in the absence of replication, spontaneous mutations occur through deamination and other base changes. It is often necessary to propagate suspect organisms in plants before genomic analyses. Such propagation is often done in laboratory hosts whose selective environment is sufficiently different from the original that multiple adaptive mutations occur. Such genome changes are particularly well documented for many phytopathogens. Many plants, particularly perennials, can harbor multiple microbes and multiple strains of individual phytopathogens. For example, grapevines carry multiple strains distinguishable by restriction (36). The population composition of such mixtures changes drastically during as few as three propagation cycles. Even triply cloned isolates of bacterial phytopathogens can alter their genomes drastically during prolonged passage (37).

# INVESTIGATION

Forensic investigation of a plant disease outbreak requires careful assessment of disease characteristics, sample collection, identification of the pathogen, identification of likely pathogen sources, and attribution or exclusion of pathogen(s) as the causal agent (33).

"First detectors" on the scene of a deliberate plant pathogen introduction are likely to be growers, crop consultants, Master Gardeners, extension agents, or other local, nongovernmental personnel. "First responders," authorized to take action after a potential deliberate introduction, generally arrive later, after notification by first detectors. Timely and effective management of a crime scene requires that both of these groups be able to recognize that a crime has occurred and to react appropriately. A National Plant Diagnostic Network laboratory (NPDN; http://npdn.ppath.cornell.edu) may become involved if tissue samples are sent there for diagnosis (38).

Initial disease assessment should be done prior to any field disturbance and should include the pattern of disease occurrence and any relevant or unusual field characteristics. Forensically relevant SOPs may include collection of whole plants, plant parts, plant swabs, soil, insect vectors, water, air samples, and/ or biological samples, such as alternate weed hosts. Documentation should include an administrative log, a sample log, complete chain of custody, collection site map(s), and detailed information on the crop, field history, and environment. Photographs, GPS, and other aids are useful supplements to this documentation.

What constitutes a "good" sample depends on the disease incidence, the pathogen, and the host. Samples should be collected from a representative number of disease foci (see earlier discussion), from outside the focal areas, and from different plants and plant parts (39). Pooling samples from several sources allows a larger proportion of the plant population to be tested and improves the detection limit (40); positives can then be tested individually if appropriate. Sampling of necrotic lesions is from their edges, as the centers may be invaded by saprophytic microbes. Seeds are a good source for seed-borne pathogens, whereas underground stems and tubers are suitable for other pathogens. Specialized pathogen structures, such as galls or tumors, may also be collected.

Sampling for pathogen detection (i.e., presence or absence) requires different sampling patterns and sample size than that to determine disease incidence or severity. Presence–absence data can be more important than incidence or severity data for forensic purposes, for example, to determine the geographical extent of the disease, or to decide whether a field should be quarantined. In such cases, sampling can concentrate on high-risk areas in a field, such as borders or wet areas, depending on the pathogen. In most forensic applications, disease incidence or severity data will be needed to develop spatial disease intensity maps to identify the potential point(s) of inoculation.

Sample integrity and security must be preserved during collection, movement, storage, and analysis (41). Storage conditions must be documented, and chain-of-custody records must reflect all aspects of exposure to the environment and records of individuals having access.

## ROLES AND RESPONSIBILITIES

A successful response to a plant health event involving a criminal investigation requires extensive collaboration, coordination, and communication between numerous agencies and organizations at the local, state, federal, and potentially international level. Because the primary interests and goals of the agricultural and law enforcement communities differ in some signficant ways (Table 7.1), it is important that the groups are able to work in a coordinated manner. Most states have laws requiring the reporting of any diseases of regulatory significance to regulatory officials.

At the state level, the State Plant Regulatory Official (SPRO) is the highest level plant health official and serves the State Secretary of Agriculture or State Agriculture Commissioner. In most states, the State Department of Agriculture (SDA) has the authority to conduct an agriculture investigation in the field. Most SDAs have investigative services units that will investigate cases where plant health regulatory statutes and laws may have been violated. The SDA also has the authority to implement a 90-day stop movement on plant materials and to implement quarantines with the assistance of local law enforcement and/or the National Guard.

**Table 7.1** Comparison of Objectives of Agricultural (Ag) and Law Enforcement (LE) Specialists in a Plant Disease Emergency

| Ag Objectives | LE Objectives |
| --- | --- |
| Damage assessment | Security |
| Economic impact | Investigation |
| Potential for spread | Perimeter control |
| Impact to market/populations | Surveillance |
| Delimited area | Profiling |
| Trace in/out or forward/back | Trace in/out or forward/back |
| Personal safety, responders and public | Catch the perpetrator |
| Outreach, education, public information | |
| Containment/control | |
| Evidence security/collection | |
| Stop the epidemic | |

The federal plant regulatory authority belongs to the U.S. Department of Agriculture (USDA) Animal and Plant Health Inspection Service (APHIS) Plant Protection and Quarantine (PPQ). In each state, the State Plant Health Director (SPHD) is an APHIS employee and has the highest level of federal authority for that state. The SPHD and the SPRO work together to leverage state and federal roles and authorities in a complementary manner to respond optimally to an event. The SPHD has the authority to implement a local mitigation measure (quarantine, crop destruction, sanitation, etc.) as an Emergency Action Notice.

Agriculture diagnostic laboratory testing is conducted by university plant clinics or SDA laboratories, usually members of the National Plant Diagnostic Network (NPDN), which coordinates and collaborates with the APHIS National Identification Service, the national confirmatory authority. However, evidentiary samples collected by law enforcement will be analyzed by those laboratories that have been vetted to handle evidence.

Initially, the intent of the criminal act may be unknown, therefore requiring multiple law enforcement agency participation until the lead agency can be identified. The Federal Bureau of Investigation (FBI) is designated as the lead authority for the investigation of domestic terrorism, as outlined in Homeland Security Presidential Directive/HSPD-5. However, the USDA's Office of the Inspector General will be the lead agency for criminal acts involving agriculture of a nonterrorism nature. The Department of Homeland Security (DHS) agencies Customs and Border Protection (CBP), Immigrations and Customs Enforcement, and Coast Guard were assigned authority for incident management and resource coordination. Regardless of which agency serves as the investigative lead, coordination with the response and recovery agencies will be crucial for the preservation of evidence, both microbial and traditional.

## EDUCATION AND OUTREACH

The discipline of microbial forensics was purposefully expanded, following dissemination of the anthrax letters in 2001, with incorporation of new and more discriminatory scientific technologies. The U.S. Homeland Security community recognized the need for a broad capability in forensic microbiology, including pathogens of humans, animals, and plants (7). Because new homeland security initiatives require capable, well-trained professionals to carry them out, that capability must include provision for the education of young scientists and for training of those already working in homeland security roles.

New career roles require scientists trained and experienced in both agricultural and forensic sciences, and both knowledgeable and appreciative of the concerns of homeland security. Traditional academic units (i.e., departments of plant pathology and similar disciplines) at several U.S. universities have developed new coursework at the graduate and/or undergraduate levels on biosecurity, agricultural biosecurity, plant health, and plant biosecurity. Although these programs introduce students to important new issues in plant health, they are limited in coverage of related security areas. An ideal training program for agricultural forensics would provide both a strong footing in agricultural sciences (available in existing, traditional strong programs) and substantive new coursework and applications in forensic sciences and homeland security. Because new security-focused careers in the FBI, the DHS, the Central Intelligence Agency, and even in the USDA's regulatory agency, APHIS, are unfamiliar to students, it is important also to provide opportunities for them to learn about these careers through (i) interactions with agency personnel at meetings and seminars and (ii) internships in which students experience agency operations and receive hands-on experience. A program that incorporates all of these elements has been established at the National Institute for Microbial Forensics & Food and Agricultural Biosecurity (NIMFFAB) at Oklahoma State University (http://www.ento.okstate.edu/nimffab/).

In addition to targeted educational programs for students, training and outreach to career specialists who might be first on the scene or involved in the response are also critical. Specific training on recognition of intentional pathogen introductions and on the appropriate conduct of a criminal investigation (sampling, chain of custody, and site preservation) will facilitate attribution and assure that justice is done. Audiences targeted by NIMFFAB for training exercises include agricultural specialists, plant disease diagnosticians, extension educators, Master Gardeners, and security and law enforcement officers of the FBI and the DHS, as well as state and local law enforcement officers, regulatory officials, and others.

## RESOURCES AND INFRASTRUCTURE

Preparedness for a criminal event involving a plant pathogen includes *prevention, detection and diagnostics, response*, and *recovery* (42,43). The responsibility for protecting U.S. crops, rangelands, forests, and other plant resources from introduced pathogens and pests is shared by the USDA (especially the APHIS-PPQ), the DHS (through CBP), and the NBFAC, the FBI, and local law enforcement. In a *prevention* strategy, focus is on agents having a high probability of introduction and establishment. Because threat characterizations and determinations of vulnerability to a specific plant pathogen and, ultimately, the risk, are imprecise, prioritization is based on perceived potential to cause persistent, wide-scale damage.

Because huge numbers and volumes of plants and plant products move through our ports and borders we cannot completely exclude the introduction of new agents that arrive accidentally or intentionally, and we must be prepared at all times to respond to the introduction of pathogens that threaten our plant systems. The principal capabilities of the United States in *plant pathogen identification* and *disease diagnostics* center in the NPDN, an interconnected network of plant disease diagnostic laboratories, generally one per state. In 2002, these formerly independent laboratories, affiliated either with the state's land grant university or SDA, were organized by the USDA into a highly effective and coordinated network that works with APHIS to monitor, diagnose, and report plant diseases in the United States (38).

Our *surveillance and detection systems* vary significantly with the plant system, target pathogen or pest, and geographic region. Because surveillance usually targets specific agents of concern, programs are concentrated in "at-risk" areas. For some plant systems, industry also conducts effective surveillance programs and provides data to APHIS.

*Diagnosis* is provided primarily by the NPDN, which has developed a triage system for rapid and accurate diagnosis of introduced plant pathogens and insect pests (38). The NPDN sends diagnostic data collected at network laboratories to a national database; tools for data and syndromic analyses are currently under development to enhance the usefulness of the collected data.

*Response* to plant disease outbreaks resulting from new pathogen introductions is a responsibility of USDA APHIS, which provides leadership for a coordinated response that often includes APHIS-led rapid deployment teams, SDAs, industry, and NPDN laboratories. Response elements include surveillance, epidemic delimitation, application of disease control or management strategies, and other actions to minimize both spread and damage.

*Forensic capability* is another important response element in cases in which intentional introduction is suspected. Bioforensic analyses for a number of human and animal high-consequence biological agents have been developed, but few similar bioforensic analyses/assays exist for plant pathogens. The need for this capability is now well recognized and efforts are moving forward through the development of new assays by APHIS and Agricultural Research Service (ARS), the DHS NBFAC, and the NIMFFAB at Oklahoma State University (44).

*Recovery* is intended to restore pre-event status or establish a new, but stable, status. Effective recovery, which must include both short- and long-term plans, generally focuses on local and system-level issues and considers ecological impacts, production declines, and downstream effects on transportation systems, trade, market reentry, and replacement systems. The National Plant Disease Recovery System (NPDRS), mandated by Homeland Security Presidential Directive 9 (HSPD-9), is managed by the USDA ARS. NPDRS has involved other federal agencies [e.g., APHIS and Cooperative State Research, Education, and Extension Service (now National Institute of Food and Agriculture)], SDAs, scientific societies, and universities in the development of national response plans for the select agents and other plant pathogens of high consequence.

## GAPS

Forensic plant pathologists may arise not only from within the discipline of plant pathology, but also from related disciplines such as microbiology, molecular biology, and genetics. These scientists must accommodate the needs and stringent requirements of forensic science while adapting some of the existing tools, knowledge, and resources in plant pathology, which were developed for peaceful purposes and natural disease outbreaks, as well as by developing targeted new technologies. It is not enough to identify a pathogen to genus and species; we also must discriminate among highly similar pathogen strains. We need to know the confidence levels of our tests. For many plant pathogens, detection and identification tools are not optimized, standardized, or validated. Some still-used traditional methods, such as host range studies and use of sets of "differential" cultivars, are tedious. Tools based on DNA typing and genomics are highly promising, but new, rigorous, and reliable analytical methods are needed. Priority should be given for development of technologies applicable to high-priority plant pathogens, such as those on the "Select Agent" list, for multiplex tests, and for assays that are portable and rapid. We need to better understand the mutation rates of threatening pathogens in natural settings and in culture and how they affect a forensic investigation. It is important also to better understand the microbial communities that make up natural environments and influence sample characterization.

There continues to be a need for education and training at several levels. Bright, well-trained scientists having experience in both plant pathology and forensic sciences are needed to fill new positions in federal agencies, yet few graduate programs provide coursework relevant to both disciplines. Although existing training programs for plant disease diagnosticians and for extension personnel and law enforcement officials are excellent, few address law enforcement issues. Security and law enforcement training, similarly, rarely provides exposure to agricultural issues and threats. More training opportunities are needed in which law enforcement and agricultural experts are brought together to address not only the scientific aspects of an incident but also the unique roles and responsibilities of various agencies and responders so that actions at the crime scene are seamless and that appropriate follow-up occurs.

## SUMMARY

Forensic plant pathology combines elements of a host of disciplines. The targeted stakeholders of forensic plant pathology are members of the law enforcement and security communities, whose immediate goals are to identify the source of a criminally introduced pathogen and to attribute responsibility to the perpetrator(s) so that they are brought to justice. For this emerging discipline to function optimally, the law enforcement community must communicate their needs to plant pathologists effectively. Similarly, forensic plant pathologists must design their work based on regular interaction and communication with members of the security community so as to assure its relevance and utility in solving real problems.

## REFERENCES

[1] R. Casagrande, Biological terrorism targeted at agriculture: the threat to U.S. national security, The Nonproliferation Rev. Fall-Winter (2000) 92–105. Available from: <http://cns.miis.edu/pubs/npr/vol07/73/73casa.pdf>.

[2] L. Madden, M. Wheelis, The threat of plant pathogens as weapons against U.S. crops, Annu. Rev. Phytopathol. 41 (2003) 155–176.

[3] S.M. Whitby, Biological Warfare Against Crops, Palgrave, Basingstoke, UK, 2002.

[4] American Phytopathological Society Public Policy Board. 2002. The American Phytopathological Society, first line of defense. APSnet. <http://www.apsnet.org>.

[5] Food and Drug Administration, U.S. Department of Agriculture, and the Federal Bureau of Investigation. Undated. Criminal Investigation Handbook for Agroterrorism, U.S. Government Publication.

[6] S.M. Rogers, R. Hunger, J. Fletcher, An agricultural biosecurity decision tool: Is it natural or intentional? Phytopathology 99 (2009) S110.

[7] J. Fletcher, C.L. Bender, B. Budowle, W.T. Cobb, S.E. Gold, C.A. Ishimaru, et al., Plant pathogen forensics: capabilities, needs and recommendations, Microbiol. Mol. Biol. Rev. 70 (2006) 450–471.

[8] R. Harris, J. Paxman, A Higher Form of Killing, Random House, Inc., New York, 2002.

[9] D.M. Huber, Anti-crop bioterrorism (Chapter 7), in: S.A. Amas (Ed.), The Science of Homeland Security, vol. 1, Purdue University Press, W. Lafayette, IN, 2006.

[10] M. Wheelis, R. Casagrande, L.V. Madden, Biological attack on agriculture: Low-tech, high impact bioterrorism, Bioscience 52 (2002) 569–576.

[11] B. Budowle, R. Chakraborty, Genetic considerations for interpreting molecular microbial forensic evidence, in: C. Doutremepuich, N. Morling (Eds.), Progress Forensic Genet. 10, Elsevier, Amsterdam, 2004, pp. 56–58.

[12] B. Budowle, M.D. Johnson, C.M. Fraser, T.J. Leighton, R.S. Murch, R. Chakraborty, Genetic analysis and attribution of microbial forensics evidence, Crit. Rev. Microbiol. 31 (2005) 233–254.

[13] B. Budowle, R.S. Murch, R. Chakraborty, Microbial forensics: The next forensic challenge, Int. J. Leg. Med. 119 (2005) 317–330.

[14] M.A. Cooper, Label-free screening of bio-molecular interactions, Anal. Bioanal. Chem. 377 (2003) 834–842.

[15] P.G. Jones, A. Gladkov, FloraMap. A Computer Tool for Predicting the Distribution of Plants and Other Organisms in the Wild, Centro Internacional de Agricultura Tropical (CIAT), Cali, Colombia, 1999.

[16] F. Kaffarnik, P. Muller, M. Leibundgut, R. Kahmann, M. Feldbrugge, PKA and MAPK phosphorylation of Prf1 allows promoter discrimination in Ustilago maydis, EMBO J. 22 (2003) 5817–5826.

[17] J. Fletcher, The need for forensic tools in a balanced national agricultural security program, in: Crop Biosecurity: Assuring Our Global Food Supply, Proceedings of a NATO Project, Springer Science + Business Media B.V., 2008, pp. 93–101.

[18] J. Fletcher, U. Melcher, D. Luster, J.L. Sherwood, Microbial Forensics and Plant Pathogens: Attribution of Agricultural Crime, in: Handbook of Science & Technology for Homeland Security, Ed. J. G. Voeller, John Wiley & Sons, Inc. 2008.

[19] B. Budowle, J. Burans, R.G. Breeze, M.R. Wilson, R. Chakraborty, Microbial forensics, in: R.G. Breeze, B. Budowle, S.E. Schutzer (Eds.), Microbial Forensics, Elsevier Academic Press, San Diego, CA, 2005, pp. 1–26.

[20] B. Budowle, Defining a new forensic discipline: Microbial forensics, Profiles in DNA 6 (2003) 7–10. Available from: <http://www.promega.com/profiles/601/ProfilesInDNA_601_07.pdf>.

[21] B. Nyvad, Diagnosis versus detection of caries, Caries Res. 38 (2004) 192–198.

[22] F.M. Ochoa-Corona, J. Tang, B.S.M. Lebas, L. Rubio, A. Gera, B.J.R. Alexander, Diagnosis of Broad bean wilt virus 1 and Verbena latent virus in *Tropaeolum majus* in New Zealand, Australasian Plant Pathol. 39 (2) (2010) 120–124.

[23] F.M. Ochoa Corona, B.S.M. Lebas, D.R. Elliott, J.Z. Tang, B.J.R. Alexander, New host records and new host family range for Turnip mosaic virus in New Zealand, Australasian Plant Dis. Notes 2 (2007) 127–130.

[24] R.G. Grogan, The science and art of plant-disease diagnosis, Annu. Rev. Phytopathol. 19 (1981) 333–351.

[25] B.S.M. Lebas, F.M. Ochoa-Corona, Impatiens necrotic spot virus, in: G.P. Rao, C. Bragard, B.S.M. Lebas (Eds.), Characterization, Diagnosis and Management of Plant Viruses, vol. 4, Studium Press LLC, Houston, TX, 2007, pp. 241–243.

[26] R.R. Martin, J. Delano, A.C. Lévesque, Impacts of molecular diagnostic technologies on plant disease management, Annu. Rev. Phytopathol. 38 (2000) 207–239.

[27] S.A. Miller, R.R. Martin, Molecular diagnosis of plant disease, Annu. Rev. Phytopathol. 26 (1988) 409–432.

[28] N.W. Schaad, R.D. Frederick, J. Shaw, W.L. Schneider, R. Hickson, M.D. Petrillo, et al., Advances in molecular-based diagnostics in meeting crop biosecurity and phytosanitary issues, Annu. Rev. Phytopathol. 41 (2003) 305–324.

[29] N. Boonham, J. Tomlinson, R. Mumford, Microarrays for rapid identification of plant viruses, Annu. Rev. Phytopathol. 45 (2007) 307–328.

[30] M.F. Clark, Immunosorbent assays in plant pathology, Annu. Rev. Phytopathol. 19 (1981) 83–106.

[31] E.L. Halk, S.H. De Boer, Monoclonal antibodies in plant-disease research, Annu. Rev. Phytopathol. 23 (1985) 321–350.

[32] J.M. Henson, R. French, The polymerase chain reaction and plant disease diagnosis, Annu. Rev. Phytopathol. 31 (1993) 81–109.

[33] F.W. Nutter Jr., Developing forensic protocols for the post-introduction attribution of threatening plant pathogens, Phytopathology 94 (2004) S77.

[34] F.W. Nutter Jr., L.V. Madden, Plant pathogens as biological weapons against agriculture, in: L.I. Lutwick, S.M. Lutwick (Eds.), Beyond Anthrax: The Weaponization of Infectious Disease, Springer, New York, 2008, pp. 335–363.

[35] F.W. Nutter Jr., N.S. Holah, S.K. Eggenberger, E. Byamukama, D.L. Wright, J. Marois, N. Van Rij, Emerging GPS, GIS, and remote sensing technologies for improved crop biosecurity, in: D.M. Gadory, R.C. Seem, M.M. Moyer, W.E. Fry (Eds.), Proceedings of the 10th International Epidemiology Workshop, New York Agricultural Experiment Station, Geneva, NY, 2009, pp. 116–117.

[36] P. Naraghi-Arani, S.D. Daubert, A. Rowhani, Quasispecies nature of the grapevine fanleaf virus genome, J. Gen. Virol. 82 (2001) 1791–1795.

[37] F. Ye, U. Melcher, J.E. Rascoe, J. Fletcher, Extensive chromsome aberrations in *Spiroplasma citri* strain BR3, Biochem. Genet. 34 (1996) 269–286.

[38] J. Stack, K. Cardwell, R. Hammerschmidt, J. Byrne, R. Loria, K. Snover-Clift, W. Baldwin, G. Wisler, H. Beck, R. Bostock, C. Thomas, E. Luke, The National Plant Diagnostic Network, Plant Dis. 90 (2006) 128–136.

[39] F.W. Nutter Jr., Post-introduction mapping of new and emerging agricultural pathogens in real-time using GPS and GIS technologies, Phytopathology 94 (2004) S130 (Abstract).

[40] G. Hughes, T.R. Gottwald, Survey strategies for citrus tristeza virus disease assessment, Phytopathology 88 (1998) 715–723.

[41] FBI Scientific Working Group on Forensic Analysis on Chemical, Biological, Radiological, and Nuclear Terrorism. Best practices for the collection of chemical, biological, radiological or nuclear evidence, Forensic Sci. Commun. (in press).

[42] J.P. Stack, J. Fletcher, Plant biosecurity infrastructure for disease surveillance and diagnostics, in: Microbial Threats, National Academy of Sciences, Institute of Medicine, Washington, DC, 2007.

[43] J. Fletcher, J. Stack, Agricultural Biosecurity: Threats and Impacts for Plant Resources, in: Microbial Threats, National Academy of Sciences, Institute of Medicine, Washington, DC, 2007.

[44] J. Fletcher, D. Luster, R. Bostock, J. Burans, K. Cardwell, T. Gottwald, et al., Emerging infectious plant diseases (in press), in: J. Hughes (Ed.), Emerging Infectious Diseases, ASM Press, 2010.

# 2

# Emerging Infections

# CHAPTER 8

# Influenza Forensics

**Robin M. Bush**

*Department of Ecology and Evolutionary Biology, University of California, Irvine, California*

## INTRODUCTION

To our knowledge, the influenza virus has never been used as a weapon. Nonetheless, every influenza outbreak investigation resembles a forensics analysis in that the characterization, analysis, and interpretation of evidence for attribution are its primary goals (1). Attribution in the case of influenza is more likely to indicate an animal or animal link than an ill-intentioned human; rather than an arrest, the goals are those of basic investigatory science. We seek an understanding of the context within which, and mechanisms whereby, a new influenza virus comes to infect a susceptible host population. Yet the potential for intentional harm from influenza does exist.

The most oft-mentioned concern is the release or escape of the recently reconstructed "Spanish flu" H1N1 influenza virus, which killed tens of millions of people between 1918 and 1920 (2). The gene sequences of this virus have been deposited in GenBank, making reconstruction for malicious purposes theoretically possible by those with ill intent. Also of particular concern are the highly pathogenic H5N1 avian influenza viruses, should they evolve, or be genetically manipulated, to become easily transmissible among humans (3–5).

However, an influenza virus need not be unusually deadly to cause harm. The current 2009 H1N1 pandemic serves as an example of the social and economic disruption that can ensue from any outbreak of a novel influenza strain. The H1N1 2009 pandemic tested our capacity for influenza outbreak investigation at a time when a relative wealth of forensic technology was at our disposal. Nonetheless, the origin of the H1N1 2009 virus remains unknown, as are the exact origins of all past influenza pandemic strains.

This chapter focuses on the biological features of influenza that have repeatedly allowed outbreaks to emerge undetected. It first briefly reviews influenza genetics and evolution and then reviews those aspects of human and swine influenza evolution during the past century required to frame our current

Microbial Forensics. DOI: 10.1016/B978-0-12-382006-8.00008-6

knowledge of the origin of the 2009 H1N1 influenza pandemic. The chapter then provides an overview of symptoms, diagnostics, surveillance, and techniques of molecular evolution and phylogenetics required to study the origin and spread of new strains and finishes by considering how current forensic techniques would suffice in the investigation of a criminal release.

## THE INFLUENZA VIRUS

The influenza viruses comprise three genera in the family Orthomyxoviridae. The genera are, in practice, referred to as "types" A, B, and C. However, in terms of their potential to cause harm to humans, the influenza viruses can be divided into two categories.

The first category includes the very few influenza strains that circulate in humans and to which most of the human population has some immunity. The second category includes the great majority of influenza viruses—those that currently circulate in other animal hosts. We typically have little or no immunity to these zoonotic viruses, and they thus pose the greatest potential for human harm.

The zoonotic influenza viruses are all of type A—a diverse group that, for the most part, are asymptomatic inhabitants of the guts of aquatic waterfowl such as ducks, gulls, and shorebirds. Type A influenza viruses occasionally infect but rarely become established in nonavian animals. A few lineages of influenza A, all originally of avian origin, currently circulate in humans, swine, equines, and dogs. Reviews of the diversity of influenza in animals can be found elsewhere (6–8).

Human-adapted influenza viruses include all strains of types B and C and a few strains of influenza A. The familiar winter epidemics of influenza in the temperate zones are caused by influenza types A and B. These so-called seasonal human influenza strains also cause disease in the tropics on a more temporally irregular basis (9,10). Influenza C, a distant relative of types A and B, infects children primarily and typically causes only mild symptoms.

The majority of this chapter addresses influenza type A, as the forensic techniques needed to investigate outbreaks of types B and C, which lack a zoonotic reservoir, would be similar but simpler. General reviews of the biology of types B and C can be found elsewhere (11,12).

## INFLUENZA A GENOMICS

The influenza A genome is composed of eight negative-sense, single-stranded RNA segments and is around 14 kb in size. Each of the eight RNA segments encodes one or two proteins; the genome consists of a total of 10 or 11 genes depending on the viral strain. These genes evolve at different rates depending on functional constraints and on the amount of selective pressure they experience from the host immune system.

The most informative genes for fine-scale analysis are those coding for the antigenic surface glycoproteins hemagglutinin (HA, segment 4) and neuraminidase (NA, segment 6). These genes evolve rapidly in humans due to strong selective pressure by the humoral immune system favoring antigenic variants not recognized by host antibodies (13). Hemagglutinin is involved in binding to host cell surface receptors and fusion with the host endosomal membrane. Neuraminidase is an enzyme required for viral release from host cells. While HA is the primary target for neutralizing antibodies, antibodies against NA also may reduce occurrence and severity of illness, and possibly prevent infection if present at high titers (11,14).

Hemagglutinin and NA evolve more slowly in nonhuman animals. In birds, the influenza virus typically does not cause disease and is thus presumably under little selective pressure from the immune system. Influenza rarely causes severe illness in swine, and although they do mount an antibody response to the virus, there is little opportunity for immune selection because the lifespan of domestic swine is generally short due to agricultural practice. Thus, generally speaking, the rates of HA and NA evolution are most rapid in humans, less rapid in swine, and even less rapid in avians (15,16). These host-specific rate differences are important to forensic analysis, as we often find ourselves trying to time unobserved events from the past using evolutionary rates estimated from gene sequence data.

More slowly evolving genes are used in outbreak investigations as targets of reverse transcription polymerase chain reaction (RT-PCR) protocols used for quick assignment of viruses to type, subtype, and host-specific lineages, as described later. These include genes encoding the three polymerase proteins: PB2, PB1, PA (segments 1, 2, and 3); three structural proteins NP (segment 5), M1, and M2 (segment 7); and two nonstructural proteins involved in nuclear export, NS1 and NS2 (segment 8). An eleventh open reading frame, PB1-F2, is encoded by the +1 reading frame of the PB1 gene in some influenza A viruses (17).

Genetic diversity in influenza viruses is derived primarily by two distinct mechanisms. The first mechanism is the generation of point mutations as a consequence of errors made by the RNA polymerase complex during viral replication (18,19). Indels are rarely seen in influenza, although a few have become fixed in human seasonal H1N1 influenza A and in influenza B. Recombination may occur but if so it is very rare (20).

The most notable effect of point mutations is their contribution to the ongoing serial replacement of amino acids on the distal surface and around the receptor-binding site of the hemagglutinin (21). These changes, many of

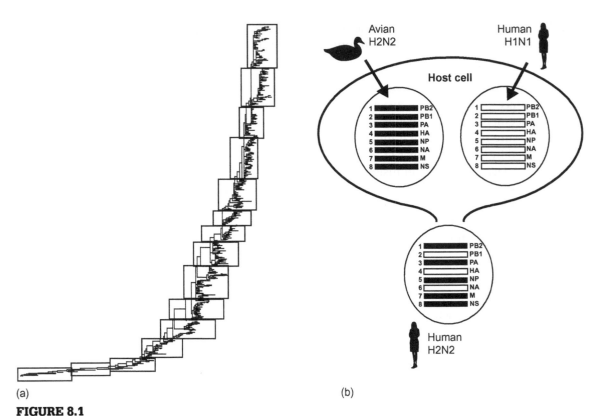

**FIGURE 8.1**

(a) Antigenic drift caused by point mutations in the hemagglutinin gene of H3N2 influenza A. Boxes superimposed on the phylogenetic tree are (non-accurate) cartoons grouping antigenically similar genetic variants. (b) Antigenic shift resulting from reassortment between avian and human-adapted viruses.

which occur in known antibody-binding sites, eventually allow the virus to escape from antibodies raised by prior infection or vaccination (22–24). Change in antigenicity due to the accumulated effects of small mutations is referred to as "antigenic drift."

We typically visualize antigenic drift by drawing boxes over the sections of phylogenetic trees that contain antigenically similar viruses (Figure 8.1a); however, the genetic basis of antigenic drift is not well understood. Experimental manipulation of archived viruses suggests that antigenic drift is influenced more heavily by some individual amino acid replacements than others (25–27) but these experiments present a picture that is simpler than expected based on epidemiological observation.

Reassortment is the second mechanism by which genetic diversity is generated in influenza viruses. Reassortment, only possible in viruses with segmented genomes, is the production of mixed genome offspring within host cells that have been coinfected by more than one virus (Figure 8.1b). Reassortment occurs within but not between types A, B, and C.

Reassortment is typically detected by constructing separate phylogenetic trees for each of the eight gene segments and then visually examining sets of trees for noncongruent branching patterns, although more sophisticated techniques have been developed (28). The rate of reassortment in nature is unknown. Reassortment is most likely to occur during periods of high viral transmission, but it is at this time when cocirculating viruses are most likely to be similar to one another, thus making analysis problematic.

Reassortment within evolutionary lineages likely plays a role in disease dynamics; however, it is most famously known for its role in generating the 1957 H2N2 and 1968 H3N2 human pandemic viruses. In both cases, human-adapted seasonal influenza type A viruses obtained gene segments from avian viruses encoding surface antigens to which most humans living at the time had no prior immunity (Figure 8.2). The establishment of such viruses in the human population is called an "antigenic shift." although this term can also be used to describe the wholesale jump of a novel virus, with all of its segments intact, to humans.

The goal of this chapter is to leave the reader with an understanding of how the 2009 H1N1 pandemic virus evolved and the forensic (or epidemiologic) approaches taken to discover its origin. This requires an understanding of how the 2009 H1N1 strain differs from other H1N1 viruses, such as the 1918 "Spanish flu" virus, human seasonal H1N1 influenza, and the "classic" swine H1N1 virus. It also requires an understanding of the evolutionary history of the 1957 H2N2 and 1968 H3N2 human pandemic strains as well.

Before providing a brief overview of those events, we first discuss three aspects of influenza nomenclature that have potential for causing confusion in influenza outbreak investigations: the practice of referring to evolutionary lineages by their subtype classification, the convention used for naming individual viral isolates, and the use of colloquial expressions such as "swine flu."

# INFLUENZA NOMENCLATURE

## Influenza A Subtypes

Type A influenza viruses are classified into subtypes based on the antigenic differentiation of their HA and NA alleles. Genes coding for these antigenically distinct alleles are also quite distinct genetically, differing from one another on average by about 50% at the amino acid level. Thus far influenza viruses bearing 16 HA alleles and 9 NA alleles have been isolated from birds.

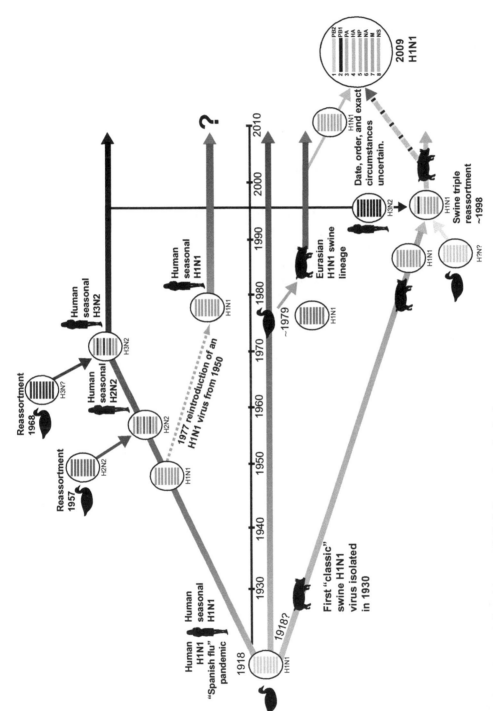

**FIGURE 8.2** Evolution of the 2009 H1N1 influenza A pandemic virus.

The human 1918 "Spanish flu" and "classic" swine influenza A H1N1 viruses probably evolved from a single avian-adapted ancestor (left side of figure). Since 1918, genetic variation has accumulated in both human and swine influenza A lineages as a result of reassortment (explicitly shown in the figure, see bottom right for key to gene segments) and gradually, via point mutation (suggested by gradual color transitions on the lines that represent individual lineages). The 2009 H1N1 pandemic virus appears to have been derived through the reassortment of several viruses currently known to circulate in swine. (See Color Insert.)

Subtype classification of the four pandemics of the past century are 1918 H1N1, 1957 H2N2, 1968 H3N2, and 2009 H1N1.

Referring to a strain by its subtype serves as a convenient shorthand only when a single distinct evolutionary lineage of that subtype is in circulation in a host species. This was the case until recently, when the 2009 H1N1 pandemic strain emerged during a time period in which a human-adapted H1N1 seasonal strain was circulating. Nomenclature adopted and abandoned before settling on "2009 H1N1" included (H1N1)swl, indicating "swine lineage," S-OIV for "swine-origin influenza virus," n(H1N1) for "novel," and (H1N1)v for "variant" and (H1N1) pdm for pandemic. One way to prevent such confusion in the future might include extension of a recently invented hierarchical numbering system for H5N1 avian influenza lineages to other influenza A subtypes (29).

An additional source of confusion with respect to subtypes concerns the numbering of the amino acids in viral proteins. Position number will vary depending on whether one is considering a protein before or after aligning it with a reference strain, which may be from another subtype. For example, oseltamivir resistance is conferred by an H274Y (histidine to tyrosine) amino acid replacement where 274 refers to the position in the N2 neuraminidase protein. One may see reference to this mutation in studies of the H1N1 2009 pandemic virus as well, despite the fact that the equivalent change is in position 275 in the N1 neuraminidase (30).

## Isolate Names

The current format for influenza isolate names is type/location/isolate number/year. Thus A/California/4/2009 is an influenza A virus that was isolated in California in 2009 and assigned the sample number 4 in the laboratory that characterized it. This name does not convey that this is an H1N1 (as opposed to an H3N2) influenza A virus nor does it tell you that it is from the swine-derived 2009 H1N1 pandemic lineage (as opposed to the human H1N1 seasonal lineage). There is no standard convention for including these additional distinctions in an isolate name.

Names of viruses isolated from nonhuman animals should, but do not always, indicate the host. For example, A/swine/Belgium/1/98 or A/sw/Belgium/1/98 is a virus isolated from a pig. The words equine (eq) and canine similarly indicate isolation from horses and dogs, respectively. Avian isolate names typically indicate the common rather than the scientific name of the avian host. This is unfortunate; while common names used sometime indicate a particular species (A/northern pintail/Alberta/11701/2005), in many cases they do not (A/duck/Jiangsu/022/2009).

## Colloquial Strain Names

Colloquial names such as "bird flu" and "swine flu" refer to influenza A lineages that are circulating in populations of those animal hosts. They do not refer to the host from which a particular virus was isolated. For example, if a farmer infects a pig with a seasonal human influenza virus, the pig is not said to have "swine flu." However, if that pig infects other pigs and the resulting viral lineage becomes established in swine herds, it will eventually be referred to as a swine-adapted strain or "swine flu." There are no formal rules governing the time frame over which this change in colloquial nomenclature happens.

The 2009 H1N1 pandemic virus is a prime example of how forensic investigations can potentially be compromised by the use of colloquial names. The genome of this reassortment virus is composed of segments genetically most similar to those of viruses currently known from circulation in swine. Thus the 2009 H1N1 pandemic virus was originally (and often still is) referred to as a "swine flu" despite the fact that no virus with this particular genetic configuration has, to date, been recovered from swine, with the exception of those who had contracted the virus from an infected farmer (http://www .promedmail.org archive 20091106.3834, http://www.ncagr.gov/paffairs/ release/2009/12-09pigsconfirmed.htm).

Unfortunately, use of the term "swine flu" generated public fear and resulted in financial loss for the pork industry. Similarly, referring to an outbreak based on geographic location (such as "Mexican flu") carries with it the potential stigma that the outbreak was caused by poor agricultural practice. Although the first known outbreak of the 2009 H1N1 pandemic occurred in Mexico, it is unclear where the 2009 H1N1 virus actually originated. It was not found to be circulating in swine near the region of human outbreaks in Mexico, and indeed may have been imported to Mexico from elsewhere in the world (31).

# TIME: THE MAJOR FOE OF INFLUENZA FORENSICS

Type A influenza caused four human influenza pandemics during the past century. We do not know the exact sequence of events that led to the emergence of any of these pandemic strains. In each case the passage of time between emergence and detection of the outbreak led to an irretrievable loss of critical data.

There are several reasons why time is so critical to the investigation of an influenza outbreak. First, the symptoms of influenza are wide ranging and nonspecific; many of those infected do not seek medical help or, if they do, their symptoms may not initially be seen as unusual. Second, rapid viral transmission among humans and rapid viral clearance within humans can quickly obscure the geographical origin of an outbreak.

Finally, the RNA genome of influenza is highly mutable and has the potential to evolve rapidly, particularly when it jumps to a new host (32). As time passes, the genetic relationship between the newly emerged virus and its progenitors becomes increasingly difficult to discern. Depending on the genetic configuration of the virus and the infection history of the host population, serology may be of some use in a retrospective study of an outbreak. Indeed, "seroarcheological" evidence obtained from the elderly in the mid-1900s was used to infer the influenza strains responsible for influenza pandemics in the late 1800s (33). Serology does not, however, offer the precision of genetic data for pinpointing the exact origin of a new strain. Thus, this chapter focuses on the use of genetic analysis in influenza forensics.

## A BRIEF HISTORY OF PANDEMIC INFLUENZA A IN HUMANS

Our ability to understand the origin of the 2009 H1N1 pandemic strain and to anticipate the emergence of future pandemic strains requires familiarity with the evolutionary history of influenza A in both humans and swine. The processes of mutation and reassortment and of antigenic shift and drift all come into play. Here we begin at the point at which our scientific knowledge of influenza, for all practical purposes, begins—with the 1918 H1N1 pandemic. The chronology is illustrated in Figure 8.2.

### 1918 H1N1 "Spanish Flu" Pandemic

The 1918 H1N1 pandemic occurred in waves of increasingly virulent disease that started in the spring of 1918 and continued through the subsequent winter, resulting in the deaths of tens of millions of people worldwide. Although called "Spanish flu," the geographic origin of this virus remains unknown and somewhat controversial (34). The 1918 pandemic had a much higher per-case mortality rate than seen in seasonal influenza, with deaths primarily occurring in young adults rather than, as typical of seasonal influenza, among the elderly and infants.

The manner in which deaths occurred was also unusual. Influenza-related mortality typically occurs a week or two postinfection and is associated with secondary bacterial pneumonia or other complications. During the "Spanish flu" epidemic, many people died within just a few days from acute respiratory distress syndrome (35).

To better understand this difference in pathology, Taubenberger, Reid, and colleagues undertook a forensic investigation that resulted in the sequencing and reconstruction of the 1918 H1N1 strain using viruses preserved in the lung tissue of two army soldiers and from an Alaskan Inuit woman frozen in permafrost—all victims of the 1918 pandemic (36,37). Subsequent experiments using these genetic resources suggest that the 1918 H1N1 virus was able to interfere with host immunity via mechanisms such as upregulating genes coding for cytokines (38,39).

An outbreak of respiratory disease also occurred in swine in the United States in 1918. Serological tests of the first influenza viruses isolated from swine in 1930 suggested that the human and swine H1N1 lineages had descended from a common ancestor. These results were confirmed later using protein sequencing and then gene sequencing, as reviewed elsewhere (40). This "classic" H1N1 swine lineage circulates in swine to this day. Although we will not address swine influenza in any more detail here, in addition to the classic H1N1 and Eurasian H1N1 domestic strains, swine also harbor human H1N1 and H3N2 viruses and various reassortments of these various lineages (41–46).

The relationship between the 2009 H1N1 pandemic strain and human seasonal H1N1 influenza A is a major source of confusion in the lay community. The H1 hemagglutinin allele of the 2009 H1N1 human pandemic virus was derived from the classic H1N1 swine lineage, which, over the years, has diverged genetically and antigenically from the more rapidly evolving human seasonal H1N1 lineage. Humans in older age groups apparently retain immune memory to the viruses they were originally exposed to, back when seasonal H1N1 viruses were more similar to the classic swine strain (47).

## 1957 H2N2 "Asian Flu" and 1968 H3N2 "Hong Kong Flu" Pandemics

The 1918 H1N1 pandemic strain evolved, within a few years' time, into a virus with the virulence and epidemiological characteristics of a seasonal influenza virus. This lineage circulated in humans until it was displaced during the 1957 pandemic by an H2N2 reassortant that consisted of an H1N1 seasonal virus that had obtained avian segments coding for H2, N2, and PB1 alleles. The H2N2 pandemic virus circulated as a seasonal influenza strain until being displaced by an H3N2 reassortant composed of a seasonal H2N2 virus with avian H3 and PB1 genes (48).

The paucity of archived viral isolates from human, swine, and avian hosts may prevent us from ever determining the sequence of events culminating in the 1918 pandemic. No influenza viruses isolated from modern birds are particularly similar to the 1918 pandemic strain, and sequences from viruses isolated from waterfowl that were collected in 1917 and preserved in alcohol in the American Museum of Natural History did little to resolve this mystery (49).

Unfortunately, almost as little is known about the avian and swine viruses in circulation at the time of the 1957 and 1968 pandemics. This lack of knowledge points to the importance to forensics in preserving existing museum collections of animals from which viral RNA can be extracted and for expanded surveillance of extant zoonotic strains.

The serial replacement of one influenza A subtype by another in 1957 and again in 1968 suggested [for reasons that are presumably related to immune memory but remain poorly understood (50)] that only a single influenza A subtype could circulate in humans at one time. However, an event in 1977 proved this not to be so.

## The 1977 "Russian Flu" Reemergence of Seasonal H1N1

In 1977, 20 years after it had last circulated in humans, human seasonal H1N1 influenza reappeared in northern China (51). The reemergent viruses were identical to H1N1 viruses that had been isolated from humans in 1950 and are thought to have escaped from a laboratory somewhere near the Soviet–Chinese border (6,52,53). The reemerged H1N1 strain spread rapidly in the human population, but caused relatively mild disease. Illness occurred almost exclusively among those under 20 years of age, as older individuals retained humoral immune memory from exposure prior to 1957 (54).

Both the (reemergent) H1N1 and H3N2 subtypes of seasonal influenza A cocirculated in humans from 1977 until the emergence of the 2009 H1N1 pandemic strain. During cocirculation, reassortment between the two strains occurred repeatedly; however, consistent with previous observations of reassortment between human-adapted strains (21), the reassortant strains were not persistent (55,56). It is not clear at the time of this writing whether the seasonal H1N1 strain became extinct during the 2009 H1N1 pandemic.

## 2009 H1N1 PANDEMIC

The 2009 H1N1 pandemic is ongoing from a public health point of view. Forensically, however, it seems that, as in past influenza pandemics, the events leading to its emergence were over well before we knew anything was happening. Here we review the initial time course of the pandemic, which has been described in much more detail elsewhere (31,57–61) from a forensic point of view.

In March and early April of 2009, several regions in Mexico experienced outbreaks of respiratory disease. This outbreak appeared at an unusual time of year and, as reviewed later, the symptoms in some people, but certainly not all, were unusually severe. On April 17, 2009, a case of atypical pneumonia in the La Gloria area of Oaxaca State prompted enhanced surveillance throughout Mexico (62).

In the meantime, viruses isolated from two children in southern California that tested positive for influenza A during a routine screen but could not be assigned to subtype were sent to the Centers for Disease Control and Prevention (CDC) for assessment. On April 21, 2009, the CDC announced existence of the 2009 H1N1 pandemic strain (63). Two days later five more cases—two from Texas and three from southern California—were reported (62). By late April there were reports of outbreaks in schools in New York and Nova Scotia, and it soon became clear that the pandemic had spread to Europe as well (64).

Gene sequencing revealed that the 2009 H1N1 pandemic strain is the result of reassortment between an H1N1 virus from the "Eurasian" swine lineage and an H1N1 virus from the "triple reassortant" swine lineage (Figure 8.2), despite the fact that these lineages are not known to circulate on the same continent (58). The Eurasian swine lineage originated via transmission of an avian virus to a pig sometime around 1979 and contributed segments 6 and 7 to the pandemic strain. The "triple reassortant" viruses contain segments from an avian virus of unknown subtype (segments 1 and 3), a human seasonal H3N2 virus (segment 2), and a virus from the "classic" swine H1N1 lineage (segments 4–8), which is a direct descendant of the swine viruses first isolated in 1930 and thus presumably from the same 1918 outbreak that caused the "Spanish flu" pandemic in humans.

The 2009 H1N1 pandemic strain has an H1N1 subtype designation but its H1 and N1 alleles came from different evolutionary lineages: the H1 allele is from a swine H1N1 triple reassortant virus, while the N1 allele is from the Eurasian swine linage and thus descended from viruses that were, until around 1979, circulating in birds. It is interesting but of unknown biological significance that the PB1 gene (segment 2) of the 2009 H1N1 pandemic virus had recently circulated in a human-adapted lineage; segment 2 reassorted from avian viruses to human-adapted strains in the formation of both the 1957 and the 1968 pandemic strains. The sequences in GenBank that were most similar to those of the 2009 H1N1 virus were from viruses isolated anywhere from 5 to 17 years ago, depending on the segment.

The exact circumstances surrounding the origin of the 2009 H1N1 virus remain a mystery. It has been reported (31) that there was no outbreak of influenza in either swine or farm workers in the La Gloria area of Mexico, where the first outbreaks in humans were seen. Searching for the origin of the 2009 H1N1 virus using serological surveys of swine herds is probably fruitless. The HA of the pandemic strain was derived from the classic H1N1 swine lineage (via the triple reassortant), and most domestic swine have been either infected with or vaccinated with the classic H1N1 strain (65). Lopez-Cervantes and colleagues (31) suggested that the outbreak may have stemmed from transmission to humans by travelers, as this area of Mexico has high human migration rates and also because the first cases with viral confirmation (the two children in southern California) had no reported contact with swine.

The influenza surveillance community responded to the outbreak by searching archived viruses in the hope of finding strains that would help bridge the events leading to the origin of the pandemic strain. Smith and colleagues (60) reported extensive reassortment among the three main swine influenza A lineages (classic, Eurasian, and triple reassortant; Figure 8.2). One virus of interest, A/Sw/HK/915/04, was similar to the pandemic strain in seven of eight gene segments, but phylogenetic analysis did not place this virus on a direct evolutionary lineage leading to the 2009 H1N1 outbreak (60).

An exhaustive search of GenBank produced two reassortant viruses carrying M and NA alleles similar to those in the 2009 H1N1 strain, but again, based on analysis of all segments, these viruses did not appear to have stemmed from the same evolutionary lineage as the H1N1 2009 pandemic virus (66). This work, along with that of Smith and colleagues (60), clearly illustrates that much undetected reassortment has been occurring in swine lineages without resulting in outbreaks of human disease.

Forensic analysis of any outbreak, whether intentional or not, should include a screen for characters associated with traits such as host range specificity and virulence. In influenza, only a few such characters are known.

The PB2 protein of avian and equine influenza A viruses typically has a glutamic acid in position 627 while a lysine is found in human influenza A viruses. An E627K replacement in an avian influenza A virus has been shown to increase the virulence of avian H5N1 and equine H7N7 influenza A viruses (67,68), and as such one suspects an increased propensity to spread and perhaps more severe outcomes with 627K. Swine may have either lysine or glutamic acid; the H1N1 2009 virus had an E, with this segment being derived from an avian virus that reassorted in 1998 to form the swine "triple reassortant" virus (69). Replacement of the 627E with K in the 2009 H1N1 virus was not found to be associated with increased virulence in mice (70), perhaps due to compensatory mutations at other positions in the PB2 gene, such as 701 (71).

Another genomic area associated with, but not diagnostic of, host range and virulence is the length and amino acid composition of the neuraminidase stalk (72). The 2009 H1N1 virus was similar to other swine and human-adapted viruses in this respect.

The PB1-F2 protein has been associated with increased pathogenicity of the 1918 flu and highly pathogenic H5N1 avian influenza (73,74); however, the PB1-F2 gene of the H1N1 2009 pandemic virus is truncated and nonfunctional (58).

The influenza NS1 protein is involved in inhibition of the antiviral interferon response (75); substituting the last four C-terminal residues from the 1918 H1N1 and from an H5N1 highly pathogenic avian influenza virus into a seasonal influenza virus caused a number of negative effects on the host by mechanisms that remain to be elucidated (76). However, the NS1 of the 2009 H1N1 pandemic virus was truncated at the stop codon and did not contain these residues (58).

Amino acids at several hemagglutinin positions (such as 190 and 225) are known to be involved in host-cell receptor binding; these differ between subtypes and between avian and mammalian-adapted viruses (77) but are not precisely diagnostic. The binding site can change very rapidly once in a new host or laboratory culture (78). A comparison between the 1979 avian-origin H1N1 European swine lineage and classic H1N1 swine viruses failed to find differences that could be considered diagnostic (79). The receptor-binding pocket of the 2009 H1N1 pandemic virus resembled that of the classic swine lineage and showed no obvious signs of adaptation to humans (58).

In summary, despite our best efforts, the exact origin and evolutionary history of the 2009 H1N1 virus remain unclear. The evolutionary gap between this virus and its possible progenitors is similar to gaps seen in previous pandemics, and while it is not impossible that this virus was created intentionally and released intentionally or accidentally (80), there is nothing about it that makes it more mysterious than any prior pandemic strain.

## OUTBREAK DETECTION: SYMPTOMS, DIAGNOSTICS, AND SURVEILLANCE

As reviewed elsewhere (1), epidemiological evidence that a disease outbreak may have been caused intentionally include increased severity of symptoms or unusual symptoms, circulation out of season, an unusual age distribution, or rapid spread of disease. As shown later, the range of presentation of natural influenza outbreaks is so great with respect to these factors that it is hard to imagine an intentional outbreak that would make one suspect ill intent.

### Human Seasonal Influenza

The symptoms of infection by a human-adapted seasonal influenza virus range from none to lethal, varying both by viral strain and by health, genotype, and immune status of the individual host. Typically, however, they include abrupt onset of fever, cough, sore throat, myalgia, arthralgia, chills, headache, and fatigue. Gastrointestinal symptoms are sometimes seen in children, but are infrequent in adults. For reviews, see Wright and Webster (11) and Cox and associates (12).

Most people with influenza-like symptoms do not seek medical care; when they do, they are generally not tested for a disease-causing organism unless hospitalization is required. This can cause a delay in the realization that an outbreak is occurring. Many cases of influenza-like illness are not even caused by influenza; rather they can be attributed to rhinoviruses, respiratory syncytial viruses, and other respiratory viruses (81).

Mortality due to seasonal influenza is most often seen in the very old, via a secondary bacterial pneumonia, and the very young, whose immune systems are not yet well developed. This age pattern is not, however, diagnostic of an influenza outbreak. The 2009 H1N1 pandemic, for example, caused a disproportionate amount of disease in young people because they lacked the preexisting immunity seen in older cohorts (82–84), apparently due to preservation of a B-cell epitope from the 1918 H1N1 progenitor (47). A similar age pattern was observed with reintroduction of a human-adapted seasonal H1N1 strain in 1977, after being out of circulation in humans for 20 years (52).

## Swine Influenza Infection of Humans

Swine farmers are occasionally infected by swine-adapted influenza viruses. A wide range of symptoms and attack profiles can result, but these infections are typically not problematic (85) and, even in the case of the infamous Fort Dix 1976 "swine flu outbreak" [reviewed elsewhere (86)], rarely spread among humans. Further study is needed to reveal why the 2009 H1N1 pandemic virus was different.

## Avian Influenza Infection of Humans

Human influenza to date has been restricted to viruses carrying a very small subset of the possible avian HA and NA alleles (H1, H2, H3, N1, and N2). Given the vast number of influenza A viruses extant in birds, the rarity of type A pandemics in humans reflects the fact that avian-adapted influenza A viruses do not infect or transmit easily among humans. The reasons for this are not clear; host-cell receptor-binding specificity has been implicated but does not seem to be the only factor involved (87).

When avian influenza viruses do infect humans, the symptoms range from none to lethal, as with seasonal influenza. Most of our knowledge of avian influenza comes from human infections with the avian H5N1 virus, which typically does not involve mortality from secondary bacterial pneumonia but instead is due to acute respiratory distress syndrome associated with virus-induced cytokine dysregulation (88–90). Interference with host immunity also seems to have been a factor in the high mortality of the 1918 H1N1 human pandemic virus (39) and in some deaths from the 2009 H1N1 pandemic virus as well (61).

Release of a "bird flu" bioweapon would probably not escape notice due to many years of media exposure. Avian-adapted strains can also, however, present with symptoms that most people would not associate with influenza, such as cases of human conjunctivitis seen in infections associated with outbreaks of highly pathogenic avian H7N7 influenza in the Netherlands in 2003 (91) and H7N3 infection in Canada in 2004 (92).

Although there have been few serological surveys of animal workers for exposure to nonhuman-adapted influenza viruses, some (85) suggest that hunting and occupational exposure to poultry present some risk to humans, as evidenced by seropositive human responses to H5 and H7 avian alleles.

## SURVEILLANCE

The time lag between emergence of a novel strain and its detection depends greatly on the intensity of surveillance, which varies from quite minimal in parts of Africa and other developing areas of the world to well-organized systems in countries participating in the WHO Global Influenza Programme (http://www.who.int/csr/disease/influenza).

Surveillance has been increasing in both geographic range and sophistication in recent years and thus will likely have changed before this chapter is published. Briefly, in the United States, influenza surveillance is a collaborative effort between the CDC and state and local health departments, clinical laboratories, vital statistics offices, hospitals, and health care providers. Information is based on data on viral surveillance, outpatient and hospital illness, patterns of mortality, and geographical spread of disease. Information about animal influenza is coordinated through various agencies, including the World Organization for Animal Health (OIE) and the Food and Agriculture Organization (FAO) and Offlu, a network of experts coordinated by the OIE/FAO and the WHO.

### Syndromic Surveillance

Syndromic surveillance has been a source of interest ever since the 2001 anthrax letter incidents, but as yet has not proved to be of much help in outbreak detection. For example, Google Flu Trends, which uses search engine query data to detect deviations from a baseline, showed a spike in mid-April of 2009 in Mexico due to only two data points, April 19 and 26, which correspond to media announcements about the H1N1 outbreak. After that inquiries declined, despite the fact that the outbreak was spreading rapidly.

## DIAGNOSTICS

Many of the problems related to microbial typing outlined by Budowle and colleagues (93), such as choosing particular parts of large genomes to type, are not of significant impact in an influenza investigation because of the small size (14 kb) of its genome. The current gold standard for influenza identification is RT-PCR, followed by viral culture and sequencing (64). Sets of RT-PCR amplification primers designed to distinguish between types A and B typically use the highly conserved matrix gene. Until 2009, all currently circulating swine-adapted influenza A lineages could be distinguished from human H1N1 lineages by their nucleoprotein gene, which is of classical swine H1N1 origin. Seasonal H1 and pandemic H1 alleles can be differentiated using the hemagglutinin gene. These RT-PCR protocols are updated continually and made available on the WHO (94) and CDC Web sites.

For isolates of particular interest, RT-PCR is followed by viral isolation and whole genome sequencing. Gene sequence data, particularly of the rapidly evolving HA and NA genes, allow determination of evolutionary relationships using genetic similarity measures. These are determined most easily by comparison with the almost 15,000 influenza gene sequences and over 5800 complete genomes in GenBank using the NCBI BLAST tool (http://blast.ncbi.nlm.nih.gov/). Phylogenetic analysis for assignment to evolutionary lineage follows, along with a comparison of individual gene trees to check for reassortment.

The evolutionary scenario shown in Figure 8.1b was based on phylogenetic analysis of human H3 HA gene sequence data; the (nonaccurate cartoon) overlay of boxes indicates antigenic relationships based on the hemagglutination inhibition assay, the standard antigen–antibody assay used in influenza surveillance. Figure 8.2 is an attempt to provide an overview of the events leading to formation of the 2009 H1N1 virus without showing the evolutionary trajectory of every gene in each viral lineage, something bravely attempted by Smith and colleagues (60). This chapter does not detail the methodology used in phylogenetic reconstruction. Influenza sequence data are essentially free of indels and variation due to recombination; thus alignment is straightforward. Satisfactory phylogenetic trees can be constructed using any of the usual neighbor-joining, maximum likelihood, or Bayesian algorithms (95).

There has to date only been one in-depth analysis of the evolutionary events involved in generation of a human pandemic strain. Figure 8.2 suggests that evolution of the 1968 H3N2 pandemic strain involved a single reassortment event. Genome sequencing of archived viruses revealed that the human H3N2 strain was, in fact, the single survivor of multiple reassortment lineages that cocirculated during the early years of that pandemic (32). The sorting of gene segments during this time, reminiscent of introgression during plant hybridization in nature, is at this point simply an observation. We do not understand the mechanistic basis for the fact that only one lineage remained by 1971.

Lacking such samples for other outbreaks, we are forced to resort to molecular evolution analyses to estimate past divergence times. These analyses have as yet added little to our knowledge of the origin of the 2009 H1N1 pandemic virus over what was known from simply blasting the first segments sequenced against GenBank and noting the isolation dates of the closest relative of each segment (58,96–98).

It has been hypothesized that swine served as intermediate hosts in which the reassortment events of 1957 and 1968 took place; however, there is as yet no genetic evidence in support of this hypothesis (99). The avian H2 allele related most closely to the H2 allele in the 1957 pandemic virus is from a duck isolated in 1973 in Germany; the surface-exposed (HA1) domain of these HAs differs at 22 of 328 amino acid positions. The H3 HA of the 1968 H3N2 pandemic strain differs from a number of avian alleles isolated between 1972 and 1980 by 8 to 12 fixed changes in the HA; the genetic distance being independent of the year of isolation. Genetic differences of this magnitude are similar to those that can take years, if not decades, to accumulate via antigenic drift, with the time course depending on the host in which the virus resides. It is very difficult to estimate the time passed, as the rate of antigenic drift can vary considerably over time even within individual lineages of human seasonal influenza (100).

Before going on to discuss how influenza might be used as a bioweapon, we note that although various types of quick tests, called "rapid influenza diagnostic tests" and "direct immunofluorescence assays," are available for diagnosing influenza, they lack the sensitivity needed to rule out an influenza infection, especially in specimens with low viral titers. Their use in typing an unusual outbreak is also limited by their scope—many can only distinguish influenza A from B, whereas others can determine that a sample is influenza type A but cannot distinguish subtype (101). Discovery of the 2009 H1N1 pandemic was delayed due to these viruses showing up only as "untypeable" influenza A on such test kits.

Microbial forensic investigations may also include examination of isotopes or other artifacts of the manufacturing process in cases of intentional release. While these activities might compose part of an influenza forensics investigation, they are outside the scope of this review. For more information on diagnostic testing, see reviews by Fan and colleagues (102) and Poon and colleagues (103). An excellent discussion of the advantages and disadvantages of different types of diagnostic tests can be found elsewhere (104,105).

## INFLUENZA AS A BIOWEAPON

Even among the general public the influenza virus is well known for its capacity to transmit rapidly and to evolve quickly into forms against which humans lack immunity. It is also widely known that the virus can evolve quickly to evade immunity provided by well-matched vaccines. Clearly, anyone developing an influenza bioweapon would know these issues too and thus be acting without regard to the survival of self or family.

Some influenza viruses are listed as select agents by the CDC (106). The Select Agent Program specifically lists the reconstructed 1918 H1N1 virus as an HHS Select agent. There is also concern that influenza could be used as an agricultural bioweapon (105); accordingly, highly pathogenic avian influenza is listed as a Select Agent by the U.S. Department of Agriculture (http://www .selectagents.gov).

What would an influenza bioweapon look like and how would we respond to an attack? The latter question is the easier: as outlined elsewhere (106), the basic procedures for dealing with influenza bioterrorism differ from those already in place in the extensive and ever-expanding literature on pandemic planning by the involvement of law enforcement.

What would an influenza bioweapon look like? An influenza bioweapon could come in a variety of forms, but it would have to meet two major requirements. First, it must have the ability to transmit among humans. We know very little about the biological determinants of viral transmission (or of virulence) except that these are obviously not simply determined traits (11,76). Engineering these traits into a virus would probably be accomplished most easily through trial and error, such as by serial passage in animals. There is no record of this ever having been attempted, so it is hard to say if it would work or what animal model might be required.

The second requirement for an influenza bioweapon is a sufficient lack of immunity in the human population. Rather surprisingly, this is another poorly understood area of influenza biology. The human population experiences yearly outbreaks of seasonal influenza viruses whose surface proteins have evolved into forms not well recognized by antibodies formed in response to previous infection or vaccination. On a per-individual basis, however, it is not clear how much antigenic drift is required to avoid reinfection or how this varies among the different influenza lineages. As discussed earlier, when a seasonal H1N1 virus was reintroduced to the human population in 1977, after 20 years' time, there was clearly evidence of some long-term immune protection of older age cohorts, despite significant antigenic drift during the intervening years.

This pattern was also seen during the 2009 H1N1 pandemic, where older cohorts of humans derived some protection due to exposure to seasonal H1N1 viruses that had not yet undergone many rounds of genetic drift (47). Together, residual immunity in older cohorts, along with newly derived immunity via exposure to the 2009 H1N1 virus, may provide sufficient breadth of protection against the most feared influenza bioweapon, a reconstruction of the 1918 H1N1 "Spanish flu" virus.

In theory, any desired influenza bioweapon can be constructed in the laboratory using techniques similar to those used by Tumpey and associates (2) in reconstruction of the 1918 H1N1 pandemic virus. The A/WSN/33 (H1N1) backbone plasmids (107,108) used in these protocols have been shared with laboratories worldwide, making attribution problematic.

An optimized influenza bioweapon would be likely to have been engineered for resistance to antiviral drugs, such as via the S31N mutation in the M2 protein. The M2 protein is the target of M2 ion channel blocker drugs (adamantanes) that include amantadine and rimantadine. However, because mutations resulting in antiviral resistance are known to occur in nature in the absence of selective pressure by antiviral treatment, their presence would not necessarily be indicative of genetic engineering.

Influenza strains currently most capable of transmission and spread, and obtained most easily, are the archived H2N2 seasonal human influenza A viruses. As shown in Figure 8.2, the H2 HA allele only circulated in humans from 1957 to 1968, and thus anyone currently under 42 years of age has not been exposed to the H2 antigen. It is probably only a matter of time until we experience a reemergence of H2N2, which is still under study in many laboratories, including those interested in developing vaccines should the H2N2 virus reemerge (109).

In 2005, human H2N2 viruses were distributed accidentally to over 4000 clinical laboratories as parts of laboratory proficiency test kits. This error was realized prior to spread, and the test kits were destroyed (http://www.cdc.gov/flu/h2n2backgroundqa.htm)(http://www.who.int/csr/disease/influenza/h2n2_2005_04_12/en/). As noted (110), laboratory contamination of patient material by lab strains can cause a "pseudo-outbreak" [an outbreak in viral cultures, not people (111)] and should be considered when genetic tests suggest infection by archival influenza isolates.

The continued use of archival viruses for both basic research and vaccine development is essential but also problematic, as this makes it impossible to rule out laboratories as sources of future outbreaks, whether intentional or accidental. Experimental attempts to discover the genetic determinants of transmission using reassortant viruses composed of gene segments from H5N1 avian flu and human seasonal H3N2 viruses have not provided simple biological answers to these questions, but do illustrate our capacity to construct potentially problematic viruses in the laboratory (112–115). In 2008, live H5N1 avian influenza viruses were shipped accidentally to other laboratories in Europe, a fact only realized upon the death of ferrets in a receiving laboratory (http://www.promedmail.org archive 20090226.0801). Fortunately, these viruses failed to spread among humans, as have H5N1 viruses in nature to date.

## MOVING FORWARD

As demonstrated in the 2009 H1N1 influenza pandemic, the main impediment to influenza forensics is the absence of data due to the passage of time between emergence of a new virus into the human population and its detection. As a review by Snacken and colleagues (116), now 12 years old, aptly outlines, steps needed to close this gap are clear but difficult and expensive: improved capacity for international response, enhanced human and veterinary surveillance accompanied by a more broadly representative sequence database, better low-cost surveillance techniques, improved laboratory safety, enhanced electronic communications, enhanced vaccine production capacity, and improved access to vaccines and antivirals via efficient distribution systems. In addition, reconstruction of evolutionary lineages linking outbreak strains to their progenitors requires increased collection, sequencing, and deposition of sequences, particularly from nonhuman species that are sorely underrepresented in online databases, and continual updating of comprehensive reviews by specialists that integrate these sequence data with host animal ecology (16,43,45).

## ACKNOWLEDGMENTS

R.M. Bush is funded by the National Institute of General Medical Sciences Models of Infectious Disease Agent Study (MIDAS). She also thanks the Santa Fe Institute for support.

## REFERENCES

[1] S.A. Morse, B. Budowle, Microbial forensics: Application to bioterrorism preparedness and response, Infect. Dis. Clin. North Am. 20 (2006), xi, 455–473.

[2] T.M. Tumpey, C.F. Basler, P.V. Aguilar, H. Zeng, A. Solorzano, D.E. Swayne, et al., Characterization of the reconstructed 1918 Spanish influenza pandemic virus, Science 310 (2005) 77–80.

[3] P.A. Sharp, 1918 flu and responsible science, Science 310 (2005) 17.

[4] S. Miller, M.J. Selgelid, Ethical and philosophical consideration of the dual-use dilemma in the biological sciences, Sci. Eng. Ethics 13 (2007) 523–580.

[5] M.J. Selgelid, Governance of dual-use research: An ethical dilemma, Bull. World Health Organ. 87 (2009) 720–723.

[6] R.G. Webster, W.J. Bean, O.T. Gorman, T.M. Chambers, Y. Kawaoka, Evolution and ecology of influenza A viruses, Microbiol. Rev. 56 (1992) 152–179.

[7] R.J. Webby, R.G. Webster, J.A. Richt, Influenza viruses in animal wildlife populations, Curr. Top. Microbiol. Immunol. 315 (2007) 67–83.

[8] M.F. Ducatez, R.G. Webster, R.J. Webby, Animal influenza epidemiology, Vaccine 26 (Suppl 4) (2008) D67–D69.

[9] A.W. Hampson, Epidemiological data on influenza in Asian countries, Vaccine 17 (Suppl 1) (1999) S19–S23.

[10] L. Yang, C.M. Wong, E.H. Lau, K.P. Chan, C.Q. Ou, J.S. Peiris, Synchrony of clinical and laboratory surveillance for influenza in Hong Kong, PLoS One 3 (2008) e1399.

[11] P.F. Wright, R.G. Webster, Orthomyxoviruses, in: D.M. Knipe, P.M. Howley (Eds.), Fields Virology, Lippincott Williams and Wilkins, Philadelphia, PA, 2001, pp. 1533–1579.

[12] N.J. Cox, G. Neumann, R.O. Donis, Y. Kawaoka, Orthomyxoviruses: Influenza, in: B.W.J. Mahy, L. Collier (Eds.), Topley & Wilson's Microbiology and Microbial Infections, Wiley-Blackwell, London, 2009, pp. 634–698.

[13] R.M. Bush, W.M. Fitch, C.A. Bender, N.J. Cox, Positive selection on the H3 hemagglutinin gene of human influenza virus A, Mol. Biol. Evol. 16 (1999) 1457–1465.

[14] J.L. Schulman, The role of antineuraminidase antibody in immunity to influenza virus infection, Bull. World Health Organ. 41 (1969) 647–650.

[15] D.L. Suarez, Evolution of avian influenza viruses, Vet. Microbiol. 74 (2000) 15–27.

[16] C.W. Olsen, The emergence of novel swine influenza viruses in North America, Virus Res. 85 (2002) 199–210.

[17] W. Chen, P.A. Calvo, D. Malide, J. Gibbs, U. Schubert, I. Bacik, et al., A novel influenza A virus mitochondrial protein that induces cell death, Nat. Med. 7 (2001) 1306–1312.

[18] R.G. Webster, W.G. Laver, Determination of the number of nonoverlapping antigenic areas on Hong Kong (H3N2) influenza virus hemagglutinin with monoclonal antibodies and the selection of variants with potential epidemiological significance, Virology 104 (1980) 139–148.

[19] J.D. Parvin, A. Moscona, W.T. Pan, J.M. Leider, P. Palese, Measurement of the mutation rates of animal viruses: Influenza A virus and poliovirus type 1, J. Virol. 59 (1986) 377–383.

[20] M.F. Boni, Y. Zhou, J.K. Taubenberger, E.C. Holmes, Homologous recombination is very rare or absent in human influenza A virus, J. Virol. 82 (2008) 4807–4811.

[21] N.J. Cox, C.A. Bender, The molecular epidemiology of influenza viruses, Sem. Virol. 6 (1995) 359–370.

[22] D.C. Wiley, I.A. Wilson, J.J. Skehel, Structural identification of the antibody-binding sites of Hong Kong influenza haemagglutinin and their involvement in antigenic variation, Nature 289 (1981) 373–378.

[23] G.W. Both, M.J. Sleigh, N.J. Cox, A.P. Kendal, Antigenic drift in influenza virus H3 hemagglutinin from 1968 to 1980: Multiple evolutionary pathways and sequential amino acid changes at key antigenic sites, J. Virol. 48 (1983) 52–60.

[24] I.A. Wilson, N.J. Cox, Structural basis of immune recognition of influenza virus hemagglutinin, Annu. Rev. Immunol. 8 (1990) 737–771.

[25] D.J. Smith, A.S. Lapedes, J.C. De Jong, T.M. Bestebroer, G.F. Rimmelzwaan, A.D. Osterhaus, et al., Mapping the antigenic and genetic evolution of influenza virus, Science 305 (2004) 371–376.

[26] H. Jin, H. Zhou, H. Liu, W. Chan, L. Adhikary, K. Mahmood, et al., Two residues in the hemagglutinin of A/Fujian/411/02-like influenza viruses are responsible for antigenic drift from A/Panama/2007/99, Virology 336 (2005) 113–119.

[27] N.J. McDonald, C.B. Smith, N.J. Cox, Antigenic drift in the evolution of H1N1 influenza A viruses resulting from deletion of a single amino acid in the haemagglutinin gene, J. Gen. Virol. 88 (2007) 3209–3213.

[28] R. Rabadan, A.J. Levine, M. Krasnitz, Non-random reassortment in human influenza A viruses, Influenza Other Respir. Viruses 2 (2008) 9–22.

[29] W.O.F.H.N.E.W. Group, Toward a unified nomenclature system for highly pathogenic avian influenza virus (H5N1). Emerging infectious diseases, 2008. Available from: http://www.cdc.gov/EID/content/14/17/e11.htm.

[30] L. Guo, R.J. Garten, A.S. Foust, W.M.X. Sessions, M. Okomo-Adhiambo, L.V. Gubareva, et al., Rapid identification of oseltamivir-resistant influenza A (H1N1) viruses with H274Y mutation by RT-PCR/restriction fragment length polymorphism assay, Antiviral Res. 82 (2009) 29–33.

[31] M. Lopez-Cervantes, A. Venado, A. Moreno, R.L. Pacheco-Dominguez, G. Ortega-Pierres, On the spread of the novel influenza A (H1N1) virus in Mexico, J. Infect. Dev. Ctries. 3 (2009) 327–330.

[32] S.E. Lindstrom, N.J. Cox, A. Klimov, Genetic analysis of human H2N2 and early H3N2 influenza viruses, 1957–1972: Evidence for genetic divergence and multiple reassortment events, Virology 328 (2004) 101–119.

[33] W.R. Dowdle, Influenza A virus recycling revisited, Bull. World Health Organ. 77 (1999) 820–828.

[34] A. Trilla, G. Trilla, C. Daer, The 1918 "Spanish flu" in Spain, Clin. Infect. Dis. 47 (2008) 668–673.

[35] J.K. Taubenberger, A.H. Reid, T.G. Fanning, The 1918 influenza virus: A killer comes into view, Virology 274 (2000) 241–245.

[36] J.K. Taubenberger, A.H. Reid, A.E. Krafft, K.E. Bijwaard, T.G. Fanning, Initial genetic characterization of the 1918 "Spanish" influenza virus, Science 275 (1997) 1793–1796.

[37] A.H. Reid, T.G. Fanning, J.V. Hultin, J.K. Taubenberger, Origin and evolution of the 1918 "Spanish" influenza virus hemagglutinin gene, Proc. Natl. Acad. Sci. USA 96 (1999) 1651–1656.

[38] J.K. Taubenberger, The virulence of the 1918 pandemic influenza virus: Unraveling the enigma, Arch. Virol. Suppl (2005) 101–115.

[39] R. Billharz, H. Zeng, S.C. Proll, M.J. Korth, S. Lederer, R. Albrecht, et al., The NS1 protein of the 1918 pandemic influenza virus blocks host interferon and lipid metabolism pathways, J. Virol. 83 (2009) 10557–10570.

[40] R.G. Webster, Wet markets: A continuing source of severe acute respiratory syndrome and influenza? Lancet 363 (2004) 234–236.

[41] B.R. Murphy, R.G. Webster, Orthomyxoviruses, in: B.N. Fields, D.M. Knipe, P.M. Howley (Eds.), Fields Virology, Lippincott-Raven, Philadelphia, PA, 1996, pp. 1397–1445.

[42] I.H. Brown, The epidemiology and evolution of influenza viruses in pigs, Vet. Microbiol. 74 (2000) 29–46.

[43] B. Olsen, V.J. Munster, A. Wallensten, J. Waldenstrom, A.D. Osterhaus, R.A. Fouchier, Global patterns of influenza a virus in wild birds, Science 312 (2006) 384–388.

[44] K. Van Reeth, I.H. Brown, R. Durrwald, E. Foni, G. Labarque, P. Lenihan, et al., Seroprevalence of H1N1, H3N2 and H1N2 influenza viruses in pigs in seven European countries in 2002–2003, Influenza Other Respi. Viruses 2 (3) (2008) 99–105.

[45] C. Brockwell-Staats, R.G. Webster, R.J. Webby, Diversity of influenza viruses in swine and the emergence of a novel human pandemic influenza A (H1N1), Influenza Other Respi. Viruses 3 (2009) 207–213.

[46] G. Kuntz-Simon, F. Madec, Genetic and antigenic evolution of swine influenza viruses in Europe and evaluation of their zoonotic potential, Zoonoses Public Health (2009).

[47] R. Xu, D.C. Ekiert, J.C. Krause, R. Hai, J.E. Crowe Jr., I.A. Wilson, Structural basis of preexisting immunity to the 2009 H1N1 pandemic influenza virus, Science 328 (2010) 357–360.

[48] Y. Kawaoka, S. Krauss, R.G. Webster, Avian-to-human transmission of the PB1 gene of influenza A viruses in the 1957 and 1968 pandemics, J. Virol. 63 (1989) 4603–4608.

[49] T.G. Fanning, R.D. Slemons, A.H. Reid, T.A. Janczewski, J. Dean, J.K. Taubenberger, 1917 avian influenza virus sequences suggest that the 1918 pandemic virus did not acquire its hemagglutinin directly from birds, J. Virol. 76 (2002) 7860–7862.

[50] N.M. Ferguson, A.P. Galvani, R.M. Bush, Ecological and immunological determinants of influenza evolution, Nature 422 (2003) 428–433.

[51] F.L. Raymond, A.J. Caton, N.J. Cox, A.P. Kendal, G.G. Brownlee, The antigenicity and evolution of influenza H1 haemagglutinin, from 1950-1957 and 1977–1983: Two pathways from one gene, Virology 148 (1986) 275–287.

[52] A.P. Kendal, G.R. Noble, J.J. Skehel, W.R. Dowdle, Antigenic similarity of influenza A (H1N1) viruses from epidemics in 1977-1978 to "Scandinavian" strains isolated in epidemics of 1950-1951, Virology 89 (1978) 632–636.

[53] K. Nakajima, U. Desselberger, P. Palese, Recent human influenza A (H1N1) viruses are closely related genetically to strains isolated in 1950, Nature 274 (1978) 334–339.

[54] N. Cox, H. Regnery, Global influenza surveillance: Tracking a moving target in a rapidly changing world, in: L.E. Brown, Q.W. Hampson, R.G. Webster (Eds.), Options for the Control of Influenza III, Elsevier Science Publishers B. V., 1996, pp. 591–598.

[55] Y.J. Guo, X.Y. Xu, N.J. Cox, Human influenza A (H1N2) viruses isolated from China, J. Gen. Virol. 73 (1992) 383–387.

[56] V. Gregory, M. Bennett, M. Orkhan, S. Al Hajjar, N. Varsano, E. Mendelson, et al., Emergence of influenza A H1N2 reassortant viruses in the human population during 2001, Virology 300 (2002) 1–7.

[57] Centers for Disease Control and Prevention (CDC), Intensive-care patients with severe novel influenza A (H1N1) virus infection—Michigan, June 2009, MMWR Morb. Mortal. Wkly. Rep. 58 (2009) 749–752.

[58] R.J. Garten, C.T. Davis, C.A. Russell, B. Shu, S. Lindstrom, A. Balish, et al., Antigenic and genetic characteristics of swine-origin 2009 A (H1N1) influenza viruses circulating in humans, Science 325 (2009) 197–201.

[59] G. Neumann, T. Noda, Y. Kawaoka, Emergence and pandemic potential of swine-origin H1N1 influenza virus, Nature 459 (2009) 931–939.

[60] G.J. Smith, D. Vijaykrishna, J. Bahl, S.J. Lycett, M. Worobey, O.G. Pybus, et al., Origins and evolutionary genomics of the 2009 swine-origin H1N1 influenza A epidemic, Nature 459 (2009) 1122–1125.

[61] K.K. To, I.F. Hung, I.W. Li, K.L. Lee, C.K. Koo, W.W. Yan, et al., Delayed clearance of viral load and marked cytokine activation in severe cases of pandemic H1N1 2009 influenza virus infection, Clin. Infect. Dis. 50 (2010) 850–859.

[62] Centers for Disease Control and Prevention (CDC), Outbreak of swine-origin influenza A (H1N1) virus infection—Mexico, March-April 2009, MMWR Morb. Mortal. Wkly. Rep. 58 (2009) 467–470.

[63] Centers for Disease Control and Prevention (CDC), Swine influenza A (H1N1) infection in two children—Southern California, March-April 2009, MMWR Morb. Mortal. Wkly. Rep. 58 (2009) 400–402.

[64] F.S. Dawood, S. Jain, L. Finelli, M.W. Shaw, S. Lindstrom, R.J. Garten, Novel Swine-Origin Influenza A (H1N1) Virus Investigation Team, et al., Emergence of a novel swine-origin influenza A (H1N1) virus in humans, N. Engl. J. Med. 360 (2009) 2605–2615.

[65] C.S. Kyriakis, C.W. Olsen, S. Carman, I.H. Brown, S.M. Brookes, J.V. Doorsselaere, et al., Serologic cross-reactivity with pandemic (H1N1) 2009 virus in pigs, Europe, Emerg. Infect. Dis. 16 (2010) 96–99.

[66] C. Kingsford, N. Nagarajan, S.L. Salzberg, 2009 Swine-origin influenza A (H1N1) resembles previous influenza isolates, PLoS One 4 (2009) e6402.

[67] E.K. Subbarao, W. London, B.R. Murphy, A single amino acid in the PB2 gene of influenza A virus is a determinant of host range, J. Virol. 67 (1993) 1761–1764.

[68] G. Gabriel, M. Abram, B. Keiner, R. Wagner, H.D. Klenk, J. Stech, Differential polymerase activity in avian and mammalian cells determines host range of influenza virus, J. Virol. 81 (17) (2007) 9601–9604.

[69] N.N. Zhou, D.A. Senne, J.S. Landgraf, S.L. Swenson, G. Erickson, K. Rossow, et al., Genetic reassortment of avian, swine, and human influenza A viruses in American pigs, J. Virol. 73 (1999) 8851–8856.

[70] H. Zhu, J. Wang, P. Wang, W. Song, Z. Zheng, R. Chen, et al., Substitution of lysine at 627 position in PB2 protein does not change virulence of the 2009 pandemic H1N1 virus in mice, Virology 401 (1) (2010) 1–5.

[71] J. Steel, A.C. Lowen, S. Mubareka, P. Palese, Transmission of influenza virus in a mammalian host is increased by PB2 amino acids 627K or 627E/701N, PLoS Pathog. 5 (2009) e1000252.

[72] J. Banks, E.S. Speidel, E. Moore, L. Plowright, A. Piccirillo, I. Capua, et al., Changes in the haemagglutinin and the neuraminidase genes prior to the emergence of highly pathogenic H7N1 avian influenza viruses in Italy, Arch. Virol. 146 (2001) 963–973.

[73] D. Zamarin, M.B. Ortigoza, P. Palese, Influenza A virus PB1-F2 protein contributes to viral pathogenesis in mice, J. Virol. 80 (2006) 7976–7983.

[74] G.M. Conenello, D. Zamarin, L.A. Perrone, T. Tumpey, P. Palese, A single mutation in the PB1-F2 of H5N1 (HK/97) and 1918 influenza A viruses contributes to increased virulence, PLoS Pathog. 3 (2007) 1414–1421.

[75] A. Garcia-Sastre, A. Egorov, D. Matassov, S. Brandt, D.E. Levy, J.E. Durbin, et al., Influenza A virus lacking the NS1 gene replicates in interferon-deficient systems, Virology 252 (1998) 324–330.

[76] D. Jackson, M.J. Hussain, D. Hickman, D.R. Perez, R.A. Lamb, A new influenza virus virulence determinant: The NS1 protein four C-terminal residues modulate pathogenicity, Proc. Natl. Acad. Sci. USA 105 (2008) 4381–4386.

[77] M. Matrosovich, A. Tuzikov, N. Bovin, A. Gambaryan, A. Klimov, M.R. Castrucci, et al., Early alterations of the receptor-binding properties of H1, H2, and H3 avian influenza virus hemagglutinins after their introduction into mammals, J. Virol. 74 (2000) 8502–8512.

[78] J.S. Robertson, An overview of host cell selection, Dev. Biol. Standard. 98 (1999) 7–11, discussion 73-14.

[79] E.J. Dunham, V.G. Dugan, E.K. Kaser, S.E. Perkins, I.H. Brown, E.C. Holmes, et al., Different evolutionary trajectories of European avian-like and classical swine H1N1 influenza A viruses, J. Virol. 83 (2009) 5485–5494.

[80] A.J. Gibbs, J.S. Armstrong, J.C. Downie, From where did the 2009 "swine-origin" influenza A virus (H1N1) emerge? Virol. J. 6 (2009) 207.

[81] Centers for Disease Control and Prevention (CDC), Considerations for distinguishing influenza-like illness from inhalational anthrax, MMWR Morb. Mortal. Wkly. Rep. 50 (2001) 984–986.

[82] K. Hancock, V. Veguilla, X. Lu, W. Zhong, E.N. Butler, H. Sun, et al., Cross-reactive antibody responses to the 2009 pandemic H1N1 influenza virus, N. Engl. J. Med. 361 (2009) 1945–1952.

[83] J.C. Krause, T.M. Tumpey, C.J. Huffman, P.A. McGraw, M.B. Pearce, T. Tsibane, et al., Naturally occurring human monoclonal antibodies neutralize both 1918 and 2009 A (H1N1) pandemic influenza viruses, J. Virol. 84 (6) (2010) 3127–3130.

[84] J.A. McCullers, L.A. Van De Velde, K.J. Allison, K.C. Branum, R.J. Webby, P.M. Flynn, Recipients of vaccine against the 1976 "swine flu" have enhanced neutralization responses to the 2009 novel H1N1 influenza virus, Clin. Infect. Dis. 50 (11) (2010) 1487–1492.

[85] G.C. Gray, T. McCarthy, A.W. Capuano, S.F. Setterquist, M.C. Alavanja, C.F. Lynch, Evidence for avian influenza A infections among Iowa's agricultural workers, Influenza Other Respi. Viruses 2 (2) (2008) 61–69.

[86] E.D. Kilbourne, Influenza pandemics of the 20th century, Emerg. Infect. Dis. 12 (2006) 9–14.

[87] T. Ito, Y. Kawaoka, Host-range barrier of influenza A viruses, Vet. Microbiol. 74 (2000) 71–75.

[88] M.D. de Jong, C.P. Simmons, T.T. Thanh, V.M. Hien, G.J. Smith, T.N. Chau, et al., Fatal outcome of human influenza A (H5N1) is associated with high viral load and hypercytokinemia, Nat. Med. 12 (10) (2006) 1203–1207.

[89] S.M. Lee, J.L. Gardy, C.Y. Cheung, T.K. Cheung, K.P. Hui, N.Y. Ip, et al., Systems-level comparison of host-responses elicited by avian H5N1 and seasonal H1N1 influenza viruses in primary human macrophages, PLoS One 4 (2009) e8072.

[90] J.S. Peiris, C.Y. Cheung, C.Y. Leung, J.M. Nicholls, Innate immune responses to influenza A H5N1: Friend or foe? Trends Immunol. 30 (2009) 574–584.

[91] M. Koopmans, B. Wilbrink, M. Conyn, G. Natrop, H. van der Nat, H. Vennema, et al., Transmission of H7N7 avian influenza A virus to human beings during a large outbreak in commercial poultry farms in the Netherlands, Lancet 363 (2004) 587–593.

[92] D.M. Skowronski, S.A. Tweed, M. Petric, T. Booth, Y. Li, T. Tam, Human illness and isolation of low-pathogenicity avian influenza virus of the H7N3 subtype in British Columbia, Canada, J. Infect. Dis. 193 (2006) 899–900.

[93] B. Budowle, M.D. Johnson, C.M. Fraser, T.J. Leighton, R.S. Murch, R. Chakraborty, Genetic analysis and attribution of microbial forensics evidence, Crit. Rev. Microbiol. 31 (2005) 233–254.

[94] WHO. CDC protocol of realtime RTPCR for influenza A(H1N1). Revision 2, October 6, 2009.

[95] B.G. Hall, Phylogenetic Trees Made Easy: A How-to Manual, Sinauer Associates, Inc., 2007.

[96] C. Fraser, C.A. Donnelly, S. Cauchemez, W.P. Hanage, M.D. Van Kerkhove, T.D. Hollingsworth, et al., WHO Rapid Pandemic Assessment Collaboration. Pandemic potential of a strain of influenza A (H1N1): Early findings, Science 324 (2009) 1557–1561.

[97] A. Rambaut, E. Holmes, The early molecular epidemiology of the swine-origin A/H1N1 human influenza pandemic, PLoS Curr. Influenza (2009) RRN1003.

[98] G.J. Smith, J. Bahl, D. Vijaykrishna, J. Zhang, L.L. Poon, H. Chen, et al., Dating the emergence of pandemic influenza viruses, Proc. Natl. Acad. Sci. USA 106 (2009) 11709–11712.

[99] C. Scholtissek, Pigs as "mixing vessels" for the creation of new pandemic influenza A viruses, Med. Principles Practice 2 (1990) 65–71.

[100] R.M. Bush, Influenza as a model system for studying the cross-species transfer and evolution of the SARS coronavirus, Philos. Trans. R. Soc. Lond. B Biol. Sci. 359 (2004) 1067–1073.

[101] Centers for Disease Control and Prevention (CDC), Evaluation of rapid influenza diagnostic tests for detection of novel influenza A (H1N1) Virus–United States, 2009, MMWR Morb. Mortal Wkly. Rep. 58 (2009) 826–829.

[102] J. Fan, A.J. Kraft, K.J. Henrickson, Current methods for the rapid diagnosis of bioterrorism-related infectious agents, Pediatr. Clin. North Am. 53 (2006) vii–viii, 817–842.

[103] A.F. Poon, S.L. Kosakovsky Pond, P. Bennett, D.D. Richman, A.J. Leigh Brown, S.D. Frost, Adaptation to human populations is revealed by within-host polymorphisms in HIV-1 and hepatitis C virus, PLoS Pathog. 3 (2007) e45.

[104] B. Hoffmann, M. Beer, S.M. Reid, P. Mertens, C.A. Oura, P.A. van Rijn, et al., A review of RT-PCR technologies used in veterinary virology and disease control: Sensitive and specific diagnosis of five livestock diseases notifiable to the World Organisation for Animal Health, Vet. Microbiol. 139 (2009) 1–23.

[105] M.L. Perdue, Molecular diagnostics in an insecure world, Avian Dis. 47 (2003) 1063–1068.

[106] K.F. Gensheimer, M.I. Meltzer, A.S. Postema, R.A. Strikas, Influenza pandemic preparedness, Emerg. Infect. Dis. 9 (2003) 1645–1648.

[107] G. Neumann, T. Watanabe, H. Ito, S. Watanabe, H. Goto, P. Gao, et al., Generation of influenza A viruses entirely from cloned cDNAs, Proc. Natl. Acad. Sci. USA 96 (1999) 9345–9350.

[108] E. Hoffmann, G. Neumann, Y. Kawaoka, G. Hobom, R.G. Webster A DNA transfection system for generation of influenza A virus from eight plasmids, Proc. Natl. Acad. Sci. USA 97 (2000) 6108–6113.

[109] G.L. Chen, E.W. Lamirande, H. Jin, G. Kemble, K. Subbarao, Safety immunogenicity, and efficacy of a cold-adapted A/Ann Arbor/6/60 (H2N2) vaccine in mice and ferrets, Virology 398 (1) (2010) 109–114.

[110] C.D. Salgado, B.M. Farr, K.K. Hall, F.G. Hayden, Influenza in the acute hospital setting, Lancet Infect. Dis. 2 (2002) 145–155.

[111] L.D. Budnick, M.E. Moll, H.F. Hull, J.M. Mann, A.P. Kendal, A pseudo-outbreak of influenza A associated with use of laboratory stock strain, Am. J. Public Health 74 (1984) 607–609.

[112] T.R. Maines, L.M. Chen, Y. Matsuoka, H. Chen, T. Rowe, J. Ortin, et al., Lack of transmission of H5N1 avian-human reassortant influenza viruses in a ferret model, Proc. Natl. Acad. Sci. USA 103 (32) (2006) 12121–12126.

[113] L.M. Chen, C.T. Davis, H. Zhou, N.J. Cox, R.O. Donis, Genetic compatibility and virulence of reassortants derived from contemporary avian H5N1 and human H3N2 influenza A viruses, PLoS Pathog. 4 (2008) e1000072.

[114] S. Jackson, N. Van Hoeven, L.M. Chen, T.R. Maines, N.J. Cox, J.M. Katz, et al., Reassortment between avian H5N1 and human H3N2 influenza viruses in ferrets: A public health risk assessment, J. Virol. 83 (2009) 8131–8140.

[115] C. Li, M. Hatta, C.A. Nidom, Y. Muramoto, S. Watanabe, G. Neumann, et al., Reassortment between avian H5N1 and human H3N2 influenza viruses creates hybrid viruses with substantial virulence, Proc. Natl. Acad. Sci. USA 107 (2010) 4687–4692.

[116] R. Snacken, A.P. Kendal, L.R. Haaheim, J.M. Wood, The next influenza pandemic: Lessons from Hong Kong, 1997, Emerg. Infect. Dis. 5 (1999) 195–203.

# CHAPTER 9

# Keeping Track of Viruses

**Catherine L. Murray, Thomas S. Oh, and Charles M. Rice**

*Laboratory of Virology and Infectious Disease, Center for the Study of Hepatitis C,*
*The Rockefeller University, New York*

## INTRODUCTION

Viruses are the most abundant biological entities on earth (1). These obligate parasites infect every form of life, from archaea and eubacteria to fungi, plants, and animals; even viruses can be affected by a coinfecting satellite species (2). Viruses play key roles in global ecology—they form a vast reservoir of genetic diversity, influence the composition and evolution of host populations, and affect the cycling of chemical compounds through the environment (3). While research has focused on the tiny fraction that causes disease in humans, domestic animals, and crops, sequencing surveys have suggested that the majority of viruses are completely unknown (1). The ability of viruses to jump species barriers, move between habitats, and circle the globe rapidly underscores the importance of continued vigilance for naturally emerging or deliberately engineered outbreaks. This chapter reviews methods of isolating, identifying, and tracking viruses with potential applications to microbial forensic investigations.

## WHAT IS A VIRUS?

Viruses are extremely simple "life" forms without metabolic capacity, organelles, translational machinery, or autonomous replicative potential (4). Virus particles constitute a minimal set of components, primarily those required to deliver the genome to the target cell and initiate replication. Consequently, virus particles (or virions) are extremely small, most in the range of 20 to 200 nm in diameter. A notable exception is a recently discovered "giant virus," termed mimivirus, for "mimicking microbe," which has a particle diameter of 400 nm, comparable to a small bacterium (5). Virions are not only diverse in size, but also in composition, morphology, and genome

Microbial Forensics. DOI: 10.1016/B978-0-12-382006-8.00009-8

characteristics. Virus particles may be irregular in shape or possess a distinct symmetry, such as helical or icosahedral. Particles may be surrounded by a host-derived membrane(s), termed "enveloped," or a tight protein shell, termed "nonenveloped." Inside the virion, the genome is associated with nucleic acid-binding proteins; some viruses carry additional factors, such as enzymes required to initiate replication. While bacteria, fungi, parasites, plants, and animals use exclusively deoxyribonucleic acid (DNA) as their genetic material, a viral genome may be composed of either DNA or ribonucleic acid (RNA). The genome may be single stranded (ss) or double stranded (ds), circularized or linear, consist of a single nucleic acid strand, or be "segmented" on multiple molecules. Viruses do not share any characteristic sequence that is conserved across all families, as are ribosomal (r)RNAs in cellular organisms. Virus genomes also vary greatly in size. The ssRNA genomes of poliovirus and human immunodeficiency virus (HIV) are 7.4 and 9.2 kb, respectively, whereas the dsDNA genome of mimivirus is approximately 800 kbp.

## VIRUS LIFE CYCLE

### Virus Attachment and Entry

Viruses must enter a target cell in a way that does not do excessive damage to the host or alert immune defenses (6). This is generally accomplished by hijacking normal cellular processes, including receptor–ligand binding, endocytosis, and nuclear import. The virion attaches by binding to a protein, lipid, and/or carbohydrate displayed on the cell surface. Envelope glycoproteins, or the spikes and indentations of the nonenveloped virus shell, participate in these initial interactions. The specific cellular molecule to which a virus binds is termed its "receptor." Some viruses, such as HIV and hepatitis C virus (HCV), bind to several receptors and coreceptors, which perform distinct roles in complex multistep uptake pathways. Receptor binding initiates internalization of the virus particle, transport to the appropriate cellular compartment, and uncoating of the genome. Enveloped virus glycoproteins are triggered to mediate fusion of the viral and host membranes during uptake. Delivery of a replication-competent viral genome to a permissive intracellular site is the first step in establishing a productive infection.

### Replication Strategies

The diversity of viral genomes necessitates a variety of replication strategies. Viruses are divided into seven groups based on genetic material, polarity, and messenger (m)RNA synthesis (7). Polarity refers to the protein-coding capacity of a nucleic acid strand, where positive (+) strand nucleic acid has a 5′→ 3′ polarity, identical to mRNA, and negative (−) strand nucleic acid has a 3′→5′ polarity, complementary to mRNA (Figure 9.1).

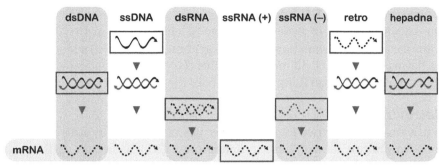

**FIGURE 9.1**
Replication strategies of viruses. All virus genomes must be used to produce mRNA in order for the viral proteins to be expressed inside the cell. The schematic represents the seven classes of viruses, according to the Baltimore classification, and the intermediates through which mRNA is produced. The nucleic acid character of the viral genome is indicated by a box. Black, positive-strand nucleic acid; gray, negative-strand nucleic acid. Arrows on nucleic strands indicate their directionality, pointing from 5′ to 3′ ends. DNA is indicated as solid lines, RNA as dashed lines. The partially double-stranded nature of the hepadnavirus DNA genome is indicated by a gap.

## Double-Stranded DNA Viruses

Viruses with dsDNA genomes may replicate in the nucleus or cytoplasm, with transcription of the genome into mRNA by host or viral RNA polymerases, respectively. Variola major, the causative agent of smallpox, is an example of an enveloped virus with a linear dsDNA genome.

## Single-Stranded DNA Viruses

These viruses have an ssDNA genome of (+) polarity. The genome is converted to dsDNA in the nucleus and is subsequently transcribed and translated by host machinery to produce viral proteins. Parvoviruses, which cause rash in children and often-fatal infection in dogs, are members of this group.

## Double-Stranded RNA Viruses

Viruses in this class contain segmented dsRNA genomes. mRNA is synthesized by a virally encoded RNA-dependent RNA polymerase (RdRp). Because most eukaryotic cells do no encode this type of enzyme, the virus must import its own RdRp within the incoming virion. Rotavirus, a common etiologic agent of severe infectious diarrhea in children, has a dsRNA genome.

## Positive-Strand RNA Viruses

The ssRNA genomes of these viruses are translated directly by host ribosomes in the cytoplasm. The virally encoded RdRp then replicates the genome through a complementary (−) strand intermediate. Examples of viruses in this class include poliovirus, West Nile virus, and HCV.

### Negative-Strand RNA Viruses

The ssRNA genomes of these viruses may be either segmented or continuous. Some are ambisense, with portions of the genome acting as (+) strands and others having (−) polarity. All members of this class, which includes influenza and Ebola viruses, import an RdRp that transcribes the viral genome into mRNA.

### Retroviruses

Retroviruses package two identical molecules of (+) polarity ssRNA. A virally encoded enzyme termed "reverse transcriptase (RT)" generates dsDNA from the RNA templates. The name "retrovirus" reflects the fact that this replicative cycle is retrograde (RNA→DNA→mRNA→protein) relative to the central dogma of modern biology (DNA→mRNA→protein). HIV, the virus that causes acquired immune deficiency syndrome (AIDS), is a retrovirus.

### Hepadnaviruses

Members of this group have a partially double-stranded DNA genome and replicate via an RNA intermediate, similar to retroviruses. mRNA is packaged into immature particles before conversion to DNA by the virally encoded RT. An example of this group is hepatitis B virus, an important etiologic agent of chronic liver disease.

## Assembly and Release

Transport of the amplified genome to a new permissive host requires the production of infectious viral particles. This is a complicated process that is well understood for only a few viruses. Similar to replication, virion assembly takes place at defined intracellular locations, such as in the nucleus, at membranous cytoplasmic organelles, or at the cell surface. Virions can be released from the cell by noncytopathic budding or through host cell lysis.

## HOW DO YOU IDENTIFY A VIRUS?

Sudden emergence of an infectious disease demands methods to rapidly and accurately identify the infectious agent, diagnose patients, and explain routes of transmission. A "staged" approach is often employed, in which epidemiology, pathology, and serological assays suggest candidate pathogens, which can be confirmed by nucleic acid-based methods. In the absence of suspects, microarray, next-generation sequencing, or subtractive cloning can be informative (8,9).

## Culture and Cytopathic Effect

Patient samples may be directly infectious to immortalized cell lines, allowing the pathogen to be isolated, quantified, and amplified. Some viruses do not grow well in cultured cells, but may be coaxed to replicate in primary cells, embryonated eggs, or experimental animals. If culture is successful, the phenotype of the infected cells can reveal valuable clues about the identity of the virus (4). Infection may kill the cells, creating a characteristic cytopathic effect (CPE). Cells may appear to be rounded or growing in grape-like clusters, indicating adenovirus or herpes simplex virus. Cells may fuse into multinucleated "syncytia," suggesting influenza, mumps, or measles. Poxviruses create foci of fused cells, whereas positive-strand RNA viruses induce proliferation of membranes in the cytoplasm. Many other viruses do not cause noticeable cell damage. Observance of CPE was a critical factor in identifying the causative agent of acute fever with encephalitis among pig farmers on the Malay Peninsula in 1998–1999. Multinucleate syncytia were seen in Vero cells inoculated with cerebral spinal fluid obtained from fatal infections, implicating a paramyxovirus (10). The new pathogen was named "Nipah virus," and the outbreak was stopped after culling over one million pigs.

## Electron Microscopy (EM)

Infected cultures or amplified virus can serve as material for visualization by EM. Electron microscopes use a beam of electrons, rather than visible light, to form an image at extremely high magnification (up to 1,000,000×). Staining with an electron-rich "negative stain" and thin sectioning of the specimen can increase contrast and enhance visualization of internal features. Virions can appear ribbon like (rabies), rod shaped (measles), spherical (poliovirus), or filamentous (Ebola) (Figure 9.2). Some viruses show irregular shapes or multiple morphologies, referred to as "pleomorphic." Visualization of particles by EM was a defining step in identification of a novel virus responsible for severe acute respiratory syndrome (SARS), a disease that spread rapidly around the globe in 2003. The virus was isolated by inoculation of cell lines with an oropharyngeal specimen obtained from a fatally infected patient. Cultures were subjected to thin section and negative stain EM, and particles showing the distinctive halo of *Coronaviridae* were detected (Figure 9.2). The characteristic particles enabled researchers to focus swiftly on a specific virus group, and numerous other tests validated and characterized the novel SARS-coronavirus (SARS-CoV) (11).

## Serological Assays

The specificity and high affinity of antibody–antigen recognition is widely used in virus diagnostics. Antibodies may be isolated from sera of infected or recovering individuals or be generated experimentally in animals immunized with viral antigens. Serological assays can discover similarities between a novel pathogen and a known virus through antibody cross-reactivity.

### Enzyme-Linked Immunosorbent Assay (ELISA)

The ELISA is a rapid and versatile method of detecting antigens or antibodies. Viral proteins or virus-specific antibodies are adsorbed to the surface of a microtiter plate, allowing specific capture of the cognate antibody or antigen

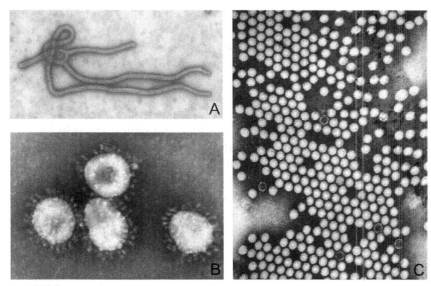

**FIGURE 9.2**

Electron micrographs of virus particles. (A) Ebola virus particles showing filamentous morphology. Courtesy of the CDC/C.Goldsmith. (B) SARS coronavirus showing characteristic "corona-like" morphology. Courtesy of the CDC/Fred Murphy. (C) Polio virus particles showing spherical morphology; courtesy of the CDC/Fred Murphy and Sylvia Whitfield.

from patient serum. Complexes are detected with labeled secondary antibodies, followed by a colorimetric readout. ELISA can distinguish between different classes of antibodies, such as those indicative of recent infection (IgM) or previous exposure and vaccination (IgG). This assay was used to investigate cases of encephalitis and/or profound muscle weakness in Queens, New York, in 1999. IgM-capture ELISA was used to survey antibodies against common encephalitic viruses. Results implicated St. Louis encephalitis (SLE) virus, a mosquito-borne flavivirus. Sequencing later revealed the agent was not SLE but the related West Nile virus—a pathogen never before detected in the Western hemisphere. The ability of ELISA to detect cross-reactive flavivirus antibodies meant that the appropriate vector control measures could be implemented quickly (12).

### Neutralization and Hemagglutination Inhibition Assays

Antibodies produced during infection often have the ability to interfere with the native properties of virus particles. Mixing dilutions of antibodies with a virus sample, followed by measurement of the decrease in virion activity, can be used to identify viruses and to classify them into serotypes. Neutralization assays measure the ability of antibodies to block viral entry. Hemagglutinin inhibition (HI) assays detect antibodies that can block the ability of some viruses to aggregate red blood cells. In May 2009, neutralization and HI assays were used to demonstrate that previous vaccination against seasonal influenza offered little protection against the novel H1N1 pandemic strain, indicating that the new swine-origin virus was substantially different from those that had circulated in recent years (13).

### Immunostaining

Antibodies can be used to detect viral proteins in patient tissues or infected cultures. Specific binding can be visualized by secondary staining using an antibody conjugated to a fluorescent dye (immunofluorescence) or enzyme (immunohistochemistry). Immunohistochemistry was used to investigate a disease cluster in the region bordering Arizona, Colorado, New Mexico, and Utah in spring 1993. Symptoms included fever, headache, and cough that progressed rapidly to respiratory distress and death (14). Patient sera were found to contain antibodies targeting members of the genus *Hantavirus*. Immunohistochemistry of autopsy tissues indicated the presence of hantavirus antigens in endothelial cells of the lung and other involved organs (15). Nucleic acid sequencing confirmed the diagnosis, and the new pathogen was named "Sin Nombre virus."

## Polymerase Chain Reaction (PCR)

Development of the PCR in 1987 ushered in a new era of nucleic acid-based pathogen detection systems (16). Amplification of viral genomes allows rapid, specific, and sensitive detection and analysis without the need for *in vitro* culture or quality antibodies. PCR allows a dsDNA target to be amplified exponentially using a pair of oligonucleotide primers designed to flank the region of interest. The PCR product, or amplicon, can be detected by a variety of methods, such as nucleic acid staining. PCR is a sensitive way to confirm the presence of a suspected virus in patients or environmental samples.

### Real-Time PCR

Because methods for detecting the final PCR product can be laborious and time-consuming, an alternative strategy is to monitor amplicon synthesis in real time (17). This technique provides a wealth of information, including accurate quantification of the starting template, and is termed "real-time" or "quantitative" PCR. Real-time PCR depends on the emergence of a signal as the amplification reaction proceeds. The simplest form uses reporter molecules that fluoresce when bound nonspecifically to dsDNA. Alternatively, sequence-specific detection can be achieved using an ssDNA "probe" designed to bind within the amplified region. The probe is labeled with a fluorophore in close proximity to a quencher, which dissipates the fluorescence energy until annealing occurs. Real-time PCR is a rapid and effective method for assessing the presence of candidate viruses, distinguishing between genotypes, and measuring viral load.

### Multiplex PCR

Considerable savings in time, effort, and sample volume can potentially be achieved by combining multiple PCR reactions in a single tube, termed "multiplex PCR" (17–19). Development of multiplex assays, however, can require significant optimization, and primers must be designed carefully to work well without interference (18). Furthermore, the number of targets that can be distinguished is limited—for example only a few fluorophores are available (17,20). To increase the capacity of multiplex PCR, new methods of amplicon differentiation have been established. One is MassTag PCR, which uses primers labeled with a tag of a known, unique molecular weight. After PCR, the identity of the incorporated tags can be determined by mass spectrometry. This method has been used to multiplex up to 22 respiratory pathogens in a single reaction (21) and for simultaneous detection of viral hemorrhagic fever agents (22). In an alternative method, the precise weight of the amplicon can be measured directly, allowing multiple microbes to be detected in a single complex sample based on amplification of conserved sequences (23–25). These systems, termed "TIGER" or "Ibis T5000" (Ibis Biosciences), identify virus(es) using primers to amplify broadly conserved regions from large groupings of species, followed by electrospray ionization mass spectrometry and analysis of the nucleic acid base composition (i.e., the number of adenosines, cytidines, guanosines, and thymidines in the amplicon) (23). This technology, which is used currently at the Centers for Disease Control (CDC) and the National Bioforensic Analysis Center, has been used to uncover a novel encephalitis virus (26) and detected the second case of novel H1N1 (swine flu) in the United States (26a).

## Microarray and Virus Chips

Although PCR is well suited for sensitive detection of a small number of candidate viruses, the technique is inherently biased by primer and/or probe design. Recently, the application of microarray technology has provided a more impartial approach to pathogen discovery. Microarrays, or "chips," consist of short oligonucleotide probes immobilized as spots on a solid support (20,27). Isolated DNA or RNA is labeled with a fluorescent dye and hybridized to the chip; the bound spot(s) indicates the presence of sequences in the target sample. Microarrays have been used extensively since the early 1990s to investigate cell biology (27), but have only recently been adapted for the detection of infectious agents. Initial success was achieved with arrays targeting a few dozen pathogens (28,29). The subsequently developed "virus chip" included probes to the most highly conserved regions of 140 virus genomes (30) and was later expanded to over 1000 different species (31). Species and serotypes not represented explicitly on the chip can also be detected, as they form unique signatures based on hybridization to conserved sequences. The "GreeneChip" is a similar platform that includes probes for bacteria, fungi, and parasites, as well as viruses (19). The GreeneChip was used in a postmortem diagnosis of a health care worker who had succumbed to fever and liver failure during a Marburg virus outbreak in Angola. Multiplex PCR had failed to detect Marburg or other hemorrhagic fever viruses. Hybridization to the GreeneChip revealed the presence of *Plasmodium*, a parasite that causes malaria (19). Although chip technologies are limited by the need for updates as emerging, mutating, or engineered viruses occur, they are nonetheless an important tool for epidemiologic or microbial forensic investigation.

## Next-Generation Sequencing

Unbiased sequencing of the entire microbial population in an environmental sample or diseased tissue has become a real possibility with the advent of next-generation sequencing technologies—termed "high-throughput (HTS)," "deep," "massively parallel," or "Next Gen" sequencing. 454 (Roche), SOLiD (Applied Biosystems), and Illumina (Solexa) represent several of the most widely used platforms (32). HTS uses sheared DNA as a template for millions of parallel amplification reactions. 454 amplification occurs on beads, which are then arrayed in individual wells of a picotiter plate for parallel "pyrosequencing"—using the pyrophosphate released by each nucleotide incorporation to trigger a reporter signal. SOLiD sequencing also begins with amplification on beads, followed by "sequencing by ligation" in which short labeled probes, only two bases in length, bind to and reveal the target sequence over multiple rounds of annealing. Illumina sequencing begins with amplification of DNA clusters on a glass support. Fluorescently labeled nucleotides are then incorporated one at a time, through a series of blocking and unblocking steps, and images of each cycle record the sequence (32). HTS reads are short, but typically sufficient to query a database and discover pathogens with even low similarity to known sequences (9). With new HTS technologies developing at a rate so rapid as to inspire the colloquial term "Now Gen," a major challenge is the need for bioinformatic analysis to keep pace.

High-throughput sequencing was used to unravel the mystery of colony collapse disorder (CCD), a phenomenon that began devastating the honeybee industry between 2004 and 2006. CCD is characterized by the very rapid disappearance of the entire adult bee population of a hive. The observation that irradiated, but not untreated, hives could be repopulated suggested an infectious etiology. In an attempt to identify the agent, total RNA extracted from diseased or healthy colonies was analyzed by pyrosequencing. A number of bacteria, fungi, parasites, and viruses were revealed as candidates, and real-time PCR assays were developed to assess the distribution of each agent. One pathogen correlated most strongly with the occurrence of CCD: Israeli acute paralysis virus, a positive-strand RNA virus that had not been found previously in the United States (33).

## Subtractive Cloning

A novel virus may display an unusual or chronic pathology that complicates epidemiology, occurs in conjunction with other microbes, or has a genome that is completely unknown. Subtractive cloning is a classical technique that comprehensively surveys differences between samples and can still be useful in revealing otherwise undetectable pathogens (9). Typically, a cDNA probe derived from infected material is used to identify sequences specific to diseased, but not normal, tissue by sequential rounds of hybridization and amplification. Borna disease virus, a pathogen implicated in a range of behavioral and neuropsychiatric pathologies in animals, and possibly in humans, was discovered by this method (34,35). Disease-specific cDNA clones can also be expressed as proteins and selected by binding to patient, but not healthy, sera. This technique was used to identify HCV, a chronic liver pathogen for which previous detection of antigens, antibodies, nucleic acid, and virus particles had been unsuccessful (36).

# VIRAL DIVERSITY AND PHYLOGENETICS

During the investigation of a disease outbreak or suspected biocrime, it may be critical to determine not only the species of virus involved, but how the infection is moving through a population—here the often dramatic diversity generated during viral replication can be highly informative.

## Viral Evolution

In most organisms, evolution takes place over very long timescales—much longer than could be observed in a laboratory experiment or in a criminal investigation. Viruses, however, evolve rapidly enough to make the study of genetic change a very relevant and powerful tool for the forensic scientist. What accounts for the remarkable speed of viral evolution? Two important features set viruses apart from other organisms with regard to rates of change: high mutation frequencies and replicative potential. During replication of the genetic sequence of any organism, copying errors are inevitably made, leaving the new sequence with differences from the original. However, whereas such misincorporations occur perhaps once every billion bases in most living cells, in some viruses errors are made as often as once every thousand bases copied. This results from the use of enzymes without proofreading activity (RdRp or RT) rather than higher fidelity DNA polymerases. Not only do mutations occur more frequently, many copies are made and very quickly. While a single cell cycle results in two progeny, a virus might be copied hundreds or even thousands of times in a single life cycle. Because of their inherent simplicity and small size, viruses can be assembled quickly and very large populations can be supported. The number of viruses in a single infected person may be in the billions. The large numbers of genomes and high error rates result in a diverse and rapidly changing population. In addition to mutation, reassortment and recombination are two other important mechanisms of virus diversity. Reassortment occurs in species with segmented genomes, when two related viruses infect the same cell. As progeny are assembled, each genome segment that is packaged may be derived from either of the two original viruses, producing a virus with a combination of genes. Similarly, genetic material can be recombined from two viruses or even between a virus and its host cell, as events in the process of replicating DNA or RNA result in a new strand that is partially copied from one source and partly from another. In each case, it is the tremendous scale of viral replication that allows these seemingly rare events to have a significant impact on the process of viral evolution.

## Phylogenetic Analysis

Generally, the process of genetic variation through mutation, reassortment, and recombination cannot be observed directly; however, resulting viral sequences can. By examining the sequences of many samples, it is often possible to reconstruct a family tree (or phylogeny) of a set of sequences and to infer what series of events occurred, and in what order, to produce that set. This sort of reconstruction is known as phylogenetic analysis. Figure 9.3 shows three nine-base sequences, which differ at two positions. How might these sequences be related? One possibility is that virus A acquired a mutation, giving rise to C, which in turn mutated into B. Alternatively, C may be a common ancestor to both A and B, each differing from C by one mutation. Both scenarios seem quite plausible. A third possibility is that sequence A acquired two mutations

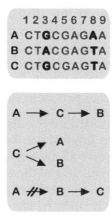

**FIGURE 9.3**

Phylogenetic analysis. How are these sequences related? (Top) Three nine-base sequences each differing at two positions (bold). (Bottom) Three scenarios that may relate the sequences. The probability that each scenario is correct can be calculated and used to construct a phylogenetic history of the sequence set.

(which may or may not have arisen simultaneously), producing B, and then B mutated once more to become C. This third scenario is certainly possible, but is less likely than the first two because it involves either a double mutation or sequence C arising twice—once in between A and B and again after B. Given a model of how sequences evolve, it is possible to calculate statistically how probable each scenario is relative to the others and then to determine the most likely sequence of events. While the example may seem trivial, as the length and number of samples increase, this type of analysis grows in both complexity and power. The Schmidt case, described below, was decided in large part on the basis of a phylogenetic analysis involving over a hundred viral sequences of nearly 2000 bases in length. Phylogenetic trees are used to identify the origins of new outbreaks, to determine the transmission path from one person to another, and to shed light on the evolutionary history of an unknown virus.

## SOLVING A BIOCRIME

### The State of Louisiana Versus Richard Schmidt

On August 4, 1994, a physician from Lafayette, Louisiana, gave an intramuscular injection to a former mistress who had recently broken off their affair. He told her he was administering a vitamin B shot, but when she became ill, suspicions mounted (4). A few months later, the victim tested positive for HIV, and Richard Schmidt was accused of using blood from a patient under his care to deliberately infect his former girlfriend—but could it be proven? Multiple viral sequences from the victim, the patient, and infected individuals in the community unrelated to the case were obtained. At first glance, it might be supposed a transmission event could be established by determining whether the patient and the victim carried identical viral sequences; the rapid rate of HIV evolution, however, makes this expectation overly simplistic. Likewise, a measurement of sequence similarity might be enough to establish a relationship between the viruses, but the direction of spread would be unknown. Phylogenetic evidence was needed to unravel the allegation of deliberate virus transmission from patient to victim.

Phylogenetic analysis was conducted on two regions of the HIV genome with different rates of evolution—an important consideration for extracting useful data. If evolution occurs too slowly, all the sequences would be similar, including those from the victim, the patient, and the unrelated controls. If the rate of change is too high, differences between the sequences may be so large that it becomes impossible to determine their relationship. Two portions of the HIV genome were analyzed to help achieve the appropriate range: the envelope (gp120) and RT coding regions. gp120, a structural component of the virion, is relatively plastic and evolves rapidly to escape host immune responses; RT performs conserved enzymatic functions and is less amenable to sequence change. Analysis of gp120 revealed that all the patient sequences formed one cluster, while all victim sequences formed another cluster; these clusters were related by a common ancestor, which was not shared by any of the other HIV sequences analyzed (Figure 9.4A). This ancestor could be a sequence that existed in the patient before the transmission event and had since disappeared or be a virus from a third individual that infected both the patient and the

**A**

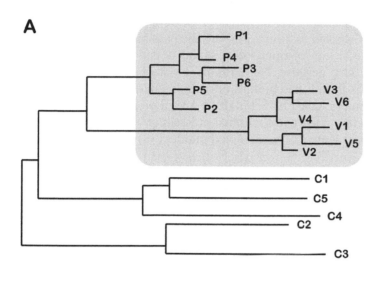

**B**

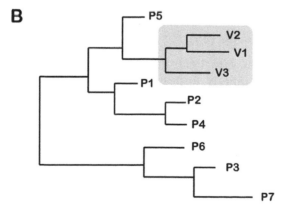

**FIGURE 9.4**

Schematic examples of phylogenetic trees. Phylogenetic analysis was used to link the source of HIV infection to the victim in the State of Louisiana versus Richard Schmidt case (38). Schematics illustrate the types of sequence clustering that were observed. (A) gp120 sequences from the victim and the patient (grey) shared a common ancestor not shared by unrelated controls. (B) RT sequences from the victim (grey) clustered entirely within the patient sequences. Individual sequences representing those from the patient (P), victim (V), or unrelated controls (C) are indicated with numbers.

victim. The gp120 phylogeny, therefore, shows a relationship but cannot distinguish the direction of transfer. In contrast, the RT coding sequences of the patient and victim clustered together, reflecting the slower evolution of this region (Figure 9.4B). Furthermore, all the RT sequences from the victim shared a single common ancestor contained within the patient sequences. This was a strong indication that transmission took place from the patient to the victim.

Importantly, phylogenetics analysis was used as only one piece of the puzzle, with traditional detective investigation helping to build the case (37). Although the victim worked as a nurse, raising the possibility of occupational exposure, she had a history of unexcluded blood donation and her past sexual partners were HIV negative. Furthermore, a vial of blood from the HIV-infected patient was found in Schmidt's office, a highly unusual occurrence. The State of Louisiana vs Richard J. Schmidt set a legal precedent for the admissibility of phylogenetic evidence in a criminal proceeding, and the defendant is now serving a 50-year sentence for attempted murder (38).

## The Case of Kristen D. Parker

Spreading an infectious agent may not be the deliberate intention, but a by-product of criminal activity. Between October 2008 and June 2009, a surgical technician in Denver, Colorado, began stealing pain medication from patients under her care. Kristen D. Parker removed syringes of fentanyl from the operating room and replaced them with her own—often previously used—syringes, filled with saline solution. In the process, the technician exposed hundreds of patients to HCV, a virus she had apparently contracted through her history of injection drug use (39).

When these allegations came to light, the Colorado Department of Public Health and Environment quickly instigated systematic testing of thousands of potential victims for exposure to the virus. To maximize the possibility of effective treatment, it was important to determine rapidly which patients were positive for HCV; in building a case against Ms. Parker, it would also be critical to link the infections epidemiologically and genetically to the suspect. Out of over 5000 patients at two Denver medical centers tested for the virus, almost 70 were positive (40), but how had they been infected? The extensive genetic diversity of HCV provided a quick method to rule out unrelated cases. HCV is classified into six major genotypes, which are further divided into a large number of subtypes showing significant divergence in genome sequences. Patients presenting with viral genotypes other than "1b" were therefore unlikely to have been infected by Ms. Parker. Additional sequencing was conducted on genotype 1b viruses isolated from the suspect and the remaining patients, and the relatedness of the genomes was determined. Overall, more than 20 patients showed strong genetic and/or epidemiological evidence of a transmission link to Ms. Parker. The suspect pled guilty to charges of tampering with a consumer product and theft of a controlled substance and is now facing at least 20 years behind bars (39).

## CHALLENGES THAT REMAIN

Identification and analysis of a toxic agent are critically dependent on sufficient and appropriately collected sample material (41). This may be especially difficult in the case of viruses. Infectious virions can be labile, and harsh treatments such as extremes of pH and temperature should be avoided. Genomic material composed of RNA, unique to viruses, may also be challenging to acquire—RNA is more sensitive to degradation than DNA and must typically be prepared from fresh tissue treated to inhibit ubiquitous nucleases. Finally, the infectious material may have been cleared from the body by the time symptoms become apparent. It is therefore important to collect samples using a variety of different preservation methods, from multiple locations throughout the body, and as early as possible in the infection. Once material has been secured, unbiased identification techniques may be attractive tools for pathogen discovery; however, these methods can implicate hundreds of microorganisms, many of which may have no etiologic relationship to the disease (33). Extensive work may be required to identify a strong candidate for pathogenesis, and formalizing a causal relationship between microbe and disease may not be trivial (8). This diversity, however, may have tremendous value for forensic analysis —it is easy to see that the total microbial composition or phylogeny of multiple species within a sample might facilitate the identification of source material involved in an intentional attack. If the suspected virus is a novel species, appropriate reagents and standards may not be available, and rapid detection or diagnosis may necessitate assays that have not been validated completely (42). Finally, while classical methods, including cell culture, can be simple and highly informative, powerful new techniques require a significant time commitment, as well as specialized equipment and expertise. Because containment of a biological threat, as well as the initiation of a microbial forensic investigation, often requires a rapid point-of-care response, the challenge remains to reduce the time and ease of detection so that an accurate diagnosis can be made by the clinician while maintaining the integrity of the evidence (29,41).

## ACKNOWLEDGMENTS

Work in the laboratory of Charles M. Rice is supported by Public Health Service grants from the National Institutes of Health, the Northeast Biodefense Center (U54 AI057158 subcontract 2925-01) and the Greenberg Medical Research Institute. The authors acknowledge Jack Hietpas, Laura K. McMullan, Holly L. Hanson, and David P. Mindell for material derived from the first edition of this chapter.

# REFERENCES

[1] R.A. Edwards, F. Rohwer, Viral metagenomics, Nat. Rev. Microbiol. 3 (2005) 504–510.

[2] B. La Scola, C. Desnues, I. Pagnier, C. Robert, L. Barrassi, G. Fournous, et al., The virophage as a unique parasite of the giant mimivirus, Nature 455 (2008) 100–104.

[3] C.A. Suttle, Marine viruses—major players in the global ecosystem, Nat. Rev. Microbiol. 5 (2007) 801–812.

[4] J. Hietpas, L.K. McMullan, H.L. Hanson, C.M. Rice and D.P. Mindell, In R. Breeze, B. Budowle, S.E. Schutzer (Eds.), Microbial Forensics, Elsevier/Academic Press, 2005.

[5] B. La Scola, S. Audic, C. Robert, L. Jungang, X. de Lamballerie, M. Drancourt, et al., A giant virus in amoebae, Science 299 (2003) 2033.

[6] A.E. Smith, A. Helenius, How viruses enter animal cells, Science 304 (2004) 237–242.

[7] D. Baltimore, Expression of animal virus genomes, Bacteriol. Rev. 35 (1971) 235–241.

[8] W.I. Lipkin, Pathogen discovery, PLoS Pathog. 4 (2008) e1000002.

[9] W.I. Lipkin, G. Palacios, T. Briese, Diagnostics and discovery in viral hemorrhagic fevers, Ann. N.Y. Acad. Sci. 1171 (Suppl 1) (2009) E6–11.

[10] K.B. Chua, W.J. Bellini, P.A. Rota, B.H. Harcourt, A. Tamin, S.K. Lam, et al., Nipah virus: A recently emergent deadly paramyxovirus, Science 288 (2000) 1432–1435.

[11] T.G. Ksiazek, D. Erdman, C.S. Goldsmith, S.R. Zaki, T. Peret, S. Emery, et al., A novel coronavirus associated with severe acute respiratory syndrome, N. Engl. J. Med. 348 (2003) 1953–1966.

[12] CDC, Outbreak of West Nile-like viral encephalitis—New York, MMWR Morb. Mortal. Wkly. Rep. 48 (1999) 845–849.

[13] CDC, Serum cross-reactive antibody response to a novel influenza A (H1N1) virus after vaccination with seasonal influenza vaccine, MMWR Morb. Mortal. Wkly. Rep. 58 (2009) 521–524.

[14] CDC, Outbreak of acute illness--southwestern United States, MMWR Morb. Mortal. Wkly. Rep. 42 (1993) 421–424.

[15] S.R. Zaki, P.W. Greer, L.M. Coffield, C.S. Goldsmith, K.B. Nolte, K. Foucar, et al., Hantavirus pulmonary syndrome. Pathogenesis of an emerging infectious disease, Am. J. Pathol. 146 (1995) 552–579.

[16] K.B. Mullis, F.A. Faloona, Specific synthesis of DNA in vitro via a polymerase-catalyzed chain reaction, Methods Enzymol. 155 (1987) 335–350.

[17] I.M. Mackay, K.E. Arden, A. Nitsche, Real-time PCR in virology, Nucleic Acids Res. 30 (2002) 1292–1305.

[18] E.M. Elnifro, A.M. Ashshi, R.J. Cooper, P.E. Klapper, Multiplex PCR: Optimization and application in diagnostic virology, Clin. Microbiol. Rev. 13 (2000) 559–570.

[19] G. Palacios, P.L. Quan, O.J. Jabado, S. Conlan, D.L. Hirschberg, Y. Liu, et al., Panmicrobial oligonucleotide array for diagnosis of infectious diseases, Emerg. Infect. Dis. 13 (2007) 73–81.

[20] N. Boonham, J. Tomlinson, R. Mumford, Microarrays for rapid identification of plant viruses, Annu. Rev. Phytopathol. 45 (2007) 307–328.

[21] T. Briese, G. Palacios, M. Kokoris, O. Jabado, Z. Liu, N. Renwick, et al., Diagnostic system for rapid and sensitive differential detection of pathogens, Emerg. Infect. Dis. 11 (2005) 310–313.

[22] G. Palacios, T. Briese, V. Kapoor, O. Jabado, Z. Liu, M. Venter, et al., MassTag polymerase chain reaction for differential diagnosis of viral hemorrhagic fever, Emerg. Infect. Dis. 12 (2006) 692–695.

[23]  D.J. Ecker, R. Sampath, C. Massire, L.B. Blyn, T.A. Hall, M.W. Eshoo, et al., Ibis T5000: A universal biosensor approach for microbiology, Nat. Rev. Microbiol. 6 (2008) 553–558.

[24]  R. Sampath, T.A. Hall, C. Massire, F. Li, L.B. Blyn, M.W. Eshoo, et al., Rapid identification of emerging infectious agents using PCR and electrospray ionization mass spectrometry, Ann. N.Y. Acad. Sci. 1102 (2007) 109–120.

[25]  R. Sampath, S.A. Hofstadler, L.B. Blyn, M.W. Eshoo, T.A. Hall, C. Massire, et al., Rapid identification of emerging pathogens: Coronavirus, Emerg. Infect. Dis. 11 (2005) 373–379.

[26]  M.W. Eshoo, C.A. Whitehouse, S.T. Zoll, C. Massire, T.T. Pennella, L.B. Blyn, et al., Direct broad-range detection of alphaviruses in mosquito extracts, Virology 368 (2007) 286–295.

[26a] D.J. Faix, S.S. Sherman, S.H. Waterman. Rapid-test sensitivity for novel swine-origin influenza A (H1N1) virus in humans. N Engl J Med 361(7) (2009) 728–9.

[27]  M. Dufva, Introduction to microarray technology, Methods Mol. Biol. 529 (2009) 1–22.

[28]  W.J. Wilson, C.L. Strout, T.Z. DeSantis, J.L. Stilwell, A.V. Carrano, GL. Andersen, Sequence-specific identification of 18 pathogenic microorganisms using microarray technology, Mol. Cell Probes 16 (2002) 119–127.

[29]  C.W. Wong, C.L. Heng, L. Wan Yee, S.W. Soh, C.B. Kartasasmita, E.A. Simoes, et al., Optimization and clinical validation of a pathogen detection microarray, Genome Biol. 8 (2007) R93.

[30]  D. Wang, L. Coscoy, M. Zylberberg, P.C. Avila, H.A. Boushey, D. Ganem, et al., Microarray-based detection and genotyping of viral pathogens, Proc. Natl. Acad. Sci. USA 99 (2002) 15687–15692.

[31]  D. Wang, A. Urisman, Y.T. Liu, M. Springer, T.G. Ksiazek, D.D. Erdman, et al., Viral discovery and sequence recovery using DNA microarrays, PLoS Biol. 1 (2003) E2.

[32]  E.R. Mardis, Next-generation DNA sequencing methods, Annu. Rev. Genomics Hum. Genet. 9 (2008) 387–402.

[33]  D.L. Cox-Foster, S. Conlan, E.C. Holmes, G. Palacios, J.D. Evans, N.A. Moran, et al., A metagenomic survey of microbes in honey bee colony collapse disorder, Science 318 (2007) 283–287.

[34]  W.I. Lipkin, G.H. Travis, K.M. Carbone, M.C. Wilson, Isolation and characterization of Borna disease agent cDNA clones, Proc. Natl. Acad. Sci. USA 87 (1990) 4184–4188.

[35]  S. VandeWoude, J.A. Richt, M.C. Zink, R. Rott, O. Narayan, J.E. Clements, A borna virus cDNA encoding a protein recognized by antibodies in humans with behavioral diseases, Science 250 (1990) 1278–1281.

[36]  Q.L. Choo, G. Kuo, A.J. Weiner, L.R. Overby, D.W. Bradley, M. Houghton, Isolation of a cDNA clone derived from a blood-borne non-A, non-B viral hepatitis genome, Science 244 (1989) 359–362.

[37]  R. Harmon. In B. Budowle, R.G. Breeze, S.E. Schutzer (Eds.), Microbial Forensics, Academic Press, San Diego, 2005.

[38]  M.L. Metzker, D.P. Mindell, X.M. Liu, R.G. Ptak, R.A. Gibbs, DM. Hillis, Molecular evidence of HIV-1 transmission in a criminal case, Proc. Natl. Acad. Sci. USA 99 (2002) 14292–14297.

[39]  K. Johnson, Worker in hepatitis case is sentenced to 20 years, N.Y. Times (2009).

[40]  CDPHE, Case numbers associated with hepatitis C investigation, 2009, posting date [Online].

[41]  B. Budowle, S.E. Schutzer, J.P. Burans, D.J. Beecher, T.A. Cebula, R. Chakraborty, et al., Quality sample collection, handling, and preservation for an effective microbial forensics program, Appl. Environ. Microbiol. 72 (2006) 6431–6438.

[42]  S.E. Schutzer, P. Keim, K. Czerwinski, B. Budowle, Use of forensic methods under exigent circumstances without full validation, Sci. Transl. Med. 1 (2009) 8cm7.

# CHAPTER 10

# Microbial Forensic Analysis of Trace and Unculturable Specimens

**Mark W. Eshoo, John Picuri, David D. Duncan, and David J. Ecker**

*Ibis Biosciences Inc., a subsidiary of Abbott Molecular Inc., Carlsbad, California*

## INTRODUCTION TO BIOFORENSIC ANALYSIS OF TRACE DNA

With the advent of high-throughput sequencing technologies, genetic analysis of microbial specimens has become more than simply looking for a handful genetic signatures in bacterial or viral agents of interest. In the investigation of the anthrax-letter attacks of 2001, Sanger-based sequencing was used to identify unique genetic signatures of a minor *Bacillus anthracis* morphotype and this information was used to link the *B. anthracis* in the letters to a culture from the government laboratory at the U.S. Army Medical Research Institute for Infectious Diseases. Thus genetic information was critical for directing the investigation to the source of the attacks. The sequencing methods used for these analyses were slow and laborious, and therefore the identification and sequencing of both major and minor *B. anthracis* morphotypes in the specimens took years to complete. Furthermore, only viable cells were examined, which overlooked other possible genetic signatures present in the samples, including human DNA, that may have been useful for attribution.

Ideally, all of the DNA in a forensic sample, regardless of its source or its viability, should be analyzed. The entire genetic composition of the sample will provide a genetic fingerprint that can be used for attribution and the development of tests unique to the sample of interest. While sequencing analysis was performed on the 2001 anthrax specimens, we envision that newer sequencing technologies will allow more complete and much more rapid and comprehensive identification of genetic signatures. Sample-specific signatures may come from the bioagent itself or from other sources in the sample. For example, a purified spore preparation may contain DNA remnants from its culture medium or other contaminant DNA, such as pollen from the air and these DNA contaminants may be the unique key identifiers for the sample of interest. Characterization of these contaminating DNAs can provide a wealth of information on how and where the culture was grown and even provide

155

information as to the growth phase of the agent at the time is was processed, as described later.

Since the 2001 anthrax-letter attacks, several high-throughput whole genome sequencing technologies have become available. These "next-generation" sequencing technologies can generate over 500 megabases (millions of bases) of sequence data in a single sequencing run by performing more than a million parallel sequencing reactions. Many replicate sequencing reactions can be performed on a given sequence, providing a high level of redundancy. This repeated sequencing of the same regions is referred to as the depth of coverage or fold coverage of the sequencing reaction. Deep sequencing is sequencing of a sample with hundreds to thousands of fold depth of coverage. With increasing depth of coverage, even the DNA sequences of minor constituents can be identified. For example, with 1000-fold depth of coverage, a bacterial species constituting only 1% of the species in a sample would be fully sequenced 10 times. Deep sequencing of samples from the 2001 anthrax letters would have identified the minor morphotype *B. anthracis* and its single-nucleotide polymorphism (SNP) within weeks, provided the sequencing was performed on DNA extracted directly from the sample and not from a single colony isolate.

The obstacle to performing deep sequencing analysis is the requirement for large amounts of DNA. For instance a typical DNA sequencing run requires $1-10\mu g$ of DNA, which is the equivalent of DNA from $10^9$ *B. anthracis* cells, $3 \times 10^{10}$ smallpox viruses, or a million human cells. Figure 10.1 shows relative amounts of DNA in a single smallpox virus, a *B. anthracis* bacterium, and a human cell. Even traditional polymerase chain reaction (PCR)-based assays require 10 to 100 genome copies of template per reaction to overcome stochastic effects. It is for these reasons that a method is needed to faithfully amplify all DNA in a specimen regardless of the source to generate the DNA needed for use in whole genome deep sequencing analysis. Such a method must be able to provide more than a millionfold amplification of the DNA in a specimen without introducing bias. This means that if two bacterial genomes are present in a DNA extract at a 1-to-10 ratio, the two bacterial genomes must still be present at the same or close to the original 1-to-10 ratio following whole genome amplification (WGA). Furthermore, all of the genome from each organism must be amplified equally and the amplification approach must faithfully amplify all DNA in the sample without introducing errors into the DNA sequence.

Over the past decade, several methods have been developed for whole genome amplification of DNA. Through the use of WGA protocols it is possible to increase the amount of starting material in a very small DNA sample (a single cell) to amounts required for whole genome sequence analyses. One of these methods is degenerate oligonucleotide-primed PCR, which uses degenerate primers in a PCR reaction to amplify the DNA randomly (1,2).

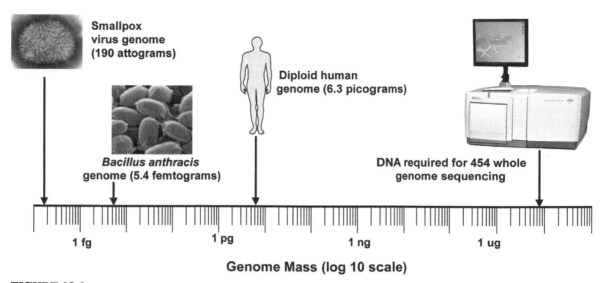

**FIGURE 10.1**

Relative genome masses in a single variola virus (smallpox), a single *Bacillus anthracis* cell, and a *Homo sapiens* cell and the amount of DNA required for a typical Next Gen sequencing run with the Genome Sequencer FLX system (shown by permission, image rights 454 Sequencing © 2009 Roche Diagnostics). Image of the variola virus-related vaccinia virus shown courtesy of the CDC/Cynthia Goldsmith. The *B. anthracis* image shown courtesy of the CDC/Laura Rose.

This method is used commonly in comparative genomic hybridization studies, but requires relatively large amounts of starting template DNA (3) and may not faithfully amplify the entire genomic DNA sample due to the inherent sequence biases of PCR amplification. Similarly, linker-adapter PCR has also been used for WGA but it too is subject to sequence-specific amplification biases of PCR and also requires DNA from hundreds of cells (4). Modified protocols based on single-primer PCR have been developed by Rubicon Genomics and are available commercially from Sigma Aldrich (St. Louis, MO), although these protocols are also limited by biases of the DNA polymerase used in the PCR.

Methods for whole genome amplification of DNA by multiple displacement amplification (MDA) have been developed that faithfully amplify all of the DNA in a specimen. MDA relies on priming the genomic DNA with exonuclease-resistant random primers and using φ29 DNA polymerase (5). φ29 DNA polymerase is a highly processive, strand-displacing polymerase with an exceptionally low error rate of 1 in $10^6$–$10^7$ nucleotides (6,7). For comparison, native *Taq* polymerase has an error rate of 3 in $10^4$ nucleotides (8) and *Pfu* polymerase, which is used commonly in high-fidelity PCR, has an error rate of 1.6 in $10^6$ nucleotides (9). Because of the high processivity and strand-displacing properties of φ29 DNA polymerase, the MDA reaction is performed under constant mesothermal conditions (30°C). The combination of high processivity and random priming makes MDA amplification an ideal approach for the amplification of DNA where there is insufficient material for standard DNA analysis technologies. Commercial kits are available for WGA by multiple displacement amplification from GE Healthcare (Pittsburgh, PA) which markets illustra GenomiPhi kits, and Qiagen (Valencia, CA) which sells the REPLI-g WGA kits. Both of these kits employ ϕ29 and are based on the multiple displacement amplification technology first described by Dean and colleagues (10). Key improvements have involved buffer formulations and the use of DNA-free reagents. These protocols for whole genome amplification can provide sufficient material for genetic analyses using PCR, DNA microarrays, Sanger sequencing, and Next Gen sequencing. This chapter lists several examples of the use of WGA by multiple displacement amplification for the microbial forensic analysis of trace DNA specimens.

## USE OF WGA FOR MICROBIAL FORENSIC ANALYSIS OF TRACE ENVIRONMENTAL SPECIMENS

The Biowatch network collects air samples on filters from several sites in the Washington, DC, area and other cities around the country. On September 24–25, 2005, environmental air monitors along the Capital Mall area in Washington, DC, signaled low-level detections of *Francisella tularensis*, the etiologic agent of *Tularemia*, which is classified as a select agent by the Centers for Disease Control and Prevention (CDC) and other federal agencies and has potential use as a bioterrorist agent (11). This organism is exceptionally virulent: exposure to as few as 10 organisms is potentially lethal to humans (11). DNA extracts from several filters tested positive for low levels of *F. tularensis* (12). Timing of the *F. tularensis* detections coincided with a well-attended antiwar protest; therefore, a large number of people were potentially exposed and subsequently traveled home to various parts of the country. On September 30, the CDC issued an official health advisory via the health alert network to alert the health care community to be aware of possible *Tularemia* exposure (Figure 10.2). Previously there have been other reports of *F. tularensis*-like organisms in air and other environmental samples (13,14). Detection of *F. tularensis* is confounded by that fact that there is a great deal of naturally occurring genetic diversity among *Francisella* species, including several subspecies of *F. tularensis* (15–18).

Ibis Biosciences, with funding from the Department of Defense, has developed a biosensor system initially called TIGER, then the Ibis T5000, and is now marketed as the PLEX-ID by Abbott Molecular, Inc. The system uses broad-range PCR and electrospray ionization mass spectrometry (PCR/ESI-MS) to identify bacteria and viruses, including unknown agents, mixtures, and novel organisms (19–23). Briefly, multiple pairs of primers

This is an official

# CDC Health Advisory

Distributed via Health Alert Network
September 30, 2005, 19:39 EDT (07:39 PM EDT)
CDCHAN-00238-05-09-30-ADV-N

## Presence of low levels of *Francisella tularensis* in the

## Washington D.C. area, September 30, 2005

CDC has become aware that from September 24th through September 25th environmental air monitors in SW Washington D.C., more specifically the Capitol Mall area, signaled the low level presence of *Francisella tularensis*, the bacterium that causes Tularemia, also known as "rabbit fever."

**At this time, public health agencies have no reports of any related human or animal illnesses. This announcement is a precautionary measure to assure that clinicians are aware of the situation and are able to recognize, test, and report any suspected cases to the appropriate medical and public health authorities.**

This is a national alert because the Capitol Mall area is a highly-trafficked tourist destination, and on Saturday, September 24th, was the site of several very well attended outdoor events.

### Clinical Presentations

The clinical presentations most likely to occur after an aerosol exposure to *F. tularensis* are pneumonic, oculoglandular and oropharyngeal. The usual incubation period is 3-5 days, but in rare instances can be longer. The disease is not communicable from person to person and can be effectively treated with readily available antimicrobials.

### Preliminary Case Definition

Onset from Monday September 26 through October 5 of an acute febrile illness associated with at least one of the following:

- conjunctivitis with preauricular lymphadenopathy (oculoglandular)
- stomatitis or pharygitis or tonsillitis and cervical lymphadenopathy (oropharyngeal)
- cough, shortness of breath, pleuritic chest pain (pneumonic),

which is not otherwise explained in a resident or visitor to the National Capitol Region on Saturday or Sunday, September 24-25.

### Human Diagnostic Specimens

Clinical specimens may include:

**FIGURE 10.2**

Copy of page one of four from the CDC Health Alert following the detection of low levels of *Francisella tularensis* in air samples collected around the Capital Mall in Washington, DC, in September 2005.

are used to amplify carefully selected regions of bacterial or viral genomes; the primer target sites are broadly conserved but the amplified regions carry information on the identity of the microbe in the nucleotide base compositions. Regions such as this appear in DNA that encodes ribosomal RNA and in housekeeping genes that encode essential proteins. Following PCR amplification, a fully automated electrospray ionization mass spectrometry analysis is performed on the PCR/ESI-MS instrument. The mass spectrometer effectively weighs the PCR amplicons, or the mixture of amplicons, with sufficient mass accuracy that the composition of A, G, C, and T can be deduced for each amplicon present. The base compositions are compared to a database of calculated base compositions derived from sequences of known organisms and to signatures from reference standards determined previously via PCR/ESI-MS.

In April 2006, a collection of DNA extracts from the Washington, DC, environmental air filters from the days around the *F. tularensis* detections were sent to Ibis for analysis on their T5000 system. Only about 20 μl of extract from each of the filters was available for analysis by Ibis scientists. Furthermore, the putative *F. tularensis* agent was reported to be present at very low levels in the extracts, necessitating that 5 μl of extract be used per PCR reaction (enough for four PCR reactions/sample). WGA by multiple displacement amplification using 25% of each specimen was employed to provide sufficient DNA for multiple PCR amplifications. This WGA DNA was then used to screen a large panel of broad-range PCR primers to identify the primers that would be most informative. Once the most resolving primers were identified, PCR/ESI-MS analysis was then repeated using the remaining original DNA extract.

One broad-range primer pair, *rpoC* primer pair (BCT354)(20), produced an amplicon with a base composition signature of A33 G32 C25 T32 from several filter extracts. This amplicon base count signature was one of several found in these specimens and was consistent with one generated by *F. tularensis*. Using WGA DNA from these putative *F. tularensis*-positive specimens, a panel of broad-range PCR primers were tested that specifically target the *Francisella* clade. Two primer pairs targeting the *asd* (BCT2328) and *galE* (BCT2332) loci best detected and identified the agent in the air samples. Other researchers have also used these two loci to characterize *Francisella* isolates (24) and in combination, the *asd* and *galE* PCR primers produced amplicons from air filter samples with base count signatures consistent with *F. tularensis* subsp. *novicida* (Table 10.1). Figure 10.3 shows mass spectra and base composition signatures obtained from one of the air samples from the day of the event. Interestingly, the *asd* primer pair BCT2328 yielded two amplicons that differed by an SNP. All of the positive specimens tested showed these two amplicons for this primer pair with one of the two observed *asd* amplicons being consistent with several subspecies of *F. tularensis*. These two amplicons may be a unique signature for this organism or may represent a mixture of more than one closely related organism in the samples.

**Table 10.1** Base Compositions of *asd* and *galE* Markers Observed with Selected *Francisella* Species and Subspecies

| Reference Isolate (strain) | BCT2328 (*asd*) | BCT2332 (*galE*) |
|---|---|---|
| *Francisella tularensis* subsp. *tularensis* SchuS4 | A16G24C10T32 | A31G21C13T36 |
| *F. tularensis* subsp. *tularensis* Wy96 | A16G24C10T32 | A31G21C14T35 |
| *F. tularensis* subsp. *mediasiatica* (FSC147) | A16G24C10T32 | A31G21C14T35 |
| *F. tularensis* subsp. *japonica* (FSC075) | A16G24C10T32 | A31G21C14T35 |
| *F. tularensis* subsp. *japonica* (FRAN043) | A16G24C9T33 | A31G21C14T35 |
| *F. tularensis* subsp. *holarctica* NAm/Eur/Asia (FSC155) | A18G22C10T32 | A31G21C14T35 |
| Washington, DC, Air sample #10 | A16G24C10T32 A16G24C9T33 | A32G20C14T35 |
| *F. tularensis* subsp. *novicida* (DPG 10B-WS) | A16G24C8T34 | A32G20C14T35 |
| *F. tularensis* subsp. *novicida* (FRAN003) | A16G24C10T32 | A32G20C14T35 |
| *F. tularensis* subsp. *novicida* (FSC156) | A16G24C10T32 | A32G19C15T35 |
| *F. tularensis* subsp. *philomiragia* | A16G24C8T34 | |

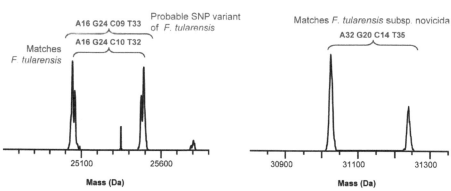

**FIGURE 10.3**

Representative mass spectra from PCR/ESI-MS analysis of air extracts collected from Washington, DC, in September 2005. Spectra show masses of forward and reverse strands of amplicons generated from sample #10 using primer pairs BCT2328 and BCT2332. Data are consistent with *F. tularensis* subsp. *novicida*. For each spectrum shown, the reverse strand corresponds to the left peak and the forward strand corresponds to the right peak. Base compositions of the amplicons are indicated.

The second primer pair targeting the *galE* locus produced a single amplicon that has a unique amplicon base count signature (A32 G20 C14 T35) that has only been observed with isolates of *F. tularensis* subsp. *Novicida,* a naturally occurring organism with generally lower virulence than *F. tularensis* subsp. *Tularensis.* The PCR reactions were subsequently repeated using the nonamplified DNA to confirm the signatures. *F. tularensis*-like organisms have been observed in a number of environmental sources, including those from air, water, and soil (13,14,25,26); thus the detections of *F. tularensis* observed in these air samples most likely represented a naturally occurring organism.

# DEEP SEQUENCE ANALYSIS *OF BACILLUS ANTHRACIS* STERNE STRAIN BY ROCHE 454 GENOME SEQUENCING TECHNOLOGY

Roche 454 whole genome sequencing is a massively multiplexed (>1 million reads per run) pyrosequencing platform that enables the sequencing of approximately 500 million high-quality bases per 10-hour run. The process consists of sample preparation (including fragmentation, library preparation, and bead-based emulsion PCR), followed by multiplexed sequencing by synthesis using sequentially applied nucleotides and chemiluminescent signal production. The luminescent signal is tracked throughout the sequencing run, and, by correlating the light production to the order of nucleotide addition, sequence data for the sample are generated (27). Pyrosequencing has been used to sequence many bacterial genomes, such as *F. tularensis* (17,28) *Escherichia coli* O157:H7 (29), *Staphylococcus aureus* (30), *and Salmonella typhi* (31). This technique, however, requires microgram quantities of template DNA, which significantly limits its utility for analysis of trace (femtogram to nanogram specimens) and uncultivable specimens unless whole genome amplification technology is used.

The use of WGA allows a relatively small amount of starting DNA to be amplified many orders of magnitude ($10^5$–$10^7$), producing the significantly larger amount of DNA needed for pyrosequencing while still maintaining the forensic signatures present in the original sample. As an example, 10 pg of *B. anthracis* Sterne strain genome ($\approx$1850 genomes) was amplified using $\phi$29-based multiple displacement amplification for a final yield of 7.75 μg of DNA ($7.75 \times 10^5$-fold amplification). This WGA DNA sample and an unamplified (native) genomic DNA sample were then whole genome sequenced on the Roche Genome sequencer FLX system and the data compared by two metrics: (i) depth of sequencing by location on the *B. anthracis* genome and (ii) SNPs unique to this specific isolate of *B. anthracis* Sterne strain.

For our analysis, the average depth of sequencing at a location on the *B. anthracis* genome was calculated as the average number of times that a location (in this specific case a "location" is a region of approximately 10 kb) was sequenced during the pyrosequencing run. By comparing the depth of sequencing at different locations, the relative abundance of each location in the original sample was determined. A plot of sequencing depth by location for the genomic DNA sample is presented in Figure 10.4a. Visually, the most noticeable feature of this plot is the "V"-shaped distribution of sequencing depth based on the location on the circular chromosome with significantly higher sequencing depth being observed the closer the location was to the origin of replication (genome location "0"). This distribution is a consequence of isolation of the *B. anthracis* genomic DNA from a logarithmically growing culture in which bacteria were dividing rapidly and hence replicating their genome rapidly. In *B. anthracis*, genome replication begins at a single origin of replication and proceeds in both directions around the circular genome until stopping roughly halfway around the circle at the replication terminus. In rapidly dividing bacteria, this replication process can be initiated a number of times before one complete genome has been replicated. This leads to a higher representation of sequences near the origin of replication than at the replication terminus. If genomic DNA used for deep sequencing is isolated from a culture in the middle of this rapid growth phase (as was done with the *B. anthracis* genomic DNA we analyzed), sequencing will reflect the higher relative abundance of sequences close to the origin of replication. With this in mind, one can use the relative sequencing depth across the genome of a specimen to elucidate information regarding the growth phase of the specimen and in turn obtain a specific signature of that specimen.

Examination of the sequencing depth by location for the WGA DNA specimen that had been amplified $7.75 \times 10^5$-fold (Figure 10.4b) shows the same

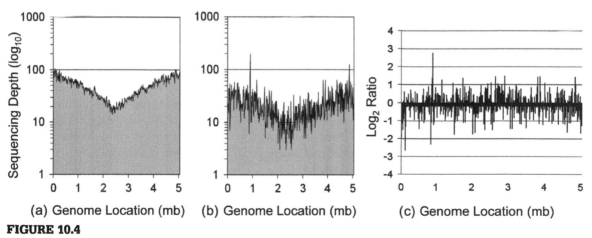

**FIGURE 10.4**

The 454 sequencing depth by location on a *B. anthracis* Sterne strain genome. (a) Data generated from 5 μg of genomic DNA (zero position represents origin of replication). (b) Data generated from 5 μg of WGA DNA obtained using 10 pg of genomic template DNA. The V shapes of the plots are characteristic of log-phase bacterial growth. (c) Plot of the $\log_2$ ratio of WGA DNA vs genomic DNA by location on the *B. anthracis* genome. Locations with negative numbers are underrepresented in WGA DNA relative to the same location in genomic DNA, whereas locations with positive numbers are overrepresented.

"V"-shaped signature typical of log phase growth as the genomic DNA specimen, albeit with a slight increase in variation. This variation is quantified in Figure 10.4c where the relative sequencing depth of the WGA DNA specimen is compared to the relative sequencing depth of the genomic DNA sample at each location along the *B. anthracis* genome. The majority of locations in the WGA DNA specimen (95%) are within 2.5-fold of what is found in the genomic DNA sample with an average difference in representation of 1.4-fold across the 500 locations measured. Overall, the level of variation observed in the WGA specimen is remarkably small considering the $7.75 \times 10^5$-fold amplification the sample underwent prior to sequencing. This is especially important considering WGA allows significantly more information (and hence more forensic signatures) to be obtained than would otherwise be possible with an unamplified sample.

The second set of signatures examined were SNPs in specimens relative to the GenBank reference sequence for the *B. anthracis* Sterne strain. An SNP is a variation in the DNA sequence at a single position where one base is either removed, added, or changed, yielding two or more different forms of the sequence. Other larger variations, such as genetic engineering events, for example, insertion of toxin-producing or antibiotic-resistance genes, are not addressed specifically in this discussion but would be observed readily by a similar analysis. The presence of specimen-specific SNPs (or any other larger genetic engineering events) provides the genetic fingerprint that can be used to determine the relationship of a bioforensic specimen to other specimens for the purpose of attribution. For example, deep pyrosequencing of the *B. anthracis* genomic DNA described earlier gave a high confidence consensus accuracy of 99.9972% when compared to the GenBank reference sequence. From this metric alone these two samples are exceptionally similar. The pyrosequencing, however, revealed 31 SNPs in the genomic DNA relative to the published GenBank sequence (GI:49183039). Although these SNPs made up only 0.0006% of the genome (31 SNP bases divided by 5.2 million total genomic bases), these changes clearly distinguished the specimen from the GenBank reference sequence. Sequence data derived from the WGA DNA specimen showed the same pattern identified in the genomic DNA. A high confidence consensus accuracy of 99.9948% was observed when compared to the GenBank reference, and each of the 31 SNPs present in the genomic DNA was detected in the WGA DNA. The presence of these SNPs makes it clear that the original 10 pg of starting DNA used in the WGA is from the same strain as the genomic DNA and that the WGA maintained these unique SNPs after $7.75 \times 10^5$-fold amplification of the template DNA.

In summary, WGA combined with highly multiplexed pyroseqeuencing is a powerful microbial forensic tool capable of revealing specimen signatures such as the growth phase of a bacterial DNA specimen, the specific strain and/or origin of a specimen (derived from the SNP profile), bioengineering events, and other trace DNA signatures such as contaminating human, plant, or bacterial DNA.

## MICROBIAL FORENSIC SEQUENCE ANALYSIS OF TRACE SPECIMENS OF *BACILLUS ANTHRACIS* BY SINGLE MOLECULE REAL-TIME SEQUENCING

Just as pyrosequencing-based Next Gen sequencing technologies described earlier have opened up new capabilities for microbial forensics, there are new "Next Next Generation Sequencing" technologies (also called "third-generation sequencing technologies") that potentially offer even faster, higher throughput sequencing at lower costs than the current Next Gen sequencing technologies. Pacific Biosciences (Menlo Park, CA) has developed single molecule real-time (SMRT™) technology that enables observation of natural DNA synthesis by DNA polymerase as it occurs in real time. Thus the technology can generate sequence data at a rate of one to three nucleotides per second across tens of thousands of separate sequencing reactions (32–35). The technology has the added advantage of not requiring a change of reagents after each nucleotide of DNA is read, thus reducing reagent costs and accelerating the sequencing process greatly. This technology generates DNA sequence data in minutes compared to hours required for current deep sequencing methods.

Single molecule real-time DNA sequencing is performed on SMRT cells, with each cell containing thousands of zero-mode waveguides (ZMWs). Produced utilizing the latest geometries available in semiconductor manufacturing, a ZMW is a hole, tens of nanometers in diameter, fabricated in a 100-nm metal film deposited on a silicon dioxide substrate. Each ZMW functions as a nano-photonic visualization chamber, providing a detection volume of 20 zepto-liters ($10^{-21}$ liters). At this volume, the activity of a single molecule can be detected among a background of thousands of labeled nucleotides.

Within each ZMW chamber, a single DNA polymerase molecule is attached to the bottom surface such that it resides permanently within the detection volume. Phospho-linked nucleotides, each base labeled with a different-colored fluorophore, are introduced into the reaction solution at concentrations that promote enzyme speed, accuracy, and processivity. As the DNA polymerase incorporates complementary nucleotides, each base is held within the detection volume for tens of milliseconds, orders of magnitude longer than the amount of time it takes a nucleotide to diffuse in and out of the detection volume. During this time, the engaged fluorophore emits fluorescent light whose color corresponds to the base identity. Then, as part of the natural incorporation cycle, the polymerase cleaves the bond holding the fluorophore in place and the dye diffuses out of the detection volume. Following incorporation, the signal immediately returns to baseline and the process repeats.

The Pacific Biosciences real-time single molecule sequencing process can sequence a variety of DNA templates. One method enables the repeated resequencing of a single molecule of DNA through the conversion of DNA fragments into single-stranded circles, which are sequenced repeatedly so that a high accuracy consensus sequence for the DNA molecule can be generated. The preparation of genomic libraries is detailed in Figure 10.5a. Briefly, the DNA sample is first fragmented to the desired size and hairpin adapters are ligated to the ends of each fragment, creating a template called a SMRTbell™ that is structurally linear but topologically circular. These templates are then bound to a polymerase and each complex is immobilized at the bottom of a ZMW for sequencing. During the sequencing reaction, a strand-displacing polymerase enzyme opens the SMRTbell into a circular template and generates linear reads composed of a concatenated SMRTbell adapter sequence, forward-strand insert sequence, SMRTbell adapter sequence, and reverse-strand insert sequence. Both forward and reverse sequences are assembled to produce a circular consensus of the insert sequence. Consensus accuracy increases linearly with each sequencing pass through the insert.

To evaluate Pacific Biosciences' single molecule real-time sequencing with trace microbial DNA specimens, we started with 500 femtograms of *B. anthracis* genomic DNA, which represents the amount of DNA from approximately 30 growing cells. The genomic DNA was subject to WGA by multiple displacement amplification, and the resulting amplified DNA was used in the Pacific Biosciences library preparation process outlined in Figure 10.5a. The library of single-stranded circular DNA was prepared using 200 to 400-bp DNA fragments and processed using a 3000 ZMW SMRT cell. Approximately 30% (1000) of prepared ZMWs had a single molecule of DNA polymerase and template and were used to generate sequence data. As the polymerase reads these single-stranded circles it reads the forward strand, then the SMRTbell adapter sequence, then the reverse strand of DNA, then the second SMRTbell adapter sequence, and back again to resequence the forward strand. Figure 10.5b shows data from one of these single molecule real-time sequencing reactions. In this example, a 161-bp DNA fragment of WGA DNA derived from 500 fg of the *B. anthracis* Sterne strain was prepared to create a single-stranded circular DNA molecule and sequenced using the Pacific Biosciences SMRT DNA sequencing system. From this template, a 1359-bp linear read was produced at a sequencing rate of 2.9 nucleotides/second. This linear read was screened for the SMRTbell adapter sequences and six being detected. Therefore, this reaction produced four forward-strand and three reverse-strand amplicon subsequences. These subsequences were assembled into a consensus sequence, which was identical to the *B. anthracis* Sterne strain. The repeating SMRTbell sequence of forward-strand insert sequence, SMRTbell adapter, reverse-strand insert sequence, SMRTbell adapter, and so on is apparent from the alignments shown in Figure 10.5b.

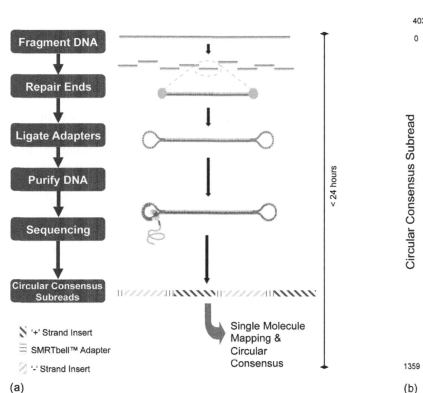
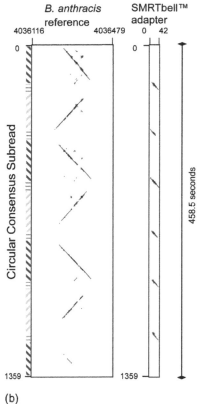

**FIGURE 10.5**

(a) Schematic of the Pacific Biosciences single molecule real-time (SMRT™) sequencing sample preparation protocol. DNA is fragmented, ends are repaired, and hairpin adapters are ligated to create SMRTbell™ single-stranded circular sequencing templates. Circular SMRTbell templates are used as substrates to generate linear reads composed of concatenated SMRTbell adapter sequence, forward-strand insert sequence, SMRTbell adapter sequence, and reverse-strand insert sequence. (b) Single molecule mapping and circular consensus of a single *B. anthracis* Sterne strain DNA molecule obtained from a 500-fg specimen. A 161-bp DNA fragment from the *B. anthracis* Sterne strain was prepared and sequenced using the Pacific Biosciences protocol. From this single molecule, a 1359-bp linear read was produced at a polymerization rate of 2.9 bp/s. Subsequences were assembled into a single molecule consensus sequence; the consensus sequence was identical to *B. anthracis* Sterne (GenBank GI:49183039). The linear SMRT read is presented on the *y* axis aligned to the *B. anthracis* and the SMRTbell adapter references. The repeating SMRTbell structure of forward-strand insert sequence, SMRTbell adapter, reverse-strand insert sequence, SMRTbell adapter, and so on is apparent.

Although our studies of single molecule real-time sequencing had a limited depth of coverage, results demonstrate the potential of WGA and third-generation real-time single molecule DNA sequencing for the analysis of trace microbial forensic samples. Single molecule real-time DNA sequencing provides long read lengths and extremely fast sequencing times and has the potential to significantly lower costs compared to Next Gen sequencing systems. This and other third-generation sequencing technologies will be especially applicable for the rapid detection of bioengineering events and emerging infectious agents where time to answer may be critical.

## CONCLUSIONS AND REMAINING CHALLENGES

This chapter demonstrated the use of whole genome amplification by $\phi29$ DNA polymerase-based multiple displacement amplification for microbial forensic PCR/ESI-MS analysis of environmental air extracts containing limited amounts of DNA. It also demonstrated the utility of WGA of trace DNA specimens coupled with either Roche 454 sequencing technology or Pacific Biosciences single molecule real-time sequencing technology. Challenges remain, however. Whole genome amplification by multiple displacement amplification will amplify all DNA in a specimen, which can create challenges for detection and characterization of an agent of interest in the presence of a large amount of background. Soil samples and clinical specimens present such challenges. For example, $1\,\mu l$ of human blood contains a 1000 copies of the human genome so identifying a pathogen in this background would necessitate >1000 human genome projects of sequencing to find the pathogen DNA. This depth of sequencing is beyond the scope of all current sequencing technologies. Fortunately, there are many methods for the separation of organisms by size that can be used to separate viruses from bacteria from eukaryotic cells prior to DNA extraction and WGA.

Multiple displacement amplification can also create chimeras (36) during the amplification process that can create issues for *de novo* sequence assemblies. It is likely, however, that microbial forensic analyses will rely on algorithms that compare individual sequence reads against a database and thus will be significantly less sensitive to these chimeric sequences. This points out the greatest challenge for microbial forensic analysis of trace and noncultivable specimens by whole genome sequence analyses: bioinformatics. Whole genome sequencing can generate enormous amounts of raw data that must be processed to generate sequence data that will be used to identify agents in the specimen. This will require development of novel algorithms that will be customized for the technology used to generate data. The results of these efforts should lead to the development of new tools to assist in microbial forensic analyses and will ultimately improve public health surveillance of emerging infectious agents.

In summary, Next Gen sequencing technologies, combined with whole genome amplification, offer the forensic investigator new capabilities for the rapid microbial forensic characterization of trace and noncultivable specimens that can identify key signatures that can lead to attribution, as well as identifying bioengineered and emerging threats.

# ACKNOWLEDGMENTS

We thank John Major and Kathryn Hunkapiller at Pacific Biosciences (Menlo Park, CA) for their technical assistance with single molecule real-time DNA sequencing. We also thank the efforts of Azad Ahmed, Garth Ehrlich, and Robert Biossy at the Center for Genomic Sciences Allegheny-Singer Research Institute (Pittsburgh, PA) for their 454 sequencing and data analysis, respectively. Funding for DNA sequencing was provided by Defense Threat Reduction Agency Contract HDTRA1-07-C-0096. Funding for the analysis of the *Francisella* air samples was provided by Department of Homeland Security contract W81XWH-05-C-0116. Opinions, interpretations, conclusions, and recommendations are those of the author and are not necessarily endorsed by the U.S. Army or the U.S. Department of Homeland Security.

# REFERENCES

[1] H. Telenius, N.P. Carter, C.E. Bebb, M. Nordenskjöld, B.A. Ponder, A. Tunnacliffe, Degenerate oligonucleotide-primed PCR: General amplification of target DNA by a single degenerate primer, Genomics 13 (3) (1992) 718–725.

[2] H. Telenius, A.H. Pelmear, A. Tunnacliffe, N.P. Carter, A. Behmel, M.A. Ferguson-Smith, et al., Cytogenetic analysis by chromosome painting using DOP-PCR amplified flow-sorted chromosomes, Genes Chromosomes Cancer 4 (3) (1992) 257–263.

[3] V.G. Cheung, S.F. Nelson, Whole genome amplification using a degenerate oligonucleotide primer allows hundreds of genotypes to be performed on less than one nanogram of genomic DNA, Proc. Natl. Acad. Sci. USA 93 (25) (1996) 14676–14679.

[4] C. Pirker, M. Raidl, E. Steiner, L. Elbling, K. Holzmann, S. Spiegl-Kreinecker, et al., Whole genome amplification for CGH analysis: Linker-adapter PCR as the method of choice for difficult and limited samples, Cytometry A 61 (1) (2004) 26–34.

[5] F.B. Dean, S. Hosono, L. Fang, X. Wu, A.F. Faruqi, P. Bray-Ward, et al., Comprehensive human genome amplification using multiple displacement amplification, Proc. Natl. Acad. Sci. USA 99 (2002) 5261–5266.

[6] J.A. Esteban, M. Salas, L. Blanco, Fidelity of phi 29 DNA polymerase: Comparison between protein-primed initiation and DNA polymerization, J. Biol. Chem. 268 (4) (1993) 2719–2726.

[7] J.G. Paez, M. Lin, R. Beroukhim, J.C. Lee, X. Zhao, D.J. Richter, et al., Genome coverage and sequence fidelity of phi29 polymerase-based multiple strand displacement whole genome amplification, Nucleic Acids Res. 32 (9) (2004) e71.

[8] K.A. Eckert, T.A. Kunkel, DNA polymerase fidelity and the polymerase chain reaction, PCR Methods Appl. 1 (1) (1991) 17–24.

[9] K.S. Lundberg, D.D. Shoemaker, M.W. Adams, J.M. Short, J.A. Sorge, E.J. Mathur, High-fidelity amplification using a thermostable DNA polymerase isolated from *Pyrococcus furiosus*, Gene 108 (1) (1991) 1–6.

[10] F.B. Dean, J.R. Nelson, T.L. Giesler, R.S. Lasken, Rapid amplification of plasmid and phage DNA using Phi 29 DNA polymerase and multiply-primed rolling circle amplification, Genome Res. 11 (6) (2001) 1095–1099.

[11] D.T. Dennis, T.V. Inglesby, D.A. Henderson, J.G. Bartlett, M.S. Ascher, E. Eitzen, et al., Working Group on Civilian Biodefense. Tularemia as a biological weapon: Medical and public health management, JAMA 285 (21) (2001) 2763–2773.

[12] P. Dvorak, Health officials vigilant for illness after sensors detect bacteria on mall, Washington Post, Washington, DC, 2005-10-02, p. 1.

[13] S.M. Barns, C.C. Grow, R.T. Okinaka, P. Keim, C.R. Kuske, Detection of diverse new *Francisella*-like bacteria in environmental samples, Appl. Environ Microbiol. 71 (9) (2005) 5494–5500.

[14] C.R. Kuske, S.M. Barns, C.C. Grow, L. Merrill, J. Dunbar, Environmental survey for four pathogenic bacteria and closely related species using phylogenetic and functional genes, J. Forensic Sci. 51 (3) (2006) 548–558.

[15] A. Johansson, J. Farlow, P. Larsson, M. Dukerich, E. Chambers, M. Byström, et al., Worldwide genetic relationships among *Francisella tularensis* isolates determined by multiple-locus variable-number tandem repeat analysis, J. Bacteriol. 186 (17) (2004) 5808–5818.

[16] K. Svensson, P. Larsson, D. Johansson, M. Byström, M. Forsman, A. Johansson, Evolution of subspecies of *Francisella tularensis*, J. Bacteriol. 187 (11) (2005) 3903–3908.

[17] M.D. Champion, Q. Zeng, E.B. Nix, F.E. Nano, P. Keim, C.D. Kodira, et al., Comparative genomic characterization of *Francisella tularensis* strains belonging to low and high virulence subspecies, PLoS Pathog. 5 (5) (2009) e1000459.

[18] P. Keim, A. Johansson, D.M. Wagner, Molecular epidemiology, evolution, and ecology of Francisella, Ann. N.Y. Acad. Sci. 1105 (2007) 30–66.

[19] R. Sampath, D.J. Ecker, Novel biosensor for infectious disease diagnostics, in: S.E. Knobler, A. Mahmoud, S. Lemon (Eds.), Forum on Microbial Threats: Learning from SARS: Preparing for the Next Disease Outbreak—Workshop Summary, National Academies Press, Washington, DC, 2004, pp. 181–185.

[20] S.A. Hofstadler, R. Sampath, L.B. Blyn, M.W. Eshoo, T.A. Hall, Y. Jiang, et al., TIGER: The universal biosensor, Int. J. Mass Spectrom. 242 (2005) 23–41.

[21] D.J. Ecker, J. Drader, J. Gutierrez, A. Gutierrez, J.C. Hannis, A. Schink, et al., The Ibis T5000 Universal Biosensor: An automated platform for pathogen identification and strain typing, J. Assoc. Lab. Automation (2006).

[22] D.J. Ecker, R. Sampath, C. Massire, L.B. Blyn, T.A. Hall, M.W. Eshoo, et al., Ibis T5000: A universal biosensor approach for microbiology, Nat. Rev. Microbiol. 6 (7) (2008) 553–558.

[23] D.J. Ecker, C. Massire, L.B. Blyn, S.A. Hofstadler, J.C. Hannis, M.W. Eshoo, et al., Molecular genotyping of microbes by multilocus PCR and mass spectrometry: A new tool for hospital infection control and public health surveillance, Methods Mol. Biol. 551 (2009) 71–87.

[24] L.E. Lindler, X.Z. Huang, M. Chu, T.L. Hadfield, M. Dobson, Genetic fingerprinting of biodefense pathogens for epidemiology and forensic investigation, in: L.E. Lindler, F.J. Lebeda, G.W. Korch (Eds.), Biological Weapons Defense: Infectious Diseases and Counterbioterrorism, Humana Press, 2005.

[25] J.M. Petersen, J. Carlson, B. Yockey, S. Pillai, C. Kuske, G. Garbalena, et al., Direct isolation of Francisella spp. from environmental samples, Lett. Appl. Microbiol. 48 (6) (2009) 663–667.

[26] J. Farlow, D.M. Wagner, M. Dukerich, M. Stanley, M. Chu, K. Kubota, et al., *Francisella tularensis* in the United States, Emerg. Infect. Dis. 11 (12) (2005) 1835–1841.

[27] GS FLX Instrument and Reagent Specifications, 2009.

[28] B. La Scola, K. Elkarkouri, W. Li, T. Wahab, G. Fournous, J.M. Rolain, et al., Rapid comparative genomic analysis for clinical microbiology: The *Francisella tularensis* paradigm, Genome Res. 18 (5) (2008) 742–750.

[29] M.L. Clawson, J.E. Keen, T.P. Smith, L.M. Durso, T.G. McDaneld, R.E. Mandrell, et al., Phylogenetic classification of *Escherichia coli* O157:H7 strains of human and bovine origin using a novel set of nucleotide polymorphisms, Genome Biol. 10 (5) (2009) R56.

[30] S.K. Highlander, K.G. Hulten, X. Qin, H. Jiang, S. Yerrapragada, E.O. Mason Jr., et al., Subtle genetic changes enhance virulence of methicillin resistant and sensitive *Staphylococcus aureus*, BMC Microbiol. 7 (2007) 99.

[31] K.E. Holt, J. Parkhill, C.J. Mazzoni, P. Roumagnac, F.X. Weill, I. Goodhead, et al., High-throughput sequencing provides insights into genome variation and evolution in *Salmonella typhi*, Nat. Genet. 40 (8) (2008) 987–993.

[32] J. Eid, A. Fehr, J. Gray, K. Luong, J. Lyle, G. Otto, et al., Real-time DNA sequencing from single polymerase molecules, Science 323 (5910) (2009) 133–138.

[33] J. Korlach, A. Bibillo, J. Wegener, P. Peluso, T.T. Pham, I. Park, et al., processive enzymatic DNA synthesis using 100% dye-labeled terminal phosphate-linked nucleotides, Nucleosides Nucleotides Nucleic Acids 27 (9) (2008) 1072–1083.

[34] P.M. Lundquist, C.F. Zhong, P. Zhao, A.B. Tomaney, P.S. Peluso, J. Dixon, et al., Parallel confocal detection of single molecules in real time, Opt. Lett. 33 (9) (2008) 1026–1028.

[35] J. Korlach, P.J. Marks, R.L. Cicero, J.J. Gray, D.L. Murphy, D.B. Roitman, et al., Selective aluminum passivation for targeted immobilization of single DNA polymerase molecules in zero-mode waveguide nanostructures, Proc. Natl. Acad. Sci. USA 105 (4) (2008) 1176–1181.

[36] R.S. Lasken, TB. Stockwell, Mechanism of chimera formation during the multiple displacement amplification reaction, BMC Biotechnol. 7 (2007) 19.

# CHAPTER 11

# Molecular Microbial Surveillance and Discovery in Bioforensics

**Evan Skowronski[a] and W. Ian Lipkin[b]**

[a]*U.S. Army Edgewood Chemical Biological Center,*
*Aberdeen Proving Ground, Maryland*
[b]*Center for Infection and Immunity, Mailman School of Public Health and College of*
*Physicians and Surgeons of Columbia University, New York*

## FUNDAMENTAL QUESTIONS: WHAT IS IT? HOW DID IT GET THERE? WHAT CAN WE DO ABOUT IT? CLIMATE IS WHAT YOU EXPECT; WEATHER IS WHAT YOU GET…HOW EXPECTATIONS ARE IMPORTANT CUES

### Location, Location, Location: Nonscientific Context Can Be Key

Perhaps the most significant challenge in microbial forensics is the ability to differentiate the natural from the intentional biological event. Mother Nature continues her relentless assault on human, animal, and plant populations with an impressive array and diversity of microbes, and situational awareness of unnatural malicious intent can be hard to come by. In some cases, the mere context of the outbreak, which agent is involved, and other nontechnical data can be incredibly telling before the first polymerase chain reaction (PCR)/immunoassay has been run, the microarrays have been read, or the first DNA sequence generated. Possibly the most pertinent example was the first case in the 2001 anthrax attacks, where a photographic editor in the American Media, Inc. (AMI) office building in Florida contracted pneumonic anthrax. While later sampling of the AMI building demonstrated widespread contamination with *Bacillus anthracis* spores (1,2), his urban workplace and the lack of occupational or recreational exposure (3–6) were salient. While the initial response from the Centers for Disease Control and Prevention (CDC) was to announce (correctly) that there were no data to suggest that the case was anything other than a natural event (7), the fact that the last pneumonic anthrax case was 26 years ago in a cattle farm worker in Texas (3,8,9) was highly suspicious and

Microbial Forensics. DOI: 10.1016/B978-0-12-382006-8.00011-6

prompted investigation and discovery of widespread contamination within the AMI building (10) and was finally linked to letters used to intentionally disseminate the agent (8). An AMI letter, if it existed, was never found.

A well-known counterexample is an incident of food poisoning of a restaurant salad bar by the Rajneeshee cult in Oregon in 1984 carried out with the intention of influencing local election results (11). While this was by far the largest intentional use of a biological agent (*Salmonella*, infecting more than 750 individuals) in recent history, it was not identified as such until nearly a year later when suspicions of the local populace were confirmed, first in the press and later when a matching strain was discovered in Rajneeshpuram (12). In this instance, the agent and context of the attack mimicked a naturally occurring event that plays out several times a year across the United States, with large, multifocal outbreaks such as the one that plagued the tomato and pepper supplies in 2008–2009, causing disease in more than 1400 people (13). The political context of the election was insufficient to provide the necessary impetus for a formal investigation of the matter in the face of such a plausible natural event. It was not until U.S. Congressman James Weaver accused the group during a speech in the House of Representatives (14) and the internal politics of the group led to self-accusations (15) that law enforcement officials were able to investigate, determine the true nature of the outbreak, and finally convict those guilty of planning and executing the attack (15).

## "*Nobody* Expects the Spanish Inquisition!"—Monty Python

With an estimated 40 million primary infectious disease hospital admissions between 1998 and 2006 (16), it is literally impossible for the etiological agent to be identified, much less to a standard that would be considered a validated fact generated by a tertiary testing facility (CDC, Laboratory Response Network, Local/State Public Health Laboratories) and that would withstand legal scrutiny. Funding for all of these investigations is limited to what is necessary for clinical treatment, often stopping short at quasi-physiological and phylogenetic descriptions such as α-hemolytic group B *Streptococcus*, influenza A, or viral encephalitis. In many cases, presumptive diagnoses and minimal screening tests are often used when clinical signs are sufficient to treat. In practice, this has been sufficient for medical care but is not adequate to address the needs of microbial forensics.

Validated pathogen identification is the first step in any microbial forensics effort, which drives the shift from diagnostics to investigation. Again, painfully learned lessons from 2001 are informative and have illustrated that even CDC category A select agents can masquerade in the guise of ubiquitous "flu-like" symptoms. The first victim of the anthrax attacks, Mr. Robert Stevens, languished for days with increasing malaise until developing meningitis, a known but uncommon clinical presentation of fulminant pneumonic anthrax (17,18). Gram staining of the cerebral spinal fluid revealed large, "boxcar"-shaped Gram-positive bacilli indicative of *B. anthracis* (one of few such organisms that cause human disease). These findings were confirmed rapidly with more selective methods (8,19). With no premonition or warning of a biological weapons attack, there was little reason to suspect infections with *B. anthracis*, and only astute work of infectious disease specialist Dr. Larry Bush allowed for correct diagnosis (19). This case would have been lost in the obscurity of clinical infectious disease statistics unless enough individuals in the same facility became ill to launch a more rigorous investigation. In fact, two other occupants of the same building were exposed: Ernesto Blanco, who went on to develop pneumonic anthrax, and Stephanie Dailey, who remained asymptomatic (8,10). In both cases, their clinical investigations were driven by Stevens's diagnosis. It is entirely possible that if Mr. Stevens's illness had never been diagnosed as anthrax, neither of the two infections would have been linked to the larger investigation that followed and additional colleagues would have been susceptible to infection from the widespread contamination within the AMI building.

## CHARACTERIZATION OF THE THREAT: WHAT KIND OF BUG IS IT?

### No School Like Old School: Use of Traditional Microbiological Techniques

Without any indication or intelligence that an especially dangerous pathogen is in play, the vast majority of microbiological testing will be done employing universally available traditional techniques that have been in use for decades. Selective and semiselective media are used to isolate and propagate bacteria from clinical samples with relatively uncomplicated (blood, cerebrospinal fluid) or complex (fecal, sputum, nasopharyngeal lavage) clinical microbiological samples (20). Viral pathogens can, in certain instances, be grown on host culture cells and identified by direct or indirect immunofluorescence (21,22). Simple measurements of bacterial metabolism and physiology such as cell wall content (Gram, polysaccharide, and capsule staining), sugar metabolism, hemolysis, antibiotic susceptibility, motility, and colony morphology are all used to confirm tentative diagnoses or categorize bacteria into broad but treatable groups (gram-positive rod-shaped bacilli, $\alpha$-hemolytic streptococci, etc.). This assumes, of course, that any such "diagnostics" of any sort are employed before palliative treatments (fluids, antipyretics, bed rest) or prophylactic antibiotics are employed, which is often the case when garden-variety infections are assumed. In the case of Robert Stevens, the presence of large gram-positive, rod-shaped bacilli was sufficient to reduce the list of likely pathogens to two (*B. anthracis* and *B. cereus*). The follow-on clinical microbiology demonstrating lack of motility and hemolytic capability was sufficient to establish a tentative diagnosis and prompt notification of the State Health Laboratory in Jacksonville, Florida, and the CDC, where findings were confirmed with more definitive and sophisticated methods (23). This process took an additional 2 days, which was soon enough to impact the clinical treatment of Ernesto Blanco (switched to the antibiotic ciprofloxacin, thought to be specifically effective against anthrax (8), potentially saving his life.

## Molecular Diagnostics
### Polymerase Chain Reaction
While traditional microbiology is often sufficient to identify a pathogen, it is a time-consuming and laborious process with significant limitations in discriminating definitively between closely related organisms (24–27). The advent of the era of genomics has provided both the technology to perform rapid molecular testing and a plethora of genetic sequences of threat agents, their near neighbors, vectors, and hosts that enable the development of sensitive agent-specific assays. Real-time PCR utilizing two amplification primers and a fluorophor quencher-tagged hydrolysis probe has been a great boon to the fields of medical diagnostics and environmental detection. "One pot" reactions with optical readout are amenable to automation, limit cross-contamination/false-positive results, and allow for high-throughput applications in times of emergency when surge capacity is needed. With the theoretical capability to detect a single copy (after sample preparation and subsampling), PCR has become the method of choice for rapid and high-throughput pathogen identification. While PCR is appropriately lauded for its sensitivity, specificity, and high-throughput capability, it has become apparent over the past decade of increased biosurveillance that even the best designed PCR assay may suffer from loss of specificity. This lesson was probably best learned from what is probably the largest continual biosurveillance effort in history, the Department of Homeland Security BioWatch program, which began in 2003 (28,55). With thousands of assays being run across the country on a daily basis, it became immediately clear that single screening assays would occasionally test positive. The vast majority of these samples was later shown not to contain the pathogen for which the screening assay was positive through additional testing with orthogonal PCR assays designed against distinct genetic targets. Reasons for this are clear when it is understood that many of these assays and others used by government, industry, and academia were developed in the mid- to late 1990s using existing available genomic sequences from target organisms, their genetic near neighbors, and a diverse background of host, vector, and environmental matrixes. These sequences could be compared by sifting

through billions of base pairs of sequences to identify candidate unique sequences that could later be further refined with laboratory screening (29). However, the phenomena of "signature erosion," coined by bioinformaticist Tom Slezak at Lawrence Livermore National Laboratory, began to be used to characterize DNA sequences shown to be less than unique when additional genetic sequencing started to show up in exponentially larger amounts. When the Salt Lake City airport was shut down for several hours based on a single PCR result [which was later shown to be negative through additional testing (30)] during the 2002 Winter Olympics, it became clear that high-consequence decision making could not depend solely upon the results of a single assay.

One solution is to merely perform multiple independent or "single-plex" PCR assays to gain a higher level of confidence in detection, with the added benefit of being able to determine agent-specific genomic architecture of interest (e.g., *B. anthracis* chromosomal and pXO1/pXO2 plasmids required for virulence, antibiotic resistance genes). If truly independent assays targeting different loci on the genome of interest were developed separately with the same rigor, probability theory using *independent* variables would indicate that three PCR assays with a $10^{-3}$ false-positive rate (i.e., *very* good assays) would have a combined false-positive rate of $10^{-9}$. If the BioWatch program tests 1000 samples nationwide each day for multiple threat agents, each agent would be predicted to give a single false-positive result every 2740 years! This kind of specificity would be acceptable for virtually any decision making, but there are several caveats to these assumptions that call this impressive number into question. First, a genomic sequence cannot be considered truly random or a statistical independent variable (31), as known biases due to conserved/convergent evolution, codon preference, repetitive elements, and other non-random elements render any such calculations invalid. Second, the cost, equipment, space, and labor of simply increasing the throughput by many-fold is often prohibitive (32). Finally, all such systems suffer from sampling-driven limitation in sensitivity based on Poisson distributions of limited genetic copies when materials are split between multiple aliquots (i.e., 50 single-plex PCR assays performed on a single sample would theoretically lower the limit of detection by 50-fold). This changes some of the fundamental assumptions of sensitivity that inform models of plume predictions, health effects modeling, and other response elements that also must be considered.

There is an obvious answer to the technical and financial shortcomings that accompany significant increases in PCR assay demand. Combining different sets of primers and probes into a single real-time PCR reaction has been demonstrated to be an effective way to "multiplex" a PCR assay (33–37). This technology has made it possible to assay for multiple genetic targets (including a positive control as necessary) simultaneously with virtually the same labor and reaction costs (33,36,38–44). Unfortunately, this approach has several design constraints and theoretical limitations that impact the widespread applicability and significant increase in throughput. When designing a multiplex real-time PCR assay, assay conditions ($Mg^{2+}$ content, pH, additives) and thermocyling parameters (annealing/melting/extension temperature, temperature ramping speed, and time spent at each temperature) must be uniform (37,45). In addition, primers must be screened against each other to minimize primer–primer hybridizations that can result in widespread amplification of nonspecific "template-less" products (46). Furthermore, sensitivity is impacted adversely by primer multiplexing. Indeed, the reduction in sensitivity is particularly problematic when more than 10 primer sets are employed in a single reaction. However, the most limiting constraint of multiplex real-time PCR reactions is the paucity of fluorophores available. So-called dark quenchers have been developed that limit the background noise of intact hydrolysis probes (47), but the wide excitation bands of each of the fluorophores requires extensive spectral overlap collection at the expense of signal and therefore sensitivity. The quantum efficiency of the fluorophores also varies greatly, making it impossible to have uniform sensitivity between assays in the same tube. At most, five separate probes can be used, with two- and three-plex assays predominating (48,49).

Other strategies have been established to enable multiplexing. PCR products can be captured by an oligonucleotide tethered to a two-color fluorescent bead and screened with a new generation of small flow cytometry devices. The process is amenable to automation, allowing for high-confidence, multi-locus detection of several threat agents (including plasmid and chromosomal targets) simultaneously (50–52). Multiplex PCR can also be pursued using primers conjugated via ultraviolet-cleavable linkers to tags that vary in mass. In this method, MassTag PCR, tags are cleaved from amplification products and detected by mass spectrometry. Panels are established for simultaneous detection of up to 30 different pathogens (53–57). The Ibis T5000™ biosensor system uses matrix-assisted laser desorption/ionization mass spectrometry to directly measure the molecular weights of PCR products obtained in an experimental sample and to compare them with a database of known or predicted product weights (58–61).

### Microarrays

Polymerase chain reaction methods have expanded significantly over the past two decades and have been established as frontline detection and characterization tools for microbiology. With great flexibility in terms of throughput, automation, and cost, PCR remains the tool of choice to screen for and confirm the presence of bacterial and viral pathogens. However, more rigorous characterization of microbes beyond evaluation of presence/absence of specific genetic loci requires a higher density of genetic information. Depending on the application, microarrays can comprise up to millions of probes that vary in length from 20 to 100+ nucleotides. Shorter probes can discriminate differences in sequence that allow speciation, detection of resistance, or virulence determinants and may have utility for forensic applications where single-nucleotide polymorphism insertions/deletions and other short unique genetic elements are important for fine discrimination. Although true discovery is not feasible (by definition as a novel agent cannot be resolved by similarity in its nucleotide sequence to that of a known agent), longer probes and variable hybridization conditions can enable detection of related but not identical targets that would be missed by shorter probes requiring an exact or near-exact match. In context of a high-density platform, this property can be exploited to survey thousands of agents across the tree of life. Preparation of sample for hybridization is key to array performance. Where the goal is fine discrimination of a limited number of targets using shorter probes (e.g., respiratory virus resequencing arrays), most platforms specify amplification by multiplex PCR followed by hybridization of the resulting products (62–65). Where the objective is broader surveillance, random amplification is required. The shift from multiplex PCR to random amplification has implications for sensitivity. Whereas multiplex PCR has a threshold of $10^2$–$10^3$ copies, random amplification typically misses template present in concentrations $<10^6$.

The GreeneChip (66) and the Virochip (67) differ in design, yet both employ random amplification strategies to allow unbiased detection of microbial targets. This is critical to exploiting the broad probe repertoire of these arrays; however, as host and microbe sequences are amplified with similar efficiency, the sensitivity for microbial detection in tissues can be problematic. Host DNA can be eliminated by enzymatic digestion; however, host ribosomal RNA remains a major confound. Thus, these platforms have been most successful with acellular template sources, such as virus cell culture supernatant, serum, plasma, cerebrospinal fluid, or urine. Although methods are described for depleting host ribosomal RNA (rRNA) prior to amplification through subtraction or use of random primers selected for lack of complementarity to rRNA (68), neither strategy has yet brought these platforms into the range achieved by consensus PCR. At present, hybridization to probes representing pathogen targets is detected by binding of fluorescent label; however, platforms are in development that will detect hybridization as changes in electrical conductance. These may enhance both ease of use and sensitivity. Elution and characterization of sequences bound to the ViroChip were helpful during the 2003 SARS outbreak (65). During a Marburg virus outbreak, the GreeneChip, a panmicrobial array, implicated *Plasmodium falciparum* in a fatal case of hemorrhagic fever that was not resolved using standard diagnostic methods (66);

a variant of the GreeneChip recently facilitated discovery of *Ebola Reston* in a porcine respiratory illness outbreak in the Philippines (62).

## High-Throughput Sequencing

High-throughput sequencing is transforming medicine and microbiology. With continued evolution of sequencing technologies, the cost for sequencing a human genome is projected to be less than $1000 by 2015. When this barrier is broken it is anticipated that whole human genome sequencing will become commonplace, facilitating a new era in personalized medicine wherein genetic analyses will become integral to understanding vulnerability to infectious diseases. Similarly, as per base costs for sequencing decrease, we anticipate a coordinate increase in efforts to sequence not only bacteria, viruses, fungi, and parasites, but also genomes of vectors that transmit them, such as bats, rodents, birds, and phlebotomus insects. Improvements in sequencing speed and the availability of databases for comparative analyses afford new opportunities to identify motifs associated with virulence and drug resistance as well as genetic fingerprints that can be used to determine the provenance of isolates for forensic applications. Another important application of high-throughput sequencing is metagenomic characterization of environmental and clinical samples. Unlike PCR or array methods wherein investigators are limited by known sequence information and must choose the pathogens to be considered in an experiment, high-throughput sequencing can be unbiased and allow an opportunity to inventory the entire tree of life.

Many high-throughput sequencing systems are in use and in development. Applications and principles for sample preparation and data analysis are similar across platforms. Sample preparation can be designed to focus on specific taxa either by treatment to eliminate nucleic acid not protected by a viral capsid or by PCR amplification that enriches for specific sequences such as 16S rRNA (e.g., analyses of gastrointestinal or skin flora) or those associated with pathogens of interest (e.g., influenza or dengue virus). As in microarray applications based on unbiased PCR amplification strategies, host nucleic acid can be a critical impediment to sensitivity. The same caveats and potential solutions also apply. After amplification and sequencing, raw sequence reads are clustered into nonredundant sequence sets. Unique sequence reads are assembled into contiguous sequences, which are then compared to databases using programs that examine homology at the nucleotide and amino acid levels using all six potential reading frames (69). To detect truly new microbes that elude alignment analyses, we have begun to use methods that examine motifs, codon bias, and nucleotide order and usage frequency. While insufficient to positively identify a sequence as microbial, these methods can facilitate genome assembly and focus additional iterative sequencing efforts that can do so.

Several instruments are coming to market that promise to deliver small-footprint, inexpensive high-throughput sequencing. As throughput increases and automation improves, the major challenge to wider application of these new sequencing platforms will be bioinformatics, that is, sequence assembly and analysis. In 2008 the ratio at the Center for Infection and Immunity of personnel devoted to sequencing versus bioinformatics was 10:1; the ratio in 2010 is 1:1.

## CONCLUSION

Over the past decade, the explosion of genetic sequence data and technologies such as PCR (and all of its variants), microarrays, high-throughput "next-gen" sequencing, and bioinformatics have yielded highly specific and sensitive and exacting techniques for detecting and characterizing pathogens. The utility for these methods is broadly applicable to public health, biosurveillance, and obviously forensics and attribution. The speed of data acquisition has increased exponentially, now challenging our ability to store, much less make sense of the implications of, the information being generated. However, our capabilities to detect, track, and assign sources to microbial pathogens are now making their way into routine epidemiology and forensic cases (70). While the power of molecular methods cannot be overstated, other old and emerging techniques will continue to play major and irreplaceable roles in forensic analysis. These include traditional microbiology, medical epidemiology, materials science, isotope ratio determination, and traditional law enforcement investigation. Molecular methods themselves will continue to be refined over time, as the following important capability gaps remain:

- Viral nucleic acid recovery and amplification from host-rich tissue samples
- Metagenomic bioinformatic analysis of complex sample types (soil, sputum, feces, etc.)
- Virulence/countermeasure resistance prediction based on genetic profile
- More complete geographical, temporal, and phylogenetic genomic data set of pathogens and their genetic near neighbors
- Discriminating natural from intentional outbreaks of infectious disease
- Detecting deliberate selection events without overt genetic modification

## ACKNOWLEDGMENTS

The authors thank the editors for the opportunity to contribute to the revised edition of this publication. We also thank Kyle Hubbard and Katrina Ciraldo for assisting with literature research and formatting of the text.

Work in the Center for Infection and Immunity is supported by the National Institutes of Health (grants AI062705, AI56118, AI51202, and AI57158 [Northeast Biodefense Center – Lipkin]), and the Transformational Medical Technologies program at the Department of Defense (grant LSI-03-514).

# REFERENCES

[1] R. Biagini, D. Sammons, J. Smith, E. Page, J. Snawder, C. Striley, et al., Determination of serum IgG antibodies to *Bacillus anthracis* protective antigen in environmental sampling workers using a fluorescent covalent microsphere immunoassay, Occup. Environ. Med. 61 (2004) 703–708.

[2] M. Traeger, S. Wiersma, N. Rosenstein, J. Malecki, C. Shepard, P. Raghunathan, et al., First case of bioterrorism-related inhalational anthrax in the United States, Palm Beach County, Florida, 2001, Emerg. Infect. Dis. 8 (2002) 1029–1034.

[3] P. Brachman, Inhalation anthrax, Ann. N.Y. Acad. Sci. 353 (1980) 83–93.

[4] C. Dahlgren, L. Buchanan, H. Decker, S. Freed, C. Phillips, P. Brachman, *Bacillus anthracis* aerosols in goat hair processing mills, Am. J. Hyg. 72 (1960) 24–31.

[5] M. Meselson, J. Guillemin, M. Hugh-Jones, A. Langmuir, I. Popova, A. Shelokov, et al., The Sverdlovsk anthrax outbreak of 1979, Science 266 (1994) 1202–1208.

[6] S. Plotki, P. Brachman, M. Utell, F. Bumford, M. Atchison, An epidemic of inhalation anthrax, the first in the twentieth century. I. Clinical features, Am. J. Med. 29 (1960) 992–1001.

[7] J. Maillard, M. Fischer, K.J. McKee, L. Turner, J. Cline, First case of bioterrorism-related inhalational anthrax, Florida, 2001: North Carolina investigation, Emerg. Infect. Dis. 8 (2002) 1035–1038.

[8] J. Jernigan, D. Stephens, D. Ashford, C. Omenaca, M. Topiel, M. Galbraith, et al., Bioterrorism-related inhalational anthrax: The first 10 cases reported in the United States, Emerg. Infect. Dis. 7 (2001) 933–944.

[9] S. Suffin, W. Carnes, A. Kaufmann, Inhalation anthrax in a home craftsman, Hum. Pathol. 9 (1978) 594–597.

[10] Centers for Disease Control and Prevention (CDC), Update: Investigation of anthrax associated with intentional exposure and interim public health guidelines, October 2001, MMWR Morb. Mortal. Wkly. Rep. 50 (2001) 889–893.

[11] C. Latkin, Seeing red: A social-psychological analysis of the Rajneeshpuram conflict, Sociol. Anal. 53 (1992) 257–271.

[12] T. Török, R. Tauxe, R. Wise, J. Livengood, R. Sokolow, S. Mauvais, et al., A large community outbreak of salmonellosis caused by intentional contamination of restaurant salad bars, JAMA 278 (1997) 389–395.

[13] Centers for Disease Control and Prevention (CDC), Outbreak of Salmonella serotype Saintpaul infections associated with multiple raw produce items—United States, 2008, MMWR Morb. Mortal. Wkly. Rep. 57 (2008) 929–934.

[14] R.O. Weaver, The Town That Was Poisoned, vol. 131 (1985) pp.4185–4189. Congressional Record (Procedures and Debates).

[15] J. Miller, S. Engelberg, W. Broad, The attack. "Germs: Biological Weapons and America's Secret War," Simon & Schuster, New York, 2001 pp. 1–34.

[16] K. Christensen, R. Holman, C. Steiner, J. Sejvar, B. Stoll, L. Schonberger, Infectious disease hospitalizations in the United States, Clin. Infect. Dis. 49 (2009) 1025–1035.

[17] J. Pile, J. Malone, E. Eitzen, A. Friedlander, Anthrax as a potential biological warfare agent, Arch. Intern. Med. 158 (1998) 429–434.

[18] S. Shafazand, R. Doyle, S. Ruoss, A. Weinacker, T. Raffin, Inhalational anthrax: Epidemiology, diagnosis, and management, Chest 116 (1999) 1369–1376.

[19] L. Bush, B. Abrams, A. Beall, C. Johnson, Index case of fatal inhalational anthrax due to bioterrorism in the United States, N. Engl. J. Med. 345 (2001) 1607–1610.

[20] P.J. Murray, E.J. Baron, J.H. Jorgensen, M.L. Landry, M.A. Pfaller (Eds.), Manual of Clinical Microbiology, ninth ed., ASM Press, 2007.

[21] M. Syrmis, D. Whiley, M. Thomas, I. Mackay, J. Williamson, D. Siebert, et al., A sensitive, specific, and cost-effective multiplex reverse transcriptase-PCR assay for the detection of seven common respiratory viruses in respiratory samples, J. Mol. Diagn. 6 (2004) 125–131.

[22] K. Templeton, S. Scheltinga, M. Beersma, A. Kroes, E. Claas, Rapid and sensitive method using multiplex real-time PCR for diagnosis of infections by influenza a and influenza B viruses, respiratory syncytial virus, and parainfluenza viruses 1, 2, 3, and 4, J. Clin. Microbiol. 42 (2004) 1564–1569.

[23] D.B. Jernigan, P.L. Raghunathan, B.P. Bell, R. Brechner, E.A. Bresnitz, J.C. Butler, et al., National Anthrax Epidemiologic Investigation Team. Investigation of bioterrorism-related anthrax, United States, 2001: Epidemiologic findings, Emerg. Infect. Dis. 8 (2002) 1019–1028.

[24] O. Fujita, M. Tatsumi, K. Tanabayashi, A. Yamada, Development of a real-time PCR assay for detection and quantification of *Francisella tularensis*, Jpn. J. Infect. Dis. 59 (2006) 46–51.

[25] S. Klee, H. Nattermann, S. Becker, M. Urban-Schriefer, T. Franz, D. Jacob, et al., Evaluation of different methods to discriminate *Bacillus anthracis* from other bacteria of the *Bacillus cereus* group, J. Appl. Microbiol. 100 (2006) 673–681.

[26] D. Lim, J. Simpson, E. Kearns, M. Kramer, Current and developing technologies for monitoring agents of bioterrorism and biowarfare, Clin. Microbiol. Rev. 18 (2005) 583–607.

[27] M. Moser, D. Christensen, D. Norwood, J. Prudent, Multiplexed detection of anthrax-related toxin genes, J. Mol. Diagn. 8 (2006) 89–96.

[28] D. Shea, S. Lister, The BioWatch Program: Detection of Bioterrorism, Congressional Research Service Report No. RL 32152, published 2003.

[29] T. Slezak, T. Kuczmarski, L. Ott, C. Torres, D. Medeiros, J. Smith, et al., Comparative genomics tools applied to bioterrorism defense, Brief Bioinform. 4 (2003) 133–149.

[30] D. Jensen, A. Welling, A brief anthrax scare, Deseret News, Deseret News Publishing Company, Salt Lake City, 2002.

[31] J. Herold, S. Kurtz, R. Giegerich, Efficient computation of absent words in genomic sequences, BMC Bioinform. 9 (2008) 167.

[32] J. Fitch, E. Raber, D. Imbro, Technology challenges in responding to biological or chemical attacks in the civilian sector, Science 302 (2003) 1350–1354.

[33] S. Bustin, Absolute quantification of mRNA using real-time reverse transcription polymerase chain reaction assays, J. Mol. Endocrinol. 25 (2000) 169–193.

[34] F. Filén, A. Strand, A. Allard, J. Blomberg, B. Herrmann, Duplex real-time polymerase chain reaction assay for detection and quantification of herpes simplex virus type 1 and herpes simplex virus type 2 in genital and cutaneous lesions, Sex Transm. Dis. 31 (2004) 331–336.

[35] M. Hayden, T. Nguyen, A. Waterman, K. Chalmers, P.C.R. Multiplex-ready, a new method for multiplexed SSR and SNP genotyping, BMC Genomics 9 (2008) 80.

[36] A. Hein, U. Bodendorf, P.C.R. Real-time, duplexing without optimization, Anal. Biochem. 360 (2007) 41–46.

[37] O. Henegariu, N. Heerema, S. Dlouhy, G. Vance, P. Vogt, Multiplex PCR: Critical parameters and step-by-step protocol, Biotechniques 23 (1997) 504–511.

[38] E. Lawrence, D. Griffiths, S. Martin, R. George, L. Hall, Evaluation of semiautomated multiplex PCR assay for determination of *Streptococcus pneumoniae* serotypes and serogroups, J. Clin. Microbiol. 41 (2003) 601–607.

[39] J. Mahony, G. Blackhouse, J. Babwah, M. Smieja, S. Buracond, S. Chong, et al., Cost analysis of multiplex PCR testing for diagnosing respiratory virus infections, J. Clin. Microbiol. 47 (2009) 2812–2817.

[40] A. Petrich, K. Luinstra, B. Page, S. Callery, D. Stevens, A. Gafni, et al., Effect of routine use of a multiplex PCR for detection of vanA- and vanB-mediated enterococcal resistance on accuracy, costs and earlier reporting, Diagn. Microbiol. Infect. Dis. 41 (2001) 215–220.

[41] J. Sachithanandham, M. Ramamurthy, R. Kannangai, H.D. Daniel, O.C. Abraham, P. Rupali, et al., Detection of opportunistic DNA viral infections by multiplex PCR among HIV infected individuals receiving care at a tertiary care hospital in South India, Indian J. Med. Microbiol. 27 (2009) 210–216.

[42] J. Sun, L. Gaidulis, M. Miller, R. Goto, R. Rodriguez, S. Forman, et al., Development of a multiplex PCR-SSP method for killer-cell immunoglobulin-like receptor genotyping, Tissue Antigens 64 (2004) 462–468.

[43] S. Yuli, L. Chengxu, M.F. Sydney. Multiplex PCR for rapid differentiation of three species in the "Clostridium clostridioforme group." FEMS Microbiol. Lett. 244 (2005) 391–395.

[44] G. Zhang, Y. Guan, H. Sheng, B. Zheng, S. Wu, H. Xiao, et al., Multiplex PCR and oligonucleotide microarray for detection of single-nucleotide polymorphisms associated with *Plasmodium falciparum* drug resistance, J. Clin. Microbiol. 46 (2008) 2167–2174.

[45] P. Markoulatos, N. Siafakas, M. Moncany, Multiplex polymerase chain reaction: A practical approach, J. Clin. Lab. Anal. 16 (2002) 47–51.

[46] E. Elnifro, A. Ashshi, R. Cooper, P. Klapper, Multiplex PCR: Optimization and application in diagnostic virology, Clin. Microbiol. Rev. 13 (2000) 559–570.

[47] K. Jinneman, K. Yoshitomi, S. Weagant, Multiplex real-time PCR method to identify Shiga toxin genes stx1 and stx2 and *Escherichia coli* O157:H7/H- serotype, Appl. Environ. Microbiol. 69 (2003) 6327–6333.

[48] B. Gröndahl, W. Puppe, A. Hoppe, I. Kühne, J. Weigl, H. Schmitt, Rapid identification of nine microorganisms causing acute respiratory tract infections by single-tube multiplex reverse transcription-PCR: Feasibility study, J. Clin. Microbiol. 37 (1999) 1–7.

[49] D. Klein, Quantification using real-time PCR technology: Applications and limitations, Trends Mol. Med. 8 (2002) 257–260.

[50] B.J. Hindson, S.M. Reid, B.R. Baker, K. Ebert, N.P. Ferris, L.F. Tammero, et al., Diagnostic evaluation of multiplexed reverse transcription-PCR microsphere array assay for detection of foot-and-mouth and look-alike disease viruses, J. Clin. Microbiol. 46 (2008) 1081–1089.

[51] W. Wilson, A. Erler, S. Nasarabadi, E. Skowronski, P. Imbro, A multiplexed PCR-coupled liquid bead array for the simultaneous detection of four biothreat agents, Mol. Cell Probes 19 (2005) 137–144.

[52] W.C. Wilson, B.J. Hindson, E.S. O'Hearn, S. Hall, C. Tellgren-Roth, C. Torres, et al., A multiplex real-time reverse transcription polymerase chain reaction assay for detection and differentiation of Bluetongue virus and epizootic hemorrhagic disease virus serogroups, J. Vet. Diagn. Invest 21 (2009) 760–770.

[53] T. Briese, G. Palacios, M. Kokoris, O. Jabado, Z. Liu, N. Renwick, et al., Diagnostic system for rapid and sensitive differential detection of pathogens, Emerg. Infect. Dis. 11 (2005) 310–313.

[54] S.R. Dominguez, T. Briese, G. Palacios, J. Hui, J. Villari, V. Kapoor, et al., Multiplex MassTag-PCR for respiratory pathogens in pediatric nasopharyngeal washes negative by conventional

diagnostic testing shows a high prevalence of viruses belonging to a newly recognized rhinovirus clade, J. Clin. Virol. 43 (2008) 219–222.

[55] D. Lamson, N. Renwick, V. Kapoor, Z. Liu, G. Palacios, J. Ju, et al., MassTag polymerase-chain-reaction detection of respiratory pathogens, including a new rhinovirus genotype, that caused influenza-like illness in New York State during 2004–2005, J. Infect. Dis. 194 (2006) 1398–1402.

[56] G. Palacios, T. Briese, V. Kapoor, O. Jabado, Z. Liu, M. Venter, et al., MassTag polymerase chain reaction for differential diagnosis of viral hemorrhagic fever, Emerg. Infect. Dis. 12 (2006) 692–695.

[57] R. Tokarz, V. Kapoor, J. Samuel, D. Bouyer, T. Briese, W. Lipkin, Detection of tick-borne pathogens by MassTag polymerase chain reaction, Vector Borne Zoonotic Dis. 9 (2009) 147–152.

[58] D. Ecker, R. Sampath, C. Massire, L. Blyn, T. Hall, M. Eshoo, et al., Ibis T5000: A universal biosensor approach for microbiology, Nat. Rev. Microbiol. 6 (2008) 553–558.

[59] S. Hofstadler, K. Sannes-Lowery, J. Hannis, Analysis of nucleic acids by FTICR MS, Mass Spectrom. Rev. 24 (2005) 265–285.

[60] R. Sampath, T. Hall, C. Massire, F. Li, L. Blyn, M. Eshoo, et al., Rapid identification of emerging infectious agents using PCR and electrospray ionization mass spectrometry, Ann. N.Y. Acad. Sci. 1102 (2007) 109–120.

[61] M. Van Ert, S. Hofstadler, Y. Jiang, J. Busch, D. Wagner, J. Drader, et al., Mass spectrometry provides accurate characterization of two genetic marker types in Bacillus anthracis, Biotechniques 37 (2004) 642–644, 646, 648 passim.

[62] R.W. Barrette, S.A. Metwally, J.M. Rowland, L. Xu, S.R. Zaki, S.T. Nichol, et al., Discovery of swine as a host for the Reston ebolavirus, Science 325 (2009) 204–206.

[63] C.Y. Chiu, A.A. Alizadeh, S. Rouskin, J.D. Merker, E. Yeh, S. Yagi, et al., Diagnosis of a critical respiratory illness caused by human metapneumovirus by use of a pan-virus microarray, J. Clin. Microbiol. 45 (2007) 2340–2343.

[64] C.Y. Chiu, S. Rouskin, A. Koshy, A. Urisman, K. Fischer, S. Yagi, et al., Microarray detection of human parainfluenzavirus 4 infection associated with respiratory failure in an immunocompetent adult, Clin. Infect. Dis. 43 (2006) e71–e76.

[65] T.G. Ksiazek, D. Erdman, C.S. Goldsmith, S.R. Zaki, T. Peret, S. Emery, et al., SARS Working Group. A novel coronavirus associated with severe acute respiratory syndrome, N. Engl. J. Med. 348 (2003) 1953–1966.

[66] G. Palacios, P.L. Quan, O.J. Jabado, S. Conlan, D.L. Hirschberg, Y. Liu, et al., Panmicrobial oligonucleotide array for diagnosis of infectious diseases, Emerg. Infect. Dis. 13 (2007) 73–81.

[67] D. Wang, L. Coscoy, M. Zylberberg, P. Avila, H. Boushey, D. Ganem, et al., Microarray-based detection and genotyping of viral pathogens, Proc. Natl. Acad. Sci. USA 99 (2002) 15687–15692.

[68] C.D. Armour, J.C. Castle, R. Chen, T. Babak, P. Loerch, S. Jackson, et al., Digital transcriptome profiling using selective hexamer priming for cDNA synthesis, Nat. Methods 6 (2009) 647–649.

[69] G. Palacios, J. Druce, L. Du, T. Tran, C. Birch, T. Briese, et al., A new arenavirus in a cluster of fatal transplant-associated diseases, N. Engl. J. Med. 358 (2008) 991–998.

[70] Amerithrax investigative summary. U.S. Department of Justice (Ed.), 2010.

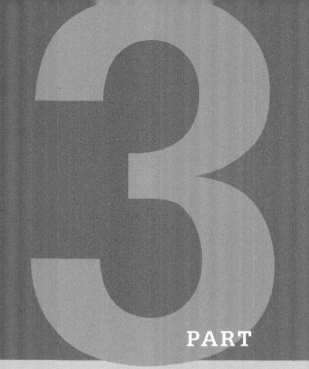

PART

3

Biosecurity

# CHAPTER 12

# Assessment of the Threat

**Jenifer A.L. Smith**

*Bioforensics Consulting, LLC, Edgewater, Maryland*

The term "threat assessment" can be broadly interpreted as evaluation of impending danger or harm by a person, group, circumstance, or set of conditions. Threat assessments can be directed against a very broad issue(s) or can be more specifically focused. This chapter examines threat assessments developed concerning the use of biological organisms as instruments to cause harm, disruption, or fear by hostile states, terrorists, or criminals. Excerpts from unclassified versions of official assessments, reports, or speeches best convey the tenor of these documents and are provided in this chapter.

## NATIONAL THREAT ASSESSMENT: ROLE OF THE INTELLIGENCE COMMUNITY

The Director of National Intelligence (DNI) serves as the head of the Intelligence Community (IC), overseeing and directing the implementation of the National Intelligence Program and acting as the principal advisor to the President, the National Security Council, and the Homeland Security Council on intelligence matters. The Office of the Director of National Intelligence is charged with:

- Integrating the domestic and foreign dimensions of U.S. intelligence so that there are no gaps in our understanding of threats to our national security
- Bringing more depth and accuracy to intelligence analysis
- Ensuring that U.S. intelligence resources generate future capabilities as well as present results (1).

The National Intelligence Council (NIC) reports to the DNI and is the center for midterm and long-term strategic thinking. Its primary functions are to:

- Support the DNI in his/her role as head of the intelligence community.
- Provide a focal point for policymakers to task the IC to answer their questions.
- Reach out to nongovernment experts in academia and the private sector to broaden the perspective of the IC.

189

Microbial Forensics. DOI: 10.1016/B978-0-12-382006-8.00012-8

- Contribute to the IC's effort to allocate its resources in response to policy makers' changing needs.
- Lead the IC's effort to produce National Intelligence Estimates (NIEs) and other NIC products (2).

National Intelligence Estimates are the DNI's most authoritative written judgments concerning national security issues. They contain the coordinated judgments of the IC regarding the probable course of future events. The goal of the NIC is to provide policy makers with the best, unvarnished, and unbiased information, regardless of whether analytic judgments conform to U.S. policy. NIEs often contain classified information that cannot be discussed openly. On occasion, unclassified versions of NIEs are released for public scrutiny. These provide valuable insights to the often closed world of intelligence and national security. There have been NIEs released that help assess the threat picture concerning the counterproliferation and/or use of weapons of mass destruction. The most recent unclassified version of an NIE that addressed biological weapons was released in 2007 and was entitled "The Terrorist Threat to the US Homeland" (3).

- We assess that al-Qa'ida's Homeland plotting is likely to continue to focus on prominent political, economic, and infrastructure targets with the goal of producing mass casualties, visually dramatic destruction, significant economic aftershocks, and/or fear among the U.S. population. The group is proficient with conventional small arms and improvised explosive devices, and is innovative in creating new capabilities and overcoming security obstacles.
- We assess that al-Qa'ida will continue to try to acquire and employ chemical, biological, radiological, or nuclear material in attacks and would not hesitate to use them if it develops what it deems is sufficient capability.

Other groups are often tasked with providing national-level threat assessments. In December 2008 in accordance with the Implementing Recommendations of the 9/11 Commission Act of 2007 (P.L. 110-53), the Commission on the Prevention of Weapons of Mass Destruction Proliferation and Terrorism submitted its report, "World at Risk" (4). That report assessed the nation's activities, initiatives, and programs to prevent weapons of mass destruction proliferation and terrorism and provided concrete recommendations to address these threats. The report provided the following threat assessment that was unanimously expressed:

> Unless the world community acts decisively and with great urgency,
> it is more likely than not that a weapon of mass destruction (WMD)
> will be used in a terrorist attack somewhere in the world by the end of
> 2013. That weapon is more likely to be biological than nuclear.

They further stated within the report:

> Biological science and technology today transcend borders. These fields
> engage a vast and expanding array of actors in the government, private,
> and commercial sectors, and they are advancing at a remarkable pace.
> The more that sophisticated capabilities, including genetic engineering
> and gene synthesis, spread around the globe, the greater the potential
> that terrorists will use them to develop biological weapons.

Less than a month after this assessment, the DNI publicly endorsed it.

Their assessment was based on four factors.

- Direct evidence that terrorists are trying to acquire weapons of mass
  destruction.
- Acquiring WMD fits the tactical profile of terrorists. They understand the
  unique vulnerability of first-world countries to asymmetrics, weapons that
  have a far greater destructive impact than the power it takes to acquire and
  deploy them.
- Terrorists have demonstrated global reach and the organizational
  sophistication to obtain and use WMD.
- The opportunity to acquire and use such weapons is growing exponentially
  because of the global proliferation of nuclear material and biological
  technologies.

The most current insight into the IC's threat assessment concerning biological
weapons can be drawn from the unclassified comments made by former DNI
Dennis C. Blair as he provided his annual threat assessment of the IC for the
Senate Select Committee on Intelligence on February 12, 2009 (5). The judgments offered to the committee were based on intelligence collected by several
agencies. The unclassified version is available and sheds light on the current
assessment of the terrorist CBRN threat.

> Over the coming years, we will continue to face a substantial threat,
> including in the US Homeland, from terrorists attempting to acquire
> biological, chemical, and possibly nuclear weapons and use them to
> conduct large-scale attacks. Conventional weapons and explosives
> will continue to be the most often used instruments of destruction in
> terrorist attacks; however, terrorists who are determined to develop
> CBRN capabilities will have increasing opportunities to do so, owing to
> the spread of relevant technological knowledge and the ability to work
> with CBRN materials and designs in safe havens.
>
> Most terrorist groups that have shown some interest, intent, or
> capability to conduct CBRN attacks have pursued only limited,

technically simple approaches that have not yet caused large numbers of casualties.

In particular, we assess the terrorist use of biological agents represents a growing threat as the barriers to obtaining many suitable starter cultures are eroding and open source technical literature and basic laboratory equipment can facilitate production. Terrorist chemical attacks ... Al-Qa'ida is the terrorist group that historically has sought the broadest range of CBRN attack capabilities, and we assess that it would use any CBRN capability it acquires in an anti-US attack, preferably against the Homeland. There also is a threat of biological or chemical attacks in the US Homeland by lone individuals.

The previously mentioned threat assessments of the national security community impact policy decisions directly. An illustration of this point is the National Strategy for Countering Biological Threats released by the National Security Council (NSC) in December 2009. This policy was designed to reduce the risks presented by the deliberate or accidental release of a biological agent (6). It describes how the U.S. government will address the challenges from the proliferation of biological weapons or their use by terrorists. It highlights the beneficial nature of advances in the life sciences and their importance in combating infectious diseases of natural, accidental, and deliberate origin, protecting the environment, expanding energy options, and enhancing agricultural production. It also outlines how risks associated with misuse and potential consequences of a biological attack require tailored actions to prevent biological threats. The strategy emphasizes the need to:

- Improve global access to the life sciences to combat infectious disease regardless of its cause
- Establish and reinforce norms against the misuse of the life sciences
- Institute a suite of coordinated activities that collectively will help influence, identify, inhibit, and/or interdict those who seek to misuse the life sciences

## THREAT CREDIBILITY ASSESSMENTS: ROLE OF LAW ENFORCEMENT AND PUBLIC HEALTH OFFICIALS

The intricate authorities, responsibilities, and actions of various agencies associated with response to a disease outbreak of known or unknown origin requiring federal assistance are addressed in the Biological Incident Annex of the U.S. National Response Plan. The scope of this annex covers the U.S. government's response to a biological terrorism event, pandemic influenza, emerging infectious disease, or novel pathogen outbreak (7). According to the annex, public health, law enforcement, and homeland security officials all share response and investigative responsibilities in either an accidental or an intentional release of a hazardous biological organism. The role of public health officials is to conduct epidemiological investigations that may be triggered by report of an outbreak or by normal surveillance systems. Their goals are clear as they are to protect the public, stop the spread of the disease, and protect public health personnel. The scope of the law and responsibility of the Federal Bureau of Investigation (FBI) has been summarized by Budowle and colleagues (7):

> "Any actual or threatened use of a disease-causing microorganism or biological material (such as a toxin) directed at humans, animals, plants, or material is regarded as a crime. The possession of a biological agent, toxin, or delivery system that cannot be justified by a prophylactic, protective *bona fide* research effort or other peaceful purpose can result in arrest, prosecution, fines or imprisonment. Moreover, it does not matter whether the perpetrator actually possesses a bioagent, the intention to obtain and use a bioagent is sufficient for arrest and prosecution. Thus, interdiction and prosecution can occur even for those attempting to develop a weapon or for those who perpetrate hoaxes"(8).

Prior to the 2001 anthrax attack, public health and law enforcement officials would likely have conducted separate investigations of suspected biological attacks. Since that event, these communities have worked in closer collaboration and have developed training methods, materials, and operational protocols to foster a greater understanding among law enforcement and public health personnel in an effort to minimize potential barriers to communication and information sharing during an actual biological event (9,10).

A bioterrorism attack may be conducted as either an "overt" (i.e., an announced attack) or a "covert" (a surreptitious release of a bioagent) operation. In an overt attack, law enforcement typically first detects the event, leads the initial response, and notifies public health officials. The FBI has established a process called the "Threat Credibility Assessment" that is initiated any time an event occurs that appears to be a potential attack using a chemical, biological, or nuclear weapon. Thus, if "any actual or threatened use of a disease-causing microorganism or biological material such as a toxin directed at humans, animals, plant, or material" appears to have occurred, the FBI will trigger the Threat Credibility Assessment. The assessment will determine whether the threat is technically feasible and operationally practical. Such assessments are done in coordination and consultation with other government experts from agencies such as the Centers for Disease Control and Prevention (CDC), U.S. Department of Agriculture, Department of Homeland Security (DHS), and

Food and Drug Administration, as well as subject matter experts from academia and industry. This threat assessment is necessary to determine whether circumstances may be the result of an intentional or criminal act, warranting law enforcement involvement. If persons are ill or preventative health services are indicated, public health will also become involved in the emergency response.

Initial consequences of a covert attack would be indistinguishable from those resulting from a natural outbreak or accidental release—the presence of ill individuals would be the first sign in such instances. Even though the covert attack would not be recognized immediately as an "attack," public health officials would still recognize the problem and initiate appropriate measures to diagnose the causative agent, provide medical care, and determine the source and extent of the outbreak. If the relevant facts raise suspicion that the outbreak or event is unusual or not from a natural event, law enforcement will be notified and the threat credibility assessment will be initiated. If the event appears to be intentional, then a joint investigative approach will be pursued.

Since the 2001 anthrax attack, there have been thousands of instances of suspicious letters or containers with potential biological threats, but the vast majority of threat credibility assessments conducted have determined that the alleged events were not bioterrorist attacks. In 2009, the FBI reported that they looked into more than 900 biological incidents from January 2007 to August 2008, the majority of which were "white powder" hoax events (11). To assist federal, state, and local law enforcement and emergency response agencies, the FBI, DHS, and CDC issued guidance concerning appropriate actions to be taken with suspicious letters or containers with a potential biological threat. The guidance detailed procedures concerning agency notification, threat credibility assessment process, and handling of suspicious items.

Some examples of cases have resulted in criminal prosecution. On February 14, 2008, Roger Von Bergendorff, who was living in an extended stay hotel in Las Vegas, Nevada, contacted emergency medical personnel because he was having trouble breathing. He was taken to a local hospital, where he was declared to be critically ill and placed on life support. Twelve days later, hotel personnel conducted an inventory of Mr. Bergendorff's property, where they discovered several weapons. They notified the Las Vegas Metropolitan Police Department, who confiscated several weapons and a silencer. The police also found castor beans, partially purified ricin, syringes, and beakers and a copy of "The Anarchist's Cookbook," which is a collection of instructions on poisons and instructions on the preparation of ricin. This discovery triggered notification to the FBI, who initiated a Threat Credibility Assessment. A team from the FBI's laboratory was sent to conduct searches of the hotel and other locations occupied previously by Mr. Bergendorff. The FBI collected castor beans, various chemicals used in the production of ricin, a respirator, filters, painter's mask, laboratory glassware, syringes, and a notebook on ricin production during a search of Salt Lake City storage units rented by Bergendorff. On March 7, 2008, tests conducted by the National BioForensic Analysis Center determined that the material recovered from Bergendorff's hotel room in Las Vegas contained 2.9% active ricin. The preparation was characterized as "crude." During the investigation, the FBI ultimately ruled out domestic terrorism as a motive for the ricin and found no evidence to indicate any intent to target any individual or individuals with the substance. On April 16, 2008, Bergendorff was discharged from the hospital in Las Vegas and was subsequently arrested by the FBI. On August 4, 2008, he pleaded guilty before U.S. District Court to one count of possession of biological toxin and one count of possession of unregistered firearms. Three months later, he was sentenced to 3 years in prison, 3 years of supervised release, and ordered to pay a $7500 fine (12,13).

Individuals who develop inhalation or cutaneous anthrax often trigger a joint public health and law enforcement response. The following case highlights the successful joint epidemiologic and environmental investigation conducted by local, state, and federal public health, animal health, and law enforcement authorities in Pennsylvania and New York City to determine the source of exposure involving a person who developed inhalation anthrax (14). On February 16, a musician had traveled from New York City to northern Pennsylvania for a performance with his dance troupe. While performing that evening, he collapsed and was admitted to a local hospital, where he reported that he had been experiencing shortness of breath, dry cough, and malaise for a few days. The next day, his condition worsened. Blood samples were provided to the Pennsylvania Department of Health laboratory, who confirmed on February 21 the presence of *Bacillus anthracis*, the causative agent of inhalation anthrax. That day, the Pennsylvania authorities notified the CDC, New York City Department of Health and Mental Hygiene (DOHMH), and local FBI that they had a case of inhalation anthrax. On February 22, the CDC conducted additional testing and confirmed the original diagnosis

Subsequently, a joint epidemiologic and environmental investigation was initiated to (i) determine the source of exposure, (ii) identify other persons who were exposed and required postexposure treatment, (iii) enhance surveillance for additional cases through outreach to the medical community, and (iv) provide frequent updates as soon as available that were consistent and appropriate messages regarding risk to the public. Interviews of the patient, his family, and his colleagues revealed that he made traditional African drums by using hard-dried African goat and cow hides. The process involved soaking hides for 1 hour in water and then scraping hair from the hides with a razor, which reportedly generated a large amount of aerosolized dust in the patient's workspace as the hides dried. He did not wear any personal protective equipment (e.g., mask or gloves) while working. After working on the hides, he usually returned home to his apartment and immediately removed his clothing and showered. He had recently returned from a trip to Côte d'Ivoire with four dried goat hides that he had been working on just before his trip to Pennsylvania. The CDC and DOHMH personnel conducted environmental sampling at the musician's workspace, home, and automobile. Environmental and epidemiologic findings suggested that the patient's primary exposure to aerosolized *B. anthracis* spores resulted from scraping a contaminated hide in his workspace. They also identified less contamination in his home and van, indicative of secondary transfer. This case was assessed and confirmed to be a case of accidental exposure to a biological organism and not a terrorist or criminal act.

## CONCLUSION

Biological weapons in the possession of hostile states or terrorists pose unique and serious threats to the safety and security of the United States. An attack with a biological agent could mimic naturally occurring disease, potentially delaying recognition of an attack and creating uncertainty about whether an intentional event has even occurred. Advances in biotechnology and life sciences present the prospect of new biological elements that would require new detection methods, preventive measures, and treatments. These developments increase the risk for surprise and make predicting future weapon threats more challenging. Anticipating such threats through intelligence efforts is made more difficult by (i) the dual-use nature of biological technologies and infrastructure and (ii) the likelihood that adversaries will use denial and deception to conceal their illicit activities. The ability to conduct appropriate biothreat assessment of whether to provide guidance to policy makers at the national level or to initiate an immediate local public health or law enforcement response has been recognized as one of the essential pillars of the U.S. government's biodefense strategy. Enhancement of threat assessment capabilities remains a priority, as evidenced by the NSC's National Strategy for Countering Biological Threats. One of the objectives of the strategy is to "obtain timely and accurate insight on current and emerging risks" by improving relevant agencies' threat identification, notification, and assessment capabilities (6). Advancements in microbial forensics will serve to enhance threat awareness and assessment capabilities of all agencies involved in this essential element of national biodefense.

## REFERENCES

[1] Office of the Director of National Intelligence. Available from: www.dni.gov/faq_about.htm.
[2] Office of the National Intelligence Council. Available from: www.dni.gov/nic/NIC_about .html.

[3] National Intelligence Estimate: The Terrorist Threat to the US Homeland, 2007. Available from: www.dni.gov/press_releases/20070717_release.pdf.

[4] World at Risk Commission on the Prevention of Weapons of Mass Destruction Proliferation and Terrorism, 2009. Available from: www.preventwmd.gov/report/.

[5] Comments from DNI Blair Annual Threat Assessment of the Intelligence Community for the Senate Select Committee on Intelligence February 12, 2009.

[6] National Strategy for Countering Biological Threats, 2009. Available from: www.whitehouse.gov/sites/default/files/National_Strategy_for_Countering_BioThreats.pdf.

[7] B. Budowle, J.A. Beaudry, N.G. Barnaby, A.M. Giusti, J.D. Bannan, P. Keim, Role of law enforcement response and microbial forensics in investigation of bioterrorism, Croat. Med. J. 48 (4) (2007) 437–449.

[8] J.C. Butler, M.L. Cohen, C.R. Friedman, M. Robert, R.M. Scripp, CG. Watz, Collaboration between public health and law enforcement: New paradigms and partnerships for bioterrorism planning and response, Emerg. Infect. Dis. 8 (10) (2002) 1152–1156.

[9] Criminal and Epidemiological Investigation Handbook, 2006. Available from: http://www2a.cdc.gov/PHLP/docs/CrimEpiHandbook2006.pdf.

[10] Biological Incident Annex of the National Response Plan, 2004. Available from: www.learningservices.us/pdf/emergency/nrf/nrp_biologicalincidentannex.pdf.

[11] M. Hall, White powder scares cost law enforcement time, money, USA Today (2008). Available from: www.usatoday.com/news/nation/2008-10-12-powder_N.htm.

[12] S. Friess, In Accord, ricin owner enters plea of guilty, N.Y. Times (2008) Available from: www.nytimes.co,/2008/08/05/us/05ricin.html.

[13] M. Manning, Man with ricin in Las Vegas hotel room sentenced, Las Vegas Sun (2008). Available from: www.lasvegassun.com/nes/2008/nov17/ran-ricin-las-vegas.

[14] Centers for Disease Control and Prevention (CDC), Inhalation anthrax associated with dried animal hides—Pennsylvania and New York City, MMWR Morb. Mortal. Wkly. Rep. 55 (10) (2006) 280–282.

References

# CHAPTER 13

# Select Agent Regulations

**Stephen A. Morse and Elizabeth Weirich**

*National Center for Emerging and Zoonotic Infectious Diseases,*
*Centers for Disease Control and Prevention, Atlanta, Georgia*

## INTRODUCTION

A number of events over the past 15 years have changed the climate in which scientists acquire and work with pathogenic microorganisms and biological toxins. These changes have occurred not only in the United States, but internationally as well. Events such as the release of the nerve agent sarin in the subway system of Tokyo on March 20, 1995, by the Aum Shinrikyo (1,2), the bombing of the Murrah Federal Building in Oklahoma City on April 19, 1995 (1), and the terrorist attacks on September 11, 2001 (1), have increased our awareness of the threat of terrorism, including bioterrorism, such as the 2001 anthrax attacks. These incidents have resulted in U.S. legislation designed to limit unauthorized access to dangerous pathogenic microorganisms and biological toxins in an effort to control the misuse of these agents. The creation of an oversight role for the federal government in the area of biological sciences has evoked controversy within the scientific community and is considered by some to actually increase societal vulnerability to biological attacks and natural epidemics (3). The purpose of this chapter is to discuss the Select Agent Regulations and other regulations that will impact the practice of microbial forensics.

Events leading up to the promulgation of the Select Agent Regulations began in May 1995 when Larry Wayne Harris, a private citizen with microbiology training and ties to white supremacist groups, fraudulently ordered three vials of lyophilized *Yersinia pestis* from the American Type Culture Collection (ATCC) (4). Mr. Harris faxed a request on falsified letterhead stationery that misled the ATCC to believe that the address was that of a legitimate laboratory. The ATCC subsequently sent this person vials of *Y. pestis*. When the vials did not arrive on time, Mr. Harris called the ATCC to complain. ATCC personnel became suspicious when he indicated the nature of his planned experiments and notified the Centers for Disease Control and Prevention (CDC).

Microbial Forensics. DOI: 10.1016/B978-0-12-382006-8.00013-X

After notification of appropriate law enforcement agencies, three unopened, intact vials were retrieved from the glove compartment of Mr. Harris's car by local law enforcement authorities using a search warrant. Shortly after this incident, a review of federal regulations was initiated by the National Security Council, which subsequently identified several regulations that restricted the possession, transfer, and use of plant and animal pathogens to qualified institutions, laboratories, and scientists. However, similar regulations that would restrict access to pathogens, toxins, and recombinant organisms dangerous to humans were not found (5). A multiagency committee was commissioned by the Secretary of the Department of Health and Human Services (HHS) to develop a regulation to address this issue and close the loophole. The regulation was to be based on the key principle of ensuring protection of the public without encumbering legitimate scientific and medical research. The committee included representatives from the following departments, agencies, and offices: the Federal Bureau of Investigation (FBI); U.S. Department of Justice; U.S. Army; National Institutes of Health (NIH); U.S. Department of Agriculture (USDA); U.S. Department of Commerce; Environmental Protection Agency; Food and Drug Administration (FDA); U.S. Postal Service; HHS Office of Emergency Preparedness; and the CDC. A framework for this regulation was developed and presented on March 5, 1996, during CDC testimony at a hearing convened by the Senate Judiciary Committee to examine concerns arising from the interstate transportation of human pathogens. This framework was subsequently incorporated into the Antiterrorism and Effective Death Penalty Act of 1996 (P.L. 104-132), which was signed into law on April 24, 1996. Section 511 of this law directed the HHS secretary to promulgate regulations to establish and maintain a list of biological agents that have the potential to pose a severe threat to public health and safety. This list ultimately became known as the select agent list. In determining whether to include an agent on this list, the following criteria were to be taken into consideration: (i) the effect on human health from exposure to the agent, (ii) the degree of contagiousness of the agent and the methods by which the agent is transferred to humans, and (iii) the availability and effectiveness of immunizations to prevent and therapies to treat any illness resulting from infection by the agent. Initially, only the transfer of agents on this list was to be regulated. In addition, the regulations were to provide for the establishment and enforcement of safety procedures for agent transfer, including measures to ensure proper training and appropriate skills for handling the select agents and proper laboratory facilities to contain and dispose of them. The regulations would also require safeguards to prevent unlawful access to the agents. Furthermore, the regulations would require the establishment of procedures to protect the public safety in the event a transfer was in violation of the safety procedures or safeguards. The statute was also clear that the regulations had to be designed to provide for the appropriate availability of these biological agents for research, education, and other legitimate purposes. The new regulations were incorporated into Part 72 of Title 42 of the Code of Federal Regulations (C.F.R. § 72), by which the CDC regulated the interstate shipment of etiologic agents. The final regulation (42 C.F.R. §§ 72.6, 72.7, and Appendix A, effective April 15, 1997) encompassed the following: (i) a list of infectious agents and biological toxins ("Select Agents") that were regarded as possible agents of interest to terrorists, (ii) a process for revising the Select Agent list when new information became available, (iii) a system of safeguards to be followed when these select agents were transported, (iv) a system for tracking the acquisition and transfer of select agents between laboratories, and (v) a process for alerting appropriate law enforcement authorities if a shipment was missing or stolen (6). The legislative requirement for ensuring the necessary skills and proper laboratory facilities for the safe handling of these agents was addressed by the incorporation by reference of the CDC/ NIH publication "Biosafety in Microbiological and Biomedical Laboratories (BMBL), 4th Edition" into the regulation. This version of the Select Agent regulations was in effect from April 15, 1997, until February 7, 2003.

On October 2, 2001, a previously healthy 63-year-old employee of American Media in Boca Raton, Florida, awoke from sleep with fever, emesis, and confusion (7). At the emergency department of a local medical center, a lumbar puncture was performed to evaluate for presumed bacterial meningitis; examination of the Gram stain of the cerebrospinal fluid sample revealed abundant polymorphonuclear white blood cells and many large gram-positive bacilli singly and in chains. On the basis of results of the Gram stain, a diagnosis of anthrax was considered. The clinical laboratory at the medical center presumptively identified the organism as *Bacillus anthracis* within 18 hours of inoculating the cerebrospinal fluid onto bacterial culture plates; this identification was confirmed by a Laboratory Response Network member laboratory within the Florida Department of Health on the following day. During the subsequent investigation, extensive environmental sampling detected the presence of *B. anthracis* spores on the patient's computer keyboard at his workplace and in the mailroom at American Media. This finding, together with the finding of *B. anthracis* spores in regional and local postal centers that processed mail destined for the American Media building, implicated one or more mailed letters or packages as the probable source of exposure. Subsequently, letters containing anthrax spores were discovered at the offices of three major television broadcasting companies in New York City (NBC, CBS, and ABC), at the newspaper headquarters of the *New York Post* (8), and in the Washington, DC, offices of Senator Patrick Leahy (8,9). In the ensuing investigation, the FBI contacted CDC's Select Agent Program and requested a list of all laboratories that possessed *B. anthracis*, in particular those that possessed the Ames strain, which was the strain isolated in all of the human anthrax cases resulting from this attack. However, only those laboratories that had transferred or received the agent since April 1997 had been required to register with the CDC Select Agent Program. This fact highlighted, from a law enforcement perspective, a weakness in the select agent regulations.

On October 26, 2001, President George W. Bush signed into law P.L. 107-56 entitled "Uniting and Strengthening America by Providing Appropriate Tools Required to Intercept and Obstruct Terrorism (USA PATRIOT) Act." The USA PATRIOT Act had important implications for handling or possessing select agents and toxins. Section 817 of the USA PATRIOT Act amended Chapter 10 (Biological Weapons) of Title 18, United States Code (U.S.C.). The first change was incorporated as a subparagraph (b) in Section 175 (18 U.S.C. § 175(b)) and stated: "Whoever knowingly possesses any biological agent, toxin, or delivery system of a type or in a quantity that, under the circumstances, is not reasonably justified by a prophylactic, protective, bona fide research, or other peaceful purpose, shall be fined under this title, imprisoned not more than 10 years, or both." The addition of subparagraph 175(b) also made clear that in this context the terms "biological agent" and "toxin" did not include a biological agent or toxin in its naturally occurring environment as long as the biological agent or toxin had not been cultivated, collected, or otherwise extracted from its natural source. The second statement in new section 175b restricted possession of Select Agents. It stated that

> "No restricted person described in subsection (b) shall ship or transport in interstate or foreign commerce, or possess in or affecting commerce, any biological agent or toxin, or receive any biological agent or toxin that has been shipped or transported in interstate or foreign commerce, if the biological agent or toxin is listed as a Select Agent in subsection (j) of section 72.6 of title 42, Code of Federal Regulations, pursuant to section 511(d)(l) of the Antiterrorism and Effective Death Penalty Act of 1996 (P.L. 104-132), and is not exempted under subsection (h) of such section 72.6 or Appendix A of part 72 Code of Federal Regulations."

The categories of restricted persons as delineated in Section 175b are listed in Table 13.1.

On June 12, 2002, President George W. Bush signed the "Public Health Security and Bioterrorism Preparedness and Response Act of 2002" (P.L. 107-188), legislation which was to have a major impact on the regulation of select agents. Title II of this act addressed enhancing controls on dangerous biological agents and toxins. Subtitle A directed the HHS Secretary to (i) establish and maintain (and review at least biennially) a list of each biological agent and each toxin that has the potential to pose a severe threat to public health and safety; (ii) provide for the regulation of transfers of listed agents and toxins; (iii) provide for the establishment and enforcement of standards and procedures governing the possession and use of listed agents and toxins; (iv) require registration with the HHS

**Table 13.1** Categories of Restricted Persons as Described in the USA PATRIOT Act of 2001

1. A person who is under indictment for a crime punishable by imprisonment for a term exceeding 1 year.
2. A person who has been convicted in any court of a crime punishable for a term exceeding 1 year.
3. A person who is a fugitive from justice.
4. A person who is an unlawful user of any controlled substance [as defined in Section 102 of the Controlled Substances Act (21 U.S.C. 802)].
5. An alien in the United States illegally or unlawfully.
6. A person who has been adjudicated as a mental defective or has been committed to any mental institution.
7. An alien (other than one admitted lawfully for permanent residence) who is a national of a country as to which the Secretary of State, pursuant to section 6(j) of the Export Administration Act of 1979 [50 U.S.C. App. 2405(j)], section 620A of chapter 1 of part M of the Foreign Assistance Act of 1961 (22 U.S.C. 2371), or section 40(d) of chapter 3 of the Arms Export Control Act [22 U.S.C. 2780(d)], has made a determination (that remains in effect) that such country[a] has repeatedly provided support for acts of international terrorism.
8. A person who has been discharged from the Armed Services of the United States under dishonorable conditions.

[a] Currently Cuba, Iran, Sudan, and Syria.

Secretary of the possession, use, and transfer of listed agents and toxins; and (v) provide appropriate safeguards and security requirements for persons possessing, using, or transferring a listed agent or toxin commensurate with the risk such agent or toxin poses to public health and safety. It also authorized the HHS Secretary to inspect the facilities of persons subject to the aforementioned requirements to ensure their compliance with such regulations.

An essential feature of this legislation was having appropriate safeguards and security requirements for persons possessing, using, or transferring a select agent. The HHS Secretary was required to establish such requirements in consultation with the Attorney General (subsequent legislation amended this requirement such that the HHS Secretary is now required to establish such requirements in collaboration with the Secretary of Homeland Security and the Attorney General). These requirements dealt with limiting access to select agents by including provisions to ensure that registered persons (i) provide access to only those individuals whom the registered person determines has a legitimate need to handle or use select agents or toxins; (ii) provide names and other identifying information (e.g., fingerprints) to both the HHS Secretary and the Attorney General for those individuals identified as needing access; (iii) deny access to select agents and toxins by individuals whom the Attorney General has identified as "restricted persons" as defined in the USA PATRIOT Act [18 U.S.C. § 175b(d)(2)]; and (iv) limit or deny access to select agents and toxins to an individual who is reasonably suspected by any federal law enforcement or intelligence agency of (a) committing a crime set forth in section 2332b(g)(5) of Title 18, U.S.C.; (b) knowing involvement with an organization that engages in domestic or international terrorism or with any other organization that engages in intentional crimes of violence; or (c) being an agent of a foreign power (as defined in section 1801 of Title 50, U.S.C.). The information provided to the Attorney General is used to search electronic databases. Results of the Attorney General's security risk assessment are then provided to the HHS Secretary (specifically to the Director of the CDC's Division of Select Agents and Toxins) to make the determination as to whether the individual is to be granted or denied access to the select agents and toxins. Revised select agent regulations are found in Part 73 of Title 42, Code of Federal Regulations (42 C.F.R. § 73) (10).

Another major change in the regulation of select agents and toxins can be found in Subtitle B of P.L. 107-188 (cited as the Agricultural Bioterrorism Protection Act of 2002), which directed the USDA Secretary to establish and maintain a list of each biological agent and toxin that the USDA Secretary determines has the potential to pose a severe threat to animal or plant health, or to animal or plant products. Criteria for the inclusion of animal pathogens included (i) availability and effectiveness of pharmacotherapies and prophylaxis to treat and prevent any illness; (ii) economic impact; inclusion on the Office International des Epizooties (OIE) A and B lists; and (iii) presence on the Australia Group List (http://www.australiagroup.net/en/biological_agents.html). Criteria for the inclusion of plant pathogens included the effect of exposure to the agent on plant health and on the production and marketability of plant products; the ability to detect the agent and diagnose the infection during its early stages; whether the agent was nonnative or exotic; and the economic importance of the host plant. The list of USDA select agents is also subject to review on a biennial basis. The USDA developed two select agent regulations that implement the provisions of Subtitle B of P.L. 107-188. Part 121 of Title 9, Code of Federal Regulations (9 C.F.R. §121) (11) regulates biological agents and toxins that have the potential to pose a severe threat to animal health or to animal products. Part 331 of Title 7, Code of Federal Regulations (7 C.F.R. §331) (12) regulates biological agents that pose a severe threat to plant health or plant products.

Those agents and toxins that are on both HHS and USDA lists are called "overlap agents and toxins" and are jointly regulated by both HHS and USDA. Agents currently identified by regulation as HHS, USDA, and overlap select agents and toxins are listed in Table 13.2. A summary of areas with significant changes to select agent regulations as a result of legislation passed in 2002 is presented in Table 13.3. Currently, the Division of Select Agents and Toxins at the CDC administers HHS select agent regulations; the Animal and Plant Health Inspection Service (APHIS) administers USDA select agent regulations.

**Table 13.2** HHS[a] and USDA[b] Select Agents and Toxins

**HHS Select Agents and Toxins**

Abrin
*Botulinum* neurotoxins
*Botulinum* neurotoxin-producing species of *Clostridium*
Cercopithecine herpesvirus 1 (herpes B virus)[c]
*Clostridium perfringens* epsilon toxin
*Coccidioides posadasii/Coccidioides immitis*
Conotoxins
*Coxiella burnetii*
Crimean-Congo hemorrhagic fever virus
Diacetoxyscirpenol (DAS)
Eastern equine encephalitis virus[c]
Ebola virus
*Francisella tularensis*
Lassa fever virus
Marburg virus
Monkeypox virus
Reconstructed replication competent forms of the 1918 pandemic influenza virus
containing any portion of the coding regions of all eight gene segments (reconstructed
1918 influenza virus)
Ricin
*Rickettsia prowazekii*
*Rickettsia rickettsii*
Saxitoxin
Shiga-like ribosome inactivating proteins
Shigatoxin
South American hemorrhagic fever viruses (Flexal, Guanarito, Junin, Machupo, Sabia)
Staphylococcal enterotoxins
T-2 toxin
Tetrodotoxin
Tick-borne encephalitis complex (flavi) viruses (Central European tick-borne encephalitis,
Far Eastern tick-born encephalitis, Kyasanur Forest disease, Omsk hemorrhagic fever,
Russian spring and summer encephalitis)[c]
*Variola* major virus (smallpox virus)
*Variola* minor (alastrim)
*Yersinia pestis*

**Overlap Select Agents and Toxins**

*Bacillus anthracis*
*Brucella abortus*
*Brucella melitensis*
*Brucella suis*
*Burkholderia mallei* (formerly *Pseudomonas mallei*)
*Burkholderia pseudomallei* (formerly *Pseudomonas pseudomallei*)
Hendra virus
Nipah virus
Rift Valley fever virus
Venezuelan equine encephalitis virus[c]

*(Continued)*

**Table 13.2** HHS[a] and USDA[b] Select Agents and Toxins (*Continued*)

**USDA Select Agents and Toxins**

African horse sickness virus
African swine fever virus
Akabane virus
Avian influenza virus (highly pathogenic)
Bluetongue virus (exotic)
Bovine spongiform encephalopathy agent
Camel pox virus
Classical swine fever virus[c]
*Ehrlichia ruminantium* (heartwater)
Foot-and-mouth disease virus[c]
Goat pox virus
Japanese encephalitis virus[c]
Lumpy skin disease virus
Malignant catarrhal fever virus (Alcelaphine herpesvirus type 1)[c]
Menangle virus
*Mycoplasma capricolum* subspecies *capripneumoniae* (contagious caprine pleuropneumonia)
*Mycoplasma mycoides* subspecies *mycoides* small colony (Mmm SC) (contagious bovine pleuropneumonia)
Peste des petits ruminants virus
Rinderpest virus
Sheep pox virus
Swine vesicular disease virus[c]
Vesicular stomatitis virus (exotic): Indiana subtypes VSV-IN2, VSV-IN3
Virulent Newcastle disease virus

**USDA Plant Protection and Quarantine (PPQ) Select Agents and Toxins**

*Peronosclerospora philippinensis* (*Peronosclerospora sacchari*)
*Phoma glycinicola* (formerly *Pyrenochaeta glycines*)
*Ralstonia solanacearum* race 3, biovar 2
*Rathayibacter toxicus*
*Sclerophthora rayssiae* var *zeae*
*Synchytrium endobioticum*
*Xanthomonas oryzae*
*Xylella fastidiosa* (citrus variegated chlorosis strain)

[a] *HHS regulation 42 C.F.R. Part 73 as updated August 31, 2009.*
[b] *USDA regulations 7 C.F.R. Part 331 and 9 C.F.R. Part 121 as updated August 31, 2009.*
[c] *Nucleic acids from these select agent viruses can produce infectious forms.*

# THE SELECT AGENT REGULATIONS

The practice of microbial forensics may necessitate working with select agents or toxins. Individuals or entities who want to possess, use, or transfer select agents or toxins should review the requirements of the select agent regulations, which can be found at http://www.selectagents.gov. Entities must register

**Table 13.3** Significant Changes to the Select Agent Regulation

|  | 1996 | 2003 |
|---|---|---|
| Agents (bacteria, fungi, virus, toxins) |  |  |
| Human (HHS) | Yes | Yes |
| Animal (USDA) | No | Yes |
| Plant (USDA) | No | Yes |
| Overlap (HHS/USDA) | No | Yes |
| Possession | No | Yes |
| Use | No | Yes |
| Transfer | Yes | Yes |
| Security risk assessment (DOJ) | No | Yes |
| Incorporation of BMBL | Yes | No |

with either the CDC or the APHIS if they plan to possess, use, or transfer select agents or toxins on the HHS or USDA list, respectively. If the select agent or toxin is an overlap agent, the entity may register with either HHS or APHIS, but is not required to register with both. Because the registration process, including the requirement for security risk assessments, may be time-consuming, an entity contemplating working with select agents or toxins should initiate the process well ahead of time. While registration is both agent and laboratory specific, the registration can be amended to include additional agents and laboratory spaces. A registration lasts for a maximum of 3 years (but may be granted for a shorter period of time), after which it must be renewed if the entity chooses to retain possession of the select agents. The HHS and USDA Select Agent Regulations are very similar in how they are structured (Table 13.4), with the exception that USDA plant select agent regulations (7 C.F.R. § 331) do not have sections that address overlap agents because there are no plant agents that affect humans directly.

## Select Agents and Toxins

The current HHS, USDA, and overlap list of select agents and toxins is shown in Table 13.2. The list is not static and is reviewed on a biennial basis by the respective select agent programs with advice and input from the Interagency Select Agents and Toxins Technical Advisory Committee (ISATTAC). The review is designed to provide an objective (i.e., quantitative) review of an agent or toxin in determining whether an agent or toxin should be added, removed, or maintained on the select agent list. Furthermore, agents and toxins can be added to (or deleted from) the list at any time by amending the regulations in accordance with the Administrative Procedure Act (5 U.S.C. § 552) by providing notification of the proposed amendment in the Federal Register and the opportunity for public comment (13,14). Currently, all regulated toxins are on the HHS list.

**Table 13.4** Structure of the Select Agent Regulations

| Component | Section | | |
|---|---|---|---|
| | 42 C.F.R. 73 | 7 C.F.R. 331 | 9 C.F.R. 121 |
| Definitions | 73.1 | 331.1 | 121.1 |
| Purpose and scope | 73.2 | 331.2 | 121.2 |
| Select agents and toxins | 73.3 | 331.3 | 121.3 |
| Overlap select agents and toxins | 73.4 | — | 121.4 |
| Exemptions | 73.5 | 331.5 | 121.5 |
| Exemptions for overlap select agents and toxins | 73.6 | — | 121.6 |
| Registration and related security risk assessments | 73.7 | 331.7 | 121.7 |
| Denial, revocation, or suspension of registration | 73.8 | 331.8 | 121.8 |
| Responsible official | 73.9 | 331.9 | 121.9 |
| Restricting access to select agents and toxins; security risk assessment | 73.10 | 331.10 | 121.10 |
| Security | 73.11 | 331.11 | 121.11 |
| Biosafety/biocontainment | 73.12 | 331.12 | 121.12 |
| Restricted experiments | 73.13 | 331.13 | 121.13 |
| Incident response | 73.14 | 331.14 | 121.14 |
| Training | 73.15 | 331.15 | 121.15 |
| Transfers | 73.16 | 331.16 | 121.16 |
| Records | 73.17 | 331.17 | 121.17 |
| Inspections | 73.18 | 331.18 | 121.18 |
| Notification of theft, loss, or release | 73.19 | 331.19 | 121.19 |
| Administrative review | 73.20 | 331.20 | 121.20 |
| Civil money penalties | 73.21 | — | — |

However, these toxins are subject to the regulation only if the aggregate amount of the toxin under the control of a principal investigator, treating physician or veterinarian, or commercial manufacturer or distributor exceeds the amount specified in the regulations (Table 13.5). These *de minimus* amounts represent an attempt to balance the requirement for regulatory oversight of toxins that have the potential to pose a severe threat to public health and safety with the need for their availability for critical research and educational purposes. The *de minimus* amounts in the regulations are based on the amount a person with a legitimate need, such as a principal investigator, could safely possess without constituting a significant threat to public health and safety.

Some native and recombinant nucleic acid molecules are also regulated. For example, single-stranded (positive strand) RNA viruses and certain

**Table 13.5** Permissible Toxin Amounts[a]

| DHHS Toxin | Amount (mg) |
| --- | --- |
| Abrin | 100 |
| Botulinum neurotoxins | 0.5 |
| *Clostridium perfringens* epsilon toxin | 100 |
| Conotoxins | 100 |
| Diacetoxyscirpenol (DAS) | 1000 |
| Ricin | 100 |
| Saxitoxin | 100 |
| Shiga-like ribosome inactivating proteins | 100 |
| Shigatoxin | 100 |
| Staphylococcal enterotoxins | 5 |
| T-2 toxin | 1000 |
| Tetrodotoxin | 100 |

[a] Toxins listed in 42 C.F.R. § 73.3(d)(3) are not regulated if the amount under the control of a principal investigator, treating physician or veterinarian, or commercial manufacturer or distributor does not exceed, at any time, the amounts indicated here.

double-stranded DNA viruses that utilize host polymerases contain nucleic acids that can produce infectious forms (e.g., Venezuelan equine encephalitis virus). Other select agent viruses that meet this criterion and fall under the same regulations are noted in Table 13.2. Recombinant nucleic acids that encode for functional form(s) of any of the toxins listed in Table 13.2 are subject to regulation if they can be expressed *in vitro* or *in vivo* or if they are in a vector or recombinant host genome and can be expressed *in vitro* or *in vivo*. The CDC and APHIS Select Agent Programs recently posted additional guidance on the regulation of select agent nucleic acids. This guidance can be found at http://www.selectagents.gov/syntheticgenomics.html. Under the current select agent regulations, the following are examples of genomic materials from select agent viruses that would not be regulated as a select agent: (i) material from regulated genomes that has been rendered noninfectious; (ii) cDNA made from genomes of regulated select agent pathogens; and (iii) complete genomes of single-stranded, negative-strand RNA viruses, double-stranded RNA viruses, and double-stranded DNA viruses that require a unique polymerase (e.g., monkeypox virus). It should be noted that the select agent regulations do not apply to variola major genetic elements; however, the World Health Organization (WHO) places significant restrictions on the possession, use, and transfer of these materials. Institutions other than the two currently recognized WHO collaborating centers may not possess genetic fragments exceeding 20% of the variola major virus genome (for further information refer    to    http://www.who.int/csr/disease/smallpox/research/en/index.html). These restrictions are followed by the CDC but are not legally binding. However, 18 U.S.C. § 175c makes it unlawful for any person to knowingly produce, engineer, synthesize, acquire, transfer directly or indirectly, receive, possess, import, export, or use, or possess and threaten to use the variola virus. This section does not apply to conduct by, or under the authority of, the HHS Secretary (e.g., research at the CDC). Genomic material from bacteria or fungi on the select agent list is not regulated as a select agent. Additionally, the nucleic acid sequence information of select agent pathogens is not regulated.

Companies that synthesize complete copies of regulated viral genomes must be registered for that agent with either the CDC or the APHIS Select Agent Program. In addition, the requestor (individual that requests the material) must also be registered for that agent. Shipment of the synthetic genome from the manufacturer to the requestor would require prior approval by the Select Agent Program.

Certain experiments with select agents or toxins are prohibited unless approved by and conducted in accordance with conditions prescribed by the Secretary (HHS or USDA). See, for example, 42 C.F.R. § 73.13 (restricted experiments). Restricted experiments are those (i) utilizing recombinant DNA that involve the deliberate transfer of a drug resistance trait to select agents that are not known to acquire the trait naturally, if such acquisition could compromise the use of the drug to control disease agents in humans, veterinary medicine, or agriculture; and (ii) involving deliberate formation of recombinant DNA containing genes for the biosynthesis of select agent toxins lethal for vertebrates at an $LD_{50}$ < 100 ng/kg body weight. Approval to conduct a "restricted experiment" requires submission of a written request to the Select Agent Program with supporting scientific information and documentation. A written decision granting or denying the request is issued by the responsible Select Agent Program (i.e., CDC or APHIS). A number of requests to introduce drug resistance genes have already been approved. However, unlike excluded attenuated strains of select agents, which once excluded are no longer subject to the regulation (as long as the attenuated strain is not manipulated to restore or enhance its virulence) and can be used by any investigator (see later), each request to conduct a restricted experiment is reviewed separately to ensure that appropriate precautions and containment will be utilized.

## Exemptions

The select agent regulatory exemptions provide that individuals or entities that may find themselves in possession of a select agent or toxin are not required to be in compliance with the select agent regulations as long as they take the specific actions required and/or meet the specific conditions proscribed by the regulations. The current exemptions in the regulations include (i) diagnostic, verification, or proficiency testing specimens in clinical or diagnostic laboratories; (ii) products licensed or otherwise approved for use by the federal government under specific statutes; (iii) investigational products approved by the federal government under specific statutes; and (iv) when either the HHS Secretary or the USDA Secretary grant specific exemptions due to a public health or agricultural emergency, respectively. A couple of examples will suffice. A clinical laboratory that isolates *B. anthracis* from a clinical specimen is exempt from requirements of the select agent regulation, provided that (i) within 7 calendar days after identification, the select agent is transferred in accordance with the regulation or destroyed on-site by a recognized sterilization process; (ii) the select agent is secured against theft, loss, or release during the period between identification and transfer or destruction; and (iii) identification of the select agent is reported to the CDC or APHIS and to other appropriate authorities when required by federal, state, or local law. Another example is that the product botox, a botulinum neurotoxin preparation approved for production and distribution under the Federal Food, Drug, and Cosmetic Act (21 U.S.C. § 310 *et seq*), is not subject to the Select Agent Regulations.

For clinical or diagnostic laboratories and other persons who possess, use, or transfer listed agents that are contained in specimens presented for diagnosis, verification, or proficiency testing, identification of such agents or toxins must be reported to the appropriate secretary, and when required under federal, state, or local law, to other appropriate authorities. Furthermore, such agents or toxins must be transferred or destroyed in a regulatory manner as set forth by the secretary in the regulations.

## Exclusions

There are certain circumstances under which the Select Agent Regulations do not apply to the possession, use, or transfer of one of the biological agents or toxins listed in the Select Agent Regulations. These exclusions include (i) any select agent or toxin that is in its naturally occurring environment, provided the select agent or toxin was not intentionally introduced, cultivated, collected, or otherwise extracted from its natural source. For example, Cercopithecine herpesvirus 1 (herpes B virus) occurring as a natural infection in a primate would be excluded from regulation. (ii) Nonviable select agents or nonfunctional toxins. For example, the purified B chain of ricin toxin is not subject to regulation. (iii) HHS toxins under the control of a principal investigator, treating physician or veterinarian, or commercial manufacturer or distributor if the aggregate amount does not, at any time, exceed amounts listed in the regulations (Table 13.5). Thus, a single investigator may possess up to 100 mg of purified ricin and not be subject to regulations.

In addition, an attenuated strain of a select agent or toxin may be excluded from requirements of the regulation based on a determination by either the CDC or the APHIS that it does not pose a severe threat to public health and safety, animal health and animal products, or to plant health and plant products. To apply for exclusion, an individual or entity must submit a written request and provide supporting scientific data to either HHS or USDA Select Agent Programs. The material is reviewed by the ISATTAC with input from appropriate subject matter experts. Based on this review, a written decision supporting or denying the request is issued by the appropriate Select Agent Program. If granted, the exclusion becomes effective upon notification of the applicant. Exclusions are listed on the Internet at http://www.selectagents .gov. The current list of excluded strains of select agents and toxins is shown in Table 13.6. If an excluded strain is subjected to any manipulation that

---

**Table 13.6** Select Agent Exclusions for Attenuated Strains[a]
Organism/toxin

- *Coccidioides posadasii* Δ*chs5* (effective 10-14-2003) —unable to form infectious arthroconidia. Also unable to form spherules *in vivo*.
- *C. posadasii* Δ*cts2*Δ*ard1*Δ*cts3* (effective 03-03-2006)—unable to form infectious arthroconidia.
- Conotoxins specifically excluded are the class of sodium channel antagonist μ-conotoxins, including GIIIA, the class of calcium channel antagonist ω-conotoxins, including GVIA, GVII, MVIIA, MVIIC, and their analogs or synthetic derivatives; the class of NMDA antagonist conotoxins, including con-G, con-R, con-T, and their analogs or synthetic derivatives; and the putative neurotensin agonist contulakin-G and its synthetic derivatives (effective 04-29-2003).
- *Yersinia pestis* Δ*pgm* (effective 03-14-2003)—strains with a deletion of a 102-kb region of the chromosome termed the *pgm* locus.
- *Bacillus anthracis* strains devoid of both virulence plasmids pX01 and pX02 (effective 02-27-2003).
- *B. anthracis* strains lacking the virulence plasmid pX02 (e.g., Sterne strain pX01[+]pX02) (effective 02-27-2003).
- *Brucella abortus* strain 19 (live vaccine strain) (effective 06-12-2003).
- *B. abortus* strain RB51 (live vaccine strain) (effective 05-07-2003).
- *Coxiella burnetii* Phase II, Nine Mile Strain, plaque-purified clone 4 (effective 10-15-2003).
- *Francisella tularensis* subspecies *novicida* strain Utah 112 (effective 02-27-2003).
- *F. tularensis* subspecies *holartica* live vaccine strain (effective 02-27-2003).
- *F. tularensis* biovar *tularensis* strain ATCC 6223 (also known as strain B38) (effective 04-14-2003).
- Rift Valley fever virus vaccine strain MP-12 (effective 02-07-2003).
- Venezuelan equine encephalitis virus vaccine candidate strain V3526 (effective 05-05-2003).
- Highly pathogenic avian influenza virus, recombinant vaccine reference strains of the H5N1 and H5N3 subtypes (effective 05-07-2004).
- Japanese encephalitis virus strain SA14-14-2 (effective 03-12-2003).

[a] Available from: http://www.selectagents.gov.

restores or enhances its virulence, the resulting select agent or toxin becomes subject to requirements of the regulations. An example of this type of manipulation is introduction of the virulence plasmid pX02 into the Sterne strain of *B. anthracis* (pX01$^+$pX02$^-$).

The current Select Agent Regulations address concerns raised by federal law enforcement agencies related to seizures (i.e., possession) of known select agent or toxins. The revisions do not authorize the seizure of a select agent or toxin by a federal law enforcement agency. Rather, they establish the conditions under which a federal law enforcement agency can conduct certain law enforcement activities (e.g., collecting evidence from a laboratory crime scene) without being in violation of the regulations.

For example, sections 73.3(f) and 73.4(f) of the HHS regulations provide that any known select agent or toxin seized by a federal law enforcement agency will be excluded from requirements of the regulation during the period between seizure of the agent or toxin and the transfer or destruction of such agent or toxin provided that (i) as soon as practicable, the federal law enforcement agency transfers the seized agent or toxin to an entity eligible to receive such agent or toxin or destroys the agent or toxin by a recognized sterilization or inactivation process; (ii) the federal law enforcement agency safeguards and secures the seized agent or toxin against theft, loss, or release and reports any theft, loss, or release of such agent or toxin; and (iii) the federal law enforcement agency reports the seizure of the select agent or toxin by submitting APHIS/CDC Form 4. In the event that a federal law enforcement agency seizes a suspected select agent or toxin or unknown material, this material will be regarded as a specimen presented for diagnosis or verification and, therefore, will not be subject to the regulations until it has been identified as a select agent or toxin.

## Transfers

With two exceptions (see later), a select agent or toxin may only be transferred to individuals or entities registered to possess, use, or transfer that agent or toxin. The transfer of the select agent or toxin must be authorized by the CDC or APHIS before transfer occurs. A transfer may be authorized if the sender (i) has an active and approved certificate of registration at the time of transfer that covers the particular select agent or toxin; (ii) meets the exemption requirements for the particular select agent or toxin (see earlier discussion); or (iii) is transferring the select agent or toxin from outside the United States and meets all import requirements. Regulations governing importation of etiologic agents of diseases of humans (42 C.F.R. § 71 Foreign Quarantine. Section 71.54); diseases of livestock, poultry, and other animals (9 C.F.R. §§ 92, 94-96, 122, and 130); and plant pests (7 C.F.R. § 330) have been presented elsewhere (15). The recipient must also have a certificate of registration that includes the particular select agent or toxin at the time of transfer. To obtain authorization for transfer, APHIS/CDC Form 2 must be submitted to either the CDC or the APHIS. An authorization for transfer is only valid for 30 calendar days after issuance. After authorization is obtained, the recipient must submit a completed APHIS/CDC Form 2 within 2 business days after the select agent or toxin is received. If the select agent or toxin has not been received within 48 hours after the expected delivery time or if the package containing select agents or toxins has been damaged to the extent that a release of the select agent or toxin may have occurred, the recipient must notify the CDC or APHIS immediately.

It is important to note that the sender must comply with all applicable laws concerning the packaging and shipping of hazardous materials. Select agents are not permitted in the U.S. postal system. Select agents are considered hazardous materials and fall under the Hazardous Materials Regulations (HMR) of the U.S. Department of Transportation (DOT). HMR are issued by the Pipeline and Hazardous Materials Safety Administration and govern the interstate transportation of materials by highway, rail, vessel, and air. They are the legally enforceable shipping regulations for hazardous materials in the United States and are enforced by the Federal Aviation Administration for air transport. For complete information, see DOT 49 C.F.R. §§171-180 at http://www.phmsa.dot.gov/hazmat/regs. The DOT HMR applies to the offering, acceptance, and transportation of hazardous materials to, from, or within the United States and to any aircraft of U.S. registry anywhere in air commerce. Select agents, infectious substances, toxins, and other dangerous goods are not allowed in the passenger compartment of aircraft. Transportation of an infectious substance by a federal, state, or local government agency or the military in a vehicle (e.g., truck, airplane) operated by a government or military employee is not subject to the HMR, which may be important, in certain circumstances, for transporting forensic evidence containing an infectious substance.

The International Civil Aviation Organization (ICAO) Technical Instructions (TI) on the Safe Transport of Dangerous Goods by Air augment the broad principles governing the international transport of hazardous materials by air contained in Annex 18 to the Convention on International Civil Aviation. The DOT HMR authorizes transport in accordance with the ICAO TI as a means of compliance with the HMR. Because both the DOT HMR and the ICAO TI are based on United Nations (UN) recommendations, the DOT allows shippers to follow the ICAO TI for domestic air transport, but additional requirements may exist (see 49 C.F.R. § 171.23, Requirements for specific materials and packages transported under the ICAO Technical Instructions, IMDG Code, Transport Canada TDG Regulations, or the IAEA Regulations).

In order to simplify the extensive and complicated requirements for transport by aircraft, the International Air Transport Association (IATA) was formed.

IATA is an international trade organization that represents approximately 230 airlines comprising 93% of scheduled international air traffic. The Dangerous Goods Regulations (DGR), which IATA publishes every year, is technically not a regulation per se, but a user-friendly "field manual" guide for the regulatory ICAO TI. Working closely with the UN Committee of Experts and other national authorities, IATA ensures that the rules and regulations governing dangerous goods transport are effective, efficient, and in complete compliance with ICAO and the UN Model Regulations. For more information, see http://www.icao.org/ and http://www.iata.org/index.htm. It is important to note that under the DGR, carriers can refuse improperly packed packages and international shipments.

The two exceptions concerning the transfer of select agents are as follow: (i) a select agent or toxin that is in a specimen for proficiency testing may be transferred without prior authorization from the CDC or APHIS provided that the sender reports to the CDC or APHIS the select agent or toxin to be transferred and the name and address of the recipient(s) at least 7 calendar days prior to the transfer and (ii) on a case-by-case basis, the Secretary of HHS or USDA can authorize a transfer of a select agent or toxin, not otherwise eligible for transfer, under conditions prescribed by the secretary.

## Records

Proper record keeping is important for the forensic analysis of microbial evidence. It is also a requirement under sections 73.17, 121.17, and 331.17 of the Select Agent Regulations (Table 13.4). Among the required records are an accurate, current inventory for each select agent (including viral genetic elements, recombinant nucleic acids, and recombinant organisms) held in long-term storage. The term "long-term storage" has been somewhat confusing but the CDC and APHIS have posted a document on the select agent Web site intended to provide guidance as to when a select agent or toxin would be considered by the Select Agent Program to be in "long-term storage" (see http://www.selectagents.gov). Long-term storage would be where the select agent or toxin meets one or more of the following criteria: (i) The material (i.e., bacteria, fungus, virus, toxin, genetic) is in a highly concentrated state and would not be used unless diluted to a less concentrated state. For example, a vial containing a high titer of virus is removed and diluted into multiple aliquots. The original vial is considered long-term storage and the aliquots "working material." (ii) It would not be used for any work by the entity for a defined period of time (e.g., no work with the material within 30 calendar days). (iii) It is not consumed within 30 calendar days of receipt or creation by the entity. For example, a vial of bacteria is received by the laboratory. There are no plans to use the contents of the vial for any work performed by the laboratory within the next 30 calendar days. (iv) It is placed in an environment that is designed to extend the viability of the material. For example, bacteria that die after 3 days under permissive growth conditions are placed in liquid nitrogen to extend their viability beyond 3 days. (v) It is placed in an environment where there is infrequent access. For example, viral suspensions are placed in a liquid nitrogen tank that is only accessed by a member of the laboratory every 2 months.

For bacteria, viruses, fungi, and genetic elements, these records must include (i) the name and characteristics (e.g., strain designation, GenBank accession number); (ii) the quantity acquired from another individual or entity (e.g., containers, vials, tubes), date of acquisition, and the source; (iii) where the material is stored (e.g., building, room, and freezer); (iv) when moved from storage and by whom and when returned to storage and by whom; (v) the select agent used and purpose of use; (vi) records created under sections 73.16, 121.16, and 331.16 (interentity transfers); (vii) intraentity transfers (where sender and recipient are covered by the same certificate of registration), the select agent, quantity transferred, date of transfer, sender, and the recipient; and (viii) records created under sections 73.19, 121.19, and 331.19 (notification of theft, loss, or release).

Records for each toxin held must include (i) name and characteristics; (ii) quantity acquired from another individual or entity; (iii) initial and current amount (e.g., milligrams, milliliters, grams); (iv) toxin used and purpose of use, quantity, date(s) of the use and by whom; (v) where stored; (vi) when moved from storage and by whom and when returned to storage and by whom including amount; (vii) interentity transfer records; (viii) intraentity transfer records; (ix) records created under sections 73.19, 121.19, and 331.19; and (x) if destroyed, the quantity of toxin destroyed, the date of such action, and by whom.

Tissue culture fluids and tissue specimens obtained from animals or plants infected experimentally with a select agent or toxin do not need to be identified as long-term storage material unless the materials were prepared with the intent to store for a long period of time or if there is no specified date established when the materials will be used. If the tissue culture fluids and tissue specimens are classified by the entity as long-term storage material, there is no requirement for a vial-by-vial inventory as required for high concentration seed stock. However, there is a requirement to label these materials with the date placed in storage, the agent contained in the fluids or tissue, and a reference identification recorded in a written record (e.g., inventory record, research notes). To facilitate inventory, vials of materials can be recorded and grouped into tamper-proof containers and audits made of intact containers rather than audits of individual vials. Any material determined to be long-term storage must be maintained in a secure location and detailed, accurate records must be kept. Any material determined to not be long-term storage does not require detailed records; however, the entity must have mechanisms in place to control the distribution of the material and to track the creation of "working material" from material in long-term storage. The entity is required to provide records, if requested, that document the stock source of production quantities of agents.

## Security

The registration process requires each laboratory to develop a written security plan that is based on a site-specific risk assessment. Regulations governing the assessment do not specify who must perform it, meaning that the assessment can be performed by officials for the laboratory itself. Each laboratory must implement a security plan that is sufficient to safeguard select agents against unauthorized access, theft, loss, or release. This assessment must provide protection based on the risk and intended use of the select agent. It includes four assessments: an agent-specific risk assessment, threat assessment, vulnerability assessment, and graded protection determination. Prior to being issued a certificate of registration, an entity must comply with all security requirements (and all other provisions of the Select Agent Regulations). In order to assist entities in developing and implementing the required written security plan, the CDC and APHIS developed a document (16) that provides possible practices and procedures that may be used in this endeavor. The plan must be reviewed and drills conducted at least annually.

## Biosafety

The registration process requires each entity to develop and implement a biosafety plan (Table 13.4) to ensure biological containment and safe handling of select agents and toxins. The biosafety plan must be based on a site-specific assessment, which provides protection commensurate with the risk of the agent and its intended use. Biosafety procedures (e.g., operational work practices and personal protective equipment) and physical containment features (e.g., facility design and engineering controls) sufficient to contain the agent must be included. The plan must be reviewed and drills conducted at least annually. Several documents are available (17–20) to help in developing this plan. These documents can be accessed at http://www.selectagents.gov.

## Incident Response

A written incident response plan must also be developed as part of the registration packet. This plan should be coordinated with all entity-wide plans and must include response procedures for biological containment and security breaches, natural disasters, and other emergencies. Information must include emergency contact information for responsible persons; roles and lines of authority; and emergency evacuation, medical treatment, first aid, and decontamination procedures. Furthermore, the plan must be reviewed and drills conducted at least annually.

## Training

The Select Agent Regulations require that each person with approved access to select agents, or any person who works in or visits areas where select agents or toxins are handled or stored, must be trained in biosafety and security principles and practices prior to having access. Training must address risks involved, procedures performed, and specific needs of the individual. Verification and documentation that the training was understood must be maintained and repeated at least annually.

## SUMMARY

The goals of the Select Agent Program are to regulate the possession, use, and transfer of select agents and toxins that have the potential to pose a severe threat to public health and safety, to animal health or animal products, and to plant health and plant products. Performing research on select agents and toxins is critical for the development of effective medical countermeasures and, ultimately, the development of effective vaccines. At the end of 2009, there were 13,171 individuals at registered facilities with approved access to select agents. This group includes scientists, technicians, and support personnel (animal care, security, facility engineering, and management staff). Approximately 400 entities are currently registered with the CDC or APHIS to possess, use, or transfer select agents. This community consists of a broad variety of institutions, including federal, state, and local research and public health laboratories, academic research institutions, and privately owned commercial research, development, production, and distribution facilities. The CDC and APHIS have developed a highly integrated regulatory oversight program, characterized by uniform regulations, a common database (the National Select Agent Registry), and a shared common Web site (http://selectagents.gov). In addition to consulting the Web site, information on this program can be obtained by sending queries via email to the CDC at lrsat@cdc.gov or the APHIS at Agricultural.Select.Agent.Program@aphis.usda.gov.

On July 2, 2010, President Barak Obama signed Executive Order 13546 (21) that when implemented will: 1) create both an Interagency Council and a Federal Experts Advisory Panel to, respectively, coordinate security policies and practices among Federal Departments and agencies that fund work on select agents and advise agency Directors on such topics as physical security and ways of ensuring the reliability of key personnel; 2) stratify the select agent list to take better account of individual agents' specific potential to cause mass casualties if deliberately misused, and issue of new rules and guidance spelling out physical security and personnel reliability practices to be applied at each tier; and 3) coordinate Federal oversight and inspections of facilities where work on select agents is underway.

## ACKNOWLEDGMENTS

The authors thank Robbin Weyant, Ph.D., Division of Select Agents and Toxins, CDC; James D. Holt, JD, Office of General Counsel, HHS; Linda M. Weigel, Ph.D., Division of Health Care Quality Promotion, CDC; and Molly A. Hughes, M.D., Ph.D., University of Virginia, for their critical comments and suggestions. The findings and conclusions in this chapter are those of the authors and do not necessarily represent the official position of the Centers for Disease Control and Prevention.

## REFERENCES

[1]  B.L. Smoak, J.A. Geiling, Mass casualty events. Lessons learned, in: M.J. Roy (Ed.), Physician's Guide to Terrorist Attack, Humana Press Inc., Totowa, NJ, 2004, pp. 3–19.

[2]  K.B. Olsen, Aum Shinrikyo: Once and future threat? Emerg. Infect. Dis. 5 (1999) 513–516.

[3]  A. Casadevall, D.A. Relman, Microbial threat lists: Obstacles in the quest for biosecurity, Nat. Rev. Microbiol. 8 (2010) 1–6.

[4]  W.S. Carus, Bioterrorism and Biocrimes: The Illicit Use of Biological Agents Since 1900, Fredonia Books, Amsterdam, The Netherlands, 2002.

[5]  Executive Summary. Report on the sale, transfer, and safe transportation of organisms in culture collections that can be used for terrorism, July 28, 1995.

[6]  M. Tipple, R.C. Knudsen, S.A. Morse, J. Foster, J.Y. Richmond, New federal regulations for transfer of infectious agents and toxins, ASM News 63 (1997) 66–67.

[7]  L.M. Bush, B.H. Abrams, A. Beall, C.C. Johnson, Index case of fatal inhalational anthrax due to bioterrorism in the United States, N. Engl. J. Med. 345 (2001) 1607–1610.

[8]  S.Y. Hunt, N.G. Barnaby, B. Budowle, S.A. Morse, Forensic microbiology, in: M. Schaechter (Ed.), The Desk Encyclopedia of Microbiology, Elsevier, Oxford, 2009, pp. 539–551.

[9]  V.P. Hsu, S.L. Lukacs, T. Handzel, J. Hayslett, S. Harper, T. Hales, et al., Opening a *Bacillus anthracis*-containing envelope, Capitol Hill, Washington, DC: The public health response, Emerg. Infect. Dis. 8 (2002) 1039–1043.

[10]  42 C.F.R. § 73: Select Agents and Toxins. Available at: http://www.Selectagents.gov/Regulations.html.

[11]  9 C.F.R. § 121: Possession, Use, and Transfer of Select Agents and Toxins. Available at: http://www.Selectagents.gov/Regulations.html.

[12]  7 C.F.R. § 331: Possession, Use, and Transfer of Select Agents and Toxins. Available at: http://www.Selectagents.gov/Regulations.html.

[13]  CDC and U.S. Department of Health and Human Services. Possession, use, and transfer of select agents and toxins—reconstructed replication competent forms of the 1918 pandemic influenza virus containing any portion of the coding regions of all eight gene segments. Interim final rule. 70 Federal Regulation 61,047 (October 20, 2005).

[14] CDC and U.S. Department of Health and Human Services. Possession, use, and transfer of select agents and toxins. Proposed addition of SARS-associated coronavirus (SARS-CoV). 74 Federal Regulation 33,401 (July 13, 2009).

[15] J.W. Ezzell, Forensic handling of biological threat samples in the lab, in: R.G. Breeze, B. Budowle, S.E. Schutzer (Eds.), Microbial Forensics, Elsevier Academic Press, San Diego, CA, 2005, pp. 213–231.

[16] CDC and APHIS. Select Agents and Toxins Security Information Document, 7 C.F.R. § 331.11, 9 C.F.R. § 121.11, 42 C.F.R. § 73.11, March 8, 2007. Available from: http://www.selectagents.gov.

[17] L.C. Chosewood, D.E. Wilson (Eds.), U.S. Department of Health and Human Services, CDC and NIH, Biosafety in Microbiological and Biomedical Laboratories, fifth edn., U.S. Govt. Printing Office, 2007. Available from: http://www.cdc.gov/od/OHS/biosfty/bmbl5/bmbl5toc.htm.

[18] Occupational Safety and Health regulation 29 C.F.R. § 1910.1450: Occupational Exposure to Hazardous Chemicals in Laboratories.

[19] Occupational Safety and Health regulation 29 C.F.R. § 1910.1200: Hazard Communication.

[20] U.S. Department of Health and Human Services, NIH. NIH Guidelines for Research Involving Recombinant DNA Molecules (NIH Guidelines). September, 2009.

# CHAPTER 14

# Biosurety in the Post-9/11 Era

**Ronald Schouten[a] and Gregory B. Saathoff[b]**

[a]Law and Psychiatry Service, Massachusetts General Hospital, Harvard Medical School,
Boston, Massachusetts
[b]Critical Incident Analysis Group, University of Virginia School of Medicine,
Charlottesville, Virginia

## INTRODUCTION: AMERITHRAX AND THE FOCUS ON BIOSURETY

The anthrax mailings in the fall of 2001 fundamentally altered the security landscape in government and private biological laboratories. Investigation of the mailings was closed with the Department of Justice's February 19, 2010, report, in which Dr. Bruce Ivins, a civilian employee at the U.S. Army Research Institute for Infectious Diseases (USAMRIID), was identified as the sole perpetrator (1). Long before the case was closed, the anthrax attacks resulted in an unprecedented demand for reexamination of security measures in government, academic, and private research facilities that handle biological select agents and toxins.

Originally feared to be a follow-on attack after the 9/11 airplane hijackings, the anthrax mailings were determined to be domestic in origin. Identifying and investigating the possible sources and perpetrators required a massive investigation, referred to as "Amerithrax," by the U.S Postal Service, Federal Bureau of Investigation (FBI), and U.S. Department of Health and Human Services. Named by the media as a possible suspect as the result of a leak, American scientist Steven Hatfill was ultimately cleared of any role. His subsequent defamation suit against the United States resulted in a settlement in his favor of $5.8 million.

By 2005, 3 years before the Hatfill case was settled, the investigation had turned to Dr. Ivins, a senior microbiologist and anthrax researcher at USAMRIID, which is housed at Ft. Detrick in Frederick, Maryland. In July 2008, approximately 1 month after the Hatfill settlement, Dr. Ivins committed suicide and his identity as the leading suspect was publicly revealed. Dr. Ivins took a lethal overdose of over-the-counter medications within a few days of his release from a psychiatric facility, where he had been hospitalized involuntarily after making

221

Microbial Forensics. DOI: 10.1016/B978-0-12-382006-8.00014-1

homicidal threats against co-workers. He died on July 29, 2008, aware that an indictment, with the potential for the death penalty, was coming.

Bruce Ivins was a respected researcher, widely viewed as highly intelligent and skilled, and an anthrax expert whose security clearance allowed him to work on biodefense projects. Dr. Ivins's motivations for the mailings have yet to be determined and may never be known with certainty. Many who knew him, as well as those who are skeptical of the government's procedures and motives, have doubted the FBI's conclusion. For some, the false accusation against Dr. Hatfill and the government's settlement of that case fueled skepticism and doubt about the investigation as a whole.

The Amerithrax case is useful as the basis for a review of the critical biosurety issues to which it brought attention. The mailings demonstrated that an attack could be carried out by a trusted insider with a select agent such as *Bacillus anthracis* and have serious consequences with 5 dead, at least 17 and as many as 68 ill (2), and significant financial costs. They also heightened demands that had begun in 1995 for increased scrutiny and regulation of those who work with biological select agents and toxins (BSAT) and the handling of those agents. The pressure for more regulations has resulted in ongoing concerns that further governmental intrusion into the privacy and professional activities of biological scientists will have a chilling effect on dual-use biological agent research in the United States.

This chapter addresses some of the issues raised by Amerithrax relating to insider threat, personnel reliability, and risk factors for a domestic scientist causing harm with biological materials, either intentionally or negligently.

## THE THREAT OF MISUSE OF BIOLOGICAL AGENTS

### Biosurety Basics

At its most basic level, the problem addressed in this chapter is domestic biosurety. Biosurety is defined as the combination of four basic elements: (i) physical security (often referred to as "guns, gates, and guards"), (ii) biosafety (appropriate handling of agents and good laboratory practice), (iii) agent accountability (keeping track of agents), and (iv) personnel reliability (3,4). Many of the underlying concepts and measures evolved in the context of handling nuclear weapons and materials, cryptographic materials, and other materials and information that require a high level of security.

The goal of any biosurety program is to decrease the risks of avoidable accidents and make it as difficult as possible for those with nefarious intent to gain access to and misuse dangerous materials. It is worthwhile to review each of the four elements before turning to a discussion of specific sources of threat.

## Physical Security

The purposes of physical security systems are to deter those who seek unauthorized access to materials and facilities and to increase the likelihood that they will be detected before they succeed. The majority of attacks on organizations and thefts are perpetrated by outsiders (5), making physical security a core component of biosurety. Insider crimes are nevertheless a concern in chemical, nuclear, and biological facilities (6). The anthrax attacks are a prime example of the possibility of such events.

In a study of insider crime prepared for the U.S. Department of Energy, Hoffman and associates (7) examined 62 insider crimes in other industries, as no such crimes had occurred in Department of Energy nuclear facilities. They concluded:

> Success in most of the incidents examined seemed to depend less on detailed planning or expert execution than on the exploitation of existing security flaws. Indeed, most of these crimes did not require sophisticated planning; they were carried out against targets of opportunity. Even those companies that were heavily secured were robbed or burgled by insiders using routine access to exploit situations where security was lax. However, it should be emphasized that none of the organizations in our database employed security equivalent to that of a nuclear facility.

Noting that no organization can function without some level of trust among employees and that security precautions can be cumbersome, they pointed out the potentially devastating consequences of a nuclear crime. As such, they concluded:

> ... adequate, or even very good, protection is not enough. Total security can never be attained, nor can insider crime ever be completely prevented. However, security officials can and must keep all possibilities in mind at all times, to avoid surprises and to be prepared at least to minimize damages.

We believe that similar levels of concern and effort are appropriate with regard to biological pathogens.

The majority of security measures are relatively simple. These include (i) identification badges, (ii) proximity card systems that control access and provide a record of entrance and exit from laboratory spaces, (iii) video recording of movement within facilities, (iv) monitoring of laboratory space utilization, (v) locks, and (vi) perimeter security.

These measures are defeated easily through noncompliance, however. "Piggybacking," in which an employee holds the door open for the employee behind him or her without checking for identification, can allow unauthorized and unrecorded access. Video recording is ineffective if the system fails or no one monitors the recordings. Similarly, unusual patterns of activity that may indicate inappropriate use of space and materials can go undetected if no one monitors access logs. Finally, perimeter security can fail if personnel are poorly trained or if credential checks are forgone, as may happen with familiar employees.

## Biosafety

Concerns about biosafety, that is, safe handling of biological agents in order to avoid contamination of laboratory facilities and the community, as well as infection of laboratory personnel and the public at large, predated the events of 2001. The Centers for Disease Control and Prevention's "Biosafety in Microbiological and Biomedical Laboratories," first published in 1984 and now in its fifth edition, is an advisory manual on safe laboratory practices (8).

Because of the risk of infections of employees and potentially their family members and others in the community, scientists and health care workers are educated in safe laboratory procedures. An example of the impact of infection of laboratory workers, community reaction, and the public health response to such an event is the 2004 *Francisella tularensis* outbreak at Boston University. In that case, three laboratory workers contracted pulmonary tularemia after working with the vaccine strain of *F. tularensis* that had become contaminated with the wild type of the organism (9).

There is a greater risk of laboratory accidents when workers suffer from certain health conditions or are taking certain medications. Occupational health programs designed to monitor overall employee health and fitness, as well as more detailed personnel reliability programs (PRPs), can be effective tools for preventing such accidents. As with physical security, however, the effectiveness of these measures can be limited if protocols are not followed and if co-workers and supervisors ignore deterioration of the physical and mental health of colleagues.

## Accountability for Agent Stocks

Biological pathogens are unique among potential agents of mass destruction in that they are self-replicating. Small amounts can pose a risk, so the quantity or volume of material that goes missing may not be the most important issue; nevertheless, larger amounts could facilitate inappropriate uses. Facilities that maintain stocks of dangerous pathogens should know what is on hand, how much they hold, and monitor for disappearance of those stocks.

For both scientific and cultural reasons, it is challenging to maintain accurate inventories of biological pathogens. U.S. governmental laboratories have occasionally been subjected to criticism and even been required to "stand down" due to their failure to produce accurate documentation of current inventories, due in part to the retirements of scientists (10). From a scientific standpoint, the dynamic nature of biological pathogens and the quantities that can be produced require active inventory monitoring.

In addition, the culture of biological research supports a respect for materials of other researchers. As a result, researchers are reluctant to dispose of materials that belong to other researchers. In well-established laboratories, containers of biological pathogens have been known to remain on site even after their originating researcher has retired. For these reasons, maintaining accurate inventories is both challenging and labor-intensive.

## Personnel Reliability

Personnel reliability is the last of the four elements of biosurety and, in many ways, the most controversial. In the aftermath of the 2001 anthrax mailings, specific and substantial concerns were raised about the policies and procedures used to ensure that those who work with these agents are reliable, safe, trustworthy, and not subject to coercion by enemies of the United States. This subject is discussed in more detail later.

## CATEGORIES OF THREATS

The potential misuse of biological agents can be divided into two broad categories: negligent (discussed earlier) and intentional.

Intentional misuse is any purposeful action by which an individual gains access to or utilizes biological materials for unauthorized purposes. This can include theft of materials for the purpose of sale and personal profit, for unauthorized experimentation, that is, biohacking, or for transfer to another party for harmful purposes such as terrorism. Biological materials and facilities can also be misused intentionally in acts of sabotage against a facility by a disgruntled employee.

Intentional misuse in furtherance of an act of terrorism has posed the greatest biosecurity concern since the anthrax mailings in 2001. Dr. Ivins's actions constitute acts of terrorism within the DOJ definition[1] as evidence shows that the mailings were, in part, an effort to influence behavior by inducing fear of anthrax infection not for political motivation but in an effort to stimulate interest and support for the anthrax vaccine program. Acts may qualify as

---

[1] The unlawful use of force or violence against persons or property to intimidate or coerce a government, the civilian population, or any segment thereof in furtherance of political or social objectives. 28 C.F.R. Section 0.85.

terrorism even if there is idiosyncratic motivation that may be the product of a mental illness or other behavioral abnormality.

## The Insider Threat

Until the 2001 anthrax mailings, there was little historical evidence that insiders pose a significant risk for intentional misuse, including terrorism (5). The occurrence and consequences of that incident, as well as incidents in which animal rights activists have infiltrated research facilities, have led to justifiable concern that acts of terrorism can emanate from research and academic facilities at the hands of insiders. These insiders may infiltrate facilities with the intent to convert them to illicit use or for the purpose of sabotage or theft. Additionally, insider threats can be posed by those who undergo radicalization, experience financial problems, or develop behavioral abnormalities after legitimately joining a facility.

With regard to terrorism, as well as other crimes, insider threats may be divided into three categories, with the possibility of some overlap among them (7):

1.  Individuals who are operating on behalf of state or nonstate groups
2.  Individuals who are acting in support of a radical ideology they wish to support
3.  Individuals who are acting for idiosyncratic reasons that may relate to mental illness, substance abuse, or personality disturbance. Individuals who fall into the third category may be further subdivided into the following groups:
    a.  Disgruntled individuals seeking to cause harm
    b.  Disgruntled individuals seeking to demonstrate ability/capacity to do harm
    c.  Individuals attempting to demonstrate weakness in the system
    d.  Individuals seeking to demonstrate their own ability and to prove their worth to an organization, community, or other individuals
    e.  Those attempting to test the bounds of science and their ability through unauthorized experimentation.

Each of these categories involves insider threat, about which there is limited information in the field of biology. Insight on this issue can be obtained from literature on the subject related to the nuclear industry and information technology. Insider threats are the focus of a 2008 report by the United Nations International Atomic Energy Agency (11). That report characterizes the insider threat as follows:

> Insider threats in particular present a unique problem for a physical protection system. Insiders could take advantage of their access rights,

complemented by their authority and knowledge of a facility, to bypass dedicated physical protection elements or other provisions such as measures for safety, material control and accountancy, and operating measures and procedures. Further, as personnel with access in positions of trust, insiders are capable of carrying out "defeat" methods not available to outsiders when confronted with protection elements and access controls. Insiders have more opportunities to select the most vulnerable target and the best time to execute the malicious act.

The report suggests that motivation of the insider "may include ideological, personal, financial and psychological factors and other forces such as coercion." It recommends a series of preventive and protective measures to reduce the risks of insider attack. Included among these preventive measures are identity verification, maintaining security of information, escort and surveillance of infrequent visitors, security awareness training, and trustworthiness assessments that include medical and psychiatric evaluations.

In their report on insider attacks for the Department of Energy, Hoffman and associates (7) had to look to other industries for the 62 cases of insider crime that they examined. The goal of the study was to identify characteristics of potential criminal actions against nuclear facilities by insiders. They organized the crimes into three categories:

- Crimes committed by insiders conspiring with outsiders
- Crimes committed by insiders conspiring with other insiders
- Crimes committed by lone insiders

They concluded that the first category was the most likely for an attack on a nuclear facility in the form of a terrorist attack, but that the threat posed by the other two categories was not negligible. Of the crimes they examined that were committed for other than financial reasons, most were committed by lone insiders who, on the whole, were less emotionally stable than other criminal insiders in the study.

One field that has had considerable experience with insider attacks is information technology. Schultz (12) notes the varied definitions of insiders and describes several models of insider attacks. Each is a variation on the "CMO model," which proposes that the inside attacker has capability, motivation, and opportunity. Among the models he discusses are those by Tuglular and Spafford (13) and the work of Shaw and associates (14). The former propose that insider misuse arises from a combination of personal characteristics, motivation, knowledge, abilities, rights and obligations, authority, and responsibility within the organization, as well as factors related to group support. They point out that insider attacks are more likely to occur in organizations in which there has been a breakdown of lines of authority. Shaw and associates (14) posit that the majority of inside attackers are introverted, are poor at handling stress or conflict, and are frustrated with work, although they note that the base rate of introversion as a personality characteristic in computer scientists is as high as 40%.

## The Problem with Profiles

The quest for a screening tool that relies on personality and other characteristics to detect employees at risk of posing an insider threat is understandable, but remains unfulfilled. Like other low incidence phenomena, such as workplace violence and suicide, inside attacker profiles suffer from the problem that even a highly sensitive test will yield an unacceptably high level of false positives. As a result, many individuals will be identified falsely as posing a risk (15). Conversely, there is a problem with false negatives when unvalidated measures are utilized to assess risk. Related problems arise due to the small size of the study sample and the difficulties that arise in testing these profiles empirically, that is, determining base rates of the proposed traits in an appropriate cohort and then testing criteria through double-blind studies.

The absence of reliable profiles allowing for prospective identification of those who have engaged or are likely to engage in nefarious acts of violence was demonstrated in studies of presidential assassins (16) and school shooters (17) conducted under the auspices of the U.S. Secret Service, as well as a study of college campus shooters (18). Shaw and colleagues (14) note that Project Slammer, a study of Americans convicted of espionage, who were mostly insiders, found an interaction of multiple factors and characteristics, none of which alone was predictive. Psychometric testing in this area, and terrorism in general, is of unproven validity and reliability (19).

## Risk Factors for Violence

Terrorism is a subset of violent behavior. As such, it is useful to look to other violence research for insights into the potential for this type of behavior. The Secret Service projects noted earlier, as well as other research on violence, have identified risk factors for violence.

Violence is often overly attributed to mental illness and it should be noted that other than in the case of lone actors, there is no indication of an association between mental illness and acts of terrorism. Consistent with other recent research on the subject, Elbogen and Johnson (20) showed that while the incidence of violence is increased among individuals with severe mental illness, that increase is statistically significant only among those mentally ill persons who also have co-occurring substance abuse or dependence. They found that severe mental illness alone did not predict violent behavior. Rather, it was associated with historical factors (past violence, juvenile detention, physical abuse, and parental arrest record), clinical factors (substance abuse and perceived threats), dispositional factors (younger age, male gender, and low income), and contextual factors (recent divorce, unemployment, and victimization).

Terrorism is targeted violence. It has been observed (16,21) that acts of targeted violence arise from an interaction among the potential perpetrator, past stressful events, situational factors, and the potential target. Individual characteristics of the potential attacker, such as trait anger, past history of difficulty handling stress, and mental illness, are considered. Situational factors, such as the type of recent stress and factors that might dissuade a potential attacker, are also taken into account. Finally, characteristics of the potential subject of attack are examined, for example, level of security, interactions with the potential perpetrator, the work situation, and the perpetrator's knowledge of the potential victim. The following risk factors were identified for such crimes: male gender; past history of violence, presence of mental illness and substance abuse, and psychopathic (antisocial) traits. The similarity between these findings and Elbogen and Johnson's is noteworthy.

Terrorism may also be considered an act of mass murder, and literature (22,23) on that subject may be informative with regard to terrorist actions by those with idiosyncratic, as opposed to purely political, motives.

In light of these observations, the risk factors of severe mental illness, substance abuse, and antisocial behavior must be considered in any threat assessment involving an identified individual. However, it must be kept in mind that the prevalence of these specific factors in the general population far exceeds the incidence of acts of violence of any type, including terrorism or insider threats. Threat assessments should be conducted by experienced individuals who are objective and able to appreciate the often subtle interactions among the various elements of risk. The focus of assessments must be on past and current behaviors and active demonstration of clearly identified risk factors rather than diagnostic labels or other arbitrary indicators.

## THE CHALLENGE OF PROTECTING AGAINST INSIDER THREATS

The authors of this chapter believe that the risk of insider attacks is real and, while small at present, likely to grow. The increasing evidence of radicalization among naturalized and native-born U.S. citizens is one source of risk from which scientists are not immune. We also know that scientists are not immune from the individual, environmental, and situational risk factors for violence or negligent handling of materials, including mental illness and substance abuse. In order to visualize the future of the scientific workplace, one need only recognize substantial changes occurring within colleges and universities. With the advent of effective treatments for severe depression, anxiety, bipolar disorder, and psychoses, academic institutions are now able to admit students whose mental illnesses might have prevented their admission in prior decades. Mental health services within universities have expanded significantly in order to service the increasing numbers of students who receive treatment and continue to function in the academic world. These students are joining the workforce and will be productive despite their illnesses. Still, the changing workplace requires greater situational awareness and an appreciation of the enhanced mental health needs of the emerging workforce.

## Current Biosecurity Programs

The events of 2001 resulted in dramatically expanded funding for research with BSAT. After the attacks, and well as prior to Dr. Ivins's identification as the perpetrator, the Department of the Army instituted the Biological Personnel Reliability Program with Army Regulation (AR) 50-X, Army Biological Surety Program. The current version of this program is AR 50-1, dated July 28, 2008 (24).

Current regulations for possessing, using, and transferring select agents and toxins working in private and university facilities are contained in three sets of regulations that govern public health, animals and animal products, and agriculture (25–27). These contain measures to ensure that those who work with BSAT are safe and do not pose a security risk, broadly referred to as "personnel reliability programs."

Pursuant to these regulations, individuals who work with BSAT must have a security risk assessment (SRA) conducted by the Department of Justice to ensure that restricted persons,[2] as defined by statute (28), are not given access to any select agent or toxin. The SRA is limited to a multiagency check for

---

[2] A restricted person is one who

(a) is under indictment for a crime punishable by imprisonment for a term exceeding 1 year
(b) has been convicted in any court of a crime punishable by imprisonment for a term exceeding 1 year
(c) is a fugitive from justice
(d) is an unlawful user of any controlled substance [as defined in Section 102 of the Controlled Substances Act (21 U.S.C. Section 802)]
(e) is an alien in the United States illegally or unlawfully
(f) has been adjudicated as a mental defective or has been committed to any mental institution
(g) is an alien (other than an alien lawfully admitted for permanent residence) who (other than an alien admitted lawfully for permanent residence) who is a national of a country as to which the Secretary of State pursuant to section 6(j) of the Export Administration Act of 1979 [50 U.S.C. App. 2405(j)], Section 620A of Chapter 1 of Part M of the Foreign Assistance Act of 1961 (22 U.S.C. Section 2371), or Section 40(d) of Chapter 3 of the Arms Export Control Act [22 U.S.C. Section 2780(d)] has made a determination (that remains in effect) that such country has repeatedly provided support for act of international terrorism;
(h) has been discharged from the Armed Services of the United States under dishonorable conditions

restricted person characteristics; it is not a full security clearance as with AR 50-1. In addition, these individuals must sign a form certifying that they do not fall within one of the categories of restricted persons.

One category of restricted persons includes those who have been "adjudicated as a mental defective," which means that "a court, board, commission, or other lawful authority has determined that he or she, as a result of marked subnormal intelligence, or mental illness, incompetency, condition, or disease, (i) is a danger to himself, herself, or others or (ii) lacks the mental capacity to contract or manage his or her own affairs." The term "adjudicated as a mental defective" explicitly includes a finding of not guilty by reason of insanity or incompetence to stand trial (29).

We note that the mental defective/civil commitment criterion is both overly broad and unduly limited for its avowed purpose. First, high-functioning, trustworthy individuals may experience an acute onset of mental illness that results in involuntary hospitalization and then subsequently return to full function. To disqualify a scientist from his or her work automatically and without further inquiry because of such an event is unnecessary. Conversely, there are many behavioral abnormalities that raise serious questions about an individual's safety, stability, and security that fall far short of resulting in a "mental defective" adjudication or involuntarily hospitalization. In addition to being a poor standard for exclusion from this line of research, the term "mental defective" is archaic and insulting to the millions of Americans who suffer from mental illness.

The SRA is limited to confirmation of identity and a multiagency check to determine that the applicant is not a restricted person. It does not capture aberrant behavior that has not resulted in an indictment or conviction. An additional problem for SRAs is that they are reliant upon the states to report adjudications regarding mental health issues. Such reporting may be slow, incomplete, or nonexistent (30).

## Advantages and Disadvantages of PRPs

Few would likely argue with the goal of maintaining biosecurity within the research community. In fact, the National Science Advisory Board for Biosecurity (NSABB) recommended in 2009 that "the culture of responsibility and accountability should be enhanced at institutions that conduct select agent research." Although some in the scientific community (31) have accepted that the Amerithrax incident will result in more stringent individual PRPs, consensus about the value and elements of mandatory programs is lacking. Indeed, the NSABB opined that a PRP similar to AR 50-1 was not necessary in non-DoD laboratories.

The movement to enforce and expand PRPs in biological research facilities has given rise to concerns that scientists will be dissuaded from working in this area by the prospect of scrutiny of their personal and professional lives, thus putting the United States at a scientific disadvantage. While the formality of such procedures has increased, the essential elements of these programs were in place prior to the most recent modifications in a number of research settings, with no apparent negative impact. All employees working with nuclear materials, including scientists, have had to undergo security background checks for many years. The same was true for scientists working at USAMRIID prior to 2001, albeit in a secure setting. In addition, those scientists were monitored for physical and mental health problems under the Special Immunization Program, under which they granted access to their medical and psychiatric records. A number of research facilities have adopted significant security measures, again with no reported adverse effects on their work (31).

Regardless of the details of personnel security programs, the hazard may lie in how those programs are executed. Review of the Amerithrax materials indicates that had all available aspects of the screening measures been applied, Dr. Ivins likely would not have been granted a security clearance before or after the mailings.

As noted, PRPs have been utilized successfully within the energy and chemical industries by researchers for many years. Research scientists within those industries have accepted the PRP process without significant attrition, and it is unclear why similar success could not be expected within the biological research community. While not advocating that nuclear and chemical industry PRPs be adopted, in his Senate testimony, Greenberger recommended the improvement of personnel reliability assessments within that community (32).

One argument for more detailed PRPs is that they provide a "paper trail" that can be used in the investigation of any incident. With that documentation, there is also increased accountability for supervisors. Detailed background investigations, medical evaluations, review of medical records, drug and alcohol testing, and interviews of character references are components of a process that yields a documented record. Because it is a part of the routine, such a process enhances situational awareness through its day-to-day administration, emphasizing the importance of biosurety for both research scientists and their supervisors.

If implementation of PRPs is experienced as onerous and "user-unfriendly," it is argued that scientists will be dissuaded from working with BSAT out of concerns that their right to privacy will be compromised. Because it could thin the ranks of researchers, it could paradoxically result in a greater biological threat to the nation—the very thing that PRPs are designed to decrease. Some have argued that the nuclear and chemical industries should not be used as models for BSAT programs (33). Furthermore, as argued in the May 2009 NSABB report, there is little evidence that PRP measures have been proven to be effective in identifying insiders within institutions who might pose a threat. As a result, the report (34) concludes that a national personnel reliability program is currently unnecessary. As noted earlier, these fears do not seem to have been realized at facilities that have employed increased security screening measures. A study of the impact of the increased regulation of BSAT research indicated that they did not have a negative impact on the amount of research in the field but did decrease both the efficiency of that research and the amount of international collaboration (35).

This ultimately begs the question: what measures should institutions adopt to ensure personnel reliability with regard to both biosafety and biosecurity? The decision to implement security measures is based on perceived risk, which can be seen as the product of consequences and probability, that is, Risk = Consequences × Probability. The measure of efficacy of any security measure is that adverse events do not happen; deterrence effects cannot be ignored. The fact that personnel security measures have not detected insider threats in biological research facilities should not be a basis for forgoing effective measures in securing these facilities. Here, as in many situations, it is worth remembering, "Absence of evidence is not evidence of absence."

In addition to thorough SRAs, we believe that the following elements should be considered in PRPs.

## Mental and Physical Health Screening

For the scientist who is found to be qualified for work with BSAT, there are ongoing needs for mental and physical health screening. Health screening is an extremely broad concept. Depending on its scope, it may or may not yield relevant and actionable information. The term "health screening" raises more questions than it answers. Who is doing the screening? Is it by self-report or administered by a health professional; if so, by whom? In addition to accuracy of the content of the screening, the frequency of administration is also important. At its simplest level, self-report screening forms require review by a qualified medical professional, with follow-up of items that are relevant to any of the four elements of biosurety. For screening to be truly useful there must be an ability to monitor the consistency of information year by year. When disparities are identified, they require follow-up by a qualified professional who does not have a treatment relationship with the subject. Still more comprehensive is the ability to contact medical and mental health professionals who have been listed as treatment providers. While this may be seen as intrusive for some, it does provide for much greater assurance that the answers are truthful. For example, a substance abuser who neglects to report a history of that condition may not be identified unless there is cross-validation by the personal physician.

## Background Checks

The term "background check" is overly broad and misunderstood in the context of BSAT. As noted earlier, standard SRAs (as opposed to security clearances) are limited to issues surrounding the restricted person standard.

More detailed background checks are required for entry into positions that require security clearances and should be considered for those working with BSAT. As with all independent background security assessments, they can entail significant expense, and privacy of health records and other sensitive information must be guarded stringently. At their most basic, background checks include automated inquiry into law enforcement databases that reflect documentation of prior criminal convictions. More extensive background checks require that an individual or individuals independently verify information provided on the initial application form. Additionally, individuals providing references may be interviewed personally with visits to current and past neighborhoods and places of business. Depending on the level of security clearance required, polygraph examinations may also be required.

Repeat background checks are required periodically for individuals to maintain security clearances (31) or when warranted due to a position change. These background checks may entail the same level of scrutiny as at entry or may be modified depending on the position and the security needed.

Given their limited nature and problems with state reporting, are current SRAs adequate? Review of the Amerithrax investigative materials suggests that investigations limited to convictions and other legal adjudications may not be sufficient to detect some who should be excluded from access. Understandably, implementation of new standards for background checks can arouse anxiety within an existing or potential workforce. The prospect of expanding current measures is controversial, yet we suggest that it, along with other measures, must be considered within the research community before harsher measures are imposed from outside.

## The Way Ahead

As we move forward in addressing the problems of biosurety, systems must address two fundamental problems. First, one of the problems with current systems is that they are plural rather than singular in nature. The presence of multiple systems engenders confusion of biosurety terms such as "background checks" and "health screening" while maintaining seams that allow for security gaps. If, for example, all BSAT systems, public and private, provide for mental and physical health screening but execute the process in radically different ways, they cannot be discussed sensibly, much less examined thoughtfully. Without a central authority to standardize systems, there is much less ability for monitoring metrics of success and to effect change.

Cost is another problem. The growth in research with these agents itself has posed its own problems and has had a significant impact on the BSAT infrastructure. Although select agents have been transported quite successfully since the adoption of the Special Agent Program (only one confirmed loss of a BSAT shipment in 20 years), a representative for the primary private carrier has expressed concern that the expense of increased security regulations may prohibit transport in the future (31).

Certainly, there is a strong argument for heterogeneity of public and private systems themselves. This diversity can provide for approaches to complex problems that are both creative and cost-effective. In order to maintain a robust environment of private and public BSAT facilities, one could argue that an umbrella biosurety approach helps ensure that private sector systems will remain viable. There is a danger in allowing the current lack of biosurety standardization to remain across BSAT facilities. In the event that the United States is faced with another insider-based attack, the controls imposed reactively could be extremely draconian in nature to the extent that private sector BSAT facilities are threatened.

Although the nuclear and chemical industries are often cited as examples for the biological research community, potential models for maintaining safety and high-performance standards exist outside of the nuclear and chemical industries. These existing systems have maintained heterogeneity while adopting central standards. As one example, the Joint Commission for Accreditation of Hospitals is composed of health professionals responsible for accreditation of a myriad of public and private medical facilities in the United States. These facilities have retained their individual identities while meeting certain evolving standards. A similar model may one day provide for enforceable standards of biosurety in public and private laboratories responsible for BSAT-related activities.

The Amerithrax case has prompted Senate testimony that consolidated oversight responsibilities within a single agency should replace current fragmented federal oversight of biosafety laboratories (32). In that testimony, Greenberger advocated a private sector model that is separate from military BSL laboratories, but allows for "strong, but appropriate and practical biosecurity procedures."

## CONCLUSION

Debate over the nature and extent of biosurety programs is not over. Regardless of their elements, however, no biosurety system can be successful without the endorsement and full cooperation of laboratory directors, principal investigators, and the research community at large. The fact that an individual does not pose a risk when hired does not mean that life and personal circumstances will not change. Initial screening of those working with BSAT should be discrete, based on valid risk factors, thorough, executed expertly, and repeated at intervals or in response to behavioral and other life changes. With an enhanced biosurety program in place, risks can be detected and addressed before adverse events occur. The Amerithrax case highlights the need for a system ensuring that co-workers, as well as supervisors, maintain a sense of responsibility for the well-being of each other and the safety and security of their laboratories.

# REFERENCES

[1]  U.S. Department of Justice. Amerithrax Investigative Summary, 2010. Available from: http://www.justice.gov/amerithrax/docs/amx-investigative-summary.pdf.

[2]  T.C. Cymet, G.J. Kerkvliet, What is the true number of victims of the postal anthrax attack of 2001? J. Am. Osteopath. Assoc. 104 (11) (2004) 452.

[3]  K. Carr, E.A. Henchal, C. Wilhelmsen, B. Carr, Implementation of biosurety systems in a department of defense medical research laboratory, J. Biosecurity Bioterrorism 2 (2004) 7–16.

[4]  G.L. Demmin, Biosurety, in: Z.F. Dembek (Ed.), Medical Aspects of Biological Warfare, Office of the Surgeon General Walter Reed Army Medical Center, Washington, DC, 2007, pp. 435–558.

[5]  M.E. Kosal, Terrorism targeting industrial chemical facilities: Strategic motivations and the implications for U.S. security, Stud. Conflict Terrorism 29 (2006) 719–751.

[6]  Commission on the Prevention of WMD Proliferation and Terrorism, World at Risk, Vintage Books, New York, 2008.

[7]  B. Hoffman, C. Meyer, B. Schwarz, J. Duncan, Insider Crime: The Threat to Nuclear Facilities and Programs, Rand Corporation, Santa Monica, CA, 1990.

[8]  Centers for Disease Control and Prevention, U.S. Department of Health and Human Services, Biosafety in Microbiological and Biomedical Laboratories, 5th ed., U.S. Government Printing Office, Washington, DC, 2007.

[9]  A.M. Barry, Report of Pneumonic Tularemia in Three Boston University Researchers, Boston Public Health Commission, Boston, 2005. Available from: http://www.bphc.org/programs/cib/environmentalhealth/biologicalsafety/Forms%20%20Documents/tularemia_report_2005.pdf.

[10] Y. Bhattacharjee, Bioweapons Lab "Stand Down" Letter Posted. ScienceInsider, February 9, 2009. Available from: http://news.sciencemag.org/scienceinsider/2009/02/bioweapons-lab.html.

[11] United Nations International Atomic Energy Agency, Preventive and Protective Measures against Insider Threats, United Nations, Vienna, 2008.

[12] E.E. Schultz, A framework for understanding and predicting insider attacks, Comput. Security 29 (6) (2002) 526–531.

[13] T. Tuglular, E.H. Spafford. A Framework for Characterization of Insider Computer Misuse. Unpublished paper, Purdue University, West Lafayette, IN, 1997.

[14] E. Shaw, K.G. Ruby, J.M. Post, The insider threat to information systems: The psychology of the dangerous insider, 2001. Available from: http://www.pol-psych.com/pubs.html.

[15] R. Schouten, Workplace violence and the clinician, in: R.I. Simon, K. Tardiff (Eds.), Violence Assessment and Management, American Psychiatric Press, Washington, DC, 2008, pp. 501–520.

[16] R.A. Fein, B. Vossekuil, Assassination in the United States: An operational study of recent assassins, attackers, and near-lethal approachers, J. Forensic Sci. 44 (2) (1999) 321–333.

[17] B. Vossekuil, M. Reddy, R. Fein, Safe School Initiative: An Interim Report on the Prevention of Targeted Violence in School, U.S. Secret Service, Washington, DC, 2000.

[18] D. Drysdale, W. Modzeleski, A. Simons, Campus Attacks: Targeted Violence Affecting Institutions of Higher Education. U.S. Secret Service, U.S. Department of Homeland Security, Office of Safe and Drug-Free Schools, U.S. Department of Education, and Federal Bureau of Investigation, U.S. Department of Justice. Washington, DC, 2010. Available from: http://www.fbi.gov/publications/campus/campus.pdf.

[19] M. Dernevik, A. Beck, M. Grann, T. Hogue, J. McGuire, The use of psychiatric and psychological evidence in the assessment of terrorist offenders, J Forens. Psychiatry Psychol. 20 (4) (2009) 508–515.

[20] E. Elbogen, S.C. Johnson, The intricate link between violence and mental disorder, Arch. Gen. Psych. 66 (2) (2009) 152–161.

[21] R. Borum, R. Fein, B. Vossekuil, J. Berglund, Threat assessment: defining an approach for evaluating risk of targeted violence, Behav. Sci. Law 17 (3) (1999) 323–337.

[22] R.M. Holmes, S.T. Holmes, Understanding mass murder: A starting point, Fed. Probation 56 (1) (1992) 53–61.

[23] J.R. Meloy, A.G. Hempel, B.T. Gray, K. Mohandie, A. Shiva, T.C. Richards, A comparative analysis of North American adolescent and adult mass murderers, Behav. Sci. Law 22 (3) (2004) 291–309.

[24] U.S. Department of the Army, Army Biological Surety Program, Department of the Army, Washington, DC, 2008 Army Regulation 50-1.

[25] 42 Code of Federal Regulations Sec. 73.

[26] 9 Code of Federal Regulations Sec. 121.

[27] 7 Code of Federal Regulations Sec. 331.

[28] 18 United States Code 175b.

[29] 27 Code of Federal Regulations Sec. 478.11.

[30] Office of the Inspector General, U.S. Department of Justice, Inspection of the FBI's Security Risk Assessment Program for Individuals Requesting Access to Biological Agents and Toxins, Department of Justice, Washington, DC, 2005, E & I Report No. I-2005-003.

[31] U.S. Department of Health and Human Services. Report of the Working Group on Strengthening the Biosecurity of the United States, 2009. Available from: http://www.hhs.gov/aspr/omsph/biosecurity/biosecurity-report.pdf.

[32] Strengthening Security and Oversight at Biological Research Laboratories. Hearings before the Subcommittee on Terrorism and Homeland Security, of the Senate Committee on the Judiciary, 111th Congress, 2009 (testimony of Michael Greenberger).

[33] Strengthening Security and Oversight at Biological Research Laboratories. Hearings before the Subcommittee on Terrorism and Homeland Security, of the Senate Committee on the Judiciary, 111th Congress, 2009 (testimony of Jean D. Reed).

[34] National Science Advisory Board for Biosecurity, Enhancing Personnel Reliability among Individuals with Access to Select Agents, Author, Washington, DC, 2009.

[35] M.B. Dias, L. Reyes-Gonzalez, F.M. Veloso, E.A. Casman, Effects of the USA PATRIOT Act and the 2002 Bioterrorism Preparedness Act on select agent research in the United States, PNAS Early Edition (2010) 1-6, available at www.pnas.org/cgi/doi/10.1073/pnas.0915002107.

References

# Forensic Public Health: Epidemiological and Microbiological Investigations for Biosecurity

**Ali S. Khan and Nicki Pesik**

*National Center for Zoonotic, Vector-Borne, and Enteric Diseases,*
*Centers for Disease Control and Prevention, Atlanta, Georgia*

## INTRODUCTION

Epidemiology is the study of how disease is distributed in populations and of the factors that influence this distribution (1). More broadly, it is the study of the distribution and determinants of health-related states or events in specified populations and application of this study to control health problems (2). Epidemiology is based on the premise that disease, illness, and ill health are not distributed randomly in a population and that individuals have certain characteristics (e.g., genetic) that interact with the environment and predispose to, or protect against, a variety of different diseases. The specific objectives of epidemiology (1) are to (i) identify the etiology or cause of a disease and the factors that increase a person's risk for disease; (iii) determine the extent of disease found in the community; (iii) study the natural history and prognosis of disease; (iv) evaluate new preventive and therapeutic measures and new modes of health care delivery; and (v) provide a foundation for developing public policy and regulations. This chapter discusses how epidemiology integrated with laboratory science can be used to identify the source of diseases caused by microorganisms or toxins.

## DYNAMICS OF DISEASE TRANSMISSION

Disease has been classically described as the result of an epidemiological triad, where disease results from the interaction of the human host, an infectious agent or toxin, and the environment that promotes the exposure (1). In some instances, an animal or an arthropod vector such as a mosquito or tick is involved to maintain or transmit the pathogen. Among the assumptions necessary for this interaction to take place is that there is a susceptible host. The susceptibility of the host is influenced by a variety of factors, including

Microbial Forensics. DOI: 10.1016/B978-0-12-382006-8.00015-3

genetic, nutritional, and immunological factors. Bacteria, viruses, prions, fungi, and parasites responsible for disease can be transmitted in either a direct or an indirect fashion (Table 15.1). Different organisms spread in different ways, and the potential of a given organism to spread and produce outbreaks depends on the characteristics of the organism and the route by which it is transmitted from person to person.

Diseases can be defined as endemic, epidemic, and pandemic. *Endemic* can be defined as either the habitual presence of a disease within a given geographical area or as the usual occurrence of a given disease within such an area. *Epidemic* can be defined as the occurrence in a community or region of disease, clearly in excess of normal expectancy, and generally derived from a common source or from a propagated source. *Epidemic* and *outbreak* are interchangeable but are used differentially to imply degrees of severity. A cluster implies an apparent excess of cases that may or may not be normal pending an epidemiological investigation. *Pandemic* refers to a worldwide epidemic. The usual or expected level of a disease is determined through ongoing *surveillance*.

Using a number of different strategies and mechanisms, microorganisms are very efficient at infecting humans. These are exemplified both by the various strategies employed by the pathogen to survive prior to infecting a host, such as spore formation or harboring in drought-resistant mosquito eggs, and by the various modes of transmission, for example, direct contact (including large droplets) or indirect contact with fomites, or by insect vectors, and airborne via small particle droplets (3). Natural experiments, however, have highlighted the true diversity in the abilities of microorganisms to infect humans and animals: *Salmonella* outbreaks due to contaminated alfalfa sprouts (4) and to ice cream made from milk contaminated in a tanker that had previously contained raw eggs (5), legionellosis associated with grocery store

---

**Table 15.1** Modes of Agent Transmission (Modified from Ref. 1)

Horizontal (transmission from one individual to another in the same generation)

Direct transmission
- Direct contact (touching, biting, sexual intercourse, etc.)
- Direct projections (large droplet spread, e.g., coughs to mucous membranes)
- Direct exposure (animals, soils)

Indirect transmission
- Vehicle borne (fomites, blood transfusion)
- Vector borne (mechanical or biological propagation)
- Airborne (droplet nuclei or dust)

Vertical (transmission from mother to offspring)

misters (6), and pneumonic tularemia on Martha's Vineyard from mowing over a rabbit (7). These few examples are a semblance of the seemingly endless list of novel ways that agents and their vectors are spread. The ability to exploit newly created biological conditions is both the hallmark and the challenge of emerging infections (8). Many factors contribute to the emergence of infectious diseases, including human susceptibility to infection, international travel and trade, microbial adaption and change, and intent to harm (9). In 2003, African rodents infected with monkeypox reached Texas, resulting in infections of persons in the United States (10). Translocation of Rift Valley fever virus from Africa to the Arabian Peninsula and West Nile virus to the United States has had a more global impact (11,12), as does chikungunya virus or the recently identified henipaviruses, which pose significant threats in southeast Asia and Oceanian (13). Many of the biological threat agents are also considered to be reemerging or emerging infectious pathogens. Viral hemorrhagic fever viruses are considered high-priority threat agents and are a concern as an emerging disease, as illustrated in ongoing outbreaks in Africa and the recent identification of a novel arenavirus in Zambia (14). Approximately 60% of emerging infections are zoonotic pathogens (diseases transmitted from animal to human), whereas vector-borne diseases account for approximately 23% of emerging infections (15). For early detection and recognition of emerging infections, it is critical that proper epidemiologic investigations are integrated with laboratory surveillance (9).

## OUTBREAK INVESTIGATION

Occurrence of a disease at more than an endemic level may stimulate an investigation during which investigators may ask three questions. *Who* has the disease? The answer to this question will help identify those characteristics of the human host that are closely related to disease risk. *When* did the disease occur? Some diseases occur with a certain periodicity. This question is also addressed by examining trends of disease incidence over time. *Where* did the cases rise? Answers to the previous questions lead to determining the *how* and *why* of an outbreak. Disease is not distributed randomly in time and place. These questions are central to virtually all outbreak investigations. Investigation of an outbreak may be primarily deductive (i.e., reasoning from premises or propositions proved antecedently), inductive (i.e., reasoning from particular facts to a general conclusion), or a combination of both. Important considerations in the investigation of acute outbreaks of infectious disease include (i) determining that an outbreak has in fact occurred, (ii) defining the population at risk, (iii) determining the method of spread and reservoir, and (iv) characterizing the agent. Steps used commonly for investigating an outbreak are shown in Table 15.2.

| **Table 15.2** Commonly Used Steps in Investigation of Infectious Disease Outbreak |
| --- |
| **Step 1. Verify the diagnosis** |
| **Step 2. Establish a case definition (person, place, and time)** |
| **Step 3. Identify cases** |
| **Step 4. Verify you have an epidemic (descriptive epidemiology)** |
| Time: Look for temporal clustering and time–place interactions |
| Place: Look for geographic clustering |
| Person: Examine the risk in subgroups of affected population according to personal characteristics: sex, age, residence, occupation, social groups, etc. |
| Look for combination (interactions) of relevant variables |
| **Step 5. Develop hypotheses based on the following:** |
| Existing knowledge (if any) of the disease |
| Analogy to diseases of known etiology |
| **Step 6. Test hypotheses** |
| Further analyze existing data (e.g., case-control studies) |
| Collect additional data, environmental samples, animal/vectors |
| **Step 7. Recommend and implement control and prevention measures** |
| Control of present outbreak |
| Prevention of future similar outbreaks |
| **Step 8. Communicate findings** |

# DELIBERATE INTRODUCTION OF A BIOLOGICAL AGENT

Deliberate dissemination of a biological agent via a number of different routes, including air, water, food, and infected vectors, presents the latest challenge to global public health. The deliberate nature of such dissemination may be obvious, as in the case of multiple mailed letters containing spores of *Bacillus anthracis*. However, some forms of bioterrorism may be more covert, for example, the deliberate contamination of salad bars in The Dalles, Oregon, in 1984 by a religious cult in an effort to test their ability to incapacitate the local population prior to an election (16). This outbreak, which sickened more than 750 persons, was specifically excluded as bioterrorism during the initial investigation and was only recognized as such following a tip from an informant (16,17). Given the natural ability of infectious agents to emerge, the Oregon outbreak serves to highlight difficulties in determining a characteristic signature for an infectious disease outbreak resulting from deliberate transmission.

Difficulties in identifying a covert dissemination of a biological agent are exemplified by the aforementioned investigation of a food-borne outbreak with a very unusual pattern and a rare strain of *Salmonella typhimurium* (16). Although the possibility of intentional contamination was considered early in the investigation, it was specifically excluded for the following reasons: (i) such an event had never been reported previously; (ii) no one claimed responsibility; (iii) no disgruntled employee was identified; (iv) no motive was apparent; (v) the epidemic curve suggested multiple exposures, which was presumed to be unlikely behavior for a saboteur; (vi) law enforcement officials failed to establish a recognizable pattern of unusual behavior; (vii) a few employees had onset of illness before the patrons, suggesting a possible inside source of infection; (viii) the outbreak was biologically plausible—even if highly unlikely; and (ix) it is not unusual to be unable to find a source in even highly investigated outbreaks. Although one of the initial reasons to exclude terrorism (i.e., no prior incidents) is no longer applicable, based on similar actions since 1984, determining if an unusual outbreak is biologically plausible will remain a challenge. In this context, it is important to remember that the first case of inhalation anthrax identified in Florida in 2001 was initially thought to be natural. It is clear from the two documented cases of bioterrorism in the United States—the 1984 Oregon *Salmonella* outbreak and the 2001 anthrax attack—that a terrorist will not necessarily announce his/her intentions or take credit for such an attack (16,18). Similarly, divining motives behind an attack should be abandoned as a public health tool to assess whether an outbreak is natural or deliberate in nature. Fortunately, a number of epidemiological clues, alone or in combination, may suggest that an outbreak is deliberate. It is essential to make this determination not only from a law enforcement standpoint to prevent future such actions, but to protect the public health. There is a very short "window of opportunity" in which to implement postexposure prophylaxis for many of the agents likely to be used for bioterrorism (19). Therefore, it is critical that all outbreaks be rapidly investigated and assessed for whether they are of deliberate origin.

A set of epidemiological clues (Table 15.3) has been proposed by the Department of Health and Human Service's Centers for Disease Control and Prevention (CDC) in collaboration with the Federal Bureau of Investigation (FBI) (20). These clues are based on distinctive epidemiology and laboratory criteria of varying specificity to evaluate whether an outbreak may be of deliberate origin. The clues focus on aberrations in the typical characterization of an outbreak by person, place, and time in addition to consideration of the microorganism. Some of the clues, such as a community-acquired case of smallpox, are quite specific for bioterrorism, whereas others, such as similar genetic typing of an organism, may simply denote a natural outbreak. A combination of clues, especially those that suggest suspicious point source outbreaks, will increase the probability that the event is likely due to terrorism. Although these clues are an important set of criteria to help evaluate outbreaks, no list will replace sound epidemiology to assess an outbreak.

**Table 15.3** Epidemiological Clues That May Signal a Biological or Chemical Terrorist Attack (Modified from Ref. 20)

1. Single case of disease caused by an uncommon agent (e.g., glanders, smallpox, viral hemorrhagic fever, inhalation, or cutaneous anthrax) without adequate epidemiologic explanation
2. Unusual, atypical, genetically engineered, or antiquated strain of agent (or antibiotic resistance pattern)
3. Higher morbidity and mortality in association with a common disease or syndrome or failure of such patients to respond to usual therapy
4. Unusual disease presentation (e.g., inhalation anthrax or pneumonic plague)
5. Disease with an unusual geographic or seasonal distribution (e.g., plague in a nonendemic area, influenza in the summer)
6. Stable endemic disease with an unexplained increase in incidence (e.g., tularemia, plague)
7. Atypical disease transmission through aerosols, food, or water in a mode suggesting sabotage (i.e., no other possible physical explanation)
8. No illness in persons who are not exposed to common ventilation systems (have separate closed ventilation systems) when illness is seen in persons in close proximity who have a common ventilation system
9. Several unusual or unexplained diseases coexisting in the same patient without any other explanation
10. Unusual illness that affects a large, disparate population (e.g., respiratory disease in a large heterogeneous population may suggest exposure to an inhaled pathogen or chemical agent)
11. Illness that is unusual (or atypical) for a given population or age group (e.g., outbreak of measles-like rash in adults)
12. Unusual pattern of death or illness among animals (which may be unexplained or attributed to an agent of bioterrorism) that precedes or accompanies illness or death in humans
13. Unusual pattern of death or illness in humans that precedes or accompanies illness or death in animals (which may be unexplained or attributed to an agent of bioterrorism)
14. Ill persons who seek treatment at about the same time (point source with compressed epidemic curve)
15. Similar genetic type among agents isolated from temporally or spatially distinct sources
16. Simultaneous clusters of similar illness in noncontiguous areas, domestic or foreign
17. Large numbers of cases or unexplained diseases or deaths

It is important to note that epidemiological clues can only be assessed in the context of a rapid and thorough epidemiological investigation. Not surprisingly, ongoing surveillance to identify increases in disease incidence is both the first step and the cornerstone of bioterrorism epidemiology. The majority of the clues described in Table 15.3 simply suggest an unusual cluster of cases. They have been reorganized by specificity to trigger increasingly broader investigations by state and federal public health officials and to alert law enforcement authorities (Tables 15.4 and 15.5). However, even the most specific of clues may signal a new natural outbreak. An epidemiological investigation should consider all potential sources and routes of both natural and potential deliberate exposure. For example, the recent community outbreak of individuals with smallpox-like lesions in the Midwest may, on first blush, have indicated the deliberate release of smallpox virus. However, a thorough integrated epidemiological and laboratory investigation identified the disease as monkeypox an exotic disease in the United States, which in itself should suggest bioterrorism (10). Instead, affected individuals were infected by prairie

**Table 15.4** Recommendations for Level of Public Health Involvement for Investigation of Potential Biologic or Chemical Terrorism (Modified from Ref. 20)

Initial investigation at local level

a. Higher morbidity and mortality than expected, associated with a common disease or syndrome
b. Disease with an unusual geographic or seasonal distribution
c. Multiple unusual or unexplained disease entities coexisting in the same patient
d. Unusual illness in a population (e.g., renal disease in a large population, which may be suggestive of toxic exposure to an agent such as mercury)
e. Ill persons seeking treatment at about the same time
f. Illness in persons suggesting a common exposure (e.g., same office building, meal, sporting event, or social event)

Continued investigation with involvement of state health department and/or Centers for Disease Control and Prevention

a. At least a single, definitively diagnosed case(s) with one of the following:
   ■ Uncommon agent or disease
   ■ Illness due to genetically altered organism
b. Unusual, atypical, or antiquated strain of agent
c. Disease with unusual geographic, seasonal, or "typical patient" distribution
d. Endemic disease with unexplained increase in incidence
e. No illness in persons not exposed to common ventilation systems
f. Simultaneous clusters of similar illness in noncontiguous areas, domestic or foreign
g. Cluster of patients with similar genetic type among agents isolated from temporally or spatially distinct sources

**Table 15.5** Considerations for Notifying Law Enforcement of Possible Biologic or Chemical Terrorism Initial Investigation at Local Level (Modified from Ref. 20)

Immediate notification of the FBI when:

a. Notification is received from individual or group that a terrorist attack has occurred or will occur
b. A potential dispersal/delivery device such as munition or sprayer or questionable material is found

Notification of the FBI as soon as possible after investigation confirms the following:

a. Illness due to unexplained aerosol, vector, food, or water transmission
b. At least a single, definitively diagnosed case(s) with one of the following:
   ■ Uncommon agent or disease occurring in a person with no other explanation
   ■ Illness due to a genetically altered organism

Notification of FBI after investigation confirms the following (with no plausible natural explanation):

a. Disease with an unusual geographic, seasonal, or "typical patient" distribution
b. Unusual, atypical, or antiquated strain of agent
c. Simultaneous clusters of similar illness in noncontiguous areas, domestic or foreign
d. Clusters of patients presenting with similar genetic type among agents isolated from temporally or spatially distinct sources
e. Infection due to novel vehicle or mode of transmission

dogs purchased as pets, which had acquired their infection while co-housed with infected giant Gambian rats that had been imported recently from Ghana, and not from deliberate dissemination. More recently in 2005, four U.S. soldiers acquired hemorrhagic fever with renal syndrome near the demilitarized zone in South Korea (21). Despite initial suspicions of deliberate infection, epidemiological and laboratory analysis ultimately linked exposures of the U.S. soldiers to rodent hosts at training sites visited by the soldiers (21). Similarly, the death of a wildlife biologist working for the National Park Service in 2007 from *Yersinia pestis* required a thorough epidemiological investigation. The wildlife biologist was found deceased at his home by colleagues and a subsequent postmortem determined cause of death as primary pneumonic plague (22). Epidemiology and an ecological and laboratory investigation concluded the biologist's source of exposure to *Y. pestis* was most likely during a necropsy on a mountain lion he performed prior to his death (22). Concerns regarding the potential deliberate use of biological agents such as *Y. pestis* and the presence of emerging infections will continue to complicate efforts to distinguish between naturally occurring disease and disease resulting from deliberate release of a biological agent.

## MOLECULAR STRAIN TYPING

The microbiology laboratory has made significant contributions to the epidemiology of infectious diseases. Repeated isolation of a specific microorganism from patients with a given disease or syndrome has helped prove infectious etiologies. In addition, isolation and identification of microorganisms from animals, vectors, and environmental sources have been invaluable in identifying reservoirs and verifying modes of transmission. In dealing with an infection, it is often necessary to identify the infecting microorganism and determine its antimicrobial susceptibilities in order to prescribe effective therapy. Many of the techniques that have evolved for such purposes are both rapid and accurate but, in general, do not provide the kind of genetic discrimination necessary for addressing epidemiological questions. Historically, the typing methods that have been used in epidemiological investigations fall into two broad categories: *phenotypic* and *genotypic*. Phenotypic methods are those methods that characterize the products of gene expression in order to differentiate strains. For example, the use of biochemical profiles to discriminate between genera and species of bacteria is used as a diagnostic method, but can also be used for biotyping. Other methods, such as phage typing, can be used to discriminate among groups within a bacterial species. Biotyping emerged as a useful tool for epidemiological investigations in the 1960s and early 1970s, while phage typing of bacteria and serological typing of bacteria and viruses have been used for decades. Today, the majority of these tests are considered inadequate for epidemiological purposes. First, they do not provide enough unrelated parameters to obtain a good

**Table 15.6** Characteristics of Phenotypic Typing Methods (Modified from Ref. 28)

| Typing System | Proportion of Strains Typeable | Reproducibility | Discriminatory Power | Ease of Interpretation | Ease of Performance |
|---|---|---|---|---|---|
| Biotyping | All | Poor | Poor | Moderate | Easy |
| Antimicrobial susceptibility patterns | All | Good | Poor | Easy | Easy |
| Serotyping | Most | Good | Fair | Moderate | Moderate |
| Bacteriophage or pyocin typing | Some | Good | Fair | Difficult | Difficult |
| MLEE[a] | All | Excellent | Excellent | Moderate | Moderate |

[a]Multilocus enzyme electrophoresis.

reflection of genotype. For example, serotyping of *Streptococcus pneumoniae* discriminates among only a limited number of groups. In addition, some viruses, such as human cytomegalovirus and measles virus, cannot be divided into different types or subtypes by serology because significant antigenic differences do not exist. Second, the expression of many genes is affected by spontaneous mutations, by environmental conditions, and by developmental programs or reversible phenotypic changes, such as high-frequency phenotypic switching. Because of this, many of the properties measured by phenotypic methods have a tendency to vary and, for the most part, have been replaced by genotypic methods. The one major exception is multilocus enzyme electrophoresis (MLEE) (23,24), which is a robust phenotypic method that performs comparably to many of the most effective DNA-based methods (25,26). Characteristics of selected phenotypic methods are presented in Table 15.6. These methods have been characterized by *typeability*, the ability of the technique to assign an unambiguous result (i.e., type) to each isolate; *reproducibility*, the ability of a method to yield the same results upon repeat testing of a bacterial strain; *discriminatory power*, the ability of the method to differentiate among epidemiologically unrelated isolates; *ease of interpretation*, the effort and experience required to obtain useful, reliable typing information using a particular method; and *ease of performance*, which reflects the cost of specialized reagents and equipment, technical complexity of the method, and the effort required to learn and implement the method.

Extremely sensitive and specific molecular techniques have recently been developed to facilitate epidemiological studies. Our ability to use these molecular techniques (genotypic methods or proteomic methods for prions) to detect and characterize the genetic variability of infectious agents (bacteria, fungi, protozoa, viruses) is the foundation for the majority of molecular epidemiological studies. The application of appropriate molecular techniques has been an aid in the surveillance of infectious agents and in determining sources of infection.

**Table 15.7** Molecular Characteristics of Genetic Diversity at Different Hierarchical Level (Modified from Ref. 29)

| Function | Purpose | Regions of DNA |
|---|---|---|
| Discrimination above level of species | Taxonomy/evolution | Highly conserved coding regions (e.g., rDNA) |
| Discrimination between species | Taxonomy/diagnosis/ epidemiology | Moderately conserved regions |
| Discrimination between intraspecific variants/strains | Population genetics | Variable regions |
| Discrimination between individual isolates/clonal lineages | "Fingerprinting"—tracking transmission of genotypes/ identifying sources of infection and risk factors | Highly variable genetic markers that are not under selection by the host |
| Genetic markers/linking phenotype and genotype | Identifying phenotypic traits of clinical significance | Genotype linked to phenotype |

The ability to link isolates to sources has direct implications for investigating both natural and deliberate outbreaks (27). These molecular techniques can be used to study health and disease determinants in animal (including human) as well as in plant populations. Molecular techniques may also be applied to clinical as well as environmental samples. It requires choosing a molecular method(s) that is capable of discriminating genetic variants at different hierarchical levels, coupled with the selection of a region of nucleic acid, which is appropriate to the questions being asked (Table 15.7).

Genotypic methods are those based on an analysis of the genetic structure of an organism. Over the past decade, a number of genotypic methods have been used to fingerprint pathogenic microorganisms (Table 15.8). The methods have been described in detail elsewhere (28–33). In general, molecular typing methods can be divided into three general categories: restriction endonuclease-based methods, amplification-based methods, and sequence-based methods (33). Among these methods, restriction fragment-length polymorphism/pulsed-field gel electrophoresis (RFLP-PFGE) and RFLP + probe and ribotyping have been the most commonly used methods for fingerprinting bacteria (30,34). Random amplification of polymorphic DNA (RAPD) and karyotyping have been used for fingerprinting fungi (30,35). MLEE, RAPD, and polymerase chain reaction (PCR)-RFLP have been used for fingerprinting parasitic protozoa (30). Multilocus variable number tandem repeat analysis (MLVA) has been used to subtype *B. anthracis*, *Y. pestis*, and *Francisella tularensis*. MLVA schemes are now available for most bioterroism agents (36). Single-nucleotide polymorphisms (SNPs) have been used to analyze strains of *B. anthracis* and several gram-negative food-borne pathogens (33,37). An assay used for scoring SNPs of *B. anthracis* has been shown to have high-throughput

**Table 15.8** Examples of Genotypic Methods Used in Epidemiologic Investigations

Restriction endonuclease-based methods
a. Restriction fragment-length polymorphism without hybridization
- Frequent cutter (4- to 6-bp recognition site) coupled with conventional electrophoresis to separate restriction fragments
- Infrequent cutter (generally 6- to 8-bp recognition site) coupled with pulsed-field gel electrophoresis to separate restriction fragments
b. RFLP with hybridization
- Frequent cutter (4- to 6-bp recognition site) coupled with conventional electrophoresis to separate restriction fragments followed by Southern transfer to nylon membrane. Power and efficacy of typing method depend on the probe.
- 16S and 23S rRNA (ribotyping)
- Insertion sequences (e.g., IS6110 of *Mycobacterium tuberculosis*)

Amplification-based methods
a. Random amplification of polymorphic DNA analysis; arbitrarily primed PCR
b. Amplified fragment-length polymorphism method
c. Repetitive element method; variable number tandem repeat fingerprinting

Sequence-based methods
a. Multilocus sequence typing
b. Electrophoretic karotyping
c. Single-nucleotide polymorphism analysis
d. Whole genome sequencing

---

capability and can be performed with small amounts of DNA (37). Select gene or complete genome characterization, as well as other molecular methods, has been used for viruses (38).

When should fingerprinting be used? Strain typing data are most effective when they are collected, analyzed, and integrated into results of an epidemiological investigation. The epidemiologist should consult the laboratory when investigating a potential outbreak of an infectious disease. Microbial fingerprinting should supplement, and not replace, a carefully conducted epidemiological investigation. In some cases, typing data can effectively rule out an outbreak and thus avoid the need for an extensive epidemiological investigation. In other cases, these data may reveal the presence of outbreaks caused by more than one strain. Data interpretation is facilitated greatly by an appreciation of the molecular basis of genetic variability of the organism being typed and the technical factors that can affect results. With the exception of whole-genome sequencing, molecular methods analyze only a small portion of the organisms' genetic complement. Thus, isolates that give identical results are classified as "indistinguishable," not "identical." Theoretically, a more detailed analysis should uncover differences in the isolates that appeared to give identical patterns but that were unrelated epidemiologically. This is unlikely to occur when a set of epidemiologically linked isolates are analyzed (28).

For this reason, only whole-genome sequencing would provide unequivocal data required for attribution.

The power of molecular techniques in epidemiological investigations is well exemplified by a few examples. PulseNet, the national molecular subtyping network for food-borne disease surveillance, was established by the CDC and several state health departments in 1996 to facilitate subtyping bacterial food-borne pathogens for epidemiological purposes. Twenty years ago, most food-borne outbreaks were local problems that typically resulted from improper food-handling practices. Outbreaks were often associated with individual restaurants or social events, and often came to the attention of local public health officials through calls from affected persons. Today, food-borne disease outbreaks commonly involve widely distributed food products that are contaminated before distribution, resulting in cases that are spread over several states or countries. The PulseNet network, which began with 10 laboratories typing a single pathogen (*Escherichia coli* O157:H7), achieved full national participation in 2001 and includes food safety laboratories of the U.S. Food and Drug Administration (FDA) and the U.S. Department of Agriculture. Sister networks have also been established internationally (34,39). Currently, PulseNet USA utilizes standardized PFGE protocols for seven organisms: *E. coli* O157: H7, *Salmonella enterica*, *Shigella* spp., *Listeria monocytogenes*, thermotolerant *Campylobacteria* spp., *Clostridium perfringens*, and *Vibrio cholera* (39). The laboratories follow a standardized protocol using similar equipment so that results are highly reproducible and DNA patterns generated at different laboratories can be compared. Isolates are subtyped on a routine basis, and data are analyzed promptly at the local level. Clusters can often be detected locally that could not have been identified by traditional epidemiological methods alone. PFGE patterns are shared between participating laboratories electronically, which serves to link apparently unrelated outbreaks and facilitates the identification of a common vehicle (40). For example, in 2006, PulseNet was critical to facilitating the identification of an *E. coli* O157:H7 outbreak affecting 26 states in the United States (41). In September 2006, the Wisconsin Division of Public Health and Wisconsin State Laboratory of Hygiene linked geographically dispersed *E. coli* O157:H7 isolates from stool samples of symptomatic patients, all of which had the same PFGE pattern. Epidemiological investigation revealed that infection was caused by ingestion of contaminated spinach. Additionally, environmental samples obtained by the FDA and California Department of Health Services from river water, cattle manure, and wild pig feces located within and surrounding fields where spinach was grown were identified as *E. coli* O157:H7 isolates with an identical PFGE pattern to those obtained from clinical samples (41). Without molecular typing, epidemiologists would have found it difficult to identify cases associated with each cluster. However, the use of PFGE subtyping as part of routine surveillance has benefits beyond outbreak detection.

For example, the temporal clustering of unrelated cases is not uncommon, and without molecular typing, valuable public health resources would be wasted investigating pseudo-outbreaks or unrelated outbreaks. Molecular genotyping of food-borne pathogens continues to evolve. PulseNet is currently evaluating other potential subtyping methodologies, including MLVA and SNP analysis (39). PulseNet is a powerful tool that can be applied for the early detection of cluster(s) of illness as a result of deliberate food-borne contamination (39).

Another example of the power of molecular techniques is demonstrated in the invaluable information provided during the 2001 bioterrorism-associated anthrax attacks. MLVA was used to subtype isolates obtained from patients, environmental samples, and powders. Information from MLVA identified the subtype of *B. anthracis* and was able to link clinical cases to environmental samples and powders, thereby providing information on possible sources of exposure (42). Molecular subtyping also allowed for confirmation that clinical cases were caused by the same strain and that suspected cases outside the United States were not linked (42). Both forensic and epidemiological investigations can result in the collection of hundreds of clinical and environmental samples for testing. In this event, MLVA assisted with the identification of potential laboratory contamination of samples as a result of the large number of samples requested to be tested (42). MLVA can reliably and rapidly genotype an isolate within 8 hours of receipt of an isolate. Molecular subtyping identified the *B. anthracis* used in the 2001 attack as the Ames strain, a strain rarely found in nature (37). This information was a critical epidemiological factor in determining that these cases were most likely the result of a deliberate release (37). Additionally, whole genomic sequencing of isolates obtained from spores indicated that the genome and plasmid sequences were identical to those of an Ames strain stored at a U.S. Army research facility (43). The utility of molecular typing methodologies is clearly demonstrated in this forensic investigation involving the deliberate use of a biological agent in the United States.

In 2006, the CDC was notified of two cases of brucellosis in microbiologists at clinical laboratories in Indiana and Minnesota (44). Because *Brucella* spp. is considered a category B agent (19), infections with *Brucella* spp. should have a thorough epidemiological investigation to determine potential sources of exposure. MLVA was utilized to assist in identifying the source of the *Brucella* infections. The CDC compared blood culture isolates from the microbiologists with the isolates they handled in the laboratory. The epidemiological investigation revealed that the clinical isolate from the infected microbiologist in Indiana had been forwarded to the clinical laboratory in Minnesota; however, investigation revealed that the second microbiologist did not handle this clinical isolate (44). Further epidemiological investigation determined that the Minnesota microbiologist had handled unidentified isolates later determined to be *Brucella* spp. on an open bench. MLVA confirmed that the source of the Minnesota microbiologist's infection was one of the isolates handled on the open bench (44). The source of the Indiana microbiologist's infection was an unidentified isolate from a referral laboratory that had requested identification of this particular specimen (44). Molecular genotyping provided critical confirmation of the source of exposure for these microbiologists and confirmed that these cases resulted from a laboratory exposure.

Additional advances in molecular laboratory techniques have begun to allow for the rapid detection of antimicrobial resistance. In one prospective study on methicillin-resistant *Staphylococcus aureus*, automated clonal alerts based on real-time subtyping were faster than traditional methods (45). At present, however, direct identification of resistance genes by PCR or similar techniques is limited because only a few resistance genes are strongly associated with phenotypic resistance (46). PCR followed by electrospray ionization mass spectrometry has been reported to be able to detect quinolone resistance in *Acinetobacter* spp. (47). However, this technique must be further evaluated and limitations acknowledged, such as whether detection of a resistance gene indicates that a resistant phenotype is always present (47). The ability to establish antimicrobial susceptibility patterns rapidly is particularly critical in order to quickly provide the appropriate antimicrobial agents for treatment or postexposure prophylaxis in a situation where the deliberate dissemination of a potentially engineered drug-resistant organism is being considered. Because there are numerous mechanisms for antimicrobial resistance in bacteria, current phenotypic methods will continue to be the basis for laboratory determination of antimicrobial susceptibility patterns for the foreseeable future (46).

## SUMMARY

With few exceptions, a careful epidemiological investigation will be required to determine whether an outbreak of infectious disease is due to intentional release of an agent or is naturally occurring. A number of molecular techniques have been developed for subtyping microbes that have been shown to complement the epidemiological investigation as well as identify related cases. For example, since the establishment of PulseNet, the routine use of molecular subtyping by PFGE has improved both the sensitivity and the specificity of epidemiological investigation of food-borne outbreaks at the state and local level (48). As current subtyping methodologies evolve, applications and uses in the public health response to deliberate releases of biologic agents must be considered and applied.

### Challenges

Unfortunately, molecular genotyping information exists within multiple databases and in a variety of formats. Although PulseNet and other systems have Web-based access, integration and sharing of data among multiple databases remains a challenge. As information and databases expand, data will also become more challenging to analyze. Therefore, there is a need to refine analytic methods to improve pattern recognition and integration of multiple streams of epidemiologic and laboratory data so that outbreaks and events can be detected quickly. Informatics capacity at local, state, and federal level requires continued investment in order to maximize the integration of epidemiology and laboratory information.

Finally, the threat of bioterrorism has initiated the development of mechanisms to quickly identify the presence of biological agents in the environment in order to rapidly initiate public health and medical response efforts. Molecular technologies allow for the rapid identification of genetic material of biological agents from collection devices such as those used for outdoor and indoor environmental monitors. Public health, forensic, and laboratory assessments must be made on the basis of material collected in a distinct area covered by the monitor or sensor. Because these detectors or devices collect only genetic material, detectors cannot indicate that a live organism was released, that individuals were exposed, or that a deliberate release occurred. As a result, it is critical that information from public health and epidemiological investigations be considered when interpreting information from environmental monitors. Public health must consider the limits of these new technologies, previous history of environmental detection of a biological agent in a given area, and environmental sampling methods. As the recent Institute of Medicine report on "Effectiveness of National BioSurveillance Systems: Biowatch and the Public Health System" indicated, the challenge is "understanding the clinical context in which disease detection and reporting occurs and the factors that shape the decision-making process for the state and local public health officials who must interpret the data" from these systems as well as that from traditional public health surveillance systems (49).

## ACKNOWLEDGMENTS

We thank Dr. Stephen Morse for his review and insights in integrating epidemiology and laboratory science. The findings and conclusions in this report are those of the authors and do not necessarily represent views of the Centers for Disease Control and Prevention.

## REFERENCES

[1] L. Gordis, Epidemiology, WB Saunders Co., Philadelphia, PA, 1996.

[2] J.M. Last, A Dictionary of Epidemiology, second edn., Oxford University Press, New York, 1988.

[3] D.L. Heymann (Ed.), Control of Communicable Diseases Manual, nineteenth edn., American Public Health Association, Washington, DC, 2008.

[4] Centers for Disease Control, Outbreak of *Salmonella* serotype Kottbus infections associated with eating alfalfa sprouts—Arizona, California, Colorado, and New Mexico, February–April, 2001, MMWR Morb. Mortal. Wkly. Rep. 51 (2001) 7–9.

[5] T.W. Hennessy, C.W. Hedberg, L. Slutsker, K.E. White, J.M. Besser-Wiek, M.E. Moen, et al., A national outbreak of *Salmonella enteritidis* infections from ice cream, N. Engl. J. Med. 334 (1996) 1281–1286.

[6] F.J. Mahoney, C.W. Hoge, T.A. Farley, J.M. Barbaree, R.F. Breiman, R.F. Benson, et al., Community-wide outbreak of Legionnaires' disease associated with a grocery store mist machine, J. Infect. Dis. 165 (1992) 736–739.

[7] K.A. Feldman, R.E. Enscore, S.L. Lathrop, B.T. Matyas, M. McGuill, M.E. Schrieffer, et al., An outbreak of pneumonic tularemia on Martha's Vineyard, N. Engl. J. Med. 345 (2001) 1601–1606.

[8] Institute of Medicine, Emerging Infections: Microbial Threats to Health in the United States, National Academy Press, Washington, DC, 1994.

[9] M.S. Smolinski, M.A. Hamburg, J. Lederberg, Microbial Threats to Health: Emergence, Detection and Response, National Academy Press, Washington, DC, 2003.

[10] Centers for Disease Control, Update: Multistate outbreak of monkeypox—Illinois, Indiana, Kansas, Missouri, Ohio, and Wisconsin, 2003, MMWR Morb. Mortal. Wkly. Rep. 52 (2003) 642–646.

[11] Centers for Disease Control, Outbreak of Rift Valley fever—Saudi Arabia, August–October, 2000, MMWR Morb. Mortal. Wkly. Rep 49 (2000) 905–908.

[12] D. Nash, F. Mostashari, A. Fine, J. Miller, D. O'Leary, K. Murray, et al., The outbreak of West Nile virus infection in the New York City area in 1999, N. Engl. J. Med. 344 (2001) 1807–1814.

[13] M.K. Lo, P.A. Rota, The emergence of Nipah virus, a highly pathogenic paramyxovirus, J. Clin. Virol. 43 (4) (2008) 396–400.

[14] T. Briese, J.T. Paweska, L. McMullan, S.K. Hutchison, C. Street, G. Palacios, et al., Genetic detection and characterization of Lujo virus, a new hemorrhagic fever-associated Arenavirus from Southern Africa, PLoS Pathog. 4 (5) (2009) e1000455.

[15] K.E. Jones, N.G. Patel, M.A. Levy, A. Storeygard, D. Balk, J.L. Gittleman, et al., Global trends in emerging infectious diseases, Nature 451 (2008) 990–993.

[16] T.J. Torok, R.V. Tauxe, R.P. Wise, J.R. Livengood, R. Sokolow, S. Mauvais, et al., A large community outbreak of salmonellosis caused by intentional contamination of restaurant salad bars, JAMA 278 (1997) 389–395.

[17] W.S. Carus, Bioterrorism and Biocrimes. The Illicit Use of Biological Agents since 1900, Fredonia Books, Amsterdam, The Netherlands, 2002.

[18] J.A. Jergnigan, D.S. Stephens, D.A. Ashford, C. Omenaca, M.S. Topiel, M. Galbraith, et al., Anthrax Bioterrorism Investigation Team. Bioterrorism-related anthrax: The first 10 cases reported in the United States, Emerg. Infect. Dis. 7 (2001) 933–944.

[19] A.S. Khan, S. Morse, S. Lillibridge, Public health preparedness for biological terrorism in the USA, Lancet 356 (2000) 1179–1182.

[20] T.A. Treadwell, D. Koo, K. Kuker, A.S. Khan, Epidemiologic clues to bioterrorism, Pub. Health Rep. 118 (2003) 92–98.

[21] J.W. Song, S.S. Moon, S.H. Gu, K.J. Song, L.J. Baek, H.C. Kim, et al., Hemorrhagic fever with renal syndrome in 4 US soldiers, South Korea, 2005, Emerg. Infect. Dis. 15 (2009) 1833–1836.

[22] D. Wong, M.A. Wild, M.A. Walburger, C.L. Higgins, M. Callahan, L.A. Czarnecki, et al., Primary pneumonic plague contracted from a mountain lion carcass, Clin Infect. Dis 49 (2009) 33–38.

[23] N. Pasteur, G. Pasteur, F. Bonhomme, J. Catalan, J.B. Davidian, Practical Isozyme Genetics, Halstead Press, New York, 1998.

[24] B.J. Richardson, P.R. Baverstock, M. Adams, Alloenzyme Electrophoresis. A Handbook for Animal Systematics and Population Studies, Academic Press, Orlando, FL, 1986.

[25] C. Pujol, S. Joly, S.R. Lockhart, S. Noel, M. Tibayrenc, D. Soll, Parity among the randomly amplified polymorphic DNA method, multilocus enzyme electrophoresis, and Southern blot hybridization with the moderately repetitive DNA probe Ca3 for fingerprinting Candida albicans, J. Clin. Micro 35 (1997) 2348–2358.

[26] M. Tibayrenc, K. Neubauer, C. Barnabe, F. Guerrini, D. Skarecky, F.J. Ayala, Genetic characterization of six parasitic protozoa: Parity between random primer DNA typing and multilocus enzyme electrophoresis, Proc. Natl. Acad. Sci. USA 90 (1993) 1335–1339.

[27] R.E. Coleman, A.J. Vogler, J.L. Lowell, K.L. Gage, C. Morway, P.J. Reynolds, et al., Fine-scale identification of the most likely source of a human plague infection, Emerg. Infect. Dis. 15 (2009) 1623–1625.

[28] F.C. Tenover, R.D. Arbeit, R.V. Goering, How to select and interpret molecular strain typing methods for epidemiological studies of bacterial infections: A reveiw for healthcare epidemiologists, Inf. Contr. Hosp. Epidemiol. 18 (1997) 426–439.

[29] R.C.A. Thompson, C.C. Constantine, U.M. Morgan, Overview and significance of molecular methods: What role for molecular epidemiology? Parasitology 117 (1998) S161–S175.

[30] D.R. Soll, S.R. Lockhart, C. Pujol, Laboratory procedures for the epidemiological analysis of microorganisms, in: P.R. Murray, E.J. Baron, J.H. Jorgensen, M.A. Pfaller, R.H. Yolken (Eds.), Manual of Clinical Microbiology, eighth edn., ASM Press, Washington, DC, 2002, pp. 139–161.

[31] T.H. Pennington, Electrophoretic typing, in: M. Sussman (Ed.), Molecular Medical Microbiology, Academic Press, San Diego, CA, 2002, pp. 535–547.

[32] M. Arens, Methods for subtyping and molecular comparison of human viral genomes, Clin. Microbiol. Rev. 12 (1999) 612–626.

[33] S.L. Foley, A.M. Lynne, R. Nayak, Molecular typing methodologies for microbial source tracking and epidemiological investigations of Gram-negative bacterial foodborne pathogens, Infect. Gen. Evol. 9 (2009) 430–440.

[34] B. Swaminathan, T.J. Barrett, S.B. Hunter, R.V. Tauxe, C.D.C. PulseNet, Task Force. PulseNet: The molecular subtyping network for foodborne bacterial disease surveillance, United States, Emerg. Infect. Dis. 7 (2001) 382–389.

[35] D.R. Soll, The ins and outs of DNA fingerprinting of infectious fungi, Clin. Microbiol. Rev. 13 (2000) 332–370.

[36] A. Van Belkum, Tracing isolates of bacterial species by multilocus variable number of tandem repeat analysis (MLVA), FEMS Immunol. Med. Microbiol. 40 (2007) 22–27.

[37] P. Keim, M.N. Van Ert, T. Pearson, A.J. Vogler, L.Y. Huynh, D.M. Wagner, Anthrax molecular epidemiology and forensics: Using the appropriate marker for different evolutionary scales. Infect, Gen. Evol. 4 (2004) 205–213.

[38] M. Arens, Methods for subtyping and molecular comparison of human viral genomes, Clin. Microbiol. Rev. 12 (1999) 612–626.

[39] P. Gerner-Smidt, K. Hise, J. Kincaid, S. Hunter, S. Rolando, E. Hyytia-Trees, et al., PulseNet USA: A five-year update, Foodborne Path Dis. 3 (2006) 9–19.

[40] Centers for Disease Control, Update: Outbreaks of Shigella sonnei infection associated with eating fresh parsley—United States and Canada, July–August, 1998, MMWR Morb. Mortal. Wkly. Rep. 48 (1999) 285–289.

[41] A.M. Wendel, D.H. Johnson, U. Sharapov, J. Grant, J.R. Archer, T. Monson, et al., Multistate outbreak of Escherichia coli O157:H7 infection associated with consumption of packaged

spinach, August–September 2006: The Wisconsin Investigation, Clin. Infect. Dis. 48 (2009) 1079–1086.

[42] A.R. Hoffmaster, C.C. Fitzgerald, E. Ribot, L.W. Mayer, T. Popovic, Molecular subtyping of *Bacillus anthracis* and the 2001 bioterrorism-associated anthrax outbreak, United States, Emerg. Infect. Dis. 8 (2002) 1111–1116.

[43] W.F. Fricke, D.A. Rasko, J. Ravel, The role of genomics in the identification, prediction and prevention of biological threats, PLoS Biol. 7 (2009) e1000217.

[44] Centers for Disease Control, Laboratory-acquired brucellosis-Indiana and Minnesota, 2006, MMWR Morb. Mortal. Wkly. Rep. 57 (2) (2008) 39–42.

[45] V. Sintchenko, B. Gallego, Laboratory-guided detection of disease outbreaks: Three generations of surveillance systems, Arch. Pathol. Lab. Med. 133 (2009) 916–925.

[46] J.H. Jorgensen, M.J. Ferraro, Antimicrobial susceptibility testing: A review of general principles and contemporary practices, Clin. Infect. Dis. 49 (2009) 1749–1755.

[47] K.M. Hujer, A.M. Hujer, A. Endimiani, J.M. Thomson, M.D. Adams, K. Goglin, et al., Rapid determination of quinolone resistance in Acinetobacter spp, J. Clin. Microbiol. 47 (2009) 1436–1442.

[48] C.W. Hedberg, J.M. Besser, Commentary: Cluster evaluation, Pulsenet, and public heath practice, Foodborne Path Dis. 3 (2006) 32–35.

[49] Institute of Medicine and National Research Council of the National Academies, Effectiveness of National Biosurveillance Systems: BioWatch and the Public Health System, National Academies Press, Washington, DC, 2009, Interim Report.

# 4

# Subject Areas

# CHAPTER 16

# Forensic Analysis in Bacterial Pathogens

Richard T. Okinaka,[a] Karen H. Hill,[b] Talima Pearson,[a]
Jeffrey T. Foster,[a] Amy J. Vogler,[a] Apichai Tuanyok,[a]
David M. Wagner,[a] and Paul S. Keim[a,c]

[a]Center for Microbial Genetics and Genomics, Northern Arizona University, Flagstaff, Arizona
[b]Los Alamos National Laboratory, Biosciences Division, Los Alamos, New Mexico
[c]Pathogen Genomics Division, Translational Genomics Research Institute,
Flagstaff, Arizona

## INTRODUCTION

Microbiology began as the study of organisms that could not be easily seen with the naked eye. It then evolved into a science focused on microbes that impacted human health and industrial processes that could also be isolated, cultured, and studied in detail. The bacterium *Bacillus anthracis*, the causative agent for anthrax, had an important role in the beginnings of medical microbiology. It was the first bacterial agent shown to have a cause-and-effect relationship for a disease (1877) and it led Robert Koch to develop his germ theory of disease (1). Koch went on to isolate the causative agents for tuberculosis (*Mycobacterium tuberculosis*) and cholera (*Vibrio cholera*) and soon thereafter other cause-and-effect relationships were established for *Brucella melitensis* and brucellosis by David Bruce (1887) and *Yersinia pestis* and plague by Alexandre Yersin (1894). It has been more than 115 years since these seminal studies helped lay the foundation for research in bacterial pathogenesis, but a highly visible anthrax letter attack case and evolution of drug-resistant markers means that these same pathogens are still at a forefront that demands significant research efforts.

Our view of taxonomic diversity within the microbial world has undergone an exponential expansion over the past few decades because of the realization that the small-subunit ribosomal RNA (SSU, a.k.a. 16S RNA) is conserved in all life forms (except viruses) and that its sequence can provide a relatively accurate phylogenetic "tree of life" (2). Equally as important has been application of the polymerase chain reaction (PCR) to analysis of the SSU from a wide array of environmental sources that allowed scientists to capture

259

Microbial Forensics. DOI: 10.1016/B978-0-12-382006-8.00016-5

glimpses of the enormous diversity of unculturable small eukaryotes (3), other bacterial groups (4), and expansion of the Archaea, the new domain first described by Carl Woese (2).

Microbiology, therefore, now consists of the study of unicellular organisms that belong to these three domains of life: eukaryotes (the domain of plants and animals), bacteria (including many of the earliest known pathogens), and Archaea [a significant, if not dominant, presence in our ecosystem's biomass, despite its relatively recent discovery (5)]. In addition to these three domains of life are small, relatively inert, "biological packets" of DNA called "viruses" that are mostly defined by the specificity of the host to which they are associated. Viruses are obligate parasites that require the metabolic capabilities of their hosts to replicate their DNA or RNA and to produce "infectious" viral particles that can restore and maintain the viral life cycle. In contrast to bacteria, viral genomes are relatively small, ranging between 5 and 250 kbp of nucleotide residues, whereas metabolically active and complex bacteria have genomes that range in size from 1 to 10 million bp of nucleotide residues.

## 16S SEQUENCES AND RAPID IDENTIFICATION OF MICROBES

Initial identification and phylogenetic reconstruction of large numbers of bacterial species have been facilitated greatly by the analysis of the sequence of the 16S small ribosomal subunit (6). These and other studies have been used to create public databases such as the "Microbial Rosetta Stone" (7) that emphasize the identification of pathogenic microorganisms and the use of accurate taxonomic nomenclature. These databases represent the tools that are necessary as a first step in the rapid evaluation of epidemics and/or bioterrorist attacks.

## FORENSIC SIGNATURES: HUMAN VERSUS BACTERIAL PATHOGENS

"DNA fingerprinting" began with the discovery in 1985 that simple tandem repeat (STR) sites in the DNA of humans could be used to differentiate certain human populations (8). In 1997 the Federal Bureau of Investigation adopted a 13 human marker STR system that produced astronomical statistics for matching a single individual to other individuals in our global population with probabilities that equaled $1 \times 10^{-15}$ or greater (9). The main scientific basis for these incredible probabilities lies in meiosis and *random* sorting of the 23 pairs of chromosomes between the sperm and the egg of the parents that yield a combination of $2^{23}$ or >8 million possibilities in the offspring.

This is not the case for bacteria. All bacteria are clonal in nature, and the degree of diversity is caused by a variation in mutation, mutation rates, horizontal gene transfer, recombination, and rates of recombination. An interesting component in the forensic analysis of a bacterium such as *Bacillus anthracis* is that all the offspring and their descendants are genetically identical to their parent until a mutation occurs in one of the descendents. Unlike the situation in humans where the progeny's genome is already a complex mixture of the two parents, it may take a thousand generations or more before a single nucleotide change would be observed in a *B. anthracis* lineage. However, mutation rates vary considerably in the landscape that makes up the bacterial genome; the following section describes the progression of discovery that has allowed *B. anthracis* to be dissected down to the level of individual isolates.

## *BACILLUS ANTHRACIS*: A MODEL SYSTEM

Several years before the 2001 anthrax-letter attacks, *B. anthracis*, the causative agent, was known to be a genetically monomorphic organism with few molecular markers that could differentiate individual isolates (10). However, in ensuing years, research on molecular genotyping of *B. anthracis* evolved rapidly as a model for developing evolutionary relationships and molecular signatures for clonally propagated microbes.

In 1997, Keim and colleagues (11), using an emerging technology, amplified fragment length polymorphism (AFLP) analyses, provided the first high-level resolution of 31 genotypes among 79 distinct *B. anthracis*. By 2000, a multiple locus variable number tandem repeat (MLVA) analysis method was published (12) using sequences generated from the original AFLP markers and the initial whole genome sequences of the plasmids pXO1 and pXO2 (12,13). These sequences were used to identify small repeat regions that would mimic the rapidly evolving genetic markers being used in human forensics (9). These analyses identified 89 genotypes within 419 *B. anthracis* isolates (12) and eventually provided the first forensic evidence that the anthrax-letter attacks were fostered by a clone of the Ames strain (14,15).

While the MLVA approach revolutionized our view of the strictly clonal population structure of *B. anthracis*, it is the use of whole genome sequencing and comparative genomic tools that promise to resolve these populations down to the level of individual isolates. *B. anthracis* was one of the first organisms to have genomes of multiple isolates sequenced in their entirety (16). Comparative analysis of five whole genome sequences of *B. anthracis* uncovered 3500 single nucleotide polymorphisms (SNP) between these isolates (16). Distribution of these SNPs among these five isolates (i.e., number of

SNPs unique to each isolate versus SNPs shared between the different isolates) created an accurate evolutionary relationship among these isolates and resulted in an SNP-based "phylogenetic tree" for *B. anthracis*.

The accuracy and conserved nature of the *B. anthracis* SNP tree was reinforced by designing genotyping assays for each of 990 SNP positions and using each of these assays to type 26 diverse *B. anthracis* isolates. These ≈25,000 data points demonstrated the conserved nature of the branches and also indicated that the 990 SNPs contained a large amount of redundant information. Two important concepts evolved from these analyses (17). First, it was suggested that a few select canonical SNP (canSNP) assays located at strategic positions within the tree could replace the original 990 SNP assays and still accurately "bin" all *B. anthracis* isolates into one of 12 phylogenetically conserved subgroups. Second, a *hierarchical* approach (*PHRANA*) was proposed (17) to genotype any new isolate of *B. anthracis* to initially define an accurate phylogenetic position and then provide the highest resolution genotype available (Figure 16.1). Figure 16.1 illustrates how canSNP assignments could be paired with more rapidly evolving markers (two different VNTR systems) in a hierarchical manner to provide the highest resolution for each *B. anthracis* isolate. These ideas were verified when 1033 worldwide isolates of *B. anthracis* were first placed into one of 12 subgroups or sublineages using only 13 canSNPs. This was followed by MLVA using 15 VNTR markers to identify 221 different genotypes (18). Furthermore, by combining canSNP groupings with MLVA15

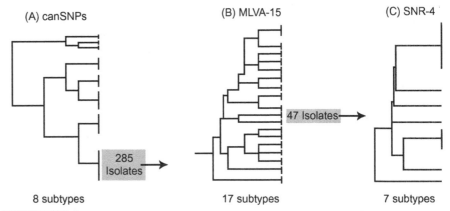

**FIGURE 16.1**

A hierarchical approach to the resolution of *Bacillus anthracis*. (A) Thirteen canonical SNPs separated 1033 isolates into 12 major phylogenetic groups; 8 of these are shown in this diagram (17,18). (B) MLVA15 analysis of a large but extremely conserved cluster of 285 isolates designated Western North America (WNA) yielded 17 unique types (17,18,68). (C) SNR-4 analyses were conducted on 47 isolates recovered from a natural anthrax outbreak in North Dakota in cattle in 2005 and revealed seven closely related subtypes (19).

and four single nucleotide repeat (SNR) assays, the number of unique genotypes could be increased to >450 (17,19).

These numbers, however, are still minuscule in comparison to the odds that can be generated in human forensics, but bacteria possess several properties that can be used to advantage in developing evidentiary material. Their relatively compact genomes and the declining costs of next generation sequencing allow the luxury of being able to generate the whole genome sequence of a pathogen from any incident as an "on the fly" operation. The value of whole genome sequencing and the comparison of strict clonal organisms were discovered in the Ames letter attacks when genomes from the Florida strain were shown to be identical (at ≈5 Mb positions) to what is believed to be an ancestral strain of the Ames isolate (20). From an evidentiary point of view, these data sets indicate that for recent and clonally derived pathogens, the sequencing and analytical tools are readily available to demonstrate that whole genomes from an incident and a source can be matched.

## MUTATION RATES IN BACTERIA

DNA sequencing approaches are now being used to study the evolution of bacteria and other microorganisms in the laboratory. In 2003, Lenski and colleagues (21) generated nearly a million base pairs of sequence (36 regions × 500 bp × 50 clones) from 12 populations of *Escherichia coli* that had undergone 20,000 population doublings. The goal of these experiments was to obtain some measure of the level of genomic evolution within their controlled laboratory experiments. Their random targeting of 36 regions for sequencing allowed them to observe rates of substitution at neutral (synonymous) sites and at sites that cause phenotypic changes (nonsynonymous). Results reinforce previous estimates (22,23) and indicate that mutation rates for substitutions in wild-type *E. coli* are very low ($1.44 \times 10^{-10}$ per bp per generation).

This information suggested that a significant challenge in the then ongoing anthrax-letters case would be that genomes from letter sources and the ancestral Ames strain would have very few, if any, differences (20,21,24). However, current comparative genome sequencing analysis of these laboratory evolution experiments (25) also addresses conceptual issues that may be common to large culture and serial transfer vessels that might be used in a production facility. Inoculums for large fermenter applications are often not pure isogenic populations because larger volumes of cells are required to "jump start" 10, 20, or 500 liter vessels. These inoculums or mixtures of inoculum are more likely to house dynamic and adaptive clones that can gain an advantage in these mass cultures and produce background "signatures" that become distinctive for a particular production. This happened in the case of the anthrax letters (15).

In the section that follows on *Yersinia pestis*, an important development focused on ascertaining the mutation rates in loci that evolve much more rapidly than genomic substitution mutations. These are regions that contain variable-number tandem repeat (VNTR) markers. More importantly, these estimations for mutation rates have been used to study the evolution of *Y. pestis* in naturally occurring animal populations and provide an example for estimating probabilities and likelihood for the source of a particular human infection of plague.

## *YERSINIA PESTIS* AND PLAGUE: ANOTHER RECENTLY EMERGED PATHOGEN

*Yersinia pestis*, the etiological agent for plague, is a relatively young species that likely arose within the past 20,000 years (26–28). As a result, there have been relatively few base substitution genetic polymorphisms that have accumulated within this species as a whole. The lack of polymorphisms in *Y. pestis* was first demonstrated in a multilocus sequence typing (MLST) analysis of six housekeeping genes that revealed no variation among 36 globally diverse strains (28). The lack of variation is especially evident in the *Y. pestis* subgroup 1.ORI (Figure 16.2), which experienced a global expansion when it was spread from coastal regions in China to Africa, Europe, North America, South America, and Australia during the third plague pandemic beginning in the mid-1850s (29). Despite its worldwide distribution, the subgroup is genetically, highly monomorphic due to its relatively recent emergence and because of a genetic bottleneck that occurred in China (29). In addition, specific populations of this particular global expansion of the 1.ORI group, such as the North American and South American populations, are even more monomorphic because they appear to have resulted from a single and more recent introduction (30).

Whole genome sequence comparisons have allowed for the identification of relatively small numbers of phylogenetically informative SNPs among multiple strains from these very recent populations, such as North American *Y. pestis* (31,32). In 2004, an international consortium published a global population structure of *Y. pestis* (27). In this study, a pair-wise comparison of the three whole genome sequences [*Yersinia pseudotuberculosis*, 91001; *Y. pestis* CO92 (orientalis); and *Y. pestis* Kim (mediasiatica)] revealed only 76 shared synonymous SNP positions in 3250 homologous gene pairs. As in the case with *B. anthracis*, the *Y. pestis* SNP distribution does not reveal inconsistent designations or homoplasy in the first 105 isolates examined using 40 of the discovered SNPs (27).

The consensus phylogenetic tree from this study indicates that there are three conserved, major branches (0, 1, and 2) in the *Y. pestis*/*Y. pseudotuberculosis*

SNP tree (Figure 16.2). The canSNP approach is being used to rapidly establish accurate phylogenetic positions for thousands of isolates for *Y. pestis.* In addition, more than 20 whole genome sequences and many others in progress promise to create many new branches along these three major branches (Wagner and Keim, personal communication). A more complete genetic population structure for *Y. pestis* will soon be resolved by analysis of these whole genome sequences and use of phylogenetic inference.

*Yersinia pestis* is also suited to a hierarchical approach whereby canSNP analysis rapidly establishes accurate phylogenetic positioning and MLVA provides high-level subtyping resolution down to the level of individual analysis. A 43 marker MLVA system for *Y. pestis* has been used to determine the natural diversity within 1565 isolates (33,34). VNTR markers evolve several orders of magnitude faster than substitution mutations, and *in vitro* individual mutation rates for these markers have been estimated for these 43 VNTR loci using serially passaged ($\approx$96,000 generations) *Y. pestis* strains (34). Mutation rate estimates for these VNTR markers are of importance because their relatively rapid evolution can be used to study the population dynamics of plague outbreaks within natural rodent populations (33). More importantly, the understanding of mutation rates of VNTR markers in *Y. pestis* and *E. coli* O157 can be used in building probabilistic models for genetic relatedness between potential sources and human infections. These provide important statistical considerations for potential attribution of disease outbreaks in epidemiological and/or forensics cases.

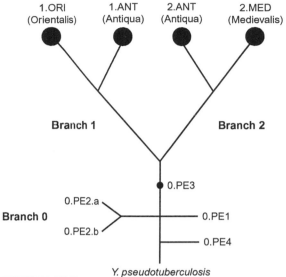

**FIGURE 16.2**

A SNP tree for *Yersinia pestis*: Inferred synonymous mutation tree for *Y. pestis* (27). SNPs along specific branches are useful for either epidemiological or forensic purposes because they can serve as molecular markers for different populations of this organism.

A classic example involves two tourists visiting New York City from New Mexico who were diagnosed with plague in 2002. Multilocus variable number tandem repeat analysis genotypes of these clinical samples were compared to genotypes of 632 isolates in an MLVA database and, although no exact matches were found, the closest matches were to isolates from the same county in New Mexico where the victims lived (35). This finding was in agreement with other epidemiological data that suggested that the victims were infected from natural sources in New Mexico and not by a bioterrorism event in New York City. In this case, mutation rate data and transmission modeling suggested that several isolates from northern New Mexico were the most likely geographic source of the human plague infections just described and that several local infection sources were possible (35).

Fleas infected with *Y. pestis* were found in the victims' backyard as well as along a hiking trail that they utilized shortly before traveling to New York City. Isolates from both locations were close but not perfect MLVA matches to the isolate obtained from one of the victims. Previous studies had determined the individual VNTR mutation rates (34) for the genetic markers in question and relative probabilities in a maximum likelihood framework could be used to evaluate the significance of the genotypic near matches. These calculations established that the couple was most likely infected in their backyard (36). These kinds of analyses provide an example of how likely sources can be determined using relatively rapidly mutating loci.

## *FRANCISELLA TULARENSIS*

*Francisella tularensis*, the causative agent of tularemia, consists of three official subspecies: *tularensis*, *holarctica*, and *mediasiatica*. *Francisella novicida*, officially a separate species, is often considered an unofficial fourth subspecies. *Francisella tularensis* subsp. *tularensis* is the most virulent and is geographically restricted to North America. *Francisella tularensis* subsp. *holarctica* causes a less severe form of disease and occurs throughout the northern hemisphere. *Francisella tularensis* subsp. *mediasiatica* has virulence similar to *F. tularensis* subsp. *holarctica* but has only been isolated from a small region in central Asia (37). *Francisella novicida* is the least virulent and has only rarely been isolated from North America (37) and once from Australia (38).

There are also unculturable genetic near neighbors to *F. tularensis* that have been identified from the soil (37,39). Other relatives include tick endosymbionts (e.g., *Wolbachia persica*) (37,40) and *F. philomiragia* (37). However, the deeper phylogenetic structure within this genus is not understood, making the development of species-specific markers problematic.

An MLVA subtyping system using 25 loci was developed to establish genetic relationships among 192 *F. tularensis* isolates representing a global population (41). This included representatives of each of the four subspecies. These analyses revealed 120 genotypes among the 192 isolates with significantly greater diversity within *F. tularensis* subsp. *tularensis*, the type A subspecies (39 genotypes in 45 isolates) than within *F. tularensis* subsp. *holarctica*, the type B subspecies (74 genotypes in 139 isolates). These studies were the first to demonstrate a distinct genetic division within the highly virulent type A strains of *F. tularensis* subsp. *tularensis* that were designated A.I and A.II.

More recently the comparative analysis of 13 diverse whole genome sequences were used to develop 23 canSNPs that could be used to genotype nearly 500 worldwide isolates of *F. tularensis* (42). This approach has been used to demonstrate a phylogeographical distribution of *F. tularensis* subsp.

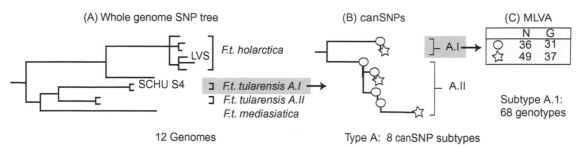

(A) Whole genome SNP tree

(B) canSNPs

(C) MLVA

12 Genomes

Type A:  8 canSNP subtypes

Subtype A.1:
68 genotypes

**FIGURE 16.3**

A hierarchical approach for *Francisella tularensis.* A whole genome SNP tree (A) was used to design 23 clade- and subclade-specifc canSNPs; these 23 canSNPs were used to genotype 496 isolates (42). The *F. tularensis* subsp. *tularensis* or the type A clade was subdivided into 8 canSNP groups (B), including the distinct A.I and A.II subclades. Eighty-five isolates belonging to the A.1 clade were split into canSNP subclades containing 36 and 49 isolates (column N in C). A MLVA-11 analysis was then able to identify 31 and 37 (column G in C) different genotypes within these two subclades.

*tularensis* and *F. tularensis* subsp. *holarctica* within vast areas of the northern hemisphere (42). From a microbial forensic perspective, the current phylogenetically stable canSNP data sets have subdivided the type A and type B subspecies of *F. tularensis* into 19 clades or subclades (42). Figure 16.3 illustrates a hierarchical approach to resolving *F. tularensis* strains using canSNPs (42) and MLVA-11 (43). The utility of next-generation sequencing also promises to further resolve the two main types of *F. tularensis* where there are currently 29 complete or draft whole genome sequences available.

## *BRUCELLA* SPP. AND BRUCELLOSIS

Brucellosis is a ubiquitous disease of livestock and wildlife and is the most common zoonotic infection worldwide. The disease is caused by a closely related group of bacteria, the *Brucella*, that infect a wide range of animals, including cattle, pigs, seals, rodents, and an ever-expanding list of animals (44,45). Despite a high degree of homology between genomes of *Brucella* spp., host (animal) specificities are associated with what are defined as individual species, for example, *B. suis, B. abortus, B. canis, B. melitensis, B. ovis,* and *B. neotomae.* The *Brucella* remain category B select agents because they are highly infectious (fewer than 10 cells cause disease), grown easily, and occur commonly throughout the world, particularly in politically unstable regions (46). *Brucella suis* was also the first species to be weaponized within the U.S. Bioagent program that ended in 1969. The bioterror risks to humans would be relatively minor compared to many other select agent pathogens but potential impacts to agriculture could be devastating. Although eradicated from most of Europe and North America, high levels of debilitating human and livestock infections remain in the Middle East and central Asia.

Several multilocus, variable number tandem repeat systems have been developed and described for *Brucella* spp. (47–50). Despite indications for homoplasy in more rapidly evolving markers (50), the selection of variable but relatively conserved markers allows for clustering to the species level in most cases. Equally important for forensic purposes, the most rapidly evolving markers, for example, hypervariable octamer oligonucleotide footprints (47), offer markers whose diversity indices approach levels of 0.9. While these markers may not be phylogenetically stable, they can provide additional resolution when comparing closely related strains.

Like *B. anthracis*, *Y. pestis*, and *F. tularensis*, the phylogenetic analysis and resolution of *Brucella* spp. have benefited greatly from the comparative analysis of whole genome sequences (45). Analysis of ≈9000 polymorphic nucleotides shared among all 13 genomes revealed an extremely low homoplasy index (0.0104). This is an indication that the sequencing error rate and/or the degree of phylogenetic inconsistencies within this analysis was very low. This also suggests that *Brucella* has a clonal population structure and would, therefore, have features that are quite similar to the previously discussed pathogens. A proposed hierarchical approach to resolving the *Brucella* is illustrated in Figure 16.4. The whole genome SNP tree in Figure 16.4A shows strong

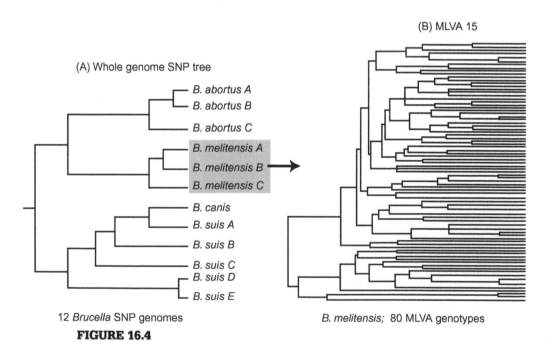

**FIGURE 16.4**

Proposed hierarchical approach to resolving *Brucella* spp. (A) The conserved SNP tree allows selection of a select number of canonical SNP that can be used to place each *Brucella* isolate into appropriate positions on the SNP tree (45). (B) The *B. melitensis* group (88 isolates) represents 80 different MLVA genotypes (48). Resolution of this species by canSNP groupings and subgroupings will provide additional resolution.

differentiation and clustering by species and subspecies level. The 12 genomes represent four different *Brucella* species (*B. abortus*, *B. melitensis*, *B. canis*, and *B. suis*), and the phylogenetic tree created by the analysis supported a tree with three major branches with *B. canis* recently emerging from *B. suis*. Although this tree is an early version of current efforts to sequence and analyze a considerably larger number of genomes from all *Brucella* spp., it illustrates a large diversity within several of these species. *Brucella suis*, for example, has significant resolution between all four of the sequenced genomes, not counting the *B. canis* subgroup, and the three *B. melitensis* genomes also appear to have significant resolution between them.

The hierarchical relationship depicted in Figure 16.4B indicates three *B. melitensis* genomes that provide a basis for the 85 isolates that are separated into 80 distinct MLVA genotypes. The precise relationship between the three sequenced genomes and the 85 isolates can be established by determining the status of the SNPs that define each of the new genomic branches of these 85 isolates. These sequencing assays help determine the branch location for each isolate and whether there are new nodes (branch points) on each of the original branches.

Molecular tools used to assist in the detection, epidemiology, and possible forensic analysis of this species are close to being in place using a hierarchical approach. The potential for significant progress in this area is also being enhanced because the *Brucella* research community also recently used next-generation sequencing technologies to sequence ≈100 additional genomes of research importance. The analysis of these genomes is now beginning to populate the *Brucella* sequence databases and allowing new gaps to be identified. From a forensic standpoint, particular emphasis has been placed on resolving the virulent species *B. melitensis*, *B. suis*, and *B. abortus* because nearly all of the human infections are caused by these three species and because they also represent the large bulk of the economic burden on agriculture.

## BURKHOLDERIA PSEUDOMALLEI

*Burkholderia pseudomallei*, the causative agent of melioidosis, is listed by the CDC as a category B select agent. *B. pseudomallei* commonly lives as a saprophyte that is endemic to soil and freshwater in the tropical regions of southeast Asia and northern Australia. Recent analysis of 33 whole genome sequences suggests that not only are the populations of southeast Asia and northern Australia distinct, but also that the southeast Asian population is a monophyletic derivative from an ancestral Australian population (51). The most dominant characteristic of *B. pseudomallei* is a high degree of recombination that obscures the overall clonality of the species. The genome contains a plethora of insertion sequence elements (52,53), genomic islands (53–55), and VNTRs (52,56). Analysis of MLST data from >1700 isolates and >600

STs suggests that recombination is 18–30 (57) times more likely than mutation to cause allelic changes.

The population dynamics of *B. pseudomallei* are extremely complex due to the high rates of genetic exchange or transfer between different lineages. As expected, the ancestral Australian population is genetically more diverse than the monophyletic southeast Asian population; but, the southeast Asian population appears to recombine more frequently (57). However, many other parameters that influence genetic exchange within this species are poorly understood. This includes identification of regions within the genome that are more likely to undergo genetic exchange and/or the frequency or rates of horizontal gene transfer and homologous recombination.

The extensive role of genetic exchange within *B. pseudomallei* may present a distinct advantage over clonally derived pathogens when attempting to develop models for estimating confidence limits for genotype "matches," "near matches," or "nonmatches." The high recombination rate causes a relatively rapid "scrambling" of even the relatively conserved housekeeping genes used in MLST analysis (57). But an even more dramatic aspect of the evolution *B. pseudomallei* has been demonstrated in a study of four cases of humans with acute forms of melioidosis (58). An MLVA-23 system (59) was used to examine 182 isolates recovered from different infected regions from these four patients. Results of this analyses indicated that despite a relatively short period of infection, all four of these patients showed significant divergence from the putative founder genotype (58). These results suggest that MLVA mutation rates and probabilistic applications can be used in tracking and identifying likely sources in the epidemiology and forensic analysis of melioidosis cases.

## CLOSTRIDIUM BOTULINUM

*Clostridium botulinum* is an anaerobic spore-forming bacterium that produces one of the most toxic substances known to man, botulinum neurotoxin (BoNT). The bacterium is endemic in soils and aquatic environments throughout the world (60). Inhalation or ingestion of bacterial cells, spores, or toxin results in a flaccid paralysis that can require mechanical ventilation and administration of botulinum antitoxin. BoNTs are listed by the CDC as category A select agents because they are relatively easy to produce, extremely potent, and exposure requires prolonged intensive hospital care (61).

Unlike most bacterial species, the primary criterion for the *C. botulinum* species designation has been based on production of the BoNT. This definition was adopted in order to prevent scientific and medical confusion regarding intoxication known as "botulism." However, this species definition has resulted in a pathogen whose genomic background encompasses at least four separate species by 16S rRNA analysis (62,63). A few isolates of two other species, *C. baratii* and *C. butyricum*, have also been identified that produce BoNTs (64,65).

In addition to genetic diversity within host bacteria, there are seven serologically distinct BoNTs designated A–G. Comparisons of the BoNT/A–G protein sequences reveal that BoNT protein identities differ by up to 70% among the seven serotypes (66). Diversity observed among the various BoNTs argues that divergence of these proteins was not recent and that their presence in diverse bacterial backgrounds (effectively six different species) appears to be the result of horizontal gene transfer.

Incongruence between phylogeny of toxin genes and host genomic background indicates a complex evolutionary pattern, but these differences can be used to develop a hierarchical approach to resolving *C. botulinum* isolates that appear to be tightly clustered. For example, Figure 16.5A depicts BoNT/A gene sequence differences in 60 different isolates belonging to the A serotype. Note that there are eight different sequence genotypes but that four of these are closely clustered as the A1 subtype. Linking these genotypes to an MLVA analysis (Figure 16.5B) of genomes that contain the BoNT/A gene yields 38 different subtypes and improves the resolution of serotype A *C. botulinum* isolates greatly (67).

From a forensics perspective, another layer of resolution could precede the MLVA when whole genome sequences and SNP discovery are coupled to these analyses to include a canonical SNP approach. The BoNT/A1 subtype

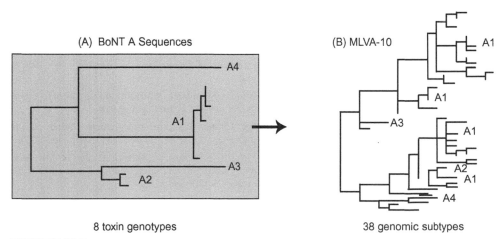

**FIGURE 16.5**

A proposed hierarchical approach to resolve serotype A *Clostridium botulinum* isolates. (A) A dendrogram depicting variation in the neurotoxin A gene. (B) These 8 genotypes can be further resolved into 38 subtypes by MLVA using 10 markers.

is currently the only example where multiple genomes are available for comparative analysis. Additional genomic sequences of isolates representing other serotypes and subtypes will provide further resolution of this species and be vital to clinical, epidemiological, and forensic analyses of *C. botulinum*.

# CONCLUSIONS

Significant progress has been made since the anthrax-letter attacks in defining molecular forensic approaches to investigate potential biocrimes. In this review the current status is described for genotyping and analysis of the bacterial pathogens *Bacillus anthracis*, *Yersinia pestis*, *Francisella tularensis*, *Brucella* spp., *Burkholderia pseudomallei*, and *Clostridium botulinum*. *C. botulinum* is unusual in that genomes represent several distinct species and the clonal nature of the organism is somewhat obscured by the diverse toxin complexes. With the exception of *B. pseudomallei*, these bacterial pathogens are basically clonal organisms that do not possess the mating properties and the independent assortment statistics that enhance human and plant forensic analyses greatly. Despite this, a hierarchical approach can now be proposed for most of these pathogens to provide the highest possible resolution for epidemiologic and forensics applications.

# REFERENCES

[1] P. Turnbull, Introduction: Anthrax history, disease and ecology, in: T.M. Koehler (Ed.), Anthrax, Springer-Verlag, Berlin, 2002, pp. 1–20.

[2] C.R. Woese, G.E. Fox, Phylogenetic structure of the prokaryotic domain: The primary kingdoms, Proc. Natl. Acad. Sci. USA 74 (11) (1977) 5088–5090.

[3] S.C. Dawson, N.R. Pace, Novel kingdom-level eukaryotic diversity in anoxic environments, Proc. Natl. Acad. Sci. USA 89 (12) (2002) 8324–9329.

[4] E.F. Delong, N.R. Pace, Environmental diversity of Bacteria and Archaea, Syst. Biol. 50 (4) (2001) 470–478.

[5] J.S. Lipp, Y. Morono, F. Inagaki, K.U. Hinrichs, Significant contribution of Archaea to extant biomass in marine subsurface sediments, Nature 454 (7207) (2008) 991–994.

[6] P. Hugenholtz, B.M. Goebel, N.R. Pace, Impact of culture-independent studies on the emerging phylogenetic view of bacterial diversity, J. Bacteriol. 180 (18) (1998) 4765–4774.

[7] D.J. Ecker, R. Sampath, P. Willett, J.R. Wyatt, V. Samant, C. Massire, et al., The Microbial Rosetta Stone Database: A compilation of global and emerging infectious microorganisms and bioterrorist threat agents, BMC Microbiol. 5 (2005) 19.

[8] A.J. Jeffreys, V. Wilson, S.L. Thein, Individual-specific "fingerprints" of human DNA, Nature 316 (6023) (1985) 76–79.

[9] P. Keim, T. Pearson, R. Okinaka, Microbial forensics: DNA fingerprinting of *Bacillus anthracis* (anthrax), Anal. Chem. 80 (13) (2008) 4791–4799.

[10] L.J. Harrell, G.L. Andersen, K.H.J. Wilson, Genetic variability of *Bacillus anthracis* and related species, J. Clin. Microbiol. 33 (1995) 1847–1850.

[11] P. Keim, A. Kalif, J. Schupp, K. Hill, S.E. Travis, K. Richmond, et al., Molecular evolution and diversity in *Bacillus anthracis* detected by amplified fragment length polymorphism markers, J. Bacteriol. 179 (3) (1997) 818–824.

[12] P. Keim, L.B. Price, A.M. Klevytska, K.L. Smith, J.M. Schupp, R. Okinaka, et al., Multiple-locus variable-number tandem repeat analysis reveals genetic relationships within *Bacillus anthracis*, J. Bacteriol. 182 (10) (2000) 2928–2936.

[13] R.T. Okinaka, K. Cloud, O. Hampton, A.R. Hoffmaster, K.K. Hill, P. Keim, et al., Sequence and organization of pXO1, the large *Bacillus anthracis* plasmid harboring the anthrax toxin genes, J. Bacteriol. 181 (20) (1999) 6509–6515.

[14] A.R. Hoffmaster, C.C. Fitzgerald, E. Ribot, L.W. Mayer, T. Popovic, Molecular subtyping of Bacillus anthracis and the 2001 bioterrorism-associated anthrax outbreak, United States, Emerg. Infect. Dis. 8 (10) (2002) 1111–1116.

[15] Keim P, Budowle B, Ravel J. Anthrax-Letter Attack. Microbial Forensics. Elsevier, 2010.

[16] T. Pearson, J.D. Busch, J. Ravel, T.D. Read, S.D. Rhoton, J.M. U'Ren, et al., Phylogenetic discovery bias in *Bacillus anthracis* using single-nucleotide polymorphisms from whole-genome sequencing, Proc. Natl. Acad. Sci. USA 101 (37) (2004) 13536–13541.

[17] P. Keim, M.N. Van Ert, T. Pearson, A.J. Vogler, L.Y. Huynh, D.M. Wagner, Anthrax molecular epidemiology and forensics: Using the appropriate marker for different evolutionary scales, Infect. Genet. Evol. 4 (2004) 205–213.

[18] M.N. Van Ert, W.R. Easterday, L.Y. Huynh, R.T. Okinaka, M.E. Hugh-Jones, J. Ravel, et al., Global genetic population structure of *Bacillus anthracis*, PLoS One 2 (2007) e461.

[19] L.J. Kenefic, J. Beaudry, C. Trim, R. Daly, R. Parmar, S. Zanecki, et al., High resolution genotyping of *Bacillus anthracis* outbreak strains using four highly mutable single nucleotide repeat markers, Lett. Appl. Microbiol. 46 (5) (2008) 600–603.

[20] J. Ravel, L. Jiang, S.T. Stanley, M.R. Wilson, R.S. Decker, T.D. Read, et al., The complete genome sequence of *Bacillus anthracis* Ames "Ancestor," J. Bacteriol. 191 (1) (2009) 445–446.

[21] R.E. Lenski, C.L. Winkworth, M.A. Riley, Rates of DNA sequence evolution in experimental populations of Escherichia coli during 20,000 generations, J. Mol. Evol. 56 (4) (2003) 498–508.

[22] J.W. Drake, Spontaneous mutation, Annu. Rev. Genet. 25 (1991) 125–146.

[23] H. Ochman, S. Elwyn, N.A. Moran, Calibrating bacterial evolution, Proc. Natl. Acad. Sci. USA 96 (22) (1999) 12638–12643.

[24] T.D. Read, S.L. Salzberg, M. Pop, M. Shumway, L. Umayam, L. Jiang, et al., Comparative genome sequencing for discovery of novel polymorphisms in *Bacillus anthracis*, Science 296 (5575) (2002) 2028–2033.

[25] J.E. Barrick, R.E. Lenski, Genome-wide mutational diversity in an evolving population of Escherichia coli, Cold Spring Harb. Symp. Quant. Biol (2009).

[26] M. Achtman, Age, descent and genetic diversity within *Yersinia pestis*, in: E. Carniel, B.J. Hinnebusch (Eds.), Yersinia: Molecular and Cellular Biology, Horizon Bioscience, Norwich, UK, 2004, pp. 432–445.

[27] M. Achtman, G. Morelli, P. Zhu, T. Wirth, I. Diehl, B. Kusecek, et al., Microevolution and history of the plague bacillus, *Yersinia pestis*, Proc. Natl. Acad. Sci. USA 101 (51) (2004) 17837–17842.

[28] M. Achtman, K. Zurth, G. Morelli, G. Torrea, A. Guiyoule, E. Carniel, *Yersinia pestis*, the cause of plague, is a recently emerged clone of Yersinia pseudotuberculosis, Proc. Natl. Acad. Sci. USA 96 (24) (1999) 14043–14048.

[29] P.S. Keim, D.M. Wagner, Humans and evolutionary and ecological forces shaped the phylo-geography of recently emerged diseases, Nat. Rev. Microbiol. 7 (11) (2009) 813–821.

[30] A.J. Vogler, E.M. Driebe, J. Lee, R.K. Auerbach, C.J. Allender, M. Stanley, et al., Assays for the rapid and specific identification of North American Yersinia pestis and the common labora-tory strain CO92, Biotechniques 44 (2) (2008) 201, 203–204, 207.

[31] R.K. Auerbach, A. Tuanyok, W.S. Probert, L. Kenefic, A.J. Vogler, D.C. Bruce, et al., *Yersinia pestis* evolution on a small timescale: Comparison of whole genome sequences from North America, PLoS One 2 (1) (2007) e770.

[32] J.W. Touchman, D.M. Wagner, J. Hao, S.D. Mastrian, M.K. Shah, A.J. Vogler, et al., North American *Yersinia pestis* draft genome sequence: SNPs and phylogenetic analysis, PLoS One 2 (2) (2007) e220.

[33] J.M. Girard, D.M. Wagner, A.J. Vogler, C. Keys, C.J. Allender, L.C. Drickamer, et al., Differential plague-transmission dynamics determine *Yersinia pestis* population genetic structure on local, regional, and global scales, Proc. Natl. Acad. Sci USA 101 (22) (2004) 8408–8413.

[34] A.J. Vogler, C.E. Keys, C. Allender, I. Bailey, J. Girard, T. Pearson, et al., Mutations, mutation rates, and evolution at the hypervariable VNTR loci of *Yersinia pestis*, Mutat Res 616 (1–2) (2007) 145–158.

[35] J.L. Lowell, D.M. Wagner, B. Atshabar, M.F. Antolin, A.J. Vogler, P. Keim, et al., Identifying sources of human exposure to plague, J. Clin. Microbiol. 43 (2005) 650–656.

[36] R.E. Colman, A.J. Vogler, J.L. Lowell, K.L. Gage, C. Morway, P.J. Reynolds, et al., Fine-scale identification of the most likely source of a human plague infection, Emerg. Infect. Dis. 15 (2009) 1623–1625.

[37] P. Keim, A. Johansson, D.M. Wagner, Molecular epidemiology, evolution, and ecology of *Francisella*, Ann. N.Y. Acad. Sci. 1105 (2007) 30–66.

[38] M.J. Whipp, J.M. Davis, G. Lum, J. de Boer, Y. Zhou, S.W. Bearden, et al., Characterization of a novicida-like subspecies of *Francisella tularensis* isolated in Australia, J. Med. Microbiol. 52 (Pt 9) (2003) 839–842.

[39] S.M. Barns, C.C. Grow, R.T. Okinaka, P. Keim, C.R. Kuske, Detection of diverse new *Francisella*-like bacteria in environmental samples, Appl. Environ. Microbiol. 71 (9) (2005) 5494–5500.

[40] M. Forsman, G. Sandstrom, A. Sjostedt, Analysis of 16S ribosomal DNA sequences of *Francisella* strains and utilization for determination of the phylogeny of the genus and for identification of strains by PCR, Int. J. Syst. Bacteriol. 44 (1) (1994) 38–46.

[41] A. Johansson, J. Farlow, P. Larsson, M. Dukerich, E. Chambers, M. Byström, et al., Worldwide genetic relationships among *Francisella tularensis* isolates determined by multiple-locus vari-able-number tandem repeat analysis, J. Bacteriol. 186 (17) (2004) 5808–5818.

[42] A.J. Vogler, D. Birdsell, L.B. Price, J.R. Bowers, S.M. Beckstrom-Sternberg, R.K. Auerbach, et al., Phylogeography of Francisella tularensis: Global expansion of a highly fit clone, J. Bacteriol. 191 (8) (2009) 2474–2484.

[43] A.J. Vogler, D. Birdsell, D.M. Wagner, P. Keim, An optimized, multiplexed multi-locus variable-number tandem repeat analysis system for genotyping *Francisella tularensis*, Lett. Appl. Microbiol. 48 (1) (2009) 140–144.

[44] A.M. Whatmore, Current understanding of the genetic diversity of *Brucella*, an expanding genus of zoonotic pathogens, Infect. Genet. Evol. 9 (6) (2009) 1168–1184.

[45] J.T. Foster, S.M. Beckstrom-Sternberg, T. Pearson, J.S. Beckstrom-Sternberg, P.S. Chain, F.F. Roberto, et al., Whole-genome based phylogeny and divergence of the genus *Brucella*, J. Bacteriol. 191 (8) (2009) 2864–2870.

[46] G. Pappas, P. Papadimitriou, N. Akritidis, L. Christou, E.V. Tsianos, The new global map of human brucellosis, Lancet Infect. Dis. 6 (2006) 91–99.

[47] B.J. Bricker, D.R. Ewalt, HOOF prints: Brucella strain typing by PCR amplification of multi-locus tandem-repeat polymorphisms, Methods Mol. Biol. 345 (2006) 141–173.

[48] L.Y. Huynh, M.N. Van Ert, T. Hadfield, W.S. Probert, B.H. Bellaire, M. Dobson, et al., Multiple locus variable number tandem repeat (VNTR) analysis (MLVA) of Brucella spp. identifies species-specific markers and insights into phylogenetic relationships NIH Volume 1, Frontiers in Research, in: V.S. Georgiev (Ed.), National Institute of Allergy and Infectious Disease, Vol. 1, Human Press Inc, Totowa, NJ, 2008, pp. 47–54.

[49] P. Le Fleche, I. Jacques, M. Grayon, S. Al Dahouk, P. Bouchon, F. Denoeud, et al., Evaluation and selection of tandem repeat loci for a Brucella MLVA typing assay, BMC Microbiol. 6 (2006) 9.

[50] A.M. Whatmore, S.J. Shankster, L.L. Perrett, T.J. Murphy, S.D. Brew, R.E. Thirlwall, et al., Identification and characterization of variable-number tandem-repeat markers for typing of Brucella spp, J. Clin. Microbiol. 44 (6) (2006) 1982–1993.

[51] T. Pearson, P. Giffard, S. Beckstrom-Sternberg, R. Auerbach, H. Hornstra, A. Tuanyok, et al., Phylogeographic reconstruction of a bacterial species with high levels of lateral gene transfer, BMC Biol. 7 (2009) 78.

[52] M.T. Holden, R.W. Titball, S.J. Peacock, A.M. Cerdeño-Tárraga, T. Atkins, L.C. Crossman, et al., Genomic plasticity of the causative agent of melioidosis, Burkholderia pseudomallei, Proc. Natl. Acad. Sci. USA 101 (39) (2004) 14240–14245.

[53] S.H. Sim, Y. Yu, C.H. Lin, R.K. Karuturi, V. Wuthiekanun, A. Tuanyok, et al., The core and accessory genomes of Burkholderia pseudomallei: Implications for human melioidosis, PLoS Pathog. 4 (10) (2008) e1000178.

[54] A. Tuanyok, B.R. Leadem, R.K. Auerbach, S.M. Beckstrom-Sternberg, J.S. Beckstrom-Sternberg, M. Mayo, et al., Genomic islands from five strains of Burkholderia pseudomallei, BMC Genomics 9 (2008) 566.

[55] S. Tumapa, M.T. Holden, M. Vesaratchavest, V. Wuthiekanun, D. Limmathurotsakul, W. Chierakul, et al., Burkholderia pseudomallei genome plasticity associated with genomic island variation, BMC Genomics 9 (2008) 190.

[56] J.M. U'Ren, H. Hornstra, T. Pearson, J.M. Schupp, B. Leadem, S. Georgia, et al., Fine-scale genetic diversity among Burkholderia pseudomallei soil isolates in northeast Thailand, Appl. Environ. Microbiol. 73 (20) (2007) 6678–6681.

[57] T. Pearson, P. Giffard, S. Beckstrom-Sternberg, R. Auerbach, H. Hornstra, A. Tuanyok, et al., Phylogeographic reconstruction of a bacterial species with high levels of lateral gene transfer, BMC Biol. 7 (2009) 78.

[58] E.P. Price, H.M. Hornstra, D. Limmathurotsakul, T.L. Max, D.S. Sarovich, A.J. Vogler, et al., Within-host evolution of Burkholderia pseudomallei in four cases of acute melioidosis, PLoS Pathog. 6 (1) (2010) e1000725.

[59] J.M. U'Ren, J.M. Schupp, T. Pearson, H. Hornstra, C.L. Friedman, K.L. Smith, et al., Tandem repeat regions within the Burkholderia pseudomallei genome and their application for high resolution genotyping, BMC Microbiol 7 (2007) 23.

[60] L.D.S. Smith, H. Sugiyama, Botulism: The Organism, Its Toxins, the Disease, second edn., Charles C. Thomas, 1998.

[61] S.S. Arnon, R. Schechter, T.V. Inglesby, D.A. Henderson, J.G. Bartlett, M.S. Ascher, et al., Botulinum toxin as a biological weapon: Medical and public health management, JAMA 285 (8) (2001) 1059–1070.

[62] M.D. Collins, A.K. East, Phylogeny and taxonomy of the food-borne pathogen *Clostridium botulinum* and its neurotoxins, J. Appl. Microbiol. 84 (1) (1998) 5–17.

[63] K.K. Hill, G. Xie, B.T. Foley, T.J. Smith, A.C. Munk, D. Bruce, et al., Recombination and insertion events involving the botulinum neurotoxin complex genes in *Clostridium botulinum* types A, B, E and F and *Clostridium butyricum* type E strains, BMC Biol. 7 (2009) 66.

[64] P. Aureli, L. Fenicia, B. Pasolini, M. Gianfranceschi, L.M. McCroskey, C.L. Hatheway, Two cases of type E infant botulism caused by neurotoxigenic *Clostridium butyricum* in Italy, J. Infect. Dis. 154 (2) (1986) 207–211.

[65] J.D. Hall, L.M. McCroskey, B.J. Pincomb, C.L. Hatheway, Isolation of an organism resembling *Clostridium barati* which produces type F botulinal toxin from an infant with botulism, J. Clin. Microbiol. 21 (4) (1985) 654–655.

[66] T.J. Smith, J. Lou, I.N. Geren, C.M. Forsyth, R. Tsai, S.L. Laporte, et al., Sequence variation within botulinum neurotoxin serotypes impacts antibody binding and neutralization, Infect. Immun. 73 (9) (2005) 5450–5457.

[67] T.E. Macdonald, C.H. Helma, L.O. Ticknor, P.J. Jackson, R.T. Okinaka, L.A. Smith, et al., Differentiation of *Clostridium botulinum* serotype A strains by multiple-locus variable-number tandem-repeat analysis, Appl. Environ. Microbiol. 74 (3) (2008) 875–882.

[68] L.J. Kenefic, T. Pearson, R.T. Okinaka, J.M. Schupp, D.M. Wagner, A.R. Hoffmaster, et al., Pre-Columbian origins for North American anthrax, PLoS One 4 (3) (2009) e4813.

# CHAPTER 17

# *Rickettsia* and *Coxiella*

**Robert F. Massung, Gregory A. Dasch, and Marina E. Eremeeva**

*Rickettsial Zoonoses Branch, Division of Viral and Rickettsial Diseases,
Centers for Disease Control and Prevention, Atlanta, Georgia*

## INTRODUCTION

Rickettsial diseases likely influenced the history of humankind for many years before scientists discovered and began to appreciate the characteristics of the various etiologic agents collectively identified as rickettsiae. Rocky Mountain spotted fever (RMSF) was first described in 1899, although it is believed that it was experienced by settlers and Indians of the northwestern Rocky Mountains as early as 1872. Epidemic typhus plagued humankind for centuries since at least the age of discovery in the 1500s (1); however, its modern history only goes back to 1909 when Charles Nicole established the role of the body louse in transmission of *Rickettsia prowazekii* (2). The earliest molecular evidence of *R. prowazekii* as a cause of epidemics was obtained following polymerase chain reaction (PCR) amplification of rickettsial DNA from body louse and dental remains recovered from the burial site of Napoleon's soldiers in Vilnius (3).

Query or "Q" fever was first described by Derrick in 1935 in Australia during an outbreak of febrile illness among abattoir workers (4). Burnet and Freeman (5) transmitted the Q fever agent experimentally to guinea pigs, mice, and nonhuman primates and used serological studies to demonstrate the disease to be distinct from known rickettsial infections. Also in 1935, researchers at the Rocky Mountain Laboratory (RML) in the United States transmitted an agent from *Dermacentor andersoni* ticks from Montana to guinea pigs (6). The agent, called the Nine Mile agent, was shown to be semi-filterable and could not be isolated using axenic culture, although Cox (7) was able to propagate the agent in embryonated chicken eggs. An RML worker became infected accidentally with the agent, and blood from that worker was used to infect guinea pigs. Subsequent studies confirmed that the Australian Q fever agent and the North American Nine Mile agent were the same species, and in 1948

277

Microbial Forensics. DOI: 10.1016/B978-0-12-382006-8.00017-7

the agent was named *Coxiella burnetii* in honor of the contributions of Herald Cox and Macfarlane Burnet (8,9).

*Rickettsia* and *Coxiella* are fastidious organisms, with an estimated generation time of >8 hours, and until recently, both were considered obligately intracellular bacteria. Classically, these agents were propagated similarly in embryonated chicken eggs or mammalian cell cultures but *C. burnetii* has been recently cultivated axenically (10). Initially thought to be related to the genus *Rickettsia*, more recent genetic studies have shown *C. burnetii* to be more closely related to *Legionella*, *Francisella*, and *Rickettsiella* species (11). *Rickettsia* and *Coxiella* are both members of the phylum Proteobacteria, although *Rickettsia* species are in the class Alphaproteobacteria and replicate in the cytosol of infected mammalian cells, whereas *C. burnetii* is in the class Gammaproteobacteria and grows in an acidified modified phagolysosome. *Coxiella burnetii* is very stable in the environment due to its ability to form small cell variants with spore-like properties (12,13). While *Rickettsia* may survive for years in a lyophilized form, their viability is rapidly compromised upon exposure to high humidity or temperature (14–16).

## ROCKY MOUNTAIN SPOTTED FEVER, EPIDEMIC TYPHUS, AND Q FEVER: THE DISEASES

The two classic rickettsioses, epidemic typhus and Rocky Mountain spotted fever, are among the most severe acute febrile diseases with mortality rates of >20% in the preantibiotic era. In contrast, the mortality rate for patients with acute Q fever is <1% even without antibiotic treatment. Each of these agents may cause an acute febrile illness, with a rash noted in approximately 80% of RMSF infections, 50% of typhus infections, but rare with Q fever. Epidemic typhus and RMSF have an incubation period of 1–2 weeks. For RMSF, initial symptoms may include fever, nausea, vomiting, severe headache, muscle pain, and lack of appetite. The rash typically develops 2–5 days after onset of fever and is often first noted on extremities, frequently on the palms and soles. Typical laboratory findings are thrombocytopenia, hyponatremia, and elevated liver enzymes. Common symptoms of epidemic typhus are high fever, headache, cough, chills, myalgia, and fatigue, with delirium and stupor in severe cases. A rash that typically begins on the trunk may develop approximately 5 days after onset. Laboratory findings include thrombocytopenia, anemia, and hyponatremia.

*Coxiella burnetii* can cause both acute and chronic forms of Q fever. Acute Q fever may present as flu-like illness with atypical pneumonia or hepatitis in more severe cases, but many infections are relatively mild or asymptomatic. The incubation period is approximately 2–3 weeks and may be dependent on dose, route of infection, or the infecting strain. Common symptoms of acute infections are fever, fatigue, headache, rigors, and myalgia. Abnormal lymphocyte and thrombocyte counts are rare, with elevated liver enzymes as the most common laboratory finding (17).

Chronic Q fever may develop months or years following acute infection and is strongly linked to predisposing conditions such as cardiac disease, valvular defects, vascular grafts, and immunodeficiencies, and pregnant women are at high risk for developing chronic disease (18). Endocarditis is the primary clinical presentation of chronic Q fever; rarely reported are cases with osteoarticular (osteomyelitis, coxitis, arthritis), hepatic, pulmonary, and renal involvement (17).

Tetracycline (doxycycline) is the drug of choice for treating RMSF, epidemic typhus, and Q fever (17,19). Co-trimoxazole is recommended to treat Q fever in patients for whom tetracyclines are contraindicated (pregnant women and children younger than 8 years) (20,21). The current recommendation for treating chronic Q fever is prolonged antibiotic treatment (>1 year) with a combination of doxycyline and hydroxychloroquine (17,22,23).

## EPIDEMIOLOGY

Rocky Mountain spotted fever is restricted to the Americas but is rare in the far north and south where its tick vectors cannot overwinter successfully. It often causes clusters of cases when transmitted by the brown dog tick, *Rhipicephalus sanguineus*, but is transmitted more frequently and sporadically by the American dog tick, *Dermacentor variabilis*, in the United States east of the Rocky Mountains and on the West Coast, and the wood tick, *D. andersoni*, in the Rocky Mountain states (24). *Amblyomma* spp. ticks (*A. cajennense*, *A. aureolatum*, and *A. oblongoguttatum*) are thought to be the primary vectors in Mexico and in Central and South America (25). Other ticks, such as *Haemaphysalis leporispalustris*, occasionally harbor *R. rickettsii* but whether this indicates a vector role or simply incidental acquisition by these species is not clear. A wide variety of animal species can be infected with *R. rickettsii* and it is likely that different species play important roles as reservoirs for the agent in different ecoregions.

Epidemic typhus is transmitted by the human body louse, *Pediculus humanus humanus*. Louse-borne epidemic typhus once had a worldwide distribution, but with improved sanitary practices is currently endemic only in parts of the mountainous regions of Africa, South America, and Asia where human body lice are common. Recovery from epidemic typhus results in nonsterile immunity, permitting persistence of *R. prowazekii* between epidemics; subsequent recrudescence decades later as Brill–Zinsser disease can initiate new epidemics in naïve populations if body lice are present. In the eastern United States,

*R. prowazekii* is found in the southern flying squirrel, *Glaucomys volans volans*, where it is passed between these social squirrels by their fleas and lice and possibly mites; it occasionally causes sylvatic typhus in humans coming in close contact with this reservoir (26). *Rickettsia prowazekii* has also been identified in association with *Amblyomma* ticks in Mexico and Ethiopia (27,28); however, evidence for tick transmission of *R. prowazekii* to humans has not been reported.

*Coxiella burnetii* is found worldwide with sheep, goats, and cattle most often described as the primary reservoirs (11). Humans become naturally infected by the inhalation of infectious materials shed during parturition of infected animals or inhalation of desiccated bacteria from milk, urine, and fecal matter. Q fever outbreaks are often associated with occupational exposures, and aerosolized infectious material may be carried by the wind or otherwise dispersed, resulting in widespread outbreaks in humans (29–34). Human-to-human transmission of *C. burnetii* has been reported but is rare, as is infection by ingestion of contaminated food products (35–37). Bacteria are quite resistant to environmental conditions such as temperature and humidity and can persist in soil or dairy products for extended periods of time. Dormancy induces replicating *C. burnetii* to condense into small cell variants, a spore-like form that is resistant to many methods of physical and chemical sterilization (12,13). The unique ability of *C. burnetii* to infect a large variety of potential reservoir species has often made it difficult to determine the source of natural outbreaks of human infections. Although infections of sheep, goats, and cattle have been studied most extensively, infections have been described in birds, reptiles, canids, rabbits, felines, marine mammals, arthropods, and numerous wild and laboratory-reared rodent species (38–41). The contribution of each of these to the maintenance of *C. burnetii* in nature remains undetermined. Also not clear are whether strains isolated from various potential natural reservoirs are equally infectious in humans, as strains collected from wild rodents in Dugway, Utah, were avirulent in laboratory rodents (42,43). Studies have also shown that several strains that could be differentiated based on restriction enzyme fragment profiles had variable clinical features and virulence in mouse and guinea pig models (44).

## DIAGNOSIS/DETECTION IN CLINICAL SAMPLES

The primary diagnostic techniques for RMSF, epidemic typhus, and Q fever have traditionally been based on serologic methods such as complement fixation, enzyme-linked immunosorbent assay, and indirect immunofluorescent assay (45–47). However, serologic methods require the evaluation of both acute and convalescent samples for diagnostic confirmation by seroconversion (fourfold increase in IgG titers) and therefore are of limited utility for the rapid detection of a natural outbreak or cases caused by intentional release. Isolation of the agents has long been the gold standard for diagnosis but it is technically difficult, requires specialized high-containment facilities, and is time-consuming. Molecular detection methods based on PCR have the advantages of being rapid, sensitive, and specific.

Molecular methods have been developed for the diagnosis of RMSF, epidemic typhus, and Q fever in clinical samples. Whole blood is the sample used most commonly for testing of suspected acute infections, although enrichment of the white cell fraction (buffy coat) may increase detection sensitivity. The utility of PCR of serum samples has also been shown (48–51). Molecular methods have limited utility for the detection of rickettsial agents, primarily because the number of rickettsiae in circulating blood is low. Rickettsial PCR has proven more useful for assessing skin biopsies, although these samples are not typically available for diagnostic evaluation. Difficulties in detecting *Rickettsia* in clinical samples have required the use of culture isolates for most genotypic analyses. The agents are more abundant in arthropod samples where PCR is a useful and sensitive method for detection and genetic analysis (115).

In contrast to rickettsial agents, the diagnosis of Q fever by molecular methods is commonly used and relatively sensitive because *C. burnetii* replicates in circulating neutrophils. PCR-based methods have proven useful for the detection of the agent in naturally occurring outbreaks (29,48,52,53). PCR was shown to be more sensitive than serology during the first 2 weeks after symptom onset and could provide an earlier diagnosis of Q fever than serology alone (49). Assays that amplify the IS1111 insertion sequence found as a multicopy gene in the *C. burnetii* genome generally provide increased sensitivity compared to assays that amplify single copy genes (*com*1, *rrs*, *icd*) (50,54–56).

## DETECTION IN ENVIRONMENTAL SAMPLES

Detection of *C. burnetii* in environmental samples has not been evaluated extensively, although a recent pilot study involved the collection and analysis of >1600 environmental samples from six areas of the United States. Sample collection methods were based on protocols developed for other pathogenic bacteria such as *Bacillus anthracis* and included use of wet and dry swabs, bulk sampling, and high-efficiency particulate air vacuum socks (57–59). PCR using a *Coxiella*-specific PCR assay detected *C. burnetii* DNA in 23.8% of the samples (60). Extraction of DNA from environmental samples may be problematic due to high levels of contamination by infectious agents other than the target agent, high levels of background DNAs, and the presence of PCR inhibitors, particularly in samples containing soil or humic material (61). Therefore, care must be taken to include controls to evaluate DNA extraction efficiency and PCR assay inhibition. In general, PCR assays developed for human diagnostic samples work well on properly extracted, inhibitor-free, environmental samples. The potentially high levels of *C. burnetii* DNA found in the environment may be problematic in determining whether the material detected is naturally occurring or from an intentional release. Environmental sampling for rickettsiae has not been described, although the background levels should be lower than noted for *C. burnetii* due to the limited natural reservoirs and focal geographic distribution of *R. rickettsii* and *R. prowazekii*.

## *RICKETTSIA* STRAIN TYPING

The primary limitation for developing forensic methods with *Rickettsia* is that isolates are obtained infrequently, despite numerous methods for cultivation in animals and cell culture. In the pre-PCR era before 1990, differentiation of isolates required a viable agent and methods based on antigens and polyclonal sera often failed to provide identification of species, let alone differentiate strains of *Rickettsia*. The most sensitive methods were thus based on variations in antibody neutralization tests *in vivo* (62) and virulence for animals. The virulence of *R. rickettsii* isolates differs in the guinea pig model of infection (63,64). Eleven isolates of *R. rickettsii* could be grouped by the amount of host cell injury measured by lactate dehydrogenase release and alteration of antioxidant levels (65). However, rickettsial virulence can be altered via passage attenuation with classic examples of the Iowa isolate of *R. rickettsii* (7,66) and strain E of *R. prowazekii* (67–71) and physiological adaptations (72–74) so this approach has proven to be of limited value at the strain level.

With the advent of monoclonal antibodies and high-resolution polyacrylamide gel electrophoresis (PAGE), variable antigenic features were observed at both the protein level and the epitope level for *R. prowazekii* (75–80) and *R. rickettsii* (81,82). Hlp strains of *R. rickettsii*, which have an attenuated phenotype in guinea pigs, also differ in their proteins, including major OmpA antigens. Differences were found in protein mobilities by sodium dodecyl sulfate (SDS)–polyacrylamide gel electrophoresis, by Western blotting, and by immunoprecipitation of rickettsial antigens; however, only Hlp was markedly differentiated by this method (81). Strains from North Carolina and Montana also differed in at least one epitope of the 120-kDa protein with monoclonal antibodies in an ELISA assay (82).

Old World louse-borne isolates of *R. prowazekii* can be differentiated from New World flying squirrel isolates by isoelectric focusing, SDS-PAGE, and with polyclonal and monoclonal antibodies (83). The attenuated E strain can be differentiated from its virulent parent isolate and its revertant by its biochemical and immunological properties (84–86).

## Chromosomal Differentiation

*Rickettsia prowazekii* isolates from flying squirrels exhibited unique chromosomal restriction patterns that differentiated them from the reference strain Breinl from Poland, and a North African isolate (Cairo) could be distinguished from two other African isolates (Addis Ababa and ZRS) (87,88). Similarly, avirulent vaccine strain Madrid E, its virulent revertant EVir, and isolate Katsinyian from Armenia have a genetic profile different from classic virulent isolates (89). Stable genetic patterns for isolates with multiple laboratory passages (Breinl, Madrid E) suggested that these differences could be used in tracing epidemics.

Pulsed field gel electrophoresis (PFGE) of restriction-digested genomic DNA was used to differentiate the Breinl and EVir isolates of *R. prowazekii*, estimate the small 1.1 Mb genome size of the chromosome, and distinguish this species from its closest relative, *R. typhi* (90). While *R. rickettsii* Smith had a 1.25 Mb genome like most other spotted fever rickettsiae, its macrorestriction pattern differentiated it from its nearest relatives, and the first gene localizations were obtained by Southern blotting (91). As with protein work, these methods require large amounts of cells and have largely been superseded by PCR analysis of genes and whole genome sequencing.

## Polymerase Chain Reaction Amplification

Polymerase chain reaction amplification of rickettsial genes, followed by restriction fragment length polymorphism (RFLP) analysis or DNA sequencing, has become the primary means for rapid characterization and speciation of isolates of rickettsiae for the past 20 years (92–98). PCR/RFLP methods targeting the *glt*A (99), *omp*A (100,101), and *omp*B (66) genes of *R. rickettsii* have been developed with an analytical sensitivity of between 1 and 10 rickettsiae. In contrast, while PCR/RFLP analysis of *omp*B, *glt*A, and 17-kDa protein genes did not detect genetic variability among the *R. prowazekii* strains tested (102,103), multilocus sequence analysis detected minor differences among strains isolated from patients with louse-borne typhus and from flying squirrels (104,105).

## Variable Loci Typing

In 1998, *R. prowazekii* Madrid E became the first *Rickettsia* genome sequenced (106). The genome sequences of *R. rickettsii* and near-neighbors *R. conorii* (agent of Mediterranean spotted fever) and *R. typhi* (murine typus) confirmed the highly syntenic relationship of these rickettsiae (107,108) and permitted the first detailed analyses of variable loci in these species. The 25 most variable intergenic spacers between *R. prowazekii* and *R. conorii* genomes were amplified and sequenced from five strains and 10 body louse samples of *R. prowazekii* (109). Two variable spacers, *rpm*E/tRNA(fMet) and *ser*S/*vir*B4, allowed identification of four *R. prowazekii* genotypes and demonstrated that different types can circulate within the limited area of an outbreak.

Primers initially developed for multispacer sequence typing (MST) of *R. conorii* (110) and some specific to *R. rickettsii* were used to characterize the intergenic regions (IGR) of 35 isolates of *R. rickettsii* of human, tick, and animal origin (111). Seven genotypes of *R. rickettsii* in four primary groups could be distinguished: isolates from Montana, isolates associated with *R. sanguineus* ticks and human infections in Arizona, other isolates from the United States where *D. variabilis* is thought to be the primary vector, and isolates associated primarily with *Amblyomma* ticks from Central and South America.

A tandem repeat finder (112) was used to identify 12 tandem repeats in *R. rickettsii*. Amplicons encompassing four sites had variability in electrophoretic mobility using the same panel of isolates used for IGR typing. Analysis of their concatenated sequences clustered all isolates studied into six groups (18 genotypes) that correlated with the specific tick vector from which the isolates were obtained or where these ticks are prevalent (human and animal isolates) (24).

The variable number tandem repeat (VNTR) typing system was used in molecular epidemiology to demonstrate the unique lineage of *R. rickettsii* isolates associated with *R. sanguineus* in Arizona (113) and a unique genotype of *R. rickettsii* found in *R. sanguineus* in southern California (114). The *R. rickettsii* responsible for the *R. sanguineus*-associated 2008–2009 RMSF outbreak in Mexicali, Mexico, is a unique genetic type that is different from strains detected in Arizonia and South and Central America (115).

## Insertion/Deletion (INDEL) Typing

The genome sequences of *R. typhi* and *R. prowazekii* were aligned to identify INDEL sites (INDELs) that distinguish these species (116). Of 110 INDEL bridging sites analyzed by PCR with 38 isolates of *R. prowazekii* and 18 of *R. typhi*, 8 of these sites exhibited sequence polymorphisms, 2 that were unique to Madrid E and 1 for *R. typhi*; these sites could be used to differentiate among flying squirrel isolates of *R. prowazekii*.

Insertion/deletion sites unique to *R. prowazekii* were also identified by comparison of the genome sequence of the *A. cajennense* tick isolate from Mexico with the Madrid E sequence (117). Six variable loci were compared by DNA sequencing of eight isolates of *R. prowazekii*, including Madrid E and EVir, two flying squirrel isolates, Breinl, and three isolates from Africa (ZRS, Cairo, and Addis Ababa). Three single nucleotide polymorphisms (SNP) were present in the INDEL regions, and only ZRS and Addis Ababa could not be distinguished among the seven genotypes identified.

Analysis of the genome sequences of *R. rickettsii* isolates Sheila Smith and Iowa identified 143 INDELs ranging from 1 to 10,585 bp and 492 SNP sites (118). Only a third of INDELs were within predicted coding regions, and 67 INDEL sites have been analyzed (24,119). Eight polymorphic sites were found, which permitted the identification of nine genetic clades within *R. rickettsii*. Most isolates from Montana, associated with *D. andersonii* ticks, were similar to isolate Sheila Smith, whereas *D. variabilis* tick-associated isolates were more similar to the Iowa isolate. Other isolates (Hlp#2 and 364D) were intermediate between Sheila Smith and Iowa isolates, a finding in agreement with previous VNTR and IGR analyses and again suggesting that these two isolates might be unique species of spotted fever group rickettsiae or distinct subspecies of *R. rickettsii* (24,111).

## Next-Generation Genome Sequencing of *Rickettsia*

Dasch and colleagues (120) described the sequencing of six additional isolates of *R. prowazekii* and seven isolates of *R. rickettsii*, including strains Hlp#2 and 364D. Selection of these isolates was based on VNTR, IGR, and INDEL data and geographic, vector associations, source, and passage history. Whole genome alignment identified all INDEL and SNP differences among the isolates, and those associated with VNTRs were determined by inspection and use of repeat identification software (112,121). Not surprisingly, these data confirmed polymorphisms between the isolates and provided a substantial number of new INDELs and SNPs whose distributions are being mapped in the collection of strains available. Strain 364D had substantially more unique sequences than Hlp#2 in comparison to the other isolates of *R. rickettsii*, thus supporting its placement in a new species, whereas Hlp#2 is described most accurately as a subspecies of *R. rickettsii*. *Rickettsia prowazekii* had far fewer variable sites among the seven available genomic sequences and should remain monophyletic. These sites are being examined by both INDEL and IGR assays as described earlier and the SNPs characterized by mismatch amplification mutation assays for both *R. typhi* and *R. prowazekii* (122,123).

## *COXIELLA* STRAIN TYPING

Early attempts to differentiate isolates of *C. burnetii* based on serological methods, protein patterns, or biological differences proved unsuccessful. Although immunological and phenotypic differences were noted after passage of isolates in cultured cells or embryonated eggs, these property changes were later shown due to an attenuated form (phase II) that is not seen in nature. The first methods used that could consistently differentiate *C. burnetii* isolates were based on analysis of plasmid types, lipopolysaccharide (LPS) profiles, and whole chromosome RFLP. LPS banding patterns were able to define three major groups among

*C. burnetii* isolates (124). Whole chromosome RFLP analysis was the first approach that clearly differentiated *C. burnetii* isolates into distinct genogroups (125); this method currently defines six genogroups. Isolates made from humans are found in genogroups I through V, while genogroup VI contains murine isolates from Utah. Several plasmid types have been described for *C. burnetii*, and a plasmid typing method initially separated chronic disease causing isolates from those that cause acute infections, but this distinction has not proven to be accurate (126,127). Still, plasmid typing remains a basic method for characterizing strains as one of three plasmid types (QpRS, QpH1, QpDV) or plasmidless, and PCR-based methods have been developed to facilitate such analyses (128).

Although the 16S rRNA gene is highly conserved among strains (129), significant phylogenetic heterogeneity was found among *C. burnetii* strains by DNA sequencing of selected genes (*icd, mucZ, com*1) (130–133) and, more recently, by whole genome sequencing (134,135) and microarray-based whole genome analysis (136). RFLP/PFGE was used to differentiate 80 isolates into 20 different pattern groups (137). The most comprehensive evaluation by Glazunova and colleagues (138) used MST to place 173 isolates into 30 different genogroups.

Most of the methods that have been reported for differentiating *C. burnetii* isolates are useful but time-consuming and labor-intensive, evaluated only on cultured strain isolates, and not amenable for rapid testing of clinical or environmental samples. Several recently developed PCR-based methods have shown promise for use on clinical and environmental samples. A multilocus variable number tandem repeat analysis (MLVA) defined nine MLVA types among 16 isolates from ticks, humans, goats, and rodents (139). Another MLVA study defined 36 genotypes among 42 isolates from Europe, Africa, North America, and Asia (140). A repetitive-element PCR assay that allows rapid differentiation of the five primary genogroups responsible for human infection (genogroups I through V) based on a series of PCR amplifications was shown effective for use on both isolates and clinical veterinary samples (141). A rapid and sensitive amplification method that has promise for use on clinical and environmental samples is based on SNP sites characterized within MST loci. Fourteen SNP assays have been used to genotype >40 isolates of *C. burnetii* and have been able to identify eight distinct genogroups. These genogroups distinguish isolates based on the region and type of source material, as well as infection and plasmid type. Interestingly, three Chinese isolates cluster on a branch with RSA331 (Henzerling) and M44, both isolates from southern Europe obtained near the end of World War II. Recently established American goat isolates cluster with several human heart valve isolates (KAV, PAV, Q228) and a previous American goat isolate (Q177). Isolates grouping with Nine Mile phase I have a nearly worldwide distribution, including strains from America, Australia, and Europe (142).

## CONCLUSIONS

The efficient application and characterization of DNA from pathology, clinical, vector, and animal reservoir samples require molecular assays comparable in sensitivity to current TaqMan, SYBR Green, and nested PCR assays. The feasibility of this approach has already been demonstrated in characterizing *Rickettsia* in samples from Arizona, California, and Mexico, confirming the identity of a new eschar-associated infection caused by *R. rickettsii* 364D (143) and evaluating *C. burnetii* in environmental and clinical samples associated with natural outbreaks. However, the fundamental problem in applying these methods in forensic situations remains in that it is unknown what portion of the population of these agents in nature is represented by the spectrum of available isolates and samples that have been characterized to date. The discovery of the presence of *R. prowazekii* in New World ticks far removed from the original report in Africa and the rediscovery of novel isolates of *R. rickettsii* in *Rhipicephalus sanguineus* ticks suggest that an answer to this question will require substantial investment in the collection and analysis of new field and clinical samples from around the world. The spectrum of *Coxiella*-like agents is also poorly defined. The generation of genome sequences for additional strains of *R. rickettsia*, *R. prowazekii*, and *C. burnetii* are needed to further evaluate strain variability and provide novel targets for detecting and differentiating strains and isolates. Current sequencing efforts have shown low genetic variation among the genomes of *R. prowazekii*, more variability among *R. rickettsii* strains, and a high degree of variation among the genomes of *C. burnetii* isolates. This may necessitate different strategies for the evaluation of these agents, with the differentiation of strains of a low variability agent such as *R. prowazekii* potentially requiring the more sensitive level of SNP assay discrimination. Alternatively, for *C. burnetii*, the high level of genome variability may simplify the development of assays that provide for differentiation of known strains, but such assays may be unable to differentiate novel strains with unique genome sequences or genomic rearrangements. Further, the high background level of *C. burnetii* in environmental samples will likely complicate the efficient detection and identification of strains used in a bioterrorism event. Nonetheless, the development of independent strategies utilizing multiple assays with different genome targets with high discriminatory power has provided effective tools for the detection and differentiation of *Rickettsia* and *Coxiella* species.

## REFERENCES

[1] H. Zinsser, Rats, Lice, and History, Little, Brown and Co., Boston, 1935.
[2] D. Raoult, T. Woodward, J.S. Dumler, The history of epidemic typhus, Infect. Dis. Clin. North Am. 18 (1) (2004) 127–140.

[3] D. Raoult, O. Dutour, L. Houhamdi, R. Jankauskas, P.E. Fournier, Y. Ardagna, et al., Evidence for louse-transmitted diseases in soldiers of Napoleon's Grand Army in Vilnius, J. Infect. Dis. 193 (1) (2006) 112–120.

[4] E.H. Derrick, "Q" fever, a new entity: Clinical features, diagnosis, and laboratory investigation, Med. J. Aust. 2 (1937) 281–299.

[5] F.M. Burnet, M. Freeman, Experimental studies on the virus of "Q" fever, Med. J. Aust. 2 (1937) 299–305.

[6] G.E. Davis, H.R. Cox, A filter-passing infectious agent isolated from ticks. I. Isolation from *Dermacentor andersoni*, reactions in animals, and filtration experiments, Public Health Rep. 53 (1938) 2259–2261.

[7] H.R. Cox, Cultivation of rickettsiae of the Rocky Mountain spotted fever, typhus and Q fever groups in the embryonic tissues of developing chicks, Science 94 (2444) (1941) 399–403.

[8] R.E. Dyer, A filter-passing infectious agent isolated from ticks, Hum. Infect. Public Health Rep. 53 (1938) 2277–2282.

[9] C.B. Philip, Comments on the name of the Q fever organism, Public Health Rep. 63 (1948) 58.

[10] A. Omsland, D.C. Cockrell, D. Howe, E.R. Fischer, K. Virtaneva, D.E. Sturdevant, et al., Host cell-free growth of the Q fever bacterium *Coxiella burnetii*, Proc. Natl. Acad. Sci. USA 106 (11) (2009) 4430–4434.

[11] D. Raoult, T. Marrie, J. Mege, Natural history and pathophysiology of Q fever, Lancet Infect. Dis. 5 (4) (2005) 219–226.

[12] S.A. Coleman, E.R. Fischer, D. Howe, D.J. Mead, R.A. Heinzen, Temporal analysis of *Coxiella burnetii* morphological differentiation, J. Bacteriol. 186 (21) (2004) 7344–7352.

[13] G.H. Scott, J.C. Williams, Susceptibility of *Coxiella burnetii* to chemical disinfectants, Ann. N.Y. Acad. Sci. 590 (1990) 291–296.

[14] C.R. Anderson, Survival of *Rickettsia prowazeki* in different diluents, J. Bacteriol. 47 (6) (1944) 519–522.

[15] M.R. Bovarnick, J.C. Miller, J.C. Snyder, The influence of certain salts, amino acids, sugars, and proteins on the stability of rickettsiae, J. Bacteriol. 59 (4) (1950) 509–522.

[16] N.H. Topping, The preservation of the infectious agents of some of the rickettsioses, United States Public Health Reports 55 (1940) 545–547.

[17] M. Maurin, D. Raoult, Q fever, Clin. Microbiol. Rev. 12 (4) (1999) 518–553.

[18] A. Stein, D. Raoult, Q fever during pregnancy: A public health problem in southern France, Clin. Infect. Dis. 27 (3) (1998) 592–596.

[19] D. Raoult, M. Drancourt, Antimicrobial therapy of rickettsial diseases, Antimicrob. Agents Chemother. 35 (12) (1991) 2457–2462.

[20] H.C. Maltezou, D. Raoult, Q fever in children, Lancet Infect. Dis. 2 (11) (2002) 686–691.

[21] D. Raoult, F. Fenollar, A. Stein, Q fever during pregnancy: Diagnosis, treatment, and follow-up, Arch. Intern. Med. 162 (6) (2002) 701–704.

[22] D. Raoult, P. Houpikian, H. Tissot Dupont, J.M. Riss, J. Arditi-Djiane, P. Brouqui, Treatment of Q fever endocarditis: Comparison of 2 regimens containing doxycycline and ofloxacin or hydroxychloroquine, Arch. Intern. Med. 159 (2) (1999) 167–173.

[23] J.M. Rolain, M.N. Mallet, D. Raoult, Correlation between serum doxycycline concentrations and serologic evolution in patients with *Coxiella burnetii* endocarditis, J. Infect. Dis. 188 (9) (2003) 1322–1325.

[24] M.E. Eremeeva, G.A. Dasch, Closing the gaps between genotype and phenotype in *Rickettsia rickettsii*, Ann. N.Y. Acad. Sci. 1166 (2009) 12–26.

[25] M.B. Labruna, Ecology of rickettsia in South America, Ann. N.Y. Acad. Sci. 1166 (2009) 156–166.

[26] A.S. Chapman, D.L. Swerdlow, V.M. Dato, A.D. Anderson, C.E. Moodie, C. Marriott, et al., Cluster of sylvatic epidemic typhus cases associated with flying squirrels, 2004-2006, Emerg. Infect. Dis. 15 (7) (2009) 1005–1011.

[27] A. Medina-Sanchez, D.H. Bouyer, V. Alcantara-Rodriguez, C. Mafra, J. Zavala-Castro, T. Whitworth, et al., Detection of a typhus group rickettsia in Amblyomma ticks in the state of Nuevo Leon, Mexico, Ann. N.Y. Acad. Sci. 1063 (2005) 327–332.

[28] R.J. Reiss-Gutfreund, Isolation of Rickettsia prowazekii strains from the blood of domestic animals in Ethiopia and from their ticks, Bull. Soc. Pathol. Exot. Filiales 48 (5) (1955) 602–607.

[29] H. Tissot-Dupont, M.A. Amadei, M. Nezri, D. Raoult, Wind in November, Q fever in December, Emerg. Infect. Dis. 10 (7) (2004) 1264–1269.

[30] G. Dupuis, J. Petite, O. Péter, M. Vouilloz, An important outbreak of human Q fever in a Swiss Alpine valley, Int. J. Epidemiol. 16 (2) (1987) 282–287.

[31] H.C. van Woerden, B.W. Mason, L.K. Nehaul, R. Smith, R.L. Salmon, B. Healy, et al., Q fever outbreak in industrial setting, Emerg. Infect. Dis. 10 (7) (2004) 1282–1289.

[32] M.M. Salmon, B. Howells, E.J. Glencross, A.D. Evans, S.R. Palmer, Q fever in an urban area, Lancet 1 (8279) (1982) 1002–1004.

[33] M.P. Carrieri, H. Tissot-Dupont, D. Rey, P. Brousse, H. Renard, Y. Obadia, et al., Investigation of a slaughterhouse-related outbreak of Q fever in the French Alps, Eur. J. Clin. Microbiol. Infect. Dis. 21 (1) (2002) 17–21.

[34] B. Schimmer, F. Dijkstra, P. Vellema, P.M. Schneeberger, V. Hackert, R. ter Schegget, et al., Sustained intensive transmission of Q fever in the south of the Netherlands, 2009, Euro Surveill. 14 (19) (2009).

[35] W.W. Benson, D.W. Brock, J. Mather, Serologic analysis of a penitentiary group using raw milk from a Q fever infected herd, Public Health Rep. 78 (1963) 707–710.

[36] D.B. Fishbein, D. Raoult, A cluster of Coxiella burnetii infections associated with exposure to vaccinated goats and their unpasteurized dairy products, Am. J. Trop. Med. Hyg. 47 (1) (1992) 35–40.

[37] E.R. Krumbiegel, H.J. Wisniewski, Q fever in the Milwaukee area. II. Consumption of infected raw milk by human volunteers, Arch. Environ. Health 21 (1) (1970) 63–65.

[38] G.H. Lang, Coxiellosis (Q fever) in animals, in: T.J. Marrie (Ed.), Q Fever, Vol. I, CRC Press, Boca Raton, FL, 1990, pp. 23–48.

[39] J.M. Lapointe, F.M. Gulland, D.M. Haines, B.C. Barr, P.J. Duignan, Placentitis due to Coxiella burnetii in a Pacific harbor seal (Phoca vitulina richardsi), J. Vet. Diagn. Invest 11 (6) (1999) 541–543.

[40] W.K. Reeves, A.D. Loftis, F. Sanders, M.D. Spinks, W. Wills, A.M. Denison, et al., Coxiella, and Rickettsia in Carios capensis (Acari: Argasidae) from a brown pelican (Pelecanus occidentalis) rookery in South Carolina, USA, Exp. Appl. Acarol. 39 (3–4) (2006) 321–329.

[41] A. Jasinskas, J. Zhong, A.G. Barbour, Highly prevalent Coxiella spp. bacterium in the tick vector Amblyomma americanum, Appl. Environ. Microbiol. 73 (1) (2007) 334–336.

[42] H.G. Stoenner, D.B. Lackman, The biologic properties of Coxiella burnetii isolated from rodents collected in Utah, Am. J. Hyg. 71 (1960) 45–51.

[43] H.G. Stoenner, R. Holdenried, D. Lackman, J.S. Orsborn Jr., The occurrence of Coxiella burnetii, Brucella, and other pathogens among fauna of the Great Salt Lake Desert in Utah, Am. J. Trop. Med. Hyg. 8 (1959) 590–596.

[44] K.E. Russell-Lodrigue, M. Andoh, M.W. Poels, H.R. Shive, B.R. Weeks, G.Q. Zhang, et al., *Coxiella burnetii* isolates cause genogroup-specific virulence in mouse and guinea pig models of acute Q fever, Infect. Immun. 77 (12) (2009) 5640–5650.

[45] O. Péter, G. Dupuis, M.G. Peacock, W. Burgdorfer, Comparison of enzyme-linked immunosorbent assay and complement fixation and indirect fluorescent-antibody tests for detection of *Coxiella burnetii* antibody, J. Clin. Microbiol. 25 (6) (1987) 1063–1067.

[46] B.L. Scola, Current laboratory diagnosis of Q fever, Semin. Pediatr. Infect. Dis. 13 (4) (2002) 257–262.

[47] P.E. Fournier, T.J. Marrie, D. Raoult, Diagnosis of Q fever, J. Clin. Microbiol. 36 (7) (1998) 1823–1834.

[48] P.M. Schneeberger, M.H. Hermans, E.J. van Hannen, J.J. Schellekens, A.C. Leenders, P.C. Wever, et al., Real-time PCR with serum samples is indispensable for early diagnosis of acute Q fever, Clin. Vaccine Immunol. 17 (2) (2010) 286–290.

[49] P.E. Fournier, D. Raoult, Comparison of PCR and serology assays for early diagnosis of acute Q fever, J. Clin. Microbiol. 41 (11) (2003) 5094–5098.

[50] F. Fenollar, P.E. Fournier, D. Raoult, Molecular detection of *Coxiella burnetii* in the sera of patients with Q fever endocarditis or vascular infection, J. Clin. Microbiol. 42 (11) (2004) 4919–4924.

[51] E.M. Mendes do Nascimento, S. Colombo, T.K. Nagasse-Sugahara, R.N. Angerami, M.R. Resende, L.J. da Silva, et al., Evaluation of PCR-based assay in human serum samples for diagnosis of fatal cases of spotted fever group rickettsiosis, Clin. Microbiol. Infect. 15 (Suppl 2) (2009) 232–234.

[52] W.M. Bamberg, W.J. Pape, J.L. Beebe, C. Nevin-Woods, W. Ray, H. Maguire, et al., Outbreak of Q fever associated with a horse-boarding ranch, Colorado, 2005, Vector Borne Zoonotic Dis. 7 (3) (2007) 394–402.

[53] M. Turra, G. Chang, D. Whybrow, G. Higgins, M. Qiao, Diagnosis of acute Q fever by PCR on sera during a recent outbreak in rural south Australia, Ann. N.Y. Acad. Sci. 1078 (2006) 566–569.

[54] H. Willems, D. Thiele, R. Frölich-Ritter, H. Krauss, Detection of *Coxiella burnetii* in cow's milk using the polymerase chain reaction (PCR), Zentralbl. Veterinarmed B 41 (9) (1994) 580–587.

[55] M. Panning, J. Kilwinski, S. Greiner-Fischer, M. Peters, S. Kramme, D. Frangoulidis, et al., High throughput detection of *Coxiella burnetii* by real-time PCR with internal control system and automated DNA preparation, BMC Microbiol. 8 (2008) 77.

[56] S.R. Klee, J. Tyczka, H. Ellerbrok, T. Franz, S. Linke, G. Baljer, et al., Highly sensitive real-time PCR for specific detection and quantification of *Coxiella burnetii*, BMC Microbiol. 6 (2006) 2.

[57] CDC. Procedures for collecting surface environmental samples for culturing *B. anthracis*. Available from: http://www.bt.cdc.gov/Agent/Anthrax/environmental-sampling-apr2002.asp.

[58] J.M. Edmonds, Efficient methods for large-area surface sampling of sites contaminated with pathogenic microorganisms and other hazardous agents: Current state, needs, and perspectives, Appl. Microbiol. Biotechnol. 84 (5) (2009) 811–816.

[59] C.F. Estill, P.A. Baron, J.K. Beard, M.J. Hein, L.D. Larsen, L. Rose, et al., Recovery efficiency and limit of detection of aerosolized *Bacillus anthracis* Sterne from environmental surface samples, Appl. Environ. Microbiol. 75 (13) (2009) 4297–4306.

[60] G.J. Kersh, T.M. Wolfe, K. Fitzpatrick, J.S. Self, L.D. Oliver, A.A. Kapasi, et al., Prevalence of *Coxiella burnetii* DNA in environmental samples acquired in the United States, in: 23rd Annual Meeting of the American Society for Rickettsiology. Hilton Head, South Carolina, 2009.

[61] F.M. Lakay, A. Botha, B.A. Prior, Comparative analysis of environmental DNA extraction and purification methods from different humic acid-rich soils, J. Appl. Microbiol. 102 (1) (2007) 265–273.

[62] E.J. Bell, H.G. Stoenner, Immunologic relationships among the spotted fever group of rickettsias determined by toxin neutralization tests in mice with convalescent animal serums, J. Immunol. 84 (1960) 171–182.

[63] R.L. Anacker, R.N. Philip, J.C. Williams, R.H. List, R.E. Mann, Biochemical and immunochemical analysis of *Rickettsia rickettsii* strains of various degrees of virulence, Infect. Immun. 44 (3) (1984) 559–564.

[64] W.H. Price, The epidemiology of Rocky Mountain spotted fever. I. The characterization of strain virulence of *Rickettsia rickettsii*, Am. J. Hyg. 58 (2) (1953) 248–268.

[65] M.E. Eremeeva, G.A. Dasch, D.J. Silverman, Quantitative analyses of variations in the injury of endothelial cells elicited by 11 isolates of *Rickettsia rickettsii*, Clin. Diagn. Lab. Immunol. 8 (4) (2001) 788–796.

[66] T. Hackstadt, R. Messer, W. Cieplak, M.G. Peacock, Evidence for proteolytic cleavage of the 120-kilodalton outer membrane protein of rickettsiae: Identification of an avirulent mutant deficient in processing, Infect. Immun. 60 (1) (1992) 159–165.

[67] V.F. Ignatovich, G.A. Penkina, N.M. Balaeva, Properties in culture and persistence in cotton rats of the *Rickettsia prowazekii* vaccine strain E and its mutants, Acta Virol. 34 (2) (1990) 171–177.

[68] O.M. Frolova, N.M. Balaeva, V.A. Genig, V.F. Ignatovich, Biological properties of antibiotic-resistant mutants of the mildly pathogenic *Rickettsia prowazekii* E strain, Z. Mikrobiol. Epidemiol. Immunobiol. 2 (1987) 12–15.

[69] J. Kazar, R. Brezina, J. Urvolgyi, Studies on the E strain of *Rickettsia prowazeki*, Bull. World Health Organ. 49 (3) (1973) 257–265.

[70] N.M. Balaeva, V.N. Nikolskaya, Increased virulence of the E vaccine strain of *Rickettsia prowazeki* passaged in the lungs of white mice and guinea pigs, J. Hyg. Epidemiol. Microbiol. Immunol. 17 (1) (1973) 11–20.

[71] J.P. Fox, A review of experience with an avirulent strain of *R. prowazeki* (strain E) as a living agent for immunizing man against epidemic typhus, Am. J. Public Health Nations Health 45 (8) (1955) 1036–1048.

[72] J.H. Gilford, W.H. Price, Virulent-avirulent conversions of *Rickettsia rickettsii* in vitro, in: N. Hanon (Ed.), Selected Papers on the Pathogenic Rickettsiae, Harvard University Press, Cambridge, MA, 1968, pp. 285–290.

[73] H.R. Spencer, R.R. Parker, Rocky Mountain spotted fever: Infectivity of fasting and recently fed ticks, Public Health Rep. 38 (1923) 333–339.

[74] S.F. Hayes, W. Burgdorfer, Reactivation of *Rickettsia rickettsii* in *Dermacentor andersoni* ticks: An ultrastructural analysis, Infect. Immun. 37 (2) (1982) 779–785.

[75] W.M. Ching, M. Carl, G.A. Dasch, Mapping of monoclonal antibody binding sites on CNBr fragments of the S-layer protein antigens of *Rickettsia typhi* and *Rickettsia prowazekii*, Mol. Immunol. 29 (1) (1992) 95–105.

[76] D. Raoult, G.A. Dasch, Line blot and western blot immunoassays for diagnosis of Mediterranean spotted fever, J. Clin. Microbiol. 27 (9) (1989) 2073–2079.

[77] W.M. Ching, H. Wang, B. Jan, G.A. Dasch, Identification and characterization of epitopes on the 120-kilodalton surface protein antigen of *Rickettsia prowazekii* with synthetic peptides, Infect. Immun. 64 (4) (1996) 1413–1419.

[78] W.M. Ching, G.A. Dasch, M. Carl, M.E. Dobson, Structural analyses of the 120-kDa serotype protein antigens of typhus group rickettsiae. Comparison with other S-layer proteins, Ann. N.Y. Acad. Sci. 590 (1990) 334–351.

[79] G.A. Dasch, J.P. Burans, M.E. Dobson, R.I. Jaffe, W.G. Sewell, Distinctive properties of components of the cell envelope of typhus group rickettsiae, in: J. Kazar (Ed.), Rickettsiae and Rickettsial Diseases, Publishing House of the Slovak Academy of Sciences, Bratislava, Slovakia, 1985, pp. 54–61.

[80] W.M. Ching, H. Wang, G.A. Dasch, Identification and characterization of linear epitopes recognized by mouse monoclonal antibodies on the surface protein antigen of *Rickettsia prowazekii*, in: D. Raoult, P. Brouqui (Eds.), Rickettsiae and Rickettsial Diseases at the Turn of the Third Millennium, Elsevier, Paris, France, 1999, pp. 16–22.

[81] R.L. Anacker, R.H. List, R.E. Mann, D.L. Wiedbrauk, Antigenic heterogeneity in high- and low-virulence strains of *Rickettsia rickettsii* revealed by monoclonal antibodies, Infect. Immun. 51 (2) (1986) 653–660.

[82] R.L. Anacker, R.E. Mann, C. Gonzales, Reactivity of monoclonal antibodies to *Rickettsia rickettsii* with spotted fever and typhus group rickettsiae, J. Clin. Microbiol. 25 (1) (1987) 167–171.

[83] G.A. Dasch, J.R. Samms, E. Weiss, Biochemical characteristics of typhus group rickettsiae with special attention to the *Rickettsia prowazekii* strains isolated from flying squirrels, Infect. Immun. 19 (2) (1978) 676–685.

[84] M.E. Eremeeva, V.F. Ignatovich, I.A. Nedvetzkaya, N.M. Balayeva, Studies of *Rickettsia prowazekii* antigens in immunoblotting with specific sera of infected white mice, Mol. Gen. Mikrobiol. Virusol. Nov-Dec (6) (1994) 29–33.

[85] M.E. Eremeeva, E.B. Lapina, N.M. Balaeva, V.F. Ignatovich, L.S. Belousova, Electrophoretic and immunochemical characteristics of proteins of *Rickettsia prowazekii* strains of various virulence, Mol. Gen. Mikrobiol. Virusol. May (5) (1989) 20–26.

[86] N.M. Balayeva, M.E. Eremeeva, V.F. Ignatovich, B.A. Dmitriev, E.B. Lapina, L.S. Belousova, Protein antigens of genetically related *Rickettsia prowazekii* strains with different virulence, Acta Virol. 36 (1) (1992) 52–56.

[87] R.L. Regnery, C.L. Spruill, Extent of genetic heterogeneity among human isolates of *Rickettsia prowazekii* as determined by restriction endonuclease analysis of rickettsial DNA, in: L. Leive, D. Schlessinger (Eds.) Microbiology-1984, American Society for Microbiology, Washington, DC, 1984, pp. 297–300.

[88] R.L. Regnery, Z.Y. Fu, C.L. Spruill, Flying squirrel-associated *Rickettsia prowazekii* (epidemic typhus rickettsiae) characterized by a specific DNA fragment produced by restriction endonuclease digestion, J. Clin. Microbiol. 23 (1) (1986) 189–191.

[89] N.M. Balayeva, E.B. Rydkina, M.I. Artemiev, V.F. Ignatovich, Restriction endonuclease analysis of the DNA of *Rickettsia prowazekii* vaccine strain E and its revertant, Acta Virol. 33 (5) (1989) 454–464.

[90] M.E. Eremeeva, V. Roux, D. Raoult, Determination of genome size and restriction pattern polymorphism of *Rickettsia prowazekii* and *Rickettsia typhi* by pulsed field gel electrophoresis, FEMS Microbiol. Lett. 112 (1) (1993) 105–112.

[91] V. Roux, D. Raoult, Genotypic identification and phylogenetic analysis of the spotted fever group rickettsiae by pulsed-field gel electrophoresis, J. Bacteriol. 175 (15) (1993) 4895–4904.

[92] M. Eremeeva, X. Yu, D. Raoult, Differentiation among spotted fever group rickettsiae species by analysis of restriction fragment length polymorphism of PCR-amplified DNA, J. Clin. Microbiol. 32 (3) (1994) 803–810.

[93] V. Roux, P.E. Fournier, D. Raoult, Differentiation of spotted fever group rickettsiae by sequencing and analysis of restriction fragment length polymorphism of PCR-amplified DNA of the gene encoding the protein rOmpA, J. Clin. Microbiol. 34 (9) (1996) 2058–2065.

[94] V. Roux, E. Rydkina, M. Eremeeva, D. Raoult, Citrate synthase gene comparison, a new tool for phylogenetic analysis, and its application for the rickettsiae, Int. J. Syst. Bacteriol. 47 (2) (1997) 252–261.

[95] V. Roux, D. Raoult, Phylogenetic analysis of members of the genus *Rickettsia* using the gene encoding the outer-membrane protein rOmpB (ompB), Int. J. Syst. Evol. Microbiol. 50 (Pt 4) (2000) 1449–1455.

[96] Z. Sekeyova, V. Roux, D. Raoult, Phylogeny of *Rickettsia* spp. inferred by comparing sequences of "gene D," which encodes an intracytoplasmic protein, Int. J. Syst. Evol. Microbiol. 51 (Pt 4) (2001) 1353–1360.

[97] P.E. Fournier, J.S. Dumler, G. Greub, J. Zhang, Y. Wu, D. Raoult, Gene sequence-based criteria for identification of new rickettsia isolates and description of *Rickettsia heilongjiangensis* sp. nov, J. Clin. Microbiol. 41 (12) (2003) 5456–5465.

[98] R.L. Regnery, C.L. Spruill, B.D. Plikaytis, Genotypic identification of rickettsiae and estimation of intraspecies sequence divergence for portions of two rickettsial genes, J. Bacteriol. 173 (5) (1991) 1576–1589.

[99] M.E. Eremeeva, R.M. Klemt, L.A. Santucci-Domotor, D.J. Silverman, G.A. Dasch, Genetic analysis of isolates of *Rickettsia rickettsii* that differ in virulence, Ann. N.Y. Acad. Sci. 990 (2003) 717–722.

[100] R.D. Gilmore Jr, T. Hackstadt, DNA polymorphism in the conserved 190 kDa antigen gene repeat region among spotted fever group rickettsiae, Biochim. Biophys. Acta 1097 (1) (1991) 77–80.

[101] M. Matsumoto, Y. Tange, T. Okada, Y. Inoue, T. Horiuchi, Y. Kobayashi, et al., Deletion in the 190 kDa antigen gene repeat region of *Rickettsia rickettsii*, Microb. Pathog. 20 (1) (1996) 57–62.

[102] R.L. Regnery, C.L. Spruill, B.D. Plikaytis, Genotypic identification of rickettsiae and estimation of intraspecies sequence divergence for portions of two rickettsial genes, J. Bacteriol. 173 (5) (1991) 1576–1589.

[103] M.E. Eremeeva, V.F. Ignatovich, G.A. Dasch, D. Raoult, N.M. Balayeva, Genetic, biological and serological differentiation of *Rickettsia prowazekii* and *Rickettsia typhi*, in: J. Kazar (Ed.), Rickettsiae and Rickettsial Diseases, Slovak Academy of Sciences, Bratislava, Slovakia, 1996.

[104] H. Ge, M. Tong, J. Jiang, G.A. Dasch, A.L. Richards, Genotypic comparison of five isolates of *Rickettsia prowazekii* by multilocus sequence typing, FEMS Microbiol. Lett. 271 (1) (2007) 112–117.

[105] C.G. Moron, D.H. Bouyer, X.J. Yu, L.D. Foil, P. Crocquet-Valdes, D.H. Walker, Phylogenetic analysis of the rompB genes of *Rickettsia felis* and *Rickettsia prowazekii* European-human and North American flying-squirrel strains, Am. J. Trop. Med. Hyg. 62 (5) (2000) 598–603.

[106] S.G. Andersson, A. Zomorodipour, J.O. Andersson, T. Sicheritz-Pontén, U.C. Alsmark, R.M. Podowski, et al., The genome sequence of *Rickettsia prowazekii* and the origin of mitochondria, Nature 396 (6707) (1998) 133–140.

[107] M.E. Eremeeva, A. Madan, C.D. Shaw, K. Tang, G.A. Dasch, New perspectives on rickettsial evolution from new genome sequences of *Rickettsia*, particularly *R. canadensis*, and *Orientia tsutsugamushi*, Ann. N.Y. Acad. Sci. 1063 (2005) 47–63.

[108] H. Ogata, S. Audic, P. Renesto-Audiffren, P.E. Fournier, V. Barbe, D. Samson, et al., Mechanisms of evolution in *Rickettsia conorii* and *R. prowazekii*, Science 293 (5537) (2001) 2093–2098.

[109] Y. Zhu, P.E. Fournier, H. Ogata, D. Raoult, Multispacer typing of *Rickettsia prowazekii* enabling epidemiological studies of epidemic typhus, J. Clin. Microbiol. 43 (9) (2005) 4708–4712.

[110] P.E. Fournier, Y. Zhu, H. Ogata, D. Raoult, Use of highly variable intergenic spacer sequences for multispacer typing of *Rickettsia conorii* strains, J. Clin. Microbiol. 42 (12) (2004) 5757–5766.

[111] S.E. Karpathy, G.A. Dasch, M.E. Eremeeva, Molecular typing of isolates of *Rickettsia rickettsii* by use of DNA sequencing of variable intergenic regions, J. Clin. Microbiol. 45 (8) (2007) 2545–2553.

[112] G. Benson, Tandem repeats finder: A program to analyze DNA sequences, Nucleic Acids Res. 27 (2) (1999) 573–580.

[113] M.E. Eremeeva, E. Bosserman, M. Zambrano, L. Demma, G.A. Dasch, Molecular typing of novel *Rickettsia rickettsii* isolates from Arizona, Ann. N.Y. Acad. Sci. 1078 (2006) 573–577.

[114] M.E. Wikswo, R. Hu, M.E. Metzger, M.E. Eremeeva, Detection of *Rickettsia rickettsii* and *Bartonella henselae* in *Rhipicephalus sanguineus* ticks from California, J. Med. Entomol. 44 (1) (2007) 158–162.

[115] M.E. Eremeeva, M.L. Zambrano, L. Beati, G.A. Dasch, S.E. Karpathy, H. Olguín, et al., Genetic analysis of *Rickettsia rickettsii* and *Rhipicephalus sanguineus* Latrielle from Mexicali, Mexico, in: The 23rd Meeting of the American Society for Rickettsiology, Hilton Head, SC, 2009. [Abstract].

[116] G.A. Dasch, H. Graddy, M. Wikswo, E. Pegg, D. Green, and M.E. Eremeeva, Stability of insertion/deletion events for rapid and simple differentiation of *Rickettsia prowazekii* and *Rickettsia typhi*, in: The 20th Meeting of the American Society for Rickettsiology, Asilomar, CA, 2006. [Abstract].

[117] Y. Zhu, A. Medina-Sanchez, D. Bouyer, D.H. Walker, X.J. Yu, Genotyping *Rickettsia prowazekii* isolates, Emerg. Infect. Dis. 14 (8) (2008) 1300–1302.

[118] D.W. Ellison, T.R. Clark, D.E. Sturdevant, K. Virtaneva, S.F. Porcella, T. Hackstadt, Genomic comparison of virulent *Rickettsia rickettsii* Sheila Smith and avirulent *Rickettsia rickettsii* Iowa, Infect. Immun. 76 (2) (2008) 542–550.

[119] C.Y. Kato, L.K. Robinson, F.H. White, K. Slater, S.E. Karpathy, M.E. Eremeeva, et al., Insertion/deletion (INDEL) typing of isolates of *Rickettsia rickettsii*, in: The 23rd Meeting of the American Society for Rickettsiology, Hilton Head, SC, 2009. [Abstract].

[120] G.A. Dasch, M.E. Eremeeva, K.G. Dirks, L.K. Robinson, F.H. White, C.Y. Kato, et al., Genomic sequence variation in the complete genome sequences of multiple isolates of *Rickettsia rickettsii*, *R. prowazekii*, and *R. typhi*, in: The 23rd Meeting of the American Society for Rickettsiology, Hilton Head, SC, 2009. [Abstract].

[121] P.E. Warburton, J. Giordano, F. Cheung, Y. Gelfand, G. Benson, Inverted repeat structure of the human genome: The X-chromosome contains a preponderance of large, highly homologous inverted repeats that contain testes genes, Genome Res. 14 (10A) (2004) 1861–1869.

[122] C.Y. Kato, I.H. Chung, J.C. Patel, L.K. Robinson, M.E. Eremeeva, D.C. Bruce, et al., Genotyping of *Rickettsia prowazekii* isolates by mismatch amplification mutation assays, in: 110th General Meeting of the American Society for Microbiology, San Diego, CA, 2010. [Abstract].

[123] C.Y. Kato, I.H. Chung, L.K. Robinson, M.E. Eremeeva, D.C. Bruce, C. Munk, et al., Mismatch amplification mutation assays for genotyping *Rickettsia typhi*, in: 8th ASM Biodefense Research Meeting, Baltimore, MD, 2010. [Abstract].

[124] T. Hackstadt, Antigenic variation in the phase I lipopolysaccharide of *Coxiella burnetii* isolates, Infect. Immun. 52 (1) (1986) 337–340.

[125] L.R. Hendrix, J.E. Samuel, L.P. Mallavia, Differentiation of *Coxiella burnetii* isolates by analysis of restriction-endonuclease-digested DNA separated by SDS-PAGE, J. Gen. Microbiol. 137 (2) (1991) 269–276.

[126] A. Stein, D. Raoult, Lack of pathotype specific gene in human *Coxiella burnetii* isolates, Microb. Pathog. 15 (3) (1993) 177–185.

[127] D. Thiele, H. Willems, Is plasmid based differentiation of *Coxiella burnetii* in "acute" and "chronic" isolates still valid? Eur. J. Epidemiol. 10 (4) (1994) 427–434.

[128] H. Willems, D. Thiele, H. Krauss, Plasmid based differentiation and detection of *Coxiella burnetii* in clinical samples, Eur. J. Epidemiol. 9 (4) (1993) 411–418.

[129] A. Stein, N.A. Saunders, A.G. Taylor, D. Raoult, Phylogenic homogeneity of *Coxiella burnetii* strains as determinated by 16S ribosomal RNA sequencing, FEMS Microbiol. Lett. 113 (3) (1993) 339–344.

[130] G.Q. Zhang, H. To, T. Yamaguchi, H. Fukushi, K. Hirai, Differentiation of *Coxiella burnetii* by sequence analysis of the gene (com1) encoding a 27-kDa outer membrane protein, Microbiol. Immunol. 41 (11) (1997) 871–877.

[131] Z. Sekeyova, V. Roux, D. Raoult, Intraspecies diversity of *Coxiella burnetii* as revealed by com1 and mucZ sequence comparison, FEMS Microbiol Lett. 180 (1) (1999) 61–67.

[132] S.V. Nguyen, K. Hirai, Differentiation of *Coxiella burnetii* isolates by sequence determination and PCR-restriction fragment length polymorphism analysis of isocitrate dehydrogenase gene, FEMS Microbiol. Lett. 180 (2) (1999) 249–254.

[133] M. Andoh, H. Nagaoka, T. Yamaguchi, H. Fukushi, K. Hirai, Comparison of Japanese isolates of *Coxiella burnetii* by PCR-RFLP and sequence analysis, Microbiol. Immunol. 48 (12) (2004) 971–975.

[134] R. Seshadri, I.T. Paulsen, J.A. Eisen, T.D. Read, K.E. Nelson, W.C. Nelson, et al., Complete genome sequence of the Q-fever pathogen, *Coxiella burnetii*. Proc. Natl. Acad. Sci. USA 100 (9) (2003) 5455–5460.

[135] P.A. Beare, S.F. Porcella, R. Seshadri, J.E. Samuel, R.A. Heinzen, Preliminary assessment of genome differences between the reference Nine Mile isolate and two human endocarditis isolates of *Coxiella burnetii*, Ann. N.Y. Acad. Sci. 1063 (2005) 64–67.

[136] P.A. Beare, J.E. Samuel, D. Howe, K. Virtaneva, S.F. Porcella, R.A. Heinzen, Genetic diversity of the Q fever agent, *Coxiella burnetii*, assessed by microarray-based whole-genome comparisons, J. Bacteriol. 188 (7) (2006) 2309–2324.

[137] C. Jäger, H. Willems, D. Thiele, G. Baljer, Molecular characterization of *Coxiella burnetii* isolates, Epidemiol. Infect. 120 (2) (1998) 157–164.

[138] O. Glazunova, V. Roux, O. Freylikman, Z. Sekeyova, G. Fournous, J. Tyczka, et al., *Coxiella burnetii* genotyping, Emerg. Infect. Dis. 11 (8) (2005) 1211–1217.

[139] S. Svraka, R. Toman, L. Skultety, K. Slaba, W.L. Homan, Establishment of a genotyping scheme for *Coxiella burnetii*, FEMS Microbiol. Lett. 254 (2) (2006) 268–274.

[140] N. Arricau-Bouvery, Y. Hauck, A. Bejaoui, D. Frangoulidis, C.C. Bodier, A. Souriau, et al., Molecular characterization of *Coxiella burnetii* isolates by infrequent restriction site-PCR and MLVA typing, BMC Microbiol. 6 (2006) 38.

[141] A.M. Denison, H.A. Thompson, R.F. Massung, IS1111 insertion sequences of *Coxiella burnetii*: Characterization and use for repetitive element PCR-based differentiation of *Coxiella burnetii* isolates, BMC Microbiol. 7 (2007) 91.

[142] R.A. Priestley, H.M. Hornstra, T. Pearson, P. Keim, R.F. Massung, The state of the SNP: Using real-time PCR to genotype *Coxiella burnetii*, in: 23rd Annual Meeting of the American Society for Rickettsiology. Hilton Head, South Carolina, 2009.

[143] M.R. Shapiro, C.L. Fritz, K. Tait, C.D. Paddock, W.L. Nicholson, K.F. Abramowicz, et al., *Rickettsia* 364D: A newly recognized case of eschar associated rickettsiosis in California, Clin. Infect. Dis. 50 (4) (2010) 541–548.

# CHAPTER 18

# Forensics and Epidemiology of Fungal Pathogens

**David M. Engelthaler[a] and S. Arunmozhi Balajee[b]**

[a]*Translational Genomics Research Institute, Flagstaff, Arizona*
[b]*Molecular Epidemiology, Mycotic Diseases Branch,*
*Centers for Disease Control and Prevention, Atlanta, Georgia*

## INTRODUCTION

Fungi have unique characteristics that may present significant challenges for forensics and biodefense (1). Despite this, fungi are often neglected in microbial forensics and biodefense discussions. Most members of this taxon can produce large numbers of hardy spores that can be dispersed easily into the environment. Sporulation can be induced comparatively easily in the laboratory setting, thus rendering these organisms amenable to use as bioweapons. Further, these organisms have distinct biological, reproductive, and evolutionary characteristics compared to other pathogens, thus impacting genotyping approaches and/or phylogenetic analysis (2,3) for forensic and epidemiological investigations. Additionally, several fungal species can produce mycotoxins that have potentially lethal toxigenic and/or carcinogenic effects and can be considered potential biothreats.

## THE KINGDOM FUNGI

Fungi include mushrooms, rusts, smuts, puffballs, truffles, morels, molds, and yeasts, as well as many other less well-known organisms (4). Although about 100,000 fungal species have been described thus far, it is estimated that 1.5 million species may exist in nature (5). Fungi are distinct from plants and animals and possess several distinct features that include presence of a rigid cell wall composed of chitin and glucan. Fungi are also heterotrophic, meaning that they cannot produce their own food, and obtain nourishment by secreting enzymes into the extracellular matrix for digestion and absorption of food. Fungi have a basic structural unit that is either a chain of filament-like nonmotile cells denoted as hyphae or an independent single yeast cell. Multicellular hyphal-forming organisms, called "molds," increase in length as

297

Microbial Forensics. DOI: 10.1016/B978-0-12-382006-8.00018-9

a result of apical growth of the individual hypha. Primitive molds have hyphae that lack septa or cross walls, while in other fungi the hyphae can be septate. In contrast, single-cell yeasts propagate by budding out daughter cells from their surface; these buds may either be detached from the parent cell or remain attached and bud further to produce a chain of cells. Another distinguishing feature of fungi is the mode of reproduction—fungi reproduce by means of microscopic propagules called "spores" if produced sexually and conidia if produced by an asexual process. Not all fungi can reproduce sexually; the asexual stage of an organism is denoted as an "anamorph" and the sexual stage is known as the "teleomorph." These fungal spores may be released actively or passively into the environment and fill the air we breathe.

Fungi notoriously produce numerous toxic metabolites termed "mycotoxins," which, when ingested or inhaled, can produce mycotoxicosis in humans and animals. While there are no recorded instances of live fungal organisms being used for biological warfare or bioterrorism on humans, fungal toxins have been developed and, in some cases, used as bioweapons. Most notably, aflatoxins from *Aspergillus* species were produced and placed in warheads by the Iraqi government in the 1980s and 1990s (6). These toxins are known carcinogens but are not thought to be highly important for biodefense. Other mycotoxins of primary concern are tricothecenes, produced by a number of molds, including *Stachybotrys* and *Fusarium* species. Tricothecenes have been reportedly used in warfare (e.g., in Laos in the 1970s and Afghanistan in the 1980s) and can induce immediate significant external and internal toxigenic effects, including death (7). Several fungi and their toxins are important to plant health and agriculture (e.g., *Phoma glycinicola* and *Synchytrium endobioticum*). Forensic analysis of the use of toxins and plant pathogens is described in other chapters of this text and elsewhere (1,8–11) and are not discussed further here.

## PATHOGENIC FUNGI

Of the many thousands of fungi that exist, only a few hundred are recognized as fungi capable of causing systemic disease in humans. Despite this, invasive mycoses have emerged as a serious public health problem over the past two decades with increased incidence seen in immunocompromised populations including persons with AIDS, recipients of solid organ or hematopoietic stem cell transplants, hematologic malignancies, and individuals on immunosuppressive regimens (12). There has also been a shift in fungal epidemiology, with more infections caused by fungi that are often resistant to one or more antifungal drugs and are difficult to identify using traditional methods of identification. Molecular epidemiology can provide both laboratory and analytical tools that can be used to better define the etiology of an infection and to design appropriate intervention and control strategies.

Genotyping is central to understanding fungal molecular epidemiology. Broadly, tools for fungal genotyping can be used for (i) epidemiological surveillance, (ii) retrospective and prospective multicenter studies, (iii) comparative typing in outbreak investigations for source tracking, and (iv) forensic analysis for criminal cases for both a source of infectious fungi and/or a source of environmental fungi found in crime scenes. For epidemiological surveillance, molecular markers should be stable, reproducible over time and between laboratories, and allow integration into exchangeable databases; such tools should discriminate at the species and above species level and would be used to monitor species distribution, detection, and monitoring of emerging and reemerging infections in longitudinal studies. For outbreak investigations and forensic analyses, molecular markers should (i) be reproducible within that assay, (ii) be highly discriminatory (clonal versus unrelated genotypes), (iii) assess relatedness within a set of isolates, and (iv) generate isolate-specific molecular fingerprints for assessment of epidemiological relatedness in a population.

*Coccidioides immitis* and *C. posadasii* are important human pathogens and are the only fungal organisms included on the U.S. federal government's Select Agent List (13). These dimorphic fungi cause a significant amount of morbidity in endemic regions, primarily the U.S. desert southwest, where they thrive in arid, thermic soils and can be inhaled as arthroconidia, causing pulmonary disease. *Coccidioides* is therefore a major focus of this chapter and its epidemiology and genotyping are described in detail below. Other human pathogenic fungi that are causative agents of invasive fungal infections include *Histoplasma capsulatum*, *Candida* spp., *Cryptococcus* spp., Mucormycetes, *Aspergillus* spp., *Fusarium* spp., and *Pneumocystis* spp. For the purposes of this chapter, an overview of the molecular epidemiology of select human pathogenic fungi, including brief descriptions of the methods employed to genotype these fungi, are described in subsequent sections.

## *Coccidioides* and Coccidioidomycosis
### *Background and Epidemiology*

Although this genus is considered the most virulent of human and animal fungal pathogens, there have been no known uses of *Coccidioides* for criminal purposes (13). These apparently saprophytic, soil-dwelling fungi are endemic to the southwestern United States and parts of Latin America (14,15). *C. immitis* is largely confined to California and Baja California, whereas *C. posadasii* is the dominant species found in Arizona, Texas, Mexico, and South and Central America (14,16). These closely related species are haploid, filamentous ascomycetes that reproduce asexually both in the environment (using arthroconidia) and in the infected host (spherules containing endospores) (17). However, molecular diversity patterns suggest significant recombination and likely sexual reproduction are occurring. The actual recombination mechanisms are unknown and the extent, or frequency, of this phenomenon is not well characterized, although genomic evidence of sexual reproduction has been described (18).

Outbreaks of coccidioidomycosis outside the known endemic regions may or may not be suspicious for man-made outbreaks. Sporadic cases have been identified in nonendemic regions, associated primarily with recent travel to endemic regions (14,19). Additionally, clusters associated with high-risk activities (i.e., archaeology) in areas not previously known to be endemic have been reported (20).

### Detection and Identification

Detection of *Coccidioides* in the clinical laboratory is typically accomplished by direct microscopy of spherules with appropriate stains, direct culture and identification with labeled DNA probes, or serological analysis with enzyme immunoassays or immunodiffusion (21). Recent antigen detection methodologies have been published (22,23); however, as with the aforementioned detection methodologies, these tests provide only genus-level detection and provide no insight to genotype or genetic relation to other strains. Inefficient and time-consuming testing methodologies, along with undereducation of physicians in endemic regions, result in a limited number of patients being tested and subsequently diagnosed (24).

More recently, polymerase chain reaction (PCR) (including real time and nested methodologies) has been used as a method for identification and characterization of *Coccidioides* (25,26), including an assay that detects and differentiates *C. immitis* from *C. posadasii* based on single nucleotide polymorphisms (SNPs) (27). Thus, it appears that SNP interrogation in *Coccidioides* may prove useful for genotyping and forensic analysis. Additionally, a microsatellite genotyping system (see later) has been used for molecular separation of the two species (28). As molecular technologies are becoming more prevalent in clinical laboratories, the sensitivity, specificity, and speed of detection of *Coccidioides* assays should increase greatly.

### Genetic Analysis of Coccidioides

Researchers have made great strides in understanding the genetics of *Coccidioides*. This haploid eukaryotic organism reproduces asexually and, likely, sexually. The apparent presence of both extensive genetic recombination (29) and mating type loci (18) causes difficulty in understanding the population structure. While some clonality appears to be present using microsatellite analysis, as evidenced by large-scale geographical differences correlating with major clade distinction (14), previous attempts to develop finer population structures have not been successful due the use of rapidly changing genetic markers for typing and the impact of recombination (30–32).

The major success of genetic investigations into *Coccidioides* was identification of two genetically and geographically distinct species, *C. immitis* and *C. posadasii* (28). The first evidence of this species bifurcation was discovered through use of restriction length polymorphism analysis (RFLP) (33), identifying a California subspecies and a non-California subspecies. This was later confirmed through the use of microsatellites (29,34), SNPs (32), and multilocus genotyping (31). In 2002, Fisher and associates (28) published definitive evidence of a separate species, naming the non-California lineage as *C. posadasii*. It is now widely accepted that *C. posadasii* represents the dominant species outside of California.

Additional studies have explored the mechanisms of evolution, including the impacts of recombination, positive selection, and concerted evolution on the pathogenesis of *Coccidioides* (35). Most recently, a comprehensive comparative genomic analysis of *Coccidioides* and relatives has provided more insight into genomic causes of this organism's vastly different life cycle from other Ascomycetes and the infectious nature of *Coccidioides* compared to other fungi (36). This includes significant gene family changes, such as decreases in gene families associated with plant cell wall degradation (as in other related fungi) and expansion in protease gene families (e.g., keratinase). These findings would suggest that *Coccidioides* is less of a soil saprophyte and more associated with animals, both in infection of live animals and in mycelial growth in decaying carcasses (36).

## Molecular Genotyping and Forensics of *Coccidioides*
A number of typing schemes have been developed for *Coccidioides* ranging from phenotypic variation to molecular genotyping. This history has been reviewed elsewhere (2,37). Here we focus on the current gold standard of microsatellite analysis and more recent attempts to improve detection of informative markers using SNP genotyping and whole genome analysis.

### Microsatellites
Microsatellites, also referred to as "short tandem repeats" (STRs), are generally dinucleotide repeats found throughout the genome of eukaryotic and prokaryotic organisms. These repeat regions are highly mutable and have been used successfully for genotyping numerous pathogens, including *Coccidioides* (38) and other fungal organisms (2,39). For outbreak investigations, these markers have shown promise due to their high discriminatory power.

Previously, limited molecular epidemiology has been conducted for *Coccidioides*, showing the inability of microsatellites to identify informative population clades (34) beyond those identified by large-scale geographical differences (14,28). This work identified five populations of *Coccidioides*: central California, southern California, Arizona, Texas, and Mexico, with South

American isolates belonging to a clonal subpopulation of the Texas group (Figure 18.1) (14). Although statistical algorithms have been developed to assign individual isolates to these *Coccidioides* populations using microsatellite allele data (31), there have been no microsatellite studies published to date that identify reasonable local population structures (e.g., originating within a state). This prevents molecular epidemiological analysis of cases and

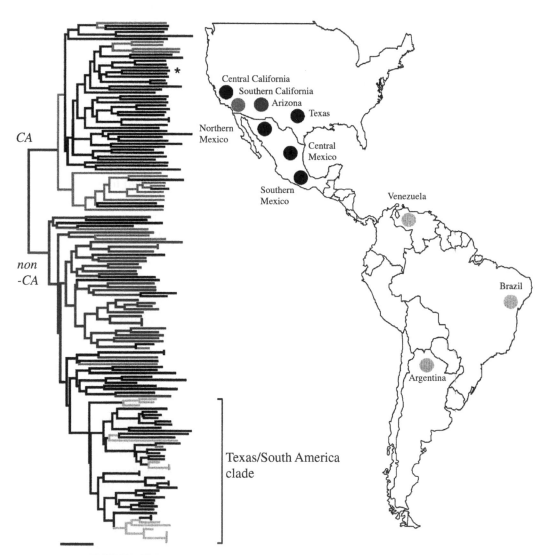

**FIGURE 18.1**

Microsatellite genotyping of *Coccidioides*: Neighbor-joining tree of pair-wise allele-sharing genetic distances calculated with the program MICROSAT (14). (Used with permission from PNAS.) (See Color Insert.)

potential outbreaks and, in turn, prevents the identification of point source locations or events (e.g., dust storms) or attribution in forensic investigations.

A recent microsatellite analysis of *Coccidioides* samples obtained within Arizona (30) provided results similar to other studies using previously published markers (38), namely, the markers allowed for genetic distinction between *C. posadasii* isolates (primarily from Arizona) and *C. immitis* isolates. They also verified that isolates coming from the same patient typically had genetically identical microsatellite profiles (as clonality would be expected within an individual infection); however, no clonality was evident from isolates collected from both relatively small and large geographic distances (30).

## Single Nucleotide Polymorphisms

Single nucleotide polymorphisms (SNPs) have been shown to be highly informative for both diagnostic identification and phylogenetic population analysis. Previous work with other microbial pathogens has led to a paradigm of nucleic acid signature identification based on evolutionary rules and comparative genomic analysis. This work was the foundation for identifying key diagnostic features called "canonical characters," which, in this case, are canonical SNPs (canSNPs) (40). A canonical diagnostic character marks a pivotal evolutionary point and, therefore, represents multiple evolutionary differences. These signatures are robust in their discrimination between target and nontarget species (27). A single canSNP is all that is required to identify a particular species, subpopulation, and/or isolate (40).

Single nucleotide polymorphism-based phylogenies of clonal microbes have been shown to be highly accurate in terms of defining population subgroups and isolate relationships (41). Synonymous, or neutral, SNPs are thought to be more evolutionarily informative than most other molecular markers due to their slow mutation rates, limited character states, and distribution across the genome (42). There are challenges, however, with using SNPs in recombining organisms (e.g., *Coccidioides*) as a result of character state conflicts, or homoplasy, arising from convergence, reversals, and/or lateral gene transfer. The challenge of homoplasy in rapidly changing multiallele loci (e.g., microsatellites) prevents development of highly informative phylogenies based on those markers. These challenges can be overcome through the use of large SNP data sets and appropriate algorithms, as has been shown in other recombining species (2,42).

Early SNP genotyping studies employing anonymous SNP loci provided initial evidence of recombination and possible sexual reproduction in *Coccidioides* (29). SNPs also provided early evidence of two separate species (32). However no viable genotyping scheme based on SNP analysis has been devised due to the challenges discussed previously, namely the effect of recombination. It has been suggested that with the increase of whole genome sequence analysis, use of large SNP data sets will provide robust phylogenetic and epidemiologic analysis of *Coccidioides* with greater resolving power than other genotyping tools (37).

In a recent study using molecular inversion probes (MIP) on a microarray (Affymetrix™) (43) targeting >500 *Coccidioides* SNPs identified from analysis of available whole genome sequences (44), a robust phylogenetic tree was developed identifying geographically linked population clades within a collection of *C. posadasii* samples originating primarily from Arizona (Figure 18.2) (45). Limited anomalies were identified within the clades; while there is known genetic recombination within species and the possibility of genetic introgression between the two species, the anomalies are more likely due to incomplete travel history or sample labeling, as has been reported previously (28). This work provides initial evidence of the ability of larger SNP data sets to provide improved population structure. In addition, this number of SNPs should provide sufficient data for epidemiological linkage and source attribution for clonally derived isolates.

## Whole Genome Sequence Typing

Perhaps the greatest resolution for genotyping and forensic epidemiology of *Coccidioides* and other fungal pathogens will come from comprehensive whole genome analysis. The advantage of what can be referred to as "whole genome sequence typing" (WGST) is that essentially all genetic differences can be discovered. This is critical for linking individual samples for outbreak tracing and source attribution, negating the need for complex statistical algorithms to assign isolates to a population. WGST also provides an unparalleled capability for defining accurate population structures. This advantage is moderated, however, by the disadvantage that comparisons are phylogenetically biased and may be relevant only to the genome sequences available (46). This bias may be mitigated by the effect of recombination and through the comparison of genomically dispersed SNPs for population analysis. Most importantly, the potential for phylogenetic bias is inversely proportional to the number of diverse and related whole genomes that are sequenced.

Both *C. immitis* and *C. posadasii* genomes have been sequenced using standard Sanger sequencing (44) under a multi-institutional project to sequence diverse isolates geographically, clinically, and environmentally (37). Population diversity of the 14 selected *Coccidioides* strains was achieved by selecting multiple samples from each of the five populations discovered by microsatellite analysis (14) (see earlier discussion). This initial work provides a critical backbone for developing a WGST system for *Coccidioides* (e.g., these sequence data were the source for the whole genome SNP array described previously). Additionally,

**FIGURE 18.2**

*Coccidioides* phylogeny (neighbor-joining tree) derived from 509 SNPs using 74 *C. posadasii* (Cp) and 24 *C. immitis* (Ci) isolates. Samples originating from Maricopa Co., Arizona (MA) and Tucson, Arizona (TU) are typically grouped within respective clades: Maricopa clade (M) and Tucson clades (T1–T3). All *C. immitis* grouped together in the outgroup (CI).

these sequences are allowing for whole genome comparative analysis studies that were not possible previously (36).

## Next-Generation Sequencing

The capabilities of "next-generation" sequencing offer great promise for high-throughput and high-coverage sequence analysis of microbes, several orders of magnitude greater than Sanger sequencing or standard pyrosequencing. This technology greatly increases the feasibility for using WGST for genotyping, molecular epidemiology, and forensics. The deep sequencing capacity of these instruments translates into a high level of coverage for much of the sequenced genomes. Additionally, with the added capacity to bar code, or index, a large number of samples in an individual run, the potential throughput capacity is unmatched by other technologies.

New bioinformatic tools are being developed rapidly (47,48) for specific next-gen applications. These tools allow for rapid alignment of millions of reads to accurately detect SNP mutations, small deletions and other features, and for other whole genome comparative analyses.

Next-generation sequencing has been used to sequence and compare several *Bacillus anthracis* strains (49), demonstrating that this technology can provide high depth of coverage across several genomes, at relatively low cost ($\approx$$1000/ genome) with high accuracy in less than 1 week. This breakthrough has paved the way for use of WGST for forensic and epidemiologic investigations. As outputs can be used for direct genotyping, there is no need for subsequent assay development (e.g., interrogation for canonical SNPs for use in SNP detection tools), and with ever-decreasing costs per genome, next-generation sequencing may be the likely molecular solution for microbial forensics in the future.

## Case Study

A transplant-related outbreak from an infected organ donor has been reported [unpublished data, U.S. Centers for Disease Control and Prevention (CDC), Atlanta, GA]. *C. immitis* isolates recovered from the infected transplant recipients (three isolates: one isolate per patient) were sequence analyzed by WGST to enhance the epidemiological investigation of the outbreak. The operating hypothesis was that isolates from the transplant recipients were from the same original source (donor samples not available) and that they would be clonal, having few to no SNP differences, as compared to isolates originating from different sources. Using the SOLiD™ sequencing system (Life Technologies, Foster City, CA), fragment libraries of each isolate provided an average over 40× coverage, covering over 94% of the fungal genome. Sequence reads were aligned and analyzed using novel bioinformatic tools. WGST was conducted by comparing whole genome shared SNPs among the three genomes, as well as comparing the isolates to all previously sequenced *C. immitis* genomes. WGST revealed that all three recipients were infected with the same strain of *C. immitis*, as only three SNPs were found among all three isolates, in contrast to finding >30,000 SNPs when compared to the reference genome (50). WGST not only allowed for definitive molecular epidemiologic linkage of isolates but also for an understanding of the placement of the suspect strains in the population of currently sequenced strains of *Coccidioides* (Figure 18.3).

## Other Pathogenic Fungi

Traditionally, fungi have been identified by their phenotypic traits where a morpho species is recognized as a group of isolates that have morphological characteristics similar to each other but distinct from other fungi. This methodology suffers from various limitations that include subjectivity, inability to identify cryptic species, and the expertise and time required to establish identification. Today, comparative sequence-based identification strategies (e.g., multilocus sequence typing, or MLST) can be considered the new gold standard for fungal species identification (51). This method is based on PCR amplification of a selected region of genomic DNA (target locus), followed by sequencing of the resulting amplicon(s) and query of the consensus sequence against a database library for evaluation for species identification. Analysis of data can be performed by generating dendrograms, examining percent similarity/percent dissimilarity, or executing more sophisticated phylogenetic analyses. Fungal phylogenetic species recognition (PSR) (2) has been used

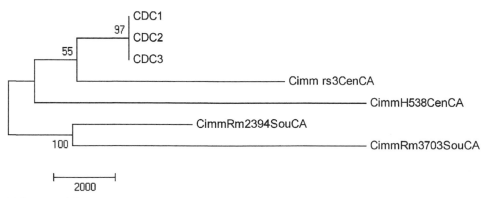

**FIGURE 18.3**

Maximum parsimony phylogenetic analysis of 32,712 single nucleotide polymorphisms shared among seven whole genome sequences of *C. immitis*. The percentage of replicate trees in which the associated taxa clustered together in the bootstrap test (1000 replicates) is shown next to the branches. The tree is drawn to scale with branch lengths calculated using the average pathway method and are in the units of the number of changes over the whole sequence. The consistency index of the tree is 0.63.

successfully to define species in *Fusarium* and *Aspergillus* (52). Once a species has been delimited by PSR using several robust loci, sequence diversity within the species is known; on the basis of this knowledge, comparative sequence analyses from a single locus can be used for rapid species identification (53).

The internal transcribed spacer (ITS) region of the rRNA operon appears to be a reasonable and commonly used choice for fungal species identification in environmental and clinical samples (54–56). Some of the limitations of this region include (i) insufficient variability to delineate the various species in the *Aspergillus* and *Fusarium* species complexes (52), (ii) recently recognized heterozygosity of this locus in *Rhizopus* species (57), and (iii) problems with reliability of the ITS sequences deposited in the reference databases (e.g., GenBank, EMBL, and DDBJ). In consideration of these limitations, a staged sequence-based identification strategy has been proposed (52). Based on this algorithm, upon receipt of an unknown fungal isolate, a morphological or molecular identification method singly or in combination can be pursued. Additional studies have shown the utility of a variety of tools to exploit the variability of the ITS region for fungal identification, including microarray hybridization (55), targeted qPCR, and pyrosequencing (56). Further resolution can be obtained from comparative sequence analyses of one or several protein-coding regions (53).

Ubiquitous in the environment, fungi have been implicated in numerous outbreaks of invasive or toxin-related disease in the community and in hospitals. In outbreak investigations, determining the source and route of transmission often requires detailed epidemiological investigation supported by appropriate laboratory strategies for determining strain relatedness. In the last decade, several fungal genomes have become available, allowing for the development of numerous fungal genotyping methods. Many large fungal outbreaks, for example, the 2006 *Fusarium* keratitis outbreak, were understood and controlled only with the use of appropriate genotyping tools.

### *Fusarium*

*Fusarium* species are filamentous fungi commonly found in the environment, particularly in soil, on plants, and in water systems and can cause a spectrum of diseases in humans ranging from superficial, invasive, and disseminated infections via inhalation, ingestion, or direct inoculation. Outbreaks of fusariosis in immunocompetent hosts are rare, and if they occur, manifest as ocular infections such as keratitis. In 2005–2006, a highly publicized keratitis outbreak spanned multiple states in the United States, concurrently affecting individuals in Hong Kong and Singapore. In the United States, the outbreak resulted in vision loss or the need for corneal transplant in over a third of the individuals affected. A case control study conducted by the CDC determined the most likely exposures associated with disease to be contact lens wear and use of a particular brand of contact lens cleaning solution. Molecular strain typing using MLST demonstrated high genotypic heterogeneity within the isolates, highlighting the presence of multiple sources of contamination and ruling out the possibility of intrinsic contamination of the contact lens solution (58). From these findings, it was hypothesized that the chemical composition of the contact lens solution allowed for the growth of fungus found naturally in these home environments, but was otherwise microbiologically sterile, resulting in discontinuation of the entire product line.

## *Aspergillus*

Members of the genus *Aspergillus* cause invasive fungal infections in immuno-compromised individuals and can often be fatal. These organisms are ubiquitous in the environment, and hospital-associated outbreaks due to these fungi, especially *A. fumigatus*, have been reported. In the setting of a nosocomial *Aspergillus* outbreak and in the presence of continued infection, the CDC recommends that an environmental assessment be undertaken to determine the point source of infection. In such investigations, appropriate strain typing methods can indicate the source of infection rapidly. Accordingly, numerous methods have been developed and validated for genotyping of *A. fumigatus*, the major etiological agent of invasive fungal infections (59). Of these, polymorphic microsatellite markers (STRAf), *Afut1* (a dispersed repetitive DNA probe), restriction fragment length polymorphism (*Afut1* RFLP), and single-locus sequence typing (CSP typing) were found to be reproducible and discriminatory methods for *A. fumigatus* genotyping (59). In another study, employing a panel of epidemiologically linked *A. fumigatus* isolates obtained from six different outbreaks of invasive aspergillosis, it was demonstrated that the STRAf assay can be a valuable molecular tool to support epidemiological investigations because it satisfied two basic tenets of a good typing system: (i) provide distinctive fingerprints from genetically unrelated isolates and identical or highly similar fingerprints from closely related organisms and (ii) establish epidemiological concordance among strains recovered from the same outbreak (60). Interestingly, this study also reported for the first time evidence of microvariation events in *A. fumigatus* populations.

## Pneumocystis

Availability of molecular tools was key to understanding the epidemiology of *Pneumocystis jirovecii* (*P. carinii* f. spp. *hominis*, formerly *P. carinii*) (61). Originally classified as a parasite, this organism was recognized as a fungus by sequence analyses of the rRNA regions. *Pneumocystis* remains an unusual member of this kingdom in that it still lacks an *ex vivo* continuous culture system, thus impeding understanding of the basic biological processes in this organism. Despite this impediment, remarkable progress has been made in understanding the molecular epidemiology of this organism. Detailed sequence-based methods and phylogenetic analyses revealed that members of this species are genotypically very diverse and that each subspecies has a specific host preference (62). For instance, *P. jirovecii* is found to colonize and infect only humans, while *P. carinii*'s host is the rat species. Molecular epidemiological studies established that this fungus could be transmitted vertically as well as from person to person, resulting in nosocomial outbreaks (63,64). Genotyping studies have also demonstrated that multiple strains of *P. jirovecii* can coinfect both immunocompetent and immunocompromised individuals (65) and that genotype switching can occur during both colonization and active infection (66). Recent restriction fragment length polymorphism analysis of the major surface glycoprotein revealed that despite genotypic heterogeneity among *P. jirovecii*, some phylogeographic structure was evident (67).

## CONCLUSIONS

This is an exciting era for molecular mycology with several whole genomes available (and many more becoming available), and numerous novel and cutting-edge molecular technologies are already in use for the detection, identification, and population structure analyses of these complex organisms. Fungal forensics and molecular epidemiology are still in their infancy and, as such, decisions made now will affect how forensic samples are analyzed in the future. The forensic, public health, mycology, and genomics communities will need to consider not only the utility of different markers, but also the advances in technology that are leading toward (i) comprehensive genotyping, (ii) exact sample matching and source attribution, (iii) improved comparative genomics and phylogenetics, and (iv) limitations on access to these technologies. Whole genome sequence typing may provide the ultimate solution for forensic and molecular epidemiologic investigations. Data generated from whole genome sequencing can be used to build robust whole genome-based phylogenies that will enable the identification of discriminatory SNP markers. From a public health perspective, identification of such global SNPs would facilitate development of targeted, discriminatory, and economical strain typing methods that can be applied to support future outbreak investigations. Many challenges need to be overcome prior to wide-scale uptake of such advanced technologies. Until then, enhancements in databases and data sharing with existing genotyping systems are critical to continue to advance the current capabilities.

## DISCLAIMER

The findings and conclusions in this chapter are those of the author(s) and do not necessarily represent the views of the CDC.

# REFERENCES

[1] R.R. Paterson, Fungi and fungal toxins as weapons, Mycol. Res. 110 (2006) 1003–1010.

[2] J.W. Taylor, D.M. Geiser, A. Burt, V. Koufopanou, The evolutionary biology and population genetics underlying fungal strain typing, Clin. Microbiol. Rev. 12 (1999) 126–146.

[3] G. Aguileta, G. Refrégier, R. Yockteng, E. Fournier, T. Giraud, Rapidly evolving genes in pathogens: Methods for detecting positive selection and examples among fungi, bacteria, viruses and protists, Infect. Genet. Evol. 9 (2009) 656–670.

[4] J.E. Stajich, M.L. Berbee, M. Blackwell, D.S. Hibbett, T.Y. James, J.W. Spatafora, et al., The fungi, Curr. Biol. 19 (2009) R840–R845.

[5] D.L. Hawksworth, A.Y. Rossman, Where are all the undescribed fungi? Phytopathology 87 (1997) 888–891.

[6] R.A. Zilinskas, Iraq's biological weapons: The past as future? JAMA 278 (1997) 418–424.

[7] R.W. Wannemacher, S.L. Wiener, Trichothecene mycotoxins, in: R. Zajtchuk, R.F. Bellamy (Eds.), Textbook of Military Medicine: Medical Aspects of Chemical and Biological Warfare, Office of the Surgeon General, Walter Reed Army Medical Center, Washington, DC, 1997, pp. 658–659.

[8] J. Fletcher, C. Bender, B. Budowle, W.T. Cobb, S.E. Gold, C.A. Ishimaru, et al., Plant pathogen forensics: Capabilities, needs, and recommendations, Microbiol. Mol. Biol. Rev. 70 (2006) 450–471.

[9] A. Quarta, G. Mita, M. Haidukowski, A. Logrieco, G. Mulè, A. Visconti, Multiplex PCR assay for the identification of nivalenol, 3- and 15-acetyl-deoxynivalenol chemotypes in *Fusarium*, FEMS Microbiol. Lett. 259 (2006) 7–13.

[10] B. Lievens, M. Rep, B.P. Thomma, Recent developments in the molecular discrimination of formae speciales of *Fusarium oxysporum*, Pest Manag. Sci. 64 (2008) 781–788.

[11] E.M. Goss, M. Larsen, G.A. Chastagner, D.R. Givens, N.J. Grunwald, Population genetic analysis infers migration pathways of *Phytophthora ramorum* in US nurseries, PLoS Pathog. 5 (2009) e1000583.

[12] D.W. Warnock, Fungal diseases: An evolving public health challenge, Med. Mycol. 44 (2006) 697–705.

[13] D.M. Dixon, *Coccidioides immitis* as a Select Agent of bioterrorism, J. Appl. Microbiol. 91 (2001) 602–605.

[14] M.C. Fisher, G.L. Koenig, T.J. White, G. San-Blas, R. Negroni, I.G. Alvarez, et al., Biogeographic range expansion into South America by *Coccidioides immitis* mirrors New World patterns of human migration, Proc. Natl. Acad. Sci. USA 98 (2001) 4558–4562.

[15] R.F. Hector, R. Laniado-Laborin, Coccidioidomycosis: A fungal disease of the Americas, PLoS Med. 2 (2005) e2.

[16] R. Bialek, J. Kern, T. Herrmann, R. Tijerina, L. Cecenas, U. Reischl, et al., PCR assays for identification of *Coccidioides posadasii* based on the nucleotide sequence of the antigen 2/proline-rich antigen, J. Clin. Microbiol. 42 (2004) 778–783.

[17] D. Pappagianis, Epidemiology of coccidioidomycosis, Curr. Top. Med. Mycol. 2 (1988) 1999–2238.

[18] M.A. Mandel, B.M. Barker, S. Kroken, S.D. Rounsley, M.J. Orbach, Genomic and population analyses of the mating type loci in *Coccidioides* species reveal evidence for sexual reproduction and gene acquisition, Eukaryot. Cell 6 (2007) 1189–1199.

[19] J.N. Galgiani, Coccidioidomycosis: A regional disease of national importance: Rethinking approaches for control, Ann. Intern. Med. 130 (1999) 293–300.

[20] CDC, Coccidioidomycosis in workers at an archeologic site—Dinosaur National Monument, Utah, June–July 2001, Morb. Mortal. Weekly Rep. 50 (2001) 1005–1008.

[21] M.A. Saubolle, Laboratory aspects in the diagnosis of coccidioidomycosis, Ann. N.Y. Acad. Sci. 1111 (2007) 301–314.

[22] T. Kuberski, R. Myers, L.J. Wheat, M. Durkin, P. Connolly, B.M. Kubak, et al., Diagnosis of coccidioidomycosis by antigen detection using cross-reaction with a Histoplasma antigen, Clin. Infect. Dis. 44 (2007) e50–e54.

[23] M. Durkin, L. Estok, D. Hospenthal, N. Crum-Cianflone, S. Swartzentruber, E. Hackett, et al., Detection of Coccidioides antigenemia following dissociation of immune complexes, Clin. Vaccine Immunol. 16 (2009) 1453–1456.

[24] D.C. Chang, S. Anderson, K. Wannemuehler, D.M. Engelthaler, L. Erhart, R.H. Sunenshine, et al., Testing for coccidioidomycosis among patients with community-acquired pneumonia, Emerg. Infect. Dis. 14 (2008) 1053–1059.

[25] M.J. Binnicker, S.P. Buckwalter, J.J. Eisberner, R.A. Stewart, A.E. McCullough, S.L. Wohlfiel, et al., Detection of Coccidioides species in clinical specimens by real-time PCR, J. Clin. Microbiol. 45 (1) (2007) 173–178.

[26] R. de Aguiar Cordeiro, R.S. Nogueira Brilhante, M.F. Gadelha Rocha, F.E. Araújo Moura, Z. Pires de Camargo, J.J. Costa Sidrim, Rapid diagnosis of coccidioidomycosis by nested PCR assay of sputum, Clin. Microbiol. Infect. 13 (2007) 449–451.

[27] K.W. Sheff, E.R. York, E.M. Driebe, B.M. Barker, S.D. Rounsley, V.G. Waddell, et al., Development of a rapid, cost-effective TaqMan real-time PCR assay for identification and differentiation of Coccidioides immitis and Coccidioides posadasii, Med. Mycol. (2009), doi:10.1080/13693780903218990.

[28] M.C. Fisher, G.L. Koenig, T.J. White, J.W. Taylor, Molecular and phenotypic description of Coccidioides posadasii sp nov., previously recognized as the non-California population of Coccidioides immitis, Mycologia 94 (2002) 73–84.

[29] A.D. Burt, A. Carter, G.L. Koenig, T.J. White, J.W. Taylor, Molecular markers reveal cryptic sex in the human pathogen Coccidioides immitis, Proc. Natl. Acad. Sci. USA 93 (1996) 770–773.

[30] K. Jewell, R. Cheshire, G.D. Cage, Genetic diversity among clinical Coccidioides spp. isolates in Arizona, Med. Mycol. 46 (2008) 449–455.

[31] M.C. Fisher, B. Rannala, V. Chaturvedi, J.W. Taylor, Disease surveillance in recombining pathogens: Multilocus genotypes identify sources of human Coccidioides infections, Proc. Natl. Acad. Sci. USA 99 (2002) 9067–9071.

[32] V. Koufopanou, A. Burt, J.W. Taylor, Concordance of gene genealogies reveals reproductive isolation in the pathogenic fungus Coccidioides immitis, Proc. Natl. Acad. Sci. USA 94 (1997) 5478–5482.

[33] C.R. Zimmermann, C.J. Snedker, D. Pappagianis, Characterization of Coccidioides immitis isolates by restriction fragment length polymorphisms, J. Clin. Microbiol. 32 (1994) 3040–3042.

[34] M.C. Fisher, G.L. Koenig, T.J. White, J.W. Taylor, A test for concordance between the multilocus genealogies of genes and microsatellites in the pathogenic fungus Coccidioides immitis, Mol. Biol. Evol. 17 (2000) 1164–1174.

[35] H. Johannesson, J.P. Townsend, C.Y. Hung, G.T. Cole, J.W. Taylor, Concerted evolution in the repeats of an immunomodulating cell surface protein, SOWgp, of the human pathogenic fungi Coccidioides immitis and C. posadasii, Genetics 171 (2005) 109–117.

[36] T.J. Sharpton, J.E. Stajich, S.D. Rounsley, M.J. Gardner, J.R. Wortman, V.S. Jordar, et al., Comparative genomic analyses of the human fungal pathogens Coccidioides and their relatives, Genome Res. 19 (2009) 1722–1731.

[37] B.M. Barker, K.A. Jewell, S. Kroken, M.J. Orbach, The population biology of coccidioides: Epidemiologic implications for disease outbreaks, Ann. N.Y. Acad. Sci. 1111 (2007) 147–163.

[38] M.C. Fisher, T.J. White, J.W. Taylor, Primers for genotyping single nucleotide polymorphisms and microsatellites in the pathogenic fungus *Coccidioides immitis*, Mol. Ecol. 8 (1999) 1082–1084.

[39] S.A. Balajee, H.A. de Valk, B.A. Lasker, J.F. Meis, C.H. Klaassen, Utility of a microsatellite assay for identifying clonally related outbreak isolates of *Aspergillus fumigatus*, J. Microbiol. Methods 73 (2008) 252–256.

[40] P. Keim, M. Van Ert, T. Pearson, A. Vogler, L. Hyunh, D.M. Wagner, Anthrax molecular epidemiology and forensics: Using different markers for the appropriate evolutionary scales, Infect. Genet. Evol. 4 (2004) 205–213.

[41] T. Pearson, R.T. Okinaka, J.T. Foster, P. Keim, Phylogenetic understanding of clonal populations in an era of whole genome sequencing, Infect. Genet. Evol. 9 (2009) 1010–1019.

[42] T. Pearson, P. Giffard, S. Beckstrom-Sternberg, R. Auerbach, H. Hornstra, A. Tuanyok, et al., Phylogeographic reconstruction of a bacterial species with high levels of lateral gene transfer, BMC Biol. 18 (2009) 78.

[43] P. Hardenbol, J. Banér, M. Jain, M. Nilsson, E.A. Namsaraev, G.A. Karlin-Neumann, et al., Multiplexed genotyping with sequence-tagged molecular inversion probes, Nat. Biotechnol. 21 (2003) 673–678.

[44] Broad Institute [*Coccidioides* Group Database]. December 10, 2009. Available from: http://www.broad.mit.edu/annotation/genome/coccidioidesgroup/-MultiHome.html.

[45] J. Schupp, J. Gillece, T. Contente, E. Driebe, K. Sheff, Pearson T., et al., Targeted large scale SNP genotyping facilitates fine scale resolution and accurate phylogeographic characterization of *Coccidioides immitis* and *C. posadasii*. 2010. Presented at the 2nd ASM Conference on Dimorphic Fungal Pathogens, Miami, FL, March 22, 2010.

[46] T. Pearson, J.D. Busch, J. Ravel, T.D. Read, S.D. Rhoton, J.M. U'Ren, et al., Phylogenetic discovery bias in *Bacillus anthracis* using single nucleotide polymorphisms from whole genome sequencing, Proc. Natl. Acad. Sci. USA 101 (2004) 13536–13541.

[47] D.W. Craig, J.V. Pearson, S. Szelinger, A. Sekar, M. Redman, J.J. Corneveaux, et al., Identification of genetic variants using bar-coded multiplexed sequencing, Nat. Methods 5 (2008) 887–893.

[48] B. Langmead, C. Trapnell, M. Pop, S.L. Salzberg, Ultrafast and memory-efficient alignment of short DNA sequences to the human genome, Genome Biol. 10 (2009) R25.

[49] C.A. Cummings, C.A. Bormann Chung, R. Fang, M. Barker, P.M. Brzoska, P. Williamson, et al., Whole-genome typing of *Bacillus anthracis* isolates by next-generation sequencing accurately and rapidly identifies strain-specific diagnostic polymorphisms, Forensic Sci. Int. Genet. Supp. Ser. 2 (2009) 300–301.

[50] D.M. Engelthaler, T. Chiller, J. Schupp, S., J.B.S. Beckstrom-Sternberg, E. Driebe, C. Robbins, et al., Use of next generation sequencing for molecular epidemiology in a suspect Coccidioides transplant-associated outbreak. 2010. Emerg. Infect. Dis. Accepted for publication.

[51] R.C. Summerbell, C.A. Levesque, K.A. Seifert, M. Bovers, J.W. Fell, M.R. Diaz, et al., Microcoding: The second step in DNA barcoding, Philos. Trans. R. Soc. Lond. Biol. Sci. 360 (2005) 1897–1903.

[52] S.A. Balajee, A.M. Borman, M.E. Brandt, J. Cano, M. Cuenca-Estrella, E. Dannaoui, et al., Sequence-based identification of *Aspergillus, Fusarium* and *Mucorales* in the clinical mycology laboratory: Where are we and where should we go from here? J. Clin. Microbiol. 47 (2009) 877–884.

[53] S.A. Balajee, J. Houbraken, P.E. Verweij, S.B. Hong, T. Yaghuchi, J. Varga, et al., *Aspergillus* species identification in the clinical setting, Studies Mycol. 59 (2007) 39–46.

[54] K.J. Martin, P.T. Rygiewicz, Fungal-specific PCR primers developed for analysis of the ITS region of environmental DNA extracts, BMC Microbiol. 5 (2005) 28.

[55] J.P. Bouchara, H.Y. Hsieh, S. Croquefer, R. Barton, V. Marchais, M. Pihet, et al., Development of an oligonucleotide array for direct detection of fungi in sputum samples from patients with cystic fibrosis, J. Clin. Microbiol. 47 (1) (2009) 142–152.

[56] J.L. Leake, S.E. Dowd, R.D. Wolcott, A.M. Zischkau, Identification of yeast in chronic wounds using new pathogen-detection technologies, J. Wound Care 18 (2009) 103–104.

[57] P.C. Woo, S.Y. Leung, K.K. To, J.F. Chan, A.H. Ngan, V.C. Cheng, et al., Internal transcribed spacer region sequence heterogeneity in *Rhizopus microsporus*: Implications for molecular diagnosis in clinical microbiology laboratories, J. Clin. Microbiol. 48 (2010) 208–214.

[58] D.C. Chang, G.B. Grant, K. O'Donnell, K.A. Wannemuehler, J. Noble-Wang, C.Y. Rao, et al., Multistate outbreak of *Fusarium* keratitis associated with use of a contact lens solution, JAMA 296 (2006) 953–963.

[59] S.A. Balajee, S.T. Tay, B.A. Lasker, S.F. Hurst, A.P. Rooney, Characterization of a novel gene for strain typing reveals substructuring of *Aspergillus fumigatus* across North America, Eukaryot. Cell 6 (2007) 1392–1399.

[60] A. van Belkum, P.T. Tassios, L. Dijkshoorn, S. Haeggman, B. Cookson, N.K. Fry, et al., Guidelines for the validation and application of typing methods for use in bacterial epidemiology, Clin. Microbiol. Infect. 13 (Suppl 3) (2007) 1–46.

[61] J.M. Beck, M.T. Cushion, *Pneumocystis* workshop: 10th anniversary summary, Eukaryot. Cell 8 (2009) 446–460.

[62] E.M. Dei-Cas, R. Chabe, I. Moukhlis, E.M. Durand-Joly, J.R. Aliouat, M. Stringer, et al., *Pneumocystis oryctolagi* sp. nov., an uncultured fungus causing pneumonia in rabbits at weaning: Review of current knowledge, and description of a new taxon on genotypic, phylogenetic and phenotypic bases, FEMS Microbiol. Rev. 30 (2006) 853–871.

[63] S. Gianella, L. Haeberli, B. Joos, B. Ledergerber, R.P. Wüthrich, R. Weber, et al., Molecular evidence of interhuman transmission in an outbreak of *Pneumocystis jirovecii* pneumonia among renal transplant recipients, Transpl. Infect. Dis. 12 (2009) 1–10.

[64] H. Yazaki, N. Goto, K. Uchida, T. Kobayashi, H. Gatanaga, S. Oka, Outbreak of *Pneumocystis jiroveci* pneumonia in renal transplant recipients: *P. jiroveci* is contagious to the susceptible host, Transplantation 88 (2009) 380–385.

[65] L.F. Nimri, I.N. Moura, L. Huang, C. del Rio, D. Rimland, J.S. Duchin, et al., Genetic diversity of *Pneumocystis carinii* f. sp. *hominis* based on variations in nucleotide sequences of internal transcribed spacers of rRNA genes, J. Clin. Microbiol. 40 (2002) 1146–1151.

[66] S.P. Keely, J.R. Stringer, Multi-locus genotype switching in *Pneumocystis carinii* sp. f. *hominis*: Evidence for reinfection, J. Eukaryot. Microbiol. 43 (1996) 50S.

[67] C. Ripamonti, A. Orenstein, G. Kutty, L. Huang, R. Schuhegger, A. Sing, et al., Restriction fragment length polymorphism typing demonstrates substantial diversity among *Pneumocystis jirovecii* isolates, J. Infect. Dis. 200 (2009) 1616–1622.

# CHAPTER 19

# Ricin Forensics: Comparisons to Microbial Forensics

**Jeffrey T. Foster,[a] Robert L. Bull,[b] and Paul S. Keim[a,c]**

[a]Center for Microbial Genetics and Genomics, Northern Arizona University, Flagstaff, Arizona

[b]FBI Laboratory, Quantico, Virginia

[c]Pathogen Genomics Division, The Translational Genomics Research Institute, Flagstaff, Arizona

## INTRODUCTION

Castor beans from the castor plant (*Ricinus communis*) are the source of a potent natural toxin, ricin. Threats posed by ricin are threefold: first, the toxin has no antidote so medical staff can only provide supportive care; second, the source plant grows throughout most tropical and temperate regions of the world so it is readily available; and third, toxin extraction is relatively easy to perform with common chemicals. Although ricin is a plant toxin and not a microbial toxin, it is included within the general area of microbial forensics because the toxin is similar to those produced by several bacteria and is studied by the same biodefense community. Forensic approaches to the toxin and the plant's DNA provide a means to compare and contrast forensic methods in microbes.

Ricin is a frequently used agent for biocrimes in the United States. Each year a handful of instances occur where someone attempts to poison another with ricin. Preparations range from crude extracts to purified toxin. Ricin purification methods abound on the Internet. In fact, a process for preparing toxin from castor beans was the subject of a U.S. patent in 1962 (1).

Detection of ricin and source attribution provides an excellent contrast to typing systems for bacteria. Typing of ricin consists of two aspects: assessment of the toxin and assessment of source plants. For ricin assessment, the process is nearly identical to procedures for toxin assessment from bacteria such as *Clostridium botulinum*. The toxin is detected by various assays, including antibody-based, enzyme-based, or polymerase chain reaction (PCR)-based tests. Verification of the presence of biologically active toxin is a key element of forensic testing because it is required for legal proof of attempted poisonings. To determine if ricin is present in an evidentiary sample, a combination of analytical methods may be

315

Microbial Forensics. DOI: 10.1016/B978-0-12-382006-8.00019-0

utilized. Each of the analytical tests detects a specific target—immunological assays and mass spectrometry for detection of the ricin protein, cellular toxicity assays or cell-free enzymatic assays for detection of ricin activity, and PCR assays for detection of residual DNA that frequently is a contaminant of the ricin toxin preparation. For clinical samples, a biomarker for ricinine is used to indicate ricin exposure (see Chapter 20). The other avenue of forensic attribution is genotyping of the plants that were the source of the ricin. Because many ricin preparations are crude extracts, DNA from the source plants is typically present. However, genotyping of plant DNA is quite different from many microbial pathogens because plants are diploid rather than haploid. Diploidly, sexual reproduction and out-crossing result in chromosomal recombination. This creates a different challenge for assessing population structure and evolutionary history from microbial pathogens. For instance, in diploid organisms, one rarely has a single-nucleotide polymorphism (SNP) that will define a particular lineage, as occurs frequently in bacteria. Thus, population genetic analyses in plants are of a statistical rather than phylogenetic nature, where diversity is partitioned in a nested fashion (e.g., Wright's statistics). Like humans, plant cytoplasmic DNA (e.g., mitochondrial and chloroplast) is frequently inherited maternally without recombination. These genetic components are conceptually similar to clonal pathogens and phylogenetic analysis is used commonly on targets in these regions.

## BACKGROUND

### History of Castor Beans

Castor beans have a long history of cultivation, and seeds have been recovered from as far back as ancient Egyptian tombs. In fact, Egypt is close to the suspected origin of the plant in East Africa, with Ethiopia the center of both diversity and its native range (2). However, the plant is now distributed widely throughout the world due to human commercial transport. The plant is not frost tolerant so although it can grow as an annual in temperate regions as far north as New England, it is largely found in tropical and subtropical climates. Castor plants have taken two paths since domestication, as a source of seeds for oil production and as a garden ornamental (3). The plant frequently escapes cultivation and can be found feral in places such as roadsides, abandoned lots, and streamsides (4).

Castor plants are fast growing and can exceed 3 meters in height in a growing season. The plant was historically a perennial shrub but has since evolved into a fast-growing annual (3), although both growth forms occur widely. The plant is characterized by morphological variation in nearly all of its characters, including color, size, and shape of leaves, stems, and seeds (Figure 19.1), seed oil content, flower and fruit size, maturation, and plant shape from dwarfed and compact to large and full. Cultivation of castor plants for oil

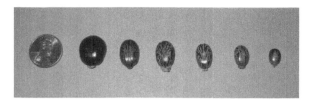

**FIGURE 19.1**
A range of sizes and coat patterns of castor bean seeds.

production is common in India, China, and Brazil (5). This high-quality oil can be used in a variety of products, including lubricants, cosmetics, soaps, paints, nylon, plastics, and other manufactured products (6). Castor oil or its derivatives are also used medicinally as a laxative and as additives in a range of drugs (7). Castor oil is derived from the seeds by either mechanical or solvent-based extraction (6). Ricin is not oil soluble so the toxin remains in the "cake," a by-product of extraction. Thus, castor oil production facilities are a potential source for large amounts of ricin; because the oil has already been removed, the toxin is much easier to extract. The cake can be used directly as a fertilizer or detoxified by heat processing and used for animal feed (8).

Castor bean plants are not true beans from the family Fabaceae (legumes) but instead belong to the Euphorbiaceae, a large family of flowering plants that includes cassava (*Manihot esculenta*), rubber tree (*Hevea brasiliensis*), and poinsettia (*Euphorbia pulcherrima*). Taxonomically, it is in the tribe Acalypheae and subtribe Ricininae. Based on analyses using several genes, the most closely related species is *Speranskia cantonensis* (9), although a more complete sampling will likely uncover other closely related species. Commonly available plants within the same tribe are members of the large and well-distributed subtribe Acalyphinae such as *Acalypha hispida* and *A. godseffiana* (chenille plants). Taxonomy is important for ricin forensics because assays must be able to distinguish *R. communis* from these genetically similar near neighbors. Thus, DNA-based assays should be screened against DNA from near neighbors to assure specificity.

## Ricin Poisoning

Ricin is a heterodimeric glycoprotein composed of two chains (subunits)—A and B. The A chain is an enzyme (*N*-glycosidase-rRNA) that inhibits protein synthesis by irreversibly inactivating eukaryotic ribosomes through removal of a single adenine residue from the 28S ribosomal RNA loop contained within the 60S subunit (10). Ricin is considered a type II ribosome-inactivating protein (11). The B chain is a lectin and binds to galactose-containing glycolipids and glycoproteins expressed on the surface of cells, facilitating entry of the A chain into the cytosol where it can function. The A chain cannot enter a cell without the B chain. Thus, the two chains work in tandem (12). Replacement of the receptor-binding subunit with another binding moiety, such as an anti-body directed against a tumor cell surface antigen, can be used to create chimeric molecules (i.e., immunotoxins), which can be used to direct ricin to tumor cells (13).

Although the amount of ricin will vary among different cultivars, toxin quantity in beans is roughly 1–5% by weight (14). The lethal dose for a person varies widely based on route of exposure. Based on studies with rodents and nonhuman primates, injection or inhalation is highly effective for administering lethal doses in a range of 3–15 µg/kg; oral ingestion is much less effective, requiring at least 20 mg/kg for a lethal dose (8). Despite its reputation as one of the most potent natural toxins, lethal dosages are only moderate when compared to toxins such as botulinum and abrin (15).

An immunological response occurs after exposure to ricin, and immunization can occur through repeated exposure to low doses, which was first described in 1891 (16). Subsequent immunization and vaccine efforts initially focused on inactivation of the toxin with chemical treatment (toxoids) and more recently have used recombinant DNA technology to produce ricin A chain lacking detectable N-glycosidase-rRNA activity (17). Most troubling for health care professionals and exposed individuals, however, is that there is currently no antidote for ricin toxicity; only supportive care can be provided.

Intentional poisonings have been far less common than those resulting from accidental ingestion of the seeds. Incidental poisoning of pets and livestock through either ingestion of the seeds or contamination of feed is the most frequent event and horses appear to be particularly susceptible (18,19). Despite these somewhat frequent occurrences, the use of ricin in a biocrime or as a bioterrorism agent remains more commonly reported and sensationalized. The assassination in London of Bulgarian writer and dissident Georgi Markov by the Bulgarian Secret Police remains the most notorious ricin event. Markov appears to have been injected in the back of the thigh with a hollow pellet containing ricin by a modified umbrella (20). In 2004, ricin was discovered along with a threatening note in a postal facility in South Carolina (21). In a possibly related event, a letter containing ricin was sent to the White House demanding changes to various federal regulations (22). These attacks and many others are detailed in a Congressional Research Service report to the U.S. Congress after ricin was found in the Dirksen Senate Office building (23). More recently, a man in Las Vegas was found in a coma and hospitalized after an apparent self-poisoning; 4 grams of ricin was found with his belongings (24). Thus, in the United States and United Kingdom a series of events involving ricin have been perpetrated by individuals and potentially state-sponsored groups. As a result of its lethality, ubiquity, and ease of preparation and dissemination, the active toxin is a considered a Select Agent by the U.S. Department of Health and Human Services (see Chapter 36).

# RICIN TOXIN DETECTION

Legally, forensic investigations must establish that biologically active ricin is present in the evidence for prosecution to be successful. Hence, sensitive assays that detect this toxin are the first step in an investigation. Highly purified forms of ricin are not necessary to prove intent and simply mashing up beans and distributing them via mail may qualify as an attempted poisoning. However, distribution of purified forms of the ricin toxoid or just one of the two chains may not qualify as a poisoning attempt. Detection of ricin toxin can be accomplished through a variety of techniques, including immunological methods, biological assays, and mass spectrometry. Each of these methods has strengths and weaknesses when applied to samples collected from a potential crime scene.

In most cases the primary tool for identification of ricin in a sample is a form of an enzyme-linked immunosorbent assay (ELISA). Formats include lateral flow and standard ELISA using a variety of conjugated reagents that facilitate the detection and possible quantification of ricin in the sample. In the ELISA, a capture antibody binds the ricin to a solid matrix and a detector antibody specific to ricin binds to the immobilized toxin, forming an antibody–toxin–antibody complex. The detector antibody may be conjugated to an enzyme (requires substrate) (25) or oligonucleotide as is the case for an immuno-polymerase chain reaction assay (26), fluorescent molecule, or any other compound that can be read visually or by an instrument (25). In other formats, an antibody specific to the detector antibody is conjugated with an enzyme or other molecule necessary for detecting the presence of ricin. In recent years, assays have been multiplexed into immunological arrays that can simultaneously screen a sample for multiple protein targets (27). Some antigen-capture ELISAs have a sensitivity <1 ng. It is critical that immunological assays be validated properly, conducted by trained and proficient staff, and include appropriate controls to ensure proper interpretation of results. All antibody-based assays detect conformational epitopes; thus the ricin protein must be in the appropriate conformation. This is critical when considering results of an antibody-based assay. If the protein that is the target of the assay is denatured, results of the assay may be negative even if ricin is present. For immunological assays, is advisable to run positive controls, negative controls, and a matrix control to ensure proper interpretation of results. It is suggested that several different assays be run. For example, positive results with the cell-free assay would not differentiate between A chain and holotoxin. In addition, some immunoassays use polyclonal antibodies and others use monoclonal antibodies for the B chain-specific capture antibody and the A chain-specific detector. This variation can be potentially problematic because recent genomic analyses have identified a family of ricin-like proteins in castor beans with enzymatic activity but with differences in reactivity with monoclonal antibodies, suggesting that ricin levels may be underestimated by these assays (10,15).

Frequently, assays for the detection of ricin activity are also utilized when examining samples, which require that the toxin not be denatured. Toxin activity can be assessed by cellular assays that measure cell death (28) or diminished protein synthesis (29) or cell-free assays that measure inhibition of protein synthesis using a rabbit reticulocyte translation system (30). As with antigen-capture assays, it is necessary for the assays to be validated properly and the competency of those that perform them established. In addition, the assays need positive, negative, and antibody-specific inhibition controls. To have confidence that the inhibition of protein synthesis is due to ricin, it is necessary to add a neutralizing antiricin antibody to the sample and show that the inhibition of protein synthesis is prevented. However, this is only relevant if the antibody recognizes all members of the ricin family.

For ricin activity assays, as well as for ELISA, establishment of a threshold is a critical component of determining the performance of the assay. With clinical samples for detection of ricinine it is possible to establish a sufficient sample size to establish a matrix background and a fixed cutoff. Samples that are frequently associated with a criminal investigation, however, are environmental in nature and likely have an undefined background. For this reason it is critical that the setting of a threshold takes into account the matrix effects of the sample and that the assay controls and background be established for each sample.

When the ricin protein is denatured or partially degraded, it may be appropriate to rely on mass spectrometry for identification. Mass spectrometry identifies the mass charge ratios of peptide fragments that are then compared against a database of potential matches. These methods are highly accurate but do have some limitations. Mass spectrometry requires more sample preparation than immunological methods and, generally, has a lower sensitivity. Even with these limitations the method has a role in the identification of ricin in some complex samples. Two technical approaches for ricin identification based on mass spectrometry are matrix-assisted laser desorption/ionization–time of flight mass spectrometry (MALDI TOF) (31) and liquid chromatography–mass spectrometry (LC MS/MS) (32). MALDI TOF requires less sample preparation and has higher sample throughput, whereas LC MS/MS can identify ricin in more complex samples and provide amino acid sequence data of specific peptides. Mass spectrometry methods are also valuable tools for determining the purification methods used in ricin preparation. Additional information can be obtained by mass spectrometry on the fatty acid, carbohydrate and protein composition, residual solvents, and stable isotope ratio in the ricin preparation of interest. These methods, which are not specific for the toxin, provide valuable information on the production methods and may provide leads as to the environment in which the seeds were grown.

Taken together, immunological, biological, mass spectrometry, and molecular analysis contribute to the investigation of samples suspected to contain ricin. It is necessary to understand the strengths and limitations of each these methods and to apply them correctly to the samples of interest. Because of the complexity of the compounds in castor beans and potential for cross reaction of nonspecific targets, extreme care must be taken when interpreting assay results. For example, $RCA_{120}$ is a lectin glycoprotein of low toxicity found in castor bean seeds that can be a confounding factor in many assays for ricin (33). Thus, multiple approaches for ricin identification are often warranted.

# CASTOR BEAN GENOTYPING

The basis of genetic differentiation of castor beans remains in its infancy, although the field has advanced rapidly in the past several years. Because most previous genetic work had focused on *R. communis* as a crop plant, research has largely centered on oil production, growth attributes, and trait heritability (34). Genetic characterization of ricin preparations for sample attribution to a source is a new issue. Forensic investigations for sample attribution of ricin rely on PCR-based methods because other characters possibly used in forensic botany cases are typically not available (35). From trace amounts of material, researchers must find genetic polymorphisms to differentiate individual plants or population-based unique genetic characteristics.

Analytic challenges with ricin preparations are threefold: first, sufficient quality DNA must be present for analysis; second, finding adequate DNA markers to differentiate samples; and third, ricin preparations are typically mixtures of seeds potentially from multiple plants. Even seeds from a particular plant probably represent genetically unique individuals due to sexual recombination. DNA quality is an issue because proteins and oils that inhibit PCR reactions can be found in ricin preparations. Furthermore, ricin purification methods can also remove much of the DNA from the sample. Even if DNA can be extracted from the samples, low genetic diversity poses an additional challenge to distinguish among potential sources. Many of the commercially available seeds, from either agricultural production or horticulture, appear genetically related (36,37). Finally, ricin preparations often come from seeds of several to dozens of plants and, hence, represent a mixture of genotypes. Forensic approaches for ricin genotyping must be sensitive to the analysis of mixtures and use a population genetics approach.

## Amplified Fragment Length Polymorphisms

Amplified fragment length polymorphisms (AFLPs) provide an effective means of genotyping, particularly when little is known about the genome or genetics of an organism. Restriction enzymes cut the DNA and adaptors are attached to the ends of the fragments. Fragments are then amplified using PCR and their varying lengths can then be visualized on gel or capillary-based platforms. AFLP is very sensitive for detecting genetic polymorphisms but requires relatively large amounts of high-quality DNA and has difficulty with mixture analysis. Thus, AFLP is not an ideal candidate for genotyping forensic ricin samples but has been used for population genetics of plants. Limited genetic diversity has been detected in a wide sampling of castor bean plants (36). Complete sequencing of the *R. communis* genome has made other geno-typing methods more desirable than AFLP.

## Simple Sequence Repeats

Simple sequence repeats (SSRs) and other repeated regions such as micro-satellites provide excellent targets and means of assessing genetic variation in samples. In general, the markers mutate rapidly so provide differentiation of even closely related samples. When nine SSR markers were screened against a worldwide collection of samples, minimal differentiation was observed (36). It appears that SSRs have low diversity, perhaps due to a historical genetic bot-tleneck. Potentially all modern domesticated castor beans were derived from a small population, which resulted in a great reduction of genetic diversity. Furthermore, SSRs are difficult to interpret with forensic samples containing large numbers of mixed genotypes. Finally, SSRs fare poorly when only trace amounts of DNA are present.

## Chloroplast DNA

DNA from chloroplasts is a potentially strong candidate for genotyping because it has a much higher copy number than nuclear DNA. Thus, in trace amounts or in degraded samples there will be much more material for assays to target (38). Despite this benefit, chloroplast DNA (cpDNA) has not proven to be highly informative. This was illustrated in a cpDNA study of eight highly diverse samples where the entire cpDNA sequence was determined. Despite this comprehensive analysis, only five cpDNA haplotypes were observed (P. Rabinowicz and A. Chan, unpublished data). Furthermore, three of these groups were separated by only two SNPs, indicating that they are closely related. Therefore, cpDNA maybe useful for determining broad-scale geo-graphical patterns and may allow for exclusion in some cases. However, it lacks the discrimination power for strong "match" or inclusionary statistics.

## Nuclear SNPs

Single-nucleotide polymorphisms from nuclear DNA provide one of the only remaining genotyping options for ricin forensics. Nuclear SNPs may also be relatively rare due to genetic bottleneck from domestication, but the genome is large with many potential SNP sites. This makes SNP identification difficult

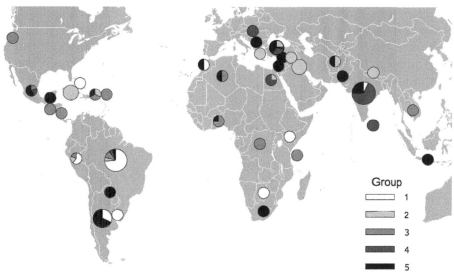

**FIGURE 19.2**

Map of distribution of five genetic groupings based on 48 single nucleotide polymorphisms. Groupings were based on a Bayesian clustering algorithm. Originally published in *BMC Plant Biology* (37).

but does not diminish their discrimination power en masse once SNPs are found. SNPs are almost always biallelic, such that their discrimination capacity as individual loci is limited, but the number of SNP loci that can be used greatly exceeds AFLP, SSR, and cpDNA. Their biallelic aspect also allows for simpler analysis of DNA mixtures because individual allele frequencies will be a simple two-component ratio. SNP genotyping assays abound with some having very high capacity and others providing single molecule sensitivity for trace evidence analysis. Analysis of a worldwide panel of >600 samples using 48 SNPs differentiated samples into five distinct groupings (Figure 19.2) (37). Similar to analyses with AFLPs and SSRs (36), however, these assays did not identify geographic structuring of plant populations.

# CHALLENGES

Reducing the availability of castor beans and castor plants as sources of the toxin is not currently a viable option, although the development of horticultural and agricultural varieties with negligible or reduced amounts of ricin is currently ongoing in agricultural plant-breeding programs. Because of its high oil content and good growth in poor soils with even low amounts of water, the plant is a potential biofuel source and increased worldwide cultivation of the plant is possible. Development of plants containing seeds with low ricin levels and high oil yields would still be useful for agriculture, while eliminating their utility as a source of ricin.

Excellent assays on a variety of platforms are available for detecting the ricin toxin and verifying its enzymatic activity, but require sophisticated equipment, the assays are challenging to run, and other compounds in castor beans may mimic ricin. Despite recent sequencing of the *R. communis* genome, large methodological gaps remain for DNA genotyping. Although there have been recent genotyping advances for assessment of castor bean populations, high-throughput methods that can assess mixtures of castor bean genotypes accurately are currently not possible. New analytical methods will likely solve this issue within a few years but the remaining challenge will be adequate reference collections of seeds. With improved genotyping methods, seeds from agriculture, horticulture, and feral populations can be compared to samples from an incident to determine likely sources. An adequate population genetic framework for analyses is therefore essential.

# REFERENCES

[1] H.L. Craig, O.H. Alderks, A.H. Corwin, S.H. Dieke, C.L. Karel, U.S. Patent 3,060,165. Preparation of toxic ricin. United States of America as represented by the Secretary of the Army, 1962.

[2] N.I. Vavilov, The Origin, Variation, Immunity and Breeding of Cultivated Plants, Chronica Botanica, Waltham, MA, 1951.

[3] A. Narain, Castor, in: J.B. Hutchinson (Ed.), Evolutionary Studies in World Crops: Change in the Indian Subcontinent, Cambridge University Press, Cambridge, UK, 1974.

[4] E. Weber, Invasive Plant Species of the World. A Reference Guide to Environmental Weeds, CABI Publishing, Wallingford, 2003.

[5] FAO, http://faostat.fao.org, 2007.

[6] D.S. Ogunniyi, Castor oil: A vital industrial raw material, Bioresource Technol. 97 (2006) 1086–1091.

[7] BASF. Technical leaflet for Cremophor EL. Available from: http://www.makeni.com.br/Portals/Makeni/prod/boletim/Cremophor%20EL.pdf.

[8] M.A. Poli, C. Roy, K.D. Huebner, D.R. Franz, N.K. Jaax, RicinChapter 15, in: Z.F. Dembek (Ed.), Medical Aspects of Biological Warfare, Borden Institute, Washington, DC, 2007.

[9] T. Tokuoka, Molecular phylogenetic analysis of Euphorbiaceae sensu stricto based on plastid and nuclear DNA sequences and ovule and seed character evolution, J. Plant Res. 120 (4) (2007) 511–522.

[10] J. Leshin, M. Danielsen, J.J. Credle, A. Weeks, K.P. O'Connell, K. Dretchen, Characterization of ricin toxin family members from *Ricinus communis*, Toxicon 55 (2–3) (2010) 658–661.

[11] L. Barbieri, M.G. Battelli, F. Stirpe, Ribosome-inactivating proteins from plants, Biochim. Biophys. Acta Rev. Biomembr. 1154 (1993) 237–282.

[12] J.M. Lord, M.R. Hartley, L.M. Roberts, Ribosome inactivating proteins of plants, Semin. Cell Biol. 2 (1991) 15–22.

[13] M.J. Lord, N.A. Jolliffe, C.J. Marsden, C.S. Pateman, D.C. Smith, R.A. Spooner, et al., Ricin. Mechanisms of cytotoxicity, Toxicol. Rev. 23 (2003) 53–64.

[14] S.M. Bradberry, K.J. Dickers, P. Rice, G.D. Griffiths, J.A. Vale, Ricin poisoning, Toxicol. Rev. 22 (1) (2003) 65–70.

[15] R.G. Darling, J.B. Woods, USAMRIID's Medical Management of Biological Casualties Handbook., U.S. Army Medical Research Institute of Infectious Diseases, Fort Detrick, MD, 2004.

[16] S. Olsnes, The history of ricin, abrin and related toxins, Toxicon 44 (4) (2004) 361–370.

[17] M.A. Olson, C.B. Millard, M.P. Byrne, R.W. Wannemacher, R.D. LeClaire. U.S. Patent 6869787. Ricin vaccine and methods of making and using thereof. The United States of America as represented by the Secretary of the Army, 2005.

[18] J.C. Albretsen, S.M. Gwaltney-Brant, S.A. Khan, Evaluation of castor bean toxicosis in dogs: 98 cases, J. Am. Anim. Hosp. Assoc. 36 (3) (2000) 229–233.

[19] M.R. Aslani, M. Maleki, M. Mohri, K. Sharifi, V. Najjar-Nezhad, E. Afshari, Castor bean (*Ricinus communis*) toxicosis in a sheep flock, Toxicon 49 (3) (2007) 400–406.

[20] B. Knight, Ricin: A potent homicidal poison, Br. Med. J. 1 (6159) (1979) 350–351.

[21] Investigation of a ricin-containing envelope at a postal facility: South Carolina, MMWR Morb. Mortal. Wkly. Rep. 52 (46) (2003) 1129–1131.

[22] D. Eggen, FBI Releases Details of Letter With Ricin Sent to White House; Postmark Was Weeks Before Discovery, The Washington Post (February 24, 2004) Washington, DC.

[23] D. Shea, F. Gottron, CRS Report for Congress, Ricin: Technical Background and Potential Role in Terrorism, Order Code RS21383, Washington, DC, February 4, 2004.

[24] K. Ritter, Man Jailed in Ricin Case Indicted on Charges in Las Vegas, Associated Press, April 22, 2008.

[25] E.A. Garber, T.W. O'Brien, Detection of ricin in food using electrochemiluminescence-based technology, J. AOAC Int. 91 (2) (2008) 376–382.

[26] C. Lubelli, A. Chatgilialoglu, A. Bolognesi, P. Strocchi, M. Colombatti, F. Stirpe, Detection of ricin and other ribosome-inactivating proteins by an immuno-polymerase chain reaction assay, Anal. Biochem. 355 (1) (2006) 102–109.

[27] B. Huelseweh, R. Ehricht, H.J. Marschall, A simple and rapid protein array based method for the simultaneous detection of biowarfare agents, Proteomics 6 (10) (2006) 2972–2981.

[28] T. Yoshida, C.H. Chen, M.S. Zhang, H.C. Wu, Increased cytotoxicity of ricin in a putative Golgi-defective mutant of Chinese hamster ovary cell, Exp. Cell Res. 190 (1) (1990) 11–16.

[29] N.J. Mantis, C.R. McGuinness, O. Sonuyi, G. Edwards, S.A. Farrant, Immunoglobulin A antibodies against ricin A and B subunits protect epithelial cells from ricin intoxication, Infect. Immun. 74 (6) (2006) 3455–3462.

[30] C.Y. Lindsey, J.D. Richardson, J.E. Brown, M.L. Hale, Intralaboratory validation of cell-free translation assay for detecting ricin toxin biological activity, J. AOAC Int. 90 (5) (2007) 1316–1325.

[31] C.S. Brinkworth, E.J. Pigott, D.J. Bourne, Detection of intact ricin in crude and purified extracts from castor beans using matrix-assisted laser desorption ionization mass spectrometry, Anal. Chem. 81 (4) (2009) 1529–1535.

[32] S.M. Darby, M.L. Miller, R.O. Allen, Forensic determination of ricin and the alkaloid marker ricinine from castor bean extracts, J. Forensic Sci. 46 (5) (2001) 1033–1042.

[33] S. Yamashiro, Y. Sano, A. Komano, H. Maruko, H. Sekiguchi, Y. Takayama, et al., Detection of proteinous toxins using the Bio-Threat Alert system, part 4. Differences in detectability according to manufactural lots and according to toxin subtypes, Forensic Toxicol. 25 (2) (2007) 80–83.

[34] S.K. Dhapke, P.W. Khorgade, M.N. Narkhede, Estimates of genetic variability in castor (*Ricinus communis* L.), Agric. Sci. Digest 12 (1992) 141–143.

[35] H.M. Coyle, Forensic Botany: Principles and Applications to Criminal Casework, CRC Press, Boca Raton, FL, 2005.

[36] G. Allan, A. Williams, P.D. Rabinowicz, A.P. Chan, J. Ravel, P. Keim, Worldwide genotyping of castor bean germplasm (*Ricinus communis* L.) using AFLPs and SSRs, Genet. Resources Crop Evol. 55 (2008) 365–378.

[37] J.T. Foster, G.J. Allan, A.P. Chan, P.D. Rabinowicz, J. Ravel, P.J. Jackson, et al., Single nucleotide polymorphisms for assessing genetic diversity in castor bean (*Ricinus communis*), BMC Plant Biol. 10 (1) (2010) 13.

[38] A.C. Hinckley, Genotyping and Bioforensics of *Ricinus communis*, Master's thesis, University of California, Davis, CA, 2006.

# CHAPTER 20

# Forensic Aspects of Biological Toxins

**James D. Marks**

*Department of Anesthesia and Pharmaceutical Chemistry, University of California, San Francisco, and San Francisco General Hospital, San Francisco, California*

## INTRODUCTION

Biological toxins are molecules produced by living organisms that are poisonous to other species, such as humans. Some Biological Toxins are so potent and relatively easy to produce that they have been classified as biothreat agents. These include botulinum neurotoxins (BoNT), ricin, staphylococcal enterotoxin B, and *Clostridium perfringens* epsilon toxin (see http://www.niaid.nih.gov/biodefense/bandc_priority.htmj for a classification of biothreat agents). These four biothreat agents are all proteins composed of amino acid building blocks. As such, they have a number of features that distinguish them from viral or bacterial threat agents. First, they are not contagious, as the threat agent is not a living organism. For the same reason, these agents cannot be cultured routinely from either patients or the environment after exposure, making forensic detection more difficult. Because proteins are composed of amino acids and not nucleic acid, it is also not generally possible to amplify and detect the presence of toxins using the polymerase chain reaction or by any type of classic DNA hybridization technology. Rather, detection typically relies on the use of antibodies to bind to and detect the presence of toxins or newer detection technologies such as mass spectrometry.

This chapter focuses on the four biothreat toxins just described and their forensic aspects. The majority of the chapter is spent on BoNTs, as these are the most poisonous substances known. The remainder of the chapter is devoted to sections on the other three toxins: ricin, staphylococcal enterotoxin B, and *C. perfringens* epsilon toxin.

## BOTULINUM NEUROTOXIN AND BOTULISM

Botulism is a rare but life-threatening disease caused by spore-forming bacteria of the genus *Clostridium*, including *Clostridium botulinum*, *C. baratii*,

Microbial Forensics. DOI: 10.1016/B978-0-12-382006-8.00020-7

**327**

and *C. butyricum* (1). BoNT is the most poisonous substance known (2). Approximately 7 pg of pure neurotoxin is the LD50 for a mouse, and it has been estimated that the human LD 50 is approximately 0.09–0.15 µg intravenously, 0.7–0.9 µg inhalationally, and 70 µg orally (3–6).

Botulism is characterized by prolonged paralysis, which, if not immediately fatal, requires prolonged hospitalization in an intensive care unit and mechanical ventilation. The potent paralytic abilities of the neurotoxin have also resulted in its development as a biowarfare and biothreat agent (7), as well as a medicine to treat a range of overactive muscle conditions, including cervical dystonias, cerebral palsy, post-traumatic brain injury, and poststroke spasticity (8). The number of medical indications for which toxin is used continues to increase and includes cosmetic uses, such as the treatment of wrinkles. At least four pharmaceutical companies now manufacture therapeutic botulinum neurotoxin and it is now manufactured in a number of countries outside of the United States.

Botulinum neurotoxins differ significantly from each other in their amino acid sequence, resulting in elicitation of different antibody responses. The different antibody responses allow classification of the neurotoxins into different serotypes; antibodies that recognize one serotype do not recognize other serotypes. There are seven BoNT serotypes (A, B, C, D, E, F, and G) (9,10), four of which (A, B, E, and F) are responsible for naturally occurring human botulism (7).

Naturally occurring botulism can result from ingestion of preformed toxin (food botulism) or from toxin produced *in situ* due to wound infection (wound botulism) or colonization of the gastrointestinal (GI) tract (infant or intestinal botulism). Botulism can also occur in exposed laboratory workers or from an overdose of therapeutic neurotoxin. In addition, BoNTs are classified by the Centers for Disease Control and Prevention (CDC) as one of the six highest risk threat agents for bioterrorism due to their extreme potency and lethality, ease of production and transport, and need for prolonged intensive care (7). Intoxication can occur via oral ingestion of toxin or inhalation of aerosolized toxin (11,12). While only four of the neurotoxin serotypes cause natural human disease, aerosolized neurotoxin serotypes C, D, and G produce botulism in primates by the inhalation route (11) and would most likely also affect humans. Thus, it is likely that any one of the seven BoNT serotypes can be used as a biothreat agent. Because of the severity of illness and potential for outbreaks, both food-borne and intentional botulism are public health emergencies.

## Types of Botulism

Five types of botulism occur in humans: food borne, wound, infant, intestinal, and inadvertent. A sixth type, intentional botulism, is likely to occur during our lifetimes. Each type is associated with different epidemiology and pathogenic mechanisms. The first recognized case of botulism in the United States occurred in 1899 and was caused by a beef tamale (13). Food botulism was the most common form of botulism in the United States prior to 1980 (1). Infant (or intestinal) botulism was first described in 1976 by two groups (14,15) and is now the most frequently reported type of botulism in the United States (1). Wound botulism was first described in the United States in 1951, with initial cases due primarily to traumatic wounds of the extremities (16). More recently, the incidence of this form of botulism has increased and has been associated with injection drug users injecting black tar heroin (17). An adult variant of infant botulism, called "botulinal autointoxication," or hidden, adult intestinal, or adult infectious botulism, was first described in 1979 (18–20). Inadvertent botulism results from unintentional exposure and typically occurs in laboratory workers (21) and in patients receiving therapeutic botulinum neurotoxin (22,23).

## Intentional Botulism

While successful use of BoNT as a bioterror agent has not occurred, it is likely only a matter of time until botulism is intentionally caused by release of toxin by terrorists. The potency and lethality of the toxin make it an ideal bioweapon. Moreover, the widespread manufacture of therapeutic toxin worldwide potentially provides a source of BoNT that could be obtained by an individual or group with ill intent without their need to produce toxin. In fact, this conundrum between the widespread manufacture of toxin for therapeutic use and its resulting availability for biothreat use was hypothesized when the toxin was first approved by the Food and Drug Administration (FDA) for cosmetic use (24). BoNT has already been released unsuccessfully by the Japanese cult Aum Shinryko (7). Both Iraq and the former Soviet Union produced BoNT for use as a weapon (25,26), and at least three additional countries (Iran, North Korea, and Syria) have developed or are believed to be developing BoNT as instruments of mass destruction. Iraq produced 19,000 liters of concentrated BoNT, of which 10,000 liters was weaponized in missile warheads or bombs (25,27).

Exposure of even a small number of civilians to botulinum neurotoxin would overwhelm the health care delivery system of any metropolitan center. Treatment of botulism requires prolonged intensive care unit (ICU) hospitalization and mechanical ventilation for up to 6 weeks. With the downsizing and closing of hospitals, most ICUs run at 80–100% occupancy. In San Francisco, for example, there are approximately 240 ICU beds, with an average occupancy rate of greater than 90%. As few as 30 cases of botulism would fill all empty ICU beds and occupy them for up to 6 weeks. This would eliminate availability of ICU beds for postoperative patients requiring ICU care, such as organ transplantation, neurosurgery, cardiac surgery, and traumatic injuries. Patients requiring such operations would represent "collateral damage," with necessary surgery postponed or transferred to outlying hospitals. Major civilian exposure to BoNT would have catastrophic effects. One study estimated that aerosol exposure of 100,000 individuals to toxin, as could occur with an aerosol release over a metropolitan area, would result in 50,000 cases with 30,000 fatalities (28). Such exposure would result in 4.2 million hospital days and an estimated cost of $8.6 billion. In this study, the most important factors reducing mortality and cost were early availability of antitoxin and mechanical ventilation (28). Such treatment could reduce deaths by 25,000 and costs by $8.0 billion.

## Microbiology of *Clostridium* spp.

Botulism is caused by a 150-kDa neurotoxin secreted by spore-forming anaerobic bacteria of the genus *Clostridium. C. botulinum* can be divided into at least four genetically and phenotypically diverse groups (I through IV) (29). While organisms in these groups are different enough to be classified as separate species, they have all been classified as *C. botulinum*, as they share the common feature of neurotoxin production. Organisms in group I are referred to as proteolytic and organisms in group II as nonproteolytic, based on their ability to digest complex proteins. All BoNT serotype A strains are in group I, serotype B and F can be produced by either group, and serotype E is produced by group II strains. Two additional species, *C. butyricum* and *C. baratii* have been found to produce neurotoxins E (30) and F (31), respectively. BoNT serotypes C and D are both produced by group III organisms. BoNT type C-producing strains are found in avian species, occurring in domestic flocks and causing massive outbreaks in wild waterfowl (32,33). Type C also occurs in other animals such as dogs, mink, and cattle. Outbreaks caused by BoNT type D are rare and are associated with cattle (34). A single human outbreak of type C and type D food botulism has been reported (35,36). Group IV was created to accommodate an organism isolated from a soil sample in Argentina that produces a unique neurotoxin (type G) that causes a flaccid paralysis in mice (37). No clinical cases of type G botulism have been reported in humans, although it has been isolated from autopsy specimens (38). The name *C. argentinense* has been proposed for group IV *Clostridia* (39). Finally, rare strains of *C. botulinum* have been reported that cause clinical disease and secrete more than one toxin, for example, A and B (Ab), A and F (Af), B and F (Bf), or B and A (Ba) (40–44).

Genome sequencing (45,46), 16S RNA analysis, and amplified fragment length polymorphism studies provide additional insights into the phylogeny and evolution of the toxin-producing species of *Clostridium* (46,47). 16S RNA analysis confirms that at least four different genetic backgrounds house the toxin genes, corresponding to the biochemical groups, each of which has acquired one or more BoNT genes independently through horizontal gene transfer. The BoNT gene can reside in the chromosome or on plasmids (48).

The presence of toxins in different genetic backgrounds suggests their move-ment both within species and between species. Indeed, analysis of genomic sequences from different strains reveals the presence of insertion sequence elements and transposon-associated proteins such as recombinases that could facilitate the horizontal transfer of BoNTs (46).

## Neurotoxin Structure and Function

Botulinum neurotoxins are secreted by *Clostridium* species as a protein com-plex with an apparent size of approximately 900 kDa (49). This complex consists of the neurotoxin and a number of proteins collectively called "neu-rotoxin-associated proteins" (NAPs). NAPs include proteins classified as hemagglutinins (50,51), due to their ability to agglutinate red blood cells, and other proteins termed "nontoxin nonhemagglutinins" (NTNH) (52,53). NTNH stabilize the toxin and protect it from environmental degradation dur-ing passage through the gastrointestinal tract (54,55).

The BoNT is secreted as a single polypeptide chain of approximately 150 kDa, which is nicked by proteases to form a 100-kDa heavy chain and a 50-kDa light chain connected by a single disulfide bond. Sequences of genes encoding neurotoxin serotypes A (56–58), B (59,60), C (61), D (62), E (63), F (53,64), and G (65) have been determined. While these toxins differ by as much as 65% at the amino acid level, it is likely that they share the same general pro-tein fold (10).

Significant sequence variability has also been observed within toxin serotypes and these different variants are termed "subtypes" (47,56,58–60). Analysis of BoNT gene sequences of 174 toxin-producing strains of *Clostridium* spp. indicated the presence of four distinct lineages (or subtypes) for BoNT/A and BoNT/B and five different lineages for BoNT/E (47). BoNTs in this study dif-fered from each other by 15% (BoNT/A), 7% (BoNT/B), and 5% (BoNT/E). More recently, additional BoNT/A and BoNT/E subtypes have been reported (66–68). Similarly, at least five BoNT/F subtypes exist, which differ from each other by 36% at the amino acid level (69). In the case of BoNT/C and BoNT/ D, there do not appear to be different subtypes, but there exist mosaic C/D and D/C toxins that contain gene segments of both BoNT/C and BoNT/D (70). The significance of the subtypes is that polyclonal and monoclonal antibodies may bind differentially, or not at all, to different subtypes (69,71,72). Because many diagnostic tests, as well as therapeutic approaches, are based on anti-bodies, failure to account for subtype sequence variation may render a test insensitive or a treatment less efficacious (69,72).

The X-ray crystal structures of BoNT types A, B, and E have been solved at high resolution (Figure 20.1) (73–75). Structural studies, combined with functional

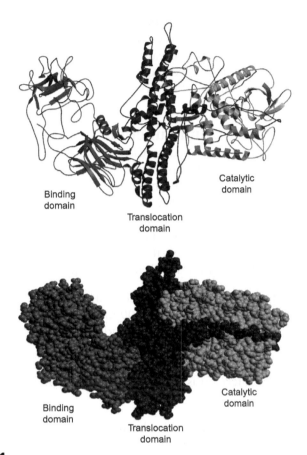

**FIGURE 20.1**

Atomic structure of botulinum neurotoxin type A. Ribbon (1A) and space-filling (1B) models of the
X-ray crystal structure of botulinum neurotoxin type A (ref. 73). The toxin consists of a binding domain,
translocation domain, and catalytic domain, as described in detail in the text.

studies, provide clear insight into how BoNTs interfere with normal release of
the neurotransmitter acetylcholine, resulting in flaccid paralysis (Figures 20.2
and 20.3). The C-terminal portion of the heavy chain (HC) comprises the
receptor-binding domain, which binds to cellular receptors on presynaptic
neurons, resulting in toxin endocytosis (76,77) (Figure 20.3). Precise determination of the cellular receptors was until recently unknown, but the presence
of two coreceptors, a protein and a sialoganglioside such as $G_{D1b}$ or $G_{T1b}$, had
been proposed (77,78). This hypothesis has been confirmed both biochemi-
cally and structurally, with identification of SV2 and synaptotagmin as protein receptors for BoNT/A and BoNT/B, respectively (79–83). Structurally,
the binding domain consists of an N-terminal subdomain consisting of a

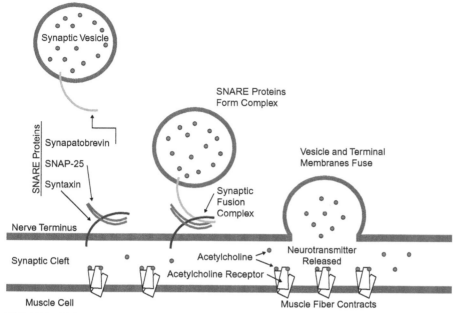

**FIGURE 20.2**

Normal neuromuscular transmission. Synaptic vesicles containing acetylcholine have the soluble
*N*-ethylmaleimide-sensitive factor attachment protein receptor (SNARE) synaptobrevin on their surfaces.
Vesicular synaptobrevin interacts with the SNARE proteins syntaxin and synaptosomal-associated protein
of 25 kDa (SNAP-25) to form a four-helix coiled coil, resulting in fusion of the synaptic vesicle with the
presynaptic membrane. Acetylcholine is released from the vesicle, diffuses across the synaptic cleft, and
binds to the acetylcholine receptor, resulting in normal muscle contraction.

jelly roll motif and a C-terminal subdomain consisting of a β-trefoil motif.
The C-terminal domain comprises both ganglioside and protein-binding sites
(73,81–83).

The N-terminal portion of the heavy chain (HN) (Figure 20.1) comprises the
translocation domain, which consists of α helices and is involved in pore formation. It is hypothesized that the lower pH of the endosome induces a conformational change in this domain, which creates a pore, allowing the light chain
to escape the endosome (Figure 20.3). The light chain (Figure 20.1) is a zinc
endopeptidase, which, depending on serotype, cleaves different members of the
soluble *N*-ethylmaleimide-sensitive factor attachment protein receptor (SNARE)
family of proteins, resulting in the blockade of neuromuscular transmission
(84,85) (Figure 20.3). SNAREs are essential for normal fusion of the synaptic
vesicle and acetylcholine release (Figure 20.3). Toxin serotypes A and E cleave distinct sites within SNAP-25 (synaptosomal-associated protein of 25 kDa) (85–87);
serotypes B, D, F, and G cleave distinct sites within vesicle-associated membrane

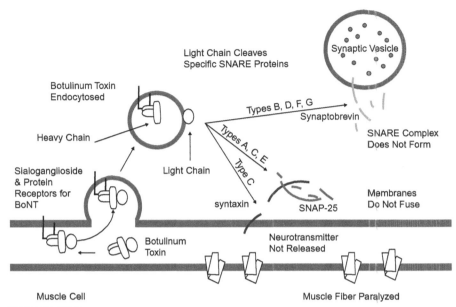

**FIGURE 20.3**

Effect of botulinum neurotoxin on normal neuromuscular transmission. Botulinum neurotoxin binds to unknown receptors on the presynaptic neuron membrane, resulting in endocytosis of the toxin. After endocytosis, the translocation domain changes conformation, resulting in release of the catalytic domain into the cytosol. Depending on the toxin serotype, the catalytic domain cleaves one or more members of the SNARE protein family. SNARE cleavage prevents formation of the SNARE complex and fusion of the vesicle with the membrane. As a result, acetylcholine is not released.

protein (also known as synaptobrevin) (84,85,88–91); and serotype C cleaves syntaxin and SNAP-25 (Figure 20.3) (92,93). These three SNARE proteins (syntaxin, SNAP-25, and synaptobrevin) interact to form a four-helix coiled coil in a step that precedes synaptic fusion (94) (Figure 20.3). Cleavage of any one of these proteins blocks fusion and acetylcholine release, leading to a flaccid paralysis.

## Clinical History, Symptoms, and Findings of Botulism

It is quite likely that the first indication of a bioterror attack with botulinum toxin will be development of clinical disease (botulism) in exposed individuals. It is thus essential that forensic personnel be familiar with the symptoms of the disease and its differential diagnosis (i.e., other diseases that may be confused with botulism). The diagnosis of botulism is made clinically, with laboratory findings and confirmation not usually immediately available. The clinical syndrome of botulism is dominated by neurological signs and symptoms resulting from blockade of neurotransmission at voluntary motor and cholinergic junctions (95–97). Patients with botulism usually present with

**Table 20.1** Symptoms and Signs of Food-Borne Botulism[a]

|  | % Cases |
|---|---|
| *Symptoms* | |
| Fatigue | 77 |
| Dizziness | 51 |
| Double vision | 91 |
| Blurred vision | 65 |
| Dysphagia | 96 |
| Dry mouth | 93 |
| Dysarthria | 84 |
| Sore throat | 54 |
| Dyspnea | 60 |
| Constipation | 73 |
| Nausea | 64 |
| Vomiting | 59 |
| Abdominal cramps | 42 |
| Diarrhea | 19 |
| Arm weakness | 73 |
| Leg weakness | 69 |
| Paresthesia | 14 |
| *Signs* | |
| Alert mental status | 90 |
| Ptosis | 73 |
| Gaze paralysis | 65 |
| Pupils dilated or fixed | 44 |
| Nystagmus | 22 |
| Facial palsy | 63 |
| Diminished gag reflex | 65 |
| Tongue weakness | 58 |
| Arm weakness | 75 |
| Leg weakness | 69 |
| Hyporeflexia or areflexia | 40 |
| Ataxia | 17 |

[a]Data from outbreaks of type A and B food-borne botulism reported in the United States in 1973–1974. Number of patients with available data varied from 35 to 55.

acute onset of weakness in muscles innervated by the cranial nerves, leading to diplopia, dysphonia, dysphagia, and dysarthria (Table 20.1). In mild cases, no other symptoms may develop. In more severe cases, symmetric weakness progresses in a descending manner, leading frequently to paralysis. If the illness is severe enough, the respiratory muscles are involved, leading to ventilatory failure and death unless intubation and mechanical ventilation are instituted. Intubation was required in 67% of cases caused by type A BoNT, 52% of type B, and 39% of type E (98). Patients may also have evidence of autonomic dysfunction, including dry mouth, blurred vision, orthostatic hypotension, urinary retention, and constipation. Sensory abnormalities are usually absent, as only motor and autonomic nerves are affected. Similarly, mental function is usually not affected.

Paralysis from botulism can be quite long lasting. Mechanical ventilation may be required for 2 to 8 weeks with food-borne botulism, with paralysis lasting as long as 7 months reported (97). Symptoms of cranial nerve dysfunction and mild autonomic dysfunction may persist for more than a year (99–101). In infants, hospital stay averages 1 month, with serotype A causing longer lasting disease (5.4-week average hospitalization) than serotype B (3.8-week average hospitalization) (102). There is experimental evidence that neurotoxin catalytic activity persists at the nerve terminal, especially for serotypes A, B, and C, and that recovery initially results from the sprouting of new neuromuscular connections (103).

## Epidemiology of Intentional Botulism and Differential Diagnosis

The intentional release of BoNT is most likely to be associated with the outbreak of a large number of cases of flaccid paralysis with prominent bulbar palsies. Other features may include an outbreak with an unusual toxin type (C, D, F, or G, vide infra), an outbreak with common geographic features but without a common dietary exposure, or multiple simultaneous outbreaks with no common source. The incubation period for intentional botulism is unknown, but is likely related to the route and degree of exposure. For oral exposure to toxin, one may obtain some idea of the incubation period from food-borne botulism literature. Food-borne botulism has an incubation period of 6 hours to 10 days (104), with the majority of cases developing between 18 and 72 hours after ingestion of contaminated food. It is difficult to know precisely the incubation period for aerosol exposure to botulinum toxin due to the paucity of data. In one study, monkeys exhibited signs of intoxication 12 to 80 hours after aerosol exposure with four to seven monkey LD50s (4). The incubation period for the three known cases of human botulism via the inhalation route was 72 hours (21).

With respect to the source of toxin, waterborne delivery is unlikely due to rapid inactivation by standard water treatments and the large innoculum of toxin required. No cases of waterborne botulism have been reported. Botulinum toxin is stable for days in untreated water and beverages, however, which could be suitable vehicles for intentional toxin delivery (7). Alternatively, toxin could be delivered via nature's way in food or food products, making it difficult to distinguish an intentional toxin release from an outbreak of food-borne botulism.

Intentional botulism must be distinguished from naturally occurring botulism and from the many diseases that may mimic botulism. Recognition of botulism depends on astute clinicians who first see intoxicated patients. Unfortunately, naturally occurring botulism is underdiagnosed and frequently misdiagnosed, usually as a polyradiculopathy (Guillain-Barré or Miller-Fisher syndrome), myasthenia gravis, or a disease of the central nervous system (Table 20.2). Common and uncommon misdiagnoses are listed in Table 20.2, along with features that distinguish botulism from these diseases. Botulism differs from other causes of flaccid paralysis in (i) its prominent cranial nerve palsies disproportionate to weakness below the neck, (ii) the symmetry of the weakness, and (iii) the absence of sensory changes, although approximately 14% of patients report paresthesias (Table 20.1).

Both intentional and naturally occurring botulism are much more likely to be associated with outbreaks (cluster of cases) than other diseases with which they may be confused, such as Guillain-Barré, poliomyelitis, or intoxications. Due to the mobility of populations and a potentially long and variable incubation time, these cases may be separated in both time and space. This fact emphasizes the importance of prompt reporting of suspected botulism cases to the public health department by first responders. In early stages, food-borne botulism is most likely confused with intentional botulism; in fact, food could be a viable route for the intentional delivery of toxin. Food-borne botulism usually occurs in outbreaks where multiple individuals ingest contaminated food. From 1899 to 1996, 921 outbreaks of food-borne botulism were reported to the CDC, with a relatively constant incidence of approximately 9.5 outbreaks/year, with an average of 2.5 cases per outbreak (1). The largest number of cases in a single food-borne outbreak was 59. Food-borne botulism is usually associated with ingestion of home canned products, most frequently foods low in acid such as vegetables, fish or marine mammals, condiments, and meat products. Fruits are rarely involved due to their high natural acidity. Outbreaks have also been reported for commercially prepared products (105) and for food prepared in restaurants (106). In the United States the incidence of food-borne botulism is highest in Alaska, where the vehicle is typically native Alaskan foods consisting of fermented or salted fish or marine mammal products.

Food-borne botulism is typically caused by BoNT types A, B, and E. Of the 1087 cases of food-borne botulism reported between 1950 and 1996, the toxin type could be determined for 786; of these, 52% were type A, 22% were

**Table 20.2** Differential Diagnosis of Botulism[a,b]

| Condition | Features That Distinguish Condition from Botulism |
|---|---|
| *Common Misdiagnoses* | |
| Guillain-Barré syndrome[c] and its variants, including Miller-Fisher syndrome | History of antecedent infection; paresthesias[d]; often ascending paralysis; early areflexia; eventual CSF protein increase; EMG findings |
| Myasthenia gravis[c] | Recurrent paralysis; EMG findings; sustained response to anticholinesterase therapy |
| Stroke[c] | Paralysis often asymmetric; abnormal CT or MRI scan |
| Intoxication with depressants (e.g., acute ethanol intoxication), organophosphates, carbon monoxide, or nerve gas | History of exposure; excessive drug levels detected in body fluids |
| Eaton-Lambert syndrome | Increased strength with sustained contraction; evidence of lung carcinoma; EMG findings similar to botulism |
| Tick paralysis | Paresthesias[d]; ascending paralysis; tick attached to skin |
| *Other Misdiagnoses* | |
| Poliomyelitis | Antecedent febrile illness; asymmetric paralysis; CSF pleocytosis |
| CNS infections, especially of the brain stem | Mental status changes; CSF and EEG abnormalities |
| CNS tumor | Paralysis often asymmetric; abnormal CT or MRI scan |
| Streptococcal pharyngitis (pharyngeal erythema can occur in botulism) | Absence of bulbar palsies; positive rapid antigen test result or throat culture |
| Psychiatric illness[c] | Normal EMG in conversion paralysis |
| Viral syndrome[c] | Absence of bulbar palsies and flaccid paralysis |
| Inflammatory myopathy[c] | Elevated creatine kinase levels |
| Diabetic complications[c] | Sensory neuropathy; few cranial nerve palsies |
| Hyperemesis gravidarum[c] | Absence of bulbar palsies and acute flaccid paralysis |
| Hypothyroidism[c] | Abnormal thyroid function test results |
| Laryngeal trauma[c] | Absence of flaccid paralysis; dysphonia without bulbar palsies |
| Overexertion[c] | Absence of bulbar palsies and acute flaccid paralysis |

[a]Common and infrequent misdiagnoses in patients with botulism (adapted from ref. 7).
[b]CSF, cerebrospinal fluid; EMG, electromyogram; CNS, central nervous system; EEG, electroencephalogram; CT, computed tomography; MRI, magnetic resonance imaging.
[c]Misdiagnoses made in a large outbreak of botulism (ref. 104).
[d]Paresthesias are reported in approximately 14% of botulism patients (Table 20.1).

type B, 25% were type E, and less than 1% were type F (1). Type A botulism is most common west of the Mississippi River, type B is most common east of the Mississippi, and type E botulism predominates in Alaska (1). This distribution corresponds to the distribution of *C. botulinum* spores in the soil (107,108). Only three outbreaks of type F botulism have been reported in the United States. These epidemiological findings might be helpful in retrospectively distinguishing intentional from food-borne botulism.

## Clinical and Laboratory Findings

The diagnosis of botulism should be based on history and physical findings, as routine laboratory tests are generally nonspecific and specific confirmation takes days. Routine laboratory tests are not particularly helpful in confirming the clinical suspicion of botulism. The complete blood count, electrolyte panel, renal and liver function tests, urinalysis, and electrocardiogram will all be normal unless complications have occurred. The cerebrospinal fluid (CSF) is typically normal in botulism, whereas the CSF protein is usually elevated in Guillain-Barré. The tensilon test is usually, but not always (104), normal in botulism and can be helpful in distinguishing botulism from myasthenia gravis. The computed tomographic (CT) scan of the head is also normal in botulism and can be used to rule out stroke or other intracranial diseases.

Patients with botulism have normal motor nerve conduction velocities and distal latencies. The electromyogram (EMG), however, may be helpful in the diagnosis of botulism and in distinguishing it from other neuromuscular diseases, such as myasthenia gravis and Guillain-Barré (109,110). The EMG of involved muscle groups reveals decreased amplitude of the muscle action potential and facilitation during rapid repetitive or post-tetanic stimulation, as can also be seen in patients with Eaton-Lambert syndrome.

Specific laboratory confirmation requires the demonstration of toxin in the blood or GI tract and/or isolation of a BoNT-producing *Clostridium* species from the GI tract (111). Currently, such testing is available only at the CDC and approximately 20 state and municipal public health laboratories (1). At the present time, the most sensitive assay for BoNT is the mouse bioassay, which is performed by injecting mice intraperitoneally with the toxin-containing sample (serum, stool, food extract, etc.) with or without polyclonal and type-specific antitoxin. The mice are observed for 4 days for the development of botulism, with mice usually dying from botulism within 6 to 96 hours. Protection by simultaneous administration of antitoxin enables determination of serotype. The mouse bioassay can detect as little as 33 pg of toxin (5), the mouse LD50. *In vitro* tests to detect BoNT, especially mass spectrometry (112,113) and variants of enzyme-linked immunosorbent assays (ELISA), are under development (114–118). Mass spectrometry methods use antibody-coupled beads to capture toxin from complex matrices such as milk, blood, or stool and then measure the substrate-specific endopeptidase activity of the toxin. Assays have been developed for all seven serotypes and can achieve sensitivities less than the mouse bioassay (119,120).

Unfortunately, current laboratory tests may not be particularly sensitive for the detection of botulism. Clostridial cultures were positive for 51% of stool specimens collected from 309 patients with clinically suspected botulism (98). Toxin testing was positive in only 37% of sera and 23% of stool specimens. At least one laboratory test was positive in 65% of patients (98). Collecting samples early in the course of disease increases the likelihood of positive results. However, large outbreaks have occurred in which no specimens, or a low percentage of specimens, gave positive results (121). It can also take days for cultures or toxin testing results to be available. It may be possible to culture *Clostridium* spp. or detect toxin in source material, if available. Because toxin prepared by terrorists is likely to be crude and unpurified, it is possible that nucleic acids of *C. botulinum* may be present, which could be amplified by polymerase chain reaction for analysis. Cultures and nucleic acid amplification assays could allow more specific classification of the precise strain of *Clostridium botulinum* used (47,122). Because specific therapy with antitoxin must be administered as rapidly as possible to be effective (see later) (123), specific toxin therapy must be based on clinical diagnosis prior to laboratory confirmation.

## Treatment

Treatment of botulism includes (i) early administration of botulinum antitoxin to prevent progression of moderate illness or reduce the duration of mechanical ventilation in patients with rapidly progressive severe botulism, (ii) close monitoring of respiratory function [vital capacity and maximal inspiratory force (MIF)], and (iii) intensive care for patients with significant paralysis and evidence of respiratory insufficiency or failure. Vital capacity (VC) should be measured as soon as the diagnosis is suspected and followed closely. In one study, 10 of 11 patients requiring mechanical ventilation had vital capacities less than 30% of predicted (124). Patients with VC less than 10 cc/kg should be monitored in an ICU for progression of respiratory failure requiring mechanical ventilation (125). Patients requiring mechanical ventilation obviously merit ICU care.

Treatment with antitoxin is the mainstay of therapy. In the United States, more than 80% of adults are treated with antitoxin. Antitoxin is most effective when administered early in the course of disease (123). Antitoxin only works on circulating toxin, as once the toxin enters the nerve terminal, antitoxin cannot bind (9). Antitoxin is an immunoglobulin preparation obtained from hyperimmunized horses (horse or equine antitoxin) from which the Fc portion has been removed enzymatically (despeciation) to reduce the incidence of side effects such as serum sickness and hypersensitivity reactions. The current licensed equine antitoxin is bivalent, having activity against serotypes A and B. For other toxin serotypes, there is an investigational equine E antitoxin for civilian use and an investigational heptavalent (A–G) antitoxin (HBAT, Cangene) under development and available from the CDC. The CDC should be contacted for information regarding these products. Efforts are under way to generate human monoclonal antibodies that could replace equine antitoxin therapy (126–129).

Despite despeciation, there is a significant (9%) risk of hypersensitivity reactions when administering equine antitoxin, including anaphylactic shock (130). Therefore, it is important that skin testing be performed prior to systemic administration according to the protocol provided on the package insert. The amount of antitoxin in one 10-ml vial of antitoxin administered intravenously is enough to neutralize toxin amounts many times in excess of those observed in patients with botulism. The CDC currently recommends a single 10-ml dose of antitoxin, unlike what is suggested in the package insert, due to the 5- to 8-day reported half-life of the antitoxin (9). Antitoxin can be obtained from the CDC by contacting the local health department and is diluted 1:10 in 0.9% saline and administered slowly intravenously. If the local health department is unavailable, the CDC can be contacted directly at 770-488-7100. Clinicians should review the package insert with public health authorities before using antitotxin.

Equine antitoxin has been administered rarely to infants with botulism because of the risk of lifelong hypersensitivity to equine antigens (131). In addition, some evidence suggests that anaphylaxis may be more severe in infants given equine antitoxin. As an alternative, S. Arnon and colleagues at the Infant Botulism Treatment and Prevention Program at the California Department of Health Services (CDHS) developed human immunoglobulin prepared from volunteers immunized with an investigational botulinum toxoid vaccine. This product, termed "botulism immune globulin" (BIG-IV), has been evaluated in a prospective randomized trial in infant botulism. Infants with a clinical diagnosis of botulism were randomized to receive either nonimmune human globulin or BIG-IV. Compared to nonimmune globulin, BIG-IV reduced the duration and cost of hospitalization significantly and the duration of mechanical ventilation and tube feedings (102,132). This benefit appeared to accrue even in patients treated as late as 5 to 7 days after the onset of symptoms, probably as a result of ongoing toxin production, and/or slow clearance of toxin from the blood (S. Arnon, personal communication). BIG-IV has been approved by the FDA for treatment of infant botulism and is available from the CDHS at 510-540-2646.

Botulism is a reportable disease and suspected cases should be reported immediately to the hospital epidemiologist or infection control practitioner, as well as the local and state health departments. The phone number of the health department can usually be found in the phone directory under government listings or via the Internet at http://www.cdc.gov/other.htm#states. If local or state health departments are not reachable, the CDC can be contacted directly at 770-488-7100. For laboratory workers who might be exposed to large amounts of botulinum toxin, an investigational pentavalent (serotypes A, B, C, D, and E) toxoid can be obtained from the CDC (133). Further details can be obtained from the National Botulism Surveillance and Reference Laboratory at 404-639-2206. A recombinant vaccine based on the toxin-binding domain for many of the serotypes is also under development (134).

# RICIN

Ricin is a plant carbohydrate-binding protein (lectin) found in high concentration in castor beans. It is also cytotoxic, with a typical human lethal dose of approximately 5 μg/kg (135). While much less toxic than botulinum toxin or SEB, ricin can be prepared in liquid or crystalline form in large quantities from castor beans with minimal sophistication or technical capacity. Access to the starting material is simple, as more than one million tons of castor beans are processed annually to produce castor oil; the waste mash contains 5% ricin by weight. Ricin is active orally or upon inhalation and thus could be used to poison food or aerosolized. Its relatively low potency makes it less likely for contaminating water supplies or large volumes of beverages. Ricin has already achieved some notoriety as a poison, being administered via an umbrella tip for a political assassination (136). A recent find of ricin and castor bean extraction equipment during a police raid of a flat in the United Kingdom indicates that this toxin has drawn the attention of terrorists (137).

## Ricin Structure and Action

Ricin consists of two polypeptide chains connected by a single disulfide bond [reviewed in Olsnes and Kozlov (138)]. The B chain of ricin folds into two globular domains (139), each of which contains a carbohydrate-binding site for β-d-galactopyranoside moieties. These carbohydrates can be present in millions of copies per cell. After binding of ricin to carbohydrate receptors on the cell surface, the toxin is endocytosed and then is translocated to the cytosol. In the cytosol, the A chain inactivates ribosomes by catalyzing removal of an adenine residue from a loop of the 28S ribosome, leading to rapid RNA hydrolysis (140). This results in failure of protein synthesis.

## Clinical Signs and Symptoms

Clinical findings of ricin intoxication vary depending on the route of exposure and are largely based on animal studies (141). After inhalation exposure, there is an incubation period of approximately 8 hours in experimental animals and 4 to 8 hours in humans after an accidental exposure. In human exposure, findings included fever, chest tightness, cough, dyspnea, nausea, and arthralgias. In animals, and presumably in humans, larger doses lead to increased permeability pulmonary edema, respiratory failure, shock, and death within 36–72 hours. Inhalational ricin poisoning could mimic a large range of diseases causing acute pulmonary disease, including pneumonias and the adult respiratory distress syndrome. By the oral route, necrosis of the gastrointestinal tract will occur, with gastrointestinal, splenic, hepatic, and renal bleeding. The key to diagnosis, as with most biothreat agents, is a high index of suspicion and recognition of a cluster of clinically related cases. Routine laboratory tests will show only nonspecific findings. A specific serum ELISA has been described, and acute and convalescent sera can be collected for detection of ricin-specific antibodies. Therapy is supportive, frequently requiring hospitalization in an intensive care unit. No specific approved antitoxin or vaccine exists. Experimental vaccines are under development (142–144).

# STAPHYLOCOCCAL ENTEROTOXIN B

Staphylococcal enterotoxins are a superfamily of proteins secreted by *Stapylococcus aureus* consisting of at least 11 members (145). Members of the family cause toxic shock syndrome and scalded skin syndrome, while staphylococcal enterotoxin is the most common cause of food poisoning in the United States. While SEB ingested orally can cause severe gastrointestinal symptoms, the low fatality rate from food poisoning suggests that aerosol release would be a greater biothreat than food or water contamination.

## Staphylococcal Enterotoxin B Structure and Action

The structure of SEB has been determined at 1.5 Å resolution (146). The protein has two binding sites, a high-affinity site for major histocompatibility II (MHC II) molecules and a low-affinity site for the Vβ T-cell receptor. As a result, SEB is capable of cross-linking up to 20% of T cells, leading to massive activation of a proinflammatory response and release of proinflammatory cytokines. The symptoms that result reflect this broad activation of the immune system. As little as 250 to 400 ng of SEB can induce symptoms (147).

## Clinical Signs and Symptoms

There is a 1- to 4-hour incubation period after oral ingestion (148). Usual symptoms include nausea, vomiting, abdominal cramping, and diarrhea. Symptoms typically last up to 20 hours. With severe intoxication, there can be profound dehydration due to loss of fluids, shock, respiratory failure, and cardiovascular collapse. Approximately 15% of patients require hospitalization, with a 5% fatality rate, usually in the very young or the very old. Inhalational SEB exposure would present with different symptoms than seen with oral ingestion (food poisoning). Cytokine activation in the lungs would lead to low-pressure pulmonary edema and acute respiratory failure. The key to diagnosis, as with other biothreat agents, is a high index of suspicion and recognition of a cluster of clinically related cases. Intentional oral release of SEB would have to be distinguished from a naturally occurring outbreak of food poisoning. Routine laboratory tests will show only nonspecific findings. A specific serum ELISA has been described that could be performed on food or environmental specimens or serum and acute and convalescent sera can be collected for detection of SEB-specific antibodies. Therapy is supportive and no specific approved antitoxin or vaccine exists; however, experimental vaccines are under development (149,150).

## *CLOSTRIDIUM PERFRINGENS* EPSILON TOXIN

*Clostridium perfringens* epsilon toxin is a potential biothreat agent; no reported cases of human intoxication have occurred. The toxin is produced by *C. perfringens* as a 311-amino acid protoxin that is cleaved into a 14-amino acid peptide, which is a potent necrotizing toxin. The toxin causes a rapidly fatal toxemia in herbivores when their gastrointestinal tracts are colonized by *C. perfringens*, leading to *in situ* toxin production (151). The toxin causes pulmonary edema, renal failure, and cardiovascular collapse. The lethal dose for rodents is 100 ng/kg, and it has been estimated that a lethal human dose would be 7 μg parenterally (152). It is thought that use of this toxin for nefarious purposes would be via the aerosol/inhalational route, leading to pulmonary edema followed by renal failure and cardiovascular collapse.

## SUMMARY

In summary, this chapter described the four major toxins—BoNT, ricin, staphylococcal enterotoxin B, and *C. perfringens* epsilon toxin—considered to be biothreat agents. The first indication of an event involving one of these toxins is likely to be the appearance of intoxicated patients at local hospitals. Rapid appreciation that an event has occurred is dependent on astute clinicians and a reporting system that can recognize the clustering of cases with a common clinical presentation consistent with intoxication. It is important to recognize that the toxins are all proteins composed of amino acid building blocks. As such, they have a number of forensic features that distinguish them from viral or bacterial threat agents. First, they are not contagious, as the threat agent is not a living organism. For the same reason, these agents cannot be cultured routinely from either patients or the environment after exposure, making forensic detection more difficult. Because proteins are composed of amino acids and not nucleic acid, it is also not possible to amplify and detect the presence of toxins using the polymerase chain reaction or by any type of classic DNA hybridization technology. Rather, detection typically relies on the use of antibodies and serological testing. Environmental and patient sampling, followed by serological testing, is likely to provide the greatest amount of forensic information.

## ACKNOWLEDGMENT

This work was partially supported by Defense Threat Reduction Agency Contract HDTRA1-07-C-0030, National Institute of Allergy and Infectious Disease Grant AI075443, and Centers for Disease Control and Prevention Contract 200-2009-30597.

## REFERENCES

[1] Centers for Disease Control and Prevention. Botulism in the United States, 1899–1998. Handbook for epidemiologists, clinicians, and laboratory workers. Atlanta, Georgia, 1998; U.S. Department of Health and Human Services, Public Health Service. Available from: http://www.bt.cdc.gov/agent/ botulism/index.asp.

[2] M.D. Gill, Bacterial toxins: A table of lethal amounts, Microbiol. Rev. 46 (1982) 86–94.

[3] B.A. Herrero, A.E. Ecklung, C.S. Streett, D.F. Ford, J.K. King, Experimental botulism in monkeys: A clinical pathological study, Exp. Mol. Pathol. 6 (1) (1967) 84–95.

[4] D.R. Franz, L.M. Pitt, M.A. Clayton, M.A. Hanes, K.J. Rose, Efficacy of prophylactic and therapeutic administration of antitoxin for inhalation botulism, in: B.R. DasGupta (Ed.), Botulinum and Tetanus Neurotoxins: Neurotransmission and Biomedical Aspects, Plenum Press, New York, 1993, pp. 473–476.

[5] E.J. Schantz, E.A. Johnson, Properties and use of botulinum toxin and other microbial neurotoxins in medicine, Microbiol. Rev. 56 (1) (1992) 80–99.

[6] A.B. Scott, D. Suzuki, Systemic toxicity of botulinum toxin by intramuscular injection in the monkey, Mov. Disord. 3 (4) (1988) 333–335.

[7] S.A. Arnon, R. Schecter, T.V. Inglesby, D.A. Henderson, J.G. Bartlett, M.S. Ascher, et al., Botulinum toxin as a biological weapon, JAMA 285 (2001) 1059–1070.

[8] N. Mahant, P.D. Clouston, I.T. Lorentz, The current use of botulinum toxin, J. Clin. Neurosci. 7 (2000) 389–394.

[9] H. Sugiyama, Clostridium botulinum neurotoxin, Microbiol. Rev. 44 (3) (1980) 419–448.

[10] D.B. Lacy, R.C. Stevens, Sequence homology and structural analysis of the Clostridial neurotoxins, J. Mol. Biol. 291 (1999) 1091–1104.

[11] J.L. Middlebrook, D.R. Franz. Botulinum toxin, in: F.R. Sidell, E.T. Takafuji, D.R. Franz (Ed.), Medical aspects of chemical and biologic warfare, Office of the Surgeon General. 1997, pp. 643–654.

[12] J.B. Park, L.L. Simpson, Inhalational poisoning by botulinum toxin and inhalation vaccination with its heavy-chain component, Infect. Immun. 71 (3) (2003) 1147–1154.

[13] K.F. Meyer, The rise and fall of botulism, Calif. Med. 118 (1973) 63–64.

[14] T.F. Midura, S.S. Arnon, Infant botulism. Identification of Clostridium botulinum and its toxins in faeces, Lancet 2 (7992) (1976) 934–936.

[15] J. Pickett, B. Berg, E. Chaplin, M.A. Brunstetter-Shafer, Syndrome of botulism in infancy: Clinical and electrophysiologic study, N. Engl. J. Med. 295 (14) (1976) 770–772.

[16] J.B. Davis, L.H. Mattman, M. Wiley, Clostridium botulinum in a fatal wound infection, JAMA 146 (1951) 934–936.

[17] S.B. Werner, D. Passaro, J. McGee, R. Schechter, D.J. Vugia, Wound botulism in California, 1951–1998: Recent epidemic in heroin injectors, Clin. Infect. Dis. 31 (4) (2000) 1018–1024.

[18] Centers for Disease Control and Prevention, Botulism-United States, MMWR Morb. Mortal. Wk. Rep. 28 (1978) 73–75.

[19] J.K. Chia, J.B. Clark, C.A. Ryan, M. Pollack, Botulism in an adult associated with food-borne intestinal infection with Clostridium botulinum, N. Engl. J. Med. 315 (4) (1986) 239–241.

[20] P.M. Griffin, C.L. Hatheway, R.B. Rosenbaum, R. Sokolow, Endogenous antibody production to botulinum toxin in an adult with intestinal colonization botulism and underlying Crohn's disease, J. Infect. Dis. 175 (3) (1997) 633–637.

[21] E. Holzer, Botulismus durch inhalation, Med. Klin. 41 (1962) 1735–1740.

[22] D.B. Cobb, W.A. Watson, M.C. Fernandez, Botulism-like syndrome after injection of botulinum toxin, Vet. Hum. Toxicol. 42 (2000) 163.

[23] B.E. Crowner, J.E. Brunstrom, B.A. Racette, Iatrogenic botulism due to therapeutic botulinum toxin a injection in a pediatric patient, Clin. Neuropharmacol. 30 (2007) 310–313.

[24] D. Kennedy, Beauty and the beast, Science 295 (2002) 1601.

[25] United Nations Security Council. Tenth report of the executive committee of the special commission established by the secretary-general pursuant to paragraph 9(b)(I) of security council resolution 687 (1991), and paragraph 3 of resolution 699 (1991) on the activities of the Special Commission. United Nations Security Council 1995.

[26] G. Bozheyeva, Y. Kunakbayev, D. Yeleukenov, Former Soviet biological weapons facilities in Kazakhstan: Past, present, and future. Center for Nonproliferation Studies, Monterey Institute of International Studies 1999.

[27] R.A. Zilinskas, Iraq's biological weapons: The past as future? JAMA 278 (1997) 418–424.

[28] R. St. John, B. Finlay, C. Blair, Bioterrorism in Canada: An economic assessment of prevention and postattack exposure, Can. J. Infect. Dis. 12 (2001) 275–284.

[29] C.L. Hatheway, Toxigenic clostridia, Clin. Microbiol. Rev. 3 (1) (1990) 66–98.

[30] P. Aureli, L. Fenicia, B. Pasolini, M. Gianfranceschi, L.M. McCroskey, C.L. Hatheway, Two cases of type E infant botulism caused by neurotoxigenic *Clostridium butyricum* in Italy, J. Infect. Dis. 154 (2) (1986) 207–211.

[31] J.D. Hall, L.M. McCroskey, B.J. Pincomb, C.L. Hatheway, Isolation of an organism resembling *Clostridium barati* which produces type F botulinal toxin from an infant with botulism, J. Clin. Microbiol. 21 (4) (1985) 654–655.

[32] E.D. Borland, C.J. Moryson, G.R. Smith, Avian botulism and the high prevalence of *Clostridium botulinum* in the Norflok Broads, Vet. Rec. 100 (6) (1977) 106–109.

[33] J.W. Galvin, T.J. Hollier, K.D. Bodinnar, C.M. Bunn, An outbreak of botulism in wild waterbirds in southern Australia, J. Wildl. Dis. 21 (4) (1985) 347–350.

[34] G.R. Smith, R.A. Milligan, *Clostridium botulinum* type D in Britain, Vet. Rec. 100 (6) (1977) 121–122.

[35] J. Demarchi, C. Mourgues, J. Orio, A.R. Prevot, Existence du botulisme de type D, Bull. Acad. Nat. Med. 142 (1958) 580–582.

[36] A.R. Prevot, J. Terrasse, J. Daumail, M. Cavaroc, J. Riol, R. Sillioc, Existence en France du botulisme humain de type C, Bull. Acad. Med. (Paris) 139 (1955) 355–358.

[37] D.F. Gimenez, A.S. Ciccarelli, Studies on strain 84 of *Clostridium botulinum*, Zentralbl. Bakteriol. 215 (2) (1970) 212–220.

[38] O. Sonnabend, W. Sonnabend, R. Heinzle, T. Sigrist, R. Dirnhofer, U. Krech, Isolation of Clostridium botulinum type G and identification of type G botulinal toxin in humans: Report of five sudden unexpected deaths, J. Infect. Dis. 143 (1) (1981) 22–27.

[39] J.C. Suen, C.L. Hatheway, A.G. Steigerwalt, D.J. Brenner, Clostridium argentinense, sp. nov: a genetically homogeneous group composed of all strains of *Clostridium botulinum* toxin type G and some nontoxigenic strains previously identified as clostridium subterminale or *Clostridium hastiforme*, Int. J. Syst. Bacteriol. 38 (1988) 375–381.

[40] C.L. Hatheway, L.M. McCroskey, G.L. Lombard, V.R. Dowell Jr., Atypical toxin variant of *Clostridium botulinum* type B associated with infant botulism, J. Clin. Microbiol. 14 (6) (1981) 607–611.

[41] C.L. Hatheway, L.M. McCroskey, Examination of feces and serum for diagnosis of infant botulism in 336 patients, J. Clin. Microbiol. 25 (12) (1987) 2334–2338.

[42] D.F. Gimenez, *Clostridium botulinum* Ba, Zentralbl. Bakteriol. Hyg. 257 (1984) 68–72.

[43] C.L. Hatheway, Botulism: The present status of the disease, Curr. Top. Microbiol. Immunol. 195 (1995) 55–75.

[44] J.A. Santos-Buelga, M.D. Collins, A.K. East, Characterization of the genes encoding the botulinum neurotoxin complex in a strain of *Clostridium botulinum* producing type B and F neurotoxins, Curr. Microbiol. 37 (5) (1998) 312–318.

[45] M. Sebaihia, M.W. Peck, N.P. Minton, N.R. Thomson, M.T. Holden, W.J. Mitchell, et al., Genome sequence of a proteolytic (Group I) *Clostridium botulinum* strain Hall A and comparative analysis of the clostridial genomes, Genome Res. 17 (7) (2007) 1082–1092.

[46] K.K. Hill, G. Xie, B.T. Foley, T.J. Smith, A.C. Munk, D. Bruce, et al., Recombination and insertion events involving the botulinum neurotoxin complex genes in *Clostridium botulinum* types A, B, E and F and *Clostridium butyricum* type E strains, BMC Biol. 7 (2009) 66.

[47] K.K. Hill, T.J. Smith, C.H. Helma, L.O. Ticknor, B.T. Foley, R.T. Svensson, et al., Genetic diversity among Botulinum neurotoxin-producing clostridial strains, J. Bacteriol. 189 (3) (2007) 818–832.

[48] T.J. Smith, K.K. Hill, B.T. Foley, J.C. Detter, A.C. Munk, D.C. Bruce, et al., Analysis of the neurotoxin complex genes in *Clostridium botulinum* A1-A4 and B1 strains: BoNT/A3, /Ba4 and /B1 clusters are located within plasmids, PLoS One 2 (12) (2007) e1271.

[49] B.R. DasGupta, D.A. Boroff, Separation of toxin and hemagglutinin from crystalline toxin of Clostridium botulinum type A by anion exchange chromatography and determination of their dimensions by gel filtration, J. Biol. Chem. 243 (5) (1968) 1065–1072.

[50] P. Balding, E.R. Gold, D.A. Boroff, T.A. Roberts, Observations on receptor specific proteins. II. Haemagglutination and haemagglutination-inhibition reactions of *Clostridium botulinum* types A, C, D and E haemagglutinins, Immunology 25 (5) (1973) 773–782.

[51] M. Sebaihia, M.W. Peck, N.P. Minton, N.R. Thomson, M.T. Holden, W.J. Mitchell, et al., Genome sequence of a proteolytic (Group I) Clostridium botulinum strain Hall A and comparative analysis of the clostridial genomes. Genome Res. 17(7) (2007) 1082–1092.

[52] R. Fujita, Y. Fujinaga, K. Inoue, H. Nakajima, H. Kumon, K. Oguma, Molecular characterization of two forms of nontoxic-nonhemagglutinin components of *Clostridium botulinum* type A progenitor toxins, FEBS Lett. 376 (1–2) (1995) 41–44.

[53] A.K. East, M. Bhandari, J.M. Stacey, K.D. Campbell, M.D. Collins, Organization and phylogenetic interrelationships of genes encoding components of the botulinum toxin complex in proteolytic *Clostridium botulinum* types A, B, and F: Evidence of chimeric sequences in the gene encoding the nontoxic nonhemagglutinin component, Int. J. Syst. Bacteriol. 46 (4) (1996) 1105–1112.

[54] I. Ohishi, S. Sugii, G. Sakaguchi, Oral toxicities of *Clostridium botulinum* toxins in response to molecular size, Infect. Immun. 16 (1) (1977) 107–109.

[55] F. Chen, G.M. Kuziemko, R.C. Stevens, Biophysical characterization of the stability of the 150-kilodalton botulinum toxin, the nontoxic component, and the 900-kilodalton botulinum toxin complex species, Infect. Immun. 66 (6) (1998) 2420–2425.

[56] A. Willems, A.K. East, P.A. Lawson, M.D. Collins, Sequence of the gene coding for the neurotoxin of *Clostridium botulinum* type A associated with infant botulism: Comparison with other clostridial neurotoxins, Res. Microbiol. 144 (7) (1993) 547–556.

[57] D.E. Thompson, J.K. Brehm, J.D. Oultram, T.J. Swinfield, C.C. Shone, T. Atkinson, et al., The complete amino acid sequence of the *Clostridium botulinum* type A neurotoxin, deduced by nucleotide sequence analysis of the encoding gene, Eur. J. Biochem. 189 (1) (1990) 73–81.

[58] T. Binz, H. Kurazono, M. Wille, J. Frevert, K. Wernars, H. Niemann, The complete sequence of botulinum neurotoxin type A and comparison with other clostridial neurotoxins, J. Biol. Chem. 265 (16) (1990) 9153–9158.

[59] S.M. Whelan, M.J. Elmore, N.J. Bodsworth, J.K. Brehm, T. Atkinson, N.P. Minton, Molecular cloning of the *Clostridium botulinum* structural gene encoding the type B neurotoxin and determination of its entire nucleotide sequence, Appl. Environ. Microbiol. 58 (8) (1992) 2345–2354.

[60] H. Ihara, T. Kohda, F. Morimoto, K. Tsukamoto, T. Karasawa, S. Nakamura, et al., Sequence of the gene for *Clostridium botulinum* type B neurotoxin associated with infant botulism, expression of the C-terminal half of heavy chain and its binding activity, Biochim. Biophys. Acta 1625 (1) (2003) 19–26.

[61] D. Hauser, M.W. Eklund, H. Kurazono, T. Binz, H. Niemann, D.M. Gill, et al., Nucleotide sequence of *Clostridium botulinum* C1 neurotoxin, Nucleic Acids Res. 18 (16) (1990) 4924.

[62] T. Binz, H. Kurazono, M.R. Popoff, M.W. Eklund, G. Sakaguchi, S. Kozaki, et al., Nucleotide sequence of the gene encoding *Clostridium botulinum* neurotoxin type D, Nucleic Acids Res. 18 (18) (1990) 5556.

[63] S.M. Whelan, M.J. Elmore, N.J. Bodsworth, T. Atkinson, N.P. Minton, The complete amino acid sequence of the *Clostridium botulinum* type-E neurotoxin, derived by nucleotide-sequence analysis of the encoding gene, Eur. J. Biochem. 204 (2) (1992) 657–667.

[64] D.E. Thompson, R.A. Hutson, A.K. East, D. Allaway, M.D. Collins, P.T. Richardson, Nucleotide sequence of the gene coding for *Clostridium barati* type F neurotoxin: Comparison with other clostridial neurotoxins, FEMS Microbiol. Lett. 108 (2) (1993) 175–182.

[65] K. Campbell, M.D. Collins, A.K. East, Nucleotide sequence of the gene coding for *Clostridium botulinum* (*Clostridium argentinense*) type G neurotoxin: Genealogical comparison with other clostridial neurotoxins, Biochim. Biophys. Acta 1216 (3) (1993) 487–491.

[66] A.T. Carter, D.R. Mason, K.A. Grant, G. Franciosa, P. Aureli, M.W. Peck, Further characterization of proteolytic *Clostridium botulinum* type A5 reveals that neurotoxin formation is unaffected by loss of the cntR (botR) promoter sigma factor binding site, J. Clin. Microbiol. 48 (3) (2010) 1012–1013.

[67] Y. Chen, H. Korkeala, J. Aarnikunnas, M. Lindstrom, Sequencing the botulinum neurotoxin gene and related genes in *Clostridium botulinum* type E strains reveals orfx3 and a novel type E neurotoxin subtype, J. Bacteriol. 189 (23) (2007) 8643–8650.

[68] N. Dover, J.R. Barash, S.S. Arnon, Novel *Clostridium botulinum* toxin gene arrangement with subtype A5 and partial subtype B3 botulinum neurotoxin genes, J. Clin. Microbiol. 47 (7) (2009) 2349–2350.

[69] T.J. Smith, J. Lou, I.N. Geren, C.M. Forsyth, R. Tsai, S.L. Laporte, et al., Sequence variation within botulinum neurotoxin serotypes impacts antibody binding and neutralization, Infect. Immun. 73 (9) (2005) 5450–5457.

[70] R.M. Curran, E. Fringuelli, D. Graham, C.T. Elliott, Production of serotype C specific and serotype C/D generic monoclonal antibodies using recombinant H(C) and H(N) fragments from *Clostridium botulinum* neurotoxin types C(1) and D, Vet. Immunol. Immunopathol. 130 (1–2) (2009) 1–10.

[71] A.M. Gibson, N.K. Modi, T.A. Roberts, C.C. Shone, P. Hambleton, J. Melling, Evaluation of a monoclonal antibody-based immunoassay for detecting type A *Clostridium botulinum* toxin produced in pure culture and an inoculated model cured meat system, J. Appl. Bacteriol. 63 (3) (1987) 217–226.

[72] C. Garcia-Rodriguez, R. Levy, J.W. Arndt, C.M. Forsyth, A. Razai, J. Lou, et al., Molecular evolution of antibody cross-reactivity for two subtypes of type A botulinum neurotoxin, Nat. Biotechnol. 25 (1) (2007) 107–116.

[73] D.B. Lacy, W. Tepp, A.C. Cohen, B.R. DasGupta, R.C. Stevens, Crystal structure of botulinum neurotoxin type A and implications for toxicity, Nat. Struct. Biol. 5 (1998) 898–902.

[74] S. Eswaramoorthy, D. Kumaran, S. Swaminathan, Crystallographic evidence for doxorubicin binding to the receptor-binding site in *Clostridium botulinum* neurotoxin B, Acta Crystallogr. D Biol. Crystallogr. 57 (Pt 11) (2001) 1743–1746.

[75] D. Kumaran, S. Eswaramoorthy, W. Furey, J. Navaza, M. Sax, S. Swaminathan, Domain organization in *Clostridium botulinum* neurotoxin type E is unique: Its implication in faster translocation, J. Mol. Biol. 386 (1) (2009) 233–245.

[76] J.O. Dolly, J. Black, R.S. Williams, J. Melling, Acceptors for botulinum neurotoxin reside on motor nerve terminals and mediate its internalization, Nature 307 (1984) 457–460.

[77] C. Montecucco, How do tetanus and botulinum toxins bind to neuronal membranes? Trends Biochem. Sci. 11 (1986) 315–317.

[78] L.L. Simpson, Kinetic studies on the interaction between botulinum toxin type A and the chloinergic neuromuscular junction, J. Pharmacol. Exp. Ther. 212 (1980) 16–21.

[79] M. Dong, F. Yeh, W.H. Tepp, C. Dean, E.A. Johnson, R. Janz, et al., SV2 is the protein receptor for botulinum neurotoxin A, Science 312 (5773) (2006) 592–596.

[80] S. Mahrhold, A. Rummel, H. Bigalke, B. Davletov, T. Binz, The synaptic vesicle protein 2C mediates the uptake of botulinum neurotoxin A into phrenic nerves, FEBS Lett. 580 (8) (2006) 2011–2014.

[81] Q. Chai, J.W. Arndt, M. Dong, W.H. Tepp, E.A. Johnson, E.R. Chapman, et al., Structural basis of cell surface receptor recognition by botulinum neurotoxin B, Nature 444 (7122) (2006) 1096–1100.

[82] R. Jin, A. Rummel, T. Binz, A.T. Brunger, Botulinum neurotoxin B recognizes its protein receptor with high affinity and specificity, Nature 444 (7122) (2006) 1092–1095.

[83] A. Rummel, T. Eichner, T. Weil, T. Karnath, A. Gutcaits, S. Mahrhold, et al., Identification of the protein receptor binding site of botulinum neurotoxins B and G proves the double-receptor concept, Proc. Natl. Acad. Sci. USA 104 (1) (2007) 359–364.

[84] G. Schiavo, F. Benfenati, B. Poulain, O. Rossetto, P.P. DeLaureto, B.R. DasGupta, et al., Tetanus and botulinum-B neurotoxins block neurotransmitter release by proteolytic cleavage of synaptobrevin, Nature 359 (1992) 832–835.

[85] G. Schiavo, O. Rossetto, S. Catsicas, P. Polverino de Laureto, B.R. DasGupta, F. Benfenati, et al., Identification of the nerve terminal targets of botulinum neurotoxin serotypes A, D, and E, J. Biol. Chem. 268 (1993) 23784–23787.

[86] G. Schiavo, A. Santucci, B.R. Dasgupta, P.P. Mehta, J. Jontes, F. Benfenati, et al., Botulinum neurotoxins serotypes A and E cleave SNAP-25 at distinct COOH-terminal peptide bonds, FEBS Lett. 335 (1) (1993) 99–103.

[87] J. Blasi, E.R. Chapman, E. Link, T. Binz, S. Yamasaki, P. De Camilli, et al., Botulinum neurotoxin A selectively cleaves the synaptic protein SNAP-25, Nature 365 (6442) (1993) 160–163.

[88] G. Schiavo, C.C. Shone, O. Rossetto, F.C. Alexander, C. Montecucco, Botulinum neurotoxin serotype F is a zinc endopeptidase specific for VAMP/synaptobrevin, J. Biol. Chem. 268 (16) (1993) 11516–11519.

[89] G. Schiavo, C. Malizio, W.S. Trimble, P. Polverino de Laureto, G. Milan, H. Sugiyama, et al., Botulinum G neurotoxin cleaves VAMP/synaptobrevin at a single Ala-Ala peptide bond, J. Biol. Chem. 269 (32) (1994) 20213–20216.

[90] S. Yamasaki, Y. Hu, T. Binz, A. Kalkuhl, H. Kurazono, T. Tamura, et al., Synaptobrevin/vesicle-associated membrane protein (VAMP) of Aplysia californica: structure and proteolysis

by tetanus toxin and botulinal neurotoxins type D and F, Proc. Natl. Acad. Sci. USA 91 (11) (1994) 4688–4692.

[91] S. Yamasaki, T. Binz, T. Hayashi, E. Szabo, N. Yamasaki, M. Eklund, et al., Botulinum neurotoxin type G proteolyses the Ala81-Ala82 bond of rat synaptobrevin 2, Biochem. Biophys. Res. Commun. 200 (2) (1994) 829–835.

[92] J. Blasi, E.R. Chapman, S. Yamasaki, T. Binz, H. Niemann, R. Jahn, Botulinum neurotoxin C1 blocks neurotransmitter release by means of cleaving HPC-1/syntaxin, EMBO J. 12 (12) (1993) 4821–4828.

[93] P. Foran, G.W. Lawrence, C.C. Shone, K.A. Foster, J.O. Dolly, Botulinum neurotoxin C1 cleaves both syntaxin and SNAP-25 in intact and permeabilized chromaffin cells: Correlation with its blockade of catecholamine release, Biochemistry 35 (8) (1996) 2630–2636.

[94] R.B. Sutton, D. Fasshauer, R. Jahn, A.T. Brunger, Crystal structure of a SNARE complex involved in synaptic exocytosis at 2.4 A resolution, Nature 395 (6700) (1998) 347–353.

[95] M.H. Merson, V.R. Dowell Jr., Epidemiologic, clinical and laboratory aspects of wound botulism, N. Engl. J. Med. 289 (19) (1973) 1105–1110.

[96] R. Wilson, J.G. Morris Jr, J.D. Snyder, R.A. Feldman, Clinical characteristics of infant botulism in the United States: A study of the non-California cases, Pediatr. Infect. Dis. 1 (3) (1982) 148–150.

[97] J.M. Hughes, J.R. Blumenthal, M.H. Merson, G.L. Lombard, V.R. Dowell Jr., E.J. Gangarosa, Clinical features of types A and B food-borne botulism, Ann. Intern. Med. 95 (4) (1981) 442–445.

[98] B.A. Woodruff, P.M. Griffin, L.M. McCroskey, J.F. Smart, R.B. Wainwright, R.G. Bryant, et al., Clinical and laboratory comparison of botulism from toxin types A, B, and E in the United States, 1975–1988, J. Infect. Dis. 166 (6) (1992) 1281–1286.

[99] J.C. Maroon, Late effects of botulinum intoxication. JAMA 238 (2) (1977) 129.

[100] J.M. Mann, S. Martin, R. Hoffman, S. Marrazzo, Patient recovery from type A botulism: Morbidity assessment following a large outbreak, Am. J. Public Health 71 (3) (1981) 266–269.

[101] H. Ehrenreich, C.G. Garner, T.N. Witt, Complete bilateral internal ophthalmoplegia as sole clinical sign of botulism: Confirmation of diagnosis by single fibre electromyography, J. Neurol. 236 (4) (1989) 243–245.

[102] S.S. Arnon, Clinical trial of human botulism immune globulin, in: B.R. DasGupta (Ed.), Botulinum and Tetanus Neurotoxins: Neurotransmission and Biomedical Aspects, Plenum Press, New York, 1993, pp. 477–482.

[103] A. de Paiva, F.A. Meunier, J. Molgo, K.R. Aoki, J.O. Dolly, Functional repair of motor end-plates after botulinum neurotoxin type A poisoning: Biphasic switch of synaptic activity between nerve sprouts and their parent terminals, Proc. Natl. Acad. Sci. USA 96 (6) (1999) 3200–3205.

[104] M.E. St. Louis, S.H. Peck, D. Bowering, G.B. Morgan, J. Blatherwick, S. Banerjee, et al., Botulism from chopped garlic: Delayed recognition of a major outbreak, Ann. Intern. Med. 108 (3) (1988) 363–368.

[105] P.A. Blake, M.A. Horwitz, L. Hopkins, G.L. Lombard, J.E. McCroan, J.C. Prucha, et al., Type A botulism from commercially canned beef stew, South Med. J. 70 (1) (1977) 5–7.

[106] J.E. Seals, J.D. Snyder, T.A. Edell, C.L. Hatheway, C.J. Johnson, R.C. Swanson, et al., Restaurant-associated type A botulism: Transmission by potato salad, Am. J. Epidemiol. 113 (4) (1981) 436–444.

[107] K.F. Meyer, B.J. Dubovsky, The distribution of the spores of C. botulinum in the United States, J. Infect. Dis. 31 (1922) 559–594.

[108] L.D.S. Smith, The occurrence of Clostridium botulinum and Clostridium tetani in the soil of the United States, Health Lab. Sci. 15 (1978) 74–80.

[109] M. Cherington, S. Ginsberg, Type B botulism: Neurophysiologic studies, Neurology 21 (1) (1971) 43–46.

[110] M. Cherington, Botulism. Ten-year experience, Arch. Neurol. 30 (6) (1974) 432–437.

[111] V.R. Dowell Jr, L.M. McCroskey, C.L. Hatheway, G.L. Lombard, J.M. Hughes, M.H. Merson, Coproexamination for botulinal toxin and clostridium botulinum. A new procedure for laboratory diagnosis of botulism, JAMA 238 (17) (1977) 1829–1832.

[112] J.R. Barr, H. Moura, A.E. Boyer, A.R. Woolfitt, S.R. Kalb, A. Pavlopoulos, et al., Botulinum neurotoxin detection and differentiation by mass spectrometry, Emerg. Infect. Dis. 11 (10) (2005) 1578–1583.

[113] A.E. Boyer, H. Moura, A.R. Woolfitt, S.R. Kalb, L.G. McWilliams, A. Pavlopoulos, et al., From the mouse to the mass spectrometer: Detection and differentiation of the endoproteinase activities of botulinum neurotoxins A-G by mass spectrometry, Anal. Chem. 77 (13) (2005) 3916–3924.

[114] J.L. Ferreira, S.J. Eliasberg, M.A. Harrison, P. Edmonds, Detection of preformed type A botulinal toxin in hash brown potatoes by using the mouse bioasssay and a modified ELISA test, J. AOAC Int. 84 (5) (2001) 1460–1464.

[115] K. Bagramyan, J.R. Barash, S.S. Arnon, M. Kalkum, Attomolar detection of botulinum toxin type A in complex biological matrices, PLoS One 3 (4) (2008) e2041.

[116] J.W. Grate, M.G. Warner, R.M. Ozanich Jr., K.D. Miller, H.A. Colburn, B. Dockendorff, et al., Renewable surface fluorescence sandwich immunoassay biosensor for rapid sensitive botulinum toxin detection in an automated fluidic format, Analyst 134 (5) (2009) 987–996.

[117] S.M. Varnum, M.G. Warner, B. Dockendorff, N.C. Anheier Jr., J. Lou, J.D. Marks, et al., Enzyme-amplified protein microarray and a fluidic renewable surface fluorescence immunoassay for botulinum neurotoxin detection using high-affinity recombinant antibodies, Anal. Chim. Acta 570 (2) (2006) 137–143.

[118] M.G. Warner, J.W. Grate, A. Tyler, R.M. Ozanich, K.D. Miller, J. Lou, et al., Quantum dot immunoassays in renewable surface column and 96-well plate formats for the fluorescence detection of botulinum neurotoxin using high-affinity antibodies, Biosens. Bioelectron 25 (1) (2009) 179–184.

[119] S.R. Kalb, J. Lou, C. Garcia-Rodriguez, I.N. Geren, T.J. Smith, H. Moura, et al., Extraction and inhibition of enzymatic activity of botulinum neurotoxins/A1, /A2, and /A3 by a panel of monoclonal anti-BoNT/A antibodies, PLoS One 4 (4) (2009) e5355.

[120] S.R. Kalb, H. Moura, A.E. Boyer, L.G. McWilliams, J.L. Pirkle, J.R. Barr, The use of Endopep-MS for the detection of botulinum toxins A, B, E, and F in serum and stool samples, Anal. Biochem. 351 (1) (2006) 84–92.

[121] W. Terranova, J.G. Breman, R.P. Locey, S. Speck, Botulism type B: Epidemiologic aspects of an extensive outbreak, Am. J. Epidemiol. 108 (2) (1978) 150–156.

[122] T.E. Macdonald, C.H. Helma, L.O. Ticknor, P.J. Jackson, R.T. Okinaka, L.A. Smith, et al., Differentiation of *Clostridium botulinum* serotype A strains by multiple-locus variable-number tandem-repeat analysis, Appl. Environ. Microbiol. 74 (3) (2008) 875–882.

[123] C.O. Tacket, W.X. Shandera, J.M. Mann, N.T. Hargrett, P.A. Blake, Equine antitoxin use and other factors that predict outcome in type A foodborne botulism, Am. J. Med. 76 (5) (1984) 794–798.

[124] W.W. Schmidt-Nowara, J.M. Samet, P.A. Rosario, Early and late pulmonary complications of botulism, Arch. Intern. Med. 143 (3) (1983) 451–456.

[125] J.M. Hughes, C.O. Tacket, "Sausage poisoning" revisited, Arch. Intern. Med. 143 (3) (1983) 425–427.

[126] P. Amersdorfer, C. Wong, S. Chen, T. Smith, S. Desphande, R. Sheridan, et al., Molecular characterization of murine humoral immune response to botulinum neurotoxin type A

binding domain as assessed using phage antibody libraries, Infect. Immun. 65 (1997) 3743–3752.

[127] P. Amersdorfer, J.D. Marks, Phage libraries for generation of anti-botulinum scFv antibodies, Methods Mol. Biol. 145 (2000) 219–240.

[128] P. Amersdorfer, C. Wong, T. Smith, S. Chen, S. Deshpande, R. Sheridan, et al., Genetic and immunological comparison of anti-botulinum type A antibodies from immune and non-immune human phage libraries, Vaccine 20 (11–12) (2002) 1640–1648.

[129] A. Nowakowski, C. Wang, D.B. Powers, P. Amersdorfer, T.J. Smith, V.A. Montgomery, et al., Potent neutralization of botulinum neurotoxin by recombinant oligoclonal antibody, Proc. Natl. Acad. Sci. USA 99 (17) (2002) 11346–11350.

[130] R.E. Black, R.A. Gunn, Hypersensitivity reactions associated with botulinal antitoxin, Am. J. Med. 69 (1980) 567–570.

[131] S.S. Arnon, Infant botulism, in: R. Feigen, J. Cherry (Eds.), Textbook of Pediatric Infectious Diseases, W.B. Saunders, Philadedlphia, 1992, pp. 1095–1102.

[132] S.S. Arnon, R. Schechter, S.E. Maslanka, N.P. Jewell, C.L. Hatheway, Human botulism immune globulin for the treatment of infant botulism, N. Engl. J. Med. 354 (5) (2006) 462–471.

[133] L.S. Siegel, Human immune response to botulinum pentavalent (ABCDE) toxoid determined by a neutralization test and by an enzyme-linked immunosorbent assay, J. Clin. Microbiol. 26 (1988) 2351–2356.

[134] M.P. Byrne, L.A. Smith, Development of vaccines for prevention of botulism, Biochimie 82 (2000) 955–966.

[135] M. Rosenbloom, J.B. Leikin, S.N. Vogel, Z.A. Chaudry, Biological and chemical agents: A brief synopsis, Am. J. Ther. 9 (2002) 5–14.

[136] B. Knight, Ricin: A potent homicidal poison, Br. Med. J. 1 (6159) (1979) 350–351.

[137] S. Mayor, UK doctors warned after ricin poison found in police raid, BMJ 326 (7381) (2003) 126.

[138] S. Olsnes, J.V. Kozlov, Ricin. Toxicon 39 (2001) 1723–1728.

[139] E. Rutenber, J.D. Robertus, Structure of ricin B-chain at 2.5 A resolution, Proteins 10 (1991) 260–269.

[140] Y. Endo, K. Mitsui, M. Motizuki, K. Tsurugi, The mechanism of action of ricin and related toxic lectins on eukaryotic ribosomes. The site and the characteristics of the modification in 28 S ribosomal RNA caused by the toxins, J. Biol. Chem. 262 (1987) 5908–5912.

[141] R.A. Greenfield, B.R. Brown, J.B. Hutchins, J.J. Iandolo, R. Jackson, L.N. Slater, et al., Microbiological, biological, and chemical weapons of warfare and terrorism, Am. J. Med. Sci. 323 (6) (2002) 326–340.

[142] J.E. Smallshaw, A. Firan, J.R. Fulmer, S.L. Ruback, V. Ghetie, E.S. Vitetta, A novel recombinant vaccine which protects mice against ricin intoxication, Vaccine 20 (27–28) (2002) 3422–3427.

[143] G.D. Griffiths, G.J. Phillips, S.C. Bailey, Comparison of the quality of protection elicited by toxoid and peptide liposomal vaccine formulations against ricin as assessed by markers of inflammation, Vaccine 17 (20–21) (1999) 2562–2568.

[144] J.H. Carra, R.W. Wannemacher, R.F. Tammariello, C.Y. Lindsey, R.E. Dinterman, R.D. Schokman, et al., Improved formulation of a recombinant ricin A-chain vaccine increases its stability and effective antigenicity, Vaccine 25 (21) (2007) 4149–4158.

[145] D.W. Dyer, J.J. Iandolo, The staphylococcal enterotoxins: A genetic overview, J. Food Safety 3 (1982) 249–264.

[146] A.C. Papageorgiou, H.S. Tranter, K.R. Acharya, Crystal structure of microbial superantigen staphylococcal enterotoxin B at 1.5 A resolution: Implications for superantigen recognition by MHC class 2 molecules and T-cell receptors, J. Mol. Biol. 277 (1998) 61–79.

[147] M.L. Evenson, M.W. Hinds, R.S. Bernstein, M.S. Bergdoll, Estimation of human dose of staphylococcal enterotoxin A from a large outbreak of staphylococcal food poisoning involving chocolate milk, Int. J. Food Microbiol. 7 (1988) 311–316.

[148] S.D. Holmberg, P.A. Blake, Staphylococcal food poisoning in the United States. New facts and old misconceptions, JAMA 251 (1984) 487–489.

[149] R.D. LeClaire, R.E. Hunt, S. Bavari, Protection against bacterial superantigen staphylococcal enterotoxin B by passive vaccination, Infect. Immun. 70 (5) (2002) 2278–2281.

[150] J.W. Boles, L.M. Pitt, R.D. LeClaire, P.H. Gibbs, E. Torres, B. Dyas, et al., Generation of protective immunity by inactivated recombinant staphlyococcal enterotoxin B vaccine in non-human primates and identification of correlates of immunity, Clin. Immun. 108 (2003) 51–59.

[151] J.L. McDonel, Toxins of *Clostridium perfringens* types A, B, C, D, and E, Pergamon Press, Oxford, 1986.

[152] O. Miyamoto, J. Minami, T. Toyoshima, S. Yamagami, S. Miyata, T. Itano, et al., Neurotoxicity of *Clostridium perfringens* epsilon-toxin for the rat hippocampus via the glutaminergic system, Infect. Immun. 66 (1998) 2501–2508.

# Methods

# CHAPTER 21

# Use of Host Factors in Microbial Forensics

**Steven E. Schutzer**

*Department of Medicine, University of Medicine and Dentistry of New Jersey,*
*New Jersey Medical School, Newark, New Jersey*

## INTRODUCTION AND BACKGROUND

Considerable advances have been made in the forensic analysis of microbes and toxins. These advances include sequencing, genomics, and microscopy. An underdeveloped and underutilized area in microbial forensics is how the host interacts with microorganisms in a way that provides unique signatures for forensic use. For investigative and forensic purposes, an immediate goal is to distinguish a potential victim and innocent person from a perpetrator and to distinguish between a naturally acquired or intentional infection. Two principal methods that are sufficiently developed are characterization of the humoral immune response and identification of vaccine-induced immunity or antibiotics that may be present in a possible perpetrator.

This chapter presents central elements of the host response in a simplified fashion and describes a few representative examples that, in the appropriate context, have a high potential of providing evidence that may aid an investigation to distinguish a perpetrator from a victim who has been exposed to a particular microorganism or by-product, such as a toxin. The chapter also presents general information about the immune system so that the interested reader can have a fuller understanding of the immune response in general.

The primary aims of a microbial forensics investigation are to identify the biological agent, its source, and the individuals responsible for the event (1). Analytic approaches will differ when the suspected biothreat agent is encountered in a container or the environment, as opposed to *in vivo* in a human, animal, or plant. Analyses of trace elements, pollens, growth media, latent fingerprints, and microbial and nonmicrobial nucleic acids are all applicable to the container and environmental sample (2). However, once the microorganism or its toxin is in the living host, it is no longer possible to analyze the preceding items except the microbial nucleic acid. However, the host's

357

Microbial Forensics. DOI: 10.1016/B978-0-12-382006-8.00021-9

response to the biological agent may be available for analytic clues. This is akin to other forensic studies where physical traces of bite marks, scratches, wound trajectories, and sizes of wounds are often surrogate evidence of the teeth, fingernails, and bullets (3). While the forensic pathologist is familiar with evidence related to determining the manner of death, including the host response, those involved with health care alone are more familiar with the host response. In the context of microbial forensics it is important to integrate all of these with intelligence information so that they may be included in the analytical data and attribution picture.

The host response to a microorganism or other foreign substance is often a well-orchestrated series of events, which may protect the individual from harm or ameliorate its effects (4). At the same time, these host responses may provide clues as to the identity of the offending microorganism or toxin, as well as a rough chronology of when it occurred and for how long it has been persisting. Emerging technologies, such as transcriptional arrays and bioinformatic analysis, will eventually be refined and methods validated to provide even greater help in delineating more of the pathways and components of the host response to an infectious agent (5,6). Other technologies are sufficiently mature to be of use today. The immune system and its components are a mainstay of our protection against infections and malignancies (4,7). Inflammation is often a side effect as the immune system contains and eradicates a microorganism or eliminates foreign tissue. Specific arms of the immune system can be used as markers in support of or against the presence of an infection. The humoral or antibody response to an invading microorganism is one example of a specific immune response that can have forensic value. Some of the antibodies produced may have a protective role together with other parts of the immune system by eradicating the pathogen or neutralizing a toxin. Other antibodies may not be as effective in this role. However, by virtue of their ability to recognize unique and specific microbial antigens, they can serve as indicators that a specific microorganism was recently present or was present in the past. In the case of vaccine-induced immunity, antibodies may recognize highly specific epitopes of one microbe versus those of a related microbe (e.g., influenza virus). This is especially so with different recombinant vaccines and could have forensic importance. Substances such as antibiotics, which can kill a pathogen rapidly, may modify the immune response by removing or reducing the stimulus for a full-scale response. As noted earlier, in clinical and veterinary medicine, measurement of the immune response helps the diagnostician decide what infection was present and how recently. In these situations, the intent is to provide treatment. The forensic scientist may exploit parts of the immune response to discover who is likely a victim of an attack and who might be responsible. This chapter discusses the basics of the host immune response that can have forensic utility.

Examples will provide a sense of what information is obtainable and what is not likely to provide highly significant clues.

Understandably, health care providers are reluctant to compromise a patient's privacy and are normally mandated to guard this privacy by Health Insurance Portability and Accountability Act regulations (8,9). However certain circumstances may compel a health care provider to reveal private information about a patient (8). Nevertheless it is important to understand how valuable information may be in the possession of the treating physician and other members of the health care team. The physician and other health care providers may be among the first to realize that a patient is a victim of a biocrime. In case of a covert attack, it may be the physician or medical examiner who first recognizes the index case. These health care workers are in key positions to preserve critical evidence and thereby contribute to the investigation (10). A number of steps should be followed when the possibility of a biological attack arises, either with the consent of the patient or because individuals are compelled by law to interact with public health and law enforcement.

A joint statement by the Federal Bureau of Investigation (FBI), the Centers for Disease Control and Prevention (CDC), and the Department of Homeland Security (DHS) advises calling the FBI and public health authorities if a suspicious situation arises (11). Some guidelines on the procedure(s) to report of suspicions of biocrimes are provided by the CDC (http://www.cdc.gov), the FBI (http://www.fbi.gov), and the DHS (http://www.dhs.gov) and are detailed in a previous article (10).

## GENERAL CONCEPTS

In response to a new exposure to a microbe, the innate immune system may be the first line of defense. Then, the immune system starts to generate a humoral immune response. Typically a phagocytic cell (i.e., macrophage) ingests and degrades some of the invading pathogens. It then presents part (antigens) of the microorganism to a helper T cell (a lymphocyte), which then directs other lymphocytes known as B cells to produce antibodies to those antigens of that particular microbe that were presented. It usually takes at least 4 days before any microbe-specific antibody can be detected (12).

Antibodies are a specific form of proteins known as "immunoglobulins" (Igs). IgM, IgG, IgA, and secretory IgA are principal classes of immunoglobulins with prime relevance to this chapter. In response to a new antigen, immunoglobulins usually appear in the order of IgM, IgG, and IgA. B cells first begin to produce IgM, and then some B cells undergo an irreversible switch to ones that produce IgG. Later some of these B cells undergo a switch to become IgA-secreting B cells. Immunoglobulins persist for varying times;

**Table 21.1** Immunoglobulin Classes and Properties

| Immunoglobulin Class | IgM | IgG | IgA | IgE | IgD |
|---|---|---|---|---|---|
| Size (kDa) | 900 | 150 | 160 | 190 | 180 |
| Serum half-life days | 5 | 21–23 | 5–6 | 1–5 | 2–8 |
| Placental transfer | No | Yes | No | No | No |
| Complement fixation | ++ | + | – | – | – |
| Percentage of serum immunoglobulin | 13 | 80 | 6 | 0.002 | 0.2 |

for example, the half-life of particular IgM antibodies is approximately 5 days, while that of IgG can be as long as 21–23 days (Table 21.1) (7).

At times in ruling in or ruling out a suspect, even specific IgE may be of value in addition to the more universal IgG and IgM responses. Those individuals unfortunate to have allergies have problems due to IgE against allergens (such as ragweed, peanut, or cat dander). In this case the IgE molecules sit on the surface of mast cells and basophils. These cells can release histamine and other allergic mediators when the offending allergen bridges two IgE molecules.

Similar to the immune response to an infection with a live microbe, vaccines can also engender an antibody response. A vaccine can be a live or attenuated microbe, a whole nonproliferating microbe or an antigenic (recombinant) component of the microbe, or a toxoid. Vaccines may contain an adjuvant (e.g., alum) to stimulate the humoral response of the host. Regardless, the intent of immunization is to engender protection, often by the generation of protective neutralizing antibodies. Although the half-life of an individual IgG molecule is less than a month, a population of antibodies of the IgG isotype form may persist for life. Memory B cells can sustain these antibodies and retain the ability to respond quickly by generating the appropriate antibodies when challenged. When the immune system encounters another infection or is subjected to a revaccination (booster), the result is an accelerated production of the particular antibody and an increase in the levels of antibodies that circulate in the blood (Figure 21.1).

Perhaps the pattern of antibody response which has the most forensic value, by providing a time frame, is the appearance of IgM first, followed by a B-cell switch to the longer lasting IgG. During the early phase of exposure, IgM predominates, as time goes on, IgG may wax and wane and IgM is no longer found (Figure 21.2).

The antibody response to a particular agent may be directed to different antigens at different times, that is, early or later after the initial exposure. That

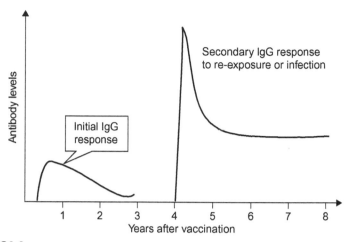

**FIGURE 21.1**

Illustration of IgG antibody response to a vaccine antigen after first immunization and subsequent exposure by natural exposure to the infectious agent or by another vaccination.

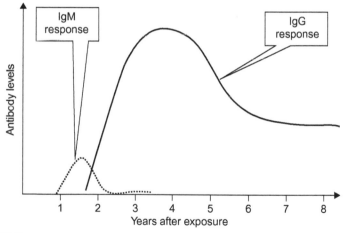

**FIGURE 21.2**

Illustration of temporal relation of IgM and IgG responses to an infection with IgM as the first and often transient response and IgG as the more sustained response.

response often involves IgM at the early stage and IgG later. Late in the course of the disease or during recovery, only IgG to particular antigens may be seen. A classic example of this is the human antibody response to Epstein-Barr virus (EBV) (13), a virus known to cause mononucleosis. During acute early disease, it is common to find high levels of IgM antibodies to the viral early antigen (EA) and viral capsid antigen (VCA). It is rare to find high levels of

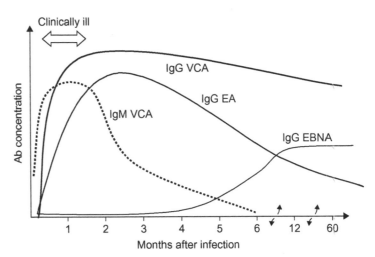

**FIGURE 21.3**
Schematic response of IgM and IgG to different antigens of EBV over an extended period of time.

IgG to the VCA or to Epstein-Barr nucleic acid (EBNA). As the patient recovers from his/her first infection with EBV, the immune response is characterized by low levels of IgM to EA or VCA, and higher or increasing levels of IgG to VCA. Antibodies to EBNA are often very low during this stage. Several months after clinical recovery, IgM to EA and VCA remain at low levels whereas IgG to VCA and EBNA are present at high levels, often for years. Table 21.3 illustrates this pattern by stage of the immune response to EBV and its particular antigens. Figure 21.3 is a graphic display of these antibody responses. For the clinician or epidemiologist, antibody responses provide a framework to determine where in the course of the infection a patient may be. Tables 21.2 and 21.3 and Figures 21.2 and 21.3 illustrate how responses to a biothreat agent or its toxin may be used to give some chronological indication of exposure. Combining the antibody response with detection of particular antigens can provide further definition as to a time frame of infection or exposure.

## ILLUSTRATIVE CONCEPTS

A controlled experiment or normal clinical event illustrates what happens when the immune system sees an infectious agent or a vaccine for the second time. The controlled experiment may be in a laboratory animal or a patient receiving a booster vaccine. The uncontrolled but normal clinical event occurs when the patient is reexposed to the infectious agent. Consider a generic antigen exposure. The first time the immune system encounters antigen X (AgX) it responds as shown in Figures 21.1 and 21.2. Initially, antibodies to AgX are

**Table 21.2** Antibody Tests for Epstein-Barr Virus

| Stage | Titers |
|---|---|
| Acute primary infection | |
| IgM EA and VCA | High |
| IgG VCA and EBNA | Low |
| Recovering from primary infection | |
| IgM EA or VCA | Lower |
| IgG VCA | Rising |
| EBNA | Low |
| After several months | |
| IgM EA and VCA | Low or normal |
| IgG VCA and EBNA | Persist at high level for several years |

**Table 21.3** Antibody Response at Different Time Points to EBV Antigens

| Disease Status | Heterophile Ab | VCA-IgM | VCA-IgG | EBNA | EA(D) |
|---|---|---|---|---|---|
| Healthy—unexposed | Negative | Negative | Negative | Negative | Negative |
| Very early infection | Possible | Possible | Possible | Negative | Negative |
| Active infection | Positive | Positive | Positive | Negative | Possible |
| Recent infection | Positive | Positive | Positive | Positive | Possible |
| Past infection | Negative | Negative | Positive | Positive | Possible |

barely discernible; then levels rise and later fall to a plateau. If a simultaneous exposure were to occur with AgX and a new AgY from another microorganism, the immune system would quickly mount a brisk response with high levels of Ab to AgX, while the course of Ab to AgY would be slow and delayed, just as it was in the response to the first exposure to AgX. This phenomenon, termed "immunological memory" or an "amnestic response," can be useful when the symptoms and signs of exposure to either X or Y are similar. This is the type of response that can occur with the early flu-like symptoms of pulmonary anthrax (14–16) and with the influenza virus itself (17–19).

Another common example is the repetitive exposure to different strains of influenza virus (17–20). As illustrated in Table 21.4, a person infected for the first time with one strain of the influenza virus has a response to most of its antigens (as a theoretical example, Ag 1, 2, 3, 4, 5, 6). Three years later, the same individual exposed to a partially similar influenza virus responds preferentially

**Table 21.4** Response to Theoretical Antigens from Different Flu Viruses at Time of Exposure (Weak vs. Strong)

| Infecting Strain and Antigen Composition | Antibody Response: Weak | Antibody Response: Strong |
|---|---|---|
| Strain A year 1 (antigens 1, 2, 3, 4, 5, 6) | 1, 2, 3, 4, 5, 6 | — |
| Strain B year 5 (antigens 1, 3, 5, 7, 8, 9) | 7, 8, 9 | 1, 3, 5 |
| Strain C year 15 (antigens 1, 3, 8, 10, 12, 13) | 10, 12, 13 | 1, 3, 8 |

to those antigens that were also present on the original influenza virus (secondary immune response). The person also has a primary antibody response to new antigens, that is, those not shared with the first virus. Ten or 20 years later, during a new flu season and exposure to a third strain of influenza, the most brisk responses would be to antigens previously recognized by the immune system. This is the scientific basis for giving the flu vaccine, which contains a variety of possible antigens common to multiple strains of the flu virus so that a rapid and protective antibody response will occur.

## UTILITY OF SEROLOGIC ANALYSIS OF PEOPLE EXPOSED TO ANTHRAX: STRENGTHS AND LIMITATIONS

Our knowledge of the humoral response to infection with biothreat microbes is limited compared with our knowledge of the kinetics of the response to common human infections. Nevertheless, in the appropriate context and with sufficient background information, detection of antibodies to a particular microbe and its antigens can have important value for a microbial forensic investigation. It may have critical probative value or it can guide investigative leads. Absence of a specific antibody response may also have value in a particular investigation. Certainly its importance is increased in the context of information of what organism could be involved, when the exposure was likely to have occurred, the route of exposure, what symptoms and signs are manifesting in the host, and other laboratory data such as presence of antigens and microbial nucleic acids (21). Other information, such as how many hosts (people or animals) have had this infection in the geographic region, what is the incidence, and background prevalence of antibody to the organism in question or a related organism, in the population being studied, is also important.

Vaccination responses can have forensic value. The current anthrax based on protective antigen (PA) vaccine contains small amounts of lethal factor (LF) and edema factor (EF), which are responsible for some of the side effects, so one might expect to see antibodies against these antigens as well as to PA. Recombinant PA is just PA so anti-LF and anti-EF would be absent in an immunized individual.

The 2001 anthrax letter attacks raised multiple questions for every person infected, potentially exposed, vaccinated, or treated. Some of these questions included how these persons were infected by spores, if at all; that is, through breaks in the skin (cutaneous anthrax), by inhalation of spores [pulmonary anthrax (22)], or by ingestion [gastrointestinal anthrax (23,24)]. Alternatively, were they among the "worried well"?

Consider the situation where a close associate comes down with symptoms compatible with inhalational anthrax after receiving a letter containing powder and that material is no longer available. Until this is shown not to be anthrax, great worry will ensue.

In several cases of documented exposure, there was not enough time for the patient to develop antibody to a specific *Bacillus anthracis* antigen, at least as probed for IgG. Serial serum samples obtained from potentially exposed individuals on November 16, 17, 18, and 19 of 2001 were tested for IgG antibody to the protective antigen (PA) component of the anthrax toxins by enzyme-linked immunosorbent assay (ELISA); all samples were negative. Serial tests for serum IgG antibody to the PA toxin of anthrax were performed on 436 workplace-exposed persons. All but one test was negative. Most of the specimens were collected on October 10 and 17 (25).

It is instructive to look at the positive antibody case in the context of the nature and duration of that individual's symptoms when he developed a positive test. Ernesto Blanco, a 73-year-old mailroom clerk (case 2) experienced fatigue on September 24. He worked in the mailroom of the AMI building and delivered mail to the index case. On September 28, he developed a non-productive cough, intermittent fever, runny nose, and conjunctivitis. These signs worsened through October 1 when he was hospitalized. In addition, he had shortness of breath with exertion, sweats, mild abdominal pain and vomiting, and episodes of confusion. His temperature was elevated to 38.5°C (101.3°F), heart rate was rapid at 109/min, respiratory rate was slightly fast at 20/min, and blood pressure was 108/61 mm Hg. He had bilateral conjunctival injection and bilateral pulmonary rhonchi. At the time of admission, his neurological examination was normal. No skin lesions were observed. The only laboratory abnormalities were low albumin, elevated liver transaminases, borderline low serum sodium, increased creatinine, and low oxygen content in the blood. Blood cultures were negative on hospital day 2, after antibiotics had been started. The chest X-ray showed a left-sided pneumonia and a small left pleural effusion but no classic mediastinal widening (26). The patient was initially given intravenous azithromycin; cefotaxime and ciprofloxacin were subsequently added. A nasal swab obtained on October 5 grew *B. anthracis* on culture. Computed tomography of the chest showed bilateral effusions and multilobar pulmonary consolidation but still no significant mediastinal lymphadenopathy. Pleural fluid aspiration was positive for *B. anthracis* DNA by polymerase chain reaction (PCR). Bacterial cultures of bronchial washings and pleural fluid were negative. Immunohistochemical staining of a transbronchial biopsy demonstrated the presence of *B. anthracis* capsule and cell-wall antigens. During hospitalization, his white blood count rose to 26,800/mm$^3$, and fluid from a second thoracentesis was positive for *B. anthracis* DNA by PCR. Immunohistochemical staining of both pleural fluid cells and pleural biopsy tissue demonstrated the presence of *B. anthracis* capsule and cell-wall antigens. Serial serum samples demonstrated >4-fold rise in serum IgG antibody to the PA component of the anthrax toxins by an ELISA assay. The patient was able to leave the hospital on October 23 and was on oral ciprofloxacin. Table 21.5 illustrates both the clinical and microbial forensic approach and the context in which to analyze such a patient. It is likely to be common to most situations where a biocrime is suspected to have potentially affected individuals. The first set of questions is directed toward whether the person is sick: does the person have any indications of not being well and is laboratory evidence indicative of an infection? The second set of questions asks whether there is any specific and objective laboratory evidence of a particular infection. A third set of questions arise if the cause of infection was an agent on the Select Agent list (27). These questions include (i) was the infection acquired naturally or was it an intentional action that led to the infection and (ii) how did the particular individual acquire it if it was not a natural

**Table 21.5** Nonspecific and Specific Indications of a Case of Anthrax

| Clinical Evidence of an Infection | Nonspecific Laboratory Evidence of an Infection | Specific Clinical Evidence of Infection with *B. anthracis* |
|---|---|---|
| Known exposure by proximity to area and infected person | Chest X-ray and computed tomography scan showing pneumonia and pleural fluid | Culture from nasal swab grew live *B. anthracis* |
| Cough, fever, shortness of breath | Elevated white blood cell count | Positive PCR for *B. anthracis* in pleural fluid on two occasions despite negative cultures |
| Sweats, abdominal pain, confusion | | Positive immunochemical staining for *B. anthracis* capsule and cell wall antigens of transbronchial biopsy, pleural fluid cells, and pleural biopsy despite negative cultures |
| Abnormal breath sounds | | Serum IgG to PA toxin component |
| Fast heart rate | | Serum IgG titer to PA toxin increased within a short time period |

infection, that is, was he the target or a bystander? An alternative possibility in the right circumstances is a laboratory-acquired infection.

This case also demonstrates that cultures may be negative at different times from different fluids and tissues because of early administration of antibiotics. However, remnants of the infection, even dead organisms, can be found by probing for antigens and DNA. This patient's response demonstrated a classic principle of infectious disease, a rising antibody titer over time. In this case it was IgG to a particular antigenic component of the anthrax toxins (28,29). The subject's antibody response may have been detected earlier if IgM to this component or to other antigens of anthrax had been sought. The case also points out the utility of integrating the detection of antibody with other indications of an anthrax infection, such as a positive culture, PCR, or antigen detection assay. These take on their greatest significance during clinical illness in someone who was possibly exposed.

Early administration of antibiotics can prevent or interfere with the isolation of a pathogen by culture (30). Of the first 10 pulmonary anthrax cases associated with the 2001 letter attacks, three patients had no isolate of *B. anthracis* from any clinical specimen; however, culture was attempted after initiation of antibiotic therapy. History of exposure in conjunction with compatible symptoms and signs of disease and objective laboratory findings were the basis for the diagnosis. *B. anthracis* was identified in pleural fluid, pleural biopsy, or transbronchial biopsy specimens by reactivity with *B. anthracis*-specific cell wall and capsular antibodies or by the detection of DNA in pleural fluid or blood by PCR (31).

An IgG-based ELISA for anti-PA illustrates the importance of understanding the limitations of an assay used in medicine or forensics (32,33). The ELISA for anti-PA was initially developed at the U.S. Army Medical Research Institute of Infectious Disease and put into operation after optimization and internal validation at the CDC (34) for functional sensitivity and specificity in detecting an antibody response to PA as a surrogate for *B. anthracis* infection. Its major limitations were that only one antigen was used and only IgG was measured. Therefore, a negative result shortly after exposure may, in effect, be a false-negative result. A gap such as this may be filled by development of an assay for antigen-specific IgM or by probing for other *B. anthracis* antigens or epitopes yet to be characterized.

The assay for anti-PA may be very useful in its present form to screen asymptomatic people for possible exposure. The study by Dewan and colleagues (26) provides a contemporary background database on a group of postal workers who may have been exposed to *B. anthracis*. Beginning on October 29, 2001, 1657 postal employees and others who had been to the Washington, DC, postal facility went to the D.C. General Hospital for antibiotics. Added to this number were those people whose treatment began on October 21,

2001. Serum samples were also obtained from the 202 individuals who had been to the Washington, DC, postal facility during the previous 2 weeks. All were negative for anti-PA IgG, including three individuals who reported a remote history of anthrax vaccination. The consistent negative findings may be explained by the fact that antibiotic therapy was initiated before serum was obtained for testing and that there were no baseline serum samples available for testing. Also, the time period from exposure to sampling was very short. Among 28 individuals in the Capitol region with culture-positive nasal swabs who received prophylactic antibiotics immediately, none had a positive culture from a nasal swab repeated 7 days later, and none developed IgG to PA 42 days after exposure. This again emphasizes the limitation and interpretation of a test in someone who had early antibiotic treatment. It does raise forensic utility considerations. Even with these easily disseminated spores, an antibody response may be aborted or modified with antibiotics by early treatment. Furthermore, antibiotics taken prior to exposure would likely be effective in preventing laboratory and clinical signs of an infection. Detection of microbial DNA, antigen, or the organism itself on a person's body, clothing, or possessions should be an indicator for exposure.

The route of infection is important in interpreting results and limitations of the assay used. The example of cutaneous anthrax in Paraguay illustrates this notion, as well as the need to search for other antigens as markers of exposure (35). In an analysis of an outbreak of 21 cases of cutaneous anthrax that followed contact with raw meat from a sick cow, sera from 12 cases and 16 colony and two nonbacterial colony controls were examined by Western blotting for antibodies to PA and LF 6 weeks after the outbreak. An ELISA was used to probe for antibodies to the poly-D-glutamic acid capsule. Of the 12 cases, 11 had antibody to PA, for a sensitivity of 91.7%; none of the 18 controls was positive. Only 6 of 12 cases had antibody to LF; all controls were negative. Anticapsule antibodies were positive in 11 of 12, but were also positive in 2 of 18 controls. Results of this study demonstrate the need to consider other antigens.

## CONSIDERATIONS AND CONCERNS RAISED BY ANALYSIS OF OTHER INFECTIONS

Some of the principles discussed earlier are highlighted by a report on severe acute respiratory syndrome (SARS). Appearance of the coronavirus responsible for this disease evoked concern of a possible terrorist origin. A report in the *Morbidity and Mortality Weekly Report* [MMWR (36)] on the "Prevalence of IgG Antibody to SARS-Associated Coronavirus in Animal Traders" discussed the need to validate and interpret tests in appropriate populations. Also discussed was the inability to date the time of infection by the IgG assay and the possibility of assay cross-reactivity to a near neighbor that might be unknown. In a

Promed bulletin, Berger looked at the same data from a different perspective and reported:

> "This week's study in MMWR indicates that animal contact may indeed promote infection; however, the most striking finding seems to have eluded the authors: 1.2–2.9% of individuals in a healthy control group of adults were also found to be seropositive! The population of the Guangdong Province is 86.42 million (2001), of whom 61.14 million are adults over age 14. If we assume that the seropositivity rates among controls is representative of the province as a whole, 734,000 to 1,773,000 adults in Guangdong have at some time been infected by the SARS virus. These figures are 87- to 211-fold the total number (8422) of SARS patients reported worldwide to date!"

This comparison is a good illustration of the advantage of open dissemination and discussion of information, as well as the need to question the methodology of acquisition of data before accepting its application in formulas or for analyses for forensics and epidemiology. It is also of value to remember that many infections with SARS coronavirus may have been asymptomatic or mildly symptomatic.

Plague is a zoonotic infection caused by *Yersinia pestis*, which occurs in the western United States with regularity and has an animal reservoir. The situation with the naturally occurring plague is in contrast to the appearance of a case of smallpox, which would be an immediate indication of a bioterrorist event (see Chapter 15). Cases need to be approached from an epidemiological standpoint first to determine whether it is a naturally acquired case or whether facts point to a deliberate introduction of the organism. Currently, analytical techniques could include genomic analysis of an isolated organism and immunological response of the host. In the new era of rapid and deep sequencing, our capacity to investigate the genomics is growing (37) (see Chapter 27). In consideration of animal reservoirs, ELISA assays were compared with other tests for detection of plague antibody and antigen in multimammate mice (*Mastomys coucha* and *M. natalensis*) (38), which were experimentally infected and then sacrificed at daily intervals. IgG ELISA was equivalent in sensitivity to passive hemagglutination and more sensitive than the IgM ELISA and complement fixation. Antibody was detectable by day 6 after infection using all four tests. IgM ELISA titers fell to undetectable levels after 8 weeks. Plague fraction 1 antigen was detected in 16 of 34 bacteremic sera from *M. coucha* and *M. natalensis*. This antibody pattern comparison shows that the principle of IgM versus IgG to this pathogen works to temporally situate the infection as an early versus late or past event. It also shows that when the information is combined with antigen detection, it engenders more confidence in the results. It should be noted that conclusions from this older reference have been substantiated with more defined antigens and assay technologies.

The context and geographic location where an infection or biothreat occurs may dictate how an infection is viewed and evaluated. An example is provided by melioidosis, which is not endemic to the United States. Melioidosis is caused by *Burkholderia pseudomallei* (39). Key clinical signs and laboratory results may raise the possibility of an infection with this pathogen. Whether it is an acute, persistent, or past infection can be determined by assessing several host responses. Often a simple nonspecific indicator such as erythrocyte sedimentation rate or C-reactive protein (CRP) can raise the clinical suspicion of an infection. In a study of 46 patients with clinical melioidosis, 35 (22 culture positive and 13 culture negative) had relatively uneventful disease courses. Initially, they had elevated serum CRP that decreased with antibiotic therapy and returned to normal as the disease resolved. In another series of patients, IgM and IgG were measured by ELISA in 95 sera from 66 septicemic cases and 47 sera from 20 cases with localized melioidosis (40). Sixty-five sera from culture-negative cases were seronegative for other endemic infections but those suspected of melioidosis were also examined. Other controls included serum from 260 nonmelioidosis cases, 169 high-exposure risk cases, and 48 healthy individuals. The IgG ELISA was 96% sensitive and 94% specific. All sera from cases with septicemic and localized infections and 61 of 63 sera from clinically suspected melioidosis cases were positive for IgG antibody. Sensitivity and specificity of the IgM ELISA were 74 and 99%, respectively. A geometric antibody index for IgM antibody in sera of melioidosis cases was significantly higher in cases compared with that of noncase controls. In another study, a rapid test for IgG and IgM was shown to have clinical utility (41). A study with the intent of evaluating the utility of an IgG assay compared with other assays illustrates how the clinical and temporal context must be integrated for interpretation (42). It also illustrates how there is room for technical improvement in tests but the best setting is often the endemic area itself or at least using samples from that area in which the infection is occurring. These tests were evaluated in the actual clinical setting in an area endemic for melioidosis. Specificity of IgG (82.5%) and IgM (81.8%) assays were significantly better than that of an indirect hemagglutination test (IHA) (74.7%). Sensitivity of the IgG assay (85.7%) was higher than that of the IHA test (71.0%) and the IgM test (63.5%). Specific IgG was found in septicemic cases (87.8%) and localized infections (82.6%). The IgG test was also better than the IgM test and the IHA test in identifying acute melioidosis cases in the first 5 days after admission. IgG antibody to a *B. pseudomallei* antigen remained high for longer than 5 years in recovered, disease-free patients. Because this is a disease that may have an incubation of days to years, an acute case may very well be detected by a rise in specific IgM if it were a matter of days from infection. Although endemic for southeast Asia, if *B. pseudomallei* was used as a biothreat agent in a different environment, its course and manifestations may not be recognized due to unfamiliarity with the disease.

The example given earlier also points out how the context in which a test is used determines is value. The concept of predictive value is instructive in determining how useful a test may be. In terms of disease detection, a high positive predictive value indicates that the test is useful in determining that the disease is present. A high negativity predictive value would indicate that the test is useful in excluding the presence of the disease.

1. Concept of sensitivity—true positives/true positives + false positives or how many with a positive test actually have the disease.
2. Concept of specificity—true negatives/true negatives + false negatives or how many with a negative test actually do not have the disease.
3. Concept of positive predictive value—how good is the test in predicting disease among a particular population or true positives/true positives + false positives.
4. Concept of negative predictive value—how good is the test in excluding the disease among a particular population under consideration or true negatives/true negatives + false negatives.
5. High + predictive values are seen where disease prevalence is high and is low where disease prevalence is low.
6. Negative predictive values are high when disease prevalence is low and lower when disease prevalence is high.

Another zoonotic agent is Rift Valley fever virus (RVFV), which can be transmitted via aerosols. One study with the intent of looking for improved tests showed the utility of IgM to determine an early exposure to RVFV (43). Two IgM ELISAs detected specific IgM antibodies to RVFV during the first 6 weeks after vaccination. Three inactivated vaccine doses were given on days 0, 6 to 8, and 32 to 34. IgM levels on days 6 to 8 were negative or in the lower range of detection; on days 32 to 34 the IgM levels were strongly positive; on days 42 to 52 they were waning; and in later collected samples were negative. The plaque reduction neutralization test was negative on days 6 to 8 and became positive in later samples. Similar to the examples shown earlier, these data suggest that three doses of RVFV vaccine induced a prolonged primary antibody response. Authors of that study concluded that ELISA IgM may be useful for early diagnosis of acute human infection. Good correlation of a neutralization test and ELISA IgG would indicate a later infection.

Taken together, these examples illustrate that an ideal test or analysis for both clinical and forensics use would incorporate endemic and incident area controls, historical contextual information, knowledge of the route of exposure, background incidence, and kinetics of transmission.

# POSSIBLE SCENARIOS OF BIOTERRORISM ATTACKS: DISTINGUISHING VICTIMS FROM PERPETRATORS

Each of these scenarios must take into account multiple factors and limitations of any analytical process to be applied. The start of the acquired immunodeficiency syndrome (AIDS) epidemic provides an example. On one extreme is the situation that occurred with the onset of AIDS from the human immune deficiency virus (HIV) in the United States. Initially, there were no cases, and therefore a precise highly sensitive and specific test with excellent positive and negative predictive values (such as exists now when a combination of tests are used) would not likely yield a positive result in an area where there was little HIV infection and disease at the onset such as, for example, Kansas. A positive test by today's methodologies from a 1970 serum sample from Kansas would be considered a probable false positive and warrant further investigation. Today, several viral and nucleic acid assays are available that would provide a definitive diagnosis in a short period of time (22). However, the same sample tested at the beginning of HIV testing could have been positive if the person had adult T-cell leukemia, which is caused by human T-cell leukemia virus-1 (HTLV-1), because original tests for what became known as AIDS involved whole viral lysates in which up to 30% of the HTLV-1 sera cross-reacted. Questions regarding interpretation of test results could be raised by knowledge of different presentations of the infection. For example, HTLV-1 can actually be used in the laboratory to immortalize cells. In the patient it actually increases the T-cell count, as is the nature with leukemia, instead of decreasing them, as with HIV infections. Other laboratory indicators such as hypercalcemia would now raise leukemia as a consideration.

Interpretation of a positive laboratory test must take into account the health status of the person being tested. This is important for the practice of medicine and can have relevance when extended to forensic analysis (10). The following examples illustrate this concept. Individuals who have syphilis, a treponemal bacterial infection, will typically have a positive fluorescent treponemal antibody test result for years, even after successful treatment. However, while infected they would have a positive venereal disease research laboratory (VDRL) test, which reverts to negative following successful antibiotic therapy. The VDRL test detects nonspecific, anticardiolipin antibodies and can produce false-positive results with other conditions (e.g., pregnancy). There are some notable exceptions related to cross-reactive epitopes or autoimmune diseases. These are readily distinguishable by history and clinical information. Similarly, individuals with active tuberculosis will likely have a positive skin test (Mantoux) or a positive interferon-$\gamma$ release assay (44), whereas the uninfected healthy person will be negative. In certain instances, a sick person with a cell-mediated immune deficiency will be anergic, that is, he/she will be negative to multiple skin tests, including common antigens such as *Candida*. The key difference here is that a great difference exists between the *healthy person* being tested and an *ill or immunocompromised individual* being subjected to the same test.

Tests may also discriminate between the length of the infection (i.e., acute or chronic); limitations of these tests may lead to different interpretations unless one is familiar with those limitations. An example of this occurred with the bacterial infection by *Borrelia burgdorferi*, which causes Lyme disease. Antibiotics can abrogate the antibody response because ELISA results are negative in 30% of patients with known disease who were treated early (45). In early cases, reactivity to a unique antigen, OspA, was also negative in serological assays, despite a demonstrable T-cell response (46). Analysis of these same sera found that there was antibody to *B. burgdorferi* but it was below the threshold of detection by conventional assays. It was detectable in its bound form in immune complexes (47,48).

Anthrax can be used as an example where investigatory leads can be generated by considering a scenario *in toto*. The elderly woman who died in Connecticut from inhalation anthrax clearly had no occupational exposure nor was she known to have had contact with anyone who had anthrax. It was possible that she had contact with cross-contaminated mail. However, if this case had occurred as the index case or out of context of the mail attacks, it would have been reasonable to question her travel history; what her work, if any, was; or if she received or used spore-contaminated products from an anthrax-endemic area. Similarly, the Vietnamese woman who died of inhalation anthrax in New York City would also have had these questions investigated. It would have been useful to search for direct or indirect evidence of anthrax by physical examinations of her contacts or close neighbors. Inspection and cultures from her workplace, apartment, and apartment complex (especially contiguous neighbors) are important for detecting the presence of *B. anthracis*. Co-workers, friends, neighbors, and other contacts could have had their serum analyzed for antibody to antigens of *B. anthracis*. These samples could have been frozen so that if one were positive it would be available for a subsequent comparison study. At a minimum, these types of studies could serve as future control data for the geographic region. With molecular methods, even trace amounts might be detectable (49), although parallel investigation using background controls would be necessary. Although hypothetical, several results could have occurred, and each will be considered separately. First example: a close contact is positive for IgM to one of the *B. anthracis* antigens, such as PA. This finding would suggest that this person had recent exposure and, if nothing else, should be treated. This individual could conceivably be the one who knowingly or unknowingly passed the spores to the patient. Given the October 26 onset of illness, which is late in the mailing sequence, it would be less likely that this individual was a perpetrator but rather a recent victim. However, if this person were IgG positive, then there are several other possibilities. Perhaps this person had past exposure in an endemic region and was treated (e.g., Haiti, where anthrax is known as "charcoal disease"); this person could have been vaccinated for bona fide reasons, such as a researcher who received it to protect against occupational exposure; or this person could have obtained the vaccine originally for legitimate or illegal purposes but was nevertheless vaccinated. The vaccine usage may have been for a clinical trial or for animal experimentation. Animal vaccines may be more obtainable without strict record keeping. This person could have loaded the mail with relative impunity if there was protective antibody generated from the vaccination. Situations similar to this one will require intelligence information regarding access, ability, and motive. In an area where recombinant vaccines are being developed or used, the antibody response would be different between someone using one type of recombinant vaccine as compared to someone using another type of vaccine. Nevertheless, finding IgG to one or more antigens of *B. anthracis* could point investigators toward such a seropositive individual, whereas an IgM finding could justify critical therapy. Where information points to a particular individual, investigation could be extended to search for ingestion or injection of antibiotics as illustrated later in the ciprofloxacin example. Questions would be raised regarding access to antibiotics, recent ingestion/injection of them, half-life of the antibiotic, half-life of the metabolites of the antibiotics, and in which body fluids or tissues can the residual be found. As illustrated from data in the earlier sections, someone with antibiotics in his/her system may be protected following exposure to a potential pathogen. This person would be antibody negative and likely antigen and microbial DNA/RNA negative, as the infection would have been eradicated before the organism could proliferate to any significant level. The widespread prophylactic use of ciprofloxacin during the period following the anthrax mailing attacks is illustrative of an understudied area. Ciprofloxacin has been increasingly associated with tendonitis and ruptured Achilles tendons (50,51). In the future, better methodology to follow pharmacokinetics of an anti-infective compound may have forensic implications. In the last example, someone who takes an antibiotic prophylactically while manipulating a lethal microbe may exhibit side effects that, in proper context of an investigation, may add to the picture of possible culpability. This area is far from established at this point in time.

Strategies can be employed to examine suspicious but possible accidental transmission of infections. This approach is illustrated by a recent study of avian influenza using a multitude of assays. Tools to determine person-to-person spread as the mode of transmission included viral culture, serologic analysis, immunohistochemical assay, reverse-transcriptase/polymerase chain reaction analysis, and genetic sequencing (52).

It is likely that future understanding of the immune system and evolving technologies such as microarrays will bring new analytic power to the field, but in the interim we can make good use of proven principles for forensic purposes.

# REFERENCES

[1] B. Budowle, S.E. Schutzer, A. Einseln, L.C. Kelley, A.C. Walsh, J.A. Smith, et al., Public health. Building microbial forensics as a response to bioterrorism, Science 301 (2003) 1852–1853.

[2] United States, Federal Bureau of Investigation, Laboratory FBI. Handbook of forensic services, 1999. Available from: http://www.fbi.gov/hq/lab/handbook/intro.htm.

[3] D.C. Averill, American Society of Forensic Odontology. Manual of Forensic Odontology. American Society of Forensic Odontology, 1991.

[4] J.B. Zabriskie, Essential Clinical Immunology, Cambridge University Press, Cambridge, 2009.

[5] C. Sala, D.C. Grainger, S.T. Cole, Dissecting regulatory networks in host-pathogen interaction using chIP-on-chip technology, Cell Host Microbe 5 (2009) 430–437.

[6] S.J. Popper, V.E. Watson, C. Shimizu, J.T. Kanegaye, J.C. Burns, D.A. Relman, Gene transcript abundance profiles distinguish Kawasaki disease from adenovirus infection, J. Infect. Dis. 200 (2009) 657–666.

[7] W.E. Paul, Fundamental Immunology, Wolters Kluwer Health/Lippincott Williams & Wilkins, Philadelphia, 2008.

[8] J.G. Hodge Jr., E.F. Brown, J.P. O'Connell, The HIPAA privacy rule and bioterrorism planning, prevention, and response, Biosecur. Bioterror. 2 (2004) 73–80.

[9] G.J. Annas, HIPAA regulations: A new era of medical-record privacy? N. Engl. J. Med. 348 (2003) 1486–1490.

[10] S.E. Schutzer, B. Budowle, R.M. Atlas, Biocrimes, microbial forensics, and the physician, PLoS Med. 2 (2005) e337.

[11] Federal Bureau of Investigation, Department of Homeland Security, Centers for Disease Control and Prevention. Guidance on Initial Responses to a Suspicious Letter/Container with a Potential Biological Threat. Available from: http://www.bt.cdc.gov/planning/pdf/suspicious-package-biothreat.pdf.

[12] T.G. Parslow, Medical Immunology, Lange Medical Books/McGraw-Hill Medical Pub. Division, New York, 2001.

[13] G.L. Mandell, R.G. Douglas, J.E. Bennett, Mandell, Douglas, and Bennett's Principles and Practice of Infectious Diseases, Churchill Livingstone, Philadelphia, 2000.

[14] B. Raymond, E. Batsche, F. Boutillon, Y.Z. Wu, D. Leduc, V. Balloy, et al., Anthrax lethal toxin impairs IL-8 expression in epithelial cells through inhibition of histone H3 modification, PLoS Pathog. 5 (2009) e1000359.

[15] G.W. Waterer, H. Robertson, Bioterrorism for the respiratory physician, Respirology 14 (2009) 5–11.

[16] L.M. Bush, B.H. Abrams, A. Beall, C.C. Johnson, Index case of fatal inhalational anthrax due to bioterrorism in the United States, N. Engl. J. Med. 345 (2001) 1607–1610.

[17] M.I. Meltzer, K.M. McNeill, J.D. Miller, Laboratory surge capacity and pandemic influenza, Emerg. Infect. Dis. 16 (2010) 147–148.

[18] B. Cao, X.W. Li, Y. Mao, J. Wang, H.Z. Lu, Y.S. Chen, et al., National Influenza A Pandemic (H1N1) 2009 Clinical Investigation Group of China. Clinical features of the initial cases of

2009 pandemic influenza A (H1N1) virus infection in China, N. Engl. J. Med. 361 (2009) 2507–2517.

[19] J. Lessler, N.G. Reich, D.A. Cummings, New York City Department of Health and Mental Hygiene Swine Influenza Investigation Team, Nair HP, Jordan HT, Thompson N. Outbreak of 2009 pandemic influenza A (H1N1) at a New York City school, N. Engl. J. Med. 361 (2009) 2628–2636.

[20] C. Janeway, Immunobiology: The Immune System in Health and Disease. Garland, New York, 2001.

[21] P.J. Jackson, M.E. Hugh-Jones, D.M. Adair, G. Green, K.K. Hill, C.R. Kuske, et al., PCR analysis of tissue samples from the 1979 Sverdlovsk anthrax victims: The presence of multiple *Bacillus anthracis* strains in different victims, Proc. Natl. Acad. Sci. USA 95 (1998) 1224–1229.

[22] G.L. Mandell, J.E. Bennett, R. Dolin, Mandell, Douglas, and Bennett's Principles and Practice of Infectious Diseases, Churchill Livingstone/Elsevier, Philadelphia, PA, 2010.

[23] D.M. Bravata, J.E. Holty, E. Wang, R. Lewis, P.H. Wise, K.M. McDonald, et al., Inhalational, gastrointestinal, and cutaneous anthrax in children: A systematic review of cases: 1900 to 2005, Arch. Pediatr. Adolesc. Med. 161 (2007) 896–905.

[24] W.D. Tutrone, N.S. Scheinfeld, J.M. Weinberg, Cutaneous anthrax: A concise review, Cutis 69 (2002) 27–33.

[25] M.S. Traeger, S.T. Wiersma, N.E. Rosenstein, J.M. Malecki, C.W. Shepard, P.L. Raghunathan, et al., Florida Investigation Team. First case of bioterrorism-related inhalational anthrax in the United States, Palm Beach County, Florida, 2001, Emerg. Infect. Dis. 8 (2002) 1029–1034.

[26] P.K. Dewan, A.M. Fry, K. Laserson, B.C. Tierney, C.P. Quinn, J.A. Hayslett, et al., Anthrax Response Team. Inhalational anthrax outbreak among postal workers, Washington, DC, 2001, Emerg. Infect. Dis. 8 (2002) 1066–1072.

[27] Department of Health and Human Services. Possession, use and transfer of select agents and toxins. 42 C.F.R. Part 73. 2003.

[28] A.M. Friedlander, S.F. Little, Advances in the development of next-generation anthrax vaccines, Vaccine 27 (Suppl 4) (2009) D28–D32.

[29] K. Cunningham, D.B. Lacy, J. Mogridge, R.J. Collier, Mapping the lethal factor and edema factor binding sites on oligomeric anthrax protective antigen, Proc. Natl. Acad. Sci. USA 99 (2002) 7049–7053.

[30] T. Kaeberlein, K. Lewis, S.S. Epstein, Isolating "uncultivable" microorganisms in pure culture in a simulated natural environment, Science 296 (2002) 1127–1129.

[31] J.A. Jernigan, D.S. Stephens, D.A. Ashford, C. Omenaca, M.S. Topiel, M. Galbraith, et al., Anthrax Bioterrorism Investigation Team. Bioterrorism-related inhalational anthrax: The first 10 cases reported in the United States, Emerg. Infect. Dis. 7 (2001) 933–944.

[32] B. Budowle, S.E. Schutzer, S.A. Morse, K.F. Martinez, R. Chakraborty, B.L. Marrone, et al., Criteria for validation of methods in microbial forensics, Appl. Environ. Microbiol. 74 (2008) 5559–5607.

[33] S.E. Schutzer, P. Keim, K. Czerwinski, B. Budowle, Use of forensic methods under exigent circumstances without full validation, Sci. Transl. Med. 1 (2009) 8cm7.

[34] C.P. Quinn, V.A. Semenova, C.M. Elie, S. Romero-Steiner, C. Greene, H. Li, et al., Specific, sensitive, and quantitative enzyme-linked immunosorbent assay for human immunoglobulin G antibodies to anthrax toxin protective antigen, Emerg. Infect. Dis. 8 (2002) 1103–1110.

[35] L.H. Harrison, J.W. Ezzell, T.G. Abshire, S. Kidd, A.F. Kaufmann, Evaluation of serologic tests for diagnosis of anthrax after an outbreak of cutaneous anthrax in Paraguay, J. Infect. Dis. 160 (1989) 706–710.

[36] Centers for Disease Control and Prevention, Prevalence of IgG antibody to SARS-associated coronavirus in animal traders, MMWR Morbid. Mortal. Wk. Rep. 52 (2003) 986–987.

[37] E.R. Mardis, Next-generation DNA sequencing methods, Annu. Rev. Genomics Hum. Genet. 9 (2008) 387–402.

[38] A.J. Shepherd, D.E. Hummitzsch, P.A. Leman, R. Swanepoel, L.A. Searle, Comparative tests for detection of plague antigen and antibody in experimentally infected wild rodents, J. Clin. Microbiol. 24 (1986) 1075–1078.

[39] L.R. Ashdown, Serial serum C-reactive protein levels as an aid to the management of melioidosis, Am. J. Trop. Med. Hyg. 46 (1992) 151–157.

[40] V. Chenthamarakshan, J. Vadivelu, S.D. Puthucheary, Detection of immunoglobulins M and G using culture filtrate antigen of *Burkholderia pseudomallei*, Diagn. Microbiol. Infect. Dis. 39 (2001) 1–7.

[41] A.J. Cuzzubbo, V. Chenthamarakshan, J. Vadivelu, S.D. Puthucheary, D. Rowland, P. L. Devine, Evaluation of a new commercially available immunoglobulin M and immunoglobulin G immunochromatographic test for diagnosis of melioidosis infection, J. Clin. Microbiol. 38 (2000) 1670–1671.

[42] T. Dharakul, S. Songsivilai, N. Anuntagool, W. Chaowagul, S. Wongbunnate, P. Intachote, et al., Diagnostic value of an antibody enzyme-linked immunosorbent assay using affinity-purified antigen in an area endemic for melioidosis, Am. J. Trop. Med. Hyg. 56 (1997) 418–423.

[43] B. Niklasson, C.J. Peters, M. Grandien, O. Wood, Detection of human immunoglobulins G and M antibodies to Rift Valley fever virus by enzyme-linked immunosorbent assay, J. Clin. Microbiol. 19 (1984) 225–229.

[44] P.K. Dewan, J. Grinsdale, S. Liska, E. Wong, R. Fallstad, L.M. Kawamura, Feasibility, acceptability, and cost of tuberculosis testing by whole-blood interferon-gamma assay, BMC Infect. Dis. 6 (2006) 47.

[45] R.J. Dattwyler, D.J. Volkman, B.J. Luft, J.J. Halperin, J. Thomas, M.G. Golightly, Seronegative Lyme disease. Dissociation of specific T- and B- lymphocyte responses to *Borrelia burgdorferi*, N. Engl. J. Med. 319 (1988) 1441–1446.

[46] A. Krause, G.R. Burmester, A. Rensing, C. Schoerner, U.E. Schaible, M.M. Simon, et al., Cellular immune reactivity to recombinant OspA and flagellin from *Borrelia burgdorferi* in patients with Lyme borreliosis. Complexity of humoral and cellular immune responses, J. Clin. Invest. 90 (1992) 1077–1084.

[47] S.E. Schutzer, P.K. Coyle, J.J. Dunn, B.J. Luft, M. Brunner, Early and specific antibody response to OspA in Lyme disease, J. Clin. Invest. 94 (1994) 454–457.

[48] S.E. Schutzer, P.K. Coyle, A.L. Belman, M.G. Golightly, J. Drulle, Sequestration of antibody to *Borrelia burgdorferi* in immune complexes in seronegative Lyme disease, Lancet 335 (1990) 312–315.

[49] R.S. Lasken, M. Egholm, Whole genome amplification: Abundant supplies of DNA from precious samples or clinical specimens, Trends Biotechnol. 21 (2003) 531–535.

[50] A.U. Akali, N.S. Niranjan, Management of bilateral Achilles tendon rupture associated with ciprofloxacin: A review and case presentation, J. Plast. Reconstr. Aesthet. Surg. 61 (2008) 830–834.

[51] S.L. Palin, S.C. Gough, Rupture of the Achilles tendon associated with ciprofloxacin, Diabet. Med. 23 (2006) 1386–1387.

[52] K. Ungchusak, P. Auewarakul, S.F. Dowell, R. Kitphati, W. Auwanit, P. Puthavathana, et al., Probable person-to-person transmission of avian influenza A (H5N1), N. Engl. J. Med. 352 (2005) 333–340.

# Collection and Preservation of Microbial Forensic Samples

**Jenifer A.L. Smith**

*Bioforensics Consulting, LLC, Edgewater, Maryland*

Microbial forensics was first defined as "a scientific discipline dedicated to analyzing evidence from a bioterrorism act, biocrime or inadvertent microorganism/toxin release for attribution purposes" (1). Although microbial forensics is most often discussed in the context of the needs of law enforcement, it was also recognized as a discipline requiring collaboration of traditionally separated communities with somewhat disparate missions (2). Coalescence of these communities into a dedicated national system became the goal of the U.S. government (USG) in recognition of the relevance of microbial forensics to the broader national security mission. The recent release of the National Strategy for Countering Biological Threats specifically mentions the need to enhance microbial forensic capabilities in order to expand the government's capability to "prevent, attribute, and apprehend" (3). The strategy states, "We must ensure that law enforcement, national security and homeland security communities have access to the full range of tools and capabilities needed to identify and disrupt the efforts of those with ill intent—preferably before they have the opportunity to conduct and attack—and apprehend and successfully prosecute all offenders". In concert with the concept of developing investigative leads based on forensic evidence, the multiplicity of applications of microbial forensic science is adeptly discussed by Thompson and Koeler in their article entitled the "Four Faces of Microbial Forensics" (4). They discuss the "potential applications of microbial forensics in the investigation of alleged use by nation-states or terrorist organizations, the assessment of biological weapons capabilities possessed by adversaries; the monitoring of nonproliferation agreements, such as the United Nations Security Council resolution mandating the elimination of Iraq's biological weapons program after the 1991 Persian Gulf War; and the verification of the Biological Weapons Convention". They proffer that the four communities of interest—law enforcement, intelligence, nonproliferation, and verification—should be able to use the same basic tools and techniques but that the specialized mission requirements of each will dictate application

379

Microbial Forensics. DOI: 10.1016/B978-0-12-382006-8.00022-0

of these methods. They cite differences in operating environment such as a controlled crime scene processed by law enforcement versus a more nonpermissive environment that might be encountered by a defense- or intelligence-led effort to detect potential signatures of a suspect terrorist facility. They emphasize that regardless of the specific application, it is important to validate techniques and protocols for sample collection and preservation to ensure that each community is using reliable and robust methods. Thus, as the field of microbial forensics advances, it is critical that the diversity in application by different communities is recognized so that all are mutually aware of existing capabilities and, equally important, so that the breadth of scientific expertise and enterprise launched against these daunting national security efforts are fully engaged.

Comprehensive microbial forensic programs are concerned with the development and implementation of validated technologies that address sample handling, collection, preservation, and technical analysis with interpretation of results (5). Simple in concept, yet complex in actual practice, the best methods of collection and preservation are highly dependent on the purpose behind the need for the samples. Also, the conditions under which the samples may be collected will often dictate the collection and preservation approaches that can be used. Finally, collection and preservation methods must allow for a variety of subsequent methods of analyses, such as microbial analysis of viable bacterial and infectious viruses, genetic analysis of DNA (plasmid and chromosomal), rRNA and mRNA analysis, ligand analysis (antibody, peptide, aptamer), visual analysis (light microscopy, electron microscopy), mass spectrometry analysis, and other emerging analytical methods.

In 2005, the Department of Homeland Security held a meeting at the Banbury Center of the Cold Spring Harbor Laboratory that focused on the collection, handling, and storage of microbial forensic samples. The group issued a report entitled "Quality Sample Collection, Handling and Preservation for an effective Microbial Forensics Program" (6). The authors noted that a critical element of successful investigation and ultimate attribution subsequent to a biological event involves the collection and preservation of vital microbial forensic evidence. A primary goal of collection is to obtain sufficient biological agent to support both species/strain or toxin identification for critical public health decisions and complete signature characterization for valuable lead information. Also, the collection of other relevant traditional forensic evidence must not be overlooked. Trace evidence, fingerprints, and other traditional evidence should be collected and preserved in order to support the attribution mission. This chapter builds upon the recommendations of this group and broadly discusses general concepts of collection and preservation of microbial forensic samples that are relevant to all microbial forensic communities. The chapter does not detail specific applications and protocols, as these should be developed by practitioners within these communities since they are best suited to tailor protocols to their missions. Instead, the chapter outlines general "best practices" involved in the collection of samples because these elements should be incorporated into standard collection operating procedures, no matter the specific application. Relevant references on the specific protocols for a variety of microbial forensic communities of interests are provided in this chapter, as well as new collection and preservation efforts subsequent to publication of the Banbury conference results. The chapter concludes with a look to the future, outlining areas of potential research and development concerning the collection and preservation of microbial forensic samples.

## GENERAL BEST PRACTICES OF COLLECTION OF FORENSIC EVIDENCE

In 1957, Paul Kirk best captured the potential "evidence" left by a perpetrator of a crime and the subsequent challenge faced by a forensic investigator charged with attribution of the event to that same perpetrator (7):

> Wherever he steps, whatever he touches, whatever he leaves, even unconsciously, will serve as a silent witness against him. Not only his fingerprints or his footprints, but his hair, the fibers from his clothes, the glass he breaks, the tool mark he leaves, the paint he scratches, the blood or semen he deposits or collects. All of these and more bear mute witness against him. This is evidence that does not forget. It is not confused by the excitement of the moment. It is not absent because human witnesses are. It is factual evidence. Physical evidence cannot be wrong, it cannot perjure itself, it cannot be wholly absent. Only human failure to find it, study it, and understand it can diminish its value.

In order to minimize "human failure to find it," the National Institute of Justice issued general crime scene investigation guidance to the law enforcement and first responder communities in order to protect, preserve, and process crime scenes (8). These guidelines, although designed for law enforcement officers, are applicable for other professionals who may be responsible for the collection of microbial forensic samples. Although there are examples of more tailored guidance for physicians (9,10), coroners (11), veterinarians, (12,13), public health investigators (14,15), and biological weapon treaty verifiers (16), it is important to ensure that these procedures address several basic elements of sample collection "best practices." Guidance intended to ensure protection, collection, and preservation of probative microbial forensic samples should involve:

- Assessment of the scene/situation
- Creation of the sampling plan of action
  - Safety of personnel

- Compliance with all regulations and legal requirements
- Prioritization of sampling
- Determination of appropriate personnel and equipment
- Timetable
- Documentation of
  - location, area, building, animal, subject
  - sample provenance and chain of possession
- Application of validated collection techniques and equipment
- Preservation and storage of samples

*Situation/scene assessment* of location, animal, and subject to be sampled involves applying all *a priori* (preliminary) information known to individuals involved in the investigation or situation. This assessment should include any relevant intelligence concerning the purpose of the collection effort. It will allow for development of a plan that coordinates identification, collection, and preservation of physical samples. This initial assessment will shape the collection plan and identify the prioritization of sample collection, necessary equipment and personnel, address safety concerns, and estimate the amount of resources and time required for collection.

*Creation of the sampling plan* prior to initiation of the sample collection is a critical step toward a successful sampling effort. Each plan should be uniquely designed to fit the circumstances. Overt collection plans such as those following a biocrime event or a mitigation effort will be very different from covert collection efforts. Medical sampling will differ from samples taken from agricultural sites. The purpose will dictate whether the overall strategy will require targeted or random sample collection. That being said, all sampling plans should address several common concerns. The plan should determine hazards involved in the collection effort to ensure that adequate protection is afforded all personnel; compliance with all regulations and legal requirements; necessary number and qualifications of the collection personnel; types and quantity of equipment, materials and reagents needed for collection and preservation, types of samples, the sampling approach to minimize contamination to ensure both the welfare of personnel and the integrity of the evidence; prioritization of areas to be sampled to ensure timely and methodical collection of evidence; an estimate of the number of samples to be taken; and the amount of "time on target" needed for the collection effort. Creation of the sampling plan is a critical step that should not be given short shrift. Exigent circumstances often require expedient action concerning the collection of samples; however, some planning prior to initiation of sampling that covers the elements just given is strongly encouraged.

*Documentation* prior to and during the collection effort will ensure integrity of the activity and provide a permanent record for later evaluation. It should include information on date and time; names of personnel present; written descriptions and or photographs (if possible) of the location, subject and/or animal; and current environmental conditions. Documentation establishing sample provenance and chain of possession must be maintained to ensure sample integrity. In a legal application this form of documentation establishes the "chain of custody." Judges must determine the authenticity of evidence prior to submission as evidence in a trial. They consider factors such as nature of the article, circumstances surrounding the preservation and custody of it, and likelihood of contamination or tampering (17). Chain of custody documentation is used to prove that integrity of the evidence has been maintained. Typical information recorded at the time of collection includes the sample's unique identifying number, the name or initials of the individual collecting the item, the date(s) the item(s) was collected and transferred, and a brief description of the item. Although not all microbial forensic samples may be collected in support of a law enforcement effort, chain of possession documentation is strongly recommended.

*Application of validated sampling techniques and procedures* reduces the risk of inefficient collection, degradation, and/or contamination during collection, mishandling, and loss during transport or storage. Retrieving sufficient quantities and maintaining the integrity of the evidence increase the chances of characterizing the material to conduct subsequent characterization and attribution analyses (18). Criteria for validation of methods in microbial forensics have been described previously as "the process that: 1. assesses the ability of procedures to obtain reliable results under defined conditions; 2. rigorously defines the conditions required to obtain results; 3. determines limitations of the procedures; 4. identifies aspects of the analysis that must be monitored and controlled; and 5. forms the basis for the development of interpretation guidelines to convey significance of the findings" (18).

*Preservation and storage of samples* must be addressed in any guidance relevant to the collection of microbial forensic evidence. Samples must be appropriately packaged, labeled, and maintained in a secure, temporary manner until final packaging and submission to secured storage or an analytical laboratory. Obtaining an analytical result can be affected by the manner and conditions under which a specimen is transported and stored. Storage conditions differ for some microorganisms. Because preservation of bacterial viability or viral integrity in specimens and samples will be affected by conditions, efforts should be made to store samples in appropriate media and at recommended temperatures. Additionally, the method of collection will affect the complexity of the storage requirements. Liquid or wet samples will need to be preserved differently from dry samples such as powders, dry surface swabs, or swipes. Additionally, consideration must be made for preservation of more traditional types of forensic evidence that may be present in the samples. Probative trace materials should also be preserved. Finally, shipping and transportation of samples must be considered, as there are numerous regulations concerning the transport of infectious substances. Organizations such as the International Civil Aviation Organization, the International Air Transport Association, and the U.S. Department of Transportation have promulgated requirements and regulations concerning the shipping and transportation of infections materials. The U.S. Department of Health and Human Services and the U.S. Department of Agriculture have lists of select agents and rules for the possession, use, and transfer of such agents. If a select agent or an item suspected of containing a select agent must be shipped or transported from one facility to another, both the shipper and the consignee must contact the appropriate state and federal authorities for guidance, instructions, and permission before such transfer occurs. In addition, the shipment must confirm that the recipients are approved for receiving select agents. Select agent rules are outlined in several references (19–21). Resources exist that provide current guidance concerning the handling, packaging, shipping, and storing of hazardous biological materials (15,22,23).

## COLLECTION STRATEGIES AND METHODS FOR MICROBIAL FORENSIC SAMPLING

Key to determination of the appropriate strategy for collection of microbial forensic samples is the underlying question to be answered or mission to be accomplished. The selection of locations, equipment, or subjects to be sampled, as well as to the methods of collection, will be determined by the purpose of the investigation. Sampling to quickly determine the presence and identity of a biological agent following a suspected attack or sampling to discern between a hoax or lethal bulk sample requires a different approach from sampling intended to determine the cleanliness of an area following decontamination after an accidental or intentional release. Sample collection strategy may be a "targeted" sampling strategy in which samples are collected from areas judged to have been most likely sites of contamination versus a "probability" sampling strategy in which samples are collected from random areas. In March of 2005, the U.S. Government Accountability Office (GAO) released a report concerning sample collection and analysis by the U.S. Postal Service (USPS), the Centers for Disease Control and Prevention (CDC), and the Environmental Protection Agency of various postal facilities subsequent to the anthrax attacks in 2001 (24). Conducted at the request of the Chairman of the Subcommittee on National Security, Emerging Threats and International Relations, of the House of Representatives Committee on Government Reform, GAO was asked to describe and assess the agencies' activities to detect anthrax in postal facilities; results of the agencies' testing; and whether the agencies' detection activities were validated. The report contains extensive information concerning the sampling and analytical efforts of the agencies. GAO concluded that results of the agencies' effort may not be totally reliable because the agencies used targeted as opposed to probability sampling strategies and did not use validated collection and analytical procedures. They reviewed the collection of environmental samples from either surfaces or air that were taken to determine the extent and degree of contamination, assess the risk of exposure, support decisions related to medical treatment or cleanup, and determine when cleanup was sufficient to allow an area to be reoccupied. The targeted approach used by the agencies was site specific and designed to sample areas and equipment believed to be contaminated. The CDC proffered in the report that they used the targeted approach because they needed to identify the source of contamination rapidly in order to institute early public health interventions. The USPS stated that they used a targeted method against the areas and equipment most likely to be contaminated because the collection was limited due to insufficient laboratory analytic capacity. The GAO countered that the targeting approach used by the various agencies was not sufficient, arguing that probability sampling would have better allowed agencies to determine, with some defined level of confidence when all results are negative, whether a building was contaminated.

When deciding whether to use a targeted or probabilistic sampling strategy, it is important to ask what the purpose of the investigation is. The targeted collection is an appropriate approach for law enforcement when use of a biological agent is suspected or when information on the source of a possible biological agent is available (6). The purpose for this sampling is different from that of a remediation effort as described within the GAO report. Following the release of a biological weapon there will be many reasons to conduct environmental sampling. First responders will need to identify and characterize the contamination to determine public health risk. Law enforcement will conduct sampling to obtain investigative leads for attribution. Environmental and occupational health professionals will sample to determine methods to remediate and decontaminate the building. The decision to use a targeted scheme based on judgment versus a random approach will be determined by the circumstances and by the missions of the various government agencies involved in postevent actions. There are circumstances in which results from both approaches can be utilized. One environmental sampling model that combines judgmental (targeted) and randomly placed collection of samples has been proposed to address the "cleanliness" of a building following a decontamination postevent. This Bayesian acceptance sampling model combines information derived from both targeted and randomly placed samples and is designed to be used primarily for clearance sampling after decontamination or to demonstrate the cleanliness of an area that is presumed not to be contaminated during the original event (25). In this model, judgment is used for sample collection, which is taken from locations believed to have been contaminated based on prior belief or knowledge and not in random fashion (i.e., targeted samples). If none of the judgment samples reveals detectable contamination, that information is used to reduce the number of random samples required to achieve the desired level of confidence that the room is clean.

Microbial forensic samples may be collected typically by three general approaches: bulk collection of an entire item, collection of a portion of an item to include vacuuming or collection of liquids, and swabbing or wiping the surfaces. Bulk collection is applied to items that are removed easily from a facility or area. These items are packaged appropriately, transported to a laboratory designed to contain hazardous organisms/toxins, and processed thoroughly under controlled laboratory conditions. Often locations that are processed contain many immovable surfaces and equipment. Collection from these surfaces can be conducted by a number of approaches using swabs, swipes, and vacuuming. The previously mentioned GAO report concluded that no validated collection methods were available to the federal agencies who conducted sampling efforts subsequent to the 2001 anthrax attacks. Since that report has been issued, numerous studies have been conducted that have tested various equipment and techniques designed to address the collection of microbial forensic samples. Several studies have determined and reported various recovery rates of swabs (cotton, macrofoam, rayon, and polyester), wipes, sponges, vacuum samples, and contact plates/films used to collect from both nonporous (glass, stainless steel, painted wallboard, metal, polycarbonate) and porous surfaces (carpet, concrete, cloth, brick) (26–35). In a comprehensive review of several of these studies, Edmonds reported that recovery efficiencies varied from 20 to 90%, which he attributed to variations in study design such as composition of the surface sampled, concentration of contaminant tested, method of disposition on testing surfaces of contaminant (i.e., aerosol dispersion vs. liquid dispersion), and amount of surface area being tested (36). General conclusions from these studies indicate that use of premoistened swabs or swipe material on both porous and nonporous surfaces is more effective than the use of dry swabs or wipes (29,30,35). Contact plates and/or adhesive tape has been shown to outperform swabs and swipes on flat nonporous, nonabsorbent as well as porous surfaces (35,36). Swab collection is most appropriate for small area sampling ($10-25\,cm^2$) with high agent concentration, but has limited value for large surface areas with low agent concentrations (32). Enhanced collection devices, such as the biological sampling kit (BiSKit), have proven effective as large surface sample collectors. In the BiSKit, a foam material is integrated into a screw-on lid, enabling either wet or dry sampling. It is designed to sample surfaces for bacteria, viruses, and toxins. Testing with the BiSKit demonstrated that both wet and dry samplings are equally efficient (28). Vacuuming, using equipment designed to take environmental samples and prevent cross contamination, is also an effective collection method and is useful when sampling large surface areas that would otherwise require numerous swipes or swabs (27,31).

The majority of validation efforts described have focused on very specific tools or devices that can be used for collection. More comprehensive "whole protocol" validation studies that involve operational applications by field operators are more difficult to find in the published literature. Beecher provides an excellent example of "real-world" sampling efforts used by the Federal Bureau of Investigation to search for mail contaminated during the 2001 anthrax attack. It was necessary to develop a unique approach to ensure that the collection of evidence was done utilizing methods to reduce additional contamination (37). Validation of the methods was conducted contemporaneously with sample collection. Another example of operational validation is a collaborative study that was conducted to validate collection methods for use by first responders who encounter visible powders suspected of being biological agents. Following the anthrax attack in 2001, there was heightened public awareness of unidentified "white powders." When responding to a visible "white powder" event, emergency personnel followed local standard operation procedures, which varied from region to region. The U.S. Department of Homeland Security (DHS) recognized that standardizing sample collection practices was needed to ensure that sufficient sample was available for laboratory and on-site testing to identify the substance expeditiously; potential criminal evidence was preserved; and chain of custody was traceable. The DHS convened a sampling standard task group representing 11 government agencies and one biodefense laboratory who worked together to develop the consensus standard "Standard Practice for Bulk Sample Collection and Swab Sample Collection of Visible Powders Suspected of Being Biological Agents from Nonporous Surfaces" as a draft American Society for Testing and Materials (ASTM) standard (38). Subsequently, they conducted a study using six teams of civil support personnel and first responders to validate the method of collection described in the draft ASTM standard. The study concluded that the sample collection procedure allows for preservation of the unadulterated bulk sample for laboratory analysis and sufficient residue remains for on-site biological analysis (39).

## LOOKING TO THE FUTURE

In 2005, several recommendations were made to help focus research and development efforts to close microbial forensic collection capability gaps (6). These recommendations included

- Compiling a database of existing U.S. government collection, storage, and transport protocols. This would allow preliminary comparison of methods to determine which are sufficiently validated; reduce duplication of effort and allow scientists to build on previous knowledge; and facilitate development of best operation plans.
- Validating collection protocols using a broad spectrum of bacterial species or strains, viruses, and toxin.
- Developing discipline-wide validation criteria. These criteria should include sensitivity, specificity, recovery efficiency, maintenance of integrity, impact on analytical assays and baseline disease and pathogen data.
- Establishing microbial forensic sample collection guidelines that can be used by multiple communities. Principles and guidance used for the collection of traditional forensic evidence can serve as a basis for this type of guidance. These guidelines should be available to each community that may be responsible for the collection of microbial forensic samples.

Since 2005, significant efforts have been made to improve and validate sampling technology. There remain areas for improvement in the standardization of evaluation criteria. Greater appreciation is needed to evaluate real-world samples. The majority of previously mentioned studies were conducted on very uniform surfaces such as glass, stainless steel, or plastic. Conducting testing on a wider variety of surfaces will be important to determine the best method of collection across a range of surface compositions. It has been suggested that it may be more beneficial to first responders and sample analysts if consistency in recovery effectiveness across a range of surface compositions is considered in the definition of recovery efficiency (36). Closer attention also needs to be given to the methods in "seeding" the surfaces to be tested. Some experiments coated the test surfaces with liquid suspensions of bacteria, allowing them to air dry, while other experiments applied bacteria to the surfaces by aerosolization. Variability in recovery efficiencies occurred as a result of test surface preparation. Thus, standardization of experimental design is needed to refine recovery efficiency estimates of various collection techniques (33).

Finally, guidance concerning the path forward to improve; sample collection, processing, preservation, and recovery and concentration of microbial pathogens and their signatures in microbial forensic samples is found in the U.S. government's recently released National Strategy to Support Research in Microbial Forensics Attribution Investigations and National Security (40).

The purpose of this strategy is "to guide and focus the research efforts of the USG to advance the discipline of microbial forensics and provide the nation with the most scientifically sound and statistically defensible capability to provide scientific data to support attribution investigations of a potential or actual biological attack" (40). Within the strategy, several actions are outlined concerning needed research and development efforts to improve sample collection, processing, preservation, recovery, and concentration of microbial pathogens and their signatures from collected samples for microbial forensic analyses. They are as follows.

1. Collect and evaluate all work conducted across the U.S. government and academic sectors that has focused on the collection of microbial samples, preservation, recovery, and concentration of microbial agents and their signatures from collected samples and use results of the evaluation to identify current research gaps and consolidated research efforts to avoid duplication.
2. Develop and improve methods for the collection, processing, preservation, and recovery of microbial agents and their signatures from microbial forensic samples that do not interfere with subsequent forensic analyses of the sample.
   a. Methods are needed to collect, process, and recover a wide range of human, animal, and agricultural microbial agents from a broad range of common surfaces, matrix types, and sample collection devices.
   b. Methods are needed for the collection of trace microbial forensic evidence applicable to the recovery of viable organisms and nonviable trace signatures from a variety of sample collection environments.
   c. An interagency working group shall be identified or, if none exists, formed to develop scientifically acceptable standards of performance and the path to validate the approaches.

Elevation of these issues within the national strategy will allow various USG agencies to continue efforts to develop and refine collection methods for all communities interested in microbial forensics. Greater collaboration among these disparate but capable communities will serve to hasten improvements in this arena.

# REFERENCES

[1] B. Budowle, S.E. Schutzer, A. Einseln, L.C. Kelley, A.C. Walsh, J.A.L. Smith, et al., Public health. Building microbial forensics as a response to bioterrorism, Science 301 (2003) 1852–1853.
[2] B. Budowle, J.A. Beaudry, N.G. Barnaby, A.M. Giusti, J.D. Bannan, P. Keim, Role of law enforcement response and microbial forensics in investigation of bioterrorism, Croat. Med. J. 48 (4) (2007) 437–449.

[3] National Strategy for Countering Biological Threats, 2009. Available from: www.whitehouse .gov/sites/default/files/National_Strategy_for_Countering_BioThreats.pdf.

[4] J.B. Tucker, G.D. Koblentz, The four faces of microbial forensics, Biosecur. Bioterror. 7 (4) (2009) 389–397.

[5] B. Budowle, S.E. Schutzer, M.S. Ascher, R.M. Atlas, J.P. Burans, R. Chakraborty, et al., Toward a system of microbial forensics: From sample collection to interpretation of evidence, Appl. Environ. Microbiol. 71 (2005) 2209–2213.

[6] B. Budowle, S.E. Schutzer, J.P. Burans, D.J. Beecher, T.A. Cebula, R. Chakraborty, et al., Quality sample collection, handling, and preservation for an effective microbial forensics program, Appl. Environ. Microbiol. 72 (2006) 6431–6438.

[7] P.L. Kirk, Crime Investigation: Physical Evidence and the Police Laboratory, Interscience Publishers, Inc., New York, 1953.

[8] Crime Scene Investigation, U.S. Department of Justice, 2000.

[9] S.E. Schutzer, B. Budowle, R.M. Atlas, Biocrimes, microbial forensics and the physician, PloS Med. 2 (12) (2005) e337.

[10] K. Yeskey, S.A. Morse, Physician recognition of bioterrorism-related disease, in: M.J. Roy (Ed.), Physician's Guide to Terrorist Attack, Humana Press, Totowa, NJ, 2003, pp. 39–46.

[11] K.D. Nolte, R.L. Hanzlick, D.C. Payne, A.T. Kroger, W.R. Oliver, Medical examiners, coroners, and biologic terrorism: A guidebook for surveillance and case management, MMWR Recomm. 53 (2004) 1–27.

[12] S.A. McEwen, T.M. Wilson, D.A. Ashford, E.D. Heegaard, T. Kuiken, B. Kournikakis, Microbial forensics for natural and intentional incidents of infectious disease involving animals, Rev. Sci. Tech. 25 (1) (2006) 329–339.

[13] B. Stacy, Handling of evidence: Maintaining admissibility. Available from: www.vet.uga.edu/ vpp/IA/SRP/vfp/admiss.html.

[14] Comprehensive Procedures for Collecting Environmental Samples for Culturing *Bacillus anthracis* CDC 2002. Available from: www.bt.cdc.gov/agent/anthrax/environmental-sampling-apr2002.asp.

[15] World Health Organization. 2000 Guidelines for collection of clinical specimens during field investigation of outbreaks, p. 6. Basic safety precautions. World Health Organization Department of Communicable Disease Surveillance and Response. Available from: http://www.who.int/csr/resources/publications/surveillance/WHO_CDS_CSR_EDC_2000_4/em/.

[16] A.J. Mohr, Biological sampling and analysis procedures for the United Nations special commission (UNSCOM) in Iraq, Politics Life Sci. 14 (2) (1995) 240–243.

[17] N.T. Kuzmack, Legal aspects of forensic science, in: R. Saferstein (Ed.), Forensic Science Handbook, Prentice-Hall, NJ, 1982, pp. 22–23.

[18] B. Budowle, S.E. Schutzer, S.A. Morse, K. Martinez, R. Chakraborty, B.L. Marrone, et al., Criteria for validation of methods in microbial forensics, Appl. Environ. Microbiol. 74 (18) (2008) 5599–5607.

[19] Animal and Plant Inspection Service, U.S. Department of Agriculture. Agricultural bioterrorism protection act of 2002: Possession, use, and transfer of biological agents and toxins; final rule 2005 (7 CFR Part 331 9 CFR Part 121). Fed. Regist. 70, 13242–13292.

[20] Centers for Disease Control and Prevention and the Office of the Inspector General, U.S. Department of Health and Human Services. Possession, use, and transfer of select agents and toxins final rule 2005 (42 CFR Parts 72 et al.). Fed. Regist. 70, 13294–13325.

[21] Centers for Disease Control and Prevention and U.S. Department of Health and Human Services. 2005. Possession, use, and transfer of select agents and toxins–reconstructed replication competent forms of the 1918 pandemic influenza virus containing any portion

of the coding regions of all eight gene segments. Interim final rule 2005 (42 CFR Part 73). Fed. Regist. 70, 61047–61049.

[22] Sentinel laboratory guidelines for suspected agents of bioterrorism and emerging infectious diseases. Available from: www.asm.org/images/pdf/PackingandShipping1-08.pdf.

[23] D.L. Sewell, Laboratory safety practices associated with potential agents of biocrime or bioterrorism, J. Clin. Microbiol. 41 (7) (2003) 2801–2809.

[24] GAO Anthrax Detection Agencies need to validate sampling activities in order to increase confidence in negative results 2005. Available from: www.gao.gov/cgi-bin/getrpt? GAO-05-251.

[25] L.H. Sego, K.K. Anderson, B.D. Matzke, K. Sieber, S. Shulman, Bennett, et al., An environmental sampling model for combining judgmental and randomly placed samples. 2007 PNNL 16636.

[26] M.S. Favero, J.J. McDade, J.A. Robertsen, F.K. Hoffman, R.W. Edwards, Microbiological sampling of surfaces, J. Appl. Bact. 31 (3) (1968) 336–343.

[27] W.T. Sanderson, M.J. Hein, L. Taylor, B.D. Curwin, G.M. Kinnes, T.A. Seitz, et al., Surface sampling methods for *Bacillus anthracis* spore contamination, Emerg. Infect. Dis. 8 (10) (2002) 1145–1151.

[28] M.P. Buttner, P. Cruz, L.D. Stetzenbach, A.K. Klima-Comba, V.L. Stevens, P.A. Emanuel, Evaluation of the biological sampling kit (BiSKit) for large area surface sampling, Appl. Environ. Microbiol. 70 (12) (2004) 7040–7045.

[29] L. Rose, B. Jensen, A. Peterson, S.N. Banerjee, M.J. Arduino, Swab materials and *Bacillus anthracis* spore recovery from nonporous surfaces, Dis. Emerg. Infect. 10 (6) (2004) 1023–1029.

[30] G.S. Brown, R.G. Betty, J.E. Brockmann, D.A. Lucero, C.A. Souza, K.S. Walsh, et al., Evaluation of a wipe surface sample method for collection of Bacillus spores from nonporous surfaces, Appl. Environ. Microbiol. 73 (3) (2007) 706–710.

[31] G.S. Brown, R.G. Betty, J.E. Brockmann, D.A. Lucero, C.A. Souza, K.S. Walsh, et al., Evaluation of vacuum filter sock surface sample collection method for Bacillus spores from porous and non-porous surfaces, J. Environ. Monit. 9 (7) (2007) 666–671.

[32] G.S. Brown, R.G. Betty, J.E. Brockmann, D.A. Lucero, C.A. Souza, K.S. Walsh, et al., Evaluation of rayon swab surface sample collection method for Bacillus spores from nonporous surfaces, J. Appl. Microbiol. 103 (4) (2007) 1074–1080.

[33] J.M. Edmonds, P.J. Collett, E.R. Valdes, E.W. Skowronski, G.J. Pellar, P.A. Emanuel, Surface sampling of spores in dry-deposition aerosols, Appl. Environ. Microbiol. 75 (1) (2009) 39–44.

[34] C.F. Estill, P.A. Baron, J.K. Beard, M.J. Hein, L.D. Larsen, L. Rose, et al., Recovery efficiency and limit of detection of aerosolized *Bacillus anthracis* Sterne from environmental surface samples, Appl. Environ. Microbiol. 75 (13) (2009) 4297–4306.

[35] D.A. Frawley, M.N. Samaan, R.L. Bull, J.M. Robertson, A.J. Mateczun, P.C. Turnbull, Recovery efficiencies of anthrax spores and ricin from nonporous or nonabsorbent and porous or absorbent surfaces by a variety of sampling methods, J. Forensic Sci. 53 (5) (2008) 1102–1107.

[36] J.M. Edmonds, Efficient methods for large area surface sampling of sites contaminated with pathogenic microorganisms and other hazardous agents: Current state, needs and perspectives, Appl. Microbial. Biotechnol. 84 (5) (2009) 811–816.

[37] D.J. Beecher, Forensic application of microbiological culture analysis to identify mail intentionally contaminated with *Bacillus anthracis* spores, Appl. Environ. Microbiol. 72 (8) (2006) 5304–5310.

[38] ASTM Standard ASTM E2458-06. Standard practice for bulk sample collection and swab sample collection of visible powders suspected of being biological agents from nonporous surfaces, 2006. Available from: www.astm.org/Standards/E2458.htm.

[39] L.E. Locascio, B. Harper, M. Robinson, Standard practice for bulk sample collection and swab sample collection of visible powders suspected of being biological agents from non-porous surfaces: Collaborative study, J. AOAC Int. 90 (1) (2007) 299–333.

[40] National Strategy to Support Research in Microbial Forensics Attribution Investigations and National Security, 2009. Available from: www.ostp.gov/galleries/NSTC%20Reports/National%20 MicroForensics%20R&DStrategy%202009%20UNLIMITED%20DISTRIBUTION.pdf.

# CHAPTER 23

# Sampling for Microbial Forensic Investigations

**Sushil K. Sharma, Hazel Bailey, and Jack Melling**
*U.S. Government Accountability Office*

## INTRODUCTION

There have been very few incidents of real biological attacks by terrorists or other criminals, as opposed to hoaxes. With real attacks, attribution to specific perpetrators—including tracing the microbial agent to particular sources—has been difficult and time-consuming. Some biological attacks may not initially be perceived as real if the resultant disease also occurs naturally. For example, before 2001, the most significant biological attack in the United States involved the 1984 contamination, with the biological agent *Salmonella typhimurium*, of several local salad bars in Oregon by the followers of Rajneesh Foundation. Through this attack, 751 people contracted salmonellosis (1). Of these, 45 were hospitalized, but there were no fatalities.

The microbial epidemiological investigation conducted by the Oregon Public Health Division and the Centers for Disease Control and Prevention (CDC) was effective in determining the actual source of the contamination, but it failed to recognize the intentional nature of the contamination of food by individuals. This outbreak was therefore treated as a normal public health outbreak. Only later, as a result of self-disclosure and without a detailed microbial forensic (criminal) investigation, were the perpetrators identified.

In contrast, the anthrax attack of 2001 was on a much wider scale than the *Salmonella* attack. In all, 22 people contracted anthrax in four states and Washington, D.C., with 5 people dying. In September and October 2001, contaminated letters containing *Bacillus anthracis* (commonly referred to as anthrax) were sent through the mail to two senators (Thomas Daschle and Patrick Leahy) and several members of the media. On October 5, 2001, the death of an American Media Inc. employee in Florida from inhalation anthrax disease triggered an investigation by several federal agencies, including the CDC, the Department of Defense (DOD), the Environmental Protection Agency (EPA), and the Federal Bureau of Investigation (FBI). Since a contaminated

393

Microbial Forensics. DOI: 10.1016/B978-0-12-382006-8.00023-2

envelope or package was not recovered in Florida, federal agencies could not initially establish how the anthrax was delivered—whether by U.S. mail or some other means, such as courier. The combination of the Florida incident and the October 15 letter to Senator Daschle established the link to the U.S. mail system. Processing of the letters containing anthrax on high-speed mail-sorting machines at the U.S. Postal Services contaminated not only the postal facilities themselves, but also cross contaminated other mail in those facilities. As a result, numerous federal and private facilities were also considered to be actually or potentially contaminated. Identifying contaminated facilities proved highly challenging for the sampling and analytical capabilities then available (2). Further, anthrax spores released in a U.S. Senate office building were reaerosolized during common office activities, which posed an additional challenge.

This contamination was clearly the result of a criminal act, given that (i) human anthrax is rare in nature and normally involves direct contact with infected animals or animal products and (ii) some of the mailed envelopes contained anthrax powder as well as letters stating this. Nevertheless, the identity of the alleged perpetrator and the source of the contaminating agent remained unknown for almost 8 years. In both the 1984 *Salmonella typhimurium* contamination and the 2001 anthrax attack, responsible authorities had to take several steps to respond to the incidents, including determining (i) what that agent was, (ii) how to treat people who had become ill from the contaminating microbial agent, and (iii) the extent of the contamination so that affected areas could be decontaminated. Two types of investigations were used: microbial epidemiological and microbial forensics, particularly in the 2001 incident. While microbial epidemiological investigation into the source of outbreaks of microbial diseases is a common practice, microbial forensic investigation is an emerging area, which has assumed more importance because of intentional incidents such as those just discussed, for which there is a need to both prosecute the perpetrator and deter future attacks. Accordingly, it is essential that microbial forensics methodologies and procedures are both validated and performed so that the evidence generated is legally robust and admissible.

## COMMONALTIES AND DIFFERENCES BETWEEN MICROBIAL EPIDEMIOLOGICAL AND MICROBIAL FORENSIC INVESTIGATIONS

Both microbial epidemiological and microbial forensic investigations have many commonalities with regard to collection and analysis of samples (3). However, there are some important differences, particularly with regard to the purpose or goal of the investigations. The goal of microbial epidemiological investigations is to effectively treat sick, as well as exposed, people and to contain the disease. This is accomplished by identifying the agent and exposed population and by determining the extent of contamination. The goal of microbial forensic investigations is to determine the source of the agent, identify the perpetrator, and present evidence to a court in order to convict the perpetrator successfully. Microbial forensic investigations therefore may require additional evidence, such as chemical and physical signatures, including by-products. Traditional forensic evidence is also required—such as fingerprints, computer records, and trace evidence—to provide clues to the identity of the perpetrator. To be legally admissible, evidence from a microbial forensic investigation must meet the standards for evidence of the scientific community as well as those required by a criminal court.

While microbial epidemiological and microbial forensic investigations have different goals, they use mostly the same tools. Both investigations start with first determining a sampling strategy, followed by selecting tools and methods for sample collection and analysis. This chapter uses the 2001 anthrax attack to highlight the process of such investigations.

## PROBLEMS ASSOCIATED WITH SAMPLING ACTIVITIES FOR BOTH MICROBIAL EPIDEMIOLOGICAL AND MICROBIAL FORENSIC INVESTIGATIONS

Microbial epidemiological investigation—involving the the CDC, EPA, and USPS—and microbiological forensic investigation—involving the FBI—sampled 286 postal facilities in 2001 to detect anthrax. To do this, a series of activities was required: (i) sampling strategy development, (ii) sample collection, (iii) transportation, (iv) extraction, and (v) analysis of samples (see Figure 23.1).

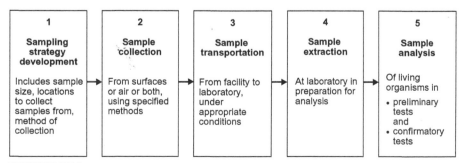

**FIGURE 23.1**
Agency sampling activities. *Source: GAO analysis of CDC, EPA, and USPS data.*

## Activity 1: Sampling Strategy

Activity 1—common to both types of investigations—involved federal investigators' developing a sampling strategy, which included deciding how many samples to collect, where to collect them from, and what collection methods to use. The investigators—based on their best technical judgment—used primarily a targeted strategy: they collected samples from specific areas considered likely to be contaminated. Such judgment can be effective in some situations, for example, in determining (i) the source of contamination in a disease outbreak investigation or (ii) whether a facility is contaminated when information on the source of potential contamination is definitive. However, in the case of a negative finding—when the source of potential contamination is not definitive—the basic question—is this building contaminated?—will remain unanswered.

In the initial sampling strategy for the 2001 investigation of the postal facilities, neither the microbial epidemiological investigators nor the microbial forensic investigators used probability sampling. In the case of a negative result, probability sampling would have allowed them to determine—with some defined level of statistical confidence—whether a facility was contaminated. A known level of confidence is needed to make informed decisions about the need for decontamination because evidence suggests that even a few anthrax spores may cause disease in susceptible individuals. In choosing targeted sampling, the microbial epidemiological investigators may have had different reasons from those of the microbial forensic investigators. For example, for microbial epidemiological investigation, targeted sampling may have been the most expeditious approach for quickly identifying contamination in facilities. Identification could support public health measures, such as decisions on the need to provide appropriate treatments for those ill or potentially exposed to the agent. In addition, the number of samples these investigators could collect was limited due to insufficient laboratory analytic capacity.

## Activity 2: Methods for Sample Collection

Activity 2 involved selecting methods for collecting samples. While some investigators used dry swabs to collect samples (the least effective method) (4), others used several methods—dry swabs, premoistened swabs, wet wipes, and a high-efficiency particulate air (HEPA) vacuum—in various combinations or alone.

However, none of the collection methods used in the investigations was evaluated and validated for anthrax detection in environmental samples. Information on the efficiency of a few sample collection methods was provided in published literature. In all the methods studied, swabs were always premoistened before samples were collected. However, according to one study, this most efficient method caused problems when used with certain analytic methods. In the absence of empirical research, agencies had no information (i) available for reliably choosing one method over another and (ii) on the limits of detection to use when evaluating negative results (5).

## Activity 3: Transporting Samples

Activity 3 involved transporting samples to laboratories for analysis. Transportation was done according to federal regulations for transporting "infectious substances"; regulations were designed primarily to prevent an inadvertent release of an infectious agent rather than maintaining the samples' biological integrity for subsequent testing (6). This dichotomy was less important in the case of the anthrax letters because anthrax spores are robust compared with other pathogenic microorganisms.

During the transportation phase, several factors could affect results. It is therefore important to know (i) specific transit times for delivering all samples to laboratories, (ii) whether sample transportation was delayed, (iii) if it was delayed, for how long, (iv) environmental conditions the samples were shipped under or when they were received at laboratories, and (v) the degree to which spores could have been exposed to varying environmental conditions (e.g., temperature, ultraviolet light, chemicals) from the time of release to the time of sample collection, which could have affected sample integrity. Whether transportation affected spore viability cannot be known because the conditions of transportation were not validated. These are important issues in the event of negative findings.

## Activity 4: Extracting Particles from Samples

Activity 4 involved laboratory personnel, using extraction fluids and procedures specified by the laboratory, extracting particles from the sample material. However, because not all laboratories used the same procedures and because no efficiency data on sample extraction were available, interpreting anthrax analytic results was problematic.

Several factors could have affected extraction efficiency. For example, the degree to which swabs or wipes can retain spores depends on their material composition. Cotton is more retentive than some artificial fibers such as rayon and may be more difficult for spore extraction. In addition, cotton swabs are characterized by a lipid matrix, which gives poor results for culture.

## Activity 5: Analyses of Environmental Samples

Final activity 5 was analysis of environmental samples. This analysis involved a variety of laboratory analytic methods and required two types of tests—preliminary and confirmatory—to generate a final result. These analytic methods, although used for detecting anthrax in clinical samples, had not been used for environmental samples. As a result, different analytic approaches were adopted for preliminary tests.

Samples deemed positive through preliminary testing were not always confirmed as positive, as was to be expected. However, actions taken based on preliminary tests could result in treatment decisions that might not have been otherwise necessary. In addition, field-based analytic methods, such as hand-held assays, were also used as a preliminary test. According to the CDC, results from hand-held assays were unreliable (http://www2a.cdc.gov/HAN/ArchiveSys/ViewMsgV.asp?AlertNum=00037). A subsequent study confirmed this (7). However, once sample extracts were subjected to the required confirmatory tests at the laboratory, a positive result was indeed a positive.

In analyzing samples, laboratories used a variety of analytic methods for preliminary and confirmatory testing. Preliminary tests included colony morphology, Gram's stain, hemolysis, and motility tests. Any culture isolates that could not be ruled out in preliminary testing were considered presumptively positive and referred for confirmatory testing. Confirmatory tests included culture analyses (traditional microbiological and biochemical analyses); $\gamma$ phage lysis (a test that identifies susceptibility of the organism to anthrax-specific viruses that create a kill zone in anthrax cultures); and direct fluorescent antibody assay, or antibody analyses, employing a two-component test that detects the cell wall and capsule, or outer covering, produced by vegetative cells of anthrax.

The test results were reported as positive—anthrax was found—or negative—anthrax was not found. Traditional microbiological analyses require 18 to 24 hours before a result can be generated, depending on the laboratory protocols and procedures. In a few instances, results were also reported as number of colony-forming units per gram of sample material. Additional tests, such as molecular subtyping, were also conducted to determine what strain of anthrax was involved.

The problems in preliminary testing in the 2001 investigation included training, quality control, and use of field-based analytic methods with limitations that were not well understood at that time. In preliminary testing, a suspect organism must first be selected; at this point, human error or quality control issues can affect the results. For example, one problem involving culture in preliminary tests is reliance on the naked human eye to identify the growth of anthrax on a petri dish. Many different types of organisms could be growing that looked like, but were not, anthrax. This is significant because when negative results were obtained during preliminary testing, no further testing was to be done.

Other problems can also affect the reliability of laboratory results.

False negatives can result from not using positive controls in performing a specific test. For example, a defective reagent can cause a test to malfunction and not reveal anthrax.

Results of public health testing in 286 postal facilities were largely negative. Public health investigators sampled facilities that processed mail from primary facilities to determine whether any other facilities had become contaminated. The majority of test results from these facilities were negative: of 283 facilities sampled, excluding the 3 primary facilities, 20 tested positive and 263 negative (see Figure 23.2).

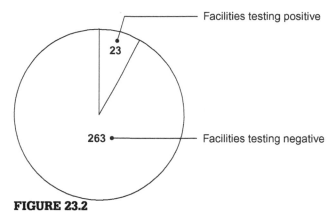

**FIGURE 23.2**

Test results were largely negative. *Source: GAO analysis of CDC, EPA, and USPS data.*

However, negative test results do not necessarily mean that a facility is free from contamination. Results can be negative if (i) samples were not collected from places where anthrax was present, (ii) the detection limit of the method was greater than the actual contamination level, (iii) not enough samples were collected, (iv) not enough spores were recovered from sample material, (v) analysis of the sample extract did not detect anthrax spores, or (vi) anthrax was not present in the facility. Of 286 facilities, 23 tested positive. For 2 of these 23 facilities, test results were negative at first, but positive on subsequent testing. However, in one of these facilities, it was not until the fourth testing that positive results were eventually obtained.

## Activities 1 to 5: Many Variables Can Affect Results

All the activities discussed previously are interdependent, and the many variables for each one can affect the results. Furthermore, problems associated with activities 1 to 5 could affect the validity of the results generated by the overall process. Given that there are so many variables, the use of different sample collection strategies, reflected in site-specific sampling plans, could yield different results. For example, three potential sampling plans could be used in one facility—plan A, using one collection method (e.g., a swab); plan B, using two methods (e.g., a swab and wipe); and plan C, using three methods (e.g., swab, wipe, and HEPA vacuum). How these collection methods are to be applied—that is, how they are physically used and how much area each sample covers—is a variable. Within each plan, sample transportation protocols could differ, involving variables such as (i) temperature: plans A and B might require transporting at ambient temperature, while plan C might require freezing temperature, (ii) moistness of the sample collection method during transport, and (iii) size and construction of the packaging.

In addition, within each sampling plan, laboratory extraction and analytic protocols used for those particular samples could differ, involving variables such as (i) different formulations of extraction fluids from different manufacturers,

(ii) different ways to physically release spores from a particular collection method (such as a swab) into the liquid extract (such as by shaking, vortexing, or sonicating), and (iii) a combination of analytic methods, such as culture or polymerase chain reaction for DNA amplification to identify anthrax. Any problems experienced with any of these variables across any of these plans could affect the final results. This is why empirical validation of the methods and the overall process is essential.

## CHALLENGES FOR MICROBIAL EPIDEMIOLOGICAL AND MICROBIAL FORENSIC INVESTIGATIONS

There are two critical challenges to investigation of a biological attack: (i) validating the methods and overall processes from sample collection through to analysis (common to both epidemiological and forensic investigation) and (ii) developing scientific techniques for attribution of an attack to a specific perpetrator (specific to forensic investigations) (8).

Regarding the first challenge, validation, as it is generally understood, is a formal, empirical process in which the overall performance characteristics of a given method are determined and certified by a validating authority as (i) meeting the requirements for the intended application and (ii) conforming to applicable standards.

Federal agencies involved in investigating the 2001 anthrax attack took some public health-related actions to respond to incidents related to bioterrorism, but they were not fully prepared for the nature of the 2001 attack. No agency activity to detect anthrax contamination in the postal facilities had been validated prior to the event. Because validation for select agents is complex and time-consuming, it was not possible for the agencies involved to perform validation studies during the emergency response itself. Therefore, because these agencies—the CDC, DOD, EPA, and FBI—did not use an empirical process to validate their testing methods, they had limited information available for reliably choosing one method over another and no information on the detection limit to use when evaluating negative results.

Without validation, sampling activities could be based on false assumptions. For example, lack of validated sample collection methods means that it is not known how many spores a particular method will collect from a surface and, thus, which method is appropriate for a given situation. Using an ineffective method or procedure could result in a finding of no contamination when in fact there is contamination—a false negative. In addition, because environmental sampling methods for anthrax are still not validated, to what extent these methods will underestimate contamination is unknown. Thus, in the case of a negative result, there would be no sound basis for taking public health measures for the occupants of a possibly contaminated facility.

Validating the overall process is important to both types of investigations because operational and health-related decisions are made on the basis of testing results generated by that process, and this information is also used as part of evidence in a criminal court. In addition, validation would offer assurance that results of using a particular method that is part of the overall process are robust enough to be reproduced, regardless of which agency, contractor, or laboratory is involved. Thus, agencies and the public could be reasonably confident that any test results generated by the process that includes that specific method would be reliable and, in particular, any negative results would mean that a sample was free from contamination (within the limits of detection of the method). However, it is important to note that validation is an expensive and time-consuming activity.

Regarding the second challenge, significant progress has been made toward attributing the 2001 anthrax attack to a specific perpetrator, leading to a major breakthrough in the investigation. Although full genome analysis of anthrax cultured from contaminated letters was performed and compared with Ames anthrax cultures from various research laboratories, those full genome sequences showed no differences at all. However, a chance observation of cultures from the contaminated letter growing on agar plates found a few colonies that had a different appearance from the majority. When the DNA from these colonies was sequenced, 10 mutations were found that differed from the normal Ames sequence. Because organisms with these mutations made up a small fraction of the total, they had not been detectable when anthrax from the contaminated letters or laboratory samples without prior colony selection were cultured and sequenced. Next, 1072 Ames anthrax samples from multiple laboratories, including the U.S. Army Medical Research Institute of Infectious Diseases (USAMRIID), were screened for 4 of the 10 mutations. Out of all 1072 samples, eight cultures showed all four mutations. One of these cultures came directly from a flask at USAMRIID, while the remaining seven subcultures had been derived from that flask. Investigators concluded that the source of anthrax in the letters was the flask at USAMRIID.

This combination—traditional microbiology in visualizing mutant colonies together with the cutting-edge genetic analysis used to characterize them—represented a major breakthrough in the investigation. However, it is not clear why it was assumed that anthrax in the letters came from the flask and not from one of the seven subcultures. Because the alleged perpetrator committed suicide before a trial could be held, it cannot be known whether the microbial forensic evidence collected by the FBI would have met the standards required by the courts and jurors to support a guilty verdict.

From the point of view of using similar methods in the future, a key challenge is validation not just of the visual selection and analytic methods used, but also of the frequency and probability with which such mutations occur. Because all the Ames anthrax originally came from a single source, it is important to know how much subculturing is required to allow such mutations to occur, whether the pattern of mutations can be replicated, and, if so, with what probability.

## CONCLUSION

An ability to determine the source, type, and extent of a biological attack reliably and accurately is critical not only to the public health response but also for successful criminal prosecution of the perpetrator. Environmental samples are one of the key elements of evidence in both epidemiological and forensic investigations. Therefore, it is critical that methods used for sample collection and analyses are validated. In terms of public health, even an unconfirmed positive result from analysis of environmental samples can usefully initiate prophylaxis, treatment, or both, as well as implementation of containment measures. These measures are based on the philosophy that prevention is better than cure.

However, for forensic investigation, a higher level of certainty is needed than that for epidemiological investigation: beyond reasonable doubt becomes the standard that all of the methods, individually and collectively (including the chain of custody), must meet. Attaining a higher level of certainty has resource implications for conducting forensic investigations, as well as requirements for methods that must meet the standards of legal acceptability.

Arguably, microbial forensics is an infant science whose development was forced because of the need to tie the 2001 anthrax attack to a specific perpetrator, by tracing the source of the biological material to a particular laboratory. It is likely that additional methods will be developed to improve the ability to pinpoint sources of microbes other than anthrax. Such improved methods will have value not only in forensic investigations, but also by enhancing the capability and robustness of epidemiological investigations.

## DISCLAIMER

The views expressed in this chapter are those of the authors and do not reflect the official position of the GAO.

# REFERENCES

[1] T.J. Török, R.V. Tauxe, R.P. Wise, J.R. Livengood, R. Sokolow, S. Mauvais, et al., A large community outbreak of salmonellosis caused by intentional contamination of restaurant salad bars, JAMA 278 (5) (1997) 389–395.

[2] GAO, Anthrax Detection: Agencies Need to Validate Sampling Activities in Order to Increase Confidence in Negative Results, GAO-05-251, Washington, DC, March 31, 2005.

[3] B. Budowle, S.E. Schutzer, M.S. Ascher, R.M. Atlas, J.P. Burans, R. Chakraborty, et al., Toward a system of microbial forensics: From sample collection to interpretation of evidence, Appl. Environ. Microbiol. 71 (5) (2005) 2209–2213.

[4] W.T. Sanderson, M.J. Hein, L. Taylor, B.D. Curwin, G.M. Kinnes, T.A. Seitz, et al., Surface sampling methods for *Bacillus anthracis* spore contamination, J. Emerg. Infect. Dis. 8 (2002) 1145–1150.

[5] M.P. Buttner, P. Cruz-Perez, L.D. Stetzenbach, Enhanced detection of surface-associated bacteria in indoor environments by quantitative PCR. Appl. Environ. Microbiol. 67 (2001) 2564–2570.

[6] Department of Transportation, 49 C.F.R. subchapter C—Hazardous Materials Regulation.

[7] FBI and CDC, Preliminary findings on the evaluation of hand-held immunoassays for *Bacillus anthracis* and *Yersinia pestis*, Forensic Sci. Commun. 5 (1) (2003).

[8] P.S. Keim, T. Pearson, B. Budowle, M. Wilson, D.M. Wagner, Microbial forensic investigations in the context of bacterial population genetics, in: Microbial Forensics, second ed. Elsevier, 2011.

# CHAPTER 24

# Toxin Analysis Using Mass Spectrometry

**Rudolph C. Johnson, Suzanne R. Kalb, and John R. Barr**

*Centers for Disease Control and Prevention, National Center for Environmental Health, Division of Laboratory Sciences, Emergency Response and Air Toxicants Branch, Atlanta, Georgia*

## INTRODUCTION

A toxin is a poison from a living source that has the potential to be used as a bioweapon (1–3). Because a toxin cannot be grown *in vitro* and does not contain DNA, microbiological techniques such as culture or polymerase chain reaction (PCR) are not applicable for toxin detection. The amount of toxin needed to cause lethality may be quite small and is often less than 1 milligram per kilogram of human body mass (4). As a result, any effective method for toxin analysis must be able to detect trace concentrations of toxin and be both highly sensitive and specific to prevent generating false-negative or -positive results. Enzyme-linked immunosorbent assays (ELISAs) are an example of a sensitive analytical method for toxins but are generally not considered a definitive analytical approach. ELISA results must be confirmed by a more specific analysis method such as mass spectrometry (5). The description of toxin analysis methods in this chapter includes an introduction to toxins, sample preparation, and mass spectrometry followed by specific methods for the analysis of saxitoxin, α-amanitin, botulinum neurotoxins (BoNT), and ricin using mass spectrometry. A discussion of strengths and weaknesses of mass spectrometry tests concludes the chapter, as well as possible future avenues of method development.

### Toxins

The chemical structures of toxins are diverse, and there are numerous methods to classify them, such as their respective biological sources, toxicity, molecular mass, and structural characteristics. For the purposes of this discussion, toxins will be divided primarily into low molecular weight alkaloids or peptides with a mass of less than 1000 Da and protein toxins with a mass of tens of thousands of daltons. Each toxin has its own unique chemical structure and

Microbial Forensics. DOI: 10.1016/B978-0-12-382006-8.00024-4

stability characteristics that are important considerations when developing a sample preparation and analytical approach. Information associated with how each toxin interacts with the human body is also useful because it can serve as the basis for measuring toxin activity, as seen with both botulinum neurotoxin and ricin. In addition, clinical symptoms following an exposure may be very characteristic for a particular toxin or class of toxins and may provide valuable insight to guide laboratory analysis efforts.

Saxitoxin (STX), with an $LD_{50\,i.p.}$ (mouse) of $10\,\mu g/kg$, is an example of a potent alkaloid toxin. Human poisoning usually results from the consumption of STX-contaminated seafood, and exposures may be recognized by the rapid onset of clinical symptoms such as tingling in the lips, gastroenteritis, respiratory paralysis, and possibly death. STX reversibly inhibits sodium channels in the body and is subsequently excreted intact in urine (6). Because STX binds reversibly and is excreted intact, it is available for direct, confirmatory mass spectrometry analysis. Sample preparations must be compatible with the high water solubility and alkaline instability of the toxin. Because STX is part of a group of more than 21 related toxins, the distribution of these toxins has been used as a selective "fingerprint" for source attribution purposes (7,8).

The peptide $\alpha$-amanitin, with an $LD_{50}$ of $300\,\mu g/kg$ (9), is an example of another potent small molecule toxin. $\alpha$-Amanitin is produced by *Amanita phalloides*, and human exposure usually occurs from accidental ingestion of these toxic mushrooms (9). Symptoms of $\alpha$-amanitin poisoning are unique and include an asymptomatic period, which occurs during the depletion of the body's protein content due to inhibition of RNA polymerase II. After the body's protein concentration reaches a critical level, severe gastroenteritis, liver failure, and death may follow. $\alpha$-Amanitin may be detected intact in human urine following poisoning and can be measured directly by mass spectrometry. $\alpha$-Amanitin is a bicyclic octapeptide with a mass of $918\,Da$. It is slightly hydrophobic, pH stable, and can be extracted readily from aqueous matrices. There are four reported forms of amanitin (5), with the $\alpha$ form being one of the most abundant. Because there is a natural distribution of amanitin toxins, it is possible to use this distribution as a toxin fingerprint for attribution purposes.

Botulinum neurotoxins are proteins (10) with a mass about 500 times greater than that of STX. The estimated toxic dose for an average adult is about $70\,\mu g$ through oral consumption (11). Human exposure to BoNTs results from eating food containing these toxins (12,13), inhalation of toxins, or through colonization by *Clostridium botulinum* or other BoNT-producing *Clostridium* species in either a wound (12) or in the gastrointestinal tract of infants or immunocompromised individuals. If bacteria colonize the human body, they continue to produce BoNT, resulting in a continuous source of toxin. Due to the high molecular weight of the toxin, BoNT is excreted primarily in stool.

This matrix contains the highest concentrations of BoNT and is the most commonly used specimen for determining human exposure to BoNT. A characteristic symptom of human exposure to BoNT is flaccid paralysis in which the patient remains aware of his/her surroundings.

Botulinum neurotoxins are fundamentally different from small molecule toxins. They are produced as single polypeptide chains of about 150 kDa that undergo proteolytic cleavage during excretion or in the extracellular milieu to generate the fully active dichain molecule consisting of a heavy 100-kDa chain and a light 50-kDa chain that are connected covalently by a disulfide bond. BoNTs are zinc metalloproteases (i.e., endopeptidases) that cleave and inactivate proteins necessary for acetylcholine release. The heavy chain is responsible for both receptor binding via its C-terminal binding domain (14,15) and for delivering the catalytic light chain to its target via its N-terminal translocation domain (16). The light chain selectively cleaves neuronal proteins required for acetylcholine release. Although the light chain accounts for specific toxicity, it requires the heavy chain to produce this toxic activity *in vivo*.

Botulinum neurotoxins are currently classified into seven serotypes (A–G), but only serotypes /A, /B, /E, and /F are associated with human disease. BoNTs are highly specific endopeptidases for members of the SNARE (SNAP-receptor) family of proteins. BoNT/A, /C, and /E cleave synaptosomal-associated protein (SNAP-25) (17–22), whereas BoNT/B, /D, /F, and /G cleave synaptobrevin-2 (VAMP-2) (23–27). Only BoNT/C is known to cleave more than one protein as it also cleaves syntaxin (17,23,24). Cleavage of any of these proteins, which comprise the SNARE complex (soluble *N*-ethylmaleimide-sensitive fusion protein attachment protein), results in the inability to form this complex, stopping nerve impulses.

Ricin is a protein toxin present in the seeds of the castor bean plant *Ricinus communis*. It has an $LD_{50}$ in humans estimated to range from 70 to 70,000 μg/kg depending on the route of exposure (28). Ricin affects the lungs or gastrointestinal tract rapidly, causing cell death by inhibiting protein synthesis. The toxin has a mass of 64 kDa and consists of two 32-kDa subunits, an A chain and a B chain, linked by a disulfide bond. The toxicity of the A chain lies in its enzymatic activity, which involves depurination of a single adenosine that is part of a GAGA tetraloop of the 28S ribosomal subunit (29,30). This depurination results in inability of the 28S ribosomal RNA to bind elongation factor 2 (EF-2), leading to an inhibition of protein synthesis (31) and resulting in the clinical symptoms associated with ricin intoxication. The B chain is a lectin and directs binding to galactose residues on cell surface glycoproteins and glycolipids (32). The B chain is heavily glycosylated, primarily mannose and glucosamine residues; this glycosylation is thought to assist in receptor binding. Both chains are needed for the toxin to show activity *in vivo*. Because ricin is extremely reactive with human cells, it is not excreted intact in urine like small alkaloid toxins but can be detected in aqueous matrices and blood.

The alkaloid ricinine is an alternative chemical target, which may be measured in lieu of ricin. Ricinine is a nontoxic component of the castor seed (164 Da) and is present at roughly 0.3–0.8% of the seed mass (33). Determining the level of ricinine present in a sample does not necessarily mean ricin is present; however, it does confirm that a sample contains components of the castor bean plant. Ricinine is extremely heat and pH stable, making it a more persistent marker than ricin. Because ricinine is a marker for a toxin, the $LD_{50}$ is not applicable (see Table 24.1).

Basic toxin analysis primarily determines the identity of the toxin, the quantity of toxin present in a sample, and whether the toxin is biologically active. Prior to undertaking these analyses it is important to consider the matrices in which a toxin may be found, as it may only be stable or present in a given matrix for a short period of time at concentrations associated with lethal or sublethal exposures. The pH of a sample can also directly impact stability of the toxin. Because the toxin may be a member of a class of toxins (e.g., STX, BoNTs), the presence or absence of other toxins in the class may be very helpful in determining the source of an exposure.

**Table 24.1** Summary of Toxin Examples and Mass Spectrometry Methods

| Name | Natural Source of Toxin | Mechanism of Toxicity | Activity Measured? | Structure and Mass[1] | Related Figures | Confirmatory or Presumptive Test |
|---|---|---|---|---|---|---|
| Saxitoxin | Marine dinoflagellate, cyanobacteria | $Na^+$ channel inhibitor in cells | No | Alkaloid 299 Da | 1, 3A | Confirmatory |
| α-Amanitin | Amanita phalloides mushroom | Inhibits RNA synthesis | No | Cyclic Peptide 918 Da | 1, 3A | Confirmatory |
| Botulinum Toxin A | Clostridium Botulinum Bacteria | Nerve synapses | Yes | A-B Toxin[2] 150,000 Da | 2, 3A, 3B | Confirmatory |
| Ricin | Ricinus Communis Plant Seeds | Inhibits RNA synthesis | Yes | A-B Toxin[2] 64,000 Da | 2, 3A, 3B | Confirmatory |
| Ricinine | Ricinus Communis Plant Seeds | n.a. | No | Alkaloid 164 Da | 1, 3A | Presumptive |

[1]Monoisotopic mass.
[2]These masses are approximate for A-B protein toxins. Ricin contains two mass chains of 32,000 Da. BoNT/A contains a light chain of 50,000 Da and a heavy chain of 100,000 Da.

Toxin activity measurements are focused primarily on high molecular weight protein toxins, which can be denatured and inactivated by heating or chemical treatment (e.g., bleach, solvent). Because there is no change in mass upon denaturation, the mass spectrometer itself cannot differentiate between inactive or active protein toxins. Thus, reaction of the protein toxin with its substrate results in cleavage or a change in the mass of the substrate, which can be measured directly in a mass spectrometer and provides the basis for determining toxin activity. Alkaloid toxins are generally not tested for activity, as the tertiary structure of the toxin does not become denatured like a protein toxin. If the presence of the alkaloid toxin is confirmed by mass spectrometry, it is assumed to be toxic.

## Sample Preparation

The goal of sample preparation is to purify the toxin from its natural matrix and reconstitute it in a matrix that is compatible with the mass spectrometer. Sample preparation approaches generally include either chemical extraction or proteomic techniques. Microbiological techniques such as culture or PCR are not useful for direct analysis of toxins as toxins are not living organisms and do not contain DNA (34). Due to the potent nature of toxins, their detection is performed by trace analysis techniques following selective sample purification. Ideally, in order to conserve the sample, preparation methodologies utilize small sample mass amounts with little toxin loss.

Solid-phase extraction (SPE) is ideally suited for removing low molecular weight toxins from a liquid matrix and can be used to remove impurities or bind, rinse, and preconcentrate toxins (35). Common SPE sorbents include $C_{18}$, $C_8$, and cation- and anion-exchange derivatized silica. SPE methods have traditionally used a polypropylene tube with a flow-through design (see Figure 24.1). SPE tubes are disposable, low cost, and frequently used for only one sample. A common sample size is about 0.1–1 ml and may require pre-extraction centrifugation to remove particulate material.

Quantitative SPE methods use internal standards that are added precisely to samples prior to any preparation steps. The fixed ratio of the toxin to the internal standard compensates for losses introduced by sample manipulation and extraction, as well as for variability introduced by instrument analysis. Stable isotopes may be incorporated into internal standards for mass spectrometry analysis because these types of internal standards are chemically identical to the target compound; however, the heavy elements can be differentiated readily by the mass spectrometer. Use of stable isotope-containing internal standards (e.g., $^{13}C$, $^2H$) is collectively referred to as "isotope dilution."

Protein toxins are removed readily from sample matrices using immunomagnetic separation (IMS). IMS offers much greater selectivity than SPE. IMS

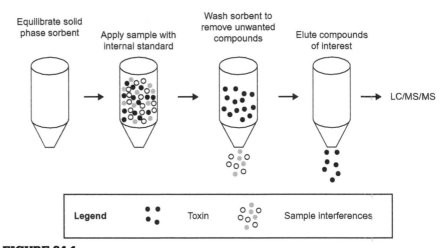

**FIGURE 24.1**

Solid-phase extraction of low molecular weight toxins such as saxitoxin and α-amanitin.

involves the use of toxin-specific antibodies conjugated to a magnetic particle and a simple magnet (36). During incubation with the antibody-conjugated iron particles, the antigen (i.e., toxin) binds to the magnetic particle. A magnetic field is then used to separate magnetic particles from the matrix, and particles are rinsed to remove nonspecific matrix components. Sample volumes generally range from 10 to 500 µl, with sample size limited by the antigen content in the sample. Following IMS, the activity of BoNT or ricin can be measured by incubating particle-bound toxins with a substrate that mimics the natural target of the toxin in the human body. The presence or absence of reaction products can be measured by mass spectrometry to determine whether the toxin is active.

Proteomic approaches to qualitative and quantitative mass spectrometric measurements include digesting the purified protein toxin with trypsin, which cleaves the protein toxin proteolytically into characteristic peptide fragments. These peptides can be compared to electronic databases of peptides to qualitatively confirm the presence of a toxin. Quantitation requires the use of internal standards, which can be added following the digestion step, and usually includes stable isotope-labeled peptides (37). Peptides are generally available from commercial suppliers and are more cost effective than generating isotopically labeled protein toxins (Figure 24.2).

## Mass Spectrometry
Mass spectrometry is a highly sensitive instrumental method, which can generate both qualitative data related to the mass and structure of a toxin and quantitative data related to its concentration in a sample (38). Stable isotope

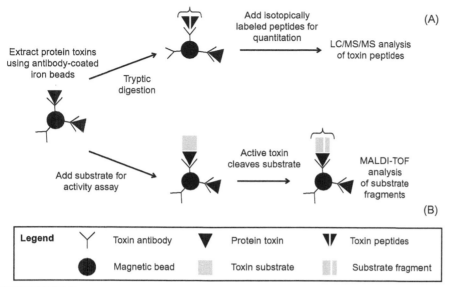

**FIGURE 24.2**

Proteomic-based sample preparation approach for botulinum toxin A and ricin. Sample extraction includes immunomagnetic separation followed by either (A) digestion and quantitation or (B) verification of toxin activity by reactivity with a selective substrate.

internal standards compensate for sample-to-sample variability resulting from autosamplers and chromatographic variation and result in an increase in analytical sensitivity. The most selective forms of mass spectrometry rely on either tandem (39,40) or high-resolution mass spectrometers (38). Tandem mass spectrometry instruments incorporate multidimensional analysis by utilizing more than one mass analyzer in a series, an example of which is the triple quadrupole. Tandem mass spectrometry dramatically increases the selectivity and sensitivity of a method by decreasing interferences. High-resolution instruments possess the ability to differentiate nominally similar ions that cannot be differentiated by low-resolution mass spectrometers. This resolution also dramatically decreases background interferences and increases method selectivity and sensitivity.

High-performance liquid chromatography (HPLC or LC) further increases the selectivity of tandem mass spectrometers (38). In addition to facilitating delivery of the sample to the mass spectrometer, HPLC further concentrates and purifies the analyte prior to analysis and sequentially delivers toxin fractions to the mass spectrometer. The separation process on the analytical column is critical to the effectiveness of the method and is based on the selective partitioning of compounds between a solid stationary phase and a liquid mobile phase. The stationary phase is typically a derivatized silica particle similar to those used in SPE but of much higher quality, more uniformity, and a smaller particle size. Because the effluent from the HPLC is a liquid and the mass spectrometer is a vacuum-based instrument, an ion source, which facilitates evaporation of the sample and solvent and ionizes the toxin, is needed prior to mass analysis. Electrospray ionization (ESI) is the interface used commonly for trace analysis of toxins; configuration of the HPLC followed by ESI and tandem mass spectrometry is commonly abbreviated LC/ESI/MS/MS or, more simply, LC/MS/MS.

Some toxins are analyzed more efficiently following dissolution into a solid matrix and direct introduction into the mass spectrometer using matrix-assisted laser desorption ionization (MALDI). Key components of MALDI are an organic matrix in which samples are mixed, a laser to ablate the matrix, and a support. The matrix is typically an unsaturated carboxylic acid, which absorbs laser radiation and is vaporized causing simultaneous vaporization of the admixed compound. The acidic matrix also donates a proton to the target analyte, causing ionization. High-resolution mass analysis is commonly applied with MALDI to compensate for the lower front-end selectivity due to absence of the HPLC separation step. A common configuration for toxin analysis is the combination of MALDI and high-resolution mass analysis using time-of-flight mass spectrometry (MALDI/TOF).

Selection of an LC/MS/MS or MALDI/TOF approach is dependent on requirements of the analytical method. LC/MS/MS is ideally suited for mixtures of compounds that would generate uninterpretable overlapping mass spectra in MALDI/TOF. A good example of such a mixture would be a tryptic digest of a protein toxin, which may contain thousands of peptides. MALDI/TOF analysis is ideally suited for high molecular weight compounds beyond the mass range of a quadrupole instrument (>3000 Da) and up to several hundred thousand daltons. MALDI/TOF can also be applied conveniently to the analysis of low molecular weight peptides for the sake of convenience or speed, assuming a very clean sample preparation using IMS (Figure 24.3).

# SPECIFIC ANALYTICAL METHODS FOR TOXINS

## Saxitoxin

Saxitoxin can be detected in liquid matrices such as water or human urine using SPE followed by LC/MS/MS analysis (6,39), which is a common approach for analysis of alkaloid toxins. Weak cation exchange (WCX) SPE has been shown to be effective for binding STX to carboxylic acids on the silica stationary phase through electrostatic attraction. Binding occurs when both the substrate (i.e., carboxylic acids) and the target ion are ionized at 2 pH units above the $pK_a$ of an acidic stationary phase and 2 pH units below

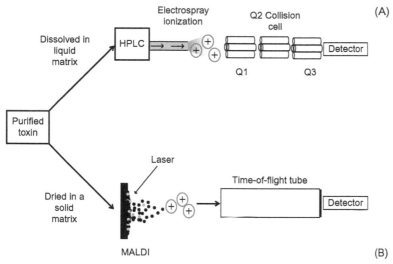

**FIGURE 24.3**

Comparison of a (A) liquid chromatography triple quadrupole mass spectrometer and (B) matrix-assisted laser desorption ionization time-of-flight mass spectrometer.

the $pK_a$ of the basic toxin. For STX, phosphate buffer, pH 6.4, facilitates efficient toxin extraction. Solvents used during sample preparation, in order of use, are methanol, then water (to wet the substrate), pH 6.4 buffer (to ionize the substrate), sample addition, water (to remove excess phosphate and matrix salts), acetonitrile (to remove neutral interferences), and 5% formic acid in methanol (pH = 1 to neutralize the stationary phase and elute the toxin). Nitrogen evaporation of the methanolic extract, using mild heating (45°C), concentrates the toxin and decreases method detection limits.

Alkaloid toxins are usually too complex to synthesize with heavy elements for isotope dilution. Instead, microorganisms that synthesize the toxin can be grown in a heavy isotope-enriched environment (e.g., $^{15}N_2$) so that the toxin incorporates the heavy elements during growth. In the case of STX, dinoflagellates (e.g., *Alexandrium* spp.) are grown in an $^{15}N_2$-enriched environment to generate the $^{15}N_7$-labeled toxin.

The LC/MS/MS method for STX uses hydrophilic interaction chromatography (39) with a high organic mobile phase for optimal retention and resolution of the polar toxin. STX can be detected to low nanograms per milliliter concentrations in urine and water. These concentrations are below the levels expected for significant human toxicity. It should be noted that if a complete fingerprint of all paralytic shellfish poisoning toxins or analysis of seafood is needed, the solid-phase extraction scheme must be altered (39). Fingerprinting of STX has been discussed in detail previously (7) and can be used to differentiate the species of dinoflagellate, source of contaminated seafood, and geographic locale. However, the generation of toxins is a transient event, and it may not be possible to definitively identify the producing organism if the conditions have changed significantly. In this case, an alternative approach would be to mimic the conditions in which a particular toxin was generated, using an assumed source organism, and determine if the toxin fingerprint was reproduced.

## α-Amanitin

α-Amanitin is a hydrophobic cyclic peptide, which is amenable to separation by traditional $C_{18}$ SPE stationary phase and traditional $C_{18}$ reversed-phase chromatography. Critical problems in analysis of α-amanitin are absence of readily available internal standards (41) and sources for growing *Amanita phalloides* in a heavy isotope environment. The issue is resolved by selecting a structurally similar surrogate cyclic peptide, which is available commercially as an internal standard.

Mass spectrometry analysis of α-amanitin is challenging because of the limited selective fragmentation of the toxin in positive ionization mode. Negative ionization mode provides the ability to fragment the toxin using tandem mass spectrometry techniques and results in an increase in method sensitivity and selectivity. A negative ionization mode also inherently reduces background noise, increasing method selectivity. The limit of detection for this compound in water or urine is 2–5 ng/ml (9,41), which is sufficient to detect toxic levels in clinical samples. While fingerprinting of the sample for all four amanitins would be ideal, there are limited sources of even the α and β forms, and a fingerprinting methodology has not yet been developed for LC/MS/MS.

## Botulinum Neurotoxins

Protein toxins, such as BoNT, can be analyzed qualitatively and quantitatively by detection of sequences that are unique to that protein and by toxin activity. IMS is integral to these methods and uses serotype-specific antibodies for BoNT to bind these toxins to a ferromagnetic particle. After the toxin-binding step, a tryptic digest of the bound toxin generates peptides that are toxin specific. These peptides are analyzed by LC/MS/MS for confirmation of the mass, and the amino acid sequence of each toxin can be used to definitively identify the toxin as BoNT and, in some cases, can identify the subtype of toxin (42) or even the strain. Subtype and strain information can be very important for epidemiological or forensic purposes. Quantitation requires the use of isotopically labeled peptides as internal standards, which are added following the digestion step. Activity measurements of BoNTs are performed in parallel or prior to tryptic digestion, as the digestion products are not enzymatically active.

For BoNT, activity measurements include incubating the IMS-bound toxin with a peptide substrate corresponding to a shortened version of the natural target of the toxin, either SNAP-25 or VAMP-2. Each BoNT cleaves the peptide substrate in a specific, toxin-dependent location, which is different for each of the BoNT serotypes (see Figure 24.4). The reaction product is then analyzed using MALDI/TOF (Figure 24.5). Detection of the peptide cleavage products corresponding to their specific toxin-dependent location indicates the presence of a particular active BoNT toxin type.

As an example, the amino acid sequence of the peptide substrate for BoNT/A is derived from the natural target of the toxin, SNAP-25. This substrate has an amino acid sequence of biotin-KGSNRTRIDEANQRATRMLGGK-biotin and a molecular mass of 2877.6 Da. The singly charged peptide substrate appears at m/z 2878.6, and the doubly charged peptide substrate appears at m/z 1440.5. When

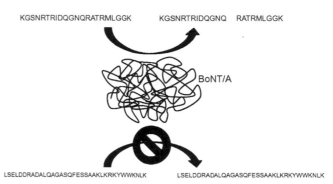

**FIGURE 24.4**

Botulinum neurotoxin, BoNT/A, will only cleave the peptide substrate designed for BoNT/A and will only cleave that substrate in a specific location.

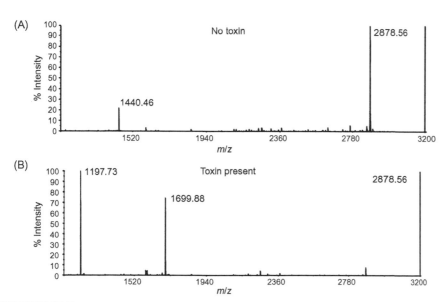

**FIGURE 24.5**

Matrix-assisted laser desorption ionization time-of-flight mass spectrometer mass spectrum (A) shows that there is no activity from botulinum neurotoxin, BoNT/A, because the substrate shown at m/z 2878.56 is intact. (B) BoNT/A is active because the substrate at m/z 2878.56 has been cleaved to form products at m/z 1197.73 and m/z 1699.88.

this peptide is incubated with BoNT/A, the toxin cleaves between the Q and the R residues. The N-terminal cleavage product, biotin-KGSNRTRIDEANQ, appears at $m/z$ 1699.9, and the C-terminal cleavage product, RATRMLGGK-biotin, appears at $m/z$ 1197.7. These cleavage products serve as biomarkers to indicate the presence of active BoNT/A in a sample, as antibodies used for IMS are directed against the heavy chain, ensuring that both heavy and light chains are present and active. Additionally, the amount of intact peptide substrate decreases upon formation of the cleavage products.

## Ricin

Ricin can also be analyzed qualitatively and quantitatively using the IMS and tryptic digestion approach described for BoNTs. Ricin activity can be determined by incubating ricin with a DNA substrate that mimics the natural target of the toxin, 28S ribosomal RNA. The DNA sequence GCGCGAGAGCGC has a molecular mass of 3699.4 Da and forms a stem loop structure with a GAGA tetraloop. When ricin contacts this GAGA tetraloop, one of the adenosines is depurinated. Depurination results in a mass shift from 3699.4 to 3581.3 Da. Detection of the depurinated DNA substrate at $m/z$ 3582.3 indicates the presence of enzymatically active ricin. Analogous to the BoNT activity test, if the DNA substrate remains unchanged, then active ricin is not present. The presence of ricin in a sample can also be determined by tryptically digesting the sample and looking for MS/MS evidence of the presence of peptides which comprise ricin (43).

Methods that analyze toxins directly by detection of peptide sequences that are unique to that protein or by their activity are considered to be confirmatory methods. In contrast, a presumptive or screening method can be valuable from a sensitivity or throughput standpoint. Ricinine is a component of the castor bean and can be monitored to confirm the presence of a castor bean product, but not ricin itself. The analytical method is identical to that used for α-amanitin, and the two targets can be extracted and analyzed simultaneously, as they have similar hydrophobicity. In contrast to the analysis of ricin, which requires 2 hours or more due to the tryptic digestion step, the biomarker ricinine can be measured much more rapidly. Ricinine is also more temperature stable, solvent resistant, pH resistant, and generally unreactive to acidic and basic pH conditions. Therefore, ricinine can be detected in matrices such as urine where the toxin has been degraded to nonspecific fragments.

## QUALITY CONTROL, VALIDATION, AND DATA REVIEW

Long-term use of a method for toxin analysis requires that the laboratory establish acceptable daily operating conditions to ensure validity of data. Positive and negative quality control materials should be incorporated into every run to verify that the method is performing properly. Analysis of toxin-enriched positive controls provides a positive instrument response confirming successful sample preparation and instrument sensitivity. The inclusion of negative controls can be used to ensure that no background interference produces false-positive signals. Positive controls with concentrations that correspond to values of greatest concern, such as those associated with regulatory limits in seafood or moderate poisoning in a clinical specimen, can be used to ensure validity of the measurement at decision points.

Establishing or validating a specific method extends beyond the characterization of quality control materials; scientists have many perspectives on what constitutes complete method validation. Common elements to method validation often include establishing analyte stability in a matrix, determining recovery from the matrix, establishing false-positive and -negative rates, testing the variability of reference materials, and comparing the method performance between different laboratories. Only validated analytical methods and corresponding sample collection routines should be used for the laboratory analysis of toxins, but this may not always be practical. In emergency situations, laboratories sometimes establish exceptions to the use of validated methods (44) and instead rely on good scientific practices and peer assistance to apply new concepts and approaches rapidly (45).

When considering toxin analysis data and whether they are valid, some key questions to consider may include (i) were positive and negative quality control materials measured within specified limits? If quality control materials failed, then none of the reported results is valid. (ii) Was the toxin measured in a previously evaluated matrix? If a new matrix is being evaluated, then toxin stability, extraction recovery, and method accuracy are not known. (iii) Are similar methods available in the peer-reviewed literature? Peer review is critical to establishing that the method uses accepted scientific principles. (iv) Were the analysts qualified to complete the method? These records are commonly retained for external auditing purposes if a laboratory is accredited.

## CURRENT LIMITATIONS TO TOXIN ANALYSIS

Primary challenges to measuring toxins are fundamental and will continue to focus on analytical sensitivity and selectivity. More sensitive methodologies are needed to detect toxins for a longer period of time after generation, especially when toxins are reduced in concentration due to environmental influences, matrix stability, or metabolism. To increase sensitivity, mass spectrometry instrumentation will continue to be developed with more efficient ion source designs, mass analyzers with rapid and precise scan rates, and improved detector response. Greater analytical selectivity is needed to ensure fewer false-positive results and will probably be addressed by the development of more selective immunoaffinity separations. Orthogonal instrument analysis techniques will also increase method selectivity by measuring different aspects of a toxin, such as molecular weight, activity, or elemental composition.

# REFERENCES

[1] J.M. Madsen, Toxins as weapons of mass destruction, Lab. Aspects Biowarfare 21 (2001) 593–605.

[2] J.R. Hancock, P.A. D'Agostino, Mass spectrometry identification of toxins of biological origin, Anal. Chim. Acta 457 (2002) 71–82.

[3] R. Russell, M. Paterson, Fungi and fungal toxins as weapons, Mycol. Res. 110 (2006) 1003–1010.

[4] L.E. Llewellyn, Saxitoxin, a toxic marine natural product that targets a multitude of receptors, Natl. Prod. Rep. 23 (2006) 200–222.

[5] M.S. Filigenzi, R.H. Poppenga, A.K. Tiwary, B. Puschner, Determination of alpha-amanitin in serum and liver by multistage linear ion trap mass spectrometry, J. Agric. Food Chem. 88 (2007) 2784–2790.

[6] R.C. Johnson, Y. Zhou, K. Statler, J. Thomas, F. Cox, S. Hall, et al., Quantification of saxitoxin and neosaxitoxin in human urine utilizing isotope dilution tandem mass spectrometry, J. Anal. Toxicol. 33 (2009) 8–14.

[7] S. Hall, G. Strichartz, E. Moczydlowski, A. Ravindran, P.B. Reichardt, The saxitoxins. Sources, chemistry, and pharmacology, ACS Symp. Ser. 418 (Mar. Toxins) (1990) 29–65.

[8] J.R. Deeds, J.H. Landsberg, S.M. Etheridge, G.C. Pitcher, S.W. Longan, Non-traditional vectors for paralytic shellfish poisoning, Mar. Drugs 6 (2008) 308–348.

[9] C. Defendenti, E. Bonacina, M. Mauroni, L. Gelosa, Validation of a high performance liquid chromatographic method for alpha amanitin determination in urine, Forensic Sci. Int. 92 (1998) 59–68.

[10] P. Databank, DOI:10.2210/pdb3boo/pdb (3BOO).

[11] B.A. Herrero, A.E. Ecklung, C.S. Street, D.F. Ford, J.K. King, Experimental botulism in monkeys: A clinical pathological study, Exp. Mol. Pathol. 6 (1967) 84–95.

[12] CDC. Botulism in the United States, 1899–1996. Atlanta, GA, 1998.

[13] G. Schiavo, M. Matteoli, C. Montecucco, Neurotoxins affecting neuroexocytosis, Physiol. Rev. 80 (2000) 717–766.

[14] S. Mahrhold, A. Rummel, H. Bigalke, B. Davletov, T. Binz, The synaptic vesicle protein 2C mediates the uptake of botulinum neurotoxin A into phrenic nerves, FEBS Lett. 580 (2006) 2011–2014.

[15] M. Dong, F. Yeh, W. Tepp, C. Dean, E.A. Johnson, R. Janz, et al., SV2 is the protein receptor for botulinum neurotoxin A, Science 312 (2006) 592–596.

[16] L. Simpson, Identification of the major steps in botulinum toxin action, Annu. Rev. Pharmacol. Toxicol. 44 (2004) 167–193.

[17] P. Foran, G.W. Lawrence, C.C. Shone, K.A. Foster, J.O. Dolly, Botulinum neurotoxin C1 cleaves both syntaxin and SNAP-25 in intact and permeabilized chromaffin cells: Correlation with its blockade of catecholamine release, Biochemistry 35 (1996) 2630–2636.

[18] T.J. Binz, S. Blasi, A. Yamasaki, A. Baumeister, E. Link, T.C. Südhof, et al., Proteolysis of SNAP-25 by types E and A botulinal neurotoxins, J. Biol. Chem. 269 (1994) 1617–1620.

[19] J. Blasi, E.R. Chapman, E. Line, T. Binz, S. Yamasaki, P. De Camilli, et al., Botulinum neurotoxin A selectively cleaves the synaptic protein SNAP-25, Nature (1993) 160–163.

[20] G. Schiavo, O. Rossetto, S. Catsicas, P. Polverino de Laureto, B.R. DasGupta, F. Benfenati, et al., Identification of the nerve terminal targets of botulinum neurotoxin serotypes A, D, and E, J. Biol. Chem. 268 (1993) 23784–23787.

[21] G. Schiavo, A. Santucci, B.R. Dasgupta, P.P. Mehta, J. Jontes, F. Benfenati, et al., Botulinum neurotoxins serotypes A and E cleave SNAP-25 at distinct COOH-terminal peptide bonds, FEBS Lett. 335 (1993) 99–103.

[22] L.C. Williamson, J.L. Halpern, C. Montecucco, J.E. Brown, E.A. Neale, Clostridial neurotoxins and substrate proteolysis in intact neurons: Botulinum neurotoxin C acts on synaptic-associated protein of 25 kDa, J. Biol. Chem. 271 (1996) 7694–7699.

[23] G. Schiavo, F. Benfenati, B. Poulain, O. Rossetto, P. Polverino de Laureto, B.R. DasGupta, et al., Tetanus and botulinum-B neurotoxins block neurotransmitter release by proteolytic cleavage of synaptobrevin, Nature 359 (1992) 832–835.

[24] G. Schiavo, C. Malizio, W.S. Trimble, P. Polverino de Laureto, G. Milan, H. Sugiyama, et al., Botulinum G neurotoxin cleaves VAMP/synaptobrevin at a single Ala-Ala peptide bond, J. Biol. Chem. 269 (1994) 20213–20216.

[25] G. Schiavo, C.C. Shone, O. Rosetto, F.C. Alexander, C. Montecucco, Botulinum neurotoxin serotype F is a zinc endopeptidase specific for VAMP/synaptobrevin, J. Biol. Chem. 268 (1993) 11516–11519.

[26] S. Yamasaki, A. Baumeister, T. Binz, J. Blasi, E. Link, F. Cornille, et al., Cleavage of members of the synaptobrevin/VAMP family by types D and F botulinal neurotoxins and tetanus toxin, J. Biol. Chem. 269 (1994) 12764–12772.

[27] S. Yamasaki, T. Binz, T. Hayashi, E. Szabo, N. Yamasaki, M. Eklund, et al., Botulinum neurotoxin type G proteolyses the Ala81-Ala82 bond of rat synaptobrevin 2, Biochem. Biophys. Res. Commun. 200 (1994) 829–835.

[28] S.M. Bradberry, K.J. Dickers, P. Rice, G.D. Griffiths, J.A. Vale, Ricin poisoning, Toxicol. Rev. 22 (2003) 65–70.

[29] Y. Endo, K. Mitsui, M. Motizuki, K. Tsurugi, The mechanism of action of ricin and related toxic lectins on eukaryotic ribosomes. The site and the characteristics of the modification in 28 S ribosomal RNA caused by the toxins, J. Biol. Chem. 262 (1987) 5908–5912.

[30] T.K. Amukele, S. Roday, V.L. Schramm, Ricin A-chain activity on stem-loop and unstructured DNA substrates, Biochemistry 44 (2005) 4416–4425.

[31] L. Montanaro, S. Sperti, A. Mattioli, G. Testoni, F. Stirpe, Inhibition by ricin of protein synthesis in vitro. Inhibition of the binding of elongation factor 2 and of adenosine diphosphate-ribosylated elongation factor 2 to ribosomes, Biochem. J. 146 (1975) 127–131.

[32] B.M. Simmons, P.D. Stahl, J.H. Russell, Mannose receptor-mediated uptake of ricin toxin and ricin A chain by macrophages. Multiple intracellular pathways for a chain translocation, J. Biol. Chem. 261 (1986) 7912–7920.

[33] R.C. Johnson, S.W. Lemire, A.R. Woolfitt, M. Ospina, K.P. Preston, C.T. Olson, et al., Quantification of ricinine in rat and human urine: A biomarker for ricin exposure, J. Anal. Toxicol. 29 (2005) 149–155.

[34] J. Audi, M. Belson, M. Patel, J. Schier, J. Osterloh, Ricin poisoning: A comprehensive review, JAMA 294 (2005) 2342–2351.

[35] M.J. Telepchak, T.F. August, G. Chaney, Forensic and Clinical Applications of Solid Phase Extraction, Humana Press, Totowa, NJ, 2004.

[36] L. Bjorck, G. Kronvall, Purification and some properties of streptococcal protein G, a novel IgG-binding reagent, J. Immunol. 133 (1984) 969–974.

[37] J. Norrgran, T.L. Williams, A.R. Woolfitt, M.I. Solano, J.L. Pirkle, J.R. Barr, Optimization of digestion parameters for protein quantification, Anal. Biochem. 393 (2009) 48–55.

[38] D.A. Skoog, J.J. Leary, Principles of Instrumental Analysis, fourth edn., Saunders College Publishing, Fort Worth, TX, 1992.

[39] C. Dell'Aversano, G.K. Eaglesham, M.A. Quilliam, Analysis of cyanobacterial toxins by hydrophilic interaction liquid chromatography-mass spectrometry, J. Chromatogr. A 1028 (2004) 155–164.

[40] Y. Seto, M. Kanamori-Kataoka, Mass spectrometric strategy for the determination of natural and synthetic organic toxins, J. Health Sci. 51 (2005) 519–525.

[41] C. Pittman, R.C. Johnson, Unpublished results, 2010.

[42] S.R. Kalb, M.C. Goodnough, C.J. Malizio, J.L. Pirkle, J.R. Barr, Detection of botulinum neurotoxin A in a spiked milk sample with subtype identification through toxin proteomics, Anal. Chem. 77 (2005) 6140–6146.

[43] S.A. Fredriksson, A.G. Hulst, E. Artursson, A.L. de Jong, C. Nilsson, B.L. van Baar, Forensic identification of neat ricin and of ricin from crude castor bean extracts by mass spectrometry, Anal. Chem. 77 (2005) 1545–1555.

[44] C.C. Chan, H. Lam, Y.C. Lee, X.M. Xue-Ming Zhang, Analytical Method Validation and Instrument Performance Verification, Wiley-Interscience, Hoboken, NJ, 2004.

[45] B. Budowle, S. Members, Quality assurance guidelines for laboratories performing microbial forensic work, Forensic Sci. Commun. 5 (2003) 4.

# CHAPTER 25

# Electron Beam-Based Methods for Bioforensic Investigations

**Joseph R. Michael,**[a] **Luke N. Brewer,**[b] **and Paul G. Kotula**[a]

[a]*Materials Characterization Department, Sandia National Laboratories, Albuquerque, New Mexico*

[b]*Computational Materials Science and Engineering, Sandia National Laboratories, Albuquerque, New Mexico*

## INTRODUCTION

Forensic investigation of biological attacks or incidents must use physical evidence, as well as genomics-based evidence, to establish the places involved and the processes used to develop and to disperse a biological-attack agent. Without a doubt, the use of accurate and precise physical characterization can produce information that is as important and informative as genomics-based tests. Generation of misinformation concerning physical evidence can be a major distraction to an investigation and provide viable defense strategies in court. In addition to validated state-of-the-art analytical methods, careful data interpretation is essential. Misinterpretation or overinterpretation of physical characterization data must be avoided.

In physical characterization studies, the microstructure and morphology of the material may be distinctive and may point to a specific production technique or to a location where the material was manufactured. The presence of additives for stabilization or dispersants may provide additional signatures that can be utilized. Other physical signatures may come from materials from the locale where the material was produced. Examples of physical signatures are environmental pollen or airborne particulates specific to a given location that potentially could allow geolocation. Tool marks present on packaging or other materials may also be used to help determine the origin of the material. It is important to consider all physical evidence that can be obtained using a variety of analytical techniques. This chapter gives an overview of electron microscopy and describes how these techniques and tools can be used by a forensic scientist for attribution.

In order to utilize these techniques and tools properly it is important to understand, at least at a basic level, how each one of the techniques is used

Microbial Forensics. DOI: 10.1016/B978-0-12-382006-8.00025-6

and the information that it can produce. For example, spatial resolution of a technique can tell you how large or small a volume can be investigated. At the same time, we need to know how sensitive the technique is in terms of concentrations so that we can understand the analytical limitation of each technique. The following sections describe in brief the techniques and their applications. This chapter also describes limitation of the techniques so that the forensic scientist can make appropriate judgments about the applicability of the technique to a given situation or study. For a more detailed discussion of the principles and limitations behind these techniques, we point the reader to some of the authoritative texts on electron microscopy and applications. This chapter also discusses sample preparation techniques that are compatible with acquiring as much as possible from the sample. These techniques are then demonstrated with some case studies.

# ELECTRON BEAM-BASED CHARACTERIZATION TECHNIQUES

## Electron Beam/Sample Interactions

The techniques discussed in this chapter use a beam of electrons or ions to interrogate the material or sample of interest. Interaction between the electron or the ion beam will provide the image or analytical signal that can be used to increase our knowledge of the sample.

In the case of electron microscopy we use energetic beams of electrons to illuminate the sample. The interaction of high-energy electrons with the sample produces many signals that can be used to characterize the sample. Figure 25.1 shows an example of signals that may be produced by interaction of the electron beam with a sample. Signals that can be used vary depending on the type of sample to be analyzed and the information needed. Interactions of the electron beam with the sample result in production of signals used for forming images of surfaces or determining elemental composition or distribution. For example, a bulk sample imaged in a scanning electron microscope (SEM) will produce electron signals used to form the image and a characteristic X-ray signal that allows qualitative (identification of elements present only) or quantitative (amounts of these elements) elemental chemistry of the sample to be measured. Different electron signals (secondary and backscattered) produce different kinds of images (1). Thin samples that are transparent (or partially transparent) to the electron beam are usually studied in a transmission electron microscope (TEM) or a scanning transmission electron microscope (STEM). In case of a thick sample, we can make use of signals that come from the entrance and exit surfaces of the sample and these signals can be used to determine the sample composition and the structure of the sample (2).

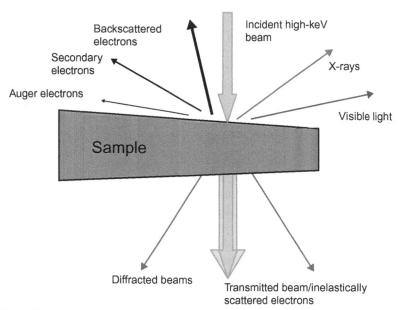

**FIGURE 25.1**

Signals for imaging and microanalysis generated by electron beams interacting with a solid sample. Signals above the sample are typically collected in the SEM, whereas those above and below the sample may be collected during a TEM or STEM analysis.

Interaction of beam electrons with tightly bound inner shell electrons produces information about the elemental makeup of the sample through the production of characteristic X-rays. The energetic beam electron ejects an inner shell electron from its electron shell, leaving the atom in an excited or ionized state with a missing inner shell electron. The atom can return to its ground state or unexcited state through a limited number of possible transitions; one of these results is production of a characteristic X-ray. The resulting X-ray has a specific or characteristic energy associated with the elemental species that was excited by the electron. Electrons may also generate continuum or Bremsstrahlung X-rays as a result of electrons losing nonquantized amounts of energy due to interactions other than those associated with inner shell electrons. Continuum X-rays are a background to the entire collected X-ray spectrum and do not provide elemental information; in fact, they limit the analytical sensitivity of X-ray microanalysis (1–3).

## Scanning Electron Microscopy

Scanning electron microscopy allows observation and characterization of organic or inorganic materials on the millimeter to nanometer scale. The SEM has proven to be extremely useful due to its capability of providing easily interpreted images of convoluted surfaces in a variety of materials.

Topographic images of surfaces are one of the main uses of SEM in organic and inorganic materials, but the SEM can also provide elemental information with spatial resolutions of about 100 nm. This combination of topographic imaging and elemental characterization makes the SEM one of the most versatile instruments for the characterization of materials, both organic and inorganic. There are currently two types of SEMs available. One type requires the sample to be placed in a high vacuum during examination. There is now the option of using low vacuum conditions in the sample chamber, allowing wet, oily, or nonconductive samples to be analyzed (1,4).

Imaging of an area of interest in the SEM is accomplished by scanning, or rastering, a finely focused beam of electrons over the sample while a desired signal is collected. Interaction of the electron beam with the sample produces a variety of signals that include secondary electrons, backscattered electrons, characteristic X-rays, and other photons with a variety of uses. Each one of these signals has a specific emission volume that sets the size of the smallest feature that can be resolved or analyzed.

The most common signal collected in the SEM for imaging of surfaces is that of secondary electron emission. Secondary electrons are low-energy electrons produced in a small region, usually limited to the beam impact area, and thus provide images with resolutions on the order of the electron beam size. Modern SEM instruments can produce images with resolutions better than 1 nm (5,6). The great advantage of using secondary electrons is the large depth of focus that exists, allowing images that appear nearly three dimensional and are intuitive in interpretation. Figure 25.2 is a typical secondary electron image obtained in the SEM from an agglomeration of bacterial spores. The detail present in these types of images is readily apparent.

Another signal used to form images in the SEM is backscattered electron emission. Backscattered electrons are primary beam electrons that are scattered back out of the sample by high-angle deflections. High-angle deflections cause some of the primary beam electrons to exit the sample surface at nearly the initial energy of the electron beam. Backscattered electron emission has the important advantage that the intensity or the number of backscattered electrons increases monotonically with atomic number. Thus, backscattered electron imaging can easily discriminate adjacent areas of the sample with differing atomic numbers. The disadvantage to backscattered electron imaging is that much of the topographic information is lost in the image and the signal can come from an area much larger than the beam impact area. This degrades the achievable resolution of this imaging technique as compared with secondary electrons (1).

The SEM can also be used to obtain elemental information from the analyzed volume through the collection of characteristic X-rays generated through

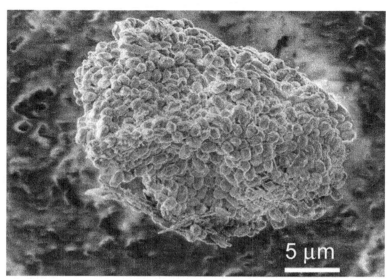

**FIGURE 25.2**
Secondary electron image of an agglomeration of *B. anthracis* spores mounted on double-stick carbon tape that demonstrates the large depth of field, enabling three-dimensional sense of size and shape.

the interaction of an energetic electron with the inner shell electrons in an atom. X-rays generated in the sample are collected with an energy-dispersive spectrometer (EDS). The EDS detects the X-rays and the associated computer system produces a plot of the intensity or number of emitted X-rays versus the energy of emitted X-rays, as shown in Figure 25.3, and can identify the peak energies and therefore the elements present (1).

Energy-dispersive spectrometer analysis cannot detect all atomic species. Modern EDS systems can now detect elements with an atomic number greater than Be (atomic number of 4). Elemental analysis with EDS has a lower detectability limit of about 0.1% by weight. This figure changes somewhat with the nature of the sample in terms of atomic number. For example, the detectability limit for a higher atomic number element in a lighter matrix is better than for a lower atomic number element in a higher atomic number matrix. Also, longer X-ray acquisition times may improve the detectability limit (1).

In the case of bulk samples, scattering of electrons within the bulk sample limits the resolution to about 1 μm in diameter and about the same in depth. The size of the interaction volume depends on the operating voltage of the SEM and the sample composition. Higher accelerating voltages and lower atomic element samples result in larger interactions volumes. However, it is not possible to continually reduce the spatial resolution by reductions in accelerating voltage due to the eventual inability of the electron beam to generate characteristic X-rays

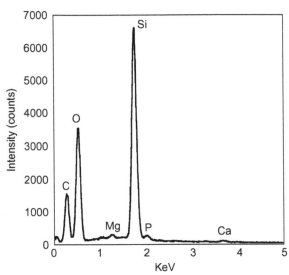

**FIGURE 25.3**

Energy-dispersive spectrometer X-ray spectrum from $SiO_2$ nanoparticles. The horizontal scale shows the energy (in keV) of the electron beam-generated X-rays, and the vertical scale shows the intensity of the X-rays. Each elemental peak in the X-ray spectrum is identified.

from the elements of interest. This is illustrated in Figure 25.4, which shows simulated electron trajectories in silicon for beam voltages of 5 and 15 kV (7). Note the large increase in the interaction volume when moving from 5 to 15 kV. This increase in interaction volume limits the spatial resolution of X-ray microanalysis.

It is possible to use the characteristic X-ray intensities to perform a quantitative analysis of the elements present in the sample. A complete discussion of this is beyond the scope of this chapter. However, quantitative analysis of inorganic material requires that the sample be flat and polished and that standards of known composition be used to provide appropriate calibration standards (1,8). Quantitative analysis of organic samples may also be accomplished using characteristic X-rays and measuring the background X-ray intensity. This technique also requires a flat and polished sample or a sample that is a thin slice (9). The accuracy of this technique must be assessed carefully for each sample composition analyzed (10). In each of these cases the most important condition that needs to be satisfied is that the sample is homogeneous over the beam interaction volume. If this condition is not met, quantitative results should be considered carefully. In materials that are inhomogeneous on a scale larger than the beam interaction volume, the compositions of each phase or constituent may be determined. If the spatial scale of the phases or constituents present is smaller than the spatial resolution, the SEM cannot be used to obtain composition of the phases. It is also very poor use of the SEM and quantitative analysis to provide a bulk chemical composition of a heterogeneous sample by scanning the beam over both phases while collecting the X-ray spectrum. This procedure violates the requirement that the sample be homogeneous over the interaction volume and can yield wildly inaccurate results.

Quantitative analysis of sample geometries that do not involve a flat, polished surface is quite difficult and may be grossly inaccurate. Typically, these alternative sample geometries involve either particles or rough surfaces. In each case, there are very real barriers to achieving accurate quantitative elemental analysis in the SEM (1,8).

This brief introduction to the SEM and elemental analysis has shown imaging signals available in the SEM. Imaging with either secondary or backscattered

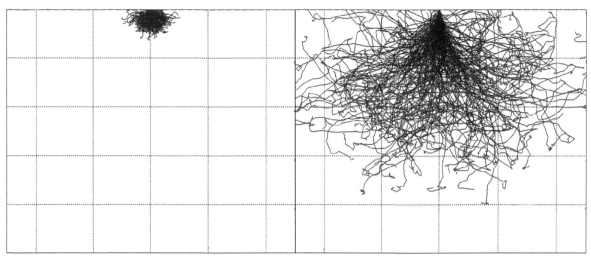

**FIGURE 25.4**

Simulation of electron interaction volumes for 5- and 15-kV electron beams interacting with silicon. The large difference in interaction volume is clearly visible in these simulations. Note that each vertical and horizontal division represents 600 nm.

electrons provides high-resolution images. Secondary electron imaging provides high resolution and high depth of field, resulting in images that appear to be nearly three dimensional. Backscattered electron imaging provides slightly lower resolution images but provides information about the relative atomic numbers of constituents in multiphase materials. X-ray microanalysis using EDS provides qualitative information about the elemental composition of the sample and, under strict analysis conditions, can be made quantitative. Certainly the mix of high-resolution imaging with elemental analysis is a very valuable and versatile combination that makes the SEM invaluable in the analysis of forensic samples.

## Transmission and Scanning Transmission Electron Microscopy

The transmission and scanning-transmission electron microscope is a powerful analytical tool for both imaging and microanalysis (2). At its heart, it is very similar to an SEM in that it has a bright electron source, a series of electron optical lenses, and image and analytical signal sensors. It differs in that electrons are typically accelerated to 100 to 300 kV, 10 or more times higher than an SEM. The second main difference is that the specimens examined are not bulk surfaces but rather thin sections of the order of 50 to 100 nm thick. The disadvantage of this with respect to SEM is that we have to sample and prepare a thin specimen of what we wish to examine. However, there are several advantages of TEM/STEM over SEM, including imaging resolution of 0.2 nm or better and practical microanalysis resolution of 1 nm typically. This section discusses the basic modes of operation of the TEM/STEM, image formation, and image and analytical signal acquisition. The following section deals with a relatively recent development in microanalysis of analytical or spectral imaging.

As the title of the section is meant to imply that there are two primary modes of operation of a modern transmission electron microscope. A complete description of the underlying physics is beyond the scope of this chapter but may be found in Williams and Carter (2). Most TEM and STEM imaging modes have potentially high spatial resolution because the samples are thin and the electron beam interacts only weakly with the specimen. Even though there is relatively little interaction, there are still extremely useful image and analytical signals that can be used to characterize the morphology and composition of our specimens, respectively.

The first mode of operation is transmission electron microscopy where the microscope is operated much like a light-optical microscope (LOM) such that a large area of the specimen is illuminated by an electron beam. In this way, looking at the transmission image of our specimen is analogous to looking at a transmission image from an LOM, albeit at significantly higher spatial resolution (2). Again analogous to an LOM, we can form a bright-field image by using an aperture (here called the "objective aperture") to stop most of the electrons that have been significantly scattered away from the direct beam (the beam traced back to the source not significantly scattered away from the optic axis). When we look at a bright-field image of a biological specimen, then, regions of higher average atomic number, such as staining agents, will appear dark and conversely lower average atomic number regions will be light. Details of biological sample preparation for TEM are discussed later (11). Figure 25.5A is a bright-field TEM image of a fluidized bacterial spore preparation simply dispersed onto a TEM support grid. The support film is less than 10 nm thick and is light, while the spore in the center of the field of view is about 1 μm thick and so looks dark in this projection image. The fluidizing agent, consisting of silicon oxide nanoparticles, is visible as fuzzy clumps surrounding the central spore. The thicker the sample locally, the darker the image as more electrons are scattered and thus cannot contribute to the final image. Figure 25.5B shows bacterial spores that have been fixed and embedded in plastic and subsequently microtomed (carefully sliced into sections of 50–80 nm thick) and stained with heavy metals to provide image contrast. The spores here are seen sectioned at random such that some look round and some look oval, reflecting the fact that the original preparation was three dimensionally dispersed; bacterial spores are almost oval in shape, and we have taken a very thin slice from that bulk preparation. More importantly is the fact that the sample is quite uniform in thickness (with the exception of some perforations in the center of the field of view) so that regions with fixative (typically osmium tetroxide) and stain (typically compounds with lead, tungsten, or uranium) scatter the electron beam more strongly and appear dark in the bright-field image. Conversely, a dark-field image, where we image with electrons scattered significantly away from the direct beam, would show higher atomic number regions (such as structures that have taken up stain) as bright and low atomic number regions such as plastic as dark. Use of an objective aperture increases the contrast of biological specimens but staining of some sort is typically used to substantially increase the contrast and aid in image interpretation by preferential absorption to specific structures in the sample. In our discussion of scanning transmission electron microscopy, we will often form dark-field images that contain the aforementioned contrast. The previous discussion did not address the case where crystalline materials may be present in the image. The contrast from crystalline materials will be significantly more complicated than noncrystalline materials and the reader is referred to a comprehensive text on the subject (2).

The second primary mode of operation of a TEM/STEM is scanning transmission electron microscopy or STEM (2,12). In this case, operation is analogous to a scanning electron microscope where we form a focused electron beam and scan it over our specimen. Unlike SEM, we primarily collect various transmitted electrons to form images as indicated in Figure 25.1. In conjunction with collecting transmitted electrons, those same electrons generate X-rays, as discussed previously, which can be used to gain elemental information from the specimen at short length scales. This is part of why STEM is so powerful, the merging of imaging and elemental analysis at such short length scales—otherwise known as "microanalysis." Although we can collect both bright-field and dark-field images with STEM, we focus on dark-field images where we use only electrons that have passed through the sample and been scattered significantly away from the direct beam

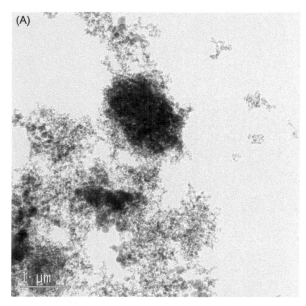

(A)

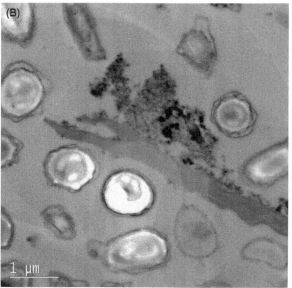

(B)

**FIGURE 25.5**

(A) Bright-field TEM image of a single spore simply dispersed on a thin carbon support film. (B) Bright-field TEM image of a microtomed section of bacterial spores.

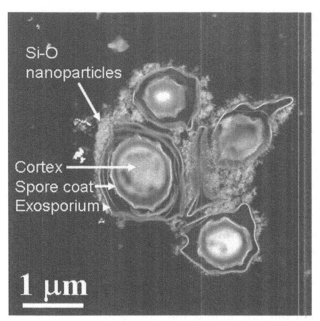

**FIGURE 25.6**

Annular dark-field STEM image of a microtomed section showing four bacterial spores coated with silicon oxide nanoparticles prior to sample preparation. This image shows high contrast due primarily to heavy metals in the fixative and stain, osmium and uranium, respectively.

to form our image. Contrast for a sample of uniform thickness in the dark-field STEM signal is usually dominated by the atomic number of the sample. Figure 25.6 is a STEM annular (annular detectors collect a range of scattered electrons subtending 360° azimuthally) dark-field image of a fixed and stained, micro-tomed specimen of bacterial spores coated with silicon oxide nanoparticles.

The brightest regions in this image represent cortexes of the spores as they have taken up the most heavy metals from fixing and staining. All of the structural features of the spores are visible and are labeled for one of the spores in the image. These bacterial spores are oblate spheroids, which are typically 1 by 1.5 μm and, when sectioned, different projections can be visualized, as seen in Figure 25.6. Also visible in this image are $SiO_x$ nanoparticles coating the outsides of the spores, attached to the exosporium. However, in order to fully understand the makeup of the materials examined in the scanning transmission electron microscope, imaging alone may not be sufficient. If we are only interested in visualizing the expected biological structures, then we need go no further. However, there may in fact be unknown and important elemental variations in samples not apparent in simple imaging. Therefore the following section describes microanalysis, where we place our focused electron beam and generate X-rays from our sample that tell us about the localized elemental makeup of our sample.

Microanalysis in the STEM consists of using the images described earlier, such as the annular dark-field signal, to locate regions of interest in the specimen. The beam is then stopped and the X-ray spectrometer acquires elemental information from small regions of the sample. The resolution of X-ray microanalysis in the SEM is limited by the size of the interaction volume of the electrons for the generation of X-rays as shown previously. This resolution is not fixed but rather depends on the material being examined. Conversely, in the STEM, because a very thin specimen is used and very high-energy electrons (100–300 keV) are used, the spatial resolution of microanalysis is limited for the most part by the electron probe size, typically 1 to 10 nm in diameter (2).

When the high-energy focused electron beam hits the specimen, two types of X-rays are generated. These include the Bremsstrahlung (X-rays produced by the deceleration of electrons in the sample) or background, and the characteristic X-rays. We can use the energy of the emitted characteristic X-rays to determine which elements are present in the sample and because we use a finely focused electron beam can furthermore determine the location of these elements. When we also take into account the relative ease or difficulty of generating and detecting X-rays from the sample, we find typical analytical sensitivities (minimum mass fraction) of 0.1–0.5% by weight (2). Elements heavier than Li can be detected with X-rays but the detection limits are much worse than those of elements heavier than Mg. Therefore, while elements such as carbon, nitrogen, and oxygen are detected readily in large amounts, their quantification is very difficult. Quantitative analysis of the beam interaction volume can be achieved through the use of a variety of approaches (2,9,10,13,14).

## CONVENTIONAL AND ADVANCED METHODS OF MICROANALYSIS, DATA ACQUISITION AND ANALYSIS

Several different methods are used for collecting microanalytical data in the SEM or STEM. The first and oldest method involves acquisition of a conventional image followed by acquisition of X-ray spectra from several points. This method requires that features of interest have contrast in the image that may not always be the case. It is also quite subjective in that the number and locations of the points chosen tend to be arbitrary. It is quite possible to miss important features in a sample simply because the number of analysis points tends to be small. The second method involves acquisition of a set of X-ray maps from a series of elements chosen in advance. This feature is most likely to be found on older instruments. The problem with this method is that the

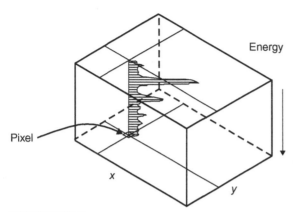

**FIGURE 25.7**

Schematic of a spectral image where a complete X-ray spectrum is at each point or pixel. A single-pixel spectrum is shown to illustrate the analysis.

elemental set chosen may not be complete. There are many examples where a map of a given element has artifacts due to the presence of another element with an X-ray line in the same location. This is referred to as "pathological overlap" and makes interpretation of simple X-ray maps problematic.

The most comprehensive method for data acquisition is full spectral imaging where a complete X-ray spectrum is acquired at each point in a two-dimensional array of points, as shown schematically in Figure 25.7. Thus, the entire area of interest is sampled and the spatial relationship between points in the array is maintained. Spectral imaging is also superior to simple X-ray mapping in that we acquire the entire spectrum at each point and can analyze data retrospectively. In such postprocessing of spectral image data we can form X-ray maps and then interrogate the underlying full spectral data to verify the presence of the mapped elements (15). While this is a powerful method for data acquisition, the following data analysis step is time-consuming and can still be subjective as data sets can consist of tens of thousands to millions of spectra and, as such, manual data analysis becomes difficult, especially when working with unknowns in terms of elements and their respective concentrations in the sample. To solve this problem, automated data analysis methods, based on multivariate statistical analysis, have been developed (15–18). By automating data analysis methods robustly, the analyst can focus then on interpreting the significance of the objective analysis rather than a hit-or-miss and time-consuming manual data analysis approach.

Depending on how our microanalysis data were acquired will determine the subsequent analysis required. If we have acquired point spectra, then our analysis consists of identifying elements present in each of the spectra acquired. While commercial X-ray acquisition and analysis software will perform this task, it is nonetheless recommended that the analyst check this result as peak identification routines can make mistakes. If, however, we have acquired simple X-ray maps, then we must verify that elements mapped are in fact present in the specimen. This can be done by selective acquisition of point spectra as a second check on the accuracy of the maps. With the advent of full spectral imaging X-ray acquisition/analysis systems, the job of data analysis becomes simpler in some ways and more difficult in others. In the first case, data are now more comprehensive than either limited point analyses or simple mapping. The spectral image can be utilized to analyze a specimen retrospectively. From the spectral image we can extract simple X-ray maps, spectra from

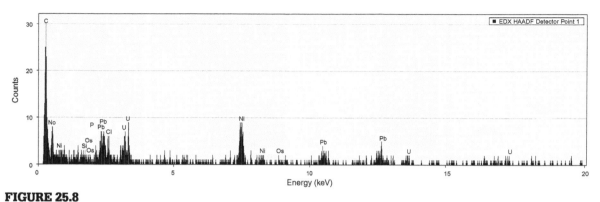

**FIGURE 25.8**

Typical spectrum, containing 1000 counts, from a spectral image. Peaks are labeled for the likely elements.

points, and spectra from selected regions. These can all be utilized to analyze the spectral image and to check the accuracy of this analysis. Although our spectral image contains as much information as possible, it is not always easy to get to it. This is especially true in the analysis of unknowns. These problems are compounded by the fact that individual spectra in spectral images are noisy and concurrently numerous. A spectral image may contain tens of thousands to millions of X-ray spectra. For this reason, a number of techniques have been developed to analyze spectral images with little operator intervention. Based on multivariate statistical analysis, these methods can reduce the amount of information needed to describe all of the unique chemical features in a spectral image, thus providing a compact and unbiased analysis for an analyst to then interpret (15–18). Generally, a spectral image can consist of more than 10,000 individual spectra each with more than 1000 energy channels and typically 100 to 300 X-ray counts in the total spectrum. Figure 25.8 shows a typical spectrum from a spectral image. As can be seen clearly, the full scale is 30 counts, meaning in any given channel, there are fewer than 30 counts. In order to deal with such noisy data, we can apply multivariate statistical analysis (MSA) methods, which first take advantage of many noisy observations (spectra) by correctly accounting for the noisy structure of data, followed by the application of efficient and robust factor analysis methods to provide a compact yet comprehensive analysis of the spectral image (19).

To demonstrate the difference between conventional mapping results and more useful MSA, consider the STEM X-ray spectral image from a region of the image shown in Figure 25.6. The sample is a microtomed section of Os-fixed and lead citrate/uranyl acetate (Pb–U heavy-metal stain)-stained *Bacillus thuringiensis* spores treated with silicon oxide nanoparticles. In this case, because we know how the sample was prepared, we can make a reasonable guess at

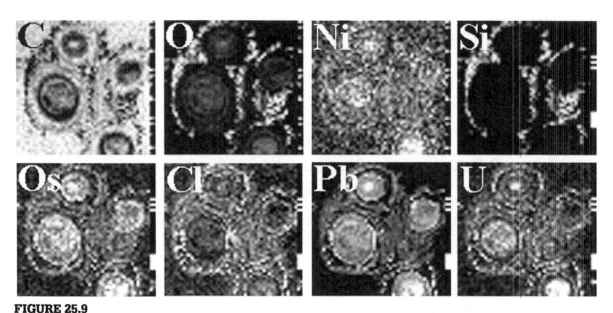

**FIGURE 25.9**

X-ray maps extracted from the spectral image acquired from the region shown in Figure 25.6. The field of view of each map is 3 μm.

the elements present and then extract X-ray maps for each of those elements. Provided we have chosen these elements correctly, no unexpected elements have been missed, and no pathological X-ray overlaps are present, we can directly see the distribution of the elements in the region analyzed. While it is relatively straightforward to perform this cursory data analysis, verifying that we have interpreted these data correctly requires a significant amount of time to go back to the raw spectral image and extract spectra summed from regions in the sample. This process in and of itself is subjective, is prone to artifacts, and tells us nothing about correlations between the X-ray maps except what we can see easily. For example, in Figure 25.9, Si and O are likely found in the same pixels, which is supported by prior knowledge of the specimen. Nickel, however, is not found in the specimen but is simply an artifact of secondary X-ray fluorescence from the Ni TEM support grid. So even though Ni is not in the sample it nonetheless contaminates the data set due to secondary fluorescence processes in the electron microscope. In many analyses, however, we have no detailed knowledge of the elemental composition of unknown materials. Therefore, multivariate statistical analysis methods as described earlier can be used to efficiently reduce the large and noisy spectral image to just its chemical essence.

If we apply MSA to the example spectral image described previously, we get four chemically significant factors to describe the data (16). These factors come in what are referred to as "component image-spectrum pairs." The component images describe where and how much of a chemical signature (described by

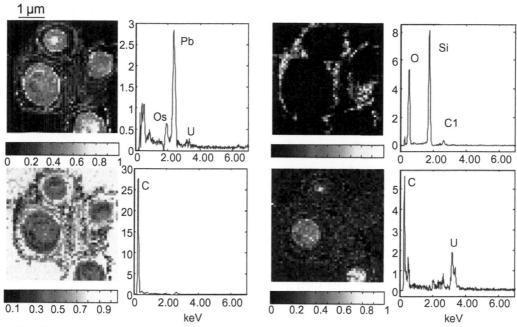

**FIGURE 25.10**

Results of multivariate statistical analysis of the spectral image that shows four chemically significant component image-spectrum pairs. Images describe where the corresponding spectral signatures are found in the microstructure. No operator input was required to calculate this result. (See Color Insert.)

the component spectrum) is found in the spectral image and therefore in the microstructure of the sample. Results of this data analysis, which took only a few seconds on a standard personal computer, are shown in Figure 25.10. Here we see four component image spectrum pairs describe all of the chemistry of the sample detected above the noise. So rather than having to analyze a series of maps and extracted spectra from regions of the sample, we can simply look at the MSA results in Figure 25.10 to see that indeed, as expected, Si and O are found on the outside of the spores. Elemental correlations have been found by MSA methodology with no prior knowledge or expectation. It is clear that the result in Figure 25.10 is much simpler for the analyst to interpret than the raw spectral image interrogated manually as in Figure 25.9. MSA software is also a commercial product and is just as applicable to STEM-EDX as well as SEM-EDX data (20).

# SAMPLE PREPARATION FOR SEM AND (S)TEM

For any technique, the preparation of samples representative of the forensic material is extremely important and critical to the success of the experiment.

This section does not attempt to give a complete discussion of sample preparation for electron beam techniques, but it tries to offer some guidelines and considerations important to a successful analysis, particularly for physical analysis of bioforensic specimens. In the case of forensics, it is important to understand how the sample preparation techniques employed may modify or alter the sample and therefore it is important to understand the techniques needed to preserve the characteristics of the samples. When considering sample preparation for SEM or TEM, we may use entirely different tools and techniques.

## Bulk Sample Preparation

One of the great advantages of the SEM is that many samples will require absolutely no sample preparation other than to fix the sample to an appropriate platform for observation within the vacuum chamber of the microscope. Generally, one will try to select a sample support or stub that is not going to interfere with the analysis of the sample. Stubs come in a variety of materials. Light materials such as Be or C are used to avoid spectral interferences when microanalysis is performed. Aluminum stubs are generally used to support the sample unless aluminum is a material of possible interest. There are many ways to attach the sample to the support, and the technique used will depend on the sample and the planned investigation. There are generally two kinds, depending on the conductive filler, of conductive adhesives utilized to attach the sample to the support. Carbon-filled adhesives are used where microanalysis will be performed, and Ag-filled adhesives are used where interference from Ag X-ray lines is not a problem. In addition, carbon-based double-stick tapes are quite useful. Care should be taken when mounting smaller particles as these tapes tend to be rough. Many samples can be imaged directly in bulk form directly after mounting. However, imaging normally requires a sample that is conductive so that the charge introduced by the electron beam can be directed to ground and does not accumulate on the sample surface. The accumulated charge can cause image distortions and sample damage and may even be large enough to deflect the electron beam away from the sample. Thus, conductive samples generally need no additional sample preparation (1,21).

In case of a nonconductive sample, something must be done to eliminate the charge buildup on the surface of the sample. If a low vacuum SEM is available, sample charging can be mitigated by removal of the electrical charge by the gas inside the sample chamber. This may be the best way to maintain sample integrity. However, if low vacuum operation is not possible, one of the simplest and most commonly applied approaches is to coat the sample with a thin layer of conductive material. Coating a sample for conductivity is achieved by either sputtering or evaporating a conductive coating onto the sample. Generally, the most common procedure is to use a sputter coating unit to place a thin (<5 nm) metal film on the sample. These metal films are most commonly an alloy of Au and Pd to produce a very fine-grained coating. Other metals commonly used are pure Au, Pt, or Cr. Proper application of the coatings should not change the surface detail visible in the microscope. Also, when conductive metal coating is used, it is important to make the coating as thin as possible so that characteristic X-ray peaks from the metal coating do not interfere with characteristic X-ray peaks from the sample. One alternative to metal conductive coating is the use of C films applied through evaporation. Generally, for inorganic samples the C coat will not produce as much spectral interference as metal films would.

Increased difficulty is encountered when soft and/or wet biological materials are prepared for SEM imaging. In the case of forensic examinations, we are interested in preserving the sample in the as-received condition. This means that we try to use methods that do not change the morphology of the sample or the chemistry of the sample. Wet or soft biological samples must be carefully dehydrated and fixed (or hardened) to allow the sample structure to be imaged in its original form. These techniques are beyond the scope of this book, and we only mention a few considerations that must be kept in mind (21). The dehydration step may seem rather straightforward and easy. However, one must remember that during dehydration, structures may be changed and water-soluble elements may migrate (21). The possibility of elemental migration must be kept in mind during subsequent microanalysis. After much of the water has been removed from the sample, it is often necessary to stabilize or fix the sample or to make it harder for subsequent handling. There are a number of ways to fix biological samples so that they retain their *in vivo* chemical distributions as well as structures. Many of the fixatives are heavy metal oxides such as osmium tetraoxide or ruthenium tetraoxide that cross-link C–C bonds (21,22). The heavy metals in these fixatives can help increase the conductivity and the contrast in images from the sample in the SEM, but they also interfere with microanalysis, as these heavy elements have many characteristic X-rays that can easily obscure elemental peaks of interest, as shown in Figure 25.11. Other methods of fixation involve a variety of aldehydes (21,22). These methods tend to be quite slow and do not protect diffusible elements, but they have the advantage of not introducing heavy metals into the stabilized sample. Other techniques that involve freeze drying, freeze substitution, or freezing of hydrated samples may provide better microanalysis fidelity, but require specialized equipment and microscope stages. It is important to note that if we are mainly concerned with inorganic materials or particles mixed with the biological material, we may not be too concerned about the changes or artifacts induced by dehydration and fixation of the sample. For a more complete discussion of sample preparation, see one of the texts listed in the references (21,22).

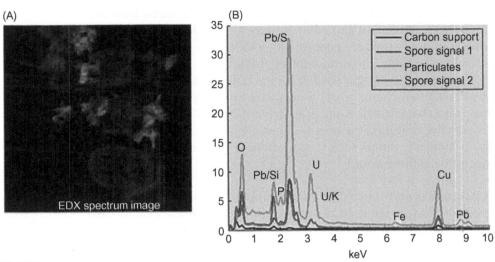

**FIGURE 25.11**

An EDX spectrum image showing features (a) and spectra (b) from microtomed TEM sections of bacterial spores stained with uranium acetate and lead citrate. Note the X-ray peak overlaps for Si, S, and K due to the staining agents. (See Color Insert.)

## Thin Sample Preparation for (S)TEM

As opposed to SEM, TEM sample preparation is generally much more extensive due to the need of electron-transparent samples in TEM (11). The thinness of the sample required depends on the voltage that the TEM is operated and the type of sample. Higher atomic number samples and lower microscope operating voltages require thinner samples. Making thin samples from bulk material can be difficult and can result in artifacts that are not representative of the original material. There are two main ways to produce thin samples. The most commonly used technique is ultramicrotomy, where the sample is sectioned mechanically (11). A newer method, not always applicable to biological samples, is the use of ion beams to produce thin section in the focused ion beam tool (23–25).

Initial sample preparation of thin samples from wet and soft biological samples requires similar dehydration and fixing procedures as discussed for bulk sample preparation. The same considerations exist in that dehydration and fixation should be done carefully so as to alter the structure or the chemistry of the sample as little as possible. Thin sectioning of the sample requires that it be embedded in a hard epoxy matrix. Once embedded, the sample is sliced into sections with the ultramicrotome. Typical thin sample biological preparation involves staining of the sample with heavy metals to increase the contrast of different structural features in the bright-field image in the TEM. Heavy metal stains typically contain Pb, W, or U, which interact with specific portions of the biological structures to make them more electron dense and thus increase the rather poor contrast of the unstained samples. Of course, these heavy metals have many characteristic lines that can interfere with many other elements of interest. Thus, for forensics work where elemental analysis is necessary, these stains should be avoided. Contrast can be obtained from unstained samples by utilizing the annular dark-field mode in the STEM.

Focused ion beam (FIB) tools can be used for the preparation of thin TEM samples from dry and hard biological materials (25–29). Use of ion beams for material sample preparation for TEM is well established (24). This technique utilizes a well-focused beam of $Ga^+$ ions to selectively remove material to produce a thin section. There are many advantages to the use of FIB tools for sample preparation: site-specific sample preparation, minimal adulteration of the sample with added dehydration or stabilization chemicals,

**FIGURE 25.12**

Focused ion beam-produced thin section for STEM analysis prepared from a particle of multiple spores. The thin section is indicated by an arrow.

and speed of sample preparation compared with ultramicrotomy (hours versus days). The ion beam is used to remove material from both sides of the desired thin sample area. An example of a TEM sample prepared using FIB from a large clump of *Bacillus anthracis* spores is shown in Figure 25.12. Final thinning of the sample results in an electron-transparent sample that is manipulated onto a suitable TEM support grid for imaging. The main disadvantage to ion-based sample preparation is limited size of the final sample of about 20 μm by 10 μm. However, due to the site specificity of the technique, this is not a big disadvantage.

## APPLICATIONS OF ELECTRON BEAM CHARACTERIZATION IN BIOFORENSICS

The techniques described previously can be illustrated by a combined SEM and (S)TEM analysis of material from a study of *B. thuringiensis*. In this study, a variety of processing methods for producing the sporulated form of this organism were analyzed for secondary particulate and compositional signatures. This discussion provides examples of SEM imaging, SEM-EDX microanalysis, STEM imaging, STEM-EDX microanalysis, and application of multivariate statistics to hyperspectral EDX data collected from these techniques (18,30).

Samples used in this study were selected from archives of simulated bioagents, produced by a variety of processing methods (termed methods A–H).

The organism involved was sporulated *Bacillus thuringiensis israelensis*, a simulant for *Bacillus anthracis*. Spore samples were deactivated by gamma ($^{60}$Co) irradiation and were subsequently cultured to demonstrate nonviability. Techniques used in this study are primarily sensitive to elemental composition, which should not be affected by irradiation.

Studies in the SEM were performed in a variable pressure instrument operated at 20 keV that was equipped with an EDS for the analytical signal. Samples were prepared by sprinkling or pressing the powders onto double-stick carbon tape on an Al stub. This process was performed inside of a glove bag to prevent cross contamination. Ceramic tweezers of known composition were used to prevent unintended incorporation of spurious particles borne by the tweezers into the samples. Spectrum images consisting of 192 × 256 pixels with 2048 X-ray channels were acquired using a dwell time of 100 ms per pixel. Each acquisition required approximately 30 minutes, and four to five fields of view were taken for each sample to ensure that sampling bias was minimized.

Studies using the scanning transmission electron microscope were performed using a TEM/STEM operated at 300 kV with EDS utilized for the analytical signal. The STEM is capable of producing and scanning a highly focused electron beam, of the order of a few nanometers, over 10 μm-sized fields of view. Samples were prepared by sprinkling powder directly onto copper grids coated with an amorphous carbon support film. Spectrum images consisting of 128 × 128 pixels with 2048 X-ray channels were collected with a dwell time of 200 ms per pixel. Each acquisition required approximately 1 hour with four to five fields of view taken for each sample.

The SEM images in Figure 25.13 show both strengths and weaknesses of electron imaging alone when searching for forensic signal in biological samples.

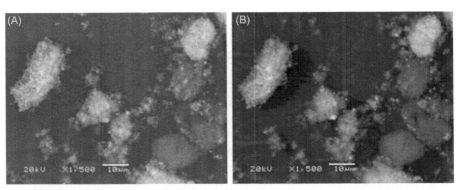

**FIGURE 25.13**
Secondary (left) and backscattered (right) electron SEM images from clumps of *Bacillus thuringiensis* spores sprinkled on carbon sticky tape.

The SEM can rapidly take images of the morphology and size of particulates and the organisms with a minimum of sample preparation using a very small volume of material, a definite advantage when examining material taken from an investigation. Images in Figure 25.13 seem to show three major features: the carbon tape background, clumps of many spores, and a collection of isolated particulates, which are particularly noticeable in the right half of each image. In the absence of any compositional information, these images are of limited use. There is no strong compositional contrast difference between secondary and backscattered electron images. It is difficult to tell how many kinds of particles there are and where they are distributed.

The inclusion of compositional data through EDS creates a more complete picture of the particulate types, but still leaves a certain amount of ambiguity mostly associated with the location of light elements and those in the support materials. Figure 25.14 shows individual EDS spectra taken from the same region as in Figure 25.13. Spectra 1 and 8, taken from clumps of spores, show high levels of calcium, phosphorous, and oxygen, as would be expected for typical bacterial spores. The carbon signals are large but not interpretable due to the carbon tape support, the large interaction volume associated with 20-keV electrons, and roughness of the sample. The silicon, chlorine, and sodium signals are interesting, but are difficult to localize because of the interaction volume and also because of the difficulty in locating the electron beam exactly for spot analysis.

By collecting an EDS spectrum image, as described previously, and then applying MSA techniques to data, we can more definitively describe the compositional signals in the spore clumps and their locations within the material (Figure 25.15). The spectrum image in Figure 25.15 was taken from approximately the same area as in Figures 25.13 and 25.14. The MSA reveals four dominant compositional signals that explain the majority of the variation in the total data set. The carbon background appears in red and shows small signals from most of the other elements present, mostly likely because of interaction volume concerns and from limitations described previously. The MSA differentiates the spore clumps into two types: one with much more silicon and less calcium (blue spectrum) than the other (green spectrum). Note that at this length scale, we cannot determine what the nature of the silicon is in the spore clumps, only that it is there and that there appear to be two types of clumps based on elemental signatures. Finally, but importantly, there is a whole collection of particles (cyan) that have very strong sodium and chlorine signals. These probable rock-salt particles make for a distinct, nonbiological attribute to this particular spore powder process method.

STEM-EDX spectrum image data show the advantage provided by the much better spatial resolution (nanometers instead of micrometers) afforded by the

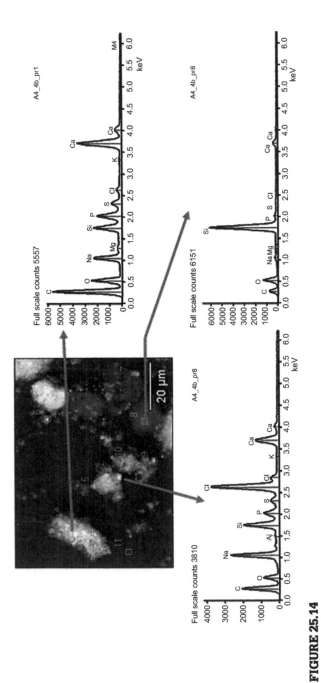

**FIGURE 25.14**

Individual EDX point spectra taken from points identified on the inset image.

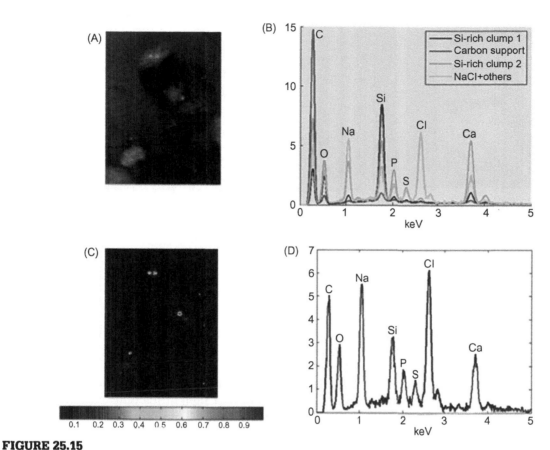

**FIGURE 25.15**

X-ray spectral image components from area in Figure 25.14. (A and B) A composite image and set of spectra from the whole area. (C and D) Features and spectrum from sodium chloride-rich particles within the field of view. (See Color Insert.)

small probe. In Figure 25.16, individual spores and much smaller particles can be seen easily. Note that a human erythrocyte (red blood cell) is usually on the order of 5–7 μm in diameter and would easily fill the entire field of view in Figure 25.16. As with SEM-EDX data, application of MSA techniques to spectrum image data allows for a systematic characterization of the compositional signals present in a fairly complex collection of structures. The spore bodies (green) have the calcium and phosphorous signals observed in SEM-EDX data. There is no silicon signal present in this collection of spores. This limited sampling is one of the key disadvantages to STEM-based measurements. There are also three kinds of particulates that are distinct only after the MSA of data. The gold–palladium particles (cyan) are clear even in the high angle annular dark-field image because of their high mass. A second mass of particles (magenta) surrounding the spores has strong calcium, manganese,

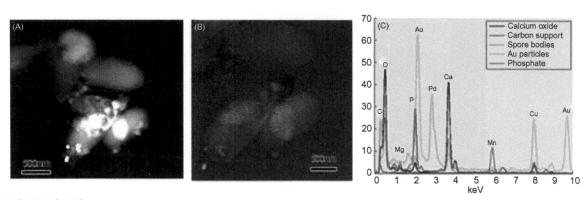

**FIGURE 25.16**

Annular dark-field (ADF) image of spores dispersed on a TEM grid. Component image overlay (B) and component spectra (C) from this region. Note that the copper signal in all spectral components is from the support grid. (See Color Insert.)

phosphorus, and oxygen signals and may be a phosphate from buffer solutions. Finally, there is a calcium- and oxygen-rich particle (blue).

Comparing the silicon-rich components from several processing routines shows the advantages of spectral imaging over simple elemental mapping. In Figure 25.17, the silicon-rich component from four different spore preparations is shown along with spatial distribution. It is clear that in each of these conditions, a high concentration of silicon is present right at the outer edge of the spore, seemingly in the spore coat. However, the ratio of silicon to other elements, sulfur, phosphorous, and so on, is not the same and may tell us something valuable about how spores take up these elements based on the process for making them and could be used to differentiate between samples collected in the field.

## SUMMARY

The electron beam analysis techniques of SEM and (S)TEM have been shown here to be invaluable tools that can be applied to the analysis of bioforensic samples. These tools can provide both morphological and elemental information with high resolution and high elemental sensitivity to elucidate important signatures for forensic attribution from biological and chemical agents.

The SEM is an important tool as samples do not require extensive preparation and thus much information about the morphology and the chemistry of the sample can be learned quickly. The SEM is capable of very high-resolution imaging (subnanometer has been demonstrated in some cases) and can provide elemental analysis of regions as small as fractions of a micrometer. These capabilities make the SEM the logical first choice for analysis of unknown solid biological or chemical materials.

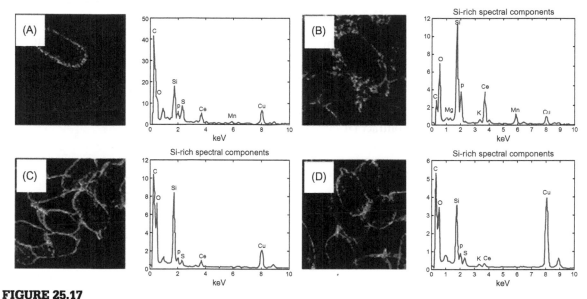

**FIGURE 25.17**

Spectral image components showing distribution of silicon-rich coating on outer portion of *Bacillus thuringiensis* spores for four different processing conditions. Note the difference in coating compositions. (See Color Insert.)

The (S)TEM is used when more detailed morphological or elemental information is needed, particularly at smaller length scales. Although typical samples for (S)TEM must be electron transparent or about 100 nm thick or less, some fine powders may be imaged directly. Preparation of thin samples for (S)TEM can be somewhat time-consuming, and the potential for contamination with fixatives and contrast stains must be constantly kept in mind. Use of thin samples in the (S)TEM allows extremely high-resolution imaging and microanalysis to be performed and important details of the sample to be revealed.

Both of these techniques can produce immense amounts of elemental data quickly through the use of EDS and spectral imaging. Spectral imaging is an important advance in that we can now visualize elemental distributions in a sample, and all elemental information about a region of a sample can be archived easily and later interrogated for other elemental signatures. The advent of MSA techniques that can extract important associations between elements in a spectrum image has reduced the difficult and time-consuming task of analyzing these huge sets of data greatly and made this type of analysis objective.

A multiple instrument approach to the analysis of biological or chemical agents for forensics and attribution provides the most information. Laboratories engaged in high-value, high-consequence analysis should be equipped with both SEM and (S)TEM capabilities so that synergisms between these techniques can be best exploited.

## ACKNOWLEDGMENTS

The authors thank the technologists at Sandia National Laboratories, Bonnie McKenzie and Michael Rye, for their expertise in SEM and FIB. Sandia National Laboratories is a multiprogram laboratory operated by Sandia Corporation, a wholly owned subsidiary of Lockheed Martin company, for the U.S. Department of Energy's National Nuclear Security Administration under Contract DE-AC04-94AL85000.

## REFERENCES

[1] J.I. Goldstein, D. Newbury, D. Joy, C. Lyman, P. Echlin, E. Lifshin, et al., Scanning Electron Microscopy and X-Ray Microanalysis, Springer, New York, 2003.

[2] D.B. Williams, C.B. Carter, Transmission Electron Microscopy: A Textbook for Materials Science, Springer, New York, 1996.

[3] D.E. Newbury, Electron beam-specimen interactions in the analytical electron microscope, in: D.C. Joy, A.D. Romig, J.I. Goldstein (Eds.), Principles of Analytical Electron Microscopy, Plenum, New York, 1986, pp. 1–28.

[4] J.B. Pawley, LVSEM in biology, in: H. Schatten, J. Pawley (Eds.), Biological Low-Voltage Scanning Electron Microscopy, Springer, New York, 2008, pp. 27–106.

[5] D.C. Joy, The aberration-corrected SEM, in: H. Schatten, J. Pawley (Eds.), Biological Low-Voltage Scanning Electron Microscopy, Springer, New York, 2008, pp. 107–129.

[6] J.R. Michael, D.C. Joy, B.J. Griffin, Challenges in achieving high resolution at low voltages in the SEM, Microsc. Microanal. 15 (Suppl 2) (2009) 660CD–661CD.

[7] N.W. Ritchie, Spectrum simulation in DTSA-II, Microsc. Microanal. 15 (2009) 454–468.

[8] K.F.J. Heinrich, Electron Beam X-Ray Microanalysis, Van Nostrand Reinhold, New York, 1981.

[9] T.A. Hall, B.L. Gupta, EDS quantitation and application to biology, in: D.C. Joy, A.D. Romig, J.I. Goldstein (Eds.), Principles of Analytical Electron Microscopy, Plenum, New York, 1986, pp. 219–248.

[10] C.E. Fiori, C.R. Swit, J.R. Ellis, A critique of the continuum normalization method used for biological X-ray microanalysis, in: D.C. Joy, A.D. Romig, J.I. Goldstein (Eds.), Principles of Analytical Electron Microscopy, Plenum, New York, 1986, pp. 413–444.

[11] M.A. Hayat, Principles and Techniques of Electron Microscopy: Biological Applications, fourth edn., Cambridge University Press, Cambridge, 2000.

[12] J.M. Cowley, Principles of image formation, in: J.J. Hren, J.I. Goldstein, D.C. Joy (Eds.), Introduction to Analytical Electron Microscopy, Plenum, New York, 1979.

[13] J.I. Goldstein, D.B. Williams, G. Cliff, Quantitative X-ray analysis, in: D.C. Joy, A.D. Romig, J.I. Goldstein (Eds.), Principles of Analytical Electron Microscopy, Plenum, New York, 1986, pp. 155–218.

[14] P.E. Champness, G. Cliff, G.W. Lorimer, Quantitative analytical electron microscopy of metals and minerals, Ultramicroscopy 8 (1982) 121–132.

[15] P.G. Kotula, M.R. Keenan, Application of multivariate statistical analysis to STEM x-ray spectral images, Microsc. Microanal. 12 (2006) 538–544.

[16] P.G. Kotula, M.R. Keenan, J.R. Michael, Automated analysis of SEM X-ray spectral images: A powerful new microanalysis tool, Microsc. Microanal. 9 (2003) 1–17.

[17] P.G. Kotula, M.R. Keenan, J.R. Michael, Tomographic spectral imaging with multivariate statistical analysis: Comprehensive 3D microanalysis, Microsc. Microanal. 12 (2006) 36–48.

[18] C.M. Parish, L.N. Brewer, Multivariate statistical applications in phase analysis in STEM-EDS spectrum images, Ultramicroscopy 110 (2010) 134–143.

[19] M.R. Keenan, P.G. Kotula, Accounting for Poisson noise in the multivariate analysis of ToF-SIMS spectrum images, Surf. Interface Anal. 36 (2004) 203–212.

[20] http://www.thermo.com/com/cda/resources/resources_detail/1,2166,13801,00.html.

[21] P. Echlin, Handbook of Sample Preparation for Scanning Electron Microscopy and X-Ray Microanalysis, Springer, New York, 2009.

[22] H. Schatten, High-resolution, low-voltage, field-emission scanning electron microscopy: Applications for cell biology and specimen preparation protocols, in: H. Schatten, J. Pawley (Eds.), Biological Low-Voltage Scanning Electron Microscopy, Springer, New York, 2008, pp. 145–171.

[23] J. Orloff, M. Utlaut, L. Swanson, High Resolution Focused Ion Beams: FIB and Its Applications, Kluwer Academic/Plenum Publishers, New York, 2003.

[24] L.A. Giannuzzi, F.A. Stevie, Introduction to Focused Ion Beams, Instrumentation, Theory, Techniques and Practice, Springer, New York, 2005.

[25] H.K. Edwards, M.W. Fay, S.I. Anderson, C.A. Scotchford, D.M. Grant, P.D. Brown, An appraisal of ultramicrotomy, FIBSEM and cryogenic FIBSEM techniques for the sectioning of biological cells on titanium substrates for TEM investigation, J. Microsc. 234 (2009) 16–25.

[26] M. Marko, C. Hsieh, R. Schalek, J. Frank, C. Mannella, Focused ion beam thinning of frozen-hydrated biological specimens for cryo-electron microscopy, Nat. Methods 4 (2007) 215–217.

[27] M. Marko, C. Hsieh, W. Moberlychan, C.A. Mannella, J. Frank, Focused ion beam milling of vitreous water: Prospects for an alternative to cryo-ultramicrotomy of frozen-hydrated biological samples, J. Microsc. 222 (2006) 42–47.

[28] M. Miliani, D. Drobne, Focused ion beam manipulation and ultramicroscopy of unprepared cells, Scanning 28 (2006) 148–154.

[29] D. Drobne, M. Miliani, V. Leser, F. Tatti, Surface damage induced by FIB milling and imaging of biological samples is controllable, Microsc. Res. Technique 70 (2007) 895–903.

[30] L.N. Brewer, J.A. Ohlhausen, P.G. Kotula, J.R. Michael, Forensic analysis of bioagents by X-ray and TOF-SIMS hyperspectral imaging, Forensic Sci. Int. 179 (2008) 98–106.

# CHAPTER 26

# Proteomics Development and Application for Bioforensics

**Karen L. Wahl, David S. Wunschel, and Brian H. Clowers**

*Pacific Northwest National Laboratory, Richland, Washington*

## INTRODUCTION

On a technical level, microbial forensics may involve "the detection of reliably measured molecular variations between related microbial strains and their use to infer the origin, relationships, or transmission route of a particular isolate" (1). Molecular variations in DNA sequence are widely used for organism and strain identification [e.g., Keim and colleagues (2)], but many other molecules and chemical species may be useful in determining microbial identity and origins; these include proteins, peptides, lipids, carbohydrates, inorganic metals, and organic metabolites. Because microbial systems produce an impressive array of molecular species, an equally broad range of analytical techniques is required to comprehensively characterize a microbial organism. Analytical methods are being developed to provide information about different sources of organisms or how a particular batch of organisms was produced to help investigators deduce what resources a perpetrator would have required, provide clues about the level of sophistication of the operation, determine whether two samples recovered in different settings came from the same source, and associate case samples with suspect media (3–11). In addition, there are efforts to integrate such analytical measurements to improve the ability to predict the production environment of unknown source microorganisms (12).

This chapter describes the role of protein and peptide analysis in the context of microbial forensics. "Proteome" is the term utilized commonly to describe the protein complement to the genome derived from an organism, while the array of measurement techniques and approaches used to analyze proteins is often referred to as "proteomics." Consequently, microbial proteomics combines a wide range of laboratory methodologies that rely on four primary scientific disciplines: microbiology, biochemistry, analytical chemistry, and computational sciences. The multidisciplinary nature of this scientific field requires a robust and rooted understanding of the uncertainties and limitations as well as the potential abilities of each discipline. In the context of

**449**

Microbial Forensics. DOI: 10.1016/B978-0-12-382006-8.00026-8

microbial forensics, proteomics seeks to identify the peptides and proteins present in a microbial isolate and compare to other isolates or to the protein sequence databases (e.g., UniProt.org). The goal is identification or sample matching related to culture environment or organism state. While gel electrophoresis is one method for analysis of proteins present in a biological sample, mass spectrometry in combination with enzymatic digestion and peptide sequencing is becoming the method of choice for protein identification for more specific and comprehensive proteomic analysis (13). This chapter discusses the strengths and challenges of mass spectrometry for proteomic analysis specifically for bioforensics.

## MICROBIOLOGY AND BIOCHEMISTRY

While the forensic importance of DNA cannot be overstated, there is much about an organism or sample that cannot be predicted or explained simply by examining the genetic code. The term "genotype" refers to the genetic complement of an organism, while the term "phenotype" broadly encompasses expressed traits of an organism's genotype. In general, DNA is read by cellular mechanisms that transform its genetic information into an intermediate state in the form of ribonucleic acid (RNA). In this intermediate state, genetic information is expressed by a second cellular mechanism by which proteins are produced. Proteins are the expression of the genotype, and the complement of proteins encoded by an organism gives rise to all of the characteristics typically associated with it. Although unique exceptions do exists, proteins are typically composed of linear chains of 20 different amino acids folded into unique three-dimensional conformations. Shorter chains of amino acids or pieces of proteins, generally fewer than 50 amino acids, are called "peptides."

Some proteins carry out the conserved functions of the bacterial cell, such as growth, division, energy utilization, and response to environmental conditions. In the case of essential cellular functions, such as protein synthesis, the proteins involved (i.e., ribosomal protein and factors required for translation) will be constitutively produced. Consequently, the identity and sequence of conserved proteins may serve as a basis of taxonomic comparison and be complementary targets for organism identification. This taxonomic role has been proposed for ribosomal and spore small acid-soluble proteins (14–17). Conversely, identities of nonconstitutive proteins in a sample may provide clues about the growth environment used to produce that sample.

Microorganisms respond dynamically to environmental conditions by altering expression of various genes. This fact implies that the suite of proteins expressed by an organism could reveal aspects of the growth environment, providing information for forensic applications. However, extracting the information on precisely what parameters were used for organism culture based on protein profile presents a challenge. A given protein may play a role in more than one cellular process and so expression may be related to more than one environmental influence. For instance, the type III secretion system in gram-negative bacteria is an example of a multifunctional set of proteins that can be expressed for different functions such as motility or virulence (18,19). As a result, caution is dictated when associating a profile of detected organism proteins to a specific culture environment based only on their predicted function without extensive testing and validation. Nevertheless, protein expression profiles can serve as a means to differentiate between samples of the same organism cultured under different conditions. For example, bacterial pathogens change their protein expression during infection to adapt or respond to the host environment (20–24). Specific factors such as ionic content, pH, and temperature are among the environmental variables that have been investigated for their effect on expression of protein virulence factors (25). *Yersinia pestis*, for example, expresses pathogenesis-related V protein and specific *Yersinia* outer-membrane proteins related to the type III secretion system at high temperatures (37°C) and low concentrations of $Ca^{2+}$ (<1 mM) (26,27). Changes in cultivation conditions such as temperature and salt concentrations affect a relatively static cellular structure such as the endospores of *Bacillus subtilis*. Producing cultures of *B. subtilis* at different temperatures alters the profile of spore coat proteins (28). In addition, culturing *Bacillus cereus* at temperatures up to 40°C resulted in a change in the expression patterns of proteins versus culturing at 30°C and also resulted in a more heat-resistant culture. Phosphate starvation of cultures also alters their protein composition with some specific proteins being identified (29). Subtle changes in medium formulation may also affect the responses of laboratory-grown organisms. The presence of specific albumin peptides in culture medium peptones were shown to affect the pigmentation phenotype of group B *Streptococci* (30). Finally, the method of cultivation has been shown to impact protein expression. Poor mass transfer kinetics of "aerobic" shake flasks is thought to occur relative to controlled aerobic conditions using chemostats (31). In a comparison of cultures produced in shaken flasks and chemostats, proteins predicted to be present during low oxygen tension (e.g., fumarate reductase) could be detected in flask cultures, but not in cells cultivated in chemostats (31).

These examples indicate that protein profiles have the potential to provide information on the environment that an organism experienced. At a high level this means that two samples of the same strain cultured under different conditions will exhibit differences in their protein composition, as has been demonstrated with profiles of small proteins in *Bacillus* spores and gram-negative bacteria (6).

Use of this information can be combined with other forensic information to provide unique insight into the growth history of a collected sample as well as matching or associating different collected samples to sources or batches.

# PROTEIN DETECTION AND IDENTIFICATION— MASS SPECTROMETRY FOR PROTEOMICS ANALYSIS

Primary tools for proteomics analysis are historically gel electrophoresis and, more recently, mass spectrometry. Gel electrophoresis is a low-resolution separation tool that provides a good indication of the size and range of proteins present in a collected sample. However, only approximate molecular weight determinations of proteins can be obtained with gel electrophoresis, and it is therefore challenging to confidently determine the identities of the proteins with a gel. Mass spectrometry, however, is a high-resolution analysis tool that can provide accurate molecular weight information indicative of a particular compound or protein.

In general, a mass spectrometer consists of an inlet system where the sample is introduced into the instrument, an ionization source to transfer the analytes into the gas phase as ions, and a mass analyzer to detect the ionized molecules (i.e., proteins or peptides) based on their mass-to-charge ratio ($m/z$).

Two primary ionization mechanisms are used in biological mass spectrometry. The first is electrospray ionization (ESI) (32) and the second is matrix-assisted laser desorption/ionization (MALDI) (33,34). The two techniques are complementary and capable of producing intact molecular ions of biological origin. There are subtle differences between the two, however, for purposes of this discussion the pertinent difference between ESI and MALDI is that ESI possesses the capacity to impart multiple charges to a single molecule and is amenable to on-line chemical separation methods, whereas the latter produces primarily singly charged ions and requires off-line chemical separation methods. Readers are directed elsewhere (32,35) for a more detailed discussion of the two ionization mechanisms. In addition, there are also many different types of mass analyzers that can be coupled to these different front end ionization methods that also vary in resolution, sensitivity, mass accuracy, and other parameters (36). Moreover, certain mass spectrometers can also perform "tandem" experiments to generate additional information such as amino acid sequence (37,38). In these tandem mass spectrometer experiments, the first step is to acquire a spectrum of the intact molecular ions present in a sample. This primary MS stage is followed by a series of secondary experiments in which an intact molecular ion is isolated from other constituents and subjected to molecular fragmentation. This secondary fragmentation step breaks the isolated, intact molecular ion (from the primary MS stage) into smaller pieces that, when examined as a whole, can provide a wealth of information regarding the composition (e.g., amino acid sequence) of the original ion. In the context of proteomics, a tandem MS experiment possesses the ability to fragment an intact peptide (or protein) ion into smaller pieces that are indicative of the sequence of amino acids comprising the original peptide. By comparing masses of the proteolytic peptides and their tandem mass spectra with those predicted from a sequence database, peptides can be identified and multiple peptide identifications assembled into a confident protein identification.

There are two basic approaches to using mass spectrometry for proteomics analysis and these are commonly referred to as "top-down" or "bottom-up" proteomics (39–41). Both have the goal of identifying the proteins present in the sample. Bottom-up proteomics involves enzymatic digestion of the complex protein sample such as a bacterial isolate to fragment all the proteins in the sample into smaller peptide pieces prior to mass spectrometric analysis. The enzyme trypsin is commonly used to cut the protein at specific amino acids along the chain, primarily at lysine or arginine residues. A small volume of this digested peptide-containing sample (typically microliter range) is injected into a separation apparatus such as a high-performance liquid chromatograph to separate the different peptides prior to mass spectrometric analysis. Individual peptides are detected in the mass spectrometer and then subjected to further fragmentation to determine the sequence of the amino acids within each peptide. This is represented in Figure 26.1. The advantage of this "bottom-up" method is that smaller peptide/protein pieces are detected more readily by mass spectrometry with better specificity and sensitivity than intact proteins. However, complete detection of all the corresponding

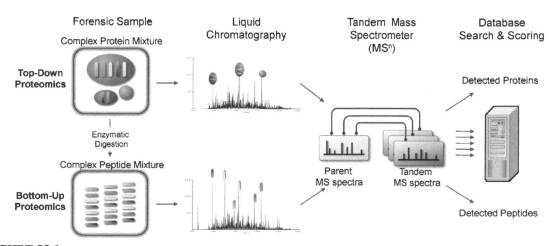

**FIGURE 26.1**

Diagram of mass spectral proteomics analysis. A complex protein mixture is separated directly or digested into peptides, followed by separations of proteins or peptides, and then subjected to mass spectrometric analysis to obtain molecular weights of the components.

digested pieces of an entire protein is rare; consequently, potential information on protein modifications such as oxidation or phosphorylation can go undetected and be lost by this method alone.

Conversely, top-down proteomics refers to performing mass spectrometric analysis on intact proteins followed by tandem MS experiments utilizing the secondary fragmentation process to produce the fragmentation/amino acid sequence information. The advantage is the potential for detecting proteins in unsequenced organisms with high homology to sequenced organisms (17) and for detecting protein modifications (via mass shifts) on the intact proteins. Disadvantages are that protein separations can be more challenging than peptide separations and that protein fragmentation/dissociation can be less efficient for large protein molecules in the gas phase and less sensitive, resulting in less peptide sequence information. Both bottom-up and top-down proteomics have value in bioforensic applications, and the choice should be driven by the specific questions at hand, the likelihood of protein sequence information for a specific organism of interest being present in the protein databases, and many other important factors.

## COMPUTATIONAL SCIENCES

As discussed previously, proteomics requires a multidisciplinary approach utilizing elements from microbiology and biochemistry, analytical chemistry, and computational sciences. The final discipline required for proteomics of microbial systems is the use of computational algorithms and rigorous statistics to maximize the confidence in interpretation of an experimental result. While the use of proteomic databases and statistics is often the final step in any experimental workflow, the importance of experimental design prior to initiating any project cannot be overstated. Without the proper experimental design, results of sample analysis may be rendered invalid due to lack of statistical rigor.

A number of computational approaches exist to parse proteomic data from mass spectrometry experiments and confidently identify proteins present in a sample (42). Perhaps the most direct approach from a conceptual point is to directly interpret spectra originating from a series of experiments. This method would use the known fragmentation behavior of peptides in a mass spectrometer to arrive at the sequence of that peptide (43). A more automated approach is often used where the accurate mass of peptides, along with fragment mass information from tandem MS experiments, is compiled into a list of masses. These peptide masses, or peak list, can then be compared to a large database containing protein sequence information from known organisms such as Uniprot or the National Center for Biotechnology Information (NCBI.nlm.nih .gov). With state-of-the-art mass spectrometry instrumentation, the molecular weight of a peptide can be determined to within three parts per million, or within 0.003 Da for a peptide with a molecular mass equal to 1000 Da. That translates to the chance that a protein database match, obtained with these accurate peptide masses, occurred at random to one in a billion (44). However, reproducibly obtaining a unique, accurate, and complete amino acid sequence directly from fragmentation data remains a formidable computational challenge. Another challenge is that the sequenced protein database(s) is rapidly expanding; consequently, the number of matches/hits per potential protein identification made for each peptide data set increases correspondingly.

Computational tools can also perform an *in silico* digestion of the sequenced proteins in the database and calculate a theoretical mass of each potential peptide to compare to actual data obtained from the sample. A comparison is made between all of the masses calculated for these peptides with measured peptide masses obtained by mass spectrometry (Figure 26.2). Possible matches for each detected peak are identified along with statistical relevance.

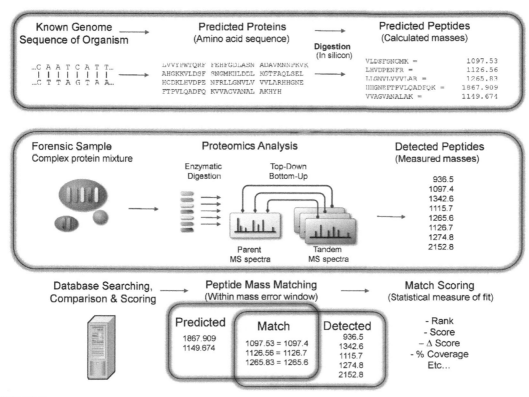

**FIGURE 26.2**
Diagram of data processing approach for proteomics analysis.

A number of proprietary and open-source tools have been developed to perform the matching function, each with its own method of calculating probability or correlation between the experimentally derived peak list and those calculated for peptides in the database (e.g., Mascot at matrixscience.com, Sequest at Sequest.com, or X!tandem at thegpm.org/TANDEM/index.html). Because of the very large sequenced database, many potential identifications are often statistically possible for each detected mass and therefore additional peptide detections are critical for increased confidence in the protein assignments. Suggestions have been made within the scientific community in setting standards for acceptable protein identifications based on the number and statistical strength of peptide data used in the analysis (45). For example, it is recommended that a single peptide match from tandem mass spectrometric data is not sufficient for protein identification. Within a proteomics publication or report, a list of all peptides identified and the protein coverage accounted for should be clearly stated to provide some indication of strength of the stated protein identification.

Knowledge of the sample and potential components also helps narrow down search criteria for peptide matching. In many proteomics studies, a known organism is digested with a known enzyme such as trypsin in a controlled laboratory setting. Tryptic peptides detected by mass spectrometry can then be searched against the protein database of that particular organism to identify which proteins are actively expressed in that sample. This greatly aids the data analysis/interpretation step by limiting the theoretical pool of peptides that must be searched and potentially increasing the statistical measures of confidence. However, analysis of forensic samples will likely require refinement of these data analysis/interpretation methods when the organism's identity may not be known and hence require searching against a much broader database to capture all possibilities. Likewise, the material being analyzed may contain protein fragments digested previously by a variety of proteases or chemical digestion processes affecting the expected resultant mass detected versus the easily predicted tryptic peptides from controlled laboratory digestion protocols often utilized in proteomics. It is also important to consider that materials for production, processing, and preservation of microorganisms may also be rich in both intact proteins and peptides. Some of these proteins and peptides may very well be present in a forensic sample and could potentially provide information as to its production history. These components are equally amenable to analysis and may add to the protein signature useful for sample matching.

## SUMMARY

Proteomics is the science associated with the study of the protein composition of a biological system. While genomic sequencing is an invaluable tool for bioforensic sample identification, proteins may offer more improved stability over DNA markers in forensic samples and have fewer or different inhibitors to analysis methods. In addition, some proteins, such as ribosomal proteins, tend to be conserved and are potentially good candidates for providing bacterial taxonomic information. Expression specificity and the relative stability of proteins with respect to genetic material make them attractive targets for microorganism identification and forensic applications to complement genomic approaches. With continuous development and improvement in mass spectrometry, protein mass spectrometric analysis is becoming an important tool for bioforensic applications. Although the field has not yet developed to the point that specific organism proteins or protein profiles can be confidently related to, for example, specific medium types or growth temperatures, protein expression patterns can be useful in a forensic application for sample differentiating or sample matching. Identification of specific proteins may be helpful in taxonomic identification as a complement to genomic identification as well as in characterizing virulence.

There are still challenges associated with data analysis and interpretation in terms of what are acceptable confident matches of identified peptides with expected proteins. One challenge is presented by rapidly expanding protein databases, which offer increasing opportunities for both correct and false matches. Another is the result of the variety of computational tools available for database matching, and the lack of standardized approaches to proteomics data analysis. However, there has been an ongoing effort within the proteomics community to standardize the reporting requirements for proteomics data. Engagement of the scientific community demonstrates the broad acceptance of proteomic methods, making them suitable for application in the legal framework required for forensic evidence. The maturity of this field is such that it is now ready for extension to microbial forensics.

# REFERENCES

[1] C.A. Cummings, D.A. Relman, Genomics and microbiology: Microbial forensics—"Cross-examining pathogens." Science 296 (5575) (2002) 1976–1978.

[2] P. Keim, T. Pearson, R. Okinaka, Microbial forensics: DNA fingerprinting of *Bacillus anthracis* (anthrax), Anal. Chem. 80 (2008) 4791–4799.

[3] J.R. Whiteaker, C. Fenselau, D. Fetterolf, D. Steele, D. Wilson, Quantitative determination of heme for forensic characterization of *Bacillus* spores using matrix-assisted laser desorption/ionization time-of-flight mass spectrometry, Anal. Chem. 76 (2004) 2836–2841.

[4] J.B. Cliff, K.H. Jarman, N.B. Valentine, S.L. Golledge, D.J. Gaspar, D.S. Wunschel, et al., Growth media differentiation of *Bacillus subtilis* spores by elemental characterization using TOF-SIMS, Appl. Environ. Microbiol. 71 (2005) 6524–6530.

[5] H.W. Kreuzer-Martin, K.H. Jarman, Stable isotope ratios and forensic analysis of microorganisms, Appl. Environ. Microbiol. 73 (2007) 3896–3908.

[6] D.S. Wunschel, E.A. Hill, J.S. Mclean, K.H. Jarman, Y.A. Gorby, N.B. Valentine, et al., Effects of varied pH, growth rate and temperature using controlled fermentation and batch culture on matrix assisted laser desorption/ionization whole cell protein fingerprints, J. Microbiol. Methods 62 (2005) 1327–1330.

[7] R.G. Breeze, B. Budowle, S.E. Schutzer, Microbial Forensics, Elsevier Academic Press, London, 2005.

[8] D. Wunschel, J. Wahl, A. Willse, N. Valentine, K. Wahl, Small protein biomarkers of culture in *Bacillus* spores detected using capillary liquid chromatography coupled with matrix assisted laser desorption/ionization mass spectrometry, J. Chromatog. B 843 (2006) 25–33.

[9] D.S. Wunschel, H.A. Colburn, A. Fox, K.F. Fox, W.M. Harley, J.H. Wahl, et al., Detection of agar, by analysis of sugar markers, associated with *Bacillus anthracis* spores, after culture, J. Microbiol. Methods 74 (2–3) (2008) 57–63.

[10] K.L. Wahl, H.A. Colburn, D.S. Wunschel, C.E. Petersen, K.H. Jarman, N.B. Valentine, Residual agar determination in bacterial spores by electrospray ionization mass spectrometry, Anal. Chem. 82 (4) (2010) 1200–1206.

[11] C.J. Ehrhardt, V. Chu, T. Brown, T.L. Simmons, B.K. Swan, J. Bannan, et al., Use of fatty acid methyl ester profiles for discrimination of *Bacillus cereus* T-strain spores grown on different media, Appl. Environ. Microbiol. 76 (2010) 1902–1912.

[12] K.H. Jarman, H.W. Kreuzer-Martin, D.S. Wunschel, N.B. Valentine, J.B. Cliff, C.E. Petersen, et al., Bayesian-integrated microbial forensics, Appl. Environ. Microbiol. 74 (11) (2008) 3573–3582.

[13] B. Domon, R. Aebersold, Mass spectrometry and protein analysis, Science 312 (5771) (2006) 212–217.

[14] H. Teeling, F.O. Gloeckner, RibAlign: A software tool and database for eubacterial phylogeny based on concatenated ribosomal protein subunits, BMC Bioinform. 7 (2006).

[15] E.R. Castanha, A. Fox, K.F. Fox, Rapid discrimination of *Bacillus anthracis* from other members of the *B. cereus* group by mass and sequence of "intact" small acid soluble proteins (SASPs) using mass spectrometry, J. Microbiol. Methods 67 (2) (2006) 230–240.

[16] F.J. Pineda, M.D. Antoine, P.A. Demirev, A.B. Feldman, J. Jackman, M. Longenecker, et al., Microorganism identification by matrix-assisted laser/desorption ionization mass spectrometry and model-derived ribosomal protein biomarkers, Anal. Chem. 75 (2003) 3817–3822.

[17] C. Wynne, C. Fenselau, P.A. Demirev, N. Edwards, Top-down identification of protein biomarkers in bacteria with unsequenced genomes, Anal. Chem. 81 (2009) 9633–9642.

[18] G.R. Cornelis, F. Van Gijsegem, Assembly and function of type III secretory systems, Annu. Rev. Microbiol. 54 (2000) 735–774.

[19] L. Journet, K.T. Hughes, G.R. Cornelis, Type III secretion: A secretory pathway serving both motility and virulence, Mol. Membr. Biol. 22 (1–2) (2005) 41–50.

[20] R. Pieper, Q.S. Zhang, P.P. Parmar, S.T. Huang, D.J. Clark, H. Alami, et al., The *Shigella dysenteriae* serotype 1 proteome, profiled in the host intestinal environment, reveals major metabolic modifications and increased expression of invasive proteins, Proteomics 9 (22) (2009) 5029–5045.

[21] R.L. Edwards, Z.D. Dalebroux, M.S. Swanson, *Legionella pneumophila* couples fatty acid flux to microbial differentiation and virulence, Mol. Microbiol. 71 (5) (2009) 1190–1204.

[22] L. Mereghetti, I. Sitkiewicz, N.M. Green, J.M. Musser, Remodeling of the *Streptococcus agalactiae* transcriptome in response to growth temperature, PLoS One 3 (7) (2008) e2785.

[23] L. Mereghetti, I. Sitkiewicz, N.M. Green, J.M. Musser, Extensive adaptive changes occur in the transcriptome of *Streptococcus agalactiae* (group B *streptococcus*) in response to incubation with human blood, PLoS One 3 (9) (2008) e3143.

[24] L. Mereghetti, I. Sitkiewicz, N.M. Green, J.M. Musser, Identification of an unusual pattern of global gene expression in group B *Streptococcus* grown in human blood, PLoS One 4 (9) (2009) e7145.

[25] K.K. Hixson, J.N. Adkins, S.E. Baker, R.J. Moore, B.A. Chromy, R.D. Smith, et al., Biomarker candidate identification in *Yersinia pestis* using organism-wide semiquantitative proteomics, J. Proteome Res. 5 (2006) 3008–3017.

[26] J.M. Fowler, R.R. Brubaker, Physiological basis of the low-calcium response in *Yersinia pestis*, Infect. Immunity 62 (12) (1994) 5234–5241.

[27] J.M. Fowler, C.R. Wulff, S.C. Straley, R.R. Brubaker, Growth of calcium-blind mutants of *Yersinia pestis* at 37 degrees C in permissive $Ca^{2+}$-deficient environments, Microbiology 155 (2009) 2509–2521.

[28] E. Melly, P.C. Genest, M.E. Gilmore, S. Little, D.L. Popham, A. Driks, et al., Analysis of the properties of spores of *Bacillus subtilis* prepared at different temperatures, J. Appl. Microbiol. 92 (6) (2002) 1105–1115.

[29] H. Antelmann, C. Scharf, M. Hecker, Phosphate starvation-inducible proteins of *Bacillus subtilis*: Proteomics and transcriptional analysis, J. Bacteriol. 182 (16) (2000) 4478–4490.

[30] M. Rosa-Fraile, A. Sampedro, J. Varela, M. Garcia-Pena, G. Gimenez-Gallego, Identification of a peptide from mammal albumins responsible for enhanced pigment production by group B *streptococci*, Clin. Diagn. Lab. Immunol. 6 (3) (1999) 425–426.

[31] D.A. Elias, S.L. Tollaksen, D.W. Kennedy, H.M. Mottaz, C.S. Giometti, J.S. McLean, et al., The influence of cultivation methods on *Shewanella oneidensis* physiology and proteome expression, Arch. Microbiol. 189 (4) (2008) 313–324.

[32] J.B. Fenn, M. Mann, C.K. Meng, S.F. Wong, C.M. Whitehouse, Electrospray ionization for mass spectrometry of large biomolecules, Science 246 (4926) (1989) 64–71.

[33] D. Heller, C. Fenselau, R. Cotter, P. Demirev, J. Olthoff, J. Honovich, et al., Mass spectral analysis of complex lipids desorbed directly from lyophilized membranes and cells, Biochem. Biophys. Res. Commun. 142 (1) (1987) 194–199.

[34] M. Karas, D. Bachmann, U. Bahr, F. Hillenkamp, Matrix-assisted ultraviolet laser desorption of non-volatile compounds, Int. J. Mass Spectrom. Ion Processes 78 (1987) 53–68.

[35] J. Peter-Katalinic, F. Hillenkamp, MALDI MS: A Practical Guide to Instrumentation, Methods and Applications, Wiley-VCH, Weinheim, 2007.

[36] J.T. Watson, O.D. Sparkman, Introduction to Mass Spectrometry: Instrumentation, Applications, and Strategies for Data Interpretation, fourth edn., Wiley, West Sussex, England, 2007.

[37] E. Mørtz, P.B. O'Connor, P. Roepstorff, N.L. Kelleher, T.D. Wood, F.W. McLafferty, et al., Sequence tag identification of intact proteins by matching tandem mass spectral data against sequence data bases, Proc. Natl. Acad. Sci. USA 93 (16) (1996) 8264–8267.

[38] N.E. Sherman, M. Kinter, Protein Sequencing and Identification Using Tandem Mass Spectrometry, John Wiley, New York, 2000.

[39] R. Aebersold, M. Mann, Mass spectrometry-based proteomics, Nature 422 (6928) (2003) 198–207.

[40] B.T. Chait, Chemistry. Mass spectrometry: Bottom-up or top-down? Science 314 (5796) (2006) 65–66.

[41] N.L. Kelleher, Top-down proteomics, Anal. Chem. 76 (11) (2004) 197A–203A.

[42] P. Hernandez, M. Müller, R.D. Appel, Automated protein identification by tandem mass spectrometry: Issues and strategies, Mass Spectrom. Rev. 25 (2) (2006) 235–254.

[43] P. Roepstorff, J. Fohlman, Proposal for a common nomenclature for sequence ions in mass spectra of peptides, Biomed. Mass Spectrom. 11 (11) (1984) 601.

[44] M. Mann, N.L. Kelleher, Precision proteomics: The case for high resolution and high mass accuracy, Proc. Natl. Acad. Sci. USA 105 (47) (2008) 18132–18138.

[45] G.K. Taylor, D.R. Goodlett, Rules governing protein identification by mass spectrometry, Rapid Commun. Mass Spectrom. 19 (2005) 3420.

# CHAPTER 27

# High-Throughput Sequencing

**Jennifer Parla, Melissa Kramer, and W. Richard McCombie**

*Cold Spring Harbor Laboratory, Cold Spring Harbor, New York*

## INITIAL DNA SEQUENCING APPROACHES

The use of DNA sequencing began in the mid-1970s. Prior to that, DNA sequencing was limited to laborious chemical analyses that would only elucidate the sequence of a few bases at a time (1–4). While these experiments yielded some significant biological insights about very specific DNA sequences, the methodology was clearly not scalable or particularly effective.

There were two roughly simultaneous advances that promoted the use of DNA sequencing, and ultimately the study of genomics, although the latter was unheard of at that time. Two methods of DNA sequencing that could be scaled considerably more than previous approaches were published in 1977. One of these, a method based on chemical cleavage of the DNA backbone at distinct nucleotides followed by separation on acrylamide gels to determine the length of DNA fragments, was published by Maxam and Gilbert (5). The other approach involved synthesizing DNA molecules in the presence of a small amount of nucleotide analogs, yielding molecules of varying length, which again could be resolved on acrylamide gels based on their size (6). Initially, the Maxam and Gilbert method predominated because the Sanger method required the use of single-stranded starting material, which was laborious to purify. However, in 1982 a method was developed to generate single-stranded DNA molecules rapidly and efficiently (7,8); once this hurdle was overcome, the Sanger method became the prevalent DNA sequencing technique. Further enhancements eliminated the need for single-stranded DNA, and the Sanger chain termination method or derivatives based on that approach were used to sequence the human genome and dominated DNA sequencing for about 30 years. As a result of the importance of this approach, it deserves further explanation.

Sanger sequencing works by taking advantage of the biological process in which DNA is synthesized (6). The process starts with a DNA template, a

**461**

Microbial Forensics. DOI: 10.1016/B978-0-12-382006-8.00027-X

primer upon which the complementary strand to the template is built, and components necessary to synthesize the new strand of DNA. Necessary components include a DNA polymerase, which is the enzyme used to synthesize the new strand, the nucleotide triphosphates needed to build the new strand, and various buffer components. However, in the case of Sanger DNA sequencing, four separate reactions were done (one for each base constituent of DNA). In addition to common reagents, each specific reaction has a small amount of an analog of the base that the reaction is designed to analyze. For instance, the A reaction has a small amount of a modified version of the nucleotide A. This modified nucleotide differs from the standard A nucleotide in that once it is incorporated in the growing chain of bases, the chain can no longer be extended, hence the term "chain terminating" for this type of sequencing. What happens in a population of molecules, which all start synthesizing from the same place (the primer), is that the modified nucleotides, present in lower quantities than the normal nucleotide, get incorporated randomly in some percentage of the new strands at every position that the given base is required. At a given ratio of normal to modified bases there may, for instance, be a 2% chance of incorporating a modified base into a new strand. So in that population of newly synthesized molecules, about 2% stop due to chain termination, at every position that an A is incorporated. This results in a population of molecules that start in the same place but contains some molecules ending at every place an A is present in the sequence. These molecules can be separated by size using electrophoresis on an acrylamide gel.

DNA sequencing is thus done by carrying out the chain termination reaction separately with all four bases. The products of the sequencing reaction are resolved using an acrylamide gel in order to create a "ladder" that readily displays the sequence of the template. A scientist looking at the autoradiogram produced from the gel sees bands in each of the four lanes showing where every chain termination event took place for a particular base. By comparing all four bases, the sequence of the template becomes apparent. The smallest molecule in any of the four bases is the site, or the base, where the first chain termination event took place. This represents the first base that can be read from the sequence. The next smallest band in any of the lanes represents the next smallest molecule, which is a chain termination event due to incorporation of the nucleotide analog used in the reaction found in that lane. So if the five smallest fragments, from smallest to largest, produce autoradiograph bands in the lanes corresponding to A, C, A, A (again), and C, the sequence of the first five incorporated bases would be ACAAC. This elegant method worked very well but was subject to some problems, particularly when efforts to scale it up were pursued.

Probably the most significant technical difficulty with the method was due to gel electrophoresis artifacts. The gels were run under fairly high voltage and generated significant heat. This heat, which was distributed unevenly across the gel, led to distortions in the resolution of similarly sized fragments. A skilled human could view the distortion and mentally compensate for it while visually scanning up the ladder. However, it proved extremely difficult to develop automation or base calling software. Nevertheless, the chemistry was very robust and chain termination technology soon became the method of choice for DNA sequencing. A series of advances in the 1980s made Sanger sequencing even more powerful. Together, these developed into the technology that was ultimately used to sequence the human genome in 2000 (9,10).

The first major advance in DNA sequencing was the use of M13 bacteriophage by Messing and colleagues to generate single-stranded DNA templates for sequencing (7,8). This was a very critical development as, until that point, it was difficult to purify suitable templates for Sanger sequencing. The next major development, which really opened floodgates of innovation and made the human genome project possible, was the development of automated, fluorescent Sanger sequencing by Leroy Hood's group at Caltech. Rather than use a single radioactive label for all four bases, Hood's group developed a method based on four separate dyes, one for each base. This provided several advantages. One, of course, was elimination of the need to handle radioactive materials. But even more important, the use of four distinct dyes meant that all four reactions (A, C, G, and T) derived from a sample could be carried out in one reaction tube and hence run in a single gel lane rather than in four separate lanes. This, paired with further innovations in DNA fragment resolution, eliminated the problem of electrophoretic distortions described earlier and made automated base calling in sequencing possible. This four-color automated Sanger sequencing was the basic method used to sequence the human genome.

# MICROBIAL GENOME SEQUENCING

## Bacteriophage λ

The most tractable way of studying DNA molecules for the purpose of determining sequence was to work with smaller DNA fragments that also did not come with extremes, such as significant GC content. Natural choices for DNA sequence study were, thus, life forms with the smallest possible genetic material, such as bacteriophages. Determining the genome sequence of bacteriophage phiX174 was a striking application of Sanger's early sequencing method, which he called the "plus and minus" method (11). Shortly after Sanger sequenced the phiX174 genome, he developed his dideoxy chain terminator sequencing method, which he then used to decode the whole genome sequence of bacteriophage λ (12). Sanger's dideoxy terminator method would then dominate in DNA sequencing procedures for the next 30 years.

While the genome of bacteriophage phiX174 consists of 5386 nucleotides, the genome of bacteriophage λ is almost 10 times longer, consisting of

48,502 bp. To produce the complete genome sequence of λ, 1454 bp of which had already been determined by other groups, Sanger worked with different sample preparation and sequencing methods. Sanger's decision to combine approaches that he as well as others had developed worked in his favor toward determining the sequence of the λ genome. The initial sample preparation technique involved the use of restriction endonucleases to fragment sample DNA prior to cloning. The restriction method is sequence specific and thus lacks randomness. Therefore, restriction-based fragmentation soon displayed its limitations by producing redundant sequence data before complete sequence data from the genome were obtained. To remediate this problem and improve the uniformity of his data, Sanger applied a random fragmentation technique using either DNase I (13) or sonication (14).

The combination of restriction-based fragmentation and random genome fragmentation allowed Sanger to identify 90% of the λ genome sequence and assemble his data into 10 to 15 contiguous regions of overlapping sequence (contigs). At this stage, Sanger had entered the "finishing" phase of the project, where he then needed to specifically produce sequence from the missing regions and build the complete assembly. These necessities inspired advances in his sequencing method, driving his read lengths from about 275 nucleotides to 500 nucleotides, as well as creative strategies such as generating sequence from the opposite end of a cloned insert, using hybridization probes to select desired clones, and subcloning fragments obtained from large inserts. Due to the large body of work produced from previous λ studies, publication of the λ genome was accompanied with a functional annotation of the sequence, showing genes and open reading frames mapped along the sequence, and a review of the organization and codon usage of the λ genome (12). Phage λ thus represents the first detailed complete genome project.

As a result of Sanger's success with sequencing the λ genome and of continual improvements to the sequencing workflow, subsequent genome sequencing projects readily employed Sanger's dideoxy sequencing method to sequence the genomes of other microbial life forms, with interest gaining in those with yet larger genomes. Advances that helped revolutionize genomics are exemplified by the invention of automated sequencing and the subsequent commercialization of this technology by Applied Biosystems (Foster City, CA), by increases in sample throughput resulting from workflow parallelization and the use of robotics, by the integration of data storage resources and management, and by the formulation of bioinformatic algorithms for sequence data manipulation and analysis.

### Haemophilus influenzae

After the genome of bacteriophage λ had been sequenced, the genomes of other viruses, as well as of endosymbionts, had also been revealed by DNA sequencing (15–17). While such genomes were up to about five times larger than the λ genome, they were not as large as other genomes of interest that scientists had decided to sequence. These larger scale projects focused on the genomes of bacterial species as well as eukaryotic species, such as *Saccharomyces cerevisiae* and *Caenorhabditis elegans*, and even *Homo sapiens*.

The first complete genome sequence from a free-living life form was produced from *Haemophilus influenzae* Rd, which is a small gram-negative bacterium capable of causing human disease (18). The genome of *H. influenzae* is a circular chromosome consisting of 1,830,138 bp, which is almost 38 times larger than the λ genome. The success of whole genome random shotgun sequencing with *H. influenzae* relied on careful and efficient random cloning of the genome using plasmid and phage vectors (18). Genome inserts for plasmid clones were carefully selected within a small and narrow size range to ensure the best clone representation possible during bacterial growth, as both large insert size and broad insert size range were expected to lead to greater variability in bacterial growth. However, phage libraries were constructed with larger inserts, where the larger insert size was not a complicating factor.

Following random library constructions with the *H. influenzae* chromosome, a combination of radiolabeled and dye terminator sequencing was carried out with the libraries, thus producing approximately sixfold coverage for both strands (18). The sequence information produced from this genome project was further strengthened by efforts taken to ensure accuracy and high quality of data. Genome regions determined to have lower coverage or accuracy relative to other regions were further analyzed by alternative sequencing approaches involving the use of different clones, by extended efforts to generate opposite end data, and by different sequencing methods. Ultimately, the work carried out to sequence the *H. influenzae* genome served to validate the random shotgun technique and improve the feasibility of more ambitious whole genome sequencing projects that were already under way or would soon follow.

## Escherichia coli

*Eschericia coli* is one of the most well-known and ubiquitous microbes in both science and general society. In the biological sciences, *E. coli* has long served as a model organism, being easy to culture, having a short generation time, and also possessing cellular biology that may be either directly applicable to human cellular biology or, rather, directly responsible for human health and disease. Therefore, the 4,639,675-bp circular chromosome of *E. coli* was one of the top favorites for complete genome sequencing, and efforts to sequence the *E. coli* genome were already under way upon the release of the *H. influenzae* genome.

The genome sequence of *E. coli* was initially determined in segments, based on genetic maps and using cloning strategies and radioactive sequencing chemistry (19–24). With integration of newer cloning and plasmid manipulation techniques, dye terminator chemistry, and automated DNA sequencing into the workflow, the rest of the *E. coli* genome was sequenced at lower cost and with greater speed and accuracy, culminating in release of the complete genome sequence in 1997 (25).

Once the bulk of the sequence data had been produced, directed efforts were required to produce data for regions that had not been covered due to lack of clone representation using the higher throughput random shotgun strategy. These missing regions included about 22.5 kb at the beginning of the *E. coli* chromosome and several gaps in the genome build. Sequencing these regions entailed going back to the *E. coli* λ clones (26) to select the appropriate clones for sequencing, as well as using long-range polymerase chain reaction (PCR) for directed sequencing, which was particularly useful for closing the sequence gaps or compensating for other problematic areas of the genome (25).

### *Bacillus anthracis* and Many More

In the decade after Sanger had sequenced the genomes of bacteriophages phiX174 and λ, just a few more complete genomes, from viruses and endosymbionts, were revealed. However, in the decade after the *H. influenzae* genome was sequenced using a random whole genome shotgun technique and automated sequencing technology, over 200 more microbial genomes were sequenced. This wave of microbial genome projects that ensued following the successful genome sequencing of *H. influenzae* and *E. coli* was spurred by the efficiency of a large-scale workflow maintained by robotics and reliable laboratory methods, the continued advances of automated sequencing technology, and the use of well-written *de novo* genome assembly algorithms.

The completion of bacterial genome sequencing projects has had a significant impact on infectious disease, agricultural, and biodefense research. Elucidation of the genome sequences of pathogenic microbes has supported improvements in diagnostic procedures and has guided the study of pathological pathways and host immunology. Given the impact of genomics on other biological fields, genome sequencing projects have been carried out with various goals in mind, and microbial genome projects can be divided readily among groupings such as human and/or agricultural pathogens, environmentally important microbes, model microbial organisms, and microbes that thrive in peculiar or extreme environments.

Microorganisms that are categorized as potential biowarfare agents are of particular interest for both scientists and the general public. Hence, the anthrax cases that surfaced following the World Trade Center attack in 2001 provided an important opportunity for genomics. With questionable tracing and ambiguous strain typing of the *B. anthracis* strains isolated from various cases and obtained from various laboratories, the clarity and comprehensiveness offered by whole genome sequencing were further illustrated by comparative genomic analyses between implicated *B. anthracis* strains (27). Because of the highly monomorphic characteristic of *B. anthracis* and the greater than 99% sequence similarity generally observed between genomes of different strains or isolates, the availability of complete genome sequences rather than incomplete genome drafts was critical for in-depth analyses between different isolates.

To facilitate detailed comparative genomic analyses between *B. anthracis* isolates, DNA from each isolate was used to generate random insert libraries, from which clones were sampled for sequencing using dye terminator chemistry (27,28). Notably, this workflow is consistent with the methodology developed during the *H. influenzae* genome project, thus illustrating the efficacy of combining the whole genome random shotgun technique with the throughput of automated sequencing technology (18). Assembling the sequence reads involved the use of the TIGR assembler (18) or the Celera assembler (29), the latter of which was used effectively to assemble whole genome shotgun reads from the almost 120 Mb euchromatic genome of *Drosophila melanogaster* (29,30). Completing the genome assemblies entailed joining contigs and closing gaps through alternative priming on selected clones and through PCR linkage (27,28).

Identification of distinct polymorphisms between *B. anthracis* isolates that were expected to be identical, based on a strain typing tool that analyzes variable number tandem repeat markers (31), revealed that accurate genome sequencing offers unparalleled sensitivity over techniques that only sample potential genomic variation at select loci. The application of DNA sequencing beyond just genomic annotation and basic variant analyses has been an important triumph in biology, and this trend has continued with the widespread use of genomics for studying all aspects of biology.

## NEXT-GENERATION SEQUENCING TECHNOLOGY

Many major advances in biology and genomics have been accomplished through application and innovation of the DNA sequencing method pioneered by Frederick Sanger (6). The establishment of automated DNA sequencing technology driven by Applied Biosystems had thus enabled the sequencing of many genomes, especially those of microbial species, and eventually the human genome and the genomes of a few plant species. Domination of the Sanger chain termination sequencing method had endured for about 30 years before a new generation of DNA sequencing technology arrived to significantly change how sequencing projects could be carried out.

Given the scientific achievements enabled by automated DNA sequencing technology based on the Sanger method, a new interest in dramatically driving down cost and driving up sample throughput and data output had begun to grow. Such changes in the nature of DNA sequencing were expected to promote the completion of projects that were stalled by limitations of the Sanger sequencing approach and to stimulate progress for projects that are more intractable for Sanger sequencing, such as epigenomics, metagenomics, and transcriptomics (32). The 454 instrument was the first next-generation sequencer to become available, which was released in 2005. The 454 instrument reflected a major change from automated Sanger sequencing technology by using a different chemistry for determining sequence and by being able to simultaneously sequence significantly more DNA templates than a traditional sequencer, without necessitating an equally large spatial increase.

## 454 Sequencing

The 454 sequencer is an advanced pyrosequencer with significant sample and data throughput (www.454.com). Fundamentally, the 454 system utilizes a sequencing-by-synthesis technique to decode DNA. 454 sequencing includes the use of beads to capture template, the use of a specialized fiber optic plate to contain template-bound beads, and the use of light as an indicator of nucleotide incorporation, which is reflected in the length of sequence reads that the instrument can reliably provide with high accuracy (33,34). The current 454 sequencing instrument is the GS FLX, which produces more than one million reads, or 400 to 600 million high-quality bases, per 10-hour sequencing run. 454 reads are 400 bases on average, which represents an advantage held over other next-generation platforms, and the technology is projected to soon advance to 1000 base reads. While the 454 system is more expensive on a per base scale compared to other next-generation systems, the ability of 454 sequencing technology to produce longer read lengths makes the system particularly applicable for projects that involve or require *de novo* sequence assembly.

To sequence a DNA sample using the 454 instrument, a compatible library must first be generated from the sample. Notably, the library preparation process is designed to simply convert the DNA sample into a collection of sequencing substrates, without the use of plasmid manipulation, plasmid cloning, and bacterial transformation and colony selection required for complete Sanger sequencing of a genome of interest. Sample conversion into a 454 library involves mechanical fragmentation of the DNA, followed by fragment end polishing and the addition of DNA oligonucleotides onto the ends of each piece of fragmented DNA. The oligonucleotides consist of sequences that are required for PCR enrichment of the library prior to sequencing, as well as for the actual sequencing of the library. Furthermore, the oligonucleotides serve to enable the attachment of library fragments to beads upon which library enrichment and sequencing occur (35). PCR enrichment of the library prior to sequencing is similar to the plasmid cloning process involved in preparing a sample for Sanger sequencing in that enrichment is used to increase the amount of library material to enable quantification and sufficient sequencing. However, error rates and biases that accompany PCR are significant concerns in next-generation sequencing due to the sensitivity of these advanced sequencers and the necessity to carry out deep sequencing for certain projects. Such concerns have also catalyzed the development of procedures that allow less or no PCR enrichment of a library (36), as well as analytical strategies with greater sensitivity for smaller amounts of DNA, and methods for generating uniform sequence data from smaller DNA inputs (37,38).

Once a sample has been converted into a 454 library, it is ready to be loaded into a 454 instrument for sequencing. Beads, each carrying clonal copies of a library fragment, are loaded into a fiber optic support called a PicoTiter plate, which consists of many channels designed to each contain one bead. During the sequencing stage, reagents necessary for sequencing, including DNA polymerase and dNTPs, are flowed across the PicoTiter plate. Only one type of base is flowed into the PicoTiter plate at a time, and base incorporation results in the release of inorganic phosphate, which is then converted into light. The 454 system utilizes light as an indicator of base incorporation, which is translated into sequence data through the imaging components of the system. One important drawback of using light detection for measuring the incorporation of deoxynucleotides is that homopolymer stretches present on the template are more likely to be difficult to measure accurately, as the system must properly quantify the magnitude of the single light signal produced by a stretch of an identical base in a template. However, a homopolymer stretch is only expected to be sequenced with significantly lower accuracy if the stretch of identical bases extends beyond about 13 of such bases.

The current functionality of the 454 system enables users to perform shotgun sequencing to generate single read data as well as paired read data. Additionally, the 454 system supports multiplex sequencing as a cost-saving and sample throughput measure, where library samples are individually bar coded using tags called MIDs prior to pooling for a sequencing run. Another throughput-enhancing strategy of the 454 system allows scientists to sequence amplicons by appending 454-specific sequences onto PCR primers, thus simplifying the sample preparation workflow prior to 454 sequencing.

## Illumina Sequencing

Following the commercial release of the 454 instrument as the first next-generation sequencer, the second advanced sequencing instrument to reach the market was released in 2006 by Solexa, which was soon acquired by Illumina (www.Illumina.com). Like the 454 instrument, the Illumina genome analyzer is a massively parallel platform that determines the sequence of DNA using a sequencing-by-synthesis approach. A key difference between the Illumina system and the 454 system, however, is read length. The early Illumina platform was capable of producing single-end data consisting of 30- to 36-base-long reads and with error rates that were hard to manage and keep below 1%. Continued developments to enhance the functionality of the Illumina system have allowed the latest platform to support read length increases up to 101 bases, with error rates below 1%. More significantly, the data output from the genome analyzer has increased dramatically, from 0.5 billion high-quality bases per single-end 36 cycle run to up to 50 billion high-quality bases per paired-end 101 cycle run, with the next data increase projected to be 95 billion high-quality bases per paired-end 151 cycle run. Illumina data output increases have been the result of continual imaging and image analysis improvements that have enabled denser packing of library fragments on the flow cell, as well as the introduction of paired-end sequencing capability in 2008. The Illumina sequencing time line has also been reduced as a result of polymerase enhancements that allow sequencing runs to complete in a shorter amount of time. The significant data output and the lower cost per base compared to 454 sequencing have been compelling reasons for scientists to employ Illumina sequencing in various genetic studies.

Illumina sequencing technology is not prone to issues with homopolymer stretches in the template, as is the case for 454 sequencing. Rather than apply the features of pyrosequencing to high-throughput DNA sequencing, Illumina technology utilizes an engineered polymerase, unique substrate nucleotides that are reversibly terminated and fluorophore labeled according to base identity, and a glass substrate called a "flow cell" that is grafted with oligonucleotides used to secure and amplify sample library fragments for sequencing. During the sequencing stage, an optical system utilizing a charge-coupled device camera detects emissions specific to the nucleotide fluorophores following each cycle of base incorporation. Imaging employed by the Illumina system is used to determine the sequence composition of each cluster of clonal DNA fragments located on the flow cell.

The Illumina sample preparation workflow for genomic sequencing is consistent with the next-generation shotgun sequencing approach, where DNA is directly modified and prepared for sequencing without the need for plasmid manipulation and bacterial cloning. To prepare DNA for Illumina sequencing, the sample is fragmented mechanically and platform-specific oligonucleotides are attached to the ends of the DNA fragments. These adapter oligonucleotides support PCR enrichment of the library prior to sequencing, which may be avoided if there is a desire to limit PCR bias and error (36) or if the user is able to quantify the library without amplification. Further, library fragments must be adapter ligated in order to be able to attach to the flow cell and be sequenced.

Prior to being loaded on the genome analyzer, an Illumina library is denatured to generate single-stranded fragments, which are then amplified on a flow cell set within a cluster station. The clonal amplification step upstream of actual sequencing ensures that the signal produced from each cluster of clonal fragments during sequencing is significant enough to be detected and allow quality assessment. Following library cluster generation, the flow cell is transferred to the genome analyzer. The sequencing polymerase is introduced during the sequencing stage, along with flows of fluorophore-labeled reversibly terminated deoxynucleotides for each cycle of incorporation. The sequencing stage is also controlled to ensure that each cycle of incorporation is tracked properly. Each instance of base incorporation is a temporarily terminal synthesis, which is followed by fluorescence imaging to record the identity of the base incorporated into each DNA cluster and subsequent terminator removal to allow the next cycle of base incorporation. To generate paired-end sequence data, a genome analyzer must be fitted with a paired-end module, which performs the functions of a cluster station. Synthesis of complementary strands from fragments on the flow cell and cluster generation with the complementary strands must occur in order to provide a template for the second sequence read and thus produce paired reads.

The genome analyzer has been particularly useful for studies that do not require long reads, such as small RNA sequencing (39) and projects that could not be addressed readily with Sanger sequencing, such as chromatin immunoprecipitation studies (40). There have been efforts to use Illumina sequence data for *de novo* assembly through the development of software tailored for short reads and the work carried out at Illumina to extend read lengths (41,42). A better understanding of the benefits and strengths of the different sequencing technologies has also resulted in projects that use data from more than one system as a way to both independently verify different data sets and combine the advantages across technologies (43,44).

## Sequencing by Oligonucleotide Ligation and Detection (SOLiD) Sequencing

Applied Biosystems had already been a leader in automated DNA sequencing technology based on the Sanger chain termination method for many years by the time next-generation sequencers became available commercially, and their continued interest in DNA sequencing was reflected by their release of the SOLiD sequencing platform in 2007 (www.appliedbiosystems.com). The current standard SOLiD platform is the SOLiD 4, which produces up to 100 Gb of high-quality sequence, or up to 1.4 billion 35-base reads per two flow cell run. The SOLiD workflow also supports various sample preparations upstream of the actual sequencing stage that provide standard shotgun or mate-pair data, or support sample multiplexing during sequencing. Overall, the SOLiD system offers advantages of an alternative sequencing chemistry based on ligation and a base calling and error correction mechanism that is integrated into the sequencing method.

The process of preparing a DNA sample for SOLiD sequencing shares similarities with the 454 shotgun sequencing sample preparation workflow, where DNA is randomly mechanically fragmented and modified with platform-specific DNA sequences appended onto the ends of each DNA fragment prior to being enriched on a bead substrate using emulsion PCR. Once the DNA library has been enriched, beads are distributed across a glass slide flow cell upon which sequencing-by-ligation chemistry (45) takes place, which is an important distinction from the sequencing-by-synthesis chemistries utilized by the 454 and Illumina platforms. The SOLiD instrument additionally allows its users to load two slides per sequencing run as a throughput-enhancing mechanism.

To initiate the sequencing stage, a primer is annealed onto an adapter sequence carried by template strands attached onto a bead, which supports the ligation of eight-base-long DNA oligonucleotides labeled fluorescently with a "two-base encoding" method used to define different dinucleotide combinations. Following ligation of the first complementary oligonucleotide, fluorescence detection is used to help determine the identity of the first two bases of the incorporated oligonucleotide. The incorporated oligonucleotide is then cleaved after its fifth base, which results in the removal of the fluorophore originally associated with the oligonucleotide and initiates the second oligonucleotide ligation cycle. These incorporation, imaging, and cleavage steps are repeated for the extent of cycles or contiguous base reads required prior to being completely reinitiated using a new primer that is exactly one base shorter at its 3′ end than the primer used in the previous set of cycles. Repriming and resequencing also occur for a number of times, up to and including the use of a sequencing primer that is four bases shorter than the original primer used at the start of the sequencing stage. Each SOLiD sequencing run thus produces five sets of independent and overlapping sequence data from each template, allowing data corroboration from a single run of a single template.

## ASSEMBLY METHODS FOR DNA SEQUENCE

The shotgun sequencing strategy is widely used in genome sequencing projects. Shotgun sequencing is a technique in which large pieces of DNA are sheared into smaller fragments that are then sequenced randomly. These random fragments must be realigned and ordered into larger contiguous pieces that are representative of the original large DNA unit. Due to the randomness of this strategy, a certain level of redundancy is required to increase the probability that the majority of the original large DNA fragment will be represented by overlapping fragments. This redundancy leads to generation of a large number of sequences. For long Sanger reads, an 8-fold redundancy is often targeted, whereas for short reads produced by a next-generation instrument, it is often 20-fold or higher, as the shorter length provides less unique information for each read. The alignment of sequence data is complicated by the fact that there is a certain amount of error inherent in any sequencing read. An assembly algorithm must be tolerant of a certain level of disagreement between sequences in order to build overlapping data sets. However, if the algorithm is too tolerant, data may be assembled incorrectly due to misalignment of high identity repeat sequences. Manual correction was often required, so many assembly tools were used in conjunction with contig editors that allowed for editing and modification of the assembly, such as GAP4 (46) and CONSED (47).

## Staden Package

The Staden assembler (48) was originally based on the consensus or majority rule method of assembly. Each read was compared to all other reads in the data set to find overlaps, and data were assembled into overlapping sections called "contigs (contiguous regions of overlapping sequence)." These contigs were then compared to find regions that might be ordered and joined together to form supercontigs. Finally, the underlying reads in each contig would be edited and used to determine the best consensus representation of data.

## PHRAP

The PHRAP assembly program (49) was different from earlier assemblers in that it used base quality scores rather than majority rule to drive the consensus. PHRAP compares sequences to each other by searching for pairs of perfectly matching "words" or sequence regions that meet certain criteria (such as a designated length). If a match of the designated word size is found, PHRAP then tries to extend the alignment. This local alignment is then scored such that matching bases are given positive value and mismatches and gaps are given negative values.

PHRAP uses PHRED (50,51) quality score information to help align and merge the random sequences. PHRED quality values range from 0 to 99 and use a logarithmic scale in which a quality value of 10 reflects a probability of 1 error in 10, a quality value of 30 reflects a probability of one error in 1000, and so on. PHRED uses information from raw sequence data such as peak height, base spacing, and level of background noise to help calculate quality values for each base call. The PHRAP contig sequences reflect a

"Golden Path"—the highest quality base call is chosen from the underlying data at each position.

## Newbler

The Newbler assembler (33) was developed to assemble pyrogram sequences produced by the 454 instrument. The three main parts of the program are the Overlapper, the Unitigger, and the Multialigner. The overlapper uses a hashing algorithm, which fragments reads into short words or k-mers, to perform an all-against-all comparison to identify overlaps. The flowgrams are compared and if the overlap exceeds a selected value, reads are flagged for overlap. Unitigger then groups the flagged reads into consistent contigs. Multialigner averages the underlying signals from reads in the unitigs and calls a consensus.

## Euler

The previously discussed assemblers used the overlap–layout–consensus approach (52). A new method of assembly has been introduced that uses the de Bruijn graph to handle the assembly of shorter reads produced by next-generation sequencing instruments. This method finds overlapping k-mers to form nodes that are connected through the graph. Repeats are only represented once in the graph. Euler then finds the best "Eulerian superpath," which is the path through the assembly that best fits all of the paths generated by the underlying reads. This method has since been used in other assemblers written for short reads such as Velvet (41) and Abyss (42).

## RESEQUENCING

Read alignment differs from *de novo* read assembly in the reads are aligned to a reference genome sequence. The reference allows reads to be mapped to an ordered standard (noting mismatches) rather than comparing all of the reads to each other and building a consensus from underlying read data. Many alignment tools have been written in the years since the advent of next-generation sequencing, but they basically fall into two main categories: aligners that use a hashing method [such as ELAND (Tony Cox, unpublished results) and MAQ (53)] and aligners that use a variation of the Burrows Wheeler transform [such as BWA (54) and BowTie (55)].

## CONCLUSION

Advances in DNA sequencing methods have had a major impact on several biological fields, with microbiology being the first field to experience the benefits of knowing the exact sequence of a genome. This advantage held by the microbiology field was a result of the more compact nature of many microbial genomes, the long-standing history of microbes as model organisms in biological research, and the initial limitations of DNA sequencing chemistry and bioinformatic methods that restricted both the scale and the scope of genomic analysis. The whole genome sequencing of microbial life forms made it feasible to perform functional annotations of genomes and strengthened the motivation of the scientific community to sequence the genomes of several species of medical, environmental, evolutionary, and/or societal importance.

With the invention of advanced DNA sequencing methods, exemplified by next-generation sequencers, biological sciences have again been presented with significant opportunities that had not been particularly feasible with more traditional methods. Microbiology has been able to benefit significantly from the functionality of next-generation sequencing. The massive data output of next-generation systems now enables scientists to use whole genome sequencing to identify a microbial isolate rapidly, as well as study microbial sequence variation without having to use various cloning and gene reporter techniques to help isolate the mutation(s) of interest. The independence of the next-generation workflow from microbiological culturing procedures has also promoted the success of metagenomic studies, the results of which have led to the reassessment of long-held scientific views regarding the diversity and interaction within microbial communities.

# REFERENCES

[1] E.B. Ziff, J.W. Sedat, F. Galibert, Determination of the nucleotide sequence of a fragment of bacteriophage phiX 174 DNA, Nat. New Biol. 241 (106) (1973) 34–37.

[2] H.D. Robertson, B.G. Barrell, H.L. Weith, J.E. Donelson, Isolation and sequence analysis of a ribosome-protected fragment from bacteriophage phiX 174 DNA, Nat. New Biol. 241 (106) (1973) 38–40.

[3] W. Gilbert, A. Maxam, The nucleotide sequence of the lac operator, Proc. Natl. Acad. Sci. USA 70 (12) (1973) 3581–3584.

[4] F. Sanger, J.E. Donelson, A.R. Coulson, H. Kossel, D. Fischer, Use of DNA polymerase I primed by a synthetic oligonucleotide to determine a nucleotide sequence in phage fl DNA, Proc. Natl. Acad. Sci. USA 70 (4) (1973) 1209–1213.

[5] A.M. Maxam, W. Gilbert, A new method for sequencing DNA, Proc. Natl. Acad. Sci. USA 74 (2) (1977) 560–564.

[6] F. Sanger, S. Nicklen, A.R. Coulson, DNA sequencing with chain-terminating inhibitors, Proc. Natl. Acad. Sci. USA 74 (12) (1977) 5463–5467.

[7] J. Messing, R. Crea, P.H. Seeburg, A system for shotgun DNA sequencing, Nucleic Acids Res. 9 (2) (1981) 309–321, PMCID: PMC326694.

[8] B. Gronenborn, J. Messing, Methylation of single-stranded DNA in vitro introduces new restriction endonuclease cleavage sites, Nature 272 (5651) (1978) 375–377.

[9] E.S. Lander, L.M. Linton, B. Birren, C. Nusbaum, M.C. Zody, J. Baldwin, et al., International Human Genome Sequencing Consortium. Initial sequencing and analysis of the human genome, Nature 409 (6822) (2001) 860–921.

[10] J.C. Venter, M.D. Adams, E.W. Myers, P.W. Li, R.J. Mural, G.G. Sutton, et al., The sequence of the human genome, Science 291 (5507) (2001) 1304–1351.

[11] F. Sanger, G.M. Air, B.G. Barrell, N.L. Brown, A.R. Coulson, C.A. Fiddes, et al., Nucleotide sequence of bacteriophage phi X174 DNA, Nature 265 (5596) (1977) 687–695.

[12] F. Sanger, A.R. Coulson, G.F. Hong, D.F. Hill, G.B. Petersen, Nucleotide sequence of bacteriophage lambda DNA, J. Mol. Biol. 162 (4) (1982) 729–773.

[13] S. Anderson, Shotgun DNA sequencing using cloned DNase I-generated fragments, Nucleic Acids Res. 9 (13) (1981) 3015–3027.

[14] S.A. Fuhrman, P.L. Deininger, P. LaPorte, T. Friedmann, E.P. Geiduschek, Analysis of transcription of the human Alu family ubiquitous repeating element by eukaryotic RNA polymerase III, Nucleic Acids Res. 9 (23) (1981) 6439–6456.

[15] A.T. Bankier, S. Beck, R. Bohni, C.M. Brown, R. Cerny, M.S. Chee, et al., The DNA sequence of the human cytomegalovirus genome, DNA Seq. 2 (1) (1991) 1–12.

[16] K. Oda, K. Yamato, E. Ohta, Y. Nakamura, M. Takemura, N. Nozato, et al., Gene organization deduced from the complete sequence of liverwort Marchantia polymorpha mitochondrial DNA. A primitive form of plant mitochondrial genome, J. Mol. Biol. 223 (1) (1992) 1–7.

[17] R.F. Massung, J.J. Esposito, L.I. Liu, J. Qi, T.R. Utterback, J.C. Knight, et al., Potential virulence determinants in terminal regions of variola smallpox virus genome, Nature 366 (6457) (1993) 748–751.

[18] R.D. Fleischmann, M.D. Adams, O. White, R.A. Clayton, E.F. Kirkness, A.R. Kerlavage, et al., Whole-genome random sequencing and assembly of Haemophilus influenzae Rd, Science 269 (5223) (1995) 496–512.

[19] D.L. Daniels, G. Plunkett 3rd, V. Burland, F.R. Blattner, Analysis of the *Escherichia coli* genome: DNA sequence of the region from 84.5 to 86.5 minutes, Science 257 (5071) (1992) 771–778.

[20] G. Plunkett 3rd, V. Burland, D.L. Daniels, F.R. Blattner, Analysis of the *Escherichia coli* genome. III. DNA sequence of the region from 87.2 to 89.2 minutes, Nucleic Acids Res. 21 (15) (1993) 3391–3398.

[21] F.R. Blattner, V. Burland, G. Plunkett 3rd, H.J. Sofia, D.L. Daniels, Analysis of the *Escherichia coli* genome. IV. DNA sequence of the region from 89.2 to 92.8 minutes, Nucleic Acids Res. 21 (23) (1993) 5408–5417.

[22] H.J. Sofia, V. Burland, D.L. Daniels, G. Plunkett 3rd, F.R. Blattner, Analysis of the Escherichia coli genome. V. DNA sequence of the region from 76.0 to 81.5 minutes, Nucleic Acids Res. 22 (13) (1994) 2576–2586.

[23] V. Burland, G. Plunkett 3rd, H.J. Sofia, D.L. Daniels, F.R. Blattner, Analysis of the *Escherichia coli* genome. VI. DNA sequence of the region from 92.8 through 100 minutes, Nucleic Acids Res. 23 (12) (1995) 2105–2119.

[24] V. Burland, G. Plunkett 3rd, D.L. Daniels, F.R. Blattner, DNA sequence and analysis of 136 kilobases of the *Escherichia coli* genome: Organizational symmetry around the origin of replication, Genomics 16 (3) (1993) 551–561.

[25] F.R. Blattner, G. Plunkett 3rd, C.A. Bloch, N.T. Perna, V. Burland, M. Riley, et al., The complete genome sequence of *Escherichia coli* K-12, Science 277 (5331) (1997) 1453–1462.

[26] Y. Kohara, K. Akiyama, K. Isono, The physical map of the whole E. coli chromosome: Application of a new strategy for rapid analysis and sorting of a large genomic library, Cell 50 (3) (1987) 495–508.

[27] T.D. Read, S.L. Salzberg, M. Pop, M. Shumway, L. Umayam, L. Jiang, et al., Comparative genome sequencing for discovery of novel polymorphisms in Bacillus anthracis, Science 296 (5575) (2002) 2028–2033.

[28] T.D. Read, S.N. Peterson, N. Tourasse, L.W. Baillie, I.T. Paulsen, K.E. Nelson, et al., The genome sequence of *Bacillus anthracis* Ames and comparison to closely related bacteria, Nature 423 (6935) (2003) 81–86.

[29] E.W. Myers, G.G. Sutton, A.L. Delcher, I.M. Dew, D.P. Fasulo, M.J. Flanigan, et al., A whole-genome assembly of Drosophila, Science 287 (5461) (2000) 2196–2204.

[30] S.E. Celniker, D.A. Wheeler, B. Kronmiller, J.W. Carlson, A. Halpern, S. Patel, et al., Finishing a whole-genome shotgun: Release 3 of the *Drosophila melanogaster* euchromatic genome sequence, Genome Biol. 3 (12) (2002) RESEARCH0079.

[31] P. Keim, L.B. Price, A.M. Klevytska, K.L. Smith, J.M. Schupp, R. Okinaka, et al., Multiple-locus variable-number tandem repeat analysis reveals genetic relationships within *Bacillus anthracis*, J. Bacteriol. 182 (10) (2000) 2928–2936.

[32] S. Fox, S. Filichkin, T.C. Mockler, Applications of ultra-high-throughput sequencing, Methods Mol. Biol. 553 (2009) 79–108.

[33] M. Margulies, M. Egholm, W.E. Altman, S. Attiya, J.S. Bader, L.A. Bemben, et al., Genome sequencing in microfabricated high-density picolitre reactors, Nature 437 (7057) (2005) 376–380.

[34] M. Droege, B. Hill, The Genome Sequencer FLX System—longer reads, more applications, straight forward bioinformatics and more complete data sets, J. Biotechnol. 136 (1-2) (2008) 3–10.

[35] D.S. Tawfik, A.D. Griffiths, Man-made cell-like compartments for molecular evolution, Nat. Biotechnol. 16 (7) (1998) 652–656.

[36] M.A. Quail, I. Kozarewa, F. Smith, A. Scally, P.J. Stephens, R. Durbin, et al., A large genome center's improvements to the Illumina sequencing system, Nat. Methods 5 (12) (2008) 1005–1010.

[37] R.A. White 3rd, P.C. Blainey, H.C. Fan, S.R. Quake, Digital PCR provides sensitive and absolute calibration for high throughput sequencing, BMC Genomics 10 (2009) 116.

[38] M. Meyer, A.W. Briggs, T. Maricic, B. Hober, B. Hoffner, J. Krause, et al., From micrograms to picograms: Quantitative PCR reduces the material demands of high-throughput sequencing, Nucleic Acids Res. 36 (1) (2008) e5.

[39] R.D. Morin, M.D. O'Connor, M. Griffith, F. Kuchenbauer, A. Delaney, A.L. Prabhu, et al., Application of massively parallel sequencing to microRNA profiling and discovery in human embryonic stem cells, Genome Res. 18 (4) (2008) 610–621.

[40] A. Barski, S. Cuddapah, K. Cui, T.Y. Roh, D.E. Schones, Z. Wang, et al., High-resolution profiling of histone methylations in the human genome, Cell 129 (4) (2007) 823–837.

[41] D.R. Zerbino, E. Birney, Velvet: Algorithms for de novo short read assembly using de Bruijn graphs, Genome Res. 18 (5) (2008) 821–829.

[42] J.T. Simpson, K. Wong, S.D. Jackman, J.E. Schein, S.J. Jones, I. Birol, ABySS: A parallel assembler for short read sequence data, Genome Res. 19 (6) (2009) 1117–1123.

[43] S. Diguistini, N.Y. Liao, D. Platt, G. Robertson, M. Seidel, S.K. Chan, et al., De novo genome sequence assembly of a filamentous fungus using Sanger, 454 and Illumina sequence data, Genome Biol. 10 (9) (2009) R94.

[44] K.E. Holt, J. Parkhill, C.J. Mazzoni, P. Roumagnac, F.X. Weill, I. Goodhead, et al., High-throughput sequencing provides insights into genome variation and evolution in Salmonella typhi, Nat. Genet. 40 (8) (2008) 987–993.

[45] J. Shendure, G.J. Porreca, N.B. Reppas, X. Lin, J.P. McCutcheon, A.M. Rosenbaum, et al., Accurate multiplex colony sequencing of an evolved bacterial genome, Science 309 (5741) (2005) 1728–1732.

[46] R. Staden, K.F. Beal, J.K. Bonfield, The Staden Package, Comput. Methods Mol. Biol. 132 (1998) 115–130.

[47] D. Gordon, C. Abajian, P. Green, Consed: A graphical tool for sequence finishing, Genome Res. 8 (3) (1998) 195–202.

[48] J.K. Bonfield, K. Smith, R. Staden, A new DNA sequence assembly program, Nucleic Acids Res. 23 (24) (1995) 4992–4999.

[49] P. Green, Documentation for PHARP, 1996. Available from: http://bozeman.mbt .washington.edu.

[50] B. Ewing, L. Hillier, M.C. Wendl, P. Green, Base-calling of automated sequencer traces using PHRED. I. Accuracy assessment, Genome Res. 8 (3) (1998) 175–185.

[51] B. Ewing, P. Green, Base-calling of automated sequencer traces using PHRED. II. Error probabilities, Genome Res. 8 (3) (1998) 186–194.

[52] J.D. Kececioglu, E.W. Myers, Combinatorial algorithms for DNA sequence assembly, Algorithmica 13 (1995) 7–51.

[53] H. Li, J. Ruan, R. Durbin, Mapping short DNA sequencing reads and calling variants using mapping quality scores, Genome Res. 18 (11) (2008) 1851–1858.

[54] H. Li, R. Durbin, Fast and accurate short read alignment with Burrows-Wheeler transform, Bioinformatics 25 (14) (2009) 1754–1760.

[55] B. Langmead, C. Trapnell, M. Pop, S.L. Salzberg, Ultrafast and memory-efficient alignment of short DNA sequences to the human genome, Genome Biol. 10 (3) (2009) R25.

# CHAPTER 28

# Genomics

**W. Florian Fricke,[a] Thomas A. Cebula,[a,b] and Jacques Ravel[a]**

[a]*Institute for Genome Sciences, University of Maryland School of Medicine, Baltimore, Maryland*
[b]*Johns Hopkins University, Baltimore, Maryland*

… and in today already walks tomorrow.

Friedrich Schiller (1)

In 1995, the "genomic revolution" began with completion of the first microbial genome sequence (2). For the first time, the genetic basis of a bacterial isolate was completely characterized. As new sequencing technologies are being developed that will facilitate access to genomic sequence information, the ongoing genomic revolution is expected to continue having its transformative effect and impact many aspects of our life, including law enforcement and the forensic sciences. The goal of this chapter is to provide an introduction to genomics and its rapidly evolving technological environment, especially as it relates to the field of microbial forensics.

Genomic research aims at revealing and analyzing DNA and RNA sequence information. Every organism on earth is defined in essence by its genome sequence. The genomic signature is considered the most specific fingerprint that can unambiguously identify most people on earth. It can help distinguish even closely related organisms, that is, those exhibiting identical phenotypes, and therefore is of special relevance to the field of microbial forensics. As an example, with knowledge of entire genome sequences, it is now possible to distinguish two bacterial strains associated with the same food-borne disease outbreak. Furthermore, these two isolates could even be classified and assigned to an evolutionary tree showing their relationship. Up until now, high sequencing costs have precluded, except for a few cases (see anthrax-letter investigation chapter 2 as an example), application of whole genome analysis as a forensic tool. However, new sequencing technologies continue to be developed that offer increasing sequence data output at decreasing costs per run, outpacing Moore's law as applied to the growth of computing resources (3). New/next-generation sequencing technologies provide the means to rapidly sequence the

479

Microbial Forensics. DOI: 10.1016/B978-0-12-382006-8.00028-1

genome of thousands of individual bacterial isolates as well as bacterial cultures (population genomics) or complex microbial communities (metagenomics). As a consequence, genomics is becoming standard not only to the research field but also to public health and microbial forensics. Genomics is not a research field in itself but is now a universal laboratory tool, and as such, once validated, it will increasingly be integrated as part of the microbial forensic investigator's toolbox.

## SEQUENCING TECHNOLOGIES

With the advent of next-generation sequencing platforms, genomics is changing rapidly, a trend that is expected to continue. Newer sequencing platforms, while producing large amounts of sequences, still have some disadvantages compared to traditional technologies (mostly decreased read length). They have, however, opened the genomics field to new research areas, such as microbial ecology (metagenomics) and single-cell biology (single-cell genomics). As sequencing costs continue to decrease, one can anticipate more applications to be developed. Today, the cost of storing, processing, and analyzing large sequence data sets already exceeds the cost of generating them.

To date, whole genome shotgun (WGS) sequencing is still the standard for sequencing a microbial genome. Next-generation sequencing technologies, combined with improved computing resources, have enabled WGS sequencing for any genome project, irrelevant of the size of the genome. For this approach, the targeted sequence is sheared randomly into multiple overlapping fragments that are then sequenced in parallel high-throughput sequencing reactions. Following the sequence generation, bioinformatics software tools assemble the random sequence fragments into larger sequence fragments ("contigs") and ultimately complete genomes *in silico*.

The most common sequencing platforms today use dye-termination electrophoretic sequencing (Sanger), sequencing by synthesis (454, Illumina, and Helicos), and sequencing by ligation (SOLiD); the last two technologies are rapidly replacing Sanger sequencing by reducing cost and increasing data generation throughput.

### Sanger Sequencing

For about 10 years after publication of the first microbial genome, *Haemophilus influenzae* in 1995 (2), genome sequencing was based entirely on Sanger sequencing (4), named after Frederick Sanger. Using this technology, not only microbes with small genome sizes (5–7), but also the first eukaryotic genomes, including those of human (8,9), mouse (10), fruit fly (11), and thale cress (12), were sequenced, albeit at a very high cost.

Electrophoretic sequencing with the Sanger method uses a modified polymerase chain reaction (PCR) amplification protocol in combination with individually fluorescently labeled dideoxynucleotides (ddNTPs). Incorporation of ddNTPs into the growing DNA strand during PCR occurs randomly, thereby terminating DNA strand extension and resulting in DNA fragments that vary in length by one nucleotide. These fragments are separated electrophoretically by size and are visualized by their nucleotide-specific fluorescent label. Base-calling software reads fluorescence emission and outputs the DNA sequence. Common challenges with this technology are low throughput and high cost on a per nucleotide basis. Further, DNA fragments are first cloned to form either small insert plasmid and/or large insert fosmid or cosmid libraries. Because not all DNA can be cloned, this step can introduce biases. However, these inserts are usually sequenced from both ends, providing so-called paired-end sequence reads separated by a piece of unknown sequence of known length. This information is useful in the downstream bioinformatics assembly of the sequence reads into contigs.

Sanger sequencing technology is still best suited to generate relatively large paired-end sequence reads. Paired-end sequence reads of at least 800 bp length, separated by up to 50-kbp insert length, are not unusual in Sanger sequencing projects. Newer sequencing platforms, thus far, have not been able to provide comparable read lengths or insert lengths of paired-end sequence reads. Due to significantly higher costs and labor per sequenced base pair, the genomics community is increasingly shifting from Sanger to next-generation sequencing platforms, which provide higher sequence output at a much lower cost.

## Next-Generation Sequencing

In 2003, the adenovirus genome was sequenced using what was then called "next-generation sequencing technology, 454 pyrosequencing" (13). This technology and other platforms use an array-based approach that allows for simultaneous detection of millions of sequencing reactions, thereby significantly increasing the number of reads per sequencing reaction, over 100,000 per run in the first-generation 454 pyrosequencers (GS20) compared with 96 per run with Sanger sequencing.

## Next-Generation Sequencing by Synthesis

All sequencing-by-synthesis approaches take advantage of photochemistry and/or fluorescence detection systems to monitor the incorporation of individual nucleotides by a DNA polymerase into a growing DNA strand. The three production-ready and currently available commercial systems—454/Roche, Illumina, and Helicos—use different approaches to amplify and immobilize or just immobilize DNA templates on a sequencing array that serves as the reaction chamber for the DNA synthesis reaction. Illumina and

454 require the addition of adapters to sheared DNA fragment to form a sequencing library. This library is then amplified by PCR amplifications, either in water–oil emulsions (454) (13) or on sequencing arrays with immobilized primers (Illumina) (14). In both cases, amplifications are intended to increase the signal intensity during the sequencing reaction. A significant limitation of the 454 system is that homopolymeric stretches consisting of more than six consecutive nucleotides of the same type are difficult to resolve and result in a high sequencing error rate. The Helicos system does not require amplification of a library and, based on the use of proprietary fluorescently labeled nucleotides, can detect directly each base pair incorporated into a DNA fragment being synthesized (15). By using immobilized single DNA molecules as templates, the Helicos system avoids any biases introduced by the DNA amplification step.

Each 4-hour run of 454/Roche pyrosequencing generates between 800,000 and 1.2 million sequence reads of about 400 bp in length, resulting in a total sequence output ranging between 300 and 480 Mbp. However, an 8- to 11-day run of Illumina sequencing has a total sequence output of 20 to 150 Gbp but in sequence reads of 50 to 100 bp in length. The shorter Illumina reads tend to have a higher accuracy than 454 reads, but represent a bioinformatics challenge for *de novo* microbial genome analysis. These new technologies are therefore currently used mainly for resequencing of organisms for which reference genomes have been deposited in the sequence databases. Another application for next-generation sequencing by synthesis, which also depends on the availability of a reference genome, is sequencing of RNA transcripts for gene expression analyses. These new technologies have reduced the cost of sequencing a microbial genome from more than $80,000 with Sanger sequencing to just a few thousand dollars or less depending on genome size.

## Sequencing by Ligation

While all other currently available sequencing technologies depend on nucleotide specificity and accuracy of DNA polymerases during the sequencing process, Applied Biosystem's SOLiD (Sequencing by Oligonucleotide Ligation and Detection) system uses DNA ligase to synthesize a complementary copy of immobilized DNA fragment templates on an array (16,17). Following emulsion PCR (similar to 454), DNA fragments are hybridized to bead-bound primers. A set of four fluorescently labeled dibase hexamer probes competes for ligation to the sequencing primer. Specificity of the dibase probe is achieved by interrogating every first and second base in each ligation reaction. Multiple cycles of ligation, detection, and cleavage are performed with the number of cycles determining the eventual read length. The template is then reset with a primer complementary to the *n*-1 position for a second round of ligation cycles. Five rounds of primer reset are completed for each sequence tag. Through the primer reset process, each base is interrogated in two independent ligation reactions by two different primers, increasing the sequencing accuracy of this technology. A total of 60–100 Gbp is generated per run (predicted to be over 300 Gbp per run by the end of 2010) with an accuracy of 99.94%.

## Future Developments

Several new sequencers are expected to become available by the end of 2010 that are likely to fundamentally change the genomics landscape. Life Sciences' 454 and other companies, such as Ion Torrent, target new niches within the growing genomics market by providing relatively inexpensive benchtop sequencers that should further open the genomics fields to smaller research laboratories. Pacific Biosciences, however, focuses on the generation of long sequence reads of several thousand nucleotides in length, using its single-molecule real-time (SMRT™) sequencing technology (18). Other platforms intend to use nanopores to directly read each nucleotide of a single DNA strand without the need for fluorescence labeling or sequence synthesis and are likely to become available in a few years. While it is difficult to predict the success of each of these new sequencing technologies, it is safe to assume that the way sequencing is performed will be different in just a few years, impacting all fields of research.

## BIOINFORMATICS SEQUENCE ANALYSIS

Sequence data can be close to meaningless without appropriate postsequencing bioinformatic analysis. As next-generation sequencing technologies increase sequence throughput, new algorithms have to be developed to perform or optimize tasks associated with sequence processing and analysis. Traditionally, genome projects involve the following three steps: (i) assembly of individual sequence reads into larger contigs and ultimately complete genome sequences; (ii) gene prediction and functional annotation based on protein and protein domain comparisons to established sequence databases; and (iii) comparative sequence analysis of single genes, DNA fragments, or entire genomes to explore gene functions and genome architectures. Downstream sequence analysis methods to identify, for example, orthologous genes from different genomes (19), define a species pan genome (20), or depict the evolutionary tree that characterizes the relationship of genes or entire chromosomes from different organisms (19,21) mostly depend on comparative sequence analysis to references from sequence databases. Relatively well-established methods have been developed to accomplish most of the analysis for each of these three steps for microbial size genome projects based on Sanger sequence data. However, with increasing amounts of sequence data being generated by next-generation sequencing platforms, bioinformatics sequence analysis is once again becoming a challenge and a major bottleneck, along with data management and storage.

# THE PREGENOMIC ERA

The transformative effect of the so-called genomic revolution is best explained by recalling limitations of microbiological research in the pregenomic era, which was characterized by the lack of absolute criteria for taxonomic classifications. Traditionally, microbiologists used to classify bacteria on the basis of observable phenotypes in order to provide microbial taxonomies. Observable phenotypes can, however, evolve at rates that are different from the rest of the organism, for example, through the acquisition of a new genotypic trait by horizontal gene transfer. Unaware of the genetic background that is responsible for a prominent bacterial phenotype, taxonomies can have difficulties in accommodating contradicting phenotype observations. In addition, microbes can evolve into lineages that are indistinguishable without complex phenotypic analyses. As an example, it became apparent during the resequencing of the original *Bacillus subtilis* 168 isolate that those laboratories that collaborated on the first *B. subtilis* genome project worked with different sublineages of the same strain, which had evolved from the original isolate over time through repeated cultivation in separate batches in different laboratories (22).

Molecular genotyping tools, such as restriction fragment length polymorphism or pulsed-field gel electrophoresis (PFGE), were developed before genomic tools became widely available. Although offering only limited phylogenetic resolution, some of these techniques continue to be applied due to their high affordability and relatively modest technical requirements. The recent food-borne outbreak of spinach-associated enterohemorrhagic *Escherichia coli* O157:H7 has demonstrated that high-resolution genome-level phylogenetic analysis can be used as a fingerprinting method in epidemiological studies. In this study, some of the associated strains were misleadingly classified as being indistinguishable using traditional fingerprinting, while genomic analyses showed substantial variation in virulence gene contents (Eppinger, Ravel and Cebula, unpublished).

After the first complete bacterial genome sequences became available, microarrays were developed that used DNA–DNA hybridization assays to screen genomic DNA of unknown composition for the presence or absence of known sequence fragments. While providing a cost- and time-efficient approach to detect a microbial genotype, for example, *E. coli* strain-specific loci, virulence, or antimicrobial resistance gene clusters, microarray-based genomic tools have a number of limitations. First, the hybridization signal depends on strong sequence homology. Microarrays therefore depend on known sequence information and are unable to detect and characterize novel sequence features, such as, for example, new virulence or resistance genes. Second, they provide only very limited information about the nature of sequence variation. Genome evolution that manifests in single-nucleotide polymorphisms, as well as genomic recombinations, can be difficult or impossible to detect through microarray hybridizations. Third, microarrays are unable to provide information about the genomic context of a detected feature. Whether or not a gene is chromosomally or plasmid encoded or whether it is predicted to originate from horizontal gene transfer has important implications for analysis of the evolutionary history of a bacterial isolate. For example, the colocalization of virulence and antimicrobial resistance genes on mobilizable plasmids implies a direct threat for the simultaneous propagation of both phenotypes, which is more critical than if both phenotypes were encoded on separate locations of the chromosome (23). All of the described limitations can be addressed through whole genome shotgun sequencing with newer sequencing technologies.

In the pregenomic era, microbial research was significantly limited by its dependence on cultivable organisms for analysis. In general, an organism had to be isolated from a sample, cultivated, and reduced to a single clonal population before it could be studied using standard molecular methods. As a consequence, it is assumed that the single cell from the original sample is representative of the original population. However, this assumption could be wrong and misleading, as it is likely that not all members of a population or culture are genomically identical. Even if, for example, individual *E. coli* strains are being isolated from a sample, for which cultivation protocols are well established, it is not clear how much variation within the original *E. coli* population is overlooked simply by concentrating on only one clone per sample in the final analysis. Newer genomic sequencing approaches, which enable the analysis of total genomic DNA isolates from samples of bacterial cultures (population genomics) or entire microbial communities (metagenomics), provide a means of overcoming this limitation and are discussed in more detail later.

## COMPARATIVE GENOMICS

To date, over 1000 microbial genomes have been completely sequenced, and thousands more are available as draft unfinished sequences (24). These numbers will only increase exponentially in the near future as next-generation sequencing technologies enable the rapid sequencing or resequencing of thousands of microbial isolates. While a wealth of new information is provided, major bioinformatics challenges are created for comparative analyses.

Soon after the first few bacterial genome sequences became available, researchers discovered the possibilities of genome comparisons for functional genome analysis and to improve our understanding of genome evolution, a key element in microbial forensic investigations. New bioinformatic algorithms had to be developed to allow for the rapid comparison of nucleotide or amino acid sequences on a genome-wide level (25–27). In addition, the number of released genome sequences had to grow to allow for the comparison of related, as well as unrelated, species. Two different approaches are commonly used in comparative genomics: (i) distantly related species are compared that share a specific phenotype and (ii) closely related species are compared that differ with respect to a specific phenotype. In the first case, shared coding features present in all compared genomes are likely candidates for identification of the genetic determinants responsible for the shared phenotype. In the second case, those genetic determinants are likely to be identified among the unique coding features that are specific for only one of the two compared groups. The first approach has been used to identify the minimal gene set of viable bacterial cells (28,29), which is of great interest in synthetic biology (30). A frequent application for the second approach is the comparison of genomes from pathogenic and non-pathogenic isolates, for example, of different commensal and pathogenic *E. coli* strains, to identify virulence factors responsible for a specific disease-causing phenotype (31). A similar method has been suggested for the identification of genes that are shared by related pathogens and encode proteins with antigenic potential that could be used for the development of vaccines. This approach has been termed "reverse vaccinology" (32). While these analyses have application in basic research, microbial forensics could make use of these technologies to determine the relatedness and/or similarities between two microbial isolates. A large number of differences would equate a distant evolutionary relationship, while very few and subtle changes would indicate recent common ancestry, and hence could link an unknown sample to its origin.

## HIGH-THROUGHPUT SCREENING ASSAYS

Comparative genomics, as discussed in the previous section, can also be used to identify marker sequences specific for a group of bacterial species or serotypes or pathogens. Using PCR primer pairs designed to specifically bind these shared marker regions, large-scale screening assays can be set up for detection, amplification, and characterization of these genotypes from various samples. Not only do these screening assays provide information about the presence or absence of a genetic feature within a sample of unknown microbial composition, if the amplified PCR product is sequenced, the information can be used for phylogenetic analyses, that is, different isolated PCR products can be aligned to generate taxonomic trees that predict the evolutionary relationships between the different isolates. The classic example for this type of screening assay is amplification of 16S ribosomal RNA genes or gene fragments, which are conversed and universally found in every bacterial genome (33). The same approach has also been used for epidemiological studies to follow disease outbreaks caused by specific bacterial pathogens (34). Multilocus sequence typing takes advantage of a limited number of conserved housekeeping genes to characterize bacterial pathogens, such as *E. coli* or *S. enterica* (http://pubmlst.org/) (35).

# METAGENOMICS

Metagenomics approaches are becoming increasingly popular in large-scale genomics applications as a way to study the taxonomic and functional composition of microbial communities from environmental, agricultural, and clinical settings. Unlike traditional single-genomics approaches, metagenomics does not rely on having to singularize individual bacterial clones from complex microbial mixtures, but catalogs by sequencing all genes and genomes from a mixed community at once (36). The single-isolate approach has been proven successful in the identification and analysis of diseases caused by essentially a single genotype. However, as practiced in epidemiological studies, selecting a single colony for sequencing might mask the possibility that a population of highly similar but still distinguishable individual genotypes may be responsible for the disease phenotype or outbreak. Moreover, if a sample such as human stool, consisting of genotypes from different bacterial, archaeal, and eukaryotic species, is to be analyzed, the one-genome-at-a-time approach would not be possible because not all microbial cells are cultivable.

In metagenomics, whole genomic DNA is prepared from samples, regardless of its microbial composition and is characterized by whole genome sequencing. The assignment of resulting DNA fragments, individual reads or assembled sequence contigs, to individual taxonomic groups or known genome sequences is carried out by sophisticated bioinformatic tools. For example, a number of tools exist that provide an overview of the species composition of metagenomic samples based on direct nucleotide sequence compositions (37,38), comparisons of conserved protein domain-coding sequences (38), identification of 16S rRNA sequences within the sample (39), or oligonucleotide frequencies (40). Other common types of WGS metagenomic sequence analyses include determination of the functional composition of a microbial community based on the assignment of protein-coding open reading frames to functional categories, such as protein domain families [Pfam (41)] or Gene Ontologies [GO (42)]. Consequently, analysis of WGS metagenomics data sets involves a large statistical component, as sequence data must be evaluated based on relative abundances rather than on absolute presence/absence data. It should be noted that the metagenomics field is still under active development, and the application of next-generation sequencing technologies is further changing the metagenomic landscape, increasing the amount of available sequence data to a point where it is now possible to fully characterize low complexity microbial communities. New types of bioinformatics sequence analyses are therefore to be expected.

# GENOME ARCHITECTURE AND EVOLUTION

A large part of our understanding of genome architecture and evolution comes from insights derived from comparative genome sequence analyses.

However, our understanding of the basic evolutionary principles that drive the emergence of new pathogen species, the spread of antibiotic resistance phenotypes, and interactions of microbial communities within and outside of their host are still largely incomplete. For example, both environmental and host-associated populations of E. coli exist, and different E. coli strains can differ largely in their pathogenic potential, from commensal inhabitants of the healthy human or animal intestine (43) to severe intestinal or extraintestinal opportunistic or nonopportunistic pathogens (44). Various virulence genes have been associated with pathogenic phenotypes, although not always is the presence of these necessarily associated with disease (31,45). As several of these virulence factors seem to have the potential for horizontal gene transfer, it might be the complete population of E. coli within a sample that determines its infectious potential. Genome projects targeting hundreds or even thousands of individual genomes, as well as entire populations, will be necessary to improve our understanding of the dynamics of genome architecture and evolution. This understanding is key to the microbial forensic investigator who is trying to establish a basic evolutionary link between two samples in order to evaluate their degree of relatedness and the possibility that one is derived from the other.

## FUTURE CHALLENGES

The availability of next-generation sequencing technologies has brought affordability to large-scale genome sequencing, which, in turn, has led to an increasing number of sequencing projects decentralized from large sequencing centers. Such developments have led to an explosive rate of sequence data acquisition and to concomitant bottlenecks regarding data storage and computational needs. As mentioned earlier, the cost of warehousing data is now rivaling the costs of generating data. Bioinformatic tools for sequence processing and analysis, developed and intended for single genome projects, currently require extensive bioinformatics hardware support to work with data generated by large-scale, next-generation sequencing projects. Genomics is thus at a critical impasse and in need of an improved infrastructure for both bioinformatics analysis and data storage.

First, many of the standard bioinformatics tools (i.e., genome sequence assembler) have to be modified or completely rewritten to allow the processing of millions of short-length sequence reads. When possible, de novo assembly of sequence reads is replaced by computationally affordable mapping of sequence reads onto a known reference. Second, new genomics applications, such as metagenomic sequencing projects, require the development of new types of sequence analysis and new bioinformatics tools. Consequently, researchers increasingly face a lack of standardized tools and vocabularies to make the results of genomics research available to the community. Third, newly generated sequence data and their processing require increasing resources for disc storage, database management, and processing power. For example, a simple sequence comparison of all proteins from 100 genomes against each other, using the standard sequence alignment tool BLAST, can easily take hours of central pprocessing unit processing time, even on complex computer grid networks. Several new approaches target the bioinformatics sequence analysis bottleneck by taking advantage of cloud computing resources available on an on-demand basis over the Internet. This allows researchers to avoid large investments in local bioinformatics infrastructure by instead using leasable Web services for their analysis

## GENOMICS AND MICROBIAL FORENSICS

Recent examples highlight the potential of genomic techniques to become new tools for the microbial forensic investigator and provide a glimpse of how genomics could be integrated into a modern microbial forensic investigation. Using whole genome shotgun sequencing on an Illumina platform, Harris and colleagues (34) analyzed the genome sequence of 63 methicillin-resistant *Staphylococcus aureus* (MRSA) isolates and demonstrated the potential of the platform for high-resolution genotyping and in-depth analysis of microevolution within a single bacterial lineage. The authors were able to trace the evolution of a single MRSA clone in health care facilities worldwide over the past decades as well as within a single hospital over a 7-month period at a level of resolution previously unachievable. This high-resolution evolutionary history was achieved by identifying a few and rare single-nucleotide polymorphisms between individual isolates. Not only did this study demonstrate the usefulness and economic feasibility of genomic tools for optimizing DNA fingerprint analysis, but also called into question current methods such as PFGE subtyping, commonly applied for investigations of food-borne pathogen outbreaks in the United States (see also chapter 3 on food-borne outbreaks).

Investigation of the 2001 anthrax mailings also relied on a combination of traditional microbiology and genomics analysis to point the police investigation to the *Bacillus anthracis* culture from which spores sent in the mail originated. The culture, like the spores from the letters, had a microbial fingerprint characterized by a consistent mixture of specific *B. anthracis* genomic variants. These variants were only detectable by whole genome sequence analysis and afforded the rapid development of high-throughput genomic assays to screen samples from more than a thousand potential sources.

More recently, Fierer and colleagues (46) demonstrated that an individual skin microbial community can be highly unique and, after touching a surface, can leave behind a characteristic fingerprint in the true sense of the word.

Microbial communities from these surfaces could be recovered and, using 16S rRNA base phylogenetic analysis, matched to a specific person, hence supporting the possibility of using skin microbiota for forensic identification. While far from being validated, this approach is novel and warrants further validation as a forensic tool (46). Thus, the relatively new and rapidly expanding metagenomics research field could provide additional scientific support in microbial forensic investigations.

While the rapidly changing genomics field raises great hopes for the microbial forensic investigator, it is important to understand that genomics, as a forensic tool, must still pass the challenges of a court of law. For this to be realized, genomics, and the attendant genome sequencing technologies and sequence analysis algorithms, must be valid, robust, and grounded with strong statistical support. However, this hurdle is readily being met with technologies, algorithms, and software that have been and promise to be developed to address the analysis of genomic data.

# REFERENCES

[1] F. Schiller. The Death of Wallenstein, 1800.

[2] R.D. Fleischmann, M.D. Adams, O. White, R.A. Clayton, E.F. Kirkness, A.R. Kerlavage, et al., Whole-genome random sequencing and assembly of *Haemophilus influenzae* Rd, Science 269 (1995) 496–512.

[3] G.E. Moore, Cramming more components onto integrated circuits, Electron. Mag. 38 (1965) 4–7.

[4] F. Sanger, S. Nicklen, A.R. Coulson, DNA sequencing with chain-terminating inhibitors, Proc. Natl. Acad. Sci. USA 74 (1977) 5463–5467.

[5] F. Kunst, N. Ogasawara, I. Moszer, A.M. Albertini, G. Alloni, V. Azevedo, et al., The complete genome sequence of the gram-positive bacterium *Bacillus subtilis*, Nature 390 (1997) 249–256.

[6] T. Kaneko, Y. Nakamura, S. Sato, E. Asamizu, T. Kato, S. Sasamoto, et al., Complete genome structure of the nitrogen-fixing symbiotic bacterium *Mesorhizobium loti*, DNA Res. 7 (2000) 331–338.

[7] J.I. Glass, E.J. Lefkowitz, J.S. Glass, C.R. Heiner, E.Y. Chen, G.H. Cassell, The complete sequence of the mucosal pathogen *Ureaplasma urealyticum*, Nature 407 (2000) 757–762.

[8] J.C. Venter, M.D. Adams, E.W. Myers, P.W. Li, R.J. Mural, G.G. Sutton, et al., The sequence of the human genome, Science 291 (2001) 1304–1351.

[9] E.S. Lander, L.M. Linton, B. Birren, C. Nusbaum, M.C. Zody, J. Baldwin, et al., International Human Genome Sequencing Consortium. Initial sequencing and analysis of the human genome, Nature 409 (2001) 860–921.

[10] R.H. Waterston, K. Lindblad-Toh, E. Birney, J. Rogers, J.F. Abril, P. Agarwal, et al., Initial sequencing and comparative analysis of the mouse genome, Nature 420 (2002) 520–562.

[11] M.D. Adams, S.E. Celniker, R.A. Holt, C.A. Evans, J.D. Gocayne, P.G. Amanatides, et al., The genome sequence of *Drosophila melanogaster*, Science 287 (2000) 2185–2195.

[12] Arabidopsis Genome Initiative, Analysis of the genome sequence of the flowering plant *Arabidopsis thaliana*, Nature 408 (2000) 796–815.

[13] M. Margulies, M. Egholm, W.E. Altman, S. Attiya, J.S. Bader, L.A. Bemben, et al., Genome sequencing in microfabricated high-density picolitre reactors, Nature 437 (2005) 376–380.

[14] D.R. Bentley, S. Balasubramanian, H.P. Swerdlow, G.P. Smith, J. Milton, C.G. Brown, et al., Accurate whole human genome sequencing using reversible terminator chemistry, Nature 456 (2008) 53–59.

[15] T.D. Harris, P.R. Buzby, H. Babcock, E. Beer, J. Bowers, I. Braslavsky, et al., Single-molecule DNA sequencing of a viral genome, Science 320 (2008) 106–109.

[16] T. Durfee, R. Nelson, S. Baldwin, G. Plunkett 3rd, V. Burland, B. Mau, et al., The complete genome sequence of *Escherichia coli* DH10B: Insights into the biology of a laboratory workhorse, J. Bacteriol. 190 (2008) 2597–2606.

[17] A. Srivatsan, Y. Han, J. Peng, A.K. Tehranchi, R. Gibbs, J.D. Wang, et al., High-precision, whole-genome sequencing of laboratory strains facilitates genetic studies, PLoS Genet. 4 (2008) e1000139.

[18] J. Eid, A. Fehr, J. Gray, K. Luong, J. Lyle, G. Otto, et al., Real-time DNA sequencing from single polymerase molecules, Science 323 (2009) 133–138.

[19] F.D. Ciccarelli, T. Doerks, C. von Mering, C.J. Creevey, B. Snel, P. Bork, Toward automatic reconstruction of a highly resolved tree of life, Science 311 (2006) 1283–1287.

[20] S. Bentley, Sequencing the species pan-genome, Nat. Rev. Microbiol. 7 (2009) 258–259.

[21] M. Wu, J.A. Eisen, A simple, fast, and accurate method of phylogenomic inference, Genome Biol. 9 (2008) R151.

[22] V. Barbe, S. Cruveiller, F. Kunst, P. Lenoble, G. Meurice, A. Sekowska, et al., From a consortium sequence to a unified sequence: The *Bacillus subtilis* 168 reference genome a decade later, Microbiology 155 (2009) 1758–1775.

[23] W.F. Fricke, P.F. McDermott, M.K. Mammel, S. Zhao, T.J. Johnson, D.A. Rasko, et al., Antimicrobial resistance-conferring plasmids with similarity to virulence plasmids from avian pathogenic *Escherichia coli* strains in *Salmonella enterica* serovar Kentucky isolates from poultry, Appl. Environ. Microbiol. 75 (2009) 5963–5971.

[24] K. Liolios, I.M. Chen, K. Mavromatis, N. Tavernarakis, P. Hugenholtz, V.M. Markowitz, et al., The Genomes On Line Database (GOLD) in 2009: Status of genomic and metagenomic projects and their associated metadata, Nucleic Acids Res. 38 (Database issue) (2010) D346–D354.

[25] D.A. Rasko, G.S. Myers, J. Ravel, Visualization of comparative genomic analyses by BLAST score ratio, BMC Bioinformatics 6 (2005) 2.

[26] S. Kurtz, A. Phillippy, A.L. Delcher, M. Smoot, M. Shumway, C. Antonescu, et al., Versatile and open software for comparing large genomes, Genome Biol. 5 (2004) R12.

[27] B. Morgenstern, DIALIGN: Multiple DNA and protein sequence alignment at BiBiServ, Nucleic Acids Res. 32 (Web Server issue) (2004) W33–W36.

[28] E.V. Koonin, Comparative genomics, minimal gene-sets and the last universal common ancestor, Nat. Rev. Microbiol. 1 (2003) 127–136.

[29] J.I. Glass, N. Assad-Garcia, N. Alperovich, S. Yooseph, M.R. Lewis, M. Maruf, et al., Essential genes of a minimal bacterium, Proc. Natl. Acad. Sci. USA 103 (2006) 425–430.

[30] C. Lartigue, S. Vashee, M.A. Algire, R.Y. Chuang, G.A. Benders, L. Ma, et al., Creating bacterial strains from genomes that have been cloned and engineered in yeast, Science 325 (2009) 1693–1696.

[31] D.A. Rasko, M.J. Rosovitz, G.S. Myers, E.F. Mongodin, W.F. Fricke, P. Gajer, et al., The pangenome structure of *Escherichia coli*: Comparative genomic analysis of *E. coli* commensal and pathogenic isolates, J. Bacteriol. 190 (2008) 6881–6893.

[32] S. Bambini, R. Rappuoli, The use of genomics in microbial vaccine development, Drug Discov. Today 14 (2009) 252–260.

[33] P. Hugenholtz, B.M. Goebel, N.R. Pace, Impact of culture-independent studies on the emerging phylogenetic view of bacterial diversity, J. Bacteriol. 180 (1998) 4765–4774.

[34] S.R. Harris, E.J. Feil, M.T. Holden, M.A. Quail, E.K. Nickerson, N. Chantratita, et al., Evolution of MRSA during hospital transmission and intercontinental spread, Science 327 (2010) 469–474.

[35] R. Urwin, M.C. Maiden, Multi-locus sequence typing: A tool for global epidemiology, Trends Microbiol. 11 (2003) 479–487.

[36] P.D. Schloss, J. Handelsman, Metagenomics for studying unculturable microorganisms: Cutting the Gordian knot, Genome Biol. 6 (2005) 229.

[37] A. Brady, S.L. Salzberg, Metagenomic phylogenetic classification with interpolated Markov models, Nat. Methods 6 (2009) 673–676.

[38] D.H. Huson, A.F. Auch, J. Qi, S.C. Schuster, MEGAN analysis of metagenomic data, Genome Res. 17 (2007) 377–386.

[39] P.D. Schloss, S.L. Westcott, T. Ryabin, J.R. Hall, M. Hartmann, E.B. Hollister, et al., Introducing mothur: Open-source, platform-independent, community-supported software for describing and comparing microbial communities, Appl. Environ. Microbiol. 75 (2009) 7537–7541.

[40] A.C. McHardy, H.G. Martin, A. Tsirigos, P. Hugenholtz, I. Rigoutsos, Accurate phylogenetic classification of variable-length DNA fragments, Nat. Methods 4 (2007) 63–72.

[41] R.D. Finn, J. Mistry, J. Tate, P. Coggill, A. Heger, J.E. Pollington, et al., The Pfam protein families database, Nucleic Acids Res. 38 (Database issue) (2010) D211–D222.

[42] The Gene Ontology's Reference Genome Project: A unified framework for functional annotation across species. PLoS Comput. Biol. 5 (2009) e1000431.

[43] O. Tenaillon, D. Skurnik, B. Picard, E. Denamur, The population genetics of commensal *Escherichia coli*, Nat. Rev. Microbiol. 8 (2010) 207–217.

[44] M.A. Croxen, B.B. Finlay, Molecular mechanisms of *Escherichia coli* pathogenicity, Nat. Rev. Microbiol. 8 (2010) 26–38.

[45] K.E. Rodriguez-Siek, C.W. Giddings, C. Doetkott, T.J. Johnson, L.K. Nolan, Characterizing the APEC pathotype, Vet. Res. 36 (2005) 241–256.

[46] N. Fierer, C.L. Lauber, N. Zhou, D. McDonald, E.K. Costello, R. Knight, Forensic identification using skin bacterial communities, Proc. Natl. Acad. Sci. USA 107 (2010) 6477–6481.

# CHAPTER 29

# Design of Genomic Signatures for Pathogen Identification and Characterization

**Tom Slezak, Shea Gardner, Jonathan Allen, Elizabeth Vitalis,
Marisa Torres, Clinton Torres, and Crystal Jaing**

*Lawrence Livermore National Laboratory, Livermore, California*

## GENOMIC SIGNATURES

This chapter addresses some of the many issues associated with the identification of signatures based on genomic DNA/RNA, which can be used to identify and characterize pathogens for biodefense and microbial forensic goals. For the purposes of this chapter, we define a "signature" as one or more strings of contiguous genomic DNA or RNA bases sufficient to identify a pathogenic target of interest at the desired resolution and that could be instantiated with particular detection chemistry on a particular platform. The target may be a whole organism, an individual functional mechanism (e.g., a toxin gene), or simply a nucleic acid indicative of the organism. The desired resolution will vary with each program's goals but could easily range from family to genus to species to strain to isolate. Resolution may not be taxonomically based but rather pan-mechanistic in nature: detecting virulence or antibiotic-resistance genes shared by multiple microbes. Entire industries exist around different detection chemistries and instrument platforms for identification of pathogens, and we only briefly mention a few of the techniques that have been used at Lawrence Livermore National Laboratory (LLNL) to support our biosecurity-related work since 2000. Most nucleic acid-based detection chemistries involve the ability to isolate and amplify the signature target region(s), combined with a technique to detect amplification.

Genomic signature-based identification techniques have the advantage of being precise, highly sensitive, and relatively fast in comparison to biochemical typing methods and protein signatures. Classic biochemical typing methods were developed long before knowledge of DNA and resulted in dozens of tests (Gram's stain, differential growth characteristics media, etc.) that could be used to roughly characterize the major known pathogens (of course, some are uncultivable). These tests could take many days to complete and precise resolution of species and strains is not always possible. In contrast, protein

493

Microbial Forensics. DOI: 10.1016/B978-0-12-382006-8.00029-3

recognition signatures composed of antibodies or synthetic high-affinity ligands offer extremely fast results but require a large quantity of the target to be present. False positives/negatives are also a problem with some protein-based techniques (home pregnancy kits use this basic approach).

## DIFFERENT TYPES AND RESOLUTIONS OF GENOMIC SIGNATURES

Genomic signatures can be intended for many different purposes and applied at multiple different resolutions. At LLNL, we have been working on signatures that can be broken out into several categories: (i) organism signatures, (ii) mechanism signatures, and (iii) method signatures.

Organism signatures are intended to uniquely identify the organism(s) involved. Mechanism signatures can be best thought of as identifying particular genes that result in functional properties such as virulence, antibiotic resistance, or host range. The primary reason to identify mechanisms, independent of organisms, is to detect potential genetic engineering. A secondary reason is that nature has shared many important mechanisms on its own over the millennia, and thus they may not be sufficiently unique to identify specific organisms. Knowledge of whether a particular isolate has the full virulence kit or possesses unusual antibiotic resistance properties and whether it is human transmissible is important for biodefense and public health responses. Method signatures present yet another dimension of analyzing pathogens: evidence of potential bacterial genetic engineering may be seen in a genome by checking for traces of the bacterial vector(s) that may have been used to insert one or more foreign genes and related components (promoters, etc.) into the genome being modified. In the future, host range signatures might indicate that an otherwise uncharacterized pathogen was potentially capable of evading or defeating the immune system of a particular host organism.

## POTENTIAL TARGET ORGANISMS

Genetic signatures can be used to identify any living organisms and viruses that contain intact DNA or RNA. Focusing on biosecurity, we are interested primarily in identifying bacteria, viruses, and fungi that could potentially be used to threaten human, animal, or plant life, to disrupt our economy, or to disturb our social order. Note that there is a wide range of genome sizes involved. RNA viruses are generally small (foot and mouth disease virus is about 8 kbp, SARS coronavirus is about 30 kbp), whereas the variola virus (causative agent of smallpox) is a large DNA virus of about 200 kbp. High-threat bacterial pathogens tend to be in the 2–5-Mbp size range (*Yersinia pestis*, causative agent of plague, is about 4 Mbp while *Bacillus anthracis* is about 5 Mbp.) Fungi can range from 10 Mbp to over 700 Mbp. As can be imagined, the sequencing databases have many more viral genomes than bacterial and many more bacterial genomes than fungal. In comparison, the human genome is about 3 Gbp and wheat is about 16 Gbp.

## SIGNATURE RESOLUTION

Organism detection signatures must be conserved sequence, reliable, and able to detect all intended organisms to minimize false negatives, and unique sequence, specific to the target organism and not detecting nontarget organisms to minimize false positives. Organism detection signatures can be at different taxonomic resolution, typically genus, species, or strain.

In biosecurity applications, high-resolution signatures are needed to precisely identify particular isolates or strains. In past years, a large distinction was drawn between identification or detection signatures and forensic signatures, where forensic signatures were typically thought of as at the strain level or below (typically thought of as substrain or isolate specific). More recently the distinction has become blurred because taxonomic distinctions have become less certain and because new signature techniques provide increased resolution levels. Using current commercially available microarray technologies that allow several millions of signatures to be designed on each chip, one can interrogate the entire resolution range (genus, species, strain, and isolate) for desired pathogen targets, providing both detection and forensic resolution. Signature design today is a combination of the desired signature purpose, our current understanding of the diversity of the organism being targeted, and the particular mission constraints that may dictate the detection chemistry and platform to be used for either biodefense or public health.

## GENOMIC SEQUENCE DATA: WHAT TO USE AND WHERE TO GET IT

There is no single resource for all genomic sequence data pertinent to signature design. The most comprehensive public source for genomic sequence data is GenBank, which is located at the National Center for Biotechnology Information (NCBI) Web site (http://www.ncbi.nlm.nih.gov/). The NCBI has reciprocal data exchange agreements with the European Molecular Biology Laboratory in the United Kingdom and the DNA Data Bank of Japan, which are equivalent databases used heavily in those parts of the world. Most authors of published sequence data usually submit a final version of their sequence data sets to GenBank. However, numerous sequence databases exist that have organism-specific data that may not be found in GenBank during the interim period of data generation and manuscript preparation and those sites would need to be probed directly to obtain the most recent and up-to-date sequence data. Some examples of these publicly available resources are the Integrated Microbial Genomics project at the Joint Genome Institute (http://img.jgi.doe.gov), the Comprehensive Microbial Resource at the JC Venter Institute (the institute formerly known as TIGR, http://cmr.jcvi.org), and the Sanger Institute in the United Kingdom (http://www.sanger.ac.uk).

Sequence data most useful for signature design fall into two major categories: finished and draft data of isolated organisms. Draft genomes are composed of multiple sets of overlapping reads, called "contigs," potentially with little or no information about the order or orientation of the contigs relative to the original genome. Draft sequence is often described by a depth factor, which is a numeric statement about the average redundancy of coverage at any base position, and thus confidence. A $3\times$ draft sequence would have, on average, at least three overlapping reads that contain each base in the genome being sequenced; $8-10\times$ depth is a common stopping point for draft genome data generation for traditional Sanger sequencing (where read lengths averaging 800+ bp are common). More recent generations of sequencing based on pyrosequencing technology yield shorter reads (100–200+ bp for Roche 454 sequencers and 32–75+ bp for machines from Illumina or Life Technologies) and may feature depths of $50\times$ or greater.

Finished whole genome microbial sequences have undergone an iterative process to assemble contigs and then use a variety of techniques to order and orient them and close any gaps. This often lengthy and costly process, when completed, produces a single string of high-quality bases from the individual and scrambled contigs of the draft sequence. Obviously, finished genomes are superior to drafts when it comes to performing annotation of gene content or other features, as well as for performing multiple sequence alignments to compare two or more genomes. In our experience at LLNL, an $8-10\times$ Sanger draft genome provides sufficient information for DNA signature design purposes (1). When you consider that finished microbial genomes can be 4–10 times as expensive as draft, it is not surprising that many microbial genomes will never be finished. Increasingly, short-read sequences are being mapped to reference genomes in lieu of a de novo assembly.

Another increasingly important category of data is the metagenomic sequence, where no attempt has been made to isolate individual organisms for sequencing. Sometimes this is because no way is known to isolate and culture the particular organism(s) of interest. Only a tiny fraction of organisms can be cultured *in vitro* and our knowledge base is greatly skewed toward those that can. At other times it is because what is desired is a sampling of an entire community of organisms. Although numerous metagenomic samples have been sequenced, it is exceedingly rare for complete assemblies of sequence from multiple organisms to result. One exception is a very small symbiotic bacterial community found living in an extremely harsh acidic environment in a mine (2). Metagenomic data are not currently of much utility for genomic signature development. A recent paper on the acid mine bacterial community is providing clues about the evolution of viral resistance (3), which illustrates the vital role metagenomic sequencing will play in expanding our systems biology knowledge at both the organism and the ecosystem level.

Searching for sequence data based on free-text queries can be problematic. For example, GenBank does not enforce consistency with sequence designation. Not all complete genomes have "complete genome" in the title, and some that do are not actually complete genomes. We have encountered complete genomes that were labeled "complete cds" (coding sequence), "complete gene," or otherwise unlabeled as a complete genome. Curation is required to validate any sequence data obtained from a public resource, and periodic in-house testing against benchmark data is necessary to maintain a database of high fidelity. A related problem is distinguishing when a new finished genome should replace a prior draft, as strain name, authors, or institutions may have changed.

## IDENTIFYING CONSERVED SEQUENCE AMONG TARGETS

Finding regions of conservation across all target genomes can be done with "alignment-based" methods and with "alignment-free" methods. The difference between methods revolves around a trade-off between time and quality.

The first issue to be faced when searching for conservation with a multiple-sequence alignment (MSA) is the amount of sequence (breadth) that an alignment method can handle. Alignments sometimes fail when input sequences are very long or when there are a large number of sequences to be aligned (depth), even if the sequences are not particularly long. Failure happens because an MSA takes impractically long to finish due to the intractable computational complexity involved or due to a lack of memory. These limitations mean the optimal alignment approach may vary depending on the breadth and depth of sequences used as input. The recent explosion of genome sequence data has resulted in a lack of MSA algorithms that can scale appropriately.

Alignment-free methods for finding consensus can be a shortcut if a complete MSA is impractical or not needed for downstream analysis. Building an alignment-free consensus relies on one sequence serving as a reference for the sequence order of the remaining sequences. This reference sequence is compared pair-wise with the remaining sequences, and the consensus is expressed in the sequence order of the reference. This is often less computationally complex than performing a complete MSA, and results are of sufficient quality to identify suitably conserved regions for potential signatures.

Another topic of concern when identifying conserved sequence regions is whether an approach can incorporate incomplete and/or draft sequences. Incomplete sequences do not cover the complete genome of the organism. Draft sequences may cover the complete genome and may be of lower quality, particularly near the ends of contigs. Increasingly, the number of genomes being finished to completion is significantly fewer than the number of genomes that will remain incomplete and in draft form. MUMmer (4) is a notable MSA program in this respect because it can align draft and complete genomes. Note that any use of incomplete genomes carries an inherent risk because regions not present in the incomplete genome(s) will not appear to be conserved and thus may not be considered for signature mining.

Finally, viruses are often highly divergent at the nucleotide level. This extreme divergence, common among many RNA viruses, can cause even alignment-free methods that rely on a pair-wise sequence search to fail at finding all shared genetic regions. Some nonviral organisms have also been observed with enough divergence to make using alignment-free methods error prone. To help overcome the hurdles of divergent targets, we have developed a novel method of signature generation, "minimal set clustering" (MSC), described later.

## IDENTIFYING SEQUENCES UNIQUE TO TARGETS

Finding regions of sequence unique to the target organism is done by searching large sequence databases. There is a trade-off in sequence search between execution time and search sensitivity. "Heuristic" algorithms (methods that take reasonable shortcuts, which may decrease sensitivity) offer the best time performance. "Nonheuristic" algorithms (methods that guarantee complete coverage within the problem space) can have more sensitive results than heuristics, but are slower and the additional sensitivity is not always significant.

Heuristics are used most commonly because they make it possible to search extremely large databases such as NCBI's NT (not nonredundant nucleotide database) quickly. The most popular of these is BLAST (5), which can scale to provide fast results with large databases by splitting the search space into many parallel processes across compute clusters. If additional limitations in search sensitivity are acceptable, other approaches, such as suffix tree-based Vmatch (http://www.vmatch.de/), can be faster. Another heuristic approach is to compute hidden Markov models that represent the sequence families of interest, such as in the program HMMER (http://hmmer.janelia.org/).

## MINING FOR SIGNATURES

After pathogen target regions that are both conserved and unique are found, they are mined for detection signatures. Signatures are found by searching for oligonucleotides with appropriate length, melting temperature, and GC ratio and by searching for oligonucleotide combinations with appropriate overall amplicon size and minimal interoligonucleotide hybridization potential. Programs such as Primer3 (6) can perform some or all of the signature selection work given a target sequence input. Primer3 can be integrated into any signature development pipeline, unlike other packages that only offer a manual graphic interface.

## HOW KPATH SIGNATURES ARE DESIGNED

This section discusses major design criteria that the LLNL KPATH (7) signature design pipeline was built around. KPATH's native signature format, which we describe, is TaqMan® PCR. Its ability to handle several other formats is not described here.

The process begins by looking across all complete target genomes for sequence conservation. We use an in-house, alignment-free, BLAST-based program for finding conservation (unpublished results).

Conserved regions of the target genomes are next screened across our complete genome database in search of potential cross-reactions. Because the oligonucleotides of TaqMan signatures are about 18 to 30 bp long, a fairly large seed length of 18 is acceptable (which means that some short perfectly matching sequences may be omitted from results). Larger seed lengths make it possible for us to search much larger databases in reasonable amounts of time. We currently use Vmatch for large database searches.

The resulting conserved and apparently unique sequence, which has no significant similarity to other known sequences, is now mined for signatures. It is important to note that we only find apparent uniqueness based on the state of the current whole genome database available to us. We anticipate that as additional pathogen targets, near-neighbor organisms, and other organisms are sequenced, our regions of conservation and uniqueness will diminish. For this reason, signature design is an iterative process and not an end point. The original KPATH system used Primer3 in a single execution to identify TaqMan signature candidates with a forward primer, reverse primer, and a hybridization probe. To let us enforce additional signature design constraints and options without ruling out potential target regions, we converted signature identification into two executions of Primer3—one for primer pairs and one for probes. Separate primer and probe results are combined with an in-house signature builder and scorer to allow us to identify the best combinations of primers and probes.

Next, signatures are filtered down so there is little or no overlap of candidate signatures within the target organism. When exhaustive signature searches are performed, many of the mathematically best signature candidates will share oligonucleotides and generally be very similar. This means that choosing the best scoring signatures for any given locus helps us remove excess redundancy from the pool of signature candidates.

The final check we typically perform is a TaqSim (http://staff.vbi.vt.edu/dyermd/publications/taqsim.html) comparison of all signature candidates against NCBI's NT database. This highly sensitive BLAST search TaqMan PCR simulator with postprocessing lets us verify that the signature candidates are conserved enough to detect all the expected targets and unique so that there are no nontarget hits. Depending on the intended uses of the signatures (e.g., environmental versus clinical samples), we may choose to do additional testing against genomes from human or other complex organisms.

We note that in recent years other DNA signature pipelines have been built that take a reverse approach. Like LLNL's minimal set clustering described, they first generate all potential valid TaqMan PCR signatures for each available genome of a target organism and then BLAST them to check for sufficient conservation and uniqueness.

## RNA VIRUSES PRESENT ADDITIONAL CHALLENGES

High rates of mutation and lack of genome repair mechanisms in many viruses generate high levels of intraspecific diversity and result in quasispecies, particularly for many single-stranded RNA viruses. Consequently, PCR-based signatures for viral detection often require high levels of degeneracy or multiplexing in order to detect all variants robustly. Large amounts of sequence data are often required to represent the range of target diversity, sometimes dozens to hundreds of genomes. As noted previously, building multiple sequence alignments with many diverse genomes taxes the capabilities of most available software. Once an alignment is built, it may reveal insufficient consensus for even a single primer, much less a pair, to detect all members of some species (e.g., human immunodeficiency virus-1 or influenza A).

One solution is to subdivide the targets into smaller or more closely related subgroups, such as clade, serotype, or phenotype, of interest (examples of phenotypes could include virulent versus vaccine, domestic versus foreign), and attempt to find signatures separately for each subgroup. This approach implies that multiple signatures will be required for species-level detection of all subgroups. One must make an assessment in advance of signature design of how best to subdivide the target sequences. A second approach is to allow degenerate or inosine bases so that a single signature will detect more diverse genomes. Specificity may suffer if some combinations of degenerate bases also pick up nontarget species. Sensitivity may decline, as the specific priming sequence for a given target is diluted in the degenerate mix. A number of tools that require a multiple sequence alignment as input are available for degenerate primer design (e.g., SCPrimer, PrimaClade, Primo, Amplicon, and HYDEN). A third approach is to forego sequence alignment altogether and to look for sets of primer-length oligomers of length k, or "k-mers," present in many targets and unique relative to nontarget sequences. Using combinatoric or greedy algorithms, one can build a signature set of k-mers such that each target contains at least two k-mers to function as forward and reverse primers. This approach demands large amounts of computing memory to store all candidate k-mers for large or many genomes, especially as k increases above 20, and may require suffix trees or other techniques for data compression.

A fourth approach employed is called MSC. Because it avoids the need for multiple sequence alignment or *a priori* subgrouping of target sequences, this method can be run blindly without expert knowledge of the target species. It begins by removing nonunique regions from consideration as primers or probes from each of the target sequences relative to a database of nontarget sequences. The remaining unique regions of each target sequence are mined for all or many candidate signatures, without regard for conservation among other targets, yet satisfying user specifications for primer and probe length, $T_m$, GC%, amplicon length, and so on. All candidate signatures are compared to all targets and clustered by the subset of targets they are predicted to detect. To predict detection, we may require that a signature's primers and probe have a perfect match to target in the correct orientation and proximity, or we may relax the match requirements to allow a limited number of mismatches, as long as $T_m$ remains above a specified threshold or those mismatches do not occur too close to the 3′ end of a primer. Signatures within a given cluster are equivalent in that they are predicted to detect the same subset of targets, so by clustering we reduce the redundancy and size of the problem to finding a small set of signatures that detect all targets. Nevertheless, finding the optimal solution of the fewest clusters to detect all targets is an "NP complete" problem, so for large data sets we use a greedy algorithm to find a small number of clusters that together should pick up all targets. LLNL has used this method to design signature sets for numerous RNA viruses, including influenza A HA serotypes, foot and mouth disease, Norwalk, Crimean–Congo hemorrhagic fever, ebola, and other divergent viruses. Figure 29.1 shows the result of an MSC computation for Crimean–Congo hemorrhagic fever performed in 2005, with the resulting signatures displayed against a whole genome phylogenetic tree of all the sequences available at that time.

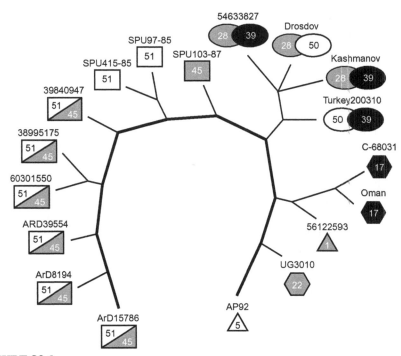

**FIGURE 29.1**
Result of minimal subset clustering signatures for Crimean–Congo hemorrhagic fever virus (CCHFV) displayed against a whole genome phylogenetic tree of available target genomes. Note that signatures 45 and 51 cover a wide range of isolates from one geographical location, whereas signatures 28, 39, and 50 cover isolates found in eastern Europe. Signatures 1, 5, 17, and 22 are required to detect some historical isolates that are not likely to be in current circulation.

# SIGNATURES OF POTENTIAL BACTERIAL GENETIC ENGINEERING

Detecting evidence for genetic engineering in bacteria is challenging when the target modification is not known and the effects of an outbreak on human health are not well understood. We may, for example, anticipate a biological outbreak that employs a bacterial host containing a foreign toxin, but the observed effects of the toxin may not implicate a known gene. Even in cases where the gene is known, it may be difficult to rule out a natural origin for the outbreak. In such cases, it may be useful to search for more direct evidence of the genetic engineering tools used to insert and express foreign genes in a bacterial host. Among the most widely used and readily available tools for genetic engineering in bacteria are artificial vector DNA molecules. Genetic engineering with artificial vectors began with efforts to improve on early work using natural plasmids for gene cloning. Natural plasmids are extrachromosomal replicons (self-replicating molecules) that come in both circular and linear form and are generally nonessential genetic material for the bacterial host but can confer important phenotypes such as virulence and drug resistance. These plasmids are mobile genetic elements that serve as a natural mechanism for the exchange of genetic material across different bacterial species (8). Artificial vectors are natural plasmid derivatives designed to improve support for the insertion and manipulation of foreign genetic elements in the carrier plasmid.

We use the term "artificial vector" to refer to replicons created through human intervention to explicitly distinguish them from their natural plasmid precursors. Sequence features designed to support genetic manipulation form the basis for methods used to distinguish artificial vector sequence from natural plasmids. The most common artificial vector-specific feature is the multiple cloning site region, which is a sequence insert containing clusters of restriction enzyme sites used to facilitate insertion of the foreign gene elements. Selection marker genes also play an important role in selecting bacteria, which maintain the artificial vector. The gene transcription control unit, which includes a promoter sequence and transcription terminator sequence for the foreign gene elements, is also an important feature, along with the origin of replication site required for maintenance of the artificial vector in the bacterial colony (9).

Detecting an artificial vector sequence in a mixed bacterial sample potentially requires testing a broad range of sequence targets. This suggests use of an assay with a high degree of multiplex capability that tests for the presence of a large number of sequences simultaneously. Microarray-based assays are a logical choice for accommodating a large number of artificial vector detection probes. The large collection of artificial vector sequences can be clustered according to exact k-mer sequence matching to find the k-mers shared among different vector sequences (10). Sequence length k corresponds to the desired probe length used in the microarray design. Each cluster of shared sequence is compared against all available sequenced natural chromosomal bacterial and viral genomes, including natural plasmids to identify which k-mers in the artificial vector sequence are distinct from the natural background. These unique k-mers are called candidate signatures. After candidate signatures are found, a probe set is created that ensures that each vector contributes a preset minimum number of candidate signatures to the final microarray probe set design. A greedy algorithm can be used to pick signatures shared by the greatest number of artificial vectors, selecting candidate signatures in decreasing order.

Additional postprocessing steps may further improve the quality of the signature probe set design to achieve the ultimate goal of sensitive detection, while maintaining a hybridization pattern on the microarray that distinguishes artificial vectors from the natural background found in a mixed sample. Once the initial probe set is designed, a BLAST search can be used to tune the probe set by replacing the candidate signatures with near matches to the background with candidates showing a greater percentage of vector unique variation. Cross-validation can be used to estimate a similarity threshold for distinguishing artificial and natural genomic sets. [An example of this approach using cross-validation is given elsewhere (10).] Another postprocessing step is to tune the probe set to ensure probes derived from each vector come from multiple functional regions. Confidence in vector detection is boosted when probes are found for multiple functional locations. Using probes from multiple regions may also provide useful forensic information on the origins and function of the detected artificial vector. Given the similarities between artificial vectors and natural plasmids, having additional probes for natural plasmids allows for direct comparison with the natural plasmid hybridization pattern, which could reduce the potential for false-positive predictions.

## VIRAL AND BACTERIAL DETECTION ARRAY

Numerous microarrays have been designed for viral discovery, detection, and resequencing (11–14). Resequencing arrays can provide sequence information for viruses closely related (>90% similarity) to sequences from which the array was designed. Discovery arrays to detect more diverse and more distantly related organisms have been built using techniques for selecting probes from regions of known conservation based on BLAST nucleotide sequence similarity (15) or profile HMM and motif indications of amino acid sequence conservation (14). Array design to span an entire kingdom on a single microarray demands substantial investment in probe selection algorithms. LLNL designed a microarray to detect all bacteria, plasmids, and viruses based on all available whole genome, whole segment, and whole plasmid sequences and is in the process of including probes for highly conserved fungal genes as well. We attempted to find probes that are unique to each viral and bacterial family, and favor probes conserved within a family. We used probes 50–65 bases long, enabling sensitive detection of targets with some sequence variation relative to the probe. We used a greedy minimal set cover algorithm to ensure that all sequences have at least 50 (for viruses) or 15 (for bacteria and plasmids) probes per sequence. We allowed some mismatches between probe and target, based on previous mismatch experiments in which we determined that probes with a contiguous match at least 29 bases long and with 85% sequence similarity between probe and target still gave a strong signal intensity. Our design should characterize unknowns to at least the family level, and in all cases tested so far, including blinded clinical samples containing multiple viruses, we are able to accurately detect and characterize all viruses contained in that sample to the species or strain level (16).

Our first-generation viral array included 36,000 probes designed from family-specific regions of all 72 viral families, and our second version included

170,000 viral probes, again from family-specific regions. There were no regions greater than 25-bp matches to human or bacteria and no regions greater than 17-bp matches to other nontarget viral families. In addition, we also included the 20,000 probes from the Virochip developed by Dr. Joseph DeRisi from University of California, San Francisco, as a control (11).

Preliminary testing using NimbleGen arrays with mixed DNA and RNA viruses and with blinded clinical samples showed accurate detection of multiple viruses in a single sample. In addition, we can identify the exact strains and isolates hybridized as a mixed sample, although the array was designed to guarantee discrimination only to family. We developed a novel statistical method that is based on likelihood maximization within a Bayesian network, incorporating a sophisticated probabilistic model of probe-target hybridization developed and validated with experimental data from hundreds of thousands of probe intensity measurements. The method is designed to enable quantifiable predictions of likelihood for the presence of each of multiple organisms in a complex, mixed sample, which is especially important in an environmental sample or one with chimeric organisms. Future detection chip designs will include probes from conserved regions of bacterial families and plasmids and fungal families. This chip will be a major platform for identification of known and unknown pathogens.

## THE FUTURE OF GENOMIC SIGNATURES

Issues related to scaling, taxonomy, and technology advances appear to be main drivers for the future of genomic signatures.

Scaling problems all stem from the exponential rate at which genomic sequence data are growing. Although it is inexpensive to buy sufficient hardware to store data physically, the current generation of bioinformatics tools was designed in an era when it was a luxury to have a handful of genomes of a particular pathogen available to work with. In recent years the Influenza Community Sequencing Project (17) has deposited many thousands of complete influenza genomes into GenBank, far exceeding the capacity of most tools to handle them. Similarly, some of the new sequencing technologies can generate billions of bases in a single run from metagenomic samples (18), but truly efficient software that takes full advantage of this information is lacking. It will likely take several years for research funding to be focused properly to close this bioinformatics tool gap. Another aspect of scaling problems is that few researchers have access to computers with large enough memories to be able to process certain classes of sequence analyses related to genomic signature design. Computer clusters optimal for physical science problems (where each node represents a point in a three-dimensional physical grid representation and almost all communication is with nearest neighbor nodes) are suboptimal for some classes of biological sequence algorithms where a large memory computer would be better.

Earlier we mentioned difficulties with the evolving taxonomy of pathogenic organisms, as classification schemes originally developed based on phenomenology are faced now with genomic inconsistencies. The current flood of metagenomic data is presenting us with an even larger problem: what exactly do concepts such as "species" and "strains" mean if it turns out that microbial life is a broad spectrum with few well-defined transitions? It is now common to refer to a "core genome" and additional distinct gene content variation that presumably is responsible for different phenotypes (19). It is possible that new concepts and terminology will be needed to map existing taxonomic categories into the genomic reality of the 21st century.

The rate of advancement in sequencing technology exceeds that even of computers, fueled by the promise of *personalized medicine* if individual drug and disease reactions can be determined and if individual genetic variation can be determined efficiently via low-cost sequencing. The field of pathogen diagnostics is riding this technology wave, too small a market to have any direct influence. Note that the read lengths of some new sequencing technologies may be too short to provide confident pathogen identification based on a single read, meaning that direct metagenomic identification of human pathogens from complex clinical or environmental samples contains some degree of uncertainty. Microarrays will have to ride their own faster/less expensive/more-information-per-chip curve if they are not to become obsolete within a few years. Alternatively, one could argue that future advances in protein detection technology could lead to breakthroughs in fast dipstick assays (similar to current home pregnancy test kits) that could provide fast, accurate, and inexpensive results for pathogen detection. In all likelihood, all these techniques will continue to compete as they evolve asynchronously.

Another technological advance is seen in the recent breakthroughs in gene and genome synthesis (20). Not only do we need to deal with emerging natural viruses from every remote corner of the planet, but now we also need to deal with the fact that for relatively modest amounts of money, it is possible to synthesize combinatorial versions of any DNA one might wish to (re)create. This potential ability to create a new class of supercharged pathogens, as well as the possibility of synthesized pathogens that do not exist in nature, puts a new urgency into ensuring that we have adequate tools to deal with these evolving biothreats.

What all this means for genomic signature design is that we will have to exist in a combination of a data avalanche, new analysis tools, and rapidly evolving new technologies. Against this background of change, we will have to deal with new missions and new challenges from adversaries equipped with the latest technologies. Fittingly for biodefense, it is indeed a very Darwinian challenge that faces us.

## DISCLAIMER

This chapter was prepared as an account of work sponsored by an agency of the U.S. government. Neither the U.S. government nor Lawrence Livermore National Security, LLC, nor any of their employees makes any warranty, expressed or implied, or assumes any legal liability or responsibility for the accuracy, completeness, or usefulness of any information, apparatus, product, or process disclosed or represents that its use would not infringe on privately owned rights. Reference herein to any specific commercial product, process, or service by trade name, trademark, manufacturer, or otherwise does not necessarily constitute or imply its endorsement, recommendation, or favoring by the U.S. government or Lawrence Livermore National Security, LLC. The views and opinions of the authors expressed herein do not necessarily state or reflect those of the U.S. government or Lawrence Livermore National Security, LLC, and shall not be used for advertising or product endorsement purposes. This work was performed under the auspices of the U.S. Department of Energy by Lawrence Livermore National Laboratory under Contract DE-AC52-07NA27344.

## REFERENCES

[1] S.N. Gardner, M.W. Lam, J.R. Smith, C.L. Torres, T.R. Slezak, Draft versus finished sequence data for DNA and protein diagnostic signature development, Nucleic Acids Res. 33 (18) (2005) 5838–5850.

[2] E.E. Allen, G.W. Tyson, R.J. Whitaker, J.C. Detter, P.M. Richardson, J.F. Banfield, Genome dynamics in a natural archaeal population, Proc. Natl. Acad. Sci. USA 104 (6) (2007) 1883–1888.

[3] J.F. Banfield, A. Andersson, Virus population dynamics and acquired virus resistance in natural microbial communities, Science 230 (2008) 1047–1050.

[4] S. Kurtz, A. Phillippy, A.L. Delcher, M. Smoot, M. Shumway, C. Antonescu, et al., Versatile and open software for comparing large genomes, Genome Biol. 5 (2) (2004) R12.

[5] S.F. Altschul, W. Gish, W. Miller, E.W. Myers, D.J. Lipman, Basic local alignment search tool, J. Mol. Biol. 215 (1990) 403–410.

[6] S. Rozen, H. Skaletsky, Primer3 on the WWW for general users and for biologist programmers, Methods Mol. Biol. 132 (2000) 365–386.

[7] T.R. Slezak, T. Kuczmarski, L. Ott, C. Torres, D. Medeiros, J. Smith, et al., Comparative genomics tools applied to bioterrorism defense, Brief Bioinform. 4 (2003) 133–149.

[8] L.S. Frost, R. Leplae, A.O. Summers, A. Toussaint, Mobile genetic elements: The agents of open source evolution, Nat. Rev. Microbiol. 3 (2005) 722–732.

[9] G.D. Solar, R. Giraldo, M.J. Ruiz-Echevarria, M. Espinosa, R. Diaz-Orejas, Replication and control of circular bacterial plasmids, Microbiol. Mol. Biol. Rev. 62 (2) (1998) 434–464.

[10] J.E. Allen, S.N. Gardner, T.R. Slezak, DNA signatures for detecting genetic engineering in bacteria, Genome Biol. 9 (3) (2008) R56.

[11] D. Wang, L. Coscoy, M. Zylberberg, P.C. Avila, H.A. Boushey, D. Ganem, et al., Microarray-based detection and genotyping of viral pathogens, Proc. Natl. Acad. Sci. USA 99 (24) (2002) 15687–15692.

[12] G. Palacios, P.L. Quan, O.J. Jabado, S. Conlan, D.L. Hirschberg, Y. Liu, et al., Panmicrobial oligonucleotide array for diagnosis of infectious diseases, Emerg. Infect. Dis. 13 (1) (2007) 73–81.

[13] B. Lin, Z. Wang, G.J. Vora, J.A. Thornton, J.M. Schnur, D.C. Thach, et al., Broad-spectrum respiratory tract pathogen identification using resequencing DNA microarrays, Genome Res. 16 (4) (2006) 527–535.

[14] O.J. Jabado, Y. Liu, S. Conlan, P.L. Quan, H. Hegyi, Y. Lussier, et al., Comprehensive viral oligonucleotide probe design using conserved protein regions, Nucleic Acids Res. 36 (1) (2008) e3.

[15] D. Wang, A. Urisman, Y.-T. Liu, M. Springer, T. Ksiazek, D.D. Erdman, et al., Viral discovery and sequence recovery using DNA microarrays, PLoS Biol. 1 (2) (2003) e2.

[16] S.N. Gardner, C. Jaing, K. McLoughlin, et al. A microbial detection array (MDA) for viral and bacterial detection. Submitted for publication.

[17] E. Ghedin, N.A. Sengamalay, M. Shumway, J. Zaborsky, T. Feldblyum, V. Subbu, et al., Large-scale sequencing of human influenza reveals the dynamic nature of viral genome evolution, Nature 437 (2005) 1162–1166.

[18] E.R. Mardis, The impact of next-generation sequencing technology on genetics, Trends Genet. 24 (3) (2008).

[19] K.E. Nelson, D.E. Fouts, E.F. Mongodin, J. Ravel, R.T. DeBoy, J.F. Kolonay, et al., Whole genome comparisons of serotype 4b and 1/2a strains of the food-borne pathogen *Listeria monocytogenes* reveal new insights into the core genome components of this species, Nucleic Acids Res. 32 (8) (2004) 2386–2395.

[20] D.G. Gibson, G.A. Benders, C. Andrews-Pfannkoch, E.A. Denisova, H. Baden-Tillson, J. Zaveri, et al., Complete chemical synthesis, assembly, and cloning of a Mycoplasma genitalium genome, Science 319 (2008) 1215–1220.

# CHAPTER 30

# Nonbiological Measurements on Biological Agents

**Stephan P. Velsko**

*Lawrence Livermore National Laboratory, Livermore, California*

## INTRODUCTION

The anthrax letters investigation engendered considerable speculation about the potential importance of nonbiological measurements on samples of biological agents (1,2). A variety of mass spectral, spectroscopic, and other instrumental techniques were applied to characterize case samples or to study the properties of surrogate agents with the hope of shedding light on morphological, trace element, isotopic, or other molecular fingerprints of the production methods used to create the anthrax powders. It was imagined by some that knowledge of the agent production method might be a key piece of evidence, supplementing that obtained from genetic analysis or other biological properties and generating valuable investigative leads (3). At the time, the generation and interpretation of chemical and physical data from biological agents (like other areas of microbial analysis utilized in the investigation) represented a new domain of forensic science.

As might be expected during this formative period, the perceived needs, forensic experience, and decidedly operational mindset of the investigators naturally led to an emphasis on standard operating procedure (SOP) development, and a number of SOPs now exist for analyzing agents by various sophisticated techniques. While these techniques could determine the composition and structural features of agents with exquisite precision and accuracy, consideration of the meaning of these data most often revealed the need for extensive exploratory research to identify and understand how specific signatures were related to case-relevant questions about the samples.

Related to this, there was a great deal of concern for analytical validation [standards, calibrations, quality assurance and quality control guidelines, etc. (4)]. Although this concern was clearly appropriate, there was less awareness of the need for a coherent framework for inferential validation (see Chapter 31). As the National Research Council (NRC) report on forensic science illustrates (5),

**509**

Microbial Forensics. DOI: 10.1016/B978-0-12-382006-8.00030-X

this need is not unique to microbial forensics, but is thought by some to apply to many traditional forensic science techniques as well.

As a result of the need to disentangle longer-term exploratory research and validation activities from near-term operational applications of certain techniques, not all directions proved immediately fruitful. Moreover, perceptions of the relative utility of this kind of evidence evolved substantially over time, and a considerably more mature perspective on many of these issues has been gained in the near decade since work began in this field. This chapter describes some examples of the progress that has been made through efforts supported by a variety of agencies and provides a view of current capabilities and future needs.

# DETERMINING MANUFACTURING METHOD

It is generally appreciated that the morphological and chemical properties of an agent reflect the methods and materials used to generate it. Thus, it is plausible that certain details about the recipe used to make a material may be deduced from chemical or physical analysis. This type of information could be of value to an investigator if it constrained the pool of suspects to those that have access to the equipment, materials, and information necessary to carry out specific recipes. Two examples of this kind of analysis involve silicon deposition and residual agar in spore preparations.

## Silicon Deposition in Spores

Early in the Amerithrax investigation, it was observed that silica ($SiO_2$) could be detected by scanning transmission electron microscopy (STEM) within the outer coat of the *Bacillus anthracis* spores found in the *N.Y. Post* and Leahy letters (6). Comparison with spore preparations made by a variety of standard growth processes led to the conclusion that the amount of $SiO_2$ observed in the case samples was unusually high, amounting to at least 1% of the spore mass by weight. At the time these observations ostensibly pointed to something unique about the process used to create the Amerithrax materials.

This led to several years of work by Weber and colleagues (7) to understand the factors that govern the concentration and distribution of silica in spores. Samples of *B. anthracis* Ames and other closely related *Bacillus* species were produced with a wide range of media, production methods, and postproduction treatments. The effect of the silica concentration in growth medium on the silica concentration in spores was determined experimentally by adding dissolved known concentrations of $SiO_2$ in different growth media whose native $SiO_2$ concentration had been determined. Postproduction experiments to test the adsorptive capacity of *B. anthracis* spores for $SiO_2$ in solution were also performed. Silicon concentrations in single spores were determined by high-resolution secondary ion mass spectrometry (nanoSIMS) analysis, and bulk silicon concentrations in growth media or spore powders were determined by inductively coupled plasma–optical emission spectrometry. Protocols and standards for quantifying the distribution of elements in single spores were developed in preliminary work (8,9).

This work demonstrated that a wide range of standard laboratory growth methods produce bacterial spores with silicon levels up to about $3000\,\mu g/g$ and that the silicon tends to accumulate preferentially in the spore coat region. This phenomenology was qualitatively consistent with the previous observations of Stewart and associates (10) and Michael (6). The addition of dissolved iron salts to the growth medium enhanced the uptake of silicon, suggesting that silicon uptake in spores might be governed by a general mechanism of silicon adsorption in bacteria reported by Davis (11), Warren and Ferris (12), Yee and colleagues (13), and Wightman and Fein (14) (referred to hereafter as DWYW). The qualitative association between iron and silicon implied by this mechanism was also consistent with the observations of Michael (6).

The average concentration of silicon in the spores was proportional to the silicon concentration in the medium, with an effective spore:solution partition coefficient as high as $\approx40:1$ (i.e., there is preferential binding of dissolved silicon oxides to the spore) at high iron concentrations. Native silicon is commonly found in growth media at concentrations well below saturation. Based on the maximum solubility of $SiO_2$ in solution [$\approx70\,ppm$ at $35\,^\circ C$ (15)] and the partition coefficient between silicon concentrations in the medium and the spores, the maximum level of silicon incorporated into spores in a normal fermentation process is $\approx3000\,\mu g/g$ (0.3%) by weight. Maximum observed silicon levels among the samples Weber and associates (7) examined were $\leq3000\,\mu g/g$ silicon per spore dry weight, consistent with this estimated upper bound, as shown in Figure 30.1.

Preferential deposition of silicon in the spore coat region had been observed previously by Stewart and colleagues (10) in *Bacillus cereus* spores. The concentration of silicon in *B. cereus* samples was estimated to be $\approx3000\,\mu g/g$ ($\approx0.3\%$) using an early X-ray spectroscopy method. In their paper, Stewart and associates (10) described this as a "surprising" concentration, but this value is at the high end of the "normal" range of silicon spore concentrations. This upper bound for normal growth methods (0.3%) is significantly lower than silicon levels inferred to be present in the Amerithrax spores ($\approx1\%$) by Michael (6). Their calculation assumed that the Leahy material is predominantly spores with little external material, which is consistent with the appearance of this material in electron micrographs (6,16). To achieve the 1% level, simple mass balance would require that the initial silica concentration

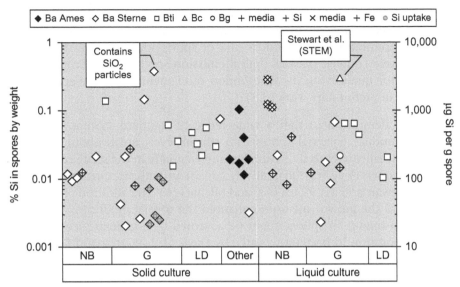

**FIGURE 30.1**

Silicon concentration in single spores determined by nanoSIMS analysis [data from Weber and colleagues (7)].

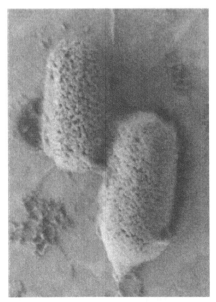

**FIGURE 30.2**

Spores coated with precipitated $SiO_2$ from a supersaturated solution.

would have to have exceeded saturation by a factor of $\approx 3\times$ even if all the silica in the growth medium had been adsorbed onto spores and cell debris in this material. The observed difference in silica concentration is too high to be explained by statistical fluctuations. Weber and collaborators (7) concluded that the silica concentrations reported for the case samples could not be explained simply by the DWYW mechanism without invoking growth conditions that involve considerable supersaturation of silica in the growth medium, some sort of "Ostwald ripening" mechanism to increase the size of the silica deposits within the spore, or some as yet unrecognized chemical process.

Two additional observations provide further support for this conclusion. First, prolonged exposure of spores to a solution of silicon at saturation was not able to increase spore silicon levels above those seen in spores produced in growth media containing added silicon. As displayed in Figure 30.2, in some of these spores, $SiO_2$ actually precipitated onto the spore outer surface during these experiments. Second, in addition to the limitation imposed by $SiO_2$ solubility, the DWYW iron–silicon adsorption mechanism implies an additional bound (Si:Fe $\approx$ 1:1) on silicon adsorption determined by saturation of all iron-binding sites within the spore. The reported silicon-to-iron ratios in the Amerithrax material (6) were approximately 10:1, significantly too high to be attributable to this mechanism.

Thus, if the estimated silicon concentrations in the Amerithrax spores are correct, they are not consistent with our current understanding of silica deposition or those materials must have indeed been produced under an unusual set of conditions. If the latter were true, the silica evidence might provide a significant bound on the credible growth and production scenarios that would be consistent with the prosecution narrative in this case.

## Presence of Agar in Spore Preparations

Another topic of interest early in the investigation was whether the Amerithrax spore preparations were grown on agar plates rather than in liquid culture. This information might aid an investigation by indicating that certain resources and specific training on agent preparation methods were available to the perpetrator. The effort focused primarily on the idea that when harvesting spores grown on agar plates, agar itself can be entrained into the samples.

Initially Fenselau (17) developed an assay that uses liquid phase extraction and gas chromatography–mass spectrometry (GC-MS) to identify trace quantities of 3,6-anhydrogalactose (AGal). This method is based on the observation that AGal is a characteristic component of agar and thus is expected to be a chemical signature indicating the presence of residual agar in a spore preparation. There are several versions of this analytical method reported in the literature (18–20). Each version uses a chemical extraction procedure that converts the AGal into a unique stable derivative that can be detected using GC-MS instrumentation. The protocol was shown to be nearly quantitative when applied to agar standards (17).

Subsequently, Wunschel and colleagues (21) found that spores grown in broth (where AGal is initially absent) yielded detectable amounts of AGal when subjected to the Fenselau protocol, leading to false-positive detections. Therefore, they developed an alternative analysis method that permitted them to avoid this problem and detected the AGal marker readily, despite its partial destruction during hydrolysis. An analysis of a wide range of spore preparations provided preliminary validation that this new protocol could be used to identify agar-grown spores with high detection probability and low false-positive probability.

Note, however, that strictly speaking, the analysis does not directly answer the question "Was the agent grown on agar medium?" Drawing this inference from detection of AGal in an agent sample would be incorrect, for example, if agar were added deliberately to the agent after it had been grown in

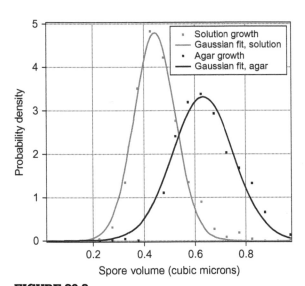

**FIGURE 30.3**

Observed distributions of spore sizes for agar and liquid cultures.

liquid culture. [Aside from the possibility that a perpetrator might do this to misdirect investigators, there could be other motivations, for example, AGal containing polysaccharides such as carrageenan can be used in certain microencapsulation methods for bacteria (22,23).] To rigorously validate such an interpretation, it would be necessary to have some independent method of analysis that could exclude the possibility of postgrowth addition of some material that contains AGal. Likewise, a negative result cannot simply be interpreted as excluding the possibility that the agent was grown on agar plates. This could happen if the agent were washed sufficiently well after growth that the residual agar concentration was below the limit of detection for the analysis. Clearly, by itself agar analysis can only reveal that AGal is present in the sample or determine that it is undetectable. At best it can provide quantitative information such as "this sample has $x \pm y$ micrograms of AGal per gram of agent" or "this sample has fewer than $x$ micrograms of AGal per gram of agent."

Several years after the AGal analysis protocols had been developed, Plomp and co-workers (24) observed measureable differences in the distributions of spore sizes from agar plate and liquid shake flask cultures. Spore dimensions could be quantified using atomic force microscopy (AFM) imaging and expressed as spore length or calculated volume. A preliminary validation study was performed with a limited number of B. anthracis Sterne samples provided from several different laboratories; results are shown in Figure 30.3. These empirical distributions are based on measurements of more than 100 individual spores from each sample.

To understand the inferential power of observing that a spore in a questioned sample has a certain volume, it is useful to plot the receiver-operating characteristic (ROC) curve corresponding to size distribution data, as is done in Figure 30.4. The slope (derivative) of the ROC curve at any point is the likelihood ratio corresponding to the value of the metric corresponding to that point. Spore volumes smaller than $\approx 0.53\,\mu m^3$ correspond to points on the ROC curve where the likelihood ratio for the liquid growth hypothesis is higher than 1 (and consequently the measurement does not favor agar growth). For a single spore measurement, likelihood ratios over the 95% probability size range for shake flask grown spores are fairly modest, in the range of 1 to 10. However, as Figure 30.4 shows, the ROC curve is considerably

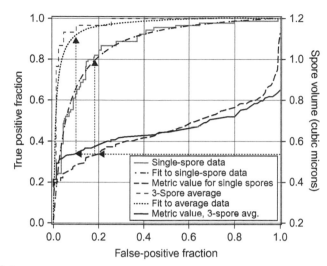

**FIGURE 30.4**

Receiver-operating characteristic for the size-based test for agar plate culture. Dashed arrows indicate the approximate position of the spore volume value (0.53 $\mu m^3$) that corresponds to a likelihood ratio value of 1. (See Color Insert.)

sharper, and likelihood ratios are considerably higher if the comparison metric is based on the average of 3 spores. For averages over 10 spores (not shown), the ROC curve is nearly a perfect classifier with an area under curve (AUC) value >0.99 (25).

These results suggested that spore size measurements might provide a viable addition to chemical analysis for determining if an agent was grown on agar plates. In addition to providing an "orthogonal" method, AFM analysis has the advantage that it does not require a bulk sample but can work on trace levels of material because it does not rely on agar as a signature—it is immune to the presence of other sources of AGal or to rigorous washing. However, it was not clear what factors actually determine the difference in size so interpretation of evidence on the basis of preliminary validation results alone demanded substantial caution. Therefore, a more extensive validation cycle using a greater variety of media and inclusion of fermentor-grown spores in addition to shake flask samples was initiated and is ongoing.

Silica and agar analyses are typical representatives of a variety of assay development efforts that were initiated after Amerithrax. Some were abandoned after initial exploratory results proved puzzling or disappointing, others moved on to SOP development, and a few were subjected to more extensive validation studies. These then form a core of capabilities and exemplars for future casework.

## SAMPLE MATCHING

Physical and chemical measurements can be used to compare evidence samples collected by investigators at different locations or at different times. Matching of sample properties can help establish the relatedness of disparate incidents, and mismatches might exclude certain scenarios or signify a more complex etiology of the events under investigation. An early report outlining research issues in the chemical and physical analysis of microbial agents for forensics states:

> Chemical and morphological analysis for sample matching has a long history in forensics and is likely to be acceptable *in principle* in court, assuming that match criteria are well defined and derived from known limits of precision of the measurement techniques in question. Thus, apart from certain operational issues (such as how to prioritize such measurements in the face of limited sample availability or how to render samples safe for handling in the analytical laboratory), instrumental analysis of biological agents for purposes of sample matching alone is unlikely to present fundamental problems that require extensive research and development investments (26).

In the years after this was written, it became apparent that there were, in fact, fundamental issues regarding inferential validation of sample-matching protocols for biological agents. This concern arose from the National Research Council's report on bullet lead analysis and the subsequent abandonment of this method by the Federal Bureau of Investigation (27). The NRC study contains some valuable insights into the way in which inferences about sample matching should be presented in court and the statistical framework in which the interpretation of such data is validated. Studies undertaken within the past few years have applied these considerations to matching biological agent samples (28,29) by means of elemental analysis and other methods.

A key feature of the assays that have been developed is that the term "match" is never used to describe test results. The term "match" itself is often difficult to define objectively and can lead to the adoption of arbitrary or subjective criteria for declaring that two samples are related. Therefore, tests that have been developed adhere to a general paradigm for evaluating and validating the inferential power of sample comparisons that does not rely on defining match criteria. This paradigm consists of

- Formulating a testable hypothesis concerning common origin of the two samples; for example, were the samples made by the same process in the same laboratory?
- Defining the population of samples for which the test is relevant; for example, the population of all bench-top processes for making dry spore preparations.

- Developing representative sampling frames for the relevant populations; for example, a list of all processes obtained from open sources and a list of laboratories that can generate surrogate (nonpathogenic) agent samples.
- Performing sampling of exemplars from those frames. Typically this involves having multiple laboratories make samples using multiple processes where the laboratories and processes were selected at random from their respective frames.
- Selecting the chemical or physical signature to be determined by analysis; for example, concentrations of a particular set of elements.
- Defining an objective metric derived from measurement of the signature in two samples that will be used as the basis for the hypothesis test. The metric expresses quantitatively how similar the signatures are in the two samples.
- Evaluating the metric for each pair of samples drawn from the exemplar set and using the metric to construct an ROC curve estimator for the likelihood ratio at each observed value of the metric.

An example of an ROC curve generated using this method is shown in Figure 30.5. In this case, the metric delta is the sum over squared differences between concentrations of 11 elements in pairs of samples drawn from a population of samples made in multiple batches by seven different processes in four different laboratories. The pair population consists of two subpopulations: one

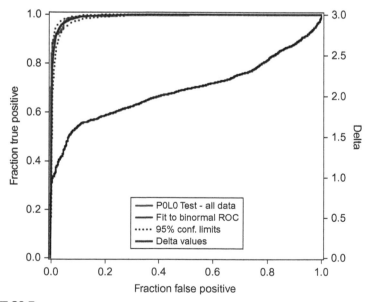

**FIGURE 30.5**

Receiver-operating characteristic curve for testing if two samples were made by the same process in the same laboratory. (See Color Insert.)

consisting of pairs made by the same process in the same laboratory and the other consisting of pairs made by different processes, or in different laboratories (or both). The ROC curve indicates that if a pair of samples has delta less than $\approx 1.5$, then the likelihood ratio is greater than 1, supporting the hypothesis that they were made by the same process in the same laboratory. Over the range of delta values observed for this subpopulation, the likelihood ratio ranges between one and several hundred.

Other hypotheses can be tested by pair comparisons as well, for example, that the two samples were made in the same laboratory but in different batches, or that they were made by the same process but in different laboratories. The testable hypotheses form a hierarchy of tests in which the corresponding ROC curves are formed from pairs of samples drawn from appropriate subpopulations. The analyst reports results of comparing two samples by quoting a likelihood ratio for the hypothesis in question.

Constructing a representative sampling frame for an agent production process requires a systematic understanding of the many different possible recipes for generating agents. Several different frames are possible, including a very general "unit process" frame in which generic recipes are broken down into their basic steps (26,28). For each unit process in the recipe, there are usually many choices of specific techniques and materials. End-to-end processes are constructed by sampling randomly from the list of specific variants for each unit process. Of course, agents used in biocrimes could be processed crudely; in some cases, no processing as such may be used. Challenges to the validity of the likelihood ratios determined by sample matching tests will ultimately rest on whether the questioned samples were really drawn from the same population as the samples used to generate the ROC curve.

## REMAINING RESEARCH AND DEVELOPMENT CHALLENGES

A number of significant research and development challenges remain associated with developing a robust forensic capability for chemical and physical analysis of biological agents. One of the most significant gaps is the ability to date agents. Determining how long before dispersal a biological agent was prepared is important for establishing timelines that are consistent with any narrative that ties together the acquisition, manufacture, transport, and deployment of a device or material. Of course, in situations where the attack agent is only recovered from infected victims, dating its preparation would not be possible in principle. However, if a bulk agent is recovered, radiocarbon assay methods provide the only established method for dating. Radiocarbon dating of biological agents using accelerator mass spectrometry

(1,30) relies on the fact that common growth medium components are derived from agricultural products whose radiocarbon content reflects that of the atmosphere at the time they were produced. The radiocarbon content of an organism grown in these media will reflect its production date, assuming that the agent was produced not too long after the medium components were manufactured. However, this technique has a resolution of $\pm 1$ year, at best (for recent materials), and is difficult to apply to complex "weaponized" agents, as these can contain petroleum-derived components that are radiocarbon depleted and can distort the estimated age of the sample.

It seems likely that other chemical or physical phenomena could be harnessed for dating agents made only months, weeks, or days prior to release, and might be applicable to single particles. One such technique that was proposed involved using nanoSIMS to map the relaxation of elemental gradients in spores and bacteria after they have been processed (31). However, relaxation rates of easily measured elements appeared to be too slow to be used this way. Therefore, the development of new agent dating capabilities remains a salient research and development gap.

There is a general need to increase the number of tests that can be applied to trace quantities of agents instead of requiring bulk material and to replace destructive tests with nondestructive alternatives. Currently, the choice of analytical method that can be applied depends critically on the amount of material available and whether it is contaminated with extraneous material that can compromise the analysis. An initial determination of these qualities has been made for many of the common analysis methods that are available (8,16,17,21,24,32–40), shown in Table 30.1. These sample requirement determinations are generalizations and can depend on the precise object of the analysis. Most of the existing protocols require uncontaminated bulk material and are destructive. Examples include protocols based on solvent extraction of a signature compound, imaging that requires sample fixation, and ablative mass spectroscopy techniques such as MALDI and SIMS.

During the Amerithrax investigation, the types of chemical and physical studies that could be considered were generally constrained by available sample size. Because several letters contained substantial quantities of powder (41), the scientific community was able to offer a number of reasonable methods for addressing important questions about this material. However, at some Amerithrax crime scenes, bulk powder evidence was not available, so these methods could not be applied. In many imaginable future attack scenarios, the only agent samples that would be available for analysis are those recovered from contaminated surfaces, ductwork or filters from building air conditioning systems, or material on the filter units used in urban air samplers. Such samples would be mixed heavily with other materials, requiring the analyst

**Table 30.1** Characteristics of Some Instrumental Analysis Methods

| Type of Analysis | Method | Destructive or Nondestructive? | Can Use Irradiated Sample? | Can Use ClO$_2$ or VHP to Decontaminate? | Minimum Amount Needed | Pristine Sample Required? |
|---|---|---|---|---|---|---|
| Elemental analysis | ICP-MS (1) | Destructive | Yes | Yes | 1 mg | Yes |
| | MXRF (2) | Nondestructive | Yes | Yes | 1 mg | No |
| | PIXE (3) | Nondestructive | Yes | Yes | 1 µg | No |
| | SIMS (4) | Destructive | Yes | Yes | Trace | No |
| Isotopic analysis | MS (5) | Destructive | Yes | ? | 1 mg | Yes |
| | SIMS | Destructive | Yes | ? | Trace | No |
| Organic analysis | GC-MS (6) | Destructive | No | ? | 1 mg | Yes |
| | ES-MS (7) | | No | | | Yes |
| | MALDI (8) | Destructive | No | ? | 1 mg | Yes |
| | IR (9) | Nondestructive | No | ? | 1 mg | Yes |
| | NMR (10) | Nondestructive | No | ? | 1 mg | Yes |
| | µ-Raman (11) | Nondestructive | No | ? | 1 µg | Yes |
| Micro-structural analysis | SIMS | Destructive | No | Yes | Trace | No |
| | STEM (12) | Destructive | No | Yes | Trace | No |

to identify and isolate the agent particles from the mixture. In the future, it will clearly be important to extend the reach of more kinds of chemical and physical analyses to such situations.

In most cases, chemical and physical analyses of case samples take place on dedicated instrumentation that is not housed in a biosafety level 3 or 4 environment. As a result, it is desirable to develop inactivation protocols to render samples safe for analysis. In many bulk analysis procedures, extraction protocols can be designed to render the analytical sample safe, and extraction can be performed in a biosafety area. However, research needs to be conducted to determine the effect of irradiation or decontamination protocols based on chlorine dioxide (ClO$_2$) (42) or vaporous hydrogen peroxide (VHP) (43) on signatures that must be determined directly from the agent particles themselves.

Little systematic development of specialized collection methods for instrumental analysis of agent samples has occurred. As more sophisticated analysis methods are developed, it will be important to establish collection and storage methods that do not inadvertently distort subtle structural signatures or alter chemical clues by contamination or allow them to degrade prior to analysis.

Finally, in many cases it is desirable to use nonpathogenic surrogates for research and development of instrumental analysis methods. Current practice has settled on the use of *B. anthracis* Sterne as a surrogate for virulent *anthracis* strains. There is no standard surrogate nonpathogenic substitute for other pathogens, as yet. The transferability of conclusions drawn from studies on surrogates to the actual agent may not always be straightforward. Therefore, it will be necessary to understand better the limitations of surrogates and how to choose the best surrogate for a given study.

## PREPARING FOR FUTURE EVENTS

The National Bioforensics Analysis Center (NBFAC) is the lead organization for conducting technical (biological, chemical, and physical) analysis of biological agents used in terrorist or criminal activities affecting U.S. persons or assets (44,45). In future cases, NBFAC may need to call upon the specialized chemical and physical analysis capabilities of other laboratories ("spoke labs") in order to carry out analyses that utilize sophisticated and expensive instrumentation and unique expertise. It is clearly prudent to have a set of standard procedures in place that can be called upon as needed. However, there will always be a need to develop certain analyses ad hoc, depending on the particular circumstances of a given case. Thus, a preexisting plan should be flexible enough to accommodate new or unusual samples or situations. In this sense, a seamless sample analysis plan involves two major activities: (i) to formalize and standardize methodologies as much as possible by anticipating the types of analyses that may be called for and developing standard operating procedures for performing them (i.e., think ahead) and (ii) to formalize a mechanism for deciding how to handle new situations for which there are no existing SOPs. When only limited or trace samples are available, a systematic procedure for choosing the most informative analytical tests becomes crucial. It will be important to formalize such considerations to expedite the development of the best analytical plan when an incident occurs. More generally, formulation of a comprehensive plan involves:

- Developing a mechanism for quickly formulating a consensus analytical plan when a new sample (or set of samples) arises
- Keeping and updating a set of standardized procedures (SOPs) and validation data for analyzing case samples
- Maintaining a set of documented guidelines, requirements, and procedures for sample preparation for each analytical procedure
- Maintaining approved procedures for handling and storing samples
- Developing standardized methods for data analysis, reporting, and presentation
- Maintaining reliable channels for sending and receiving samples
- Maintaining secure conduits for data, information, and discussion
- Developing a mechanism for formulating an on-the-fly validation plan for a new procedure

While many of these items are in place, there has been little or no systematic testing of the existing infrastructure through planned exercises. In the absence of actual casework, comprehensive exercises that include evidence collection, analysis, and expert testimony at mock trials and admissibility hearings are critical for maintaining effective preparedness.

## CONCLUDING REMARKS

The forensic utility of data that can be obtained from chemical and physical analyses of agents depends critically on establishing rigorous and defensible scientific underpinnings. While much progress has been made over the past decade, it is important to keep in mind that technical microbial forensics to date is largely untested in U.S. courtrooms and has only begun to be subjected to the scrutiny of domestic and international scientific fora. The most prominent biological terrorism case to date, the anthrax letters incident, will apparently close without a trial (46). Given the relative rarity of major biological terrorism events, there is some danger of complacency about the strength of our preparations and uncertainty about research and development priorities.

If there is any lesson to be drawn from past experience, it is that microbial forensic collection and analysis are not very effective when they are conducted as ad hoc activities, by nonspecialists, using improvised methods and on-the-fly attempts at validation, without prior review by a knowledgeable community. The utility of microbial forensic analysis rises in proportion to the extent that it is anticipatory, well planned, driven by a cadre of qualified experts, and resourced adequately. This, in turn, requires adequate attention to the need for long-term fundamental research on many aspects of biological agent composition and structure. It is encouraging that a robust microbial forensic research and development program is now recognized as an essential part of the national strategy to combat biological threats (45). With sufficient targeted investment the technical challenges associated with chemical and physical analysis can be met, and the expertise gained in this area in the past decade will continue to evolve toward an established capability.

## ACKNOWLEDGMENTS

I acknowledge my collaborators at Lawrence Livermore National Laboratory who generated data in the illustrative images used in this chapter. Figures 30.1, 30.2, and 30.5 are based on data provided by Peter Weber. Alexander Malkin provided data used in Figures 30.3 and 30.4.

## DISCLAIMER

This was work performed under the auspices of the U.S. Department of Energy by Lawrence Livermore National Laboratory under Contract DE-AC52-07NA27344. This chapter was prepared as an account of work sponsored by an agency of the U.S. government. Neither the U.S. government nor Lawrence Livermore National Security, LLC, nor any of their employees makes any warranty, expressed or implied, or assumes any legal liability or responsibility for the accuracy, completeness, or usefulness of any information, apparatus, product, or process disclosed or represents that its use would not infringe on privately owned rights. Reference herein to any specific commercial product, process, or service by trade name, trademark, manufacturer, or otherwise does not necessarily constitute or imply its endorsement, recommendation, or favoring by the U.S. government or Lawrence Livermore National Security, LLC. The views and opinions of authors expressed herein do not necessarily state or reflect those of the U.S. government or Lawrence Livermore National Security, LLC, and shall not be used for advertising or product endorsement purposes.

## REFERENCES

[1] C.M. Schaldach, J.J. DeYoreo, T. Esposito, D.P. Fergensen, E. Gard, C. Hollars, et al., Non-DNA methods for biological signatures, in: R.G. Breeze, B. Budowle, S.E. Schutzer (Eds.), Microbial Forensics, Elsevier, Amsterdam, 2005.

[2] P. Keim, Microbial Forensics: A Scientific Assessment. American Academy of Microbiology, 2003.

[3] G. Matsumoto, Anthrax powder: State of the art? Science 302 (2003) 1492–1497.

[4] B. Budowle, S.E. Schutzer, A. Finseln, L.C. Kelly, A.C. Walsh, J.A. Smith, et al., Public health: Building microbial forensics as a response to bioterrorism, Science 301 (2003) 1852.

[5] National Research Council, Strengthening Forensic Science in the United States, National Academies Press, Washington, DC, 2009.

[6] Sandia National Laboratory News Release, August 21, 2008. Available at www.sandia.gov/news/resources/2008/anthrax.html; Joseph Michael, Presentation to the National Research Council Scientific Review of the FBI Anthrax Investigation, September 25, 2009; podcast available at http://nationalacademies.org/newsroom/nalerts/20090925.html.

[7] P. Weber, B. Vianni, L. Davisson, S. Velsko, Nanometer-Scale Secondary Ion Mass Spectroscopy For Microbial Characterization, Presentation to the National Research Council Scientific Review of the FBI Anthrax Investigation, September 25, 2009.

[8] S. Ghosal, S.J. Fallon, T.J. Leighton, K.E. Wheeler, M.J. Kristo, I.D. Hutcheon, et al., Imaging and 3D elemental characterization of intact bacterial spores by high-resolution secondary ion mass spectrometry, Anal. Chem. 80 (15) (2008) 5986–5992.

[9] M.L. Davisson, P.K. Weber, J. Pett-Ridge, S. Singer, Development of Standards for NanoSIMS Analyses of Biological Materials. Lawrence Livermore National Laboratory Report LLNL-TR-406039, August 5, 2008.

[10] M. Stewart, A.P. Somlyo, A.V. Somlyo, H. Shuman, J.A. Lindsay, W.G. Murrell, Distribution of calcium and other elements in cryosectioned *Bacillus cereus* T spores, determined by high-resolution scanning electron probe x-ray microanalysis, J. Bacteriol. 143 (1) (1980) 481–491.

[11] C.C. Davis, Modeling silica adsorption to iron hydroxide, Environ. Sci. Technol. 36 (2002) 582–587.

[12] L.A. Warren, F.G. Ferris, Continuum between sorption and precipitation of Fe(III) on microbial surfaces, Environ. Sci. Technol. 32 (1998) 2331–2337.

[13] N. Yee, D.A. Fowle, F.G.A Ferris, Donnan model for metal sorption onto *Bacillus subtilis*, Geochim. Cosmochim. Acta 68 (2004) 3657–3664.

[14] P.G. Wightman, J.B. Fein, Iron adsorption by *Bacillus subtilis* bacterial cell walls, Chem. Geol. 216 (2005) 177–189.

[15] R.K. Iler, The Chemistry of Silica, Solubility, Polymerization, Colloid and Surface Properties, and Biochemistry, Wiley & Sons, New York, 1979.

[16] L.N. Brewer, J.A. Ohlhausen, P.G. Kotula, J.R. Michael, Forensic analysis of bioagents by X-ray and TOF-SIMS hyperspectral imaging, Forensic Sci. Int. 179 (2-3) (2008) 98–106.

[17] C. Fenselau, 17th Sanibel Conference on Mass Spectrometry: Forensic Science and Counterterrorism, Sanibel Island, FL, 2005.

[18] C.N. Jol, T.G. Neiss, B. Penninkhof, B. Rudolph, G.A. De Ruiter, A novel high-performance anion-exchange chromatographic method for the analysis of carrageenans and agars containing 3,6-anhydrogalactose, Anal. Biochem. 268 (1999) 213–222.

[19] T.T. Stevenson, R.H. Furneaux, Chemical methods for the analysis of sulfated galactans from red algae, Carbohydr. Res. 210 (1991) 277–298.

[20] Y. Hama, H. Nakagawa, M. Kurosawa, T. Sumi, X. Xia, K. Yamaguchi, A gas chromatographic method for the sugar analysis of 3,6-anhydrogalactose-containing algal galactans, Anal. Biochem. 265 (1998) 42–48.

[21] D.S. Wunschel, H.A. Colburn, A. Fox, K.F. Fox, W.M. Harley, J.H. Wahl, et al., Detection of agar, by analysis of sugar markers, associated with *Bacillus anthracis* spores, after culture, J. Microbiol. Methods 74 (2-3) (2008) 57–63.

[22] T.B. Hammill, R.L. Crawford, Bacterial microencapsulation with three algal polysaccharides, Can. J. Microbiol. 43 (1997) 1091–1095.

[23] E. Murano, Use of natural polysaccharides in the microencapsulation techniques, J. Appl. Ichthyol. 14 (1998) 245–249.

[24] M. Plomp, T.J. Leighton, K.E. Wheeler, A.J. Malkin, The high-resolution architecture and structural dynamics of Bacillus spores, Biophys. J. 88 (2005) 603–608.

[25] W.J. Krzanowski, D.J. Hand, ROC Curves for Continuous Data, CRC Press, Boca Raton FL, 2009.

[26] S.P. Velsko, Physical and Analytical Chemical Analysis: A Key Component of Bioforensics. Presented at the AAAS Annual Meeting, February 2005, Washington, DC, available from Lawrence Livermore National Laboratory as UCRL-CONF-209735, 2005.

[27] a. National Academies Board on Chemical Sciences and Technology, Forensic Analysis: Weighing Lead Bullet Evidence. The National Academies Press, Washington, DC, 2004; b. M.O. Finkelstein, B. Levin, Compositional analysis of bullet lead as forensic evidence. J. Law Policy 13 (1) (2005) 119–142; c. FBI response to the NRC study can be found in FBI Laboratory Announces Discontinuation of Bullet Lead Examinations." Press Release, Federal Bureau of Investigation, Washington, DC, September 1, 2005.

[28] S.P. Velsko, Bioagent Sample Matching Using Elemental Composition Data: An Approach to Validation, UCRL-TR-220803 Lawrence Livermore National Laboratory, 2006.

[29] S.P. Velsko, P. Weber, C.E. Ramon, R.E. Lindvall, M.L. Davisson, M. Robel, Bioagent Sample Matching Using Elemental Composition Data, Lawrence Livermore National Laboratory Report LLNL-TR-419683, 2009.

[30] D.H. Ubelaker, Artificial radiocarbon as an indicator of recent origin of organic remains in forensic cases, J. Forensic Sci. 46 (2001) 1285–1287.

[31] S. Ghosal, S. Fallon, T. Leighton, K. Wheeler, I. Hutcheon, P.K. Weber, Analysis of Bacterial Spore Permeability to Water and Ions Using Nano-Secondary Ion Mass Spectrometry (NanoSIMS), Lawrence Livermore National Laboratory, UCRL-ABS-229725, 2007.

[32] C.G. Worley, S.S. Wiltshire, T.C. Miller, G.J. Havrilla, V. Majidi, Detection of visible and latent fingerprints using micro-X-ray fluorescence elemental imaging, J. Forensic Sci. 51 (1) (2006) 57–63.

[33] J.S. Becker, N. Jakubowski, The synergy of elemental and biomolecular mass spectrometry: New analytical strategies in life sciences, Chem. Soc. Rev. 38 (7) (2009) 1969–1983.

[34] E.F. Garman, G.W. Grime, Elemental analysis of proteins by microPIXE, Prog. Biophys. Mol. Biol. 89 (2) (2005) 173–205.

[35] H.W. Kreuzer-Martin, K.H. Jarman, Stable isotope ratios and forensic analysis of microorganisms, Appl. Environ. Microbiol. 73 (12) (2007) 3896–3908.

[36] B.L. van Baar, Characterisation of bacteria by matrix-assisted laser desorption/ionisation and electrospray mass spectrometry, FEMS Microbiol. Rev. 24 (2) (2000) 193–219.

[37] N. Valentine, S. Wunschel, D. Wunschel, C. Petersen, K. Wahl, Effect of culture conditions on microorganism identification by matrix-assisted laser desorption ionization mass spectrometry, Appl. Environ. Microbiol. 71 (1) (2005) 58–64.

[38] M. Harz, P. Rösch, J. Popp, Vibrational spectroscopy: A powerful tool for the rapid identification of microbial cells at the single-cell level, Cytometry A 75 (2) (2009) 104–113.

[39] F. Freitas, V.D. Alves, J. Pais, N. Costa, C. Oliveira, L. Mafra, et al., Characterization of an extracellular polysaccharide produced by a Pseudomonas strain grown on glycerol, Bioresource Technol. 100 (2) (2009) 859–865.

[40] A.P. Esposito, C.E. Talley, T. Huser, C.W. Hollars, C.M. Schaldach, S.M. Lane, Analysis of single bacterial spores by micro-Raman spectroscopy, Appl. Spectrosc. 57 (7) (2003) 868–871.

[41] D.J. Beecher, Forensic application of microbiological culture analysis to identify mail intentionally contaminated with Bacillus anthracis spores, Appl. Environ. Microbiol. 72 (2006) 5304–5310.

[42] D.E. Cortezzo, K. Koziol-Dube, B. Setlow, P. Setlow, Treatment with oxidizing agents damages the inner membrane of spores of Bacillus subtilis and sensitizes spores to subsequent stress, J. Appl. Microbiol. 97 (4) (2004) 838–852.

[43] J.A. Otter, G.L. French, Survival of nosocomial bacteria and spores on surfaces and inactivation by hydrogen peroxide vapor, J. Clin. Microbiol. 47 (1) (2009) 205–207.

[44] HSPD-10. Biodefense for the 21st century, 2004. Available from: www.fas.org/irp/offdocs/nspd/hspd-10.html.

[45] National Strategy for Countering Biological Threats, November 2009. Available from: www.whitehouse.gov/the-press-office/president-obama-releases-national-strategy-countering-biological-threats.

[46] Vahid Majidi, Public briefing on the Amerithrax case, August 18, 2008. Available from: www.fbi.gov/page2/august08/anthraxscience_081808.html.

# CHAPTER 31

# Inferential Validation and Evidence Interpretation

**Stephan P. Velsko**

*Lawrence Livermore National Laboratory, Livermore, California*

## THE NEED TO VALIDATE THE INTERPRETATION OF MICROBIAL FORENSIC EVIDENCE

The field of microbial forensics is being created at a time when forensic science in general faces unprecedented skepticism. The foundations of many long-accepted forensic science methods have been questioned, and recent National Research Council studies have supported these criticisms (1,2). It is likely that both the admissibility and the evidentiary weight of microbial forensic evidence in future cases will be scrutinized closely and that *Daubert* challenges will occur. Thus, it is imperative that this new area of forensic science build sound, *Daubert*-resistant foundations by carefully considering both the framework for validation and the way in which microbial forensic evidence is conveyed in reports, hearings, and trials.

This concern is generally appreciated by the community of scientists engaged in microbial forensic research and operations, who have addressed certain important aspects of validation. In particular, guidelines for quality assurance have been formulated and published widely (3). It is also possible to find clear and useful guidance for establishing the precision and accuracy of a variety of assays of use in microbial forensics. However, as shown later, this addresses only one aspect of validation, and by itself cannot impart *Daubert* resistance to microbial forensic evidence. This is because the most salient criticisms that have been leveled at forensic science do not question data quality, but rather interpretation. This issue is best illustrated by two quotations that clearly differentiate between the validity of data and the validity of the interpretation of that data in forensic science testimony:

> Even if an instrument yields exquisitely precise measurements,
> the witness's inferences from the measurements may be badly
> flawed. As Justice Blackmun stressed in *Daubert*, it is the expert's

Microbial Forensics. DOI: 10.1016/B978-0-12-382006-8.00031-1

ultimate inference which "must be derived by the scientific method…
[and] supported by appropriate validation.…"

**Edward J. Imwinkelried in
"The Methods of Attacking Scientific Evidence" (4)**

The committee found the analytical technique used is suitable and
reliable for use in court, as long as FBI examiners apply it uniformly
as recommended. […] However, for legal proceedings, the probative
value of these findings and how the probative value is conveyed to
a jury remains a critical issue.

**From the NRC report "Forensic Analysis:
Weighing Bullet Lead Evidence" (1)**

This chapter suggests that validation of interpretation is a distinct research
and development activity that is separable from other kinds of validation and
can be formalized to a large extent. Although there is a large body of literature that discusses methods for validating data interpretation, the concept is
seldom treated as a separate activity in the development of forensic assays.
The statistical concepts and methodologies described in this chapter play
much more familiar roles in the area of medical diagnostics, where they may
be considered mainstream.

Because the term "validation" is used in various ways in the forensic context,
a short description of the various types of validation that have been described,
their interrelationships, and their connections to the *Daubert* decision (5) and
Federal Rules of Evidence (6) is provided. Next, a general scheme for inferential validation in the context of microbial forensics is described, followed by
a discussion of population and sampling issues that apply to the specific area
of chemical and physical analysis of biological agents.

## THE TAXONOMY OF VALIDATION

The term "validation" is used to describe a number of distinct activities in
forensic research and operations. To understand relationships among the different classes of validation activities it is useful to turn to the original text of
the *Daubert* decision, which noted that scientific validity rests on two factors:
reliability and relevance (5). The reliability of a technique is its ability to produce consistent, objective results with known precision and traceable accuracy. Those quality assurance procedures that assure the reliability of scientific
evidence are termed "analytical validation."

The term "relevance" refers to the fact that analytical measurements or other
scientific data usually are not of interest to the court per se, but are proffered as

**Table 31.1** Assays and Inferences in Microbial Forensics

| Test or Assay Results | Ultimate Inference |
|---|---|
| Elemental profiles of two agent samples | The two agent samples were made by the same (different) method(s). |
| Carbon 14 content of an agent sample | The agent was produced later than a certain date. |
| Presence of certain organic compounds | The agent was made using certain materials or methods. |
| Genetic sequences of two bacterial isolates A and B | Isolate A could have been derived by culturing isolate B. |

evidence to support or refute by inference a fact at issue in the trial (4). According to Federal Rules of Evidence 401 and 402, relevant evidence is that "having any tendency to make the existence of any fact [at issue in the trial] more probable or less probable than it would be without the evidence." Evidence that is not relevant in this sense is not admissible. Inferential validation is the process that establishes the strength of support (i.e., the degree of relevance, or probative value) that a given observation or other data provide for the expert's ultimate inference for which the observation or data are offered as evidence. (The term "ultimate inference" should not be confused with the legal term "ultimate issue," which refers to a question that the jury must decide, for example, the guilt or innocence of the accused.) Table 31.1 provides some examples of assays and the ultimate inferences that they may be used to support in the field of microbial forensics.

Corresponding to these two classes of validation, it is sometimes useful to distinguish between a "reporting expert witness" and an "interpreting expert witness" (4). The former testifies as to the test result and how it was obtained and seeks to assure the fact finders that results of the analysis are reliable. The latter provides an expert opinion regarding the ultimate inference to be drawn from that test result. (In practice, of course, the same expert may perform both these roles, and the distinction is useful even if the evidence is never used in court.) The reporting expert comes armed with results of analytical validation, while the interpreting expert supports his testimony with experimental results that provide inferential validation. No matter how exacting the quality assurance and quality control regime, the expert's interpretation of scientific evidence is still vulnerable to challenge, especially if the interpretation is particularly crucial to the prosecution narrative. For either expert, the ultimate product of a validation study is, in essence, the value of a statistical estimator. For analytical validation, examples of such estimators are precision values associated with measurements or a distribution of difference values relative to a known standard (7,8). For inferential validation, typical estimators are receiver-operating characteristic (ROC) curves or likelihood ratio estimators for a well-constructed hypothesis test (9). The aforementioned considerations are summarized in Table 31.2.

**Table 31.2** Relationship Between Analytical and Inferential Validation

| Daubert Criterion | Type of Expert Witness | Type of Validation | Statistical Metrics | Assurance Value |
|---|---|---|---|---|
| Reliability | Reporting | Analytical | Precision Accuracy Reproducibility | The technique produces consistent, objective results with known precision and traceable accuracy |
| Relevance | Interpreting | Inferential | ROC curves Likelihood ratios | The result supports the expert's inference |

Inferential validation is intrinsically a research activity, but there are variants of analytical validation that apply to the operational implementations of forensic assays. For example, a distinction can be made between "developmental" and "internal" validation. Following definitions given in the Scientific Working Group on Microbial Genetics and Forensics (SWGMGF) Quality Assurance Guidelines for Laboratories Performing Microbial Forensic Work (3), developmental validation is an activity carried out by the laboratory that develops the technique and thus is a research activity, while internal validation is carried out by laboratories that are implementing the technique in-house for operational use. Inferential validation need only be performed in developmental mode, as the same inferential power will apply to the same technique used in a different laboratory, assuming that internal analytical validation has been performed.

Two other terms sometimes found in the literature are "external validation" and "preliminary validation." In external validation, the performance of a technique by a laboratory is evaluated by one or more (usually blind) tests administered by an independent entity. In this regard, external validation is a species of analytical validation that provides additional assurance of the consistency and reliability of a technique by showing it to be independent of the particular laboratory or operator. In microbial forensics, preliminary validation has been defined as the acquisition of limited test data to enable evaluation of a method used to assess materials derived from a biocrime or bioterrorism event (3,10,11). Preliminary validation enables evaluation of a previously uncharacterized method used to provide investigative support (e.g., generating investigative leads.) Preliminary validation involves both analytical and inferential validation. The latter is clearly required at some level in order to evaluate the value of the test for generating investigative leads based on the ultimate inference drawn from the test. The SWGMGF Guidelines for Microbial Forensics stipulate that if results are to be used for other than investigative support, then a panel of peer experts, external to the laboratory, should be convened to assess

**Table 31.3** Required Types of Validation for the Four Validation Categories

|  | Analytical | Inferential |
|---|---|---|
| Preliminary | Yes | Yes |
| Developmental | Yes | Yes |
| Internal | Yes | No |
| External | Yes | No |

the utility of the method and to define the limits of interpretation and conclusions drawn (3,11). Table 31.3 summarizes the matrix of requirements for validation corresponding to the foregoing discussion.

## THE ROC/LR FRAMEWORK FOR INFERENTIAL VALIDATION

Human DNA analysis is generally regarded as the gold standard for forensic science, and the statistical foundation for DNA evidence is sometimes suggested as a paradigm for inferential validation of other kinds of forensic tests and assays (12). For example, the NRC studies of forensic science have consistently advocated a "likelihood" or "likelihood" ratio framework for interpreting scientific evidence (1,13,14). [Counterarguments to this notion are sometimes put forth by professional forensic scientists, who argue that other kinds of forensic assays and tests cannot be treated in the same framework (15).] This section outlines a foundation for inferential validation based on a likelihood ratio approach.

The standard likelihood equation is shown in Figure 31.1. E represents some piece of evidence, in our case some measurement or set of observations made on one or more samples of a biological agent. H is some hypothesis concerning the production or source of that agent. $O_0(H)$ are the odds that H is true in the absence of E, and $O(H|E)$ are the posterior odds. The likelihood ratio is determined by the probability that E would be observed if H were true versus if it were false ($\bar{H}$). The likelihood ratio is often considered the strength (or probative value) of evidence E with respect to the hypothesis H. Because Federal Rule 401 explicitly defines the relevance of evidence in terms of whether it makes H more probable or less probable, legal scholars have often cited the likelihood ratio (LR) as a measure of relevance, and hence admissibility (16–18). Specifically, if

$$LR(E) = P(E|H)/P(E|\bar{H}) = 1$$

the evidence is not logically relevant and thus inadmissible according to Rule 402.

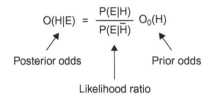

**FIGURE 31.1**
Basic equation for interpreting forensic evidence.

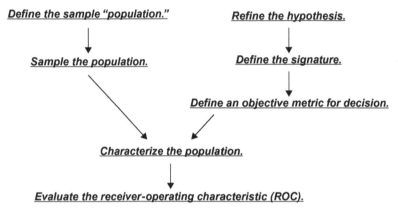

**FIGURE 31.2**
Steps for a generic inferential validation study.

Given this correspondence, the approach to demonstrating the probative value of a given test or assay with respect to a given hypothesis (e.g., those in Table 31.1) is to estimate the LR associated with the measurements or observations E produced by the test when it is applied to samples that conform to H and $\bar{\text{H}}$. When the test is applied to a questioned sample and the result E is obtained, if LR(E) > 1, E supports the hypothesis H; if LR(E) < 1, E supports $\bar{\text{H}}$. Thus, a scientist may testify that his/her measurement of a certain value of some metric for a sample provides a particular level of support to the hypothesis in question rather than stating that his values are "consistent with" the hypothesis (which is simply the statement that $P(E|H) \neq 0$) or, worse, that results make it "likely that the hypothesis is true." In many respects, the most important aspect of this approach is the change it represents in the language used to present forensic science evidence (18).

A general procedure that allows one to estimate LR is given in Figure 31.2. A critical first step is careful formulation of the hypothesis that constitutes the ultimate inference that is to be tested by the method. Referring to Table 31.4, tests can generally be classified into one of three categories.

**Table 31.4** Most Assays Can Be Placed in One of 4 Categories

| Type of Test | Purpose of Test | Examples |
| --- | --- | --- |
| Sample matching | To establish that two agent samples originate from the same batch of material or were made by the same process | Did the Leahy and Daschle samples come from the same batch? |
| Classification: Single hypothesis | To establish that a certain material or that a certain process condition was used in the manufacture of the agent | Was the sample grown on agar plates? |
| Classification: Multiple hypotheses | | Which growth medium was used? |
| Calibration | To establish bounds on some parameter associated with the agent | How old is the sample? |

*Single-hypothesis tests* seek to establish support for a "yes or no" inference. For example, did two samples of agent originate from the same batch of material or was the agent grown on agar plates? One can separate single hypothesis tests into two distinct categories: sample matching and classification. *Multiple-hypothesis tests* seek to establish support for a "one of several choices" inference. For example, what growth medium was used to culture the agent? *Calibrations* seek to establish bounds on some parameter associated with a material being tested. For example, was a biological agent produced within a certain time interval in the past?

A well-formed hypothesis is one that can be realized objectively in a set of reference samples that can be subjected to the test. For example, the hypothesis that "the two samples match" would not be well formed because declaring a match is inherently subjective, that is, a matter of definition. One can always find differences between two samples if one looks hard enough, or similarity by increasing the tolerable differences. However, the hypothesis that "the two samples were drawn from a common batch of material" would be testable because it is possible to objectively produce test samples that are drawn from the same or different batches.

Once a testable hypothesis has been determined, it is necessary to define the signature, that is, the set of molecular, chemical, or physical characteristics that provide the basis for decision (H or $\bar{H}$). In practice, this is often accomplished through an empirical, exploratory study that identifies observable (preferably quantifiable) differences between H and $\bar{H}$ samples. [It is assumed that the measurement process for characterizing the signatures has undergone prior analytical validation and has been codified as a standard operating procedure (SOP) before the inferential validation study is initiated.] Based on the

**Table 31.5** Examples of Signatures and Metrics for Some Notional Anthrax Powder Assays Based on Published Work

| Assay/Test | Signature | Metric |
|---|---|---|
| Assay for presence of residual agar (19) | Mass spectral peaks at relevant m/z values | Ion counts at each m/z value |
| Assay for presence of added silica (20) | X-Ray emission (EDX) spectrum for Si and 0 | Peak areas |
| Assay for presence of residual heme (21) | MALDI mass spectral peaks | Sum of peak heights |
| Sample matching using isotopes (22) | Stable isotope ratios for C,N,O, and H | Euclidian distance between isotopic δ values for two samples |

signature, one then defines an objective metric for decision, that is, a scalar quantity defined in terms of the signature that is used to decide H or $\bar{H}$. The objectivity of the metric is not strictly necessary, but if subjective criteria for decisions are used, then the validation procedure strictly applies only to the operator making the subjective decision and not the method in general. Table 31.5 provides some examples of signatures and possible metrics for various anthrax powder assays discussed in the literature.

In addition to careful hypothesis formulation, careful consideration of the population over which the test applies is essential. The sample set used to perform inferential validation should be representative of the population of samples for which the inference is intended, meaning that it is not a biased sampling of members of that population (23). This follows from the general principle that inferences about the questioned sample based on the properties of a set of reference samples are only valid if all samples were drawn from the same population. Therefore, understanding the relevant population and choosing a sampling strategy are key questions that arise in executing the processes outlined in Figure 31.2.

Two important general observations can be made about the concept of a population from which reference samples are drawn. First, the relevant population may be real or virtual. For analysis of materials such as fibers or drugs, samples can be drawn from a real population (i.e., materials that already exist) that is generated by commercial manufacturing activities. In contrast, because biological agents are clearly not manufactured continuously in quantity, the population of interest is actually determined by the set of possible manufacturing processes that could be used to make them. Sampling from this virtual population necessarily involves simulating the diversity in manufacturing methods by using representative recipes and laboratories to make reference samples. However, suppose we wish to validate an antibody assay that is intended to provide evidence that a person received vaccination for anthrax. Clearly the population is real: humans who have and have not received the anthrax vaccine.

Second, whether the population is virtual or real, it is ultimately defined by the types of variations one could expect among real samples. For example, in the case of chemical and physical analysis:

- The exact method of growth and production of an agent
- The exact source of materials used in the production process
- The temperature and humidity conditions under which an agent might have been stored prior to dissemination

   or, in the case of the vaccination assay,

- The immune system condition, health, and treatment history of the suspect that a blood sample was drawn from

- The type and formulation of the vaccine that might have been administered.

For a method to be applicable to a questioned sample for which factors like these are not known, the set of samples used for validation must reflect an unbiased selection from a population in which those factors are allowed to vary over their naturally occurring ranges. Thus, as a prelude to any validation exercise, it is necessary to consider the possible factors that could affect the relationship between the measured value of the metric and the hypothesis in question but cannot be controlled and would not be known about a questioned sample.

Once the population is defined, the next critical element of an inferential validation study is to develop a *sampling frame* that represents the population adequately. [A frame is basically a list or tabular representation of actual members of the population that could be sampled (23).] Obviously, the sampling frame should include samples that conform to the hypothesis H and its complement $\bar{H}$, which can be thought of as two subpopulations within the larger population of possible samples. Individual samples are drawn randomly from this list and characterized according to the SOP for the analytic method under study. The metric is computed for each sample, and the end result of the characterization process is two sets of metric values, one from H samples, and one from $\bar{H}$ samples, with their associated probability distributions.

Figure 31.3 is a notional representation of distributions of metric values observed for the H and $\bar{H}$ subpopulations, displayed as histograms. A standard way to express the performance of an assay over a population of samples is the ROC curve, which can be constructed in a straightforward way from metric value distributions (24–26). Figure 31.4 is the ROC curve representation of data in Figure 31.3. Once the population has been characterized this

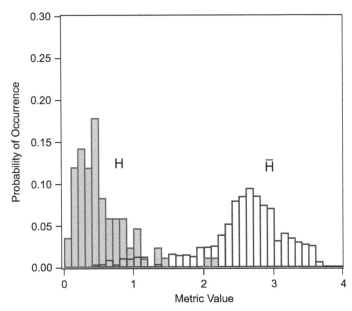

**FIGURE 31.3**

A notional example of a histogram of metric values resulting from characterizing H and $\bar{H}$ subpopulations.

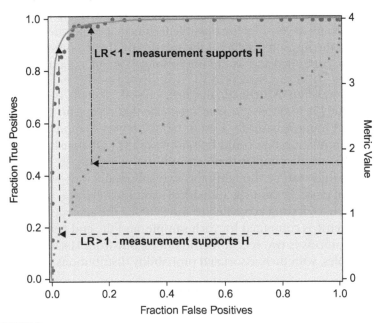

**FIGURE 31.4**

The ROC curve corresponding to data in Figure 31.3. Heavy dots, empirical ROC curve; solid curve, fitted ROC curve; light dots, metric values. Shaded regions demarcate zones of positive and negative support for hypothesis H.

way, the slope of the ROC curve can be used to estimate the likelihood ratio using the process illustrated by the dashed arrows in Figure 31.4. When a new sample is encountered, it is characterized using the same SOP and the metric value is calculated from the measurement(s). That metric value corresponds to a location on the reference ROC curve. If it lies on the rising part of the curve, the slope (LR) is greater than 1, and the observed metric provides support to hypothesis H. In this notional example, metric values smaller than 1 favor H, while values larger than 1 favor $\bar{H}$ since LR < 1.

The degree of separation between distributions of metric values for the two subpopulations is reflected in the steepness of the slope in the ROC curve. If the two subpopulations do not overlap at all, the ROC curve is perfect, with an infinitely steep slope for values of the metric smaller than the highest value found in the H population. If the two subpopulations are fully overlapping, the resulting ROC curve would have a slope of 1 and the test would have no inferential power regardless of the metric value.

There are several advantages of adopting the ROC/LR framework for inferential validation studies and expert testimony on interpretation. It uses an accepted nonparametric method for interpreting evidence that passes muster with modern evidence scholarship (18). It avoids the implicit or explicit assumption of prior odds, which may pose problems in some courts (27). Arguments about the interpretation of assay results based on the ROC/LR framework are likely to center on population, frame, and sampling issues, just as they did for human DNA forensics during its early phases (28). The issues of population definition and sampling bias are also familiar in a number of other contexts where critical decision making is dependent on test results, including clinical testing and medical diagnosis (25,26). Thus, the ROC/LR approach is a generally accepted methodology for scientific inference.

Challenges to population definition generally speak to the weight of the evidence, not admissibility, as long as the bias that might be introduced is not overwhelming, or deliberate. However, the nature of the conceptual source population and whether the samples used to construct ROC curves are truly representative could clearly be a potential point of contention. It may happen that a study that uses one explicit frame for sampling is called into question when other frames may be reasonably suggested. In this context, inferential validation can be thought of as a multiphase process, as illustrated schematically in Figure 31.5.

At an early stage of validation, or under exigent circumstances, only opportunistic or very limited sample sets may be available for testing. Results of such preliminary validation studies may only be useful for generating investigative leads (11). Test performance is subsequently evaluated on a set of samples drawn from a more carefully constructed, putatively representative sampling

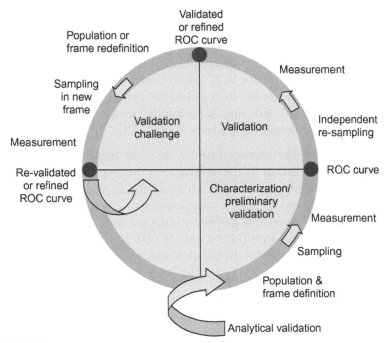

**FIGURE 31.5**
The ROC/LR approach defines a cycle for continuous improvement.

frame and subsequently validated on a completely independent set of samples drawn from an independent frame. Standard statistical methods have been developed for testing whether two independent ROC curves or their underlying distributions are drawn from the same underlying population (29). Results of two or more studies can be combined to make a composite ROC curve that is ostensibly based on a more representative overall population sample. Several cycles of evaluation and validation may occur as our understanding of the structure of the underlying population and the choice of representative frame evolve. Eventually, reasonable challenges to the population or frame definition must decline, and the ability of the test procedure to provide reliable estimates of the likelihood ratio will become accepted.

The description of the ROC/LR method provided earlier primarily considered sample matching or classification by single-hypothesis tests. The same basic framework also applies to multiple-hypothesis tests, although several precautions must be considered. First, the hypotheses encompassed by the test must constitute a complete and nonoverlapping set. That is, every possible sample that could be encountered must conform to one member of the set of hypotheses, and only one (30). Second, the inferential power associated with a multiple-hypothesis test must be reduced to account for the increased probability of assigning an unknown sample to any particular hypothesis purely by chance (31).

Calibration shares certain technical features with hypothesis testing, although it is a distinct activity. In calibrations we collect a set of data that will allow us to determine the likelihood that a certain parameter of a questioned sample lies within a certain range based on a measurement (or measurements) of some other property. The use of calibration curves in analytical validation is well known (7,8). In the context of inferential validation, calibration curves (more accurately scatter plots) are constructed by unbiased sampling over the population of sample types, just as ROC curves or other likelihood estimators are for hypothesis tests. Quantitative evidence extracted from a calibration curve can also be expressed as a likelihood ratio (9). Like hypothesis testing assays, inferential calibrations are validated by independent resampling of the sample population.

## APPLICATION TO CHEMICAL AND PHYSICAL ANALYSIS OF BIOLOGICAL AGENTS

Tables 31.1 and 31.5 and the discussion in the previous section have provided some examples of chemical and physical analysis methods that might be used to infer relationships between two agent samples or whether certain materials or process steps were involved in their manufacture. This section briefly describes some considerations about framing and sampling the population of manufacturing methods for biological agents for inferential validation studies of chemical and physical methods. Much of this discussion is based on previous experience with validating sample matching tests based on elemental analysis of agents (32,33).

The composition and morphology of a bioagent such as *Bacillus anthracis* are end results of the end-to-end process used to produce it. Certain steps influence the overall composition of the agent through the addition or removal of certain substances, and certain steps influence the physical form of the material. The relevant population for evaluating and validating chemical and physical analysis methods is therefore the set of materials that could be generated by any growth and preparation method that may be used to generate a bioagent, using starting materials from any potential sources. Thus, the population of biological agents is an imaginary construct, and the problem is how to generate a set of real samples that adequately provides a statistically representative sample of this imaginary space of possibilities. Moreover, this population must be sampled in an unbiased way, capturing all sources of possible variation: batch-to-batch variation in the same laboratory, laboratory–laboratory variation in executing the same nominal process, and vendor-to-vendor variation in starting material properties.

It is clear from this last requirement that a proper reference sample set for generating ROC curves would involve multiple laboratories making multiple batches of an agent using multiple processes. How the laboratories and processes are chosen is an important aspect of experimental design, as this choice must be representative of the kinds of laboratories and processes from which case samples are likely to originate. The key is to establish an objective frame that represents the population, that is, a specific list of all the members of the population, and then use a random selection process to perform the sampling.

One very general frame is illustrated in Figure 31.6, where each end-to-end process is broken down into unit process steps such as growth, separation of the microbe from the growth medium, washing, drying, milling, and combining with additives. The sampling frame is effectively the list of all combinations of unit processes that plausibly result in an end product. For preparation of a toxin such as ricin, a similar matrix can be constructed with columns defined by the unit process steps appropriate to the particular toxin. For each unit process, there are a number of options, including the "null" option in which that particular unit process is not carried out. Sampling from this frame would involve randomly choosing a variant for each unit process step to create an end-to-end process and then choosing a laboratory at random to execute it. It should be noted that different laboratories might implement a particular unit process in a slightly different way or use materials from different sources, and this variance can be captured by executing the process at multiple laboratories.

An alternative to the unit process frame is based on the observation that methods for making bacterial preparations are usually communicated as end-to-end recipes. Thus, a valid frame would be a list of all known end-to-end processes that have been used in the past. This is clearly a subset of the possible processes generated by the unit process frame, but arguably captures the most probable processes. Note that both unit process and end-to-end process frames explicitly connect the validation process with intelligence about terrorist

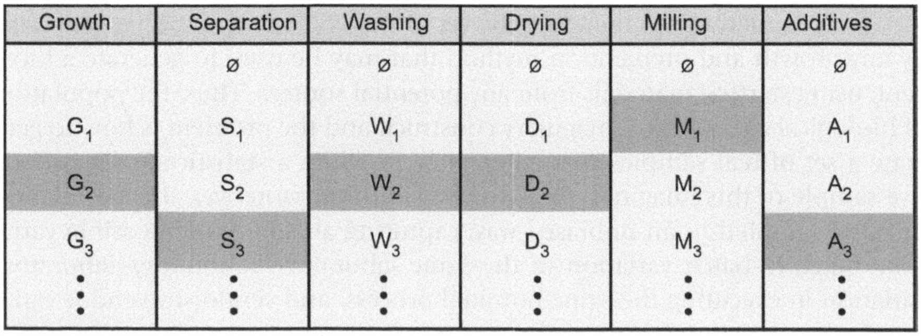

| Growth | Separation | Washing | Drying | Milling | Additives |
|---|---|---|---|---|---|
| ∅ | ∅ | ∅ | ∅ | ∅ | ∅ |
| $G_1$ | $S_1$ | $W_1$ | $D_1$ | $M_1$ | $A_1$ |
| $G_2$ | $S_2$ | $W_2$ | $D_2$ | $M_2$ | $A_2$ |
| $G_3$ | $S_3$ | $W_3$ | $D_3$ | $M_3$ | $A_3$ |
| ⋮ | ⋮ | ⋮ | ⋮ | ⋮ | ⋮ |

**FIGURE 31.6**
Unit process decomposition of biological agent production. An end-to-end process draws a process variant from each column.

interests and state program practice. Biological agent manufacturing information that a criminal or terrorist might use can come from many sources. This includes material derived from open sources such as recipes provided by underground cookbooks and Internet sites, relevant knowledge from the open scientific literature, and inadvertent leaks of sensitive (but often inaccurate) information published in the news media. In some instances, intelligence collection efforts may uncover information about the technical knowledge possessed by particular terrorist groups or foreign BW programs. Both of the frames discussed here require periodic updating and always leave open the question of whether there may be important but unknown subpopulations that have not been sampled.

Assuming that a set of processes and executing laboratories have been chosen randomly from a suitable frame, partial factorial sampling designs can be used to reduce the number of samples to a reasonable value (to control costs). An example of a design involving three processes and three laboratories is shown in Figure 31.7. The symmetric design helps ensure that the reduction in sample number does not introduce bias. Partitioning of the total number of samples per laboratory among the processes executed by each laboratory represents a degree of freedom that can be optimized for certain tests.

Such designs have been executed for validation exercises involving sample matching and other assays, for the population of bench-top scale processes for producing dry spore agent preparations [using nonpathogenic *B. anthracis* surrogates (33)]. Different frames have been constructed and sampled in order to examine the sensitivity of the resulting ROC curves. Preliminary findings indicate that the sampling frames discussed earlier provide a reasonable basis for defining the population, and the method for constructing the sample sets is defensibly unbiased. The resulting sample set provides a useful library for other studies.

|  | Lab 1 | Lab 2 | Lab 3 | # Batches per process |
|---|---|---|---|---|
| Process 1 | Batch 1 Batch 2 Batch 3 | Batch 5 Batch 6 | Batch 4 | 6 |
| Process 2 | Batch 4 | Batch 1 Batch 2 Batch 3 | Batch 4 Batch 6 | 6 |
| Process 3 | Batch 5 Batch 6 | Batch 4 | Batch 1 Batch 2 Batch 3 | 6 |
| # Batches per lab | 6 | 6 | 6 | Total # of batches = 18 |

**FIGURE 31.7**
A 3 × 3 partial factorial design for sample production.

## CONCLUSION

The ROC/LR method represents a transparent and straightforward approach to inferential validation that uses mainstream statistical concepts and leads naturally to an interpretation of microbial forensic data that does not overstate its probative value. This approach also makes it easy to compare two methods designed for the same purpose or to combine results of two independent analyses using orthogonal methods. Although the effort to apply it systematically has only begun recently, it can be applied to a large number of microbial forensic assays. Wider adoption of this methodology will help assure that the interpretation of microbial forensic evidence will meet modern scientific standards.

## DISCLAIMER

This work performed under the auspices of the U.S. Department of Energy by Lawrence Livermore National Laboratory under Contract DE-AC52-07NA27344. This document was prepared as an account of work sponsored by an agency of the U.S. government. Neither the U.S. government nor Lawrence Livermore National Security, LLC, nor any of their employees makes any warranty, expressed or implied, or assumes any legal liability or responsibility for the accuracy, completeness, or usefulness of any information, apparatus, product, or process disclosed or represents that its use would not infringe privately owned rights. Reference herein to any specific commercial product, process, or service by trade name, trademark, manufacturer, or otherwise does not necessarily constitute or imply its endorsement, recommendation, or favoring by the U.S. government or Lawrence Livermore National Security, LLC. The views and opinions of authors expressed herein do not necessarily state or reflect those of the U.S. government or Lawrence Livermore National Security, LLC, and shall not be used for advertising or product endorsement purposes.

## REFERENCES

[1] National Research Council, Committee on Scientific Assessment of Bullet Lead Elemental Composition Comparison, Forensic Analysis: Weighing Bullet Lead Evidence, National Academies Press, Washington, DC, 2004.

[2] National Research Council, Committee on Identifying the Needs of the Forensic Science Community: Strengthening Forensic Science in the United States: A Path Forward, National Academies Press, Washington, DC, 2009.

[3] Scientific Working Group on Microbial Genetics and Forensics (SWGMGF), Quality Assurance Guidelines for Laboratories Performing Microbial Forensic Work, June 20, 2003, *Science*. Supporting Online Material, doi: 10.1126/science.1090270.

[4] E.J. Imwinkelried, The Methods of Attacking Scientific Evidence, fourth edn., LexisNexis, 2004.

[5] Daubert v. Merrell Dow Pharmaceuticals, 509 U.S. 579 (1993).

[6]  Federal Rules of Evidence, December 31, 2004.

[7]  J. Mandel, The Statistical Analysis of Experimental Data, Dover Publications, 1984.

[8]  A. Fajgelj, A. Ambrus (Eds.), Principles and Practices of Method Validation, Royal Society of Chemistry, 2000.

[9]  S. Velsko, Validation Strategies for Microbial Forensic Analysis of Biological Agents: Beyond Sample Matching. Lawrence Livermore National Laboratory Technical Report UCRL-TR-229944, 2007.

[10]  B. Budowle, M.D. Johnson, C.M. Fraser, T.J. Leighton, R.S. Murch, R. Chakraborty, Genetic analysis and attribution of microbial forensic evidence, Microbiol. Mol. Biol. Rev. 70 (2) (2006) 233–254.

[11]  S.E. Schutzer, P. Keim, K. Czerwinski, B. Budowle, Use of forensic methods under exigent circumstances without full validation, Sci. Transl. Med. 1 (8) (2009) 1–3.

[12]  M.J. Saks, J.J. Koehler, The coming paradigm shift in forensic identification science, Science 309 (2005) 892–895.

[13]  National Research Council, Committee on DNA Forensic Science, The Evaluation of Forensic DNA Evidence, National Academy Press, Washington, DC, 1996.

[14]  National Research Council, Committee to Review the Scientific Evidence on the Polygraph, The Polygraph and Lie Detection, National Academy Press, Washington, DC, 2003.

[15]  M.M. Houck, Statistics and trace evidence: The tyranny of numbers, Forensic Sci. Commun. 1 (3) (1999).

[16]  R. Lempert, Modeling Relevence. Michigan Law Review Vol. 75, 1977, pp. 1021–1057.

[17]  D.H. Kaye, J.J. Koehler, The misquantification of probative value, Law Hum. Behav. 27 (2003) 645–659.

[18]  B. Robertson, G.A. Vignaux, Interpreting Evidence: Evaluating Forensic Science in the Courtroom, Wiley and Sons, Ltd., New York, 1995.

[19]  D.S. Wunschel, H.A. Colburn, A. Fox, K.F. Fox, W.M. Harley, J.H. Wahl, et al., Detection of agar, by analysis of sugar markers, associated with *Bacillus anthracis* spores, after culture, J. Microbiol. Methods 74 (2008) 57–63.

[20]  L.N. Brewer, J.A. Ohlhausen, P.G. Kotula, J.R. Michael, Forensic analysis of bioagents by X-ray and TOF-SIMS hyperspectral imaging, Forensic Sci. Int. 179 (2008) 98–106.

[21]  J.R. Whiteaker, C.C. Fenselau, D. Fetterolf, D. Steele, D. Wilson, Quantitative determination of heme for forensic characterization of bacillus spores using matrix-assisted laser desorption/ionization time-of-flight mass spectrometry, Anal. Chem. 76 (2004) 2836–2841.

[22]  H.W. Kreuzer-Martin, K.H. Jarman, Stable isotope ratios and forensic analysis of microorganisms, Appl. Environ. Microbiol. 73 (2007) 3896–3908.

[23]  W.E. Deming, Some Theory of Sampling, Dover Publications, Inc., New York, 1966.

[24]  W.J. Krzanowski, D.J. Hand, ROC Curves for Continuous Data, CRC Press, Boca Raton, FL, 2009.

[25]  M.S. Pepe, The Statistical Evaluation of Medical Tests for Classification and Prediction, Oxford Statistical Science Series 28; Oxford University Press, New York, 2003.

[26]  NCCLS. Assessment of the Clinical Accuracy of Laboratory Tests Using Receiver Operating Characteristics (ROC) Plots: Approved Guideline. NCCLS Document GP10-A, 1995.

[27]  P.I. Good, Applying Statistics in the Courtroom, Chapman and Hall/CRC, CRC Press LLC, Boca Raton, FL, 2001.

[28]  National Research Council, Committee on DNA Forensic Science, The Evaluation of Forensic DNA Evidence, National Academy Press, Washington, DC, 1996.

[29] A.D.M. Kester, F. Buntinx, Meta-analysis of ROC curves, Med. Decision Making 20 (2000) 430–439.

[30] T.Y. Young, T.W. Calvert, Classification, Estimation, and Pattern Recognition, American Elsevier Publishing Company, Inc., New York, 1974.

[31] J.P. Shafer, Multiple hypothesis testing, Annu. Rev. Psych. 46 (1995) 561–584.

[32] S. Velsko, Bioagent Sample Matching Using Elemental Composition Data: An Approach to Validation. Lawrence Livermore National Laboratory Report UCRL-TR-220803, 2006.

[33] S.P. Velsko, P. Weber, C.E. Ramon, R.E. Lindvall, M.L. Davisson, M. Robel, Bioagent Sample Matching Using Elemental Composition Data. Lawrence Livermore National Laboratory Report LLNL-TR-419683, 2009.

# CHAPTER 32

# Microbial Forensic Investigations in the Context of Bacterial Population Genetics

**Paul S. Keim,**[a,d] **Talima Pearson,**[a] **Bruce Budowle,**[b]
**Mark Wilson,**[c] **and David M. Wagner**[a]

[a]*Center for Microbial Genetics and Genomics, Northern Arizona University, Flagstaff, Arizona*
[b]*Institute of Investigative Genetics, Department of Forensic and Investigative Genetics, University of North Texas Health Science Center, Ft. Worth, Texas*
[c]*Forensic Science Program, Western Carolina University, Cullowhee, North Carolina*
[d]*Pathogen Genomics Division, The Translational Genomics Research Institute, Flagstaff, Arizona*

## INTRODUCTION AND BACKGROUND

Genetic analysis created a revolution in the field of forensics, and its application to microbial forensics will be a major part of many investigations involving a biothreat agent. The utility and importance of genetic analysis are not surprising given that genomes contain extensive and varied information content that can be exploited to precisely characterize and identify biological evidentiary material and support other investigative efforts. In human forensic DNA analysis, molecular biology tools have become incredibly powerful due to a great understanding of human biology, the human genome, and, most critically, human population-level genetics. One of the early scientific and, later, legal challenges to DNA fingerprinting was the lack of high-quality human population genetic data on the forensically relevant genetic markers. Over the past two decades these data have been generated and represent an invaluable resource to forensic analyses, as they are a point of reference against which forensic DNA profiles can be considered for weighing the significance of an observation. In contrast to the large-scale effort dedicated to generating human population data on forensically relevant genetic markers, generating, collecting, and analyzing population genetic data accurately still represent a greater challenge for microbial forensics. Each pathogen has a unique biology and population genetic structure and there is no widespread multiple laboratory effort contributing to such studies.

Genetic and genomic analyses of forensic evidence can only be interpreted properly in the context of a specific pathogen's population genetic structure,

545

Microbial Forensics. DOI: 10.1016/B978-0-12-382006-8.00032-3

diversity, and reproductive mechanisms. Genetic and genomic analyses should lead to quantitative similarity data where evidentiary materials may match, nearly match, or exclude, which represents just three points along a nondiscrete continuum of possibilities. Additionally, evidence may be inconclusive such that there are insufficient data to render a conclusion about the relationship (or degree thereof) of the evidence with a reference sample(s). The significance of a particular genetic/genomic similarity measurement cannot be assessed without an understanding of a pathogen's population genetic structure. Clearly a historical and epidemiological context can add resolution but often these data are limited (particularly the manipulations that might have induced genetic variation of the biothreat agent, which is often unknown to the microbial forensic scientist). As mode of replication (e.g., clonal) has such a large bearing on population structure, the genetic markers analyzed, statistical significance, and confidence of the estimation will be highly dependent upon each pathogen's mode of inheritance. The analysis methods can be defined broadly as either "phylogenetic" or "statistical" and should be selected based on modeling/analysis of empirical data collected from studying each particular pathogen. For example, some pathogens, such as HIV, have high rates of mutation and recombination, making an exact genotypic match unlikely. When evidentiary samples do not match exactly, the degree of relatedness can still be ascertained with empirically or theoretically derived statistical levels of confidence. Conversely, in populations with little diversity, exact genotypic matches may be widespread, decreasing the ability to attribute the source of the sample. The likelihood of sample attribution in bacterial pathogens is linked directly to fundamental biological characteristics of the species and/or source populations.

With regard to match probabilities, replication (inheritance) mode is the critical intrinsic biological characteristic of a source population. Some bacteria replicate exclusively in a clonal fashion with no recombination between different lineages. As a result, differences between strains and isolates are driven by mutational (and selection) processes with no mixing of preexisting genetic variation. Many species of bacteria, however, are notorious for horizontal transfer of genetic material. This swapping of genetic material creates new multilocus genotypes by inserting new genetic variation within a genome rapidly and effectively. The contrast in these two reproductive modalities is similar to differences observed in human genetics where mitochondrial and much of the Y chromosome is propagated clonally, but the remaining nuclear genome is not. The autosomal, the X (mostly in females), and part of the Y chromosomes undergo frequent meiotic recombination that generates novel genotypes with every new human generation. The rate of this recombination is somewhat predictable and is the basis for mendelian genetics and genetic mapping. In bacterial populations, recombination rules and frequency are less predictable. As a result, the effect of recombination upon population structure will vary greatly, even within a single species. In contrast, the evolutionary rules associated with clonality are fairly simple and well characterized, providing for more robust population genetic analysis. Hence, careful characterization of the relevant bacterial populations is desirable for microbial forensics, should lead to an understanding of their replication mode, and will ultimately dictate the appropriate analytical approach for calculating match or similarity statistics and the degree of confidence/uncertainty in the result.

Population genetic analysis of bacterial pathogens necessarily involves discovery of markers for differentiating among individual isolates. Traditionally, microbiologists used phenotypes (e.g., fermentation of sugars) or serological differentiation to discriminate among bacterial strains. Fortunately, the capacity to differentiate among bacteria has been continually refined for more than a century. Phenotypic differences such as colony morphology, microscopic characteristics, and metabolic capabilities were usually sufficient for identifying new genera and species. Otherwise identical isolates could be immunologically different, as different strains of bacteria had distinguishing antigens that could be identified with antibodies. The precise identification of bacterial pathogens was also essential for understanding disease outbreaks, identification of case clusters, and correlating cases with a common source—the basis of modern molecular epidemiology. However, phenotypic methods did not have the discrimination power to individualize isolates at the level needed for forensic attribution. Subtyping within a bacterial species for public health purposes segues into precise genotyping of bacterial genera, species, subspecies, strains, and, finally, individual isolates for forensic attribution.

The population genetics of bacteria became more robust and widespread with the advent of DNA-based methods. This revolution included many different approaches that have seemed to change continually over the past two decades, almost to the point of the absurd (1). These rapidly changing innovations all used molecular technologies to detect underlying genomic sequence differences. The plethora of approaches has led to increased data generation but with a concomitant greater confusion as to their relative usefulness, strengths, and weaknesses. The lack of data uniformity across technologies resulted in many private databases that provided limited utility among researchers. Consequently, independent validation of these population genetic data was lacking. There were, however, some successes, including 16S ribosomal RNA gene sequences, multilocus sequence typing (MLST) (2), and pulse-field gel electrophoresis (PFGE) separation of restriction fragment length polymorphisms (RFLPs). MLST and PFGE provided sufficient resolution to perform population-level genetic analysis, but 16S gene comparisons did not. Large publicly available databases were established, bringing bacterial population genetics into the scientific mainstream and the public health arena. The resolution and quality of these methods and databases, however, are less than what is needed by the forensic community where higher evidentiary standards and deeper resolution are essential for attribution.

Inspired by the use of short tandem repeat (STR) loci for human forensics and the observation that bacterial genomes also have hypervariable loci (3), high-resolution subtyping systems were developed for strain identification of bacterial pathogens (4–6). In the bacterial research community, these hypervariable loci were also named variable number tandem repeat (VNTR) loci. They proved to be multiallelic and to have relatively high mutational rates mediated by an insertion/deletion mechanism [for an experimental demonstration of this phenomenon, see Volger and colleagues (7)]. Similar to STRs, VNTR alleles were assayed easily by polymerase chain reaction and electrophoretic separation based on variation in the number of repeats contained within an amplicon. Multiple-locus VNTR analysis (MLVA) increased the potential to detect allelic differences while decreasing the probability for identical MLVA genotypes due to convergent evolution at a single locus. MLVA systems have been developed for many common pathogens (8), making this tool available for forensic analysis of many bacteria.

New genotyping technologies continue to arise and will present their own set of advantages and disadvantages for genotyping and determining the population structure of pathogens. For example, some new whole genome sequencing technologies provide a means of assaying some types of loci (e.g., SNPs), but are currently inadequate for surveying other regions of a genome (VNTRs). These technologies are also still relatively expensive and the vast amount of data is difficult to handle, making them less practical at this time for routine genotyping of a large number of isolates. In addition, whole genome databases are quite small at this time, making attribution statistics difficult to calculate with a high degree of confidence. While different technologies come and go, an understanding of the biogeography, phylogenetics, population structure, and patterns of genomic change will remain essential for forensic microbiology. With such knowledge, the natural distribution of a pathogen might be distinguished from a nefarious act, the source population can be estimated more accurately, and quantitative values can be placed on the level of relatedness between samples.

## A MICROBIAL FORENSIC PARADIGM

Based on the anthrax-letter attack investigation and our understanding of pathogen population genetics, we have devised a scheme for a microbial forensic response to a biological attack (Figure 32.1). This multistep flowchart is independent of specific genotyping methods but starts with the assumptions that attack material (e.g., spores) can be obtained by investigators and genetic

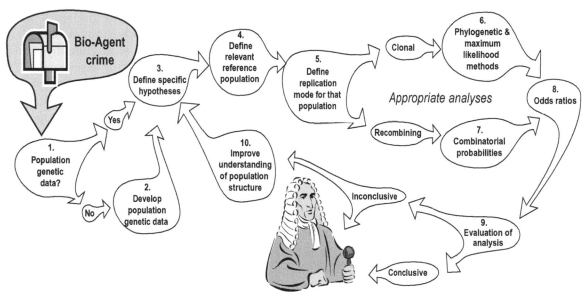

**FIGURE 32.1**

Population genetics-based confidence estimation for microbial forensics. This schema presents a logical sequence of genetic analysis activities that will follow a biological agent-based criminal event. Inevitably there will be a comparison of genotypic data to determine if crime scene samples match other evidence in the case (e.g., from a suspect's home or to a database). The genotypes may or may not match or even have different degrees of similarity. It is critical that these comparisons be quantitative and supported with confidence estimates based on statistical or probabilistic analysis.

data can be generated. This approach is applicable to data from a wide variety of genotyping methods, from the high levels of genetic resolution obtained from whole genome sequencing to low-resolution PFGE or MLST analysis of evidentiary material. Importantly, this model allows for multiple genotyping methods and technologies to be utilized to increase resolution and further refine the relevant reference population. The "attack material" genotype could then be compared to other evidentiary material (e.g., from a suspect's home or laboratory), which would result in a match, exclusion, or, perhaps, a similar genotype that is less than an exact match (or still be uninformative). The significance associated with each of these possible results requires population genetic data, regardless of the methodology or genetic resolution.

## Step 1—Availability of Population Genetic Data?

Forensic investigators will doubtlessly employ the best and most applicable genotyping technology and, as an investigation progresses, these methodologies may be improved upon. In the case of the anthrax-letter attacks, the initial genotyping system was the eight-marker MLVA system that had been developed in a university research laboratory and published in peer-reviewed scientific journals (4,5). Publication of this study made the work widely known and facilitated technology transfer to U.S. government labs at the Centers for Disease Control and Prevention, which independently verified genotyping results in a relatively rapid manner during the investigation (9). The MLVA method was capable of resolving many samples and, not surprisingly, there were still many independent isolates that were identical. The second and higher resolution approach was to generate a whole genome sequence using Sanger shotgun sequencing. This approach has the potential to differentiate any two isolates, even if they were very closely related and contingent that they harbored at least one genetic difference. But whole genome sequencing was expensive and only one other partial genome sequence was available in early 2002, making the database of whole genome sequences for *Bacillus anthracis* very small indeed. In contrast, the MLVA8 database was relatively expansive, as this system had been applied previously to a large number of isolates ($\approx$500). In this example, MLV8 genotyping data could be used to address a range of hypotheses, whereas addressing hypotheses generated from whole genome sequences was very limited.

## Step 2—Population Genetic Database Development

Population genetic database development involves generating genotypic data from a number of isolates. A larger number of isolates allows for more questions to be addressed and for greater confidence in interpreting any results. However, the total number of genotypes in a database can be very misleading due to selection biases that are likely. Thus, there will be redundancy in isolates that will not necessarily reflect the diversity of the microorganism as it applies to the case or a geographic location. The relevance of a database can be exacerbated further if there is a lack of high-quality meta-data associated with entries. While a database may be perceived to be of high quality, its utility or application may not be known until a particular hypothesis is being addressed. For example, a database with a large proportion of isolates from North America may be adequate for addressing hypotheses regarding likely origins in North America but may be inadequate for addressing similar hypotheses on a global scale. Indeed, the anthrax-letters attack was unique in that the diversity of the Ames strain was best determined by the samples collected from laboratories (because the Ames strain is rare in nature but is common in some laboratories). Most future cases will likely require different sampling strategies to develop relevant population data sets. In general, compiling data from a geographically and genotypically diverse set of isolates a priori is an important step for developing attribution capabilities for a bioterrorist event.

Following an event, the generation of additional population genetic data will be greatly dependent on the questions posed and the availability of strain collections. Archival collections from public health, academic, and/or private laboratories, as well as direct sampling of pathogen populations, may be needed. Subsequent genotyping efforts should be carried out with high-quality standards. Genotyping errors in databases could lead to inflated diversity estimates and misinterpretation of forensic data. In an ideal situation, the population genetic database should be constructed to the same quality standards as the actual evidentiary analysis. (However, this is unlikely with the current infrastructure and approaches to microbial forensics.) If no population genetic studies are available, it is essential that a study be performed before any conclusions are made concerning the evidence. Although a large and comprehensive genetic database is the theoretical goal, this may not be possible, especially prior to an unanticipated biocrime event.

## Step 3—Definition of Specific Hypotheses

Depending on the specifics of a particular case, hypotheses can be formulated concerning the evidence. With even a rudimentary population genetic database in hand, it is possible to address specific genetically based attribution hypotheses and alternatives for investigation leads. For example, this could include: (a) If the evidentiary genotype matches a source, then its match probability versus that of alternate sources can be calculated and compared. The comparison could be expressed as an odds ratio of these individual probability estimates to assess the likelihood of each source. (b) If evidentiary samples have genotypes matching each other but nothing else, the interpretation could lead to a further definition of the source. However, attribution to a specific source would not be possible in this case. (c) Evidentiary samples having a near match to other evidence would have to be assessed relative to nonevidentiary material to determine if the near match is significant. Thus, the probabilities of each of these hypothetical examples will be calculated within the context of an appropriate reference population.

A simple hypothesis used in the anthrax-letter case might have involved the similarity of all *B. anthracis* Ames strain isolates. Given that all the environmental *B. anthracis* Ames strain isolates from the postal system matched the isolates from the victims and the letters genotypically, we could hypothesize that these were all part of the same criminal event and from the same source. A population genetic database would add support that would favor this hypothesis if there were no additional known natural isolates with the same genotype. In contrast, the hypothesis would be less supported if there were numerous (unrelated or unassociated) isolates in the database with the same genotype.

## Step 4—Hypothesis Testing Needs to Be Done in Context of a Relevant Reference Population

It is important to recognize that the relevant reference population likely will not be all available data in a database. Indeed, probabilities of observing evidence for the hypotheses are highly dependent on the reference population, which may include all or perhaps just a portion of total genetic data available. The uniqueness of genotype profiles could be tested globally, regionally, or even locally. Doubtlessly, any hypothesis would necessarily include an initial assessment of all (known or available) genetic diversity within the entire species. However, if the hypotheses involve alternate sources in the same geographic region, such as New Mexico, the relevant population may be New Mexican isolates (10). The global diversity is of secondary importance to this particular set of hypotheses and should not be the basis for confidence estimation. Inclusion of African isolates, for example, might inflate the rarity of a New Mexican genotype even though the alternate hypotheses being tested are based solely in New Mexico. This does not rule out the relevance of non-New Mexico strains, especially vis-à-vis other hypotheses that may have a more global context or provide some inference about the diversity of the species.

In the anthrax-letter investigation, the relevant population varied with the particular hypotheses or questions. For example, additional sampling was carried out near the geographic origin of the Ames strain (11) to develop a relevant population to test for the Ames strain's natural variation. Additional natural isolates were obtained, but could be distinguished from the laboratory strain with whole genome-based SNP analysis. This was not the case for isolates obtained from different laboratories, particularly because all laboratory isolates derive from a 1981 isolate. In this case, the investigation returned to step 2 to further expand upon existing genetic population databases, not by adding more isolates, but by developing new genetic markers that could distinguish subpopulations that were identified from morphological variants within individual cultures [see Keim and colleagues (12)]. Thus, for particular hypotheses, relevant populations differed from the clinical and laboratory isolates of the Ames strain to address the relatedness of evidentiary samples to isolates collected from the natural environment to determine the natural variation of the Ames strain. In the Amerithrax investigation of a laboratory source, the collection of all laboratory samples/isolates did, in fact, result in a highly representative genetic database, with nearly all U.S. Ames cultures represented (or at least the data set was fairly representative).

## Step 5—Define the Replication Mode
Probability estimates will be calculated differently depending in part, on, the mode of inheritance of the genetic markers within the pathogen. Determination of the replication biology of different pathogen species is generally possible from most types of population genetic data by calculating linkage among loci and by sequence and evolutionary biology analyses. For example, phylogenetic modeling is sensitive to recombination, and its effect on character distribution is generally discoverable from genetic data sets.

However, it is important to realize that although some species may exhibit recombination, particular subpopulations within these same species could be clonal. Indeed, while some populations will be completely clonal, other populations will exhibit varying degrees of recombination, making it important to recognize that the level of recombination is a continuous rather than a discrete variable. Therefore, the mode of inheritance needs to be determined for the relevant reference population (step 4) within a species and should not be assumed from other populations but rather determined empirically from data.

## Step 6—Inheritance Mode: Clonal Replication

Clonal species do not exchange DNA across lineages; rather, diversity is driven solely by mutational processes with mutant alleles inherited by the daughter cells, drift, and selection forces. Phylogenetic analyses (e.g., maximum parsimony, maximum likelihood) are highly appropriate for clonal (and nearly clonal) pathogens where even single allelic differences, or allelic matches, and can be powerful under certain circumstances. Probability estimation can be performed using mutation rates and maximum likelihood ratios, which work well for comparing alternate hypotheses. Coleman and colleagues (10) and Vogler and colleagues (13) used VNTR mutational rates to calculate relative probabilities for alternate scenarios. Because interlocus allelic variation in clonal populations is highly correlated due to complete linkage, combinatorial probabilities ("product rule") based on allelic frequency are inappropriate. However, the "counting method" has been used for mitochondrial DNA analysis in humans, which is also inherited clonally. In this case, a population genetic database is developed and rarity of the evidentiary genotype is based on how many times it has been observed in a reference database(s). The strength of this approach is affected greatly by the genetic database and whether it has sufficiently sampled relevant populations.

## Step 7—Inheritance Mode: Recombining or Nonclonal Replication

Recombination involves the transfer of alleles across genetic lineages. This genetic mixing is reminiscent of sexual reproduction in humans, where phylogenetic methods are not always suitable for understanding relationships among individuals. High rates of recombination must be sufficient for identifying loci as independent, for practical purposes, so allele frequencies can be combined using the product rule to calculate a random match probability. Consideration of the degree of recombination is important as the likelihood of allelic linkage decreases with an increase in recombination. The product rule assumes (for practical purposes) complete linkage equilibrium (zero disequilibrium), and serious violations of this assumption make this approach suspect. The counting method can also be used but requires larger genetic databases that sufficiently sample the relevant populations. The power of any conclusions will scale with database size and its sampling of populations directly relevant to particular hypotheses.

## Step 8—Likelihood Ratios

The comparison of estimates for the probability of particular hypotheses can be accomplished through likelihood ratio calculations using relative probabilities of the alternative hypotheses derived in either step 6 for clonal populations or step 7 for recombining populations. Likelihood ratios are simply a measure of the likelihood of one hypothesis over another (10).

## Step 9—Evaluation of Analyses

Microbial forensics data will be interpreted within the context of the crime. The evaluation will be better based with the proper use of statistical and probabilistic analyses, but nevertheless a subject area expert will be essential for communicating the strengths and weaknesses of any result to the investigators, to the legal system, and to decision and policy makers. For example, if the probability of one hypothesis is significantly greater than the probabilities of other hypotheses, these data will be combined with other information so that conclusions can be made. Conversely, the lack of a strong probability difference between hypotheses may be indicative of a lack of genetic resolution between samples and/or a lack of understanding of population structure and provide little support for conclusions.

Results leading to a conclusive interpretation may lead investigators to a particular source(s) (inclusion) or, perhaps, eliminate a source (exclusion) from consideration. Some conclusive interpretations may be very narrow in scope and only eliminate a gross alternative hypothesis, without strongly supporting particular informative hypothesis. Often, exclusion of a source from consideration will be highly supported (for practical purposes—absolute) by population genetic analysis, while an inclusion conclusion will likely not be absolute. In addition, the weight of the microbial forensic evidence may be weak in relationship to other more traditional investigative lead data or other forensic data, even if microbial forensic data provide a conclusive match between a crime and reference source material. The strength or weakness of a particular conclusion is best represented in a quantitative manner but may be presented in court by nonquantitative testimony. Results leading to an inconclusive interpretation may be due to lack of genetic resolution and quality of the evidence; additional analyses with existing methodologies may not further attribution efforts. Results must be presented as is or, perhaps, an improved knowledge of population structure through further sampling of isolates and/or use of more genetic markers must be obtained. Better technologies may be developed to better exploit samples. More statistical power or perhaps additional hypotheses for testing (step 3) might develop that, in turn, could lead to different probabilities for hypotheses.

### Step 10—Improve Understanding of Population Structure

A successful forensic analysis may require multiple iterations of refining hypotheses based upon ever-increasing knowledge and new questions. This will be highly dependent on the level of population genetic analysis available at the time that an event occurs or developed subsequent to an event, as well as, biological characteristics intrinsic to the population and species being studied relevant to the biological agent.

## CALCULATING MATCH PROBABILITIES

Calculating the probability of selecting another individual randomly from a given population and finding the same genotype depends on the mode of inheritance, the allele/haplotype frequencies in a population database(s), and statistical assumptions. In clonal populations, interlocus allelic variation will not be independent but rather hierarchically associated inside the phylogenetic structure of the population. This clearly precludes the use of combinatorial probabilities based on allele frequency at different loci and calls instead for phylogenetic analysis methods or a simple counting method approach. It is important to realize that although some species may exhibit recombination, subpopulations within these taxa could still be clonal and not all genomic regions will be recombining equally. Perhaps in such situations a range of probabilities may be provided based on a priori assumptions of recombination.

For clonal populations, phylogenies can be used to determine relative levels of relatedness. Isolates that share a common ancestor are more closely related to each other than they are to isolates that do not share that common ancestor. The phylogenetic location of the common ancestor is indicative of how closely related its descendents are. Common ancestors display recent divergence as indicated by shared bifurcation points (nodes) that approach the terminal ends of the phylogeny. Thus, isolates that are related most closely to each other relative to all other isolates in the phylogeny will share the most terminal node in a phylogenetic reconstruction of the diversity. Therefore, phylogenetic reconstructions can be valuable tools in an assessment of diversity within a forensic context.

Phylogenies can also be used to estimate relatedness for clonal populations. Once the most closely related isolates are identified, genomic differences can be used to quantify levels of relatedness, with the ultimate goal of estimating the amount of evolutionary time that separates isolates (e.g., number of generations). Accuracy of mutation/substitution rate data will directly affect the accuracy of calculations on the number of intervening generations. If the amount of time needed for a given number of generations is known, then an estimate of the amount of time separating isolates can be made (under the assumption that there have been no external stresses on the microorganism). This level of quantification may be much more difficult for species such as *B. anthracis* that have quiescent stages (spores) of varying lengths where no reproduction occurs. Other factors that can create nonuniform evolutionary rates include ecological differences that alter the generations per year rates, environmental conditions that result in higher or lower mutation rates, and mutator variants defective in DNA repair, resulting in very high mutation rates.

Finally, for clonal populations, phylogenies can be used to establish the context of evidentiary material to a particular reference population. A phylogenetic approach would consider an association as two or more samples that are contained within the same phylogenetic clade and, hence, have a close evolutionary relationship to each other. Given a cladistic perspective on the question of inclusion, one relevant question could be: At what node in the cladogram should one differentiate between those isolates that are excluded as originating from the same source, from those that are included? As an investigation develops, further phylogenetic data will refine the phylogenetic relationships around a particular location within a cladogram, and hence provide a level of detail that is absent without such targeted typing efforts. All genetic data are relevant to certain questions that may be considered in these contexts, but the detailed structure of a phylogenetic tree more fully defines the particular placement of samples.

## REFERENCE DATABASE

Weight assessments must be derived from relevant population data. This may be defined globally as the entire species or related more regionally to the crime under investigation. At the beginning of an investigation, the local population (phylogenetically and spatially) will likely be unknown, necessitating inclusion of a broad range of isolates. Local populations may be difficult to define. For example, the local population could be where the sample was generated, where it was disseminated, or where individuals that are ill reside. Later, as more information is compiled and the reference population becomes more refined, carefully chosen isolates closely related to the evidence will become the focus of further phylogenetic analyses. However in some cases, it may never be possible to refine the reference population and that limitation should be stated.

The size of the reference database will affect the power of the match probability. Genotypic match results will include a high degree of uncertainty if based on a small database, as the denominator in the point estimate will be small. A large and comprehensive genetic database is the theoretical goal, but in most cases will not be possible, especially prior to a biocrime event. It may never be possible to sample all diversity in the world or even from a defined geographical location. An established database with accompanying phylogenetic structure could be instrumental for directing an investigation immediately after a biocrime event.

The extent to which the reference database reflects the natural population will affect the accuracy of the match probability. Pathogenic organisms likely to be found in a biocrime or terrorist attack will generally not have population databases constructed for the purpose of assessing the weight of a forensic association. To both place the microorganism in its evolutionary context and observe subtle changes that could assist in the question of forensic attribution, forensic investigators will generally require as much of the available genetic data as possible on a pathogenic organism. Data must be used carefully, however, because it is likely that the original data creation was for different purposes (e.g., molecular epidemiology).

The quality of data in the reference database will affect the accuracy of the match probability. Although less likely, sequencing or genotyping errors can result in near matches appearing identical. It is more likely that errors will cause identical isolates to appear different. Therefore, poor-quality genetic databases will more likely result in an overestimation of match probabilities. Genetic data quality is not the only factor affecting match probability, as errors in geographical and epidemiological source data may also cause isolates to be incorrectly included or excluded from a particular reference population.

## DISCUSSION

Technologies will change over time, and although different technologies can have a profound effect on the efficiency, quantity, and quality of data collection, the end result of a thorough understanding of population genetics remains paramount. This aspect of pathogen biology is vital for framing how data are analyzed at different points in an investigation. Understanding population structure requires a collection of isolates, the size and origin of which will have a direct bearing on the association and statistics generated from a forensic investigation.

Perhaps the most important aspect of pathogen biology relating to determining association statistics is the mode of inheritance, as it dictates the statistical/bioinformatic methods that should be employed. Genetic markers from clonal populations (or portions of the genome) can be analyzed phylogenetically, whereas genetic markers from recombining populations (or genome regions) might be analyzed combinatorially. All populations of B. anthracis,

*Yersinia pestis*, *Brucella* spp., and *Francisella tularensis* are probably completely clonal, but populations of *Burkholderia pseudomallei*, *Escherichia coli*, and *Clostridium botulinum* can be expected to have different levels of recombination. Although a species may be generally known to recombine, as the population becomes defined more narrowly, the population structure may become clonal and analytical methods will have to reflect that change. Understanding how levels of recombination differ among populations will have a direct bearing on how forensic data are analyzed and compared.

Another key aspect in calculating match statistics is the relative determination of how closely related two samples are. In general, samples that share more alleles can be assumed to be more closely related than samples with fewer shared alleles. The widely accepted principle of parsimony (Occam's razor) invokes the simplest explanation for a data set. In cases where there is little difference between the number of shared alleles, consideration of the likelihood of different types of mutations can add a finer level of resolution. Extensive studies of natural populations and laboratory-generated populations have led to a greater understanding of VNTR mutational rules, rates, and products (7,13,14), allowing more precise comparisons of the levels of relatedness of samples at a finer scale (10).

As match probabilities are based on the frequency of alleles or haplotypes in a population, definition and identification of that population are critical. Large collections of isolates with extensive genotype data are ideal as they can help jump start an investigation by providing immediate direction for further forensic sampling. While much can be done to prepare for a possible bioattack, it is impossible to predict what species/strains will be used or whether it will be synthesized *de novo*. As such, much sampling will have to be performed post hoc, the direction of which will be determined as genotype results and further knowledge of the sample are gained.

Although not necessary for legal admissibility, the strength of a forensic association should be communicated in a quantitative fashion. Importantly, we believe that scientists must clearly formulate the forensic questions and, where appropriate, consider alternate hypotheses when appropriate.

## REFERENCES

[1] M. Achtman, A surfeit of YATMs? J. Clin. Microbiol. 34 (7) (1996) 1870.

[2] M.C. Maiden, J.A. Bygraves, E. Feil, G. Morelli, J.E. Russell, R. Urwin, et al., Multilocus sequence typing: A portable approach to the identification of clones within populations of pathogenic microorganisms, Proc. Natl. Acad. Sci. USA 95 (6) (1998) 3140–3145.

[3] G.L. Andersen, J.M. Simchock, K.H. Wilson, Identification of a region of genetic variability among Bacillus anthracis strains and related species, J. Bacteriol. 178 (2) (1996) 377–384.

[4] P. Keim, A.M. Klevytska, L.B. Price, J.M. Schupp, G. Zinser, K.L. Smith, et al., Molecular diversity in *Bacillus anthracis*, J. Appl. Microbiol. 87 (2) (1999) 215–217.

[5] P. Keim, L.B. Price, A.M. Klevytska, K.L. Smith, J.M. Schupp, R. Okinaka, et al., Multiple-locus variable-number tandem repeat analysis reveals genetic relationships within *Bacillus anthracis*, J. Bacteriol. 182 (10) (2000) 2928–2936.

[6] A.M. Klevytska, L.B. Price, J.M. Schupp, P.L. Worsham, J. Wong, P. Keim, Identification and characterization of variable-number tandem repeats in the *Yersinia pestis* genome, J. Clin. Microbiol. 39 (9) (2001) 3179–3185.

[7] A.J. Vogler, C. Keys, Y. Nemoto, R.E. Colman, Z. Jay, P. Keim, Effect of repeat copy number on variable-number tandem repeat mutations in *Escherichia coli* O157:H7, J. Bacteriol. 188 (12) (2006) 4253–4263.

[8] G. Vergnaud, C. Pourcel, Multiple locus variable number of tandem repeats analysis, Methods Mol. Biol. 551 (2009) 141–158.

[9] A.R. Hoffmaster, C.C. Fitzgerald, E. Ribot, L.W. Mayer, T. Popovic, Molecular subtyping of *Bacillus anthracis* and the 2001 bioterrorism-associated anthrax outbreak, United States, Emerg. Infect. Dis. 8 (10) (2002) 1111–1116.

[10] R.E. Colman, A.J. Vogler, J.L. Lowell, K.L. Gage, C. Morway, P.J. Reynolds, et al., Fine-scale identification of the most likely source of a human plague infection, Emerg. Infect. Dis. 15 (10) (2009) 1623–1625.

[11] L.J. Kenefic, T. Pearson, R.T. Okinaka, W.K. Chung, T. Max, C.P. Trim, et al., Texas isolates closely related to *Bacillus anthracis* Ames, Emerg. Infect. Dis. 14 (9) (2008) 1494–1496.

[12] P. Keim, B. Budowle, J. Ravel, Microbial forensic investigation of the anthrax-letter attacks. Microbial Forensics, Elsevier, 2011, pp. 17–27.

[13] A.J. Vogler, C.E. Keys, C. Allender, I. Bailey, J. Girard, T. Pearson, et al., Mutations, mutation rates, and evolution at the hypervariable VNTR loci of *Yersinia pestis*, Mutat. Res. 616 (1−2) (2007) 145–158.

[14] J.M. Girard, D.M. Wagner, A.J. Vogler, C. Keys, C.J. Allender, L.C. Drickamer, et al., Differential plague-transmission dynamics determine *Yersinia pestis* population genetic structure on local, regional, and global scales, Proc. Natl. Acad. Sci. USA 101 (22) (2004) 8408–8413.

# CHAPTER 33

# Population Genetic Considerations in Statistical Interpretation of Microbial Forensic Data in Comparison with Human DNA Forensic Standard

**Ranajit Chakraborty and Bruce Budowle**

*Institute of Investigative Genetics, Department of Forensic and Investigative Genetics, University of North Texas Health Science Center, Fort Worth, Texas*

## INTRODUCTION

From initial discussions on interpretation of microbial forensic data, it was realized that information from molecular genetics, genomics, and informatics would be central to identification, virulence determination, pathogenicity characterization, and source attribution of any microbial agent used in a bioterrorist act or biocrime (1). Thus, microbial genetic information, in terms of the biology and population genetic structure of the pathogenic agent and epidemiologic implications, is critical for statistical interpretation of microbial forensic data (2). This chapter provides a brief review of some of the considerations that need to be employed to build a statistical paradigm for interpreting microbial forensic data. There are many differences in the genetic characteristics of microorganisms and these will determine to what extent a statistical paradigm can be put in place for interpretation of microbial forensic data. There remain substantive knowledge gaps on ecological diversity and maintenance of genetic diversity of microbial pathogens. This lack of knowledge will increase the uncertainty of the statistical analyses associated with microbial forensic evidence. This chapter briefly addresses comparison of human and microbial DNA forensics scenarios, from which the logic of statistical interpretation principles applicable for microbial forensics is deduced. Due to the success of human DNA forensic analyses, it can be expected that requirements of microbial forensics will be modeled on the human DNA experience (and particularly so in the legal and policy-making sectors). There are some features that should be similar and others that will be dissimilar. The chapter ends with the notion that while general outlines of interpretation of microbial forensic data can be formulated, specific recommendations

Microbial Forensics. DOI: 10.1016/B978-0-12-382006-8.00033-5

encompassing all possible microbial forensic case work currently cannot be readily made. Areas of specific knowledge gaps are identified for building a realistic statistical paradigm of interpreting microbial forensic data.

## BACKGROUND OF SOME MAJOR PATHOGENS

Microbial genetics has traditionally been a field of basic science research as microorganisms offer several features that facilitate the study of evolutionary processes. Short generation time, haploid genome, ease of culturing, and their abundance facilitate such studies. However, they offer some complexities as well; clonal and asexual propagation, recombination, and gene conversion tend to make evolutionary inference(s) a daunting task (3). Further, because single-cell analyses of such organisms remain uncommon, limits of genomic assays are also issues, as most analyses examine microorganisms at the population or quasi-population level. As microbial agents are involved in many human diseases, implications of both microbial and human genetic variations have also been productive areas for research (4). Because of the impact of disease, many molecular biological tools have developed from microbial studies that also are being used for studies in other organisms, including human (5,6).

Bioterrorism or biocrime is defined as the intentional use of a bacteria, virus, or toxin to cause fear, disruption, physical harm, or economic harm to an individual, individuals, and/or populations. While one could assert that bioterrorism and biocrime have different definitions, for the purposes of this chapter we define them as essentially the same. Basic questions for the forensic investigation of microbial evidence are (i) what is the pathogen used, (ii) where was it originally introduced, and (iii) what is the source of the introduction? Detection and source attribution are the ultimate goals of using forensic tools in microbial forensic investigations. To begin developing a strategy for placing significance on a genetic analysis of microbial forensic evidence and addressing the objectives of microbial forensics, brief comments on the biology, molecular genetics, and genetic variation of selected microorganisms should be helpful in identifying gaps in our knowledge.

Some of the microorganisms that are potential threats include *Bacillus anthracis, Yersinia pestis, Francisella tularensis, Clostridium botulinum* (and specifically its toxin), *Burkholderia mallei, Escherichia coli* O157:H7, influenza virus, and foot-and-mouth disease virus. Many of these are discussed in detail in other chapters of this book (for additional information, also see refs. 11–80). For this chapter, it is important to emphasize that these microorganisms exhibit very different modes of inheritance [clonal, sexual, horizontal gene transfer (HGT), recombination, etc.] of all or parts of their genomes. Thus, it is evident that the approach for statistical weighting of attribution of microbial forensic genetic evidence will differ. With the exception of some viruses (e.g., influenza A), diversity is unknown. Describing diversity is daunting and may rarely be possible (the Ames strain of *B. anthracis* used in the anthrax-letter attack may be an exception where most samples derive from one ancestor and descendants reside in laboratories). The microbial world is vast and variable. However, one thing shared by all major threat agents (excluding a few viruses) is that very few population data are available on the genome level to enable evaluation of the variation that could be forensically useful.

The pathogen(s) used in an event (bioterrorism or biocrime) may differ in their (i) life history and endemicity (i.e., geographic distribution, ecology, and natural reservoirs); (ii) host–pathogen interaction; (iii) genomic characteristics; and (iv) mechanism, rate, and pattern of evolution. Each of these characteristics will influence the inference of source of origin of a specific strain found in an evidentiary sample. Further, considerable knowledge gaps exist regarding the diversity and endemicity of these agents. Hence, for microbial forensics to be on a sounder footing, such knowledge gaps need to be filled, best achieved perhaps by building extensive annotated databases of these agents.

## STATISTICAL PARADIGM OF SOURCE ATTRIBUTION OF MICROBIAL AGENTS

Because it is likely that some individuals may desire to drive the field of microbial forensics along the same path as that of the human DNA forensic discipline, major differences between human and microbial DNA forensics are provided to indicate why a variant paradigm would be needed for microbial forensics. Indeed, many consider human DNA forensics to be the gold standard of forensics and one can anticipate, particularly the legal arena and policy makers, to compare the two disciplines. There are benefits in doing so, but we caution that a wholesale adoption of human DNA forensic practices could lead to erroneous interpretations. Notable differences need to be accommodated properly in the respective disciplines.

In a typical microbial forensic case, the investigation starts with collection of one or more evidentiary samples, which are subsequently subjected to laboratory analyses. Such analyses may encompass genomic as well as nongenomic assays. The notion is to generate data to enable comparison of the evidence with that of a known repository or reference samples of microorganisms to aid detection, identification, and source attribution. In this sense, it is reasonable to use the parallelism of human DNA identification and microbial forensic objectives to develop a statistical paradigm of source attribution for microbial agents. For example, the molecular biology principles and technology used in microbial forensics [e.g., extraction of DNA/RNA, polymerase chain reaction (PCR)-based methods for amplifying target regions of the genome, genotyping specific loci, or sequencing a specific fragment of DNA/RNA] are similar to those practiced in human DNA forensic applications. Both human forensic and microbial forensic practices use similar quality assurance/quality control (QA/QC) protocols, use the same terminologies of qualitative comparisons (e.g., match/inclusion, exclusion, or inconclusive), and rely on population databases for statistical evaluations. However, a temptation to follow the human DNA forensic protocol in its entirety would be erroneous in a microbial forensic context. Obvious differences that should be taken into account include database size and composition, statistical interpretation methods, and confidence in outcome of an interpretation. Table 33.1 depicts major similarities and dissimilarities of human and microbial forensics.

## Qualitative Observations Summarizing Microbial Forensic Casework Data and What They Mean

Analogous to human DNA forensics, when comparing an evidence DNA (or RNA) profile with that of a reference sample(s), the concepts of inclusion (or match/similarity), exclusion (dissimilarity), and inconclusiveness (not exclusion, but not definitely a match/similar either) have been suggested for microbial forensics. Although both human and microbial forensics use the same or similar terminologies of qualitative summary findings for casework data comparisons, the protocols, as well as meaning of the descriptive summaries, are not necessarily the same. For microbial forensics, relatively few cases will result in a definitive attribution conclusion, as conclusiveness is intrinsically related to the depth of the evidence. However, based on principles of molecular biology and mechanisms of having the character state (based on which

**Table 33.1** Comparison of Human DNA Forensics and Microbial Forensics

| Similarities | Dissimilarities |
|---|---|
| ■ Terminologies for qualitative conclusions of laboratory test results | ■ Meaning of qualitative conclusions |
| ■ Need of databases | ■ Significance of match/inclusion/same lineage |
|  | ■ Database size and composition |
|  | ■ Statistical interpretation methods |
| ■ QA/QC issues | ■ Confidence of outcome interpretation |
|  | ■ Consideration of limits of assay methods |
| ■ Need for general acceptance of protocols | ■ Use of meta-data to augment confidence of inference[a] |

[a] There could be some similarities with missing persons identifications and use of meta-data.

"match" or a similar term is defined), operational definitions of these concepts have to be reformulated for microbial forensics so that they may be subjected to further quantitative or semiquantitative analyses.

Once these concepts are defined, a tool has to be developed for comparing profiles of evidence samples with those of known reference sources (repositories, or collected reference specimens). In fact, at this stage a microbial forensic statistics paradigm starts to differ from that of human DNA forensics. The main reason is that barring the scenarios in the human context of missing person identification (which often does not have a direct reference sample of the missing person), a DNA match between an evidence sample and a reference sample is often judged in relation to whether it can be due to a direct transfer of DNA from the reference source to the crime sample. Direct transfer may not apply in many microbial forensic attribution analyses. It will be rare to find such evidence; but one can envision direct comparisons possibly in some cases involving, for example, ricin where the material found in a letter could be compared to the material found in a flask in a suspect's garage. In the context of microbial forensics, a variety of questions may arise depending on the context of DNA profile comparisons. For example, are samples from the same source (possibly if direct comparison is considered)? What might be inferred about the source of the evidentiary sample (for indirect comparisons)? Are the genetic differences between them too few to conclude that they derive from different sources (or different lineages)? Are these differences sufficient to consider that the samples are from different sources? Is it possible that the two samples have a recent common ancestor and, if so, can we determine the age of the most recent common ancestor (MRCA)? Even with in-depth knowledge of the genetics of the pathogen under investigation, statistical confidence of answers of each of these questions depends on the context of the specific case. For example, laboratory stocks, being maintained in controlled environments, may show less diversity than in nature. Thus, one or a few genetic differences between evidentiary and reference samples may be significant if the weapon originated from a laboratory-maintained culture. Alternatively, the stock could have been manipulated genetically (such as exposing it to a mutagenic agent) so that observations of a few differences would be expected even for a very recent MRCA. The number of possible passages a culture goes through after it was obtained from a source should also be a factor in such observations. Some bacterial genes (e.g., *E. coli* and *Salmonella*) have elevated mutation rates (up to a factor of $\approx$100-fold) due to defects of mismatch repair, mutator genes, and so on (81–86). Thus, it is expected that two isolates, even with a shared recent common ancestor, may differ at these rapidly evolving sites. RNA viruses are notable examples of fast evolving agents (83). This rationalizes the need of interpreting similarities or dissimilarities of observations on stable versus rapidly evolving genetic sites. Thus, even a qualitative match determination protocol in microbial agents must take into account general source (e.g., laboratory, nature), stability of genetic sites or elements, storage conditions, mutagenic treatments, and so on, which are of less consequence in direct comparison analyses with human DNA forensic data.

## Significance of Inclusion/Match/Same Lineage

Once an "inclusion/match/same lineage" description is inferred according to a protocol established (taking the aforementioned aspects into account), just as in human DNA identification, placing some significance or weight on the observation is desirable. Of course, some inferences in microbial cases may not require any quantitative assessment and a qualitative statement may suffice, such as declaring the strain of *B. anthracis* seen in some high-profile cases (87–90). When a quantitative inference is attempted, care must be taken so that such interpretation tools do not overemphasize data, nor is it performed without consideration of the limits of the assay. It may be argued that the same caution applies for human DNA forensics as well; appropriate protocols for generating human DNA forensic statistics are in place that account for such limits of detection methods and population data. Currently, there are few established statistical interpretation guidelines for microbial forensic DNA analysis. Microbes (most bacteria and viruses) are haploid and propagate largely asexually (and hence population dynamics is clonal). The majority of microbial forensic genetic evidence may only infer common lineage instead of identity. Hence, like mitochondrial DNA (mtDNA) and Y chromosome haplotype evidence of human DNA forensics, a lineage-based approach is needed. Use of a lineage-based approach extends perhaps only to the extent that the entire haplotype/sequence data have to be used and its transmission from ancestral to descendant cultures has to be recognized (which can result in a match, barring mutations and recombination—generally caused by HGT). Because the relationship between evidence and reference sample may only be that of an ancestor and progenitor, match/inclusion statistics should be formulated in terms of evolutionary relationships, which may be answered in a number of different ways. For example, one could ask: which is the nearest neighbor (evolutionarily) of the evidence sample? Are differences of the evidence sample from other reference samples significantly greater than that from the nearest neighbor? In contrast, in human DNA forensics, questions relate to how likely it is to observe such a match (with a reference sample) should they be of unrelated (maternally for mtDNA and paternally for Y chromosome haplotype) lineages. Thus, in the human context, in most cases it is not relevant how different the two haplotypes/sequences are; only that they differ is important (91) (unless differences are built into the match determination protocol itself). In other words, in the human DNA forensic context, the question is match/no match and not the degree of differences between compared profiles. In microbial forensics, determination of nearest neighbor or significance testing of distances of different clusters of strains/isolates starts with the extent of mismatches in haplotype/sequence comparisons. Such issues are addressed by phylogenetic analysis of (haplotype/sequence) data and bootstrap evaluations of significance of branch lengths of relevant segments of the phylogenetic tree (92–94).

While such approaches have been used in inferring the source of the pathogen found in evidentiary samples (94), some cautionary notes of using these interpretation tools are in order. Molecular phylogenetic analyses for determining topological relationships of operational taxonomic units (OTUs) are quite varied, and analyses use various implicit as well as explicit assumptions. For example, proper alignment of sequence/haplotype data has to be used to calibrate the extent of mismatches between haplotypes/sequences of contrasting OTUs. In segments of a microbial genome, with an extended stretch of HGT, sequence alignment may be cumbersome, having an impact on calibration of extent of mismatches, as well as inference from phylogenetic trees/networks. Second, the mechanism of production of mismatches has to be defined precisely (generally through a mutation model), and assumptions regarding the relative rate and pattern of mutations at different sites are needed to calibrate the mismatch in units of generation or chronological time of evolutionary separation of different OTUs. For example, markers used in multilocus VNTR analysis (MLVA) are known to mutate faster than single nucleotide polymorphism (SNP) sites; hence, the absolute extent of mismatches of these two markers is not readily comparable for two strains/isolates of any pathogen. Evidence of pathogen-specific differences of these factors has been already described, suggesting that the same method of evolutionary-tree construction and significance testing of the branch nodes may not hold true. Methods that are robust against violation of the assumptions are preferred rather than ones that are more accurate with the assumption, but are rather error-prone when they are violated. For example, in the presence of the possibility of HGT, the median-joining network (95) for inferring intraspecific phylogenies may be preferred as compared to the neighbor-joining method (96) where all site-specific mismatches are assumed to be due to mutational changes alone. The use of median-joining networks, however, may create ambiguities in inferring MRCA, as there can be more than one via different routes of reticulations in such network diagrams. For neighbor-joining trees, the definition of MRCA is always unambiguous.

## Database Size and Composition
While need of a database is a common element for both human and microbial forensics, database size and composition requirements for them can be

drastically different and will likely be case dependent. In the human context, because source attribution relates to members of only one species, with generation time sufficiently long, intra- and interpopulation polymorphism in contemporary scale is the only major issue of construction of database size and composition. Principles of gene frequency differences within and between populations and empirical data on interpopulation genetic distances of human populations govern database size and composition requirements (97,98). For human populations, ethnohistory and geographic migration of modern humans provided a solid rationale as to what populations are to be sampled to have a fairly accurate representation of interpopulation diversity. Consequently, contemporary population dynamics of human populations drove the principles behind the construction, composition, and size of the human DNA forensic databases, where adequate determination of allele frequencies was the focus of construction of the databases (98,99). For mtDNA and Y chromosomal databases, the rationale was slightly different; nonetheless, it was still the population substructure of contemporary populations that dictated the size and composition of forensic haplotype databases of lineage markers (91).

For microbial forensics, considerations of size and composition of databases would depend on contemporary as well as evolutionary diversity of the pathogens, their strains, and even isolates. Adequate sampling is an almost impossible task to define or achieve for microbial diversity. As a consequence, microbial databases can never be claimed to be exhaustive or comprehensive.

## Statistical Interpretation Methods

Statistical interpretation methods (for source attribution or narrowing down the possible sources) can be described generically as the (i) frequency-based approach, (ii) likelihood approach, and (iii) Bayesian approach. While there are claims that the likelihood approach (or the Bayesian approach, when it can be adopted) is the best and most relevant (100), one should note that these three approaches address three different types of questions and, furthermore, are in sequence built on each other. The frequency-based approach estimates probabilities of certain observations under specific hypotheses, which are subsequently used to generate ratios in the likelihood approach. Finally, likelihood ratios (LR) are translated into posterior probabilities by the Bayes' theorem by using prior probabilities (that are often subjective and ill-defined). In the microbial context, the frequency-based approach is the least desirable approach, as the population is most often not well defined. The phylogeny-based likelihood should be the preferred method. Some authors call this approach a method of odds ratio (OR) computations (101). However, LR, as such, is not exactly an OR; it is simply the ratio of probabilities of a set of observations under two distinct (and mutually exclusive) hypotheses (102).

Although LR can be translated to an OR, and subsequently to a posterior probability of the validity of a hypothesis and has been used in the context of microbial source tracking (103), translating an LR or OR into a posterior probability requires knowledge of the prior probabilities of the two hypotheses being contrasted in the LR computation. This is often beyond the scope of genetic data and, hence, necessitates validation of any meta-data to form the prior probability.

In summation, the statistical interpretation paradigm for microbial data will most often involve phylogenetic inference and formation of likelihoods (and their ratios) based on that inference. Estimates of likelihoods would depend on (i) detection of clusters of OTUs used in the analyses, (ii) estimates of branch lengths distinguishing the clusters, and (iii) time estimates of the MRCA. Molecular evolutionary models employed in such methods generally indicate that the confidence limits of such estimates are intrinsically wide and, hence, any translation of them to OR, or posterior probability, is likely to yield only a moderate degree of support of any assertion for source attribution (104,105). Clearly, the astronomical values often reached in the human DNA forensic context will rarely be attained for microbial forensic genetic evidence.

## Considerations of Limits of Assays

Molecular assay of a biological specimen involves certain limits of detection of true genotypes. In the human context, there were discussions on this issue of the possible effects of (a) a false-positive match or error rate (106,107) and (b) allele dropout (108). QA/QC practices of human DNA forensics are capable of detecting most common sources of laboratory errors causing a false-positive match, and recommendations are in place to address these without any statistical adjustment in the strength of DNA evidence (109). Development of adjustments for the possibility of allelic dropout for low template DNA analysis is still ongoing (110), but evaluations of parameters of the processes that generate allelic dropout would require more empirical as well as statistical studies, particularly when the evidence sample is a mixture (111).

In the microbial context, both false-positive and false-negative detection of genetic signatures are possible, particularly so since genotyping assays are currently done from multiple individual organisms, and hence low levels of certain genotypes may not be detectable in the cultures used for the assay. In the Amerithrax case, false-negative results were a concern for typing the morphological variants that were present at low levels and thus were subject to stochastic effects and dropout. Literature on the rates of such error limits of assays is quite sparse in the field of microbial forensics, and this area needs attention. Even with some defined limits, case-specific features will impact the false-negative and false-positive rates (such as those constraints that occurred with analyses of the anthrax morphotypes).

Should such data be available, bioinformatic tools to handle such limits of detection methods may be developed using the concept of hidden Markov models (HMM). To illustrate how HMM may play a role in classification and source attribution of microbial agents, some analytical details of HMM would be instructive. Consider the scenario that a sample of pathogens collected from a biocrime scene yielded genotype data, represented by $Y_1, Y_2, \ldots,$ $Y_n$, where $n$ is the number of loci and $Y_i$ is the genotype of the sample at the $i$th locus. The agent released is suspected to have come from one of $m$ laboratories or repositories. The goal is to identify the source from which the agent under investigation arose. A genotype databank/database from each laboratory for the agent under investigation is essential to begin data analyses. Because microbial agents evolve over generations, using the given data $Y_1, Y_2, \ldots,$ $Y_n$, the goal is to identify its ancestors. The problem straddles two areas of research: evolutionary biology and pattern recognition. With the possibilities of genotyping errors, the data of $Y_1, Y_2, \ldots, Y_n$ are not necessarily the true genotypes of the agent. Let $X_1, X_2, \ldots, X_n$ be the true genotypes of sample bacteria. We need to find a way to determine $X_1, X_2, \ldots, X_n$ using given data $Y_1, Y_2, \ldots,$ $Y_n$ before finding its ancestors or identifying the potential laboratory from which it was obtained. Three basic steps involved for solving the problem are (i) determine the true genotypes $X_1, X_2, \ldots, X_n$; (ii) develop a measure of closeness of given bacteria to ancestral bacteria available in a reference laboratory; and (iii) formulate a protocol for classifying sample bacteria into one of the laboratories based on the closeness measure, genotype data $X_1, X_2, \ldots, X_n$ of the sample, and genotype data of bacteria in the reference laboratories.

## Step 1

An HMM may be used to determine true genotypes. Assume, for simplicity, that each locus is biallelic. Denote the genotypes generically by *AA*, *Aa*, and *aa*. Suppose the probabilities of genotyping errors are known. Typically, these probabilities should be known for each of the genotyping laboratories from which the reference genotype data are generated. Let the genotyping error probabilities be given in the following table:

| Correct and Error Genotyping Probabilities | | | | |
|---|---|---|---|---|
| **True Genotype** | **Genotypes** | | | **Sum** |
| | **AA** | **Aa** | **aa** | |
| *AA* | $\alpha_{11}$ | $\alpha_{12}$ | $\alpha_{13}$ | 1 |
| *Aa* | $\alpha_{21}$ | $\alpha_{22}$ | $\alpha_{23}$ | 1 |
| *aa* | $\alpha_{31}$ | $\alpha_{32}$ | $\alpha_{33}$ | 1 |

Diagonal entries in this table are correct genotyping probabilities, and other entries are all error probabilities. Assume that the true genotypes $X_1, X_2, \ldots, X_n$

form a homogeneous Markov chain with state space $S = \{AA, Aa, aa\}$, which is hidden. Let the transition probability matrix be given by

$$P = \begin{pmatrix} p_{11} & p_{12} & p_{13} \\ p_{21} & p_{22} & p_{23} \\ p_{31} & p_{32} & p_{33} \end{pmatrix},$$

where $1 = AA$, $2 = Aa$, $3 = aa$, and $p_{ij}$ is the one-step conditional probability that the genotype is $j$ at location $t + 1$, given that the genotype is $i$ at location $t$. With the homogeneity assumption of the Markov chain, these one-step transition probabilities may be treated as independent of location $t$.

Using given genotype data $Y_1, Y_2, \ldots, Y_n$ on the sampled agent, the objective would be to predict the hidden genotypes at the loci. A likelihood principle may be implemented, described schematically as follows:

> Given: Data: $Y_1 = i_1, Y_2 = i_2, \ldots, Y_n = i_n$.
> A possible prediction of true genotypes: $X_1 = j_1, X_2 = j_2, \ldots, X_n = j_n$.
> Initial distribution of $X_1$: $Pr(X_1 = 1) = p_1, Pr(X_1 = 2) = p_2$, $Pr(X_1 = 3) = p_3$.
> Likelihood of data $= L(i_1, i_2, \ldots, i_n, j_1, j_2, \ldots, j_n)$
> $= Pr(Y_1 = i_1, Y_2 = i_2, \ldots, Y_n = i_n, X_1 = j_1, X_2 = j_2, \ldots, X_n = j_n)$
> $= Pr(X_1 = j_1, Y_1 = i_1) * Pr(X_2 = j_2, Y_2 = i_2 \mid X_1 = j_1) * Pr(X_3 = j_3, Y_3 = i_3 \mid X_2 = j_2)$
> $* \ldots$ etc.
> $= (pj_1 * \alpha j_1 i_1) * (pj_1 j_2 * \alpha j_2 i_2) *$ etc.

The next step is to maximize $L$ over all possibilities of $X_1 = j_1, X_2 = j_2, \ldots, X_n = j_n$. The totality of all possible guesses is $3^n$, which is an astronomical number even for moderate values of $n$. Two cases arise.

## Case 1

Initial distribution, transition probability matrix, and correct and error genotyping probabilities are all known. One can use a dynamic programming approach to calculate the likelihood for each choice of $X_1 = j_1, X_2 = j_2, \ldots, X_n = j_n$ in the log space. The Baum–Welch algorithm is specially tailored to handle such huge optimization problems (112,113).

## Case 2

One or more of the entities (initial distribution, transition probability matrix, and correct and error genotyping probabilities) are unknown. One may use the EM algorithm or a variation of it in solving the optimization problem. In the course of its execution, unknown entities will also be estimated.

### Step 2

Once the true genotypes $X_1$, $X_2$, ... , $X_n$ are estimated, alignment techniques can be employed using sample bacteria and possibly ancestral bacteria obtained from a reference laboratory. An alignment score $\delta j$ between the sample bacteria and its ancestral bacteria has to be developed, where $j$ refers to the $j$th lab.

### Step 3

Using alignment scores and genotype data of bacteria, one has to develop a classification protocol to classify the given bacteria into one of the laboratories.

## Levels of Confidence and Its Augmentation

Even with the considerations just discussed, the levels of confidence in microbial source attribution are likely to be far less probative than that generally obtained in the human DNA forensic context. Legal authorities, as well as scientific experts, should be educated in this regard to avoid confusion and/or inadmissibility of valid and reliable microbial forensic evidence. However, levels of confidence may be improved through use of multiple classification protocols in conjunction with the types of technologies used (e.g., PCR-based SNP analyses; multilocus sequence typing, MLST; pulse field gel electrophoresis, PFGE), such as those suggested for microarray gene expression data (114) and next-generation sequencing (see Chapters 3, 27, 28 in this book).

Levels of confidence of microbial data may also be somewhat boosted by use of concepts other than a simple LR or OR. Epidemiological concepts of sensitivity, specificity, and positive predictive value have been used in microbial source tracking inference (103), and these may be extended by employing the concept of receiver-operating characteristics analysis (115) (also see Chapter 31).

## Use of Meta-Data

Finally, in the human context, unless there is definitive circumstantial evidence, a simple DNA match does not necessarily place a person at the crime scene or at the time of the crime. In a microbial context, because pathogens shared by laboratories have their attendant material transfer agreement (MTA) or Select Agent transfer records, these records may provide strong evidence, in some cases, of when and for what purposes the laboratories shared the pathogen of interest and whether genetic alterations could have been accumulated in the laboratories for their continual use and culture. Such data may aid in further narrowing down the source even when the evidence sample is not particularly strong or is found to provide a genetic match with strains obtained from several repositories. MTA records may indicate their common source due to one-way transfer between laboratories and/or possibilities of *in situ* evolution due to multiple passages, whereas others may have remained dormant (by not being used at the source site). In other words, microbial source tracking can be attempted by combining genetic information with epidemiological or regulatory investigations.

In summation, discussions herein indicate that while approaches exist to assist in providing statistical strength of microbial forensic evidence data for source attribution purposes, areas of further improvements and research for making bioinformatic tools more effective are needed. They are:

- Robust algorithms(s) for sequence/haplotype alignment [robust against rate and pattern homogeneity of site-specific mutations/substitutions, the presence of repeat elements (of varied lengths), recombination and gene conversion, HGT]
- Phylogenetic algorithm(s) based on different types of markers/loci/ sequence data (co-optimized with robust algorithms for alignment, stated earlier)
- Methods for quantitatively interpreting results from case analysis (e.g., estimates of time/age of most recent common ancestor, and classification of nearest sets of neighbors, with their confidence limits specified)
- More effective algorithm(s) for detecting genomic signatures of pathogenicity/virulence and antibiotic resistance of microbial agents
- Capabilities of relating diversity to function
- Comprehensive comparative and functional genomics under varied environmental conditions
- User-friendly interfaces for interpretation and visualization of data from genomic analyses

## CONCLUSIONS

In microbial forensics, one may consider attribution solely to be the "fingerprinting" of a pathogenic agent. However, because of the clonal nature of many microorganisms and, on a case-by-case basis, lack of population and phylogenetic data unique identification of a microorganism may rarely be possible. More importantly, the ultimate goal of attribution is identification of persons who committed the bioterrorist act or biocrime, intentionally or inadvertently. Therefore, and most likely, in addition to microbiological analytical tools, traditional forensic analyses, such as human DNA analysis, dermatoglyphic patterns, analytical chemistry, tool marks, computer data, and other techniques will be used to analyze a bioterrorist event or biocrime evidence. Epidemiological data from various documents may also be a valuable set of information to infer the chronological times of transfer of strains between laboratories that are likely contenders of being the nearest genetic neighbor of the evidentiary samples.

*A strain repository* to house pathogens and other appropriate near-neighbor microorganisms must also be developed. It will facilitate research studies, assay development, validation, QA/QC requirements, and investigations. The near-neighbor concept is intimately related with methods used for detection and identification. Some methods of identification that are robust may lead to a broad class of near neighbors, while sophisticated ones may define near neighbors that are narrower. Well-characterized samples must be available to enable good-quality assay development. Assays cannot be validated adequately without proper samples and reference material. Bioinformatic interpretation of analytical results from evidence samples may be more limited without properly defined samples and controls. New analytical methods and some existing methods need to be validated properly. Such validation tasks are not limited to laboratory procedures; they equally apply to tools to be used for data interpretation and statistical assessment issues. Moreover, some biocrimes and acts of bioterrorism may require analysis by methodologies that may not have undergone the rigorous review process of that of standard operating protocols and should at least follow recommendations for preliminary validation practices to ensure that limitations and proper expectations are defined (116).

## REFERENCES

[1] B. Budowle, S.E. Schutzer, M.S. Ascher, R.M. Atlas, J.P. Burans, R. Chakraborty, et al., Toward a system of microbial forensics: From sample collection to interpretation of evidence, Appl. Environ. Microbiol. (2005) 2209–2213.

[2] P. Keim, Microbial Forensics: A Scientific Assessment, American Academy of Microbiology, Washington, DC, 2003.

[3] A. Moya, E.C. Holmes, F. González-Candelas, The population genetics and evolutionary epidemiology of RNA viruses, Nat. Rev. Microbiol. 2 (2004) 279–288.

[4] L.H. Taylor, S.M. Latham, M.E. Woolhouse, Risk factors for human disease emergence, Phil. Trans. R. Soc. Lond. B Biol. Sci. 356 (2001) 983–989.

[5] A. Van Belkum, M.C. Erasmus, Newer methods for bacterial strain typing, Clin. Microbiol. Newslett. 30 (9) (2008) 63–69.

[6] S.A. Morse, B. Budowle, Microbial forensics: Application to bioterrorism preparedness and response, Infect. Dis. Clin. North Am. 20 (2006) 455–473.

[7] R. Okinaka, K. Clod, O. Hampton, A.R. Hoffmaster, K.K. Hill, P. Keim, et al., Sequence and organization of pXO1, the large *Bacillus anthracis* plasmid harboring the anthrax toxin genes, J. Bacteriol. 181 (1999) 6509–6515.

[8] R. Okinaka, K. Cloud, O. Hampton, A. Hoffmaster, K. Hill, P. Keim, et al., Sequence, assembly and analysis of pXO1 and pXO2, J. Appl. Microbiol. 87 (1999) 261–262.

[9] T.D. Read, S.N. Peterson, N. Tourasse, L.W. Baillie, I.T. Paulsen, K.E. Nelson, et al., The genome sequence of *Bacillus anthracis* Ames and comparison to closely related bacteria, Nature 421 (2003) 81–86.

[10] T.D. Read, S.L. Salzberg, M. Pop, M. Shumway, L. Umayam, L. Jiang, et al., Comparative genome sequencing for discovery of novel polymorphisms in *Bacillus anthracis*, Science 296 (2002) 2028–2033.

[11] G.T. Vilas-Boas, A.P.S. Peruca, G.M.N. Arantes. Biology and taxonomy of *Bacillus cereus*, *Bacillus anthracis*, and *Bacillus thuringiensis*, Can. J. Microbiol. 53 (2007) 673–687.

[12] S.L. Welkos, T.J. Keener, P.H. Gibbs, Differences in susceptibility of inbred mice to *Bacillus anthracis*, Infect. Immun. 51 (1986) 795–800.

[13] I. Uchida, S. Makino, C. Sasakawa, M. Yoshikawa, C. Sugimoto, N. Terakado, Identification of a novel gene, dep, associated with depolymerization of the capsular polymer in *Bacillus anthracis*, Mol. Microbiol. 9 (1993) 487–496.

[14] P. Keim, M.W. Van Ert, T. Pearson, A.J. Vogler, L.Y. Huynh, D.M. Wagner, Anthrax molecular epidemiology and forensics: Using the appropriate marker for different evolutionary scales. Infect, Genet. Evol. 4 (2004) 205–213.

[15] K.S. Blackwood, C.Y. Turenne, D. Harmsen, A.M. Kabani, Reassessment of sequence-based targets for identification of *Bacillus* species, J. Clin. Microbiol. 42 (4) (2004) 1626–1630.

[16] L. Radnedge, P.G. Agron, K.K. Hill, P.J. Jackson, L.O. Ticknor, P. Keim, et al., Genome differences that distinguish *Bacillus anthracis* from *Bacillus cereus* and *Bacillus thuringiensis*, Appl. Environ. Microbiol. 69 (5) (2003) 2755–2764.

[17] A.R. Hoffmaster, J. Ravel, D.A. Rasko, G.D. Chapman, M.D. Chute, C.K. Marston, et al., Identification of anthrax toxin genes in a *Bacillus cereus* associated with an illness resembling inhalational anthrax, Proc. Natl. Acad. Sci. USA 101 (22) (2004) 8449–8454.

[18] M.N. Van Ert, W.R. Easterday, L.Y. Huynh, R.T. Okinaka, M.E. Hugh-Jones, J. Ravel, et al., Global genetic population structure of *Bacillus anthracis*, PLoS One 5 (2007) e46.

[19] D.S. Guttman, D.E. Dykhuizen, Clonal divergence in *Escherichia coli* as a result of recombination, not mutation, Science 266 (1994) 1380–1383.

[20] M. Achtman, K. Zurth, G. Morelli, G. Torrea, A. Guiyoule, E. Carniel, *Yersinia pestis*, the cause of plague, is a recently emerged clone of *Yersinia* pseudotuberculosis, Proc. Natl. Acad. Sci. USA 96 (1999) 14043–14048.

[21] D. Zhou, Y. Han, R. Yang, Molecular and physiological insights into plaque transmission, virulence, and etiology, Microb. Infect. 8 (2006) 273–284.

[22] B.W. Wren, The Yersiniae: A model genus to study the rapid evolution of bacterial pathogens, Nat. Rev. Microbiol. 1 (2003) 55–64.

[23] D. Zhou, Y. Han, Y. Song, P. Huang, R. Yang, Comparative and evolutionary genomics of *Yersinia pestis*, Microbes Infect. 6 (2004) 1226–1234.

[24] J. Parkhill, B.W. Wren, N.R. Thomson, R.W. Titball, M.T. Holden, M.B. Prentice, et al., Genome sequence of *Yersinia pestis*, the causative agent of plague, Nature 413 (2001) 523–527.

[25] W. Deng, V. Burland, G. Plunkett 3rd, A. Boutin, G.F. Mayhew, P. Liss, et al., Genome sequence of *Yersinia pestis* KIM, J. Bacteriol. 184 (2002) 4601–4611.

[26] Y. Song, Z. Tong, J. Wang, Z. Guo, Y. Han, J. Zhang, et al., Complete genome sequence of *Yersinia pestis* strain 91001, an isolate avirulent to humans, DNA Res. 11 (2004) 179–197.

[27] D.Z. Zhou, Y. Tong, Y. Song, Y. Han, D. Pei, X. Pang, et al., Genetics of metabolic variations between *Yersinia pestis* biovars and the proposal of a new biovar, microtus, J. Bacteriol. 186 (2004) 5147–5152.

[28] P.S. Chain, E. Carniel, F.W. Larimer, P.O. Stoutland, W.M. Regala, A.M. Georgescu, et al., Insights into the evolution of *Yersinia pestis* through whole-genome comparison with *Yersinia pseudotuberculosis*, Proc. Natl. Acad. Sci. USA 101 (38) (2004) 13826–13831.

[29] D.T. Dennis, T.V. Inglesby, D.A. Henderson, J.G. Bartlett, M.S. Ascher, E. Eitzen, et al., Working Group on Civilian Biodefense. Tularemia as a biological weapon: Medical and public health management, J. Am. Med. Assoc. 285 (2001) 2763–2773.

[30] P.C. Oyston, A. Sjostedt, R.W. Titball, Tularaemia: Bioterrorism defence renews interest in *Francisella tularensis*, Nat. Rev. Microbiol. 2 (2004) 967–978.

[31] P. Keim, A. Johansson, D.M. Wagner, Molecular epidemiology, evolution, and ecology of Francisella, Ann. N.Y. Acad. Sci. 1105 (2007) 30–66.

[32] F.R. McCrumb, Aerosol infection in man with *Pasteurella tularensis*, Bacteriol. Rev. 25 (1961) 262–267.

[33] B. Christenson, An outbreak of tularemia in the northern part of central Sweden, Scand. J. Infect. Dis. 16 (1984) 285–290.

[34] S. Dahlstrand, O. Ringertz, B. Zetterberg, Airborne tularemia in Sweden, Scand. J. Infect. Dis. 3 (1971) 7–16.

[35] A.M. Hood, Virulence factors of *Francisella tularensis*, J. Hyg. (Lond.) 79 (1977) 47–60.

[36] G. Sandstrom, S. Lofgren, A. Tarnvik, A capsule-deficient mutant of *Francisella tularensis* LVS exhibits enhanced sensitivity to killing by serum but diminished sensitivity to killing by polymorphonuclear leukocytes, Infect. Immun. 56 (1988) 1194–1202.

[37] V.M. Sorokin, N.V. Pavlovich, L.A. Prozorova, Francisella tularensis resistance to bactericidal action of normal human serum, FEMS Immunol. Med. Microbiol. 13 (1996) 249–252.

[38] G. Sandstrom, A. Sjostedt, T. Johansson, K. Kuoppa, J.C. Williams, Immunogenicity and toxicity of lipopolysaccharide from *Francisella tularensis* LVS, FEMS Microbiol. Immunol. 5 (1992) 201–210.

[39] A. Sjostedt, Virulence determinants and protective antigens of *Francisella tularensis*, Curr. Opin. Microbiol. 6 (2003) 66–71.

[40] E.V. Vinogradov, A.S. Shashkov, Y.A. Knirel, N.K. Kochetkov, N.V. Tochtamysheva, S.F. Averin, et al., Structure of the O-antigen of *Francisella tularensis* strain 15, Carbohydr. Res. 214 (1991) 289–297.

[41] C.M. Lauriano, J.R. Barker, S.S. Yoon, F.E. Nano, B.P. Arulanandam, D.J. Hassett, et al., MglA regulates transcription of virulence factors necessary for *Francisella tularensis* intraamoebae and intramacrophage survival, Proc. Natl. Acad. Sci. USA 101 (2004) 4246–4249.

[42] F.E. Nano, N. Zhang, S.C. Cowley, K.E. Klose, K.K. Cheung, M.J. Roberts, et al., A *Francisella tularensis* pathogenicity island required for intramacrophage growth, J. Bacteriol. 186 (2004) 6430–6436.

[43] M. Telepnev, I. Golovliov, T. Grundstrom, A. Tärnvik, A. Sjöstedt, *Francisella tularensis* inhibits Toll-like receptor-mediated activation of intracellular signalling and secretion of TNF-$\alpha$ and IL-1 from murine macrophages, Cell. Microbiol. 5 (2003) 41–51.

[44] P. Larsson, P.C. Oyston, P. Chain, M.C. Chu, M. Duffield, H.H. Fuxelius, et al., The complete genome sequence of *Francisella tularensis*, the causative agent of tularemia, Nat. Genet. 37 (2005) 153–159.

[45] J.F. Petrosino, Q. Xiang, S.E. Karpathy, H. Jiang, S. Yerrapragada, Y. Liu, et al., Chromosome rearrangement and diversification of *Francisella tularensis* revealed by the type B (OSU18) genome sequence, J. Bacteriol. 188 (2006) 6977–6985.

[46] L.S. Anthony, P.A. Kongshavn, Experimental murine tularemia caused by *Francisella tularensis*, live vaccine strain: A model of acquired cellular resistance, Microb. Pathog. 2 (1987) 3–14.

[47] H.T. Eigelsbach, C.M. Downs, Prophylactic effectiveness of live and killed tularemia vaccines. I. Production of vaccine and evaluation in the white mouse and guinea pig, J. Immunol. 87 (1961) 415–425.

[48] J.J. Su, D. Yang, T.H. Zhao, T.H. Kawula, J.A. Banas, J.R. Zhang, Genome-wide identification of *Francisella tularensis* virulence determinants, Infect. Immun. 75 (2007) 3089–3101.

[49] J.R. Barker, K.E. Klose, Molecular and genetic basis of pathogenesis in *Francisella tularensis*, Ann. N.Y. Acad. Sci. 1105 (2007) 138–159.

[50] J.F. Petrosino, Q. Xiang, S.E. Karpathy, H. Jiang, S. Yerrapragada, Y. Liu, et al., Chromosome rearrangement and diversification of *Francisella tularensis* revealed by the type B (OSU18) genome sequence, J. Bacteriol. 188 (2006) 6977–6985.

[51] M.P. Dempsey, J. Neitfeldt, J. Ravel, S. Hinrichs, R. Crawford, A.K. Benson, Paired-end sequence mapping detects extensive genomic rearrangement and translocation during divergence of *Francisella tularensis* subsp. *tularensis* and *Francisella tularensis* subsp. *holarctica*, J. Bacteriol. 188 (2006) 5904–5914.

[52] H. Guerra, The Brucellae and their success as pathogens, Crit. Rev. Microbiol. 33 (2007) 325–331.

[53] B.G. Mantur, S.K. Amarnath, R.S. Shinde, Review of clinical and laboratory features of human *Brucellosis*, Indian J. Med. Microbiol. 25 (3) (2007) 188–210.

[54] D.O. Sanchez, R.O. Zandomeni, S. Cravero, R.E. Verdún, E. Pierrou, P. Faccio, et al., Gene discovery through genomic sequencing of *Brucella abortus*, Infect. Immun. 69 (2001) 865–868.

[55] V.G. DelVecchio, V. Kapatral, R.J. Redkar, G. Patra, C. Mujer, T. Los, et al., The genome sequence of the facultative intracellular pathogen *Brucella melitensis*, Proc. Natl. Acad. Sci. USA 99 (2002) 443–448.

[56] I.T. Paulsen, R. Seshadri, K.E. Nelson, J.A. Eisen, J.F. Heidelberg, T.D. Read, et al., The *Brucella suis* genome reveals fundamental similarities between animal and plant pathogens and symbionts, Proc. Natl. Acad. Sci. USA 99 (2002) 13148–13153.

[57] S. Michaux, J. Paillisson, M.J. Carles-Nurit, G. Bourg, A. Allardet-Servent, M. Ramuz, Presence of two independent chromosomes in the *Brucella melitensis* 16M genome, J. Bacteriol. 175 (1993) 701–705.

[58] E. Jumas-Bilak, C. Maugard, S. Michaux-Charachon, A. Allardet-Servent, A. Perrin, D. O'Callaghan, et al., Study of the organization of the genomes of *Escherichia coli*, *Brucella melitensis* and *Agrobacterium tumefaciens* by insertion of a unique restriction site, Microbiology 141 (1995) 2425–2432.

[59] C.E. Rigby, A.D. Fraser, Plasmid transfer and plasmid-mediated genetic exchange in *Brucella abortus*, Can. J. Vet. Res. 53 (1989) 326–330.

[60] A. Cloeckaert, N. Vizcaino, J.Y. Paquet, R.A. Bowden, P.H. Elzer, Major outer membrane proteins of *Brucella* spp.: Past, present and future, Vet. Microbiol. 90 (2002) 229–247.

[61] G. Rajashekara, L. Eskra, A. Mathison, E. Petersen, Q. Yu, J. Harms, Splitter G. Brucella: Functional genomics and host-pathogen interactions, Anim. Health Res. Rev. 7 (2006) 1–11.

[62] R.A. Hutson, D.E. Thompson, M.D. Collins, Genetic interrelationships of saccharolytic *Clostridium botulinum* types B, E and F and related clostridia as revealed by small-subunit rRNA gene sequences, FEMS Microbiol. Lett. 108 (1993) 103–110.

[63] M. Lindström, H. Korkeala, Laboratory diagnostics of botulism, Clin. Microbiol. Rev. 19 (2) (2006) 298–314.

[64] M. Sebaihia, M.W. Peck, N.P. Minton, N.R. Thomson, M.T. Holden, W.J. Mitchell, et al., Genome sequence of a proteolytic (Group I) *Clostridium botulinum* strain Hall A and comparative analysis of the clostridial genomes, Genome Res. 17 (2007) 1082–1092.

[65] H. Brüggemann, Genomics of clostridial pathogens: Implication of extrachromosomal elements in pathogenicity, Curr. Opin. Microbiol. 8 (5) (2005) 601–605.

[66] H.D. Shukla, S.K. Sharma, *Clostridium botulinum*: A bug with beauty and weapon, Crit. Rev. Microbiol. 31 (2005) 11–18.

[67] D.A. Dance, Ecology of *Burkholderia pseudomallei* and the interactions between environmental *Burkholderia* spp. and human-animal hosts, Acta Trop. 74 (2000) 159–168.

[68] D. Godoy, G. Randle, A.J. Simpson, D.M. Aanensen, T.L. Pitt, R. Kinoshita, et al., Multilocus sequence typing and evolutionary relationships among the causative agents of melioidosis and glanders, *Burkholderia pseudomallei* and *Burkholderia mallei*, J. Clin. Microbiol. 41 (2003) 2068–2079.

[69] W.C. Nierman, D. DeShazer, H.S. Kim, H. Tettelin, K.E. Nelson, T. Feldblyum, et al., Structural flexibility in the *Burkholderia mallei* genome, Proc. Natl. Acad. Sci. USA 101 (2004) 14246–14251.

[70] M.A. Schell, L. Lipscomb, D. DeShazer, Comparative genomics and an insect model rapidly identify novel virulence genes of *Burkholderia mallei*, J. Bacterial 190 (7) (2008) 2306–2313.

[71] J. Gilad, *Burkholderia mallei* and *Burkholderia pseudomallei*: The causative micro-organisms of glanders and melioidosis, Recent Patents Anti-infective Drug Disc. 2 (3) (2007) 233–241.

[72] A. Leelarasamee, Recent development in melioidosis, Curr. Opin. Infect. Dis. 17 (2) (2004) 131–136.

[73] MMWR, Importance of culture confirmation of Shiga toxin-producing *Escherichia coli* infection as illustrated by outbreaks of gastroenteritis—New York and North Carolina, 2005, MMWR Morbid. Mortal. Wkly. Rep. 55 (38) (2006) 1042–1045.

[74] M. Oda, M. Morita, H. Unno, Y. Tanji, Rapid detection of *Escherichia coli* O157:H7 by using green fluorescent protein-labeled PP01 bacteriophage, Appl. Environ. Microbiol. 70 (1) (2004) 527–534.

[75] K. Kogure, E. Ikemoto, Wide occurrence of enterohemorrhagic *Escherichia coli* O157:H7 in natural fresh water, Nippon Saikingaku Zasshi 52 (1997) 601–607.

[76] T.S. Whittam, I.K. Wachsmuth, R.A. Wilson, Genetic evidence of clonal descent of *Escherichia coli* O157:H7 associated with hemorrhagic colitis and hemolytic uremic syndrome, J. Infect. Dis. 157 (1988) 1124–1133.

[77] N.T. Perna, G. Plunkett III, V. Burland, B. Mau, J.D. Glasner, D.J. Rose, et al., Genome sequence of enterohaemorrhagic *Escherichia coli* O157:H7, Nature 409 (6819) (2001) 529–533.

[78] V. Burland, Y. Shao, N.T. Perna, G. Plunkett, H.J. Sofia, F.R. Blattner, The complete DNA sequence and analysis of the large virulence plasmid of *Escherichia coli* O157:H7, Nucleic Acids Res. 26 (1998) 4196–4204.

[79] K. Makino, K. Ishii, T. Yasunaga, M. Hattori, K. Yokoyama, C.H. Yutsudo, et al., Complete nucleotide sequences of 93-kb and 3.3-kb plasmids of an enterohemorrhagic *Escherichia coli* O157:H7 derived from Sakai outbreak, DNA Res. 5 (1998) 1–9.

[80] J.Y. Lim, H. Sheng, K. Seok Seo, Y.H. Park, C.J. Hovde, Characterization of an *Escherichia coli* O157:H7 plasmid O157 deletion mutant and its survival and persistence in cattle, Appl. Environ. Microbiol. 73 (7) (2007) 2037–2047.

[81] E.W. Brown, M.L. Kotewicz, T.A. Cebula, Detection of recombination among *Salmonella enterica* strains using the incongruence length difference test, Mol. Phylogen. e-24 (2002) 102–120.

[82] J.W. Drake, B. Charlesworth, D. Charlesworth, J.F. Crow, Rates of spontaneous mutation, Genetics 148 (1998) 1667–1686.

[83] M.L. Kotewicz, E.W. Brown, J.E. LeClerc, T.A. Cebula, Genomic variability among enteric pathogens: The case of the mutS-ropS intergenic region, Trends Microbiol. 11 (2003) 2–6.

[84] J.E. LeClerc, B.G. Li, W.L. Payne, T.A. Cebula, High mutation frequencies among *Escherichia coli* and *Salmonella* pathogens, Science 274 (1996) 1208–1211.

[85] E.R. Moxon, P.B. Rainey, M.A. Nowak, R.E. Lenski, Adaptive evolution of highly mutable loci in pathogenic bacteria, Curr. Biol. 4 (1994) 24–33.

[86] P.D. Sniegowski, P.J. Gerrish, R.E. Lenski, Evolution of high mutation rates in experimental populations of *Escherichia coli*, Nature 387 (1997) 703–705.

[87] P. Keim, L.B. Price, A.M. Klevytska, K.L. Smith, J.M. Schupp, R. Okinaka, et al., Multiple-locus variable-number tandem repeat analysis reveals genetic relationships within *Bacillus anthracis*, J. Bateriol. 182 (2000) 2928–2936.

[88] P. Keim, K.L. Smith, C. Keys, H. Takahashi, T. Kurata, A. Kaufmann, Molecular investigation of the Aum Shinrikyo anthrax release in Kameido, Japan, J. Clin. Microbiol. 39 (2001) 4566–4567.

[89] C.A. Cummings, D.A. Relman, Microbial forensics: Cross-examining pathogens, Science 296 (2002) 1976–1978.

[90] R.S. Lancoitti, J.T. Roehrig, V. Deubel, J. Smith, M. Parker, K. Steele, et al., Origin of the West Nile virus responsible for an outbreak of encephalitis in the northeastern United States, Science 286 (1999) 2333–2337.

[91] B. Budowle, J. Ge, X.G. Aranda, J.V. Planz, A.J. Eisenberg, R. Chakraborty, Texas population substructure and its impact on estimating the rarity of Y-STR haplotypes from DNA evidence, J. Forensic Sci. 54 (5) (2009) 1016–1021.

[92] A.P. Martin, Phylogenetic approaches for describing and comparing diversity of microbial communities, Appl. Environ. Microbiol. 68 (8) (2002) 3673–3682.

[93] Mes THM, Microbial diversity: Insights from population genetics, Environ. Microbiol. 10 (1) (2008) 251–264.

[94] K.E. Robbins, P.J. Weidle, T.M. Brown, A.M. Saekhou, B. Coles, S.D. Holmberg, et al., Molecular analysis in support of an investigation of a cluster of HIV-I infected women, AIDS Res. Hum. Retroviruses 18 (15) (2002) 1157–1161.

[95] H.-J. Bandelt, P. Forster, A. Röhl, Median-joining networks for inferring intraspecific phylogenies, Mol. Biol. Evol. 16 (1) (1999) 37–48.

[96] N. Saitou, M. Nei, The neighbor-joining method: A new method for reconstructing phylogenetic trees, J. Mol. Evol. 4 (1987) 406–425.

[97] R. Chakraborty, R. Deka, L. Jin, R.E. Ferrell, Allele sharing at six VNTR loci and genetic distances among three ethnically defined populations, Am. J. Hum. Biol. 4 (1992) 387–397.

[98] R. Chakraborty, Sample size requirements for addressing the population genetic issues of forensic use of DNA typing, Hum. Biol. 64 (2) (1992) 141–159.

[99] B. Budowle, K. Monson, R. Chakraborty, Estimating minimum allele frequencies for DNA profile frequency estimates for PCR-based loci, Int. J. Leg. Med. 108 (1996) 173–176.

[100] J. Buckleton, C.M. Triggs, S.J. Walsh (Eds.), Forensic DNA Interpretation, CRC Press, Boca Raton, FL, 2005.

[101] R.E. Colman, A.J. Vogler, J.L. Lowell, K.L. Gage, C. Morway, P.J. Reynolds, et al., Fine-scale identification of the most likely source of a human plague infection, Emerg. Infect. Dis. 15 (10) (2009) 1623–1625.

[102] A. Agresti, Categorical Data Analysis, Wiley, New York, 1990, pp. 48–49.

[103] B.J. Kildare, C.M. Leutenegger, B.S. McSwain, D.G. Bambic, V.B. Rajal, S. Wuertz, 16S rRNA-based assays for quantitative detection of universal, human-, cow-, and dog-specific fecal Bacteroidales: A Bayesian approach, Water Res. 41 (2007) 3701–3715.

[104] S. Kumar, K. Tamura, M. Nei, MEGA: Molecular evolutionary genetics analysis, version 2, Pensylvania State University, University Park, and Arizona State University, Tempe, AZ, 2005.

[105] M. Nei, S. Kumar, Molecular Evolution and Phylogenetics, Oxford University Press, New York, 2000.

[106] W.C. Thompson, F. Taroni, C.G.G. Aitkin. How the probability of a false positive affects the value of DNA evidence, J. Forensic Sci. 48 (1) (2003) 47–54.

[107] F. Pompanon, A. Bonin, E. Bellemain, P. Taberlet, Genotying errors: Causes, consequences and solutions, Nat. Rev. Genet. 6 (2005) 847–856.

[108] T. Tvedebrink, P.S. Eriksen, H.S. Mogensen, N. Morling, Estimating the probability of allelic-drop-out of STR alleles in forensic genetics, Forensic Sci. Int. Genet. 3 (2009) 222–226.

[109] NRC, National Research Council, The Evaluation of Forensic DNA Evidence, Academy Press, National Washington, DC, 1996.

[110] P. Gill, C.H. Brenner, J.S. Buckleton, A. Carracedo, M. Krawczak, W.R. Mayr, et al., DNA Commission of the International Society of Forensic Genetics: Recommendations on the interpretation of mixtures, Forensic Sci Int. 160 (2-3) (2006) 90–101.

[111] B. Budowle, R. Chakraborty, A. van Daal, Authors' response—letter to the editor, J. Forensic Sci. 55 (1) (2010) 269–272.

[112] R. Durbin, S. Eddy, A. Krogh, G. Mitchison, Biological Sequence Analysis: Probabilistic Models of Proteins and Nucleic Acids, Cambridge University Press, Cambridge, 1998.

[113] R.C. Doenier, S. Tavare, M.S. Waterman, Computational Genome Analysis, Springer, New York, 2005.

[114] J.W. Lee, J.B. Lee, M. Park, S.H. Song, An extensive comparison of recent classification tools applied to microarray data, Comput. Stat. Data Anal. 48 (1) (2005) 869–885.

[115] J.A. Hanley, B.J. McNeil, The meaning and use of the area under a receiver operating characteristics (ROC) curve, Radiology 143 (1982) 29–36.

[116] S. Schutzer, P. Keim, J. Czerwinski, B. Budowle, Use of forensic methods under exigent circumstances prior to full validation, Sci. Transl. Med. (8cm7) (2009) 1–3.

# CHAPTER 34

# Biorepositories and Their Foundations— Microbial Forensic Considerations

**Terrance Leighton[a] and Randall Murch[b]**
*[a]Children's Hospital Research Institute, Oakland, California*
*[b]Center for Technology, Security and Policy, Virginia Polytechnic Institute and State University, Alexandria, VA*

## INTRODUCTION: ROLE OF BIOREPOSITORIES AND SYSTEMATICS IN MICROBIAL FORENSICS

### Biorepository Status and Gaps

Biorepositories are integral components of a credible microbial forensics infrastructure that spans research, development, test, evaluation, independent validation, verification, biosurety, analysis, and investigation. The existing biorepository landscape is highly heterogeneous and populated with individual laboratory collections, private collections, organism-specific collections, government collections, select-agent collections, user-group collections, geographical collections, and large-scale public or commercial collections. Many of these repositories are not publically accessible. Commercial repository sample purchase costs can also limit access to portions of their collections. Restrictions on international shipments are also problematic.

Biorepository procedures and processes for deposit, accession meta-data, accession validation, culture expansion (optimal media and growth conditions), preservation, maintenance, validation testing, contamination testing, sample documentation, quality control, quality assurance, recipient distribution data, staff training, biosecurity, and bioinformatics data can be variable and patchy. Accession meta-data such as country of origin, name of collector, date/time/geographic location of isolation, photographic documentation of the collection site, taxonomic identification methods and data, phenotypic/immunotypic/pathotypic/genotypic strain properties, bibliographic references, known distribution restrictions, and accession chain of custody may be incomplete or unavailable.

Data gaps and procedural variations (or variances) can limit or compromise the utility of repository samples in bioforensic test, validation and verification programs, and in "questioned vs. known" investigations for which forensic

Microbial Forensics. DOI: 10.1016/B978-0-12-382006-8.00034-7

science contributes to the quest for attribution. For these reasons it is desirable that uniform procedures, protocols, standards, and practices are implemented by the biorepository community. The bioforensics community also requires phylogenetic characterization of repository strains at the highest possible level of resolution.

## Need for Ultraresolution Microbial Systematics
### What Is "Ultraresolution Bacterial Systematics"?

Bacterial systematics addresses all levels of the taxonomic hierarchy, from distantly related phyla to closely related strains of a single species. We define ultraresolution bacterial systematics as a system for identifying and characterizing the most closely related taxonomic entities, namely, those at the subspecies level, at the deepest level of discrimination possible. As discussed later, this area of systematics requires a new paradigm that enables fine-grained taxonomic accuracy, reliability, and applicability to the most challenging "questioned vs. known" investigations.

### Why Is This Needed?

Microbial systematics has been based primarily on the analysis of small-subunit ribosomal RNA (16S rRNA) sequences (1). This approach, when introduced almost three decades ago, revolutionized our understanding of the phylogenetic relationships among microbes and provided the framework for a natural taxonomy that reflected these relationships. Use of this molecular chronometer revealed the inability of traditional phenetic (morphological and biochemical) methodologies to accurately reconstruct the natural genetic history of the microbial world. When applied to mixed communities of uncultivated microbes, 16S rRNA sequencing revealed an astounding diversity manifested at all taxonomic levels, from phylum to phylotype (1).

However, relationships among taxa at the extremities of the taxonomic spectrum (i.e., distantly related phyla or highly related species and subspecies) have, in many cases, proved recalcitrant to disaggregation by this approach (2). There is often insufficient information within the highly conserved 16S rRNA molecule to resolve deep relationships between phyla (resulting in "star radiation") and, due to its slow rate of evolution, among strains at the species and subspecies level (3). The latter problem is particularly evident in the case of several well-known pathogens that are recent clonal emergences where, for example, strains of the biothreat agent *Bacillus anthracis* cannot be distinguished by 16S rRNA gene sequencing from saprophytic and near-neighbor species such as *B. cereus* and *B. thuringiensis* (4). Species and subspecies of other more divergent genera show sufficient levels of sequence variation such that 16S rRNA serves as an adequate tool, with the result that the species and subspecies systematics landscape is a mosaic of rRNA-based taxonomy and orthogonal data from other methods—serotyping, plasmid typing, virulence gene typing, multilocus sequence typing, variable number tandem repeat typing, and so on (4).

Microbiologists and forensic scientists require a rigorous intellectual framework in which to organize and stratify the microbial world. Current microbial classification systems employ pragmatic, arbitrary, and ad hoc methods embodied within a formalized polyphasic structure to assign species names to microorganisms (2–4). Inflexible rules and arbitrary species boundaries defined by methodologies that are distantly connected to the forces that create microbial diversity are inadequate to delineate the natural order and structure of microbial life. Molecular phylogeny has great potential to connect microbial taxonomy to the diversity engine from which microbiota radiated. However, in traditional taxonomy, change occurs slowly and molecular cladistic approaches have not yet achieved widespread acceptance within the traditional taxonomy community.

Genome sequences are thought of and treated as "the Rosetta stone" for microbial classification, but they have also revealed the limitations of current microbial species concepts. Microorganisms can engage in intraspecies genetic recombination, horizontal interspecies gene flow, phage and plasmid exchange, transposon, and insertion element exchange. There are frequently no bright lines for demarcation of microbial species. Fortunately, there are many well-defined microbial species derived from nonrecombining clonal origins (5).

Advances in microbial systematics will depend on developing tools that are able to stratify related microorganisms into genetically distinct and ecologically cohesive groupings. Current data supporting such groupings are ambiguous due to the collision of recombination and clonality generating evolutionary forces. Improvements in the ability to extract and characterize robust phylogenetic signals are pivotal to progress in high-resolution microbial systematics. Community genomics investigations are also needed to refine our understanding of ecotype and geotype clustering, taxa-area boundaries, and biogeography (6).

A microbial species concept is a way to organize and structure extant diversity, but it is a controversial topic (7). Species categories are of obvious interest to microbial systematists; however, the most engaged parties are users of the species designations: researchers, medical and clinical professionals, forensic scientists, and applied microbiologists. These practitioners need to disentangle bacterial diversity in order to understand and manage ecosystems, human disease, and illegitimate use of bacterial pathogens. Species categories facilitate understanding microbial evolutionary structure based on natural genetic history. There is a lack of compelling data to support the deployment of either a theory-based model or an operationally based model of microbial species. Expanded population genetics and phylogenetics data sets are needed to determine which paradigm is more appropriate for particular microbial taxa or clades.

Current statistical approaches underestimate microbial diversity (8). Linear metrics, in particular, can produce very large errors because they fail to account for population size, age of clades, and variation in rates of nucleotide substitution. Microbial diversity is vast and undersampled (8), but currently available techniques for estimating the extent of this diversity are inadequate and unreliable.

Competing approaches for subspecies designation have not been reconciled, and the limits of subspecies discrimination have not been determined. The theory base for disaggregating biotypes (a population within a species that has a distinct genetic variation, biological structure, or physiological function), serotypes (immunologically distinguishable strains), geotypes (geographically distinct and bounded strain types), pathotypes (strains distinguishable by their host specificity or virulence), and ecotypes (strains distinguishable by their ecological distribution including population genetic structure or function) does not exist. Further, the limits of uncertainty imparted by genome dynamics and gene flow (pan-genomics) at the subspecies level of identification are not known with any degree of confidence (9,10).

Systematics at the subspecies level struggle to support fields and applications (e.g., molecular epidemiology and microbial forensics), which require rigorous, precise, and high-resolution assignments, intercomparisons, and unequivocal bases upon which identifications and comparisons are made (9). Ultraresolution systematics could provide the foundation for resolving and unifying subspecies identification into a single harmonized and consistent system of taxonomy, phylogenetic reconstruction, and nomenclature that would apply across taxa at the species level and below.

### What Fields Would Be Significantly Advanced by This Research?

Improvement in our ability to stratify subspecies would clearly have effects on the field of systematics itself—standardization and reconciliation of various approaches, definition of methods, and resolution limits for differentiation and taxonomy. The interaction between systematics and other subdisciplines of microbiology suggests strongly that such a revision would have wide-reaching effects. Comprehensive capabilities would inform our assessment of biological diversity by providing a universal framework for "measuring" and "visualizing" subspecies diversity. This, in turn, would have collateral ramifications for the fields of microbial ecology, molecular epidemiology (consistent, ultraresolution systematics combined with source tracking of infectious agents), and microbial forensics (sample identification, characterization, discrimination, and source attribution, i.e., "matching," association and exclusion). The emerging fields of microbial genomics and metagenomics could also benefit enormously given current concerns about the effect of multiple closely related strains on genome assembly kinetics and reliability (11).

Ultraresolution systematics is central to ensuring that biorepository collections are representative with respect to geography, ecology, pathobiology, and population genetics. A path forward to ultraresolution systematics is discussed later.

## OPERATIONAL BIOREPOSITORY COLLECTION VERIFICATION AND VALIDATION CONCERNS

Infectious disease samples can be retained for decades with minimal annotation and curation. There have been a number of instances of laboratory sample cross contamination with human immunodeficiency virus (12), simian immunodeficiency virus (13), poliovirus (14,15), and influenza viruses (16). Archival vaccine stocks have been shown to be contaminated with hepatitis B (17), SV40 (17), pseudocowpox (18), and bluetongue (19) viruses. The dimensions of biorepository sample cross contamination, misclassification, or misannotation are unknown and underestimated due to the lack of broadband pathogen assessment of these collections.

Use of misidentified biorepository samples for diagnostic and proficiency testing has contributed to biosecurity risks and erroneous test data. Beginning in September 2004 and continuing through March 2005, the College of American Pathologists (CAP) distributed a yearly test panel of influenza A and B viruses to 3747 public health laboratories in 18 countries. Contained in this panel was a misidentified influenza A virus—the H2N2 strain—which caused the 1957–1958 Asian pandemic that killed over one million people worldwide (20). The H2N2 contamination of the CAP panels was discovered fortuitously on March 26, 2005, by cross-contamination of a patient sample within a laboratory biosafety cabinet with the H2N2 containing test sample (21). The source of the H2N2 virus was a commercial culture collection repository that ships over 150,000 biological samples each year. The "influenza A" test samples were assembled and shipped by a CAP proficiency testing contractor. A massive global effort was launched on April 12, 2005, by the World Health Organization (WHO), Centers for Disease Control and Prevention (CDC), Department of State, and Health and Human Services (HHS) to notify the laboratories that received the H2N2 containing test panels and to provide guidance for verification of the destruction of the H2N2 samples (22). On April 27, 2005, CAP reported that all 3747 laboratories had destroyed the H2N2 samples (23). HHS subsequently discovered that other proficiency testing providers had also sent H2N2-containing samples to additional laboratories in the United States (20). The disposition of these samples was unclear. As a result of these events, the CDC raised the biosafety level of the H2N2 virus from BSL2 to BSL3 (24).

The H2N2 virus continued to circulate until 1968. Persons born after 1968 would be expected to have limited, if any, immunity to H2N2. H2N2 antigens are not contained in any contemporary influenza vaccine. The human transmissibility of H2N2 and a large unprotected population highlight the substantial biorisks associated with the accidental release of hazardous bioagents even within advanced and well-developed medical diagnostic infrastructures.

The WHO declared poliovirus eradicated in the European region in June 21, 2002, and certified that the European region was free of indigenous wild poliovirus transmission by destruction or containment of infectious materials. In 2005 the Russian Federation completed a similar wild poliovirus destruction and biosecurity inventory protocol. In late 2004 the Moscow WHO regional polio reference laboratory received samples of poliovirus type 1 from a collaborating non-WHO laboratory (15). Isolates were reported as originating from samples collected on different days and from different locations. The submitting laboratory reported in 1997 that it had destroyed all wild poliovirus materials. An investigation of the laboratory revealed opportunities for virus cross contamination and identified a 7-year-old mislabeled Sabin vaccine strain sample that contained a wild poliovirus 1 Mahoney strain. This incident resulted in the ordered destruction of all poliovirus materials within the Russian Federation labeled as vaccine or Sabin strains at 2355 laboratories and their replacement with authenticated Sabin stocks from the Moscow WHO regional reference laboratory. As in the case of distribution of the pandemic H2N2 stain, the mislabeled poliovirus 1 Mahoney strain was detected fortuitously by laboratory cross contamination of samples submitted to a reference laboratory.

These examples of inadequate biorepository validation and verification control systems, and their belated serendipitous discovery, illustrate the need for strengthened pathogen verification, validation, and sample release procedures and practices. The following sections discuss examples of technologies that can improve biorepository quality control, quality assurance, and biosurety.

## Broadband and Fine-Grained Sample Characterization Technologies
### Triangulation Identification for Genetic Evaluation of Risks (TIGER)
The TIGER broadband pathogen detection system (25) was initially developed by IBIS Therapeutics and Science Applications International Corporation

through a Defense Advanced Research Projects Agency (DARPA)-funded program beginning in 2000. The goal of this program was to develop a single-pass, DNA-based detection system capable of identifying all known biological warfare threats with high sensitivity and low false alarm levels. An additional requirement was to be able to identify unknown threats. The initial platform used a high-performance electrospray Fourier transform ion cyclotron resonance mass spectrometer (MS) to provide base counts for each of the four nucleotide bases contained in a short (<140 nucleotides) DNA amplicon. A more compact and automated version of the TIGER system (T5000) uses a tabletop Bio-TOF as the MS detector (26). TIGER does not evaluate the DNA amplicon nucleotide sequence, but only the numbers of each base (base counts). For threat detection or forensics applications, samples are initially assayed with a set of broad-range polymerase chain reaction (PCR) primers designed to amplify relatively conserved sequences (containing internal variable regions) spanning a broad-range of pathogens. Each DNA amplicon is unique due to its characteristic nucleotide counts, and this constrains the possible range of organisms that could be present. A second set of confirmation PCR primers focuses more precisely on a limited target clade. High-resolution genotyping of specific bacterial and viral species is accomplished by utilizing species-specific primers that interrogate regions of high intraspecies variability to distinguish closely related strains, pathotypes, and biotypes (27). The second "drill-down" step is followed by a mass deconvolution of the nucleotide counts obtained by mass spectrometry to provide species- or strain-level genotypic information.

To assess the landscape of bacteria present in diagnostic or clinical samples, a set of 16 broad-range surveillance primers are used that allow PCR amplification and quantitative identification of many different bacterial pathogens in a single assay (27). These broadband primers were chosen by computational analysis of sequence alignments of all available ribosomal DNA operons and 160 broadly conserved protein-encoding housekeeping genes. The ribosomal DNA-targeted primers have the broadest range of bacterial coverage. For example, four designed primer pairs targeted to 16S ribosomal DNA match 98% of the bacterial sequences in the Ribosomal Database Project when allowing for two to three mispairings under permissive PCR cycling conditions. Primers targeted to protein-encoding housekeeping genes have breadth of coverage at the level of major bacterial subdivisions (e.g., β proteobacteria, bacilli, streptococci, staphylococci) (28). Although any single primer target region might have an overlap of base compositions with other species, combined information from multiple primer pairs provides unambiguous organism-specific signatures for all major bacterial pathogens. Pan-microbial microarrays offer an alternative approach to PCR-based unbiased pathogen discovery and surveillance (29).

Triangulation identification for genetic evaluation of risks can resolve the quantitative composition of mixed DNA samples. For example, it can identify a plasmid, or a microbial contaminant, and quantify the ratio of amplicons to that of the reference genome. This capability makes TIGER more useful than conventional genotyping approaches that are typically only semiquantitative. An additional advantage of TIGER when compared to qPCR methods is that TIGER is robust to detection of unknown organisms and is able to phylogenetically position these unknowns by proximity to known organisms in its database, which is updated frequently from GenBank and multilocus sequence typing (MLST) data sources.

RNA viruses can be analyzed with an initial reaction using reverse transcriptase (e.g., filoviruses, SARS, and influenza virus strains) (30). TIGER assays for direct broadband PCR characterization of DNA viruses (e.g., adenovirus and orthopoxvirus) (30) have also been developed.

Variants recognized by these advanced detection systems can subsequently be fully characterized by large-scale genome sequencing (31) and ultrahigh-throughput DNA sequencing techniques (32).

### Multilocus Sequencing Typing

An ideal subspecies typing method unambiguously characterizes sequence differences across a population and thereby enables digital identification of an organism and discriminates among phylogenetic-meaningful groupings. Multilocus sequence typing adapts the concept of allelic profiles used originally in multilocus enzyme electrophoresis typing (MLEE) (33). It was observed that particular combinations of MLEE alleles (electrophoretic types, ETs) were found significantly more often than would be expected for recombining organisms. These results suggested that microbial populations could have clonal population structures. MLEE had similar problems to amplified fragment length polymorphisms (AFLP) and multilocus variable number tandem repeat analysis (MLVA) in that data were difficult to reproduce between laboratories and one had to compare each new strain to all other defined ETs.

Multilocus sequence typing directly determines the actual DNA sequence of the compared genes, as opposed to examining indirect properties associated with sequence differences. MLST data sets are defined rigorously by DNA sequences and associated data, which are deposited in reference databases (34,35). These databases can be queried with sequence information from any new isolate and the phylogenetic association of the queried strain determined directly by reconstruction algorithms. MLST has superior precision and discriminatory power when compared to other available typing methods (36).

Multilocus sequence typing schemes index variation in slowly evolving and nonrecombining housekeeping genes to identify and differentiate among organisms. Housekeeping genes encode proteins that are involved in essential metabolic processes and are, in most cases, selectively neutral. Those genes appropriate for use in MLST are not subject to diversifying selection and are not generally involved in horizontal gene transfer. Moreover, they are not located near genes that are subject to such evolutionary processes. Ideally, MLST genes are flanked by other housekeeping genes. In the most optimal cases, candidate housekeeping genes are identified and selected for inclusion into an MLST typing system through detailed analysis of a fully sequenced reference genome of the same or closely related microorganism.

In order for an MLST typing system to be meaningful and effective, the same regions of the targeted genes must be sequenced across a phylogenetically relevant population. Rather than sequence the entire protein-encoding sequence, which is laborious, expensive, and time-consuming, MLST systems typically focus on a 450- to 500-bp sequence in a conserved portion of the gene. Although examining less sequence diversity theoretically decreases the discriminatory power of a typing system, MLST sequence types are contained in conserved genes where the quantity of sequence examined will not significantly affect the discriminatory power of the system. The utility of a typing system is based on the ability to interrogate the same gene set across a population; primers used in the sequencing reaction must support PCR amplification and DNA sequencing reactions from all relevant pathotypes, ecotypes, geotypes, and so on. As in the case of 16S phylogeny, the targeted gene set regions must be sufficiently diverse to allow fine-grained discrimination among phylogenetic groups. This critical requirement substantially limits the combinations of genes that are appropriate for inclusion in a typing scheme.

Multilocus sequence typing analysis involves the abstraction of DNA sequence information into allelic profiles. Each sequence variant at a locus is considered an allele and is assigned an allelic number (an integer). Isolates with the same sequence at a locus are assigned the same allelic number. Loci with greater numbers of allelic variants will have more resolving power. The combination of alleles at all loci in an MLST typing system is referred to as the "allelic profile." An allelic profile is a representation of DNA sequence differences across the set of targeted loci. In MLST systems, "clones" (phylogenetic grouping) are defined operationally as a unique allelic profile. Clonality is only defined within the context of the MLST gene set used and may not be supported by other gene sets entrained in ecological or pathogenesis adaptation.

The discriminatory power and the clonality assignments provided by a gene set are functions of the number of genes and the diversity of those genes within the population. Discriminatory power is defined as the probability of obtaining (at random) an isolate with the most common allele at each locus. Lower probabilities correspond to higher discriminatory power. The discriminatory power of an MLST system resides in the diversity of the housekeeping genes included in the system. In clonal populations, which lack diversity, particularly in conserved sequences such as housekeeping genes, MLST is not likely to be discriminatory below the subspecies level and may not be congruent with traditional species concepts, particularly in microbial groups that do not have clonal structures (37).

The power of MLST in epidemiological investigations has been demonstrated repeatedly. The significant advantages of MLST are the repeatability and reliability of DNA assays, interpretability of results, and digital precision of sequencing data. The phylogenetic signal originating from clonal bacterial populations that exhibit linkage disequilibrium and regular diversity purging can be recovered from sampling nucleotide variation at a limited number of housekeeping gene sites. These DNA sequence data can also be used to reconstruct the evolutionary history of questioned isolates and establish their phylogenetic relationship to other members of the population. A phylogeny can then be used as a framework for association with other characteristics, such as pathogenicity, serology, host specificity, biogeography, and ecology. The ability to reconstruct phylogenies from single gene or concatenated MLST sequence data provides a rigorous means for phylogenetically positioning unknown isolates with high levels of statistical support (i.e., parsimony, likelihood, Bayesian). These phylogenetic analysis tools are able to produce statistically supported population genetic frameworks that are not obtained easily from AFLP or MLVA data.

One weakness of MLST is its discriminatory power for nonclonal populations where recombination or horizontal gene flow plays a major role in determining population structure (32,34). MLST is an extremely effective typing technology and warrants further development for biorepository and bioforensic applications.

The power of MLST is in subspecies, genotype, pathotype, and ecotype discrimination. The limitations of MLST are that frequently MLST cannot provide interspecies linkages and connections with higher order phylogenetic clustering. The next section discusses the use of molecular graticules that allow the positioning of MLST sequence types within a larger phylogenetic context.

## BEYOND MLST—EXTENDED RANGE ULTRARESOLUTION TYPING

Unambiguous and precise genetic classification of microorganisms is of pivotal importance to microbial forensics, particularly in questioned vs. known comparisons. One of the most challenging clades of microbial forensic interest is the *B. anthracis* group (2–4), which is also of significant interest for use as a biological weapon. Taxonomic relationships between *B. anthracis* and other closely related spore-forming species such as *B. cereus*, *B. thuringiensis*, *B. mycoides*, *B. pseudomycoides*, and *B. weihenstephanensis* have been controversial (4). We have explored the utility of spore structural constituents as tools for investigating the molecular evolution of the *B. anthracis* group. Among the most phylogenetically informative molecules identified was the γ-type small acid-soluble spore core protein, SspE (38). This gene is unexpectedly useful for high-resolution genotyping, as it appears to have arisen within the *Bacillus* genus, has a different sequence in ecologically distinct populations, and has a rapid rate of sequence evolution that provides fine-grained phylogenetic discrimination.

We have described previously the use of this spore structural gene for PCR assays that discriminate between *B. anthracis* clade species and subspecies (38). Phylogenetic analyses of *sspE* DNA sequence data sets from 224 *B. anthracis* clade strains available from publicly accessible biorepositories suggest that this clade is more phylogenetically complex than had been inferred by traditional classification methods (39). Forty-one *sspE* genotypes and 21 protein phylotypes were identified among the *B. cereus*, *B. thuringiensis*, *B. anthracis*, *B. mycoides*, *B. pseudomycoides*, and *B. weihenstephanensis* strains analyzed. The most extensive genetic diversity was seen in *B. thuringiensis* strains representing 78 serovar classes. In most cases, strains within a particular *B. thuringiensis* serovar showed identical *sspE* sequence, suggesting a significant correlation of *sspE* sequence clustering with serotyping. SspE phylogeny suggests that *B. cereus* and *B. thuringiensis* are more closely related to each other than to *B. anthracis* and *B. mycoides*. These results suggest that the *sspE* gene set possesses emergent DNA sequence properties, which enable the systematic study of natural variation within the *B. anthracis* group. A particularly powerful aspect of SspE phylogenetic reconstruction is the ability to generate self-ordering systematic data properties that naturally position *B. anthracis* taxa into stratifying groups. These attributes allow the creation of an internally consistent interpretative framework for the phylogenetic analysis of *B. anthracis* group microorganisms. These studies also extend the previously demonstrated utility of protein-coding genes that are entrained in the speciation process for the phylogenetic separation of closely related groups (3).

We summarize here further development of a phylogenetic positioning methodology that reliably and unambiguously identifies and stratifies members of the *Bacillus* genus that are of bioforensic, ecological, and commercial interest. This typing system, which incorporates the *sspE* gene and a seven-gene MLST set (40), has been used to extend the phylogenetic range of MLST to span several important bacillus clades (*B. anthracis* and *B. subtilis* groups) and reliably reconstruct the natural genetic history of over 380 *Bacillus* isolates that we have analyzed (39).

The orthogonal combination of these *sspE* and MLST methods provides a powerful means for identifying ecologically distinct bacterial populations. The *sspE*/MLST typing ultraresolution assay can be thought of as a digital identifier or zip code for *Bacillus* spp., where *sspE*, or a gene with similar resolving capability, is the equivalent of the first digit 9 in the Oakland, California, zip code. Genotyping by MLST, or other comparable multigene schemes, would be similar to the "4609" portion of the zip code—without the first digit, latter digits may not be decisively informative. Only with combined informative data in the proper order of descending importance, 94609, do we identify Oakland, California, in which one author's laboratory (TL) is located.

An example of the application of the *sspE*/MLST typing method to the *B. thuringiensis* group is shown in Table 34.1 (39). Note that strains highlighted in gray, which are apparently misidentified, were accessions obtained from publicly available culture collections. These results suggest that the *sspE*/MLST typing method provides a graticule coordinate system for phylogenetic parsing of biorepository accessions that can identify strains that may be improperly annotated or misidentified. These results suggest that accession data gaps and validation variations can limit or compromise the utility of repository samples in bioforensic test, validation, and verification programs and inject uncertainty into questioned vs. known investigations. For these reasons, it is desirable that ultraresolution procedures, protocols, standards, and practices are implemented by the biorepository community to phylogenetically characterize collections at the highest possible level of granularity and discrimination.

# CHALLENGES

## Biorepository Quality Assurance and Quality Control

Forensic scientists are acutely aware of the adverse impacts of misidentified or contaminated evidence on investigational procedures, processes, and conclusions. The credibility, reliability, and transferability of biorepository samples are critical to the full spectrum of microbial forensic activities. A larger vision of biorepository essentiality, criticality, and functionality is required to support this emerging field of science. Unmet bioforensic gaps and needs have been discussed in previous sections of this chapter. These issues await further attention.

## Ultraresolution Systematics

The *B. anthracis* clade has been used as an example of the challenges confronting the current resolving power of biosystematics within a microbial forensic context. *B. anthracis* is a well-studied human and animal pathogen (4). There are, however, significant knowledge gaps regarding its natural geographical distribution, population biology, and reservoirs in environments outside

**Table 34.1** *Bacillus anthracis* Group Strains Organized by *sspE*/MLST Classifier

| Classifier[a] | SspE aa Group[b] | *sspE* nt Group[c] | SspE Size (AA)[d] | MLST ST[e] | Commercial/Insecticidal Utility[f] |
|---|---|---|---|---|---|
| A1a | A | A1 | 93 | 8 | **kurstaki** (**Lepidoptera**[1,6,7,14,17,20,29,31,36,41,45,47,51,55,57,58,80–84,87,92,98]) and (**Diptera**[36,47,55,58,77,98,99]); **galleriae** (**Lepidoptera**[2,20,29,102]); **entomocidus/subtoxicus** (**Lepidoptera**[20,31,32,36,41,60,87]) |
| A1b | A | A1 | 93 | 13 | **kenyae** (**Lepidoptera**[29,31,36,41,94]) and (**Diptera**[77]); **aizawai/pacificus** (misidentified) |
| A1c | A | A1 | 93 | 15 | **aizawai/pacificus** (**Lepidoptera**[8,20,22,24,29,31,36,46,47,51,55,83]) and (**Diptera**[22,24,47,77,89]); **colmeri** (**Diptera**[23,100]) and (**Lepidoptera**[23]) |
| A1d | A | A1 | 93 | 25 | **galleriae** (**Lepidoptera**[2,20,29,102]); *wuhanensis*; ATCC 29730 |
| A1e | A | A1 | 93 | 29 | *kurstaki* (misidentified[9]) |
| A1f | A | A1 | 93 | 34 | ATCC 11778 |
| A1g | A | A1 | 93 | 138 | **dakota**[40,42,61]; B-21619[107]; *asturiensis* |
| A1h | A | A1 | 93 | 225 | *londrina*[21] |
| A1i | A | A1 | 93 | 232 | **coreanensis**[21,61] |
| A1j | A | A1 | 93 | 238 | *yosoo* |
| A1k | A | A1 | 93 | 241 | *indiana* |
| A1l | A | A1 | 93 | 251 | *jinghongiensis*[21] |
| A1m | A | A1 | 93 | 263 | **japonensis** (**Lepidoptera**[29,96,97]) and (**Coleoptera**[33,62,79]) |
| A2a | A | A2 | 93 | 226 | **nigeriae** and **nigeriensis** (**Lepidoptera**[36,69]) |
| A2b | A | A2 | 93 | 244 | **nigeriae** (**Lepidoptera**[36,69]) |
| B1a | B | B1 | 93 | 221 | **entomocidus/subtoxicus** (**Lepidoptera**[20,31,32,36,41,60,87]) |
| B1b | B | B1 | 93 | 239 | **entomocidus/subtoxicus** (**Lepidoptera**[20,31,32,36,41,60,87]) |
| C1a | C | C1 | 93 | 22 | **tolworthi** (**Lepidoptera**[20,29,69,92]) and (**Coleoptera**[15,74,88]) and (**Diptera**[77]) |
| D1a | D | D1 | 93 | 255 | ATCC 13472 |
| F1a | F | F1 | 93 | 213 | **fukuokaensis** (**Diptera**[21,39,63,73,77,78,101]) and (**Lepidoptera**[63,96,97]); **sumiyoshiensis** (**Lepidoptera**[36,96,97]) |
| F2a | F | F2 | 93 | 33 | *pirenaica*[71]; *B. licheniformis* NRRL B-571 (misidentified) |
| F2b | F | F2 | 93 | 59 | **kumamtoensis** (**Coleoptera**[74]) |
| F3a | F | F3 | 93 | 50 | **canadensis** (**Diptera**[21,39,73,77]) |
| F3b | F | F3 | 93 | 224 | **mexicanensis** (**Diptera**[77]) |

(Continued)

**Table 34.1** *Bacillus anthracis* Group Strains Organized by *sspE*/MLST Classifier (*Continued*)

| Classifier[a] | SspE aa Group[b] | *sspE* nt Group[c] | SspE Size (AA)[d] | MLST ST[e] | Commercial/Insecticidal Utility[f] |
|---|---|---|---|---|---|
| **F4a** | F | F4 | 93 | 4 | ATCC 14579[T] |
| **F4b** | F | F4 | 93 | 17 | *pakistani*[21] |
| **F4c** | F | F4 | 93 | 142 | *iberica*[71] |
| **F4d** | F | F4 | 93 | 220 | *vazensis; rongseni* |
| **G1a** | G | G1 | 93 | 212 | ***yunnanensis*** (**Isoptera**[12]) |
| **H1a** | H | H1 | 93 | 111 | Pey9 and 3466-8.1—no serotype, natural isolates |
| **H1b** | H | H1 | 93 | 218 | *xiaguangiensis* |
| **H1c** | H | H1 | 93 | 223 | 2A6 and 2C1—no serotype, natural isolates |
| **H1d** | H | H1 | 93 | 249 | Pey8—no serotype, natural isolate |
| **H2a** | H | H2 | 93 | 208 | ***amagiensis*** (**Diptera**[77]) and (**Lepidoptera**[36]) |
| **H2b** | H | H2 | 93 | 209 | *cameroun*[21] |
| **H2c** | H | H2 | 93 | 227 | ***kyushuensis*** (**Diptera**[21,39,48–50,73,77,101]) |
| **H2d** | H | H2 | 93 | 228 | ***neoleonensis*** (**Diptera**[72,103]) and (**Anti-cancer**[61]) |
| **H2e** | H | H2 | 93 | 233 | ***shandongiensis*** (**Anti-cancer**[53,54,61,66]) and (**Diptera**[39,77]) |
| **H2f** | H | H2 | 93 | 158 | *seoulensis* |
| **H2g** | H | H2 | 93 | 258 | Pey6—no serotype, natural isolate |
| **H3a** | H | H3 | 93 | 206 | ATCC 53522; ATCC 55609[104] |
| **H3b** | H | H3 | 93 | 210 | *silo*[21] |
| **H3c** | H | H3 | 93 | 242 | ***tohokuensis*** (**Diptera**[77]) |
| **H3d** | H | H3 | 93 | 243 | ***ostriniae*** (**Lepidoptera**[69]) |
| **H4a** | H | H4 | 93 | 10 | ***thuringiensis*** (**Lepidoptera**[4,20,21,29,31,36,41,47,83,92]) and (**Coleoptera**[4,36]) and (**Diptera**[38]); *kurstaki* (misidentified) |
| **H4b** | H | H4 | 93 | 204 | *B. megaterium* ATCC 55000[105] (misidentified) |
| **H4c** | H | H4 | 93 | 229 | ***sooncheon*** (**Isoptera**[12]) |
| **H4d** | H | H4 | 93 | 236 | *kim* |
| **H4e** | H | H4 | 93 | 256 | ***thuringiensis*** (misidentified) (**Lepidoptera**[4,20,21,29,31,36,41,47,83,92]) and (**Coleoptera**[4,36]) and (**Diptera**[38]) |
| **H5a** | H | H5 | 93 | 12 | ***sotto/dendrolimus*** (**Lepidoptera**[20,31,86]) and (**Diptera**[65]); ***alesti*** (**Lepidoptera**[20,29,92]) and (**Diptera**[65]); *palmanyolensis* |
| **H5b** | H | H5 | 93 | 16 | ***israelensis*** (**Diptera**[3,9–11,13,18,21,34,43,44,49,50,70,78,83,90–93,95,98]); *malayensis*; *Bacillus* spp. BGSC 18A1 (reclassified) |
| **H5c** | H | H5 | 93 | 23 | ***morrisoni*** (**Coleoptera**[15,21,29,36,56,57,74,85]) and (**Lepidoptera**[19,21,29,36,41,67,69]) and (**Diptera**[18,19,21,51,67,73]); ***thompsoni*** (**Diptera**[21,59,72,73]) and (**Lepidoptera**[5]) |

(Continued)

**Table 34.1** *Bacillus anthracis* Group Strains Organized by *sspE*/MLST Classifier (*Continued*)

| Classifier[a] | SspE aa Group[b] | *sspE* nt Group[c] | SspE Size (AA)[d] | MLST ST[e] | Commercial/Insecticidal Utility[f] |
|---|---|---|---|---|---|
| H5d | H | H5 | 93 | 56 | *darmstadiensis* (**Diptera**[16,21,39,49,68,72,73,77,101,103]) and (**Lepidoptera**[38,96,97]) |
| H5e | H | H5 | 93 | 230 | *leesis* (**Diptera**[21,28]) |
| H5f | H | H5 | 93 | 197 | *sotto/dendrolimus* (**Lepidoptera**[20,21,31,86]) and (**Diptera**[65]) |
| H5g | H | H5 | 93 | 264 | *poloniensis* |
| H5h | H | H5 | 93 | 265 | *zhaodongensis* |
| I1a | I | I1 | 93 | 257 | *morrisoni* (misidentified) (**Lepidoptera**[19,21,29,36,41,67,69]) and (**Diptera**[18,19,21,51,67,73]) and (**Coleoptera**[36,74]) |
| J1a | J | J1 | 93 | 231 | *B. mycoides* ATCC 19647 (misidentified) |
| E1a | E | E1 | 93 | 26 | ATCC 15816 |
| E1b | E | E1 | 93 | 164 | Bc ATCC 13061; *canadensis* (misidentified) |
| E1c | E | E1 | 93 | 205 | *B. subtilis* ATCC 55675[106] (misidentified) |
| E1d | E | E1 | 93 | 266 | BGSC 6A9 |
| E2a | E | E2 | 93 | 171 | *finitimus*[21] |
| E2b | E | E2 | 93 | 246 | *Bacillus* spp. ATCC 51912 (reclassified) |
| E3a | E | E3 | 93 | 211 | *konkukian* (**Diptera**[100]) |
| E4a | E | E4 | 93 | 75 | DM55—no serotype, natural isolate |
| E4b | E | E4 | 93 | 108 | BGSC 6E1; BGSC 6E2 |
| E4c | E | E4 | 93 | 109 | 003, IB, BuIB, III, III-BL, III-BS, IV—no serotypes, natural isolates |
| E4d | E | E4 | 93 | 163 | S8553/2—no serotype, natural isolate |
| E5a | E | E5 | 93 | 219 | *graciosensis* |
| E6a | E | E6 | 93 | 234 | *chanpaisis* |
| E7a | E | E7 | 93 | 104 | *tochigiensis* |
| E8a | E | E8 | 93 | 38 | ATCC 4342 |
| E8b | E | E8 | 93 | 103 | *roskildiensis* (**Isoptera**[12]) |
| E9a | E | E9 | 93 | 32 | ATCC 10987 |
| E10a | E | E10 | 93 | 78 | strain G9241[30] |
| E11a | E | E11 | 93 | 268 | strain ZK (E33L[25]) |
| K1a | K | K1 | 93 | 247 | *guiyangiensis*[21] |
| K2a | K | K2 | 93 | 106 | *brasiliensis* |
| K2b | K | K2 | 93 | 110 | *pulsiensis* |
| K2c | K | K2 | 93 | 112 | *pondicheriensis* |
| K2d | K | K2 | 93 | 113 | *konkukian* strain 97-27[25,26] |
| K2e | K | K2 | 93 | 214 | *higo* (**Diptera**[35,58,64,75,76,78]); *oswaldocruzi*[21] |
| K2f | K | K2 | 93 | 237 | *sylvestriensis* |

(Continued)

**Table 34.1** *Bacillus anthracis* Group Strains Organized by *sspE*/MLST Classifier (*Continued*)

| Classifier[a] | SspE aa Group[b] | *sspE* nt Group[c] | SspE Size (AA)[d] | MLST ST[e] | Commercial/Insecticidal Utility[f] |
|---|---|---|---|---|---|
| **K2g** | K | K2 | 93 | 254 | *azorensis* |
| **K3a** | K | K3 | 93 | 216 | *wratislaviensis*; *pingluonsis* |
| **K3b** | K | K3 | 93 | 250 | *argentinensis* |
| **K3c** | K | K3 | 93 | 262 | *balearica*[37] |
| **L1a** | L | L1 | 93 | 207 | *toguchini*[21,52] |
| **M1a** | M | M1 | 93 | 217 | *muju* |
| **M1b** | M | M1 | 93 | 245 | I2—no serotype, natural isolate |
| **N1a** | N | N1 | 92 | 107 | *monterrey*[21] |
| **O1a** | O | O1 | 95 | 1 | *B. anthracis* |
| **O1b** | O | O1 | 95 | 2 | *B. anthracis* |
| **O1c** | O | O1 | 95 | 3 | *B. anthracis* |
| **P1a** | P | P1 | 95 | 1 | *B. anthracis* (strain Western NA) |
| **Q1a** | Q | Q1 | 93 | 115 | *B. weihenstephanensis* DSM 11821[T] |
| **Q1b** | Q | Q1 | 93 | 116 | *B. mycoides* ATCC 6462[T] |
| **Q1c** | Q | Q1 | 93 | 215 | *novosibirsk* (misidentified) |
| **Q1d** | Q | Q1 | 93 | 235 | *navarrensis*[37] (misidentified) |
| **Q1e** | Q | Q1 | 93 | 248 | *B. mycoides* ATCC 11986 |
| **R1a** | R | R1 | 93 | 222 | *B. mycoides* ATCC 23258 |
| **S1a** | S | S1 | 92 | 267 | *B. mycoides* ATCC 21929 |
| **T1a** | T | T1 | 95 | 259 | *B. mycoides* ATCC 10206 |
| **T1b** | T | T1 | 95 | 260 | *B. mycoides* ATCC 31101 |
| **T1c** | T | T1 | 95 | 261 | *B. mycoides* ATCC 31102 |
| **U1a** | U | U1 | 95 | 114 | *B. pseudomycoides* DSM 12442[T] |

*Note: For references in Table 34.1, please see Wheeler and Leighton (39).*

[a]*Classifiers are **bold typed** and describe species, subspecies, and serovars of the B. thuringiensis clade by combined sspE (capital letter and number) and MLST [lowercase letter corresponds to a sequence type (ST)] within a particular sspE type.*

[b]*Translated nucleic acid sequence of the sspE gene produces SspE proteotype groups A–U.*

[c]*Nucleic acid sequences of the sspE gene are assigned to genotypes A1-x through U1-x, where the **letter** corresponds to the SspE proteotype and the **number** corresponds to a unique nucleic acid sequence of that proteotype. For example, there is only one genotype identified for proteotype U and five genotypes identified for SspE proteotype H (thus, the five H genotypes all have silent mutations with respect to each other). sspE sequence data from this study have been deposited in the GenBank nucleotide sequence database with accession numbers AF359764–AF359821, AF359823–AF359843, AF359845, AF359847–AF359860, AF359862–AF359934, AF359936–AF359938, and DQ146892–146926.*

[d]*Length of the SspE protein (92–95 amino acids, Bc group).*

[e]*The MLST sequence type (**ST**) is a **number** assigned to a unique allelic profile from nucleotide sequences of seven housekeeping gene fragments. Genes used in this scheme are glpF, gmk, ilvD, pta, purH, pycA, and tpiA, and information, including primer sequences, allelic profiles and STs, allele sequences, and isolate information, is available at pubmlst.org/bcereus. Allelic profiles for STs "267" and "268" have not yet been uploaded to the pubmlst/bcereus Web site.*

[f]*Serovars currently used commercially as insecticides or that are registered for use with the USEPA or that are described in scientific literature as insecticidal are indicated in **bold italic** font and target order insects (or other commercial application). Species or serovars that are misidentified or misclassified are highlighted in gray.*

[g]*This "kurstaki" isolate was likely misidentified by the researchers who isolated it. The culture collection agrees that, based on methods used to isolate this strain and that it has no reaction to any known Bt antisera, it is very likely B. cereus.*

of infected hosts. *B. anthracis* virulence factor gene flow among *B. cereus/ thuringiensis* near-neighbor strains has produced transitional pathogens that cause primate (41) and human diseases (42) very similar to *B. anthracis* (4) etiologies. These *B. cereus/thuringiensis* mammalian pathotypes would not be recognized by traditional *B. anthracis* (4) typing methods.

This chapter has discussed the need for and development of an ultraresolution graticule coordinate system combining MLST and spore structural gene tools for the phylogenetic parsing of this challenging clade (39). These results and similar results of others (43) suggest that the appropriate orthogonal combination of phylogenetically informative technologies can extend the range, discrimination power, applicability, stability, and consistency of microbial forensic typing methodologies. These ultraresolution tools will also find use in the emerging field of microbial trace evidence investigations due to the environmentally ubiquitous distribution of spore-forming bacteria. Similar tool sets are required for the other major clades of microbial forensic importance. It is unclear what genetic equivalents of bacillus spore structural genes (which are entrained in the ecogenetic process of speciation) can be identified that will extend the coverage range of MLST typing systems to near-neighbor taxa that must be resolved from select agents and biothreats (4). Comparative genomics based on ultraresolution methodologies has considerable potential to contribute to addressing these unmet needs, as has been demonstrated by epidemiolgcal and microevolutionary studies of a globally dominant strain of methicillin-resistant *Staphylococcus aureus* (MRSA) (44). Using ultrahigh-throughput DNA sequencing and single-nucleotide polymorphism typing methods, Harris and colleagues (44) were able to reveal the global geographic population structure and intercontinental epidemiological transmission mechanisms controlling the spread of MRSA ST239 over four decades.

## Bioagent Match Criteria

At present there is no fully defined, validated, standardized set of criteria to determine biothreat agent identity. "Sameness" or "difference" has not been adequately defined conceptually or experimentally.

Biorepositories and individual investigators may establish identity based on empirical guidelines, "criteria," "standards," or even biases. This approach may not account for inherent analytical variation or error (which may not be identified or known). Furthermore, no validated criteria or standards exist to determine a match—which is proven (or disproven) as a result of side-by-side comparisons of organisms—analyzed under identical conditions using a carefully controlled methodology (for which the resolving power of the analysis has been defined precisely).

The analytical approach employed to determine a match, resolving power of the system, repeatability of analyses, quality and precision of resulting data, and its interpretability and meaning are all determinative factors. Wherever possible, these criteria should be statistically based (45). Addressing and resolving these issues fully and robustly are fundamental to bioforensic analysis and should be developed, tested, and validated before being applied to evidentiary samples. These match/identity issues are at the core of the validity and meaning (including the assignment, value, or weight) of source exclusion (could not have originated from), inclusion (could not exclude as having originated from, including the likelihood or probability), or attribution (did absolutely originate from, to the exclusion of all other sources, and the probability or likelihood of this occurring). The lack of standardized repository collections and databases is a pivotal constraining factor in determining questioned vs. known microbial sample match or identity.

## The Political Landscape

The scientific challenges that must be addressed to harmonize biorespository structure, function, and standards of forensic value are complex, but addressable. Accomplishing the harmonization of microbial systematics across the diverse topology of communities of interest is daunting and may represent a "grand challenge" for the life sciences in the 21st century. However, even more challenging is the multidimensional, multilevel political and resource topology that has to be aligned, focused, and sustainably engaged to support a robust microbial forensic repository system. Strategic leadership is required across all of the domains of the microbial forensics community to address these challenges.

# ACKNOWLEDGMENTS

Terrance Leighton is grateful to members of his laboratory (Katie Wheeler, Ehab El-Helow, Kijeong Kim, Babak Oskouian, Esperanza Nunez, John Pool, Jackie Ho, Chrissie Chew, Jenny Lin, and Christina Lin) who contributed to the results described here and to DARPA for research support.

# REFERENCES

[1] J.T. Staley, Universal species concept: Pipe dream or a step toward unifying biology?, J. Ind. Microbiol. Biotechnol. 36 (11) (2009) 1331–1336.

[2] D. Gevers, F.M. Cohan, J.G. Lawrence, B.G. Spratt, T. Coenye, E.J. Feil, et al., Re-evaluating prokaryotic species, Nat. Rev. Microbiol. 3 (2005) 733–739.

[3] F.M. Cohan, What are bacterial species?, Annu. Rev. Microbiol. 56 (2002) 457–487.

[4] A. Casadevall, D.A. Relman, Microbial threat lists: Obstacles in the quest for biosecurity?, Nat. Rev. Microbiol. 8 (2) (2010) 149–154.

[5] F.M. Cohan, Concepts of bacterial biodiversity for the age of genomics, in: C.M. Fraser, T.D. Read, K.E. Nelson (Eds.), Microbial Genomes, Humana Press, Totowa, NJ, 2004, pp. 175–194.

[6] M. Buckley, R.J. Roberts (Eds.), Reconciling Microbial Systematics and Genomics, American Academy of Microbiology, Washington, DC, 2007.

[7] D. Gevers, P. Dawyndt, P. Vandamme, A. Willems, M. Vancanneyt, J. Swings, et al., Stepping stones towards a new prokaryotic taxonomy, Philos. Trans. R. Soc. Lond. B Biol. Sci. 361 (1475) (2006) 1911–1916.

[8] S.J. Bent, L.J. Forney, The tragedy of the uncommon: Understanding limitations in the analysis of microbial diversity, ISME J. 2 (7) (2008) 689–695.

[9] N. Galtier, V. Daubin, Dealing with incongruence in phylogenomic analyses, Philos. Trans. R. Soc. Lond. B Biol. Sci. 363 (1512) (2008) 4023–4029.

[10] V. Kunin, L. Goldovsky, N. Darzentas, C.A. Ouzounis, The net of life: Reconstructing the microbial phylogenetic network, Genome Res. 15 (7) (2005) 954–959.

[11] E. Cardenas, J.M. Tiedje, New tools for discovering and characterizing microbial diversity, Curr. Opin. Biotechnol. 19 (6) (2008) 544–549.

[12] C. Mulder, Virology: A case of mistaken non-identity, Nature 331 (1988) 562–563.

[13] H.W. Kestler 3rd, Y. Li, Y.M. Naidu, C.V. Butler, M.F. Ochs, G. Jaenel, et al., Comparison of simian immunodeficiency virus isolates, Nature 331 (1988) 619–622.

[14] M. Davies, C. Bruce, K. Bewley, M. Outlaw, V. Mioulet, G. Lloyd, et al., Poliovirus type 1 in working stocks of typed human rhinoviruses, Lancet 361 (2003) 1187–1188.

[15] World Health Organization, Progress towards wild poliovirus containment in Russian Federation, WHO Wkly. Epidemiol. Rec. 80 (49–50) (2005) 426–428.

[16] M. Enserink, Infectious diseases. Experts dismiss pig flu scare as nonsense, Science 307 (2005) 1392.

[17] P.R. Krause, Adventitious agents and vaccines, Emerg. Infect. Dis. 7 (3 Suppl) (2001) 562.

[18] C. Chastel, Adventitious viruses and smallpox vaccine, Emerg. Infect. Dis. 11 (2005) 1789.

[19] N.J. MacLachlan, 55th Annual Meeting of the American College of Veterinary Pathologists (ACVP) & 39th Annual Meeting of the American Society of Clinical Pathology (ASVCP), ACVP and ASVCP (Eds.), Bluetongue: A Review and Global Overview of the Only OIE List Disease That Is Endemic in North America. American College of Veterinary Pathologists & American Society for Veterinary Clinical Pathology, Middleton, WI, 2004.

[20] EINet. International response to the distribution of a H2N2 influenza virus for laboratory testing: Risk considered low for laboratory workers and the public, 2005. Available from: http://depts.washington.edu/einet/?a=printArticle&print=363.

[21] Recombinomics Commentary, April 14, 2005. Mislabeled 1957 H2N2 Pandemic Flu Discovered via Contamination. Available from: http://www.recombinomics.com/News/04140501/1957_H2N2_Discovery.html.

[22] USINFO.STATE.GOV, April 15, 2005. Health Officials Destroy Asian Flu at Nearly 3,800 Labs Worldwide. Potentially dangerous strain distributed in proficiency exercise. Available from: http://usinfo.state.gov/eap/Archive/2005/Apr/15-442413.html.

[23] Colleague of American Pathologists, April 27, 2005. H2N2 Specimens from CAP PT Panels Confirmed Destroyed. Available from: http://www.cap.org/apps/docs/statements/h2n2_specimens_destroyed.html.

[24] CDC, October 6, 2005. Interim CDC-NIH Recommendation for Raising the Biosafety Level for Laboratory Work Involving Noncontemporary Human Influenza Viruses. Available from: http://www.cdc.gov/flu/h2n2bsl3.htm.

[25] D.J. Ecker, R. Sampath, L.B. Blyn, M.W. Eshoo, C. Ivy, J.A. Ecker, et al., Rapid identification and strain-typing of respiratory pathogens for epidemic surveillance, Proc. Natl. Acad. Sci. USA 102 (2005) 8012–8017.

[26] D.J. Ecker, R. Sampath, C. Massire, L.B. Blyn, T.A. Hall, M.W. Eshoo, et al., New tools for discovering and characterizing microbial diversity, Curr. Opin. Biotechnol. 19 (6) (2008) 544–549.

[27] D.J. Ecker, C. Massire, L.B. Blyn, S.A. Hofstadler, J.C. Hannis, M.W. Eshoo, et al., Molecular genotyping of microbes by multilocus PCR and mass spectrometry: A new tool for hospital infection control and public health surveillance, Methods Mol. Biol. 551 (2009) 71–87.

[28] D.M. Wolk, L.B. Blyn, T.A. Hall, R. Sampath, R. Ranken, C. Ivy, et al., Pathogen profiling: Rapid molecular characterization of *Staphylococcus aureus* by PCR/electrospray ionization-mass spectrometry and correlation with phenotype, J. Clin. Microbiol. 47 (10) (2009) 3129–3137.

[29] G. Palacios, P.L. Quan, O.J. Jabado, S. Conlan, D.L. Hirschberg, Y. Liu, et al., Panmicrobial oligonucleotide array for diagnosis of infectious diseases, Emerg. Infect. Dis. 13 (2007) 73–81.

[30] R. Sampath, K.L. Russell, C. Massire, M.W. Eshoo, V. Harpin, L.B. Blyn, et al., Global surveillance of emerging influenza virus genotypes by mass spectrometry, PLoS One 2 (5) (2007) e489.

[31] J.C. Obenauer, J. Denson, P.K. Mehta, X. Su, S. Mukatira, D.B. Finkelstein, et al., Large-scale sequence analysis of avian influenza isolates, Science 311 (2006) 1576–1580.

[32] M.L. Metzker, Emerging technologies in DNA sequencing, Genome Res. 15 (2006) 1767–1776.

[33] M.C. Maiden, Multilocus sequence typing of bacteria, Annu. Rev. Microbiol. 60 (2006) 561–588.

[34] D.M. Aanensen, B.G. Spratt, The multilocus sequence typing network: mlst.net, Nucleic Acids Res. 33 (Web Server issue) (2005) W728–W733.

[35] K.A. Jolley, Internet-based sequence-typing databases for bacterial molecular epidemiology, Methods Mol. Biol. 551 (2009) 305–312.

[36] A.B. Ibarz Pavón, M.C. Maiden, Multilocus sequence typing, Methods Mol. Biol. 551 (2009) 129–140.

[37] C. Fraser, E.J. Alm, M.F. Polz, B.G. Spratt, W.P. Hanage, The bacterial species challenge: Making sense of genetic and ecological diversity, Science 323 (5915) (2009) 741–746.

[38] K. Kim, J. Seo, K. Wheeler, C. Park, D. Kim, S. Park, et al., Rapid genotypic detection of *Bacillus anthracis* and the *Bacillus cereus* group by multiplex real-time PCR melting curve analysis, FEMS Immunol. Med. Microbiol. 43 (2) (2005) 301–310.

[39] K. Wheeler, T.J. Leighton, Methods and compositions for classifying bacillus bacteria. U.S. Patent Application (2009) 20090220951.

[40] K. Kim, E. Cheon, K.E. Wheeler, Y. Youn, T.J. Leighton, C. Park, et al., Determination of the most closely related bacillus isolates to *Bacillus anthracis* by multilocus sequence typing, Yale J. Biol. Med. 78 (1) (2005) 1–14.

[41] S.R. Klee, M. Ozel, B. Appel, C. Boesch, H. Ellerbrok, D. Jacob, et al., Characterization of *Bacillus anthracis*-like bacteria isolated from wild great apes from Cote d'Ivoire and Cameroon, J. Bacteriol. 88 (15) (2006) 5333–5344.

[42] A.R. Hoffmaster, K.K. Hill, J.E. Gee, C.K. Marston, B.K. De, T. Popovic, et al., Characterization of *Bacillus cereus* isolates associated with fatal pneumonias: Strains are closely related to *Bacillus anthracis* and harbor *B. anthracis* virulence genes, J. Clin. Microbiol. 44 (9) (2006) 3352–3360.

[43] C. Callahan, E.R. Castanha, K.F. Fox, A. Fox, The *Bacillus cereus* containing sub-branch most closely related to *Bacillus anthracis*, have single amino acid substitutions in small acid-soluble

proteins, while remaining sub-branches are more variable, Mol. Cell Probes 22 (3) (2008) 207–211.

[44] S.R. Harris, E.J. Feil, M.T. Holden, M.A. Quail, E.K. Nickerson, N. Chantratita, et al., Evolution of MRSA during hospital transmission and intercontinental spread, Science 327 (5964) (2010) 469–474.

[45] S.P. Velsko, Resolution in forensic microbial genotyping. UCRL-TR-215305, Lawrence Livermore National Laboratory, Livermore, CA, 2005.

**PART**

**6**

# National-Level Capabilities

# CHAPTER 35

# A U.S. Research Strategy for Microbial Forensics: From Genesis to Implementation

Peter T. Pesenti

*Chemical and Biological Division, Science and Technology Directorate,*
*Department of Homeland Security*

## INTRODUCTION

The 2001 terrorist attack on the World Trade Center, quickly followed by the anthrax-letter mailings, forever shook the nation's concept of domestic security. The aftermath of these attacks led to dramatic policy shifts in Washington as to the nature of the terrorist threat and the required steps necessary to meet the challenge. No longer were terrorists and their complex organizations viewed as rogue disorganized bands of criminals. This shift in security policy swiftly moved the United States to address the terrorist threat as a national security priority. Critical to the nation's response was the formulation of both public and scientific policy to address the multiple challenges posed by these newly emboldened terrorists and their organizations.

The impact of these policy changes quickly rippled throughout the federal bureaucracy starting in 2001. The U.S. Congress was the first to act by formulating new laws directly aimed at countering terrorism's threat to internal domestic security. These actions were paralleled by the executive branch (Office of the President) establishing the Homeland Security Counsel (HSC) to work in concert with the National Security Counsel and by instituting Presidential Homeland Security Directives (HSPD) to address the most serious threats posed by Weapons of Mass Destruction (WMD).

The first significant new law focusing on the domestic security problem was the USA Patriot Act of 2001 (Public Law 107-56) (1), which was passed only 45 days after the 9/11 attack. The Patriot Act focuses on enhanced surveillance provisions and expanded investigative powers of law enforcement agencies. On the heels of the Patriot Act, security and defense planners were busy on the next steps required to counter the threat to domestic security. To address the nation's strategic requirements, a planning office was quickly formed in the West Wing of the White House to study the government's options for

Microbial Forensics. DOI: 10.1016/B978-0-12-382006-8.00035-9

organizational structure, levels of preparedness, and response capabilities. The Office of Homeland Security became the precursor to the newly formed Department of Homeland Security (DHS). The U.S. Congress officially established the new cabinet-level agency in November 2002 with passage of the Homeland Security Act of 2002 (Public Law 107-296) (2). DHS was the first major government restructuring since the National Security Act of 1947. The department integrated over 22 agencies spread across multiple existing cabinet departments. DHS is now assigned the primary mission for the prevention, interdiction, and response to terrorist attack. New to the DHS structure was formation of a dedicated Science and Technology Directorate (S&T) to apply technical solutions to the terrorist challenge. The scientific research and operational functions associated with the microbial forensic program are executed by DHS S&T, in coordination with the FBI and other federal agencies, and are discussed later in this chapter.

As noted, establishment of the HSC in 2002 focused security leadership to begin formulating policy direction to counter the threat of WMD. Effective security policy in the federal government can be approached via three distinct routes: (i) legislative—legislation codified in the form of U.S. laws; (ii) executive orders—national security presidential policy decisions that are promulgated in the form of official presidential directives; the current Obama administration calls these Presidential Policy Directives (PPD) with Homeland Security Presidential Directives remaining in force; and (iii) agency rule making—this is in the form of procedural regulations established by federal agencies to provide the detail and implementation to laws set by Congress. The rule-setting process follows an established rule-making procedure with final publication in the Federal Register. The most significant policy directives affecting the field of microbial forensics are found in the HSPDs.

The initial policy directive concerning microbial forensics, HSPD 10— Biodefense for the 21st Century—was promulgated in the spring of 2004 (3). This was a defining document that cut across interagency lines to create a biodefense program focused on the prevention and response to a domestic bioterrorist event. HSPD-10 provides the necessary policy direction associated with a focused approach to the forensics problem. Based partly on the lessons of the Amerithrax[1] case, the directive established the National Bioforensics Analysis Center (NBFAC) as the lead federal center for the dedicated analysis of all evidence resulting from a bioterrorist event or biocrime and established the requirement for a supporting microbial forensic research program. The NBFAC is discussed at Chapter 36.

---

[1]FBI case name for October 2001 anthrax-letter mailings.

# THE PILLARS OF BIODEFENSE

HSPD-10 provides a comprehensive framework for addressing the biological threat to the nation. Critical to the framework is the organization of biodefense process into four essential pillars of the National Biodefense Policy of the United States. The four pillars consist of (i) threat awareness, (ii) prevention and protection, (iii) surveillance and detection, and (iv) response and recovery (3). As part of the surveillance and detection pillar, attribution is specifically discussed with respect to biological attacks and can also be used pertaining to an investigation of a planned attack. It is the term "attribution" that runs to the core of the objectives of microbial forensics. Tucker and Koblentz (4) define attribution as "identifying the country, group, or individual responsible for the use of a biological weapon in order to pursue legal prosecution or military retaliation. Budowle and colleagues (5) go further, stating, "… attribution is the information obtained regarding the identification or source of a material to the degree it can be ascertained. The goal of attribution is the identification of those involved in the perpetration of the event, which is necessary for criminal prosecution, or for actions that may be taken as a result of national policy decisions."

Attribution, therefore, supports the investigative process by which the U.S. government links the identity of a perpetrator or perpetrators of illicit activity and the pathway leading to criminal activity. Making a determination of attribution for a covertly planned or actual biological attack would be culmination of a complex investigative process drawing on many different sources of information, including technical forensic analysis of material evidence collected during the course of an investigation of a planned attack or material evidence resulting from an attack (6).

The biological analysis will be coupled with traditional investigative techniques during the course of an investigation (7). These sources of information would generate many investigative leads and help draw connections among places, events, and a possible pool of suspects. In addition to the traditional types of forensic evidence such as fingerprints, hair and fibers, and human DNA, forensic material collected as part of a biological attribution investigation will yield unique types of microbiological evidence specific to the nature of the potential attack or the attack itself. Examples of such microbiological evidence could include viable samples of the microbial agent, protein toxins, nucleic acids, clinical specimens from victims, laboratory equipment, dissemination devices and their contents, environmental samples, contaminated clothing, or trace evidence specific to the process that produced and/or weaponized the biological agent (6).

A statistically sound scientific foundation supports the forensic capabilities used in traditional criminal investigations to generate investigative leads, determine inclusion, exclusion, or inconclusiveness for questioned samples when compared to known references, and establish identity (8). Current capabilities to forensically characterize microbiological evidence in support of an attribution investigation are limited primarily to detection and identification, which, while important, only begin to scratch the surface in terms of forensic requirements for detailed characterization and comparative analyses (7). Limitations of detection and superficial identification have led the microbial forensics community to seek a more long-term research strategy to fill these challenging knowledge gaps.

## SETTING THE FOUNDATION FOR A RESEARCH STRATEGY

In 2002 the Federal Bureau of Investigation (FBI) formed a collaborative interagency organization called the Scientific Working Group on Microbial Genetics and Forensics (SWGMGF) (9). The goal of the SWGMGF was to provide an avenue for government, academia, and private sector scientists to develop guidelines related to the implementation of a microbial forensics program. This program included both operational and research needs. The FBI had previously established such scientific working groups for other forensic disciplines, one of the most notable being the Scientific Working Group on DNA Analysis Methods, whose success can be seen by the common use of human DNA analysis in crime laboratories, the establishment of standards, and the widespread acceptance of DNA analysis in the courts (10).

Likewise, the SWGMGF was established to contribute to the microbial forensic infrastructure and development of a research agenda. Recommendations of the SWGMGF would help shape research and develop efforts leading to analytical method acceptance in the forensics context and implementation of the methods by the NBFAC. Over the course of the next 5 years the SWGMFG successfully set the baseline requirements for a comprehensive program. The fundamental attributes developed (10) included:

- *Detection and identification* are keys to thwart bioterrorism. To carry out attribution effectively, robust analytical techniques need to be developed and implemented. Assay development to enhance sensitivity and specificity to expand detection capabilities must be promoted. Analytical solutions need to be developed quickly and effectively. These include DNA-based systems, as well as analytical chemistry and physical analyses (i.e., nonbiological evil characterization), culture, immunoassays, and use of bioassays in a tissue culture.
- *Information databases* will play an important role in the microbial forensics endeavor. The quality and accessibility of rapidly expanding evolving databases, such as those that contain a bioagent genomic sequence, need

to improve. To achieve this, national databases on pathogen gel and other biological data (and to include nonbiological evidence) need to be created. A relational database is needed on those who have access to pathogens so that threats can be deterred or traced back effectively to possible sources. Security measures are already being enhanced to restrict control access to select pathogens and toxins. While such a database deters some individuals from participating in microbiology research, the structure will more likely protect the legitimate user so that exchange of scientific information can proceed for the betterment of society (such as developing therapeutics and better diagnostic assays). Another database needed is one that is encyclopedic in nature. There are many sources that contain scientific information, including, but not limited to, publications, presentations, Web sites, and genomic databases. It is difficult to access all sites in an effective and rapid manner. Being able to place at one's fingertips all microbiology data and data on associated nonmicrobial forensic materials will enhance the investigative capabilities of the microbial forensic scientist greatly.

- A *strain repository* to house pathogens and other appropriate near-neighbor microorganisms must be developed. The near-neighbor concept is intimately related with the methods used for detection and identification. Some methods of identification that are robust may lead to a broad class of near neighbors, whereas sophisticated methods may define near neighbors that are narrower. Well-characterized samples must be available to enable good quality assay development. Assays cannot be validated adequately without proper samples and reference material. Bioinformatic interpretation of analytical results from evidence samples may be more limited without properly defined samples and controls. In addition, better control of access and dissemination of select agents for research and development can be executed.
- New analytical methods and some existing methods need to be *validated* properly. Such validation tasks are not limited to laboratory procedures; they apply equally to tools to be used for data interpretation and statistical assessment issues. Moreover, some biological crimes may require analysis by methodologies that may not have undergone the rigorous review process of that of standard operating protocols. A preliminary review process for such assays must be implemented (see quality assurance discussion next).
- *Quality assurance* guidelines for microbial forensic laboratories must be established (some are already enacted due to public health regulations). One must employ high-quality practices to ensure that reliable results are obtained and to maintain public confidence.

These attributes continued to provide the foundation for the development of a research program that would follow a more formal path to implementation.

## IMPLEMENTING A FORMAL GOVERNMENT RESEARCH STRATEGY

The SWGMGF served a vital role in establishing baseline requirements; however, the group's mandate was restricted to developing best practices and quality assurance guidelines, not a directive research strategy (4). Hence, in early 2006 a small group of SWGMGF members recognized that a successful microbial forensics research strategy would require direction via a more official approach: an approach that had the power to direct the federal bureaucracy. The solution was straightforward: a government agency champion was required. This champion would need the power to formally promulgate the strategy in some type of policy directive. It should be noted in official Washington that federal government agencies are driven by laws, regulations, and policies that provide mission responsibilities and the funding resources to execute their assigned missions (11). Without this directive power, committees are study groups issuing reports for prompt filing on the electronic bookshelf. With this backdrop, a small meeting of major microbial forensics stakeholders was held in Langley, Virginia, in the summer of 2006. From this ad hoc group a champion emerged—after internal staffing with senior leadership the National Counter Proliferation Center (NCPC) led the charge to draft the first microbial research strategy. NCPC quickly formed a smaller planning group in 2007 to support the effort, drawing representatives from the FBI, DHS, Department of Defense (DOD), and the intelligence community. Over the course of the next 18 months, the group hammered out a comprehensive outline for a robust research program. On the heels of the NCPC study group wrapping up a draft, microbial forensics research strategy by the Commission on the Prevention of WMD Proliferation and Terrorism released their report—World at Risk (December 2008) (12). Among a number of high-priority recommendations the commission focused on biological threats as posing the highest immediate risk of catastrophic consequences. In doing so it spotlighted the need for a more aggressive bioforensics program, including research recommending, "The United States should undertake a series of mutually reinforcing domestic measures to prevent bioterrorism: ... develop a national strategy for advancing bioforensic capabilities (12)...." The commission report goes on to detail critical milestones and program attributes for such a research program:

> By the end of 2009, the U.S. government must develop a national strategy for acquiring a state-of-the-art capability for microbial forensics. Such a national strategy should (i) facilitate the development and maintenance of a comprehensive library of pathogen reference strains; (ii) establish a government-wide set of standard procedures for collecting, processing, and analyzing samples to improve consistency and quality, and identify both a lead agency to direct this effort and

the roles and responsibilities of support agencies; and (iii) fund basic research to support the further development of microbial forensic techniques (12).

The NCPC realized the need to now move the research strategy to the formal interagency coordination process. Thus, in coordination with the White House Office of Science and Technology Policy (OSTP), the next set of tasks to formalize the microbial forensics strategy was transitioned to OSTP in January 2009. OSTP moved forward rapidly with formation of a task force on microbial forensics (TFMF), co-chaired by the intelligence community, FBI, and DHS. In order to ensure the inclusion of all stakeholders in microbial forensics research or related fields, the TFMF drew on a broader set of government agencies. The task force was charged with two deliverable products: (i) a coordinated interagency microbial forensic research strategy and (ii) an implementation plan to place the strategy into action. The task force was given an aggressive schedule to complete the research strategy with interagency coordination and approval by June 2009. In the short span of only 4 months the TFMF was able to bring all parties to the table and complete a comprehensive research strategy that built upon the baseline strategy drafted by the NCPC working group. The final National Strategy to Support Microbial Forensic Research was approved and signed into effect in July 2009. The research strategy is available for public access at http://www.ostp.gov/galleries/ NSTC%20Reports/National%20MicroForensics%20R&DStrategy%202009%20 UNLIMITED%20DISTRIBUTION.pdf. The TFMF then reconvened in August 2009 to complete the final deliverable—the Microbial Forensics Implementation Plan. As of this writing the implementation plan was completed in December 2009 and is in the interagency final coordination process. The implementation plan provides further detail to the research strategy by assigning responsibility to specific government agencies for execution of the research areas spelled out in the strategy.

## A COMPREHENSIVE MICROBIAL FORENSIC STRATEGY EMERGES

The purpose of National Strategy to Support Microbial Forensic Research is to guide and focus the research efforts of the U.S. government to advance the discipline of microbial forensics and provide the nation with the most scientifically sound and statistically defensible capability to provide scientific data to support attribution investigations of a potential or actual biological attack. The strategy has three primary goals (6).

I. Develop a *strategic microbial forensic research agenda* that will produce a national microbial forensic capability that is ultimately capable of high confidence, robust detection, characterization, and comparison of biological agents in forensic samples.

II. Promote *interagency communication, coordination, and information sharing* on microbial forensic research and development efforts.

III. Develop effective interagency *education and training* on microbial forensics designed to inform policy makers and scientific and technical personnel. This strategy forms the framework of an interagency implementation plan. The detailed goals of the strategy are as follows.

## Goal I. Develop a Strategic Microbial Forensic Research Agenda That Will Produce an Enduring National Microbial Forensic Capability That Supports Sensitive Detection, Characterization, and High-Confidence Comparison of Biological Agents and/or Their Components in Forensic Samples

A directed strategic microbial forensic research agenda that meets the requirements delineated by those government agencies conducting attribution investigations will guide investment to build a microbial forensic capability consisting of both genomic and nongenomic approaches for the forensic characterization and analysis of microbiological evidence supporting an attribution investigation of a potential or actual biocrime or bioterrorism attack. An effective microbial forensic capability must include reliable, rigorous, and sensitive techniques to collect forensic samples, detect and identify forensically relevant signatures, and fully characterize forensic evidence. Additionally, a microbial forensic capability must be able to address the requirement to conduct comparative sample analyses in order to query known and questioned samples and draw inferences relating to the process used to produce a pure sample, the provenance of a sample, or relatedness between samples. Rigorous quality standards must be applied at every level of the forensic process, from sample collection through sample analysis and data analysis to the reporting and the interpretation of results. In order to meet these challenges in microbial forensics, a number of research goals will have to be achieved. In some of these areas there is ongoing work while in other areas there is very little. This strategy serves to highlight important areas for research and provide recommendations for action in those areas based on requirements of those conducting attribution investigations.

### *Objectives*

Continue to expand national microbial forensic capabilities and develop new capabilities in the following areas: (i) sample collection, processing, preservation, and recovery and concentration of microbial pathogens and signatures from collected samples; (ii) sensitive signature detection and characterization; (iii) orthogonal methods for conducting forensic comparisons between samples to include the basic scientific research, which is required to build foundational supporting data that will enable forensic comparisons to be made and interpreted; and (iv) validation of existing technologies for new application to microbial forensic problems, development of new technologies, and development of new bioinformatics analysis tools and creation of new theoretical frameworks for data analysis and interpretation.

## Goal II. Promote Interagency Communication, Coordination, and Information Sharing on Microbial Forensics Research and Development

In addition to a strong research agenda to develop microbial forensic methods, the second goal of this strategy is to facilitate interagency cooperation to spur the development of the field as a whole. Interagency communication, coordination, and collaboration will be key to developing an enduring national capability in microbial forensics to support attribution investigations. A diverse group of federal stakeholders have a shared interest in the development of microbial forensics to support a range of attribution investigations, and these shared interests and activities must translate to concerted action and cooperation. The developing field of microbial forensics faces broad scientific challenges that require sustained research and resource commitment. Only strong interagency partnerships will ensure the development of a powerful microbial forensic investigative capability to support attribution investigations. Microbial forensics stands to benefit by leveraging existing government biodefense programs wherever relevant. Assessing the areas of opportunity that could be leveraged for microbial forensic development is an important part of the interagency collaboration that is needed.

### Objectives

Improve interagency communication, coordination, and information sharing. (i) Establish and draft terms of reference and a charter for a formal Interagency Microbial Forensic Advisory Board and (ii) develop a national archive and resource for forensically important pathogen strain collections to serve microbial forensics.

## Goal III. Develop Effective Interagency Education and Training on Microbial Forensics Designed to Inform Policy Makers and Scientific and Technical Personnel

In order to better inform national security professionals, policy makers, and analytic and interested scientific communities to the complexities of microbial forensic analysis, a system of education and training is required. The educational focus will occur at two levels. The first level of education would consist of a broad overview for those individuals for whom an awareness of important forensic issues and their implications is all that is required. The next level of training shall focus, in depth, on the scientific challenges and complexities inherent to forensic analysis of microorganisms. Together both levels of education will form a core curriculum that will inform consumers and serve to produce a more informed core of analytic professionals.

### *Objectives*

Establish and develop a core curriculum of training courses that will provide (i) a high-level overview of the microbial forensic discipline and the challenges of forensic analysis and (ii) in-depth technical courses on the types of analyses used and the limitations and challenges inherent to them.

## MEETING THE RESEARCH STRATEGIC GOALS— MICROBIAL FORENSIC RESEARCH PROGRAM IN DHS S&T

One of the most critical lessons of the Amerithrax case was the need for a dedicated biocontainment laboratory for the comprehensive analysis of biological evidence resulting from biocrimes or acts of bioterrorism. The NBFAC was established for the sole purpose of supporting law enforcement in these types of investigations as promulgated in HSPD-10 (April 2004) and PPD-2 (November 2009). PPD-2 continues on the need in "establishing a national-level research and development strategy and investment plan for advancing the field of microbial forensics...." (13). Within DHS, both the NBFAC and the microbial forensic research programs fall under the responsibility of the Directorate for Science and Technology, Chemical-Biological Division (CBD). The CBD has established a robust capabilities-driven research program focused on providing advanced evidentiary analytical capabilities to the NBFAC and federal law enforcement/other government agencies with an attribution mission. These capabilities provide the tools required to conduct comprehensive analysis, characterization, and evaluation of a diverse set of biological threat agents that may be associated with a biocrime or bioterror event. This new forensics science is a supportive element in a comprehensive criminal-investigative process with the ultimate goal of attribution, apprehension, and prosecution of the perpetrator(s). For the development of a robust analytical toolkit for microbial forensics, the program supports a comprehensive research and development program to fuel the discovery of next-generation methods and techniques. This program addresses each of the steps in the evidence analytical process and supports both intramural and extramural research and development in three focus areas.

### Bioforensic Sample Management

The objective is to develop and validate operational protocols for sample management that include the collection of viable threat agents and nondenatured molecular and immunological signatures and concentration of these signatures in the presence or absence of inhibitory components. The critical technical challenge here is the need to concentrate nucleic acids and proteins in sufficient quantity to facilitate analysis protocols. Extraction and concentration of nanoliter-to-microliter quantities of nucleic acids and proteins are major goals of this area.

## Molecular Signature Analysis of Bioforensic Samples

The focus area objective is to develop and validate operational protocols for molecular-based comparative genomic assays to assist in the identification and ultimate phylogenetic characterization of the biological agent. The characterization is a drill-down analysis with respect to the degree of relatedness among organisms of the same species at the strain/isolate level. These activities focus on the deployment of molecular-based forensic assays to the NBFAC, as well as efforts to develop new assays for additional threat agents for which signatures do not yet exist. This area is also pursuing the feasibility of proteomic analysis for application to bioforensic problems. In the near term, research continues to focus on improved nucleic acid-based genotyping schemes to better answer questions regarding sample matching criteria.

With the advent of next-generation sequencing techniques for *de novo* whole genome sequencing at low cost and rapid turnaround (<5 days), this focus area has expanded to address the implications of bioinformatics and bacterial/ viral population dynamics. A recent research solicitation summed up the challenges as follows.

Current forensic analysis of biological threat agents is impeded by lack of knowledge of the underlying population genetics and ecology of the pathogens. In addition, there is a void in the availability of bioinformatics tools necessary to analyze and validate data that forensic investigators collect from incidence sites. An understanding of pathogen population genetics, including ecology, phylogeny, life cycles, genome stability, mutation rates, recombination rates, epidemiology, host preferences and interactions, geographic distribution and other source information, virulence factors, polymorphic sites, and mutation hot spots, will be necessary to interpret results from the application of multiple typing methods (e.g., MLST, VNTR, SNP, InDel, SSRs) and use of these techniques to compare the likeness of isolates associated with a bioforensic investigation. Population genetics will likely be different for each pathogen; hence the typing strategy is also likely to be different. For example, pathogens with stable genomes may be more amenable to source tracking through SNP analysis than pathogens with unstable genomes. Furthermore, as typing data on the pathogens are generated, it will need to be organized, archived, and managed in a form that is functional and accessible for forensic analysis. To further complicate the genotyping challenge is the rapid emergence of next-generation whole genome sequencing technologies, for example, Roche (GS-20 FLX/454 pyrose-quencing), Illumina (genome analyzer system), and ABI (SOLiD), which will provide faster and less expensive whole genome sequences. These platforms will potentially displace many of the traditional genotyping techniques in the near future by providing a much richer body of data associated with a whole genome sequence. This will increase by orders of magnitude the sequence data available for microorganisms. Therefore, development of statistically based methodologies to compare whole genome sequences and sort through the maze of data for match comparisons is critical to adequately make use of these powerful new methods.

In addition to an understanding of the population genetics of the patho-gens, bioinformatic tools that can bring statistical power and degrees of confidence to results will need to be developed in order to make an assess-ment of match criteria for comparative forensic purposes. In other words, can reliable inferences be made when a biological threat agent found at a crime scene is compared to a sample found at a suspect's home laboratory or place of employment? These tools should have the power to effectively describe the quantitative criteria and provide statistical support as to the likelihood or degree of similarity. These computational tools will need to be developed, standardized, and well documented and will be required to produce under-standable, explainable, and defensible results in a forensic setting.

## Physical and Chemical Analysis of Bioforensic Samples

The objective of this area is to develop and validate operational protocols for physical and chemical analysis of evidence containing biothreat agents, including threat agent matrices. This program will focus on the development and validation of standard procedures and methods for identifying inorganic and organic signatures associated with the growth, harvest, and processing conditions of a biological threat agent. The ultimate goal is to provide infor-mation on the potential production and growth conditions for the sample.

## CONCLUSION

The forensic research community has coalesced to deliver a robust and com-prehensive microbial research strategy. This strategy is now being imple-mented by DHS, the FBI, DOD, and agencies of the intelligence community in order to develop and deliver new technologies to address the complex questions associated with attribution. The chapters that follow provide an excellent overview of the science under way and efforts to close our knowl-edge gaps.

# REFERENCES

[1] USA Patriot Act of 2001. Uniting and Strengthening America by Providing Appropriate Tools Required to Intercept and Obstruct Terrorism Act of 2001, Vol. Pub. L. No. 107-156, 2001.

[2] Homeland Security Act of 2002. Pub. L. No. 107-296, 116 Stat. 2135 (Nov. 25, 2002), 2002.

[3] Homeland Security Presidential Directive (HSPD) 10. Biodefense for the 21st Century. Executive Office of the President, 2004.

[4] J.B. Tucker, G.D. Koblentz, The four faces of microbial forensics: Biosecurity and bioterrorism, Biodefense Strategy Practice Sci. 7 (4) (2009) 389–397.

[5] B. Budowle, M.D. Johnson, C.M. Fraser, T.J. Leighton, R.S. Murch, R. Chakraborty, Genetic analysis and attribution of microbial forensics evidence, Crit. Rev. Microbiol. 31 (4) (2005) 233–254.

[6] National Science and Technology Council. National Research and Development Strategy for Microbial Forensics. Office of Science and Technology Policy, 2009.

[7] S.Y. Hunt, N.G. Barnaby, B. Budowle, S. Morse, Forensic Microbiology, Encyclopedia of Microbiology, Academic Press, Oxford, 2009, pp. 22–34.

[8] B. Budowle, Genetics and attribution issues that confront the microbial forensics field, Forensic Sci. Int. 146 (Suppl 1) (2004) S185–S188.

[9] B. Budowle, S.E. Schutzer, A. Einseln, L.C. Kelley, A.C. Walsh, J.A. Smith, et al., Public health, Building microbial forensics as a response to bioterrorism. Science 301 (2003) 1852–1853.

[10] B. Budowle, J.P. Burans, R.G. Breeze, M.R. Wilson, R. Chakrabroty, Microbial forensics, in: R.G. Breeze, B. Budowle, S.E. Schutzer (Eds.), Microbial Forensics, Elsevier Academic Press, San Diego, 2005, pp. 1–25.

[11] National Science and Technology Council. The Science of Science Policy. Office of Science and Technology Policy, Washington, DC, 2008.

[12] B. Graham, J. Talent, G. Allison, S. Rademaker, T. Roemer, W. Sherman, et al., World at Risk: The Report of the Commission on the Prevention of WMD Proliferation and Terrorism, Vintage Books, New York, 2008.

[13] National Security Council. National Strategy for Countering Biological Threats. White House, 2009.

# CHAPTER 36

# The National Bioforensic Analysis Center

**James P. Burans**

*National Bioforensics and Analysis Center, Ft. Detrick, Maryland*

## INTRODUCTION

Historically, there are numerous instances where criminals and terrorists have planned or actually used biological threat agents and toxins (1-7). The National Biodefense Analysis and Countermeasures Center (NBACC) was created to prepare for and respond to those who would plan for or actually use a biological agent or toxin to cause harm to the population or economic loss to livestock and crops and other forms of agriculture. The NBACC, through the National Biological Threat Characterization Center (NBTCC) and the National Bioforensic Analysis Center (NBFAC), utilizes advanced analytical methods to provide the nation with the scientific basis for the characterization of biological threats and bioforensic analysis to support attribution of their use against the American public. At its inception in 2003 the NBACC was a federally operated program within the Department of Homeland Security (DHS). NBACC was established as a federally funded research and development center (FFRDC) and the DHS's first national laboratory in December 2006. The Battelle National Biodefense Institute, LLC (BNBI) manages and operates the NBACC FFRDC for DHS. BNBI utilizes best management and operations practices, many of which are benchmarked through communities of practice with other national FFRDCs.

## HISTORY OF THE NBFAC

The anthrax-letter attack of 2001 resulted in the realization of the need for the development of dedicated capabilities for forensic analysis of the biological agent as well as for the biothreat agent-contaminated evidence. During the early course of the Federal Bureau of Investigation's (FBI) "Amerithrax" investigation, local state public health laboratories; the Centers for Disease Control and Prevention (CDC) military biodefense laboratories, which included the

619

Microbial Forensics. DOI: 10.1016/B978-0-12-382006-8.00036-0

U.S. Army Research Institute of Infectious Diseases (USAMRIID), the Naval Medical Research Center (NMRC), and the Armed Forces Institute of Pathology (AFIP); and university laboratories such as Northern Arizona University provided containment laboratory, bacteriological, molecular biology, and electron microscopy support. These laboratories contributed significantly to the rapid identification of the *B. anthracis* Ames strain in all the mailed letters, as well as *B. subtilis*, which was present as a contaminant in letters mailed to locations in New York City (5). Most of the laboratories participating in the early stages of the Amerithrax investigation were public health or research laboratories that did not have established procedures for the processing of environmental forensic samples, the ability to support traditional forensic techniques such as fingerprint or trace fiber analysis within biocontainment, or the ability to handle large and bulky evidentiary items such as mailboxes or car seats. Additionally, most containment laboratories did not have procedures in place to strictly control for the potential of nucleic acid or antigen cross contamination in their laboratories, in a forensic context, because of the nature of their research mission. Finally, these laboratories had unique public health, biodefense, and research missions, which were significantly interrupted and were not available for the duration of a long investigation.

The NBFAC was established through Homeland Security Presidential Directive-10 (HSPD-10) (1), which states, "We have created and designated the National Bioforensic Analysis Center of the National Biodefense Analysis and Countermeasure Center, under the Department of Homeland Security, as the lead Federal facility to conduct and facilitate the technical forensic analysis and interpretation of materials recovered following a biological attack in support of the appropriate lead Federal agency." The NBFAC provides dedicated staff, containment laboratories, equipment, and procedures to conduct operational forensic analysis to support the development of scientific data that can be used by investigators for attribution analysis of planned and actual events of biocrime and bioterrorism. The NBFAC opened its doors with CDC-certified and newly renovated and equipped BSL-2 and BSL-3 laboratories within USAMRIID in May 2004—within hours it received its first samples in support of the ongoing FBI Amerithrax investigation.

Since its opening, the NBFAC has continued to provide 24/7 continuously available bioforensic analytical support to the FBI, which is the lead federal agency with investigative authority in biocrime and bioterror investigations such as the Amerithrax investigation and other biocrime investigations. To further facilitate biocrime and bioterror investigations, the NBFAC has trained a large number of FBI traditional forensic examiners from the FBI laboratory in Quantico, Virginia, to conduct their analyses directly within BSL-2 and BSL-3 containment laboratories. This eliminates the need to laboriously develop potentially destructive decontamination procedures for use with biological agent-contaminated evidence in order that it can be analyzed outside of containment. The NBFAC conducts its analyses in compliance with a very active quality management system, which incorporates International Organization for Standardization (ISO) 17025 accreditation (8,9). The NBFAC's bioforensic analyses are used in combination with traditional forensic analysis such as human DNA analysis, fingerprint, and trace fiber analysis to provide data to investigators to conduct comprehensive attribution analyses, with the objective of both deterring biological attacks and/or enabling prosecution. The NBFAC continuously maintains its ISO 17025 accredited analytical capabilities and is always expanding to provide additional and broader support to meet new and changing threats identified by the government.

## NBFAC OPERATIONAL COMPONENTS

Since its inception the NBFAC has established overlapping techniques for the complementary identification of biological agents in evidentiary samples, which significantly increases confidence in its analytical results. The NBFAC integrates sample processing, with fastidious attention to signature cross-contamination control, as a first step followed by simultaneous analysis of a sample(s) containing an unknown biological agent utilizing techniques, which enable complementary identification and characterization, including (i) bacterial culture and phenotypic characterization for free-living and obligate intracellular bacterial species; (ii) virus culture and phenotypic characterization; (iii) toxin identification by antigen capture ELISA, biological activity assays, and matrix-assisted laser desorption ionization mass spectroscopy; (iv) molecular analyses using Real-Time polymerase chain reaction (PCR) assays, sequencing, microarrays, and bioinformatic analyses; and (v) physical characterization by transmission and scanning electron microscopy.

Comparative forensic analyses, as well as evaluation of new assays for ISO 17025 accreditation and the training of staff to meet the requirements of ISO 17025, have necessitated the creation of a National Bioforensic Repository Collection, which contains a range of bacterial and viral select agents and toxins as well as environmental organisms. The collection is composed of live agents as well as nucleic acid and antigen extracts.

To ensure that bioforensic analyses conducted at the NBFAC are of the highest quality and to meet both national and international quality standards, the NBFAC has instituted an extensive quality management system (QMS), which incorporates ISO 17025 accreditation of bioforensic procedures and analytical techniques, a Web-based Laboratory Information Management (LIM), staff training, equipment and laboratory maintenance and calibration, extensive environmental monitoring of laboratories to ensure biological agent signature-free laboratories, and competency evaluation of technical staff to perform assays as well as overall laboratory proficiency evaluations.

For the conduct of bioforensic casework, analyses conducted by the NBFAC are well coordinated through FBI microbiologists who are trained forensic examiners from the Chemical, Biological Sciences Unit (CBSU) at the FBI laboratory in Quantico, Virginia. Following a determination that the FBI will open an investigative case, samples will be safely transported to the NBFAC by the FBI's Hazardous Material Response Unit (HMRU). The NBFAC then interfaces with the CBSU, microbiologist forensic examiners who liaise directly with FBI investigators at an FBI regional office to determine what samples to analyze and what assays to utilize, as well as what traditional forensic analysis is required within containment. As an example, the NBFAC can analyze samples for ricin toxin using several orthogonal assays, which include a sensitive antigen detection ELISA to detect the presence of the toxin, a luciferase-based, cell-free translation assay to characterize the biological activity of the toxin, a mass spectroscopy assay to detect the presence of the toxin as well as RT PCR to detect contaminating ricin toxin gene sequences. An investigator will choose which assays are relevant for analysis of a specific set of evidentiary samples. Based on these interactions, an analytical plan is developed and analyses are initiated following FBI CBSU approval of the plan. Following completion of analyses, the NBFAC provides the FBI CBSU with a report of the conducted analyses. All of these processes are done in accordance with NBFAC's QMS and ISO 17025 program.

From its beginning, NBFAC has added scientific depth and breadth to its capabilities through the establishment and continual evolution of a network of spoke laboratories from government, academia, and industry that provide NBACC with additional subject matter expertise, technology transfer, and surge capacity. Spoke laboratories at the Foreign Animal Disease Center at Plum Island, Oklahoma State University, CDC, and FDA provide capabilities for foreign animal disease agents, plant pathogens, BSL-3 and BSL-4 viral agents, and bacterial agents of enteric disease, respectively (5,10–13).

## NBFAC BIOTHREAT AGENT BIOFORENSIC CAPABILITY EXPANSION AND VISION FOR THE FUTURE

During it initial years of operations, the NBFAC has established a wide range of agent-based analytical capabilities for human, animal and plant, bacterial, viral, and toxin biological threat agents using a combination of agent-specific culture, phenotypic, serological, antigen detection, biological activity, RT PCR assays, and sequencing. The NBFAC has established agent-based bioforensic capabilities for the top 30 human high-consequence agents and several major foreign animal disease agents through internal assay development and collaborative technology transfer from its spoke laboratory network. DHS is also operating a robust bioforensic research and development program that has funded research to generate bioinformatic data and assays to genotype and characterize bacterial and viral isolates below the traditional species level for technology transfer to the NBFAC.

The NBFAC is now establishing a methods-based sequence approach for capability expansion in order to meet the potential investigative challenges of the future that might involve newly emerging agents, unknown–unknown agents that could be developed with genetic engineering, or new synthetic biology approaches. A methods-based sequence approach will also enable a move away from the need for advance knowledge of a biological threat agent and the need for specific reagents such as PCR and sequencing primers and antibody reagents. This approach will also enable identification and characterization of constantly mutating RNA viruses. A methods-based sequence approach that leverages new rapid sequencing technologies linked with an orthogonal approach using advanced discovery microarrays and bioinformatics analysis will enable rapid identification and characterization of known and unknown biological agents. Recent advances in molecular biology, sequencing, arrays, and bioinformatics (14–18) are enabling discovery-type approaches to be used for the identification and characterization of any biological agent in complex samples, as well as the ability to characterize rare variants in a population. Over the next 5 years the NBFAC will demonstrate a sequence-based approach to identify and characterize biological agents based on nucleic acid signatures that can be amplified through culture or whole genome amplification from a sample using rapid sequencing, microarray analysis and bioinformatic analysis.

Finally, the NBFAC will need to establish capabilities to deduce how an agent was produced, which will help provide important investigative leads to support attribution analysis. All biological agents produce products in their growth medium and are modified by temperature of cultivation, media composition, incubation process, and harvesting and postharvesting processing (19). All steps in the growth and production of a biological agent for use in a biocrime and bioterrorist event affect and may produce specific signatures that can provide clues for how an agent was produced. In order to establish a capability of production deduction, a systems biology framework has to be developed that will enable current and developmental analytical tools to associate biological agent signatures and process production methods.

## NEW NBACC LABORATORY BUILDING

The new NBACC biocontainment laboratory building is nearing the final stages of commissioning, endurance testing, and finally CDC certifications to work with select agents and operate its BSL-3 and BSL-4 laboratories. The laboratory is built on the National Interagency Biodefense Campus at Fort Detrick, Maryland. The new NBACC laboratory is approximately 160,000 square feet and was built to the highest national biocontainment safety standards. The NBACC building and its functionally designed bioforensic casework containment laboratories and bioforensic capability expansion laboratories represent a one-of-a-kind biocontainment laboratory dedicated to support bioforensic casework, as well as development, evaluation, and establishment of new techniques to support bioforensic analysis. NBFAC bioforensic casework laboratories were designed based on the analytical workflow of bioforensic samples and to control for the potential of antigen, nucleic acid and live agent cross contamination from receipt and accessioning through analysis. NBFAC casework laboratories were built with separated zones for sample processing, bacteriology, virology, toxicology, and molecular biology with change rooms for staff and air-handling units with high-efficiency particulate air (HEPA)-filtered exhaust air. Molecular biology BSL-3 casework laboratories were designed in three zones with HEPA filtration of both supply and exhaust air with one zone for reagent and master mix preparation, a second zone for sample extraction, and a third zone for PCR amplification and sequencing. Separate and distinct biocontainment laboratories built on a separate floor of the building have been designed for bioforensic capability expansion for the development, evaluation, and establishment of new bioforensic techniques and assays that can achieve ISO 17025 accreditation. The new NBACC laboratories will soon be available to provide dedicated, secure biocontainment laboratory capability to provide bioforensic analyses to support attribution investigations.

## CONCLUSION

The NBFAC has and will continue to have a significant impact in protecting the nation from biocrime and bioterrorism. As part of its long-term goals for the future, the NBFAC will establish the ability to (i) detect any biothreat agent in a sample, (ii) identify and characterize any biothreat agent, and (iii) identify how a biothreat agent was produced. Through its continuously available dedicated laboratories and staff and ongoing capability expansion to meet new, emerging, and potential unknown biological threats, the NBFAC stands ready now and in the future to rapidly support attribution investigations to successful closure.

## REFERENCES

[1] G.W. Bush, "Biodefense for the 21st Century." Homeland Security Presidential Directive 10, April 28, 2004. Available from: http://www.fas.org/irp/offdocs/nspd/hspd-10.html.

[2] M. Larkin, Microbial forensics aims to link pathogen, crime, and perpetrator, Lancet Infect. Dis. 3 (4) (2003) 180.

[3] B. Budowle, R. Harmon, HIV legal precedent useful for microbial forensics, Croat. Med. J. 46 (4) (2005) 514–521.

[4] B. Budowle, J.A. Beaudry, N.G. Barnaby, A.M. Giusti, J.D. Bannan, P. Keim, Role of Law enforcement response and microbial forensics in investigation of bioterrorism, Croat. Med. J. 48 (4) (2007) 437–449.

[5] Y. Bhattacharjee, M. Enserink, Anthrax investigation. FBI discusses microbial forensics, but key questions remain unanswered, Science 321 (5892) (2008) 1026–1027.

[6] E.L. Bahr, R. Katz, Assessing the impact of Melendez-Diaz on the investigation and prosecution of biological weapons incidents, Biosecur. Bioterror. 7 (4) (2009) 365–370.

[7] J.B. Tucker, G.D. Koblentz, The four faces of microbial forensics, Biosecur. Bioterror. 7 (4) (2009) 389–397.

[8] B. Budowle, S.E. Schutzer, J.P. Burans, D.J. Beecher, T.A. Cebula, R. Chakraborty, et al., Quality sample collection, handling, and preservation for an effective microbial forensics program, Appl. Environ. Microbiol. 72 (10) (2006) 6431–6438.

[9] B. Budowle, S.E. Schutzer, S.A. Morse, K.F. Martinez, R. Chakraborty, B.L. Marrone, et al., Criteria for validation of methods in microbial forensics, Appl. Environ. Microbiol. 74 (18) (2008) 5599–5607.

[10] P. Keim, T. Pearson, R. Okinaka, Microbial forensics: DNA fingerprinting of Bacillus anthracis (anthrax), Anal. Chem. 80 (13) (2008) 4791–4799.

[11] J. Fletcher, C.L. Bender, B. Budowle, W.T. Cobb, S.E. Gold, C.A. Ishimaru, et al., Plant pathogen forensics: Capabilities, needs and recommendations, Microbiol. Mol. Biol. Rev. 70 (2006) 450–471.

[12] J. Fletcher, The need for forensic tools in a balanced national agricultural security program, in: Crop Biosecurity: Assuring Our Global Food Supply, pp. 93–101. Proceedings of a NATO Project. Springer Science + Business Media B.V., 2008.

[13] J. Fletcher, B. Budowle, D. Luster. Attribution: Role of forensic science in solving agroterror crimes, in: Handbook of Science and Technology for Homeland Security. Wiley Publications, New York, 2008.

[14] M.L. Metzker, Sequencing technologies: The next generation, Nat. Rev. Genet. 11 (1) (2010) 31–46.

[15] W.J. Ansorge, Next-generation DNA sequencing techniques, Nat. Biotechnol. 25 (4) (2009) 195–203.

[16] J.C. Wooley, A. Godzik, I. Friedberg, A primer on metagenomics, PLoS Comput. Biol. 6 (2) (2010) e1000667.

[17] P.L. Quan, G. Palacios, O.J. Jabado, S. Conlan, D.L. Hirschberg, F. Pozo, et al., Detection of respiratory viruses and subtype identification of influenza A viruses by GreeneChipResp oligonucleotide microarray, J. Clin. Microbiol. 45 (8) (2007) 2359–2364.

[18] C.Y. Chiu, A.L. Greninger, K. Kanada, T. Kwok, K.F. Fischer, C. Runckel, et al., Identification of cardioviruses related to Theiler's murine encephalomyelitis virus in human infections, Proc. Natl. Acad. Sci. USA 105 (37) (2008) 14124–14129.

[19] D.S. Wunschel, H.A. Colburn, A. Fox, K.F. Fox, W.M. Harley, J.H. Wahl, et al., Detection of agar, by analysis of sugar markers, associated with Bacillus anthracis spores, after culture, J. Microbiol. Methods 74 (2–3) (2008) 57–63.

# CHAPTER 37

# Microbial Forensics in Australia—The Australian Federal Police Perspective

**Paul E. Roffey and James Robertson**

*Forensic and Data Centres, Australian Federal Police, Australian Capital Territory, and the Faculty of Applied Science, University of Canberra, Australian Capital Territory, Australia*

## INTRODUCTION

"Domestic Crime to International Terror: Forensic Science Perspectives"

This was the theme of the Australian and New Zealand Forensic Science Society's 19th International Symposium on the Forensic Sciences held at the Melbourne Convention Centre, October 6–9, 2008 (1). This theme aptly embodies the underlying ethos in contemporary forensic practice in Australia. Terrorism is a crime and is not new; rather, it has been elevated into public and political prominence through a series of major incidents. Examples include the sarin gas attack on the Tokyo subway in 1995, the Oklahoma City bombing in 1995, the 9/11 suicide attacks on the World Trade Center and Pentagon in the United States in 2001, and the 7/7 suicide attacks on London's public transport system in 2005. Australia has also been targeted by terrorist organizations; the Bali bombings in 2002 and 2004 and the bombing of the Australian embassy in Jakarta in 2005 are examples of successful attacks against Australian citizens and Australian interests, albeit offshore. In common with other countries, Australia has also had a number of domestic plots that have been thwarted in the planning phase.

Although devastating, the effects of terrorist acts of this nature are relatively short-lived and localized and are unlikely to impart a long-term influence on a community or economy. However, the anthrax attacks in the United States in 2001, and the subsequent hoaxes that followed in countries across the globe, demonstrated that acts of bioterrorism have the potential to do enormous harm to populations, infrastructure, and economies (2–5). Biological agents

**627**

Microbial Forensics. DOI: 10.1016/B978-0-12-382006-8.00037-2

such as *Bacillus anthracis*, the causative agent of anthrax, can persist in the environment for decades (6), thereby presenting an ongoing risk of infection that can prevent the reoccupation of contaminated areas unless costly decontamination processes are undertaken (7,8). Other biological agents, such as *Yersinia pestis*, the causative agent of pneumonic plague, are highly infectious; hence the risk to the public can spread far beyond the original site of dissemination (9). Physical harm aside, the psychological impact of bioterrorism is enormous (5,10). This is understandable given that human history is peppered with descriptions of disease epidemics that have killed millions of people (e.g., 11–14); plague and smallpox are probably the most notable of all epidemic-causing agents. Not surprisingly, these organisms feature high on government lists of potential biological agents that could be used for acts of bioterrorism (e.g., 15,16, Table 37.1). History also shows that the conduct of acts of bioterrorism and biocrime is not limited to extremist groups and international terrorist organizations but can equally be perpetrated by disgruntled nationals (17). Indeed, the anthrax attacks in the United States in 2001 are examples of how a trusted insider was likely responsible for terrorist acts (18).

Since the anthrax attacks in the United States in 2001, most developed countries have sought to embed microbiology into forensic practice. Australia is no exception. This chapter focuses on the approach taken by the Australian Federal Police

---

**Table 37.1** List of Security-Sensitive Biological Agents Regulated under SSBA National Regulatory Scheme[a]

| Tier 1 Agent | Tier 2 Agent |
|---|---|
| Abrin (reportable quantity 5 mg) | African swine fever virus |
| *Bacillus anthracis* (anthrax-virulent strains) | Capripoxvirus (sheep pox virus and goat pox virus) |
| Botulinum toxin (reportable quantity 0.5 mg) | Classical swine fever virus |
| Ebolavirus | *Clostridium botulinum* (botulism; toxin-producing strains) |
| Foot-and-mouth disease virus | *Francisella tularensis* (tularemia) |
| Highly pathogenic influenza virus, infecting humans | Lumpy skin disease virus |
| Marburg virus | Peste-des-petits-ruminants virus |
| Ricin (reportable quantity 5 mg) | *Salmonella typhi* (typhoid) |
| Rinderpest virus | *Vibrio cholerae* (cholera) (serotypes 01 and 0139) |
| SARS coronavirus | Yellow fever virus (nonvaccine strains) |
| Variola virus (smallpox) | |
| *Yersinia pestis* (plague) | |

[a]*Australian Government. Department of Health and Aging. Security Sensitive Biological Agents List. Available from: http://www.health.gov.au/SSBA#list.*

(AFP) and its role in the whole of government response to bioterrorism, to incorporate microbiology as an additional forensic discipline, as an additional form of investigation, and as an additional skill in routine forensic practice.

## THE THREAT OF BIOTERRORISM IN AUSTRALIA

On December 9, 2009, the national counterterrorism alert level for Australia was medium (19), on a four-tier scale of low, medium, high, and extreme. This indicates that a terrorist act could occur; hence risk management should take place. Australia has been at a medium level of alert since the four levels of national terrorism alert were introduced in 2003. The threat of bioterrorism is not separately published by the Australian government; nevertheless, it must be considered, as no public health or security system can guarantee complete safety from a bioterrorism attack (20).

Emergency response crews across Australia respond to potential incidents of bioterrorism on a regular basis. The bulk of responses are to false alarms or nonhazardous white powder threats and hoaxes. Like other developed countries across the globe, Australian emergency services experienced a flood of white powder callouts following the anthrax-letters attacks in the United States in 2001 (21). As public anxiety gradually reduced, the frequency of callouts reduced in parallel and, with few exceptions, have maintained at the consistent manageable level that is currently experienced. Anecdotal evidence suggests frequencies spike from time to time in accordance with emotional public events and other incidents that occur around the world.

There is no public record of any incidents of bioterrorism within Australia; however, there have been incidents where nonpathogenic endospore-containing powders have been used as threats. Not surprisingly, these have evoked a higher level of concern to authorities than incidents that involve the innocuous powders that are commonly used, such as flour and talcum powder.

Regardless of the low level of risk, threats and hoaxes are taken seriously by Australian authorities. Indeed, one of the main reasons for the establishment of the Australian Chemical, Biological, Radiological and Nuclear (CBRN) Data Center was to provide a central repository in which records of suspicious substance incidents could be collated and analyzed to provide support to law enforcement in their endeavor to track down and prosecute these public menaces.

## THE AUSTRALIAN CBRN DATA CENTER

Established in 2007, the Australian CBRN Data Center is one of three data centers in the Forensic and Data Centers Portfolio of the AFP, the other two being the Australian Bomb Data Center and the Australian Illicit Drug Data Center. The objective of the Australian CBRN Data Center is to enhance Australia's capability to prevent, prepare, and respond to the malicious use of CBRN agents within and against Australia and its interests. It does this by bringing together technical, intelligence, and law enforcement capabilities related to CBRN threats.

One of the main roles of the center is to be the national repository for CBRN incident information and intelligence and to provide technical advice to support prevention, preparedness, and response issues related to CBRN material. Specifically, the center collects and assesses data from all sources about CBRN agents, their precursors, and methodologies required to prepare and disseminate them. It assesses the availability of the agents and the feasibility and impact of their misuse in Australia, conducts trend analysis on threats, and evaluates new and emerging technology and methodology (22).

The Australian CBRN Data Center has strong links with the Australian Public Health Laboratory Network, which includes representatives from major public health diagnostic laboratories in all Australian jurisdictions, as well as health departments and agencies at the commonwealth and at state and territory levels. (Note that the Commonwealth of Australia consists of six states and two major territories, all of which are self-governing, and a number of smaller territories under the administration of the Commonwealth government.) The center conducts analysis and provides advice on biological threats as they pertain to the malicious use of biological materials or the deliberate spread of disease (22).

The center is an integral component of AFP criminal investigations and forensic operations, providing a link among the policing, forensic, and intelligence communities, allowing AFP operational portfolios to prepare for potential threats in a pre-emptive and proactive manner. It is also an integral component of emergency responses being available to provide technical advice on an around-the-clock basis.

## THE AUSTRALIAN FEDERAL POLICE AND ITS ROLE IN PREVENTING, COUNTERING, AND INVESTIGATING TERRORISM

The AFP is Australia's national policing agency, enforcing commonwealth criminal law and protecting commonwealth and national interests in Australia and overseas (23). The AFP also provides community policing services to the Australian Capital Territory, Jervis Bay, and external territories; contributes to the Australian government's international law enforcement interests such as regional peacekeeping and regional capacity building; and provides the security for major Australia airports. The AFP works closely with other law enforcement bodies at state and territory, commonwealth, and international levels to fight multijurisdictional and transnational crime and to contribute to global security. The prevention of terrorist attacks in Australia and on Australian interests overseas is a high priority for the AFP and its partner agencies.

The role of the AFP in the coordinated nationwide effort to counter terrorism and its consequences is outlined in the Australian National Counter-Terrorism Plan (24). In essence, the primary responsibility to prevent, counter, and respond to acts of terrorism resides within the jurisdictions. The nature of terrorism means its implications may cross jurisdictional boundaries and hence requires a cooperative interjurisdictional coordination of capabilities. As such, each of the states and the two self-governing territories within Australia have a capability to investigate, respond to, and prosecute acts of terrorism, which includes the capability to detect and identify biological agents that could be used for terrorist and criminal purposes. This capability is coordinated through the Australian (Counter) Bioterrorism Laboratory Network (ABLN), which contains representatives from key public health, defense, and law enforcement laboratories, as well as health departments and agencies at the commonwealth, state, and territory levels, and includes the AFP forensic laboratory and the Australian CBRN Data Center. The ABLN is administered by the Department of Health and Aging, the commonwealth government agency that administers the coordination of health systems within Australia. The role of ABLN is to advise on issues relating to the detection and analysis of security sensitive biological agents (SSBAs) and to establish, maintain, and expand collaborative links between public health and law enforcement agencies.

## THE AUSTRALIAN FEDERAL POLICE MICROBIAL FORENSICS PROGRAM

Most of the traditional forensic disciplines, such as fingerprints, chemistry, biology, and pathology, have been entrenched as forensic disciplines and are applied commonly in criminal investigations. Microbiology is also an established discipline but until recently has rarely been used in forensic application. In the quest to integrate microbiology into forensic operations, the AFP faced the reality that the law enforcement community had virtually no experience with microbiology and, conversely, the microbiology community had virtually no experience with forensic application or law enforcement. The AFP solution was to establish a basic microbiology capability in-house and then to promote partnerships with diagnostic and specialized laboratories for more intensive investigations. The AFP emphasis is on extending the capability for screening and preliminary identification of potential biological agents at the scene beyond that of first responders. The role of the AFP also includes application of normal forensic procedures from the crime scene to the laboratory.

The feasible presence of a microbiological agent does not alter, or reduce, the need to recognize, record, and recover all potential forensic materials and to maintain the integrity of the evidence from the crime scene to the laboratory and through analysis. With that said, the possible presence of a microbiological agent does have implications for these processes.

Within Australia, the expertise needed to identify and discriminate biological agents of human significance largely resides in public health diagnostic laboratories that have PC3 (i.e., BSL3) and/or PC4 (i.e., BSL4) laboratory containment facilities for culture. One notable exception in the law enforcement community is the police laboratory within the state of New South Wales, which has established a PC3 laboratory specifically for forensic analysis and culture of suspicious biological substances. Logically, partnering of law enforcement and diagnostic laboratories is an important aspect of a microbial forensic capability. In this respect the AFP and the Australian CBRN Data Center work closely with major jurisdictional public and animal health laboratories, as well as health departments and agencies at the commonwealth and at state and territory levels, to ensure that this capability is available to assist law enforcement. In a similar manner, the AFP and Australian CBRN Data Center encourage and support research conducted within academia, industry, and other government departments that have the specialized skills and knowledge required to conduct detailed characterizations of biological agents. The AFP and Australian CBRN Data Center also work closely with international law enforcement partners in the United Kingdom, Canada, and the United States. Indeed, the expertise, experience, and cooperation of international partners are important aspects of the Australian microbial forensic capability.

The forensic operations arm of the AFP itself has a basic but sound microbial forensic program. The program spans all aspects of forensic operations from sample collection to the interpretation of results from the laboratory. In this respect the program is multilayered. The AFP has a general capability across all forensic disciplines to conduct investigations and examinations in chemical and biological (CB) contaminated environments. This capability extends to all AFP personnel required to enter a scene that may contain hazardous substances. This practice is applied in a manner in which all hazards are considered—chemical, biological, radiological, nuclear, and explosives (CBRNE). With that said, the AFP requires the assistance of personnel from the Australian Nuclear Science and Technology Organization (ANSTO) and the Australian Radiation Protection and Nuclear Safety Agency (ARPANSA) to conduct investigations and examinations in environments containing radiological and nuclear (RN) hazards and to examine evidence that contains RN hazards. Crime scene personnel and laboratory staff are trained to enter CB contaminated environments in various levels of personal protective equipment to process exhibits and scenes, collect appropriate evidence, and document accordingly. AFP scientists regularly exercise with personnel from other emergency agencies, such as the fire brigade and ambulance, and from ANSTO and ARPANSA to coordinate efforts at the scene and ensure that the scene is processed logically and thoroughly in a manner that will best collect pertinent evidence and thus can withstand strict legal scrutiny. Exhibit packaging, decontamination of packaged exhibits, chain of custody, and exhibit transport, receipt, security, and storage are protocols that are all well established. Where possible these are identical to or closely follow existing practices. This is important, as in a multiagency response situation, procedures are more likely to be followed when they are already normal, embedded best practice. Where the decontamination of exhibits is required prior to laboratory examination, the AFP seeks expertise from other ABLN laboratories that have relevant experience with the decontamination of exhibits.

The AFP also has a basic capability to detect and identify biological agents. This capability provides a presumptive level of identification for a number of biological agents, including *B. anthracis* and ricin, the two biological agents reported most frequently in the media. The AFP approach is to take this analytical capability to the scene rather than bringing samples back to the laboratory for analysis. This approach provides a higher degree of certainty of identification early in the processing of a scene, thereby allowing scene examiners to process the scene in accordance with the type of agent present. This analytical capability is housed primarily within a mobile laboratory, which can be deployed to a scene or used on base. The detection capabilities used within the Mobilab are also deployable separately. The AFP also has a laboratory for the triage of potentially contaminated exhibits, thereby minimizing the risk of contamination of the main laboratory. The mobile laboratory and triage laboratory are discussed in more detail later.

Quality management is treated no differently from other forensic capabilities within the AFP. All protocols and equipment within the microbial forensic program, whether field or laboratory based, are validated, reviewed, tested, and maintained in accordance with the AFP FDC quality management framework and are accredited to International Standard ISO/IEC 17025:2005. Compliance is tested by the National Association of Testing Authorities, Australia (NATA) (25), to the aforementioned standard and supplementary requirements for accreditation in the field of forensic science. In this respect, the AFP microbial forensic program is the only forensic microbiology capability in Australia accredited under the standard NATA forensic science program. In this manner, evidence collected and analyzed within the AFP microbial forensic program meets the same standards of admissibility as traditional forensic evidence. In comparison, public health laboratories are accredited under the NATA medical testing program, and other laboratories involved in microbiological analysis of forensic evidence are accredited under the NATA veterinary testing or biological testing programs. Some of these laboratories have included the NATA forensic science module to assist with the provision of chain of custody procedures and court testimony capability.

# THE AUSTRALIAN FEDERAL POLICE MOBILE AND TRIAGE LABORATORIES

Most AFP forensic capabilities have a fundamental need to be mobile. AFP forensic operations provide support to all AFP operational areas, including those in the Australian Capital Territory, the major Australian airports, embassies and consulates within Australia, most Australian embassies, and the Christmas, Cocos, and Norfolk Islands. The AFP also provides police and forensic assistance to countries in southeast Asia, Australasia, and the Pacific. Until recently, forensic services have largely been provided through the deployment of portable detection and analysis equipment, which have been set up in temporary laboratories at, or near, the site of an incident. In 2006 the AFP commissioned a mobile forensic laboratory to complement deployable capability. The ultimate aim of the mobile laboratory is to provide a safe, secure, and clean laboratory that can be deployed by road, rail, or air virtually anywhere in the world.

The primary purpose of the mobile forensic laboratory, known as the Mobilab, is to facilitate rapid on-site screening for chemical and biological agents in cases of suspected CB incidents. With that said, it is also suitably equipped to assist the forensic investigation of other incidents that are chemical in nature, such as those that involve illicit drugs, toxic industrial chemicals, explosives, and accelerants and those that require the analysis of environmental samples. Further, the Mobilab can be equipped according to requirement and hence can enhance the deployment capability of most forensic disciplines.

The Mobilab is a large caravan fitted with two independent compartments (Figure 37.1). Compartments are accessed from the outside by external air-tight doors and between compartments with an internal air-tight door. The laboratory is powered through two shore power connections or two onboard diesel generators. Water is supplied via a town water connection through a standard hose fitting or an onboard 110-liter water tank. An identical tank holds waste water for decontamination prior to disposal. Both cabins are fitted with recycled air conditioning units, and an air management system provides the capability to establish air pressure gradients between the compartments and the compartments and the external atmosphere. Variable speed intake and exhaust fans, fitted with high-efficiency particulate air (HEPA) filters, make it possible to create negative or positive atmospheric pressures in each compartment. Positive pressure differentials are created to minimize contamination of the cabins from environmental contaminants. This is useful during investigations such as those for trace levels of explosives and is particularly useful during transport on roads where dust is a problem. Negative pressure differentials are used to contain contaminants within the laboratory. In this manner, PC3 (i.e., BSL3) laboratory-level atmospheric pressure gradients

**FIGURE 37.1**
The Mobilab.

are created for the analysis of samples from possible biological incidents. In this application the rear compartment is used as the PC3 laboratory and the forward cabin is treated as the antechamber.

The rear compartment is designed primarily for the analysis of biological samples and for initial processing of suspicious samples. It houses a large Class 3 biological safety cabinet (Figure 37.2), which has a HEPA-filtered intake duct and a series of two HEPA filters and a TEDA carbon filter on the exhaust duct. This allows both biological and chemical specimens to be examined safely. The Class 3 biological safety cabinet also has an internal power outlet and a USB port, allowing computer analysts to interrogate contaminated computers and other electronic data storage devices. It also has a nitrogen outlet so that chemists can concentrate samples prior to chemical analysis if required. Nitrogen can also be used to evacuate oxygen from the cabinet in the event of fire within the cabinet; as an inert gas, nitrogen is also used to conduct leak tests on the cabinet. The rear compartment also houses a small Class 2 biological safety cabinet, a stainless steel sink, and a small amount of bench space (Figure 37.3).

The larger forward compartment is fitted with significantly more bench and storage space. The forward compartment is used for chemical analysis, report writing, electronic communications, general laboratory techniques (Figure 37.4),

**FIGURE 37.2**
Class 3 biological safety cabinet in the rear compartment of the Mobilab.

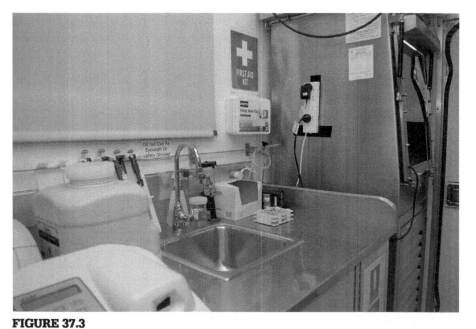

**FIGURE 37.3**
Rear compartment of the Mobilab showing sink, available bench space, and Class 2 biological safety cabinet.

**FIGURE 37.4**
Forward compartment of the Mobilab showing GC-MS/DFPD and available bench space.

**FIGURE 37.5**
Microscopy bench in forward compartment of the Mobilab.

and microscopy (Figure 37.5). A small refrigerator/freezer is located under the bench and is maintained by a bank of onboard batteries when generator power or shore power is not available. Gas lines for helium, hydrogen, nitrogen, and air span the walls of the forward compartment, all of which are plumbed into gas cylinders housed in a cylinder cupboard mounted on the drawbar of the trailer. The only instrument that is mounted permanently within the Mobilab is the combination gas chromatograph, mass spectrometer, and dual flame photometric detector for sulfur and phosphorous (GC-MS/DFPD) (Figure 37.4), which is used for the analysis of organic chemicals, including chemical warfare agents. Other equipment is imported as required.

The Mobilab has been accredited by NATA under ISO/IEC 17025:2005 as a static and mobile forensic testing laboratory. The Mobilab is also a member of the recently created Chemical Warfare Analysis Laboratory Network (CWALN). This allows the laboratory to provide presumptive-level identification of chemical warfare agents. Laboratory facilities, ventilation, protocols, and work practices are compliant with the Australian Standard AS/NZS 2243.3:2002 for a PC2 laboratory as a minimum. The laboratory can be deployed within an hour of notification and is fully operational within 2 hours of arrival; this includes cleaning, setting up, initializing, and quality checking of instruments and background checks for contamination. Laboratory procedures are tailored and practiced to return results to investigators within 60–90 minutes of

**FIGURE 37.6**
Mobilab garaged on base immediately adjacent to the ERTL.

receipt of sample. Information on the Mobilab, Mobilab procedures, and integration of the Mobilab into the investigation of hazardous materials incidents is presented in more detail in other publications (26,27).

While on base, the Mobilab is garaged immediately adjacent to the evidence recovery and triage laboratory (ERTL) (Figure 37.6) and is used as the analytical laboratory for samples taken from exhibits being processed in the ERTL. The ERTL is a stand-alone laboratory, which, as the name suggests, is used as a triage for exhibits being received into the main forensic laboratory. Triaging may be performed if it is suspected that exhibits contain, or are contaminated with, trace levels of CBRNE. This minimizes the risk of contamination of the main laboratory, ensuring that routine forensic investigations are not disrupted by investigations of this nature. The ERTL is a PC2 laboratory that in the event of a biological incident can be ramped up to mimic a PC3 laboratory level of containment. The ERTL houses a large Class 3 biological safety cabinet in the same format as that in the Mobilab (Figure 37.7), two large Class 2 biological safety cabinets, and a large fume hood that can accommodate four scientists simultaneously (Figure 37.8). Fume hood exhaust is passed through two HEPA filters and a charcoal filter prior to release into the atmosphere. When the fume hood is running, the atmospheric pressure within the ERTL is

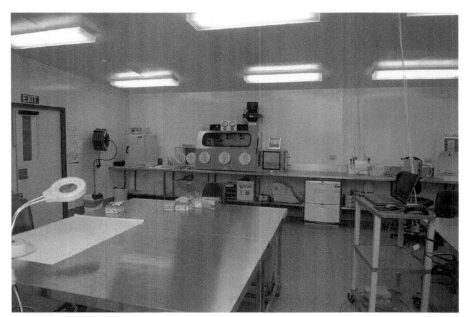

**FIGURE 37.7**
View of the ERTL showing Class 3 biological safety cabinet.

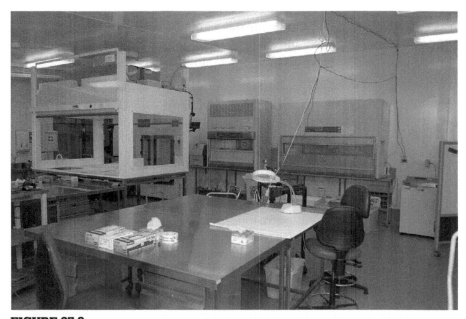

**FIGURE 37.8**
View of the ERTL showing Class 2 biological safety cabinets and fume hood.

strongly negative, which assists with the containment of potentially hazardous materials. The ERTL is also used when exhibits require analysis by scientists from a number of disciplines. The simultaneous examination of complex exhibits by multiple scientists is more efficient than consecutive examinations and also provides a forum through which plans can be developed to maximize the recovery of evidence. A third important use of the ERTL is as a surge capacity laboratory. Utilization of the ERTL for investigations that require the processing of large numbers of exhibits allows examinations to be conducted with less impact on the normal operation of the main laboratory.

## NATIONAL REGULATORY SCHEME FOR BIOLOGICAL AGENTS OF SECURITY CONCERN

In September 2007, new legislation was passed in the Australian Parliament to enhance the security of biological agents in Australia. Part 3 of the *National Health Security Act 2007* (28) introduces a national regulatory scheme for biological agents of security concern. The list of SSBAs regulated under this legislation is shown in Table 37.1. Agents on the list are derived from the Council of Australian Governments (COAG) Report on the Regulation and Control of Biological Agents, from which the COAG biological agents list was developed in 2006 (29). In the process of developing the list, nearly 200 biological agents were considered. The review included an assessment of terrorist interest, availability of the agent, ease of production, ease of dissemination for threat, morbidity/mortality of the agent, transmissibility of the agent, and difficulty to treat for consequence (30). The list was divided into two tiers—tier 1 agents, which pose the highest security risk to Australia, and tier 2 agents, which pose a high security risk. The regulatory scheme was implemented by the Australian Department of Health and Aging in January 2009 for tier 1 agents and January 2010 for tier 2 agents. An important component of the regulatory scheme is the national register of Australian entities handling SSBAs. The scheme also provides an instrument through which the government can evaluate whether entities have legitimate purpose to handle SSBAs. If approved, entities must comply with the act's regulations and relevant standards, and all entities that handle SSBAs, temporarily or otherwise, must abide by the mandatory reporting requirements.

The National SSBA Regulatory Scheme applies to all laboratories that handle SSBAs with the exception of law enforcement laboratories, which are exempt but only to carry out their functions under commonwealth, state, or territory law. Law enforcement laboratories are not exempt if they hold SSBAs for use as reference samples or controls for analytical or diagnostic purposes.

From the perspective of the AFP, implementation of the National SSBA regulatory scheme is an important milestone in the effort to reduce the risk of bioterrorism and biocrime in Australia. The exemption provided by the *National Health Security Act 2007* allows law enforcement agencies to carry out their functions under commonwealth, state, or territory law. Where possible, the AFP microbial forensic program utilizes noninfectious simulants, nonviable or nonpathogenic organisms, noninfectious products of pathogens, and inactive toxins or nontoxic subunits as controls and references; when needed, the AFP enlists assistance and expertise from ABLN laboratories.

## POTENTIAL EXPANSIONS TO THE AUSTRALIAN FEDERAL POLICE MICROBIAL FORENSIC CAPABILITY

One of the primary objectives of the Mobilab is to provide a detection and presumptive level identification capability for biological agents within 60–90 minutes of the receipt of a sample (26,27). Technologies that are currently used or are under development for use in the Mobilab include microscopy, colorimetric assay, immunoassay, capillary electrophoresis, and nucleic acid analysis. The application of GC-MS for the detection of biomarkers such as ricinine (31) and abrine (32) is also under investigation. Fourier transformed infrared spectrometry is often used when it is confirmed that a suspicious powder is not biological in nature or an explosive but the identity of the powder remains unknown.

Microbiological culture is not performed within AFP laboratories and is not being considered in the immediate future. The AFP recognizes that culture is currently considered the gold standard for the identification of biological agents; however, it has been avoided in favor of faster, albeit presumptive level identification technologies that can be taken to the scene. With that said, a new laboratory-based technology that the AFP is planning to investigate in collaboration with overseas partners is the use of multilocus polymerase chain reaction and electrospray iononization–mass spectrometry to obtain confirmatory level identification of biological agents directly from samples without the need for culture. This technology shows enormous promise, is attracting considerable interest, and has direct application in clinical and forensic disciplines (33–37).

## THE AUSTRALIAN FEDERAL POLICE IN THE WHOLE OF GOVERNMENT APPROACH TO BIOTERRORISM

The formal application of microbiology to answering questions of forensic relevance is still relatively new but has evolved considerably since the emergence of microbial forensics as a discipline. The U.S. anthrax incidents in 2001 and their extensive subsequent investigation certainly highlighted many important aspects and issues for any future major microbiological incident. Although only a small number of people died as a result of the U.S. anthrax incidents, the scale of the consequences, economic and otherwise, were almost tsunami-like in their impact and effect. The incident classically illustrated the need to treat such an incident as a whole of government (or at least multiagency) response. With respect to the forensic dimension, this also needed to be managed in a broad and holistic way.

Australia authorities have attempted to learn lessons from such incidents. In common with the experience of many countries, Australia experienced a wave of anthrax-related scares and hoaxes in the aftermath of the U.S. incident (21). This necessitated the rapid development of response protocols and procedures based on existing assets, which, in the early days following the U.S. anthrax letters, were informed and enacted by an immature response community. This is no longer the case. Australia has been proactive in putting in place appropriate protocols and procedures at a whole of government level, with the overall control residing with the Australian National Counter Terrorism Committee (NCTC). As the lead commonwealth law enforcement agency, the AFP has an important role in realizing the goals and objectives of the NCTC and the commonwealth government. As the only broad-based forensic provider for the commonwealth, forensic responsibilities also reside with the AFP. Given the federal nature of Australia, this necessarily means that the AFP works in concert with state and territory counterparts. The Australian National Institute of Forensic Sciences (NIFS) plays an important role in co-ordinating forensic science in Australia, and CWALN has been established under the NIFS framework.

The AFP has had almost a decade to evolve its model for a forensic microbiology capability. The AFP model is intended to be capable of meeting the needs of Australia while recognizing the AFP's broader role as Australia's representative on appropriate international fora. There are unique aspects of the AFP model but also elements that are similar in approach to those developed in many other countries. This is to be expected and indeed it is to be encouraged, as there is no mileage in reinventing the wheel. The challenges and issues are mostly common, and it should be no surprise that common solutions will continue to emerge.

Perhaps the most unique part of the AFP approach is interoperation of the Australian CBRN Data Center and AFP forensic operations. This is enhanced by the placement of these units under one portfolio and under one senior manager. This structure allows the AFP to bring together the specialist intelligence dimension, critical to a proper understanding of the threat, and the appropriate forensic response. The role of the Australian CBRN Data Center is to interface between forensic practitioners and highly specialized organizations such as Australia's

Defense Science and Technology Organization and the Commonwealth Scientific Industrial Research Organization with whom AFP has formal cooperation agreements in place. Many hazard response agencies and law enforcement agencies normally do not have access to security classified information where information relevant to bioterrorism threats resides. In Australia, one of the key roles of the Australian CBRN Data Center is to appropriately share information that might otherwise not be known to relevant agencies.

In developing the capability of the AFP to respond to bioterrorism, it was important for it to be appropriate to the threat and to recognize the existing roles and responsibilities of other agencies. The first responders to most incidents in Australia are the hazard response agencies, usually within fire services. Following the U.S. terrorism incidents in 2001, Emergency Management Australia, the commonwealth government agency tasked with coordinating government responses to emergency events, rolled out off-the-shelf capabilities to such agencies to conduct presumptive field testing of potential biomaterials. In the event of a positive presumptive test, a sample(s) would then go to a laboratory within the Australian health network for confirmatory testing. The policing agencies are primarily responsible for the forensic investigation. In this respect, the AFP approach differs from that followed by most other Australian policing agencies. In broad terms, the AFP fills an important forensic role that sits between the hazard response agency and the health agency but to some extent is overlapping. In this role, the AFP aims to add value to the equation but not replicate. Hence, the AFP has invested in evaluating available field equipment with a view to adding value through these being used by a microbiologist with forensic training or crime scene operatives with appropriate training. It is the view of the AFP that where an incident warrants it, this forensic intervention can play an important role in reducing disruption to normal activities. As such incidents can be in significant transport hubs, the benefit of quick action is obvious. Further, early identification of the biological agent provides a sound basis from which protective strategies can be implemented to minimize risk to personnel and from which appropriate sampling strategies can be formulated. Development of the Mobilab and associated triage center (ERTL) is aimed at balancing the need to sample the potentially active agent while protecting staff and the community from contamination but also protecting potential broader forensic evidence that may be vital in subsequent investigation. The latter is just as important in dealing with hoaxes. Of course, this approach will only be successful if roles are understood and the role of the crime scene officers and forensic microbiologist is fully recognized and incorporated into the whole of agency response plans. It is our view that, in much of the world, this broader role for forensic personnel is still immature or not well recognized. The focus of the AFP forensic response is to continue to improve field testing to provide as full an answer as possible and as quickly as possible, while recognizing the gold standard for identification remains culture.

Adaptation and overlaying of principles of microbiology across forensic and policing portfolios has been a necessary challenge for the AFP and will continue to be a challenge as the discipline evolves. A fundamental awareness has been installed across all AFP portfolios. Further, most AFP forensic disciplines have the capability to extend routine examinations to biologically contaminated scenes and exhibits. The safety of AFP personnel is an important consideration behind all AFP activities; hence the implementation of strategies to minimize risk is paramount. This includes ongoing training to maintain skills and knowledge in the use of personal protective equipment and handheld detectors for entry into potentially contaminated sites. It is also one of the primary reasons for establishment of a capacity to detect and identify biological agents at the scene. Early identification of the biological agent provides a sound basis from which protective strategies can be implemented to minimize risk and a sound basis from which strategies can be developed to maximize the evidence gathered from the scene. It also ensures that strategies can be implemented to minimize the risk of loss or contamination of biological evidence. This is particularly important given that the maintenance of the viability of biological agents is often vital for confirmatory identification and subsequent detailed physical, biochemical, metabolic, and genetic analyses.

The AFP has not developed the capacity to fully investigate a pathogenic agent. Rather, the AFP believes this is the responsibility of highly specialized laboratories. The AFP approach is to understand what is feasible and to establish collaborative arrangements with such specialized laboratories, which would be activated if necessary. The AFP adds value in this capacity by ensuring that (i) the correct forensic quality standards are applied and (ii) all potential forensic evidence is considered.

## CONCLUSION

Regrettably, the threat posed by terrorism will remain for many years ahead. Bioterrorism remains a real threat, but because actual incidents remain at a low level, complacency is a constant enemy; hence the continuing background of false alarms may indeed be beneficial, encouraging responding agencies to keep practiced and alert. The AFP model of an integrated intelligence and forensic approach is, we believe, a useful model for other nations of similar size or maturity to consider as a measured contribution to a whole of government approach to threats posed by bioterrorism.

The Australian microbial forensic capability is reliant on the combined and coordinated efforts of numerous government facilities, departments, and agencies within law enforcement and public health at the state, territory, and commonwealth levels, as well as the private sector. It is a shining example of how a coordinated effort can provide a comprehensive capability that does well to protect the Australian community. The AFP and Australian CBRN Data Center also work closely with international law enforcement partners in the United Kingdom, Canada, and the United States. The contribution of these and other international partners cannot be understated in the ongoing united effort to protect the citizens and assets of Australia and other countries against bioterrorism.

## ACKNOWLEDGMENT

The authors thank Dr. Eric Wenger, director of the Australian CBRN Data Center, for his helpful discussions and critical reading in the preparation of this manuscript.

## REFERENCES

[1] The Australian and New Zealand Forensic Science Society. The 19th International Symposium on the Forensic Sciences. Available from: http://www.anzfss2008.org.au/.

[2] G.A. Ackerman, K.S. Moran, No. 22: Bioterrorism and threat assessment, 2004. Available from: http://www.wmdcommission.org/.

[3] S.A. McEwin, T.M. Wilson, D.A. Ashford, E.D. Heegaard, T. Kuiken, B. Kournikakis, Rev. Sci. Technique 25 (1) (2006) 329–339.

[4] B. Budowle, R. Murch, R. Chakraborty, Microbial forensics: The next forensic challenge, Int. J. Legal Med. 119 (2005) 317–330.

[5] A.F. Kaufmann, M.I. Meltzer, G.P. Schmid, The economic impact of a bioterrorist attack: Are prevention and postattack intervention programs justified? Emerg. Infect. Dis. 3 (2) (1997) 83–94.

[6] R. Sinclair, S.A. Boone, D. Greenberg, P. Keim, C.P. Gerba, Persistence of Category A select agents in the environment, Appl. Environ. Microbiol. 74 (3) (2008) 555–563.

[7] P.N. Price, Anthrax sampling and decontamination: Technology trade-offs. Lawrence Berkeley National Laboratory, 2009. Available from: http://www.escholarship.org/uc/item/1mm135gt.

[8] United States of America. Capitol Hill anthrax incident. Government Accountability Office Report GAO-03-686, 2003. Available from: http://www.goa.gov/new.items/d03686.pdf.

[9] T.V. Inglesby, D.T. Dennis, D.A. Henderson, J.G. Bartlett, M.S. Ascher, E. Eitzen, et al., Plague as a biological weapon: Medical and public health management, JAMA 283 (17) (2002) 2281–2290.

[10] M.J. Hall, A.E. Norwood, R.J. Ursano, C.S. Fullerton, The psychological impacts of bioterrorism, Biosec. Bioterrorism Biodefense Strategy Practice Sci. 1 (2) (2004) 139–144.

[11] M.J. Keeling, C.A. Gilligan, Bubonic plague: A metapopulation model of a zoonosis, Proc. R. Soc. Londd. Ser. B 267 (1458) (2000) 2219–2230.

[12] N.P. Johnson, J. Mueller, Updating the accounts: Global mortality of the 1918–1920 "Spanish" influenza pandemic, Bull. Hist. Med. 76 (1) (2002) 105–115.

[13] D. Lippi, A.A. Conti, Plague, policy, saints and terrorists: A historical survey, J. Infect. 44 (4) (2002) 226–228.

[14] C.W. Potter, A history of influenza, J. Appl. Microbiol. 91 (2001) 572–579.

[15] United States of America. CDC Select Agent List. Available from: http://www.bt.cdc.gov/ agent/agentlist-category.asp.

[16] L.D. Rotz, A.S. Khan, S.R. Lillibridge, S.M. Ostroff, J.M. Hughes, Public health assessment of potential biological terrorism agents, Emerg. Infect. Dis. 8 (2) (2002) 225–230.

[17] W.S. Carus, Bioterrorism and Biocrimes: The Illicit Use of Biological Agents since 1900, Fredonia Books, Amsterdam, 2002.

[18] United States of America. Department of Justice. Transcript of Amerithax investigation press conference. Available from: http://www.justice.gov/opa/pr/2008/August/08-opa-697.html.

[19] Australian Government. Australian National Security. Available from: http://www.national-security.gov.au/.

[20] R.A. Smallwood, A. Merianos, J.D. Mathews, Bioterrorism in Australia, Med. J. Aust. 176 (2002) 251–253.

[21] A. Leask, V. Delpech, J. McAnulty, Anthrax and other suspect powders: Initial responses to an outbreak of hoaxes and scares, NSW Public Health Bull. 14 (11-12) (2002) 218–221.

[22] B. Morrish, E. Wenger, The Australian Chemical Biological Radiological and Nuclear Data Centre, Microbiol. Aust. 29 (2) (2008) 70–71.

[23] Australian Federal Police. Available from: http://www.afp.gov.au.

[24] Australian Government. Attorney-General's Department. National Counter-Terrorism Plan. Available from: http://www.ag.gov.au/agd/www/nationalsecurity.nsf/AllDocs/85A16ADB-86A23AD1CA256FC600072E6B?OpenDocument.

[25] National Association of Testing Authorities, Australia. Available from: http://www.nata.asn.au/.

[26] P. Roffey, K. Norman, D. Royds, A mobile laboratory for real time analysis during forensic operations, Microbiol. Aust. 29 (2) (2008) 91–94.

[27] K. Norman, P. Roffey, D. Royds, Application of a mobile laboratory for real time response to chemical, biological, radiological, nuclear and explosive incidents, Aust. J. Forensic Sci. 41 (1) (2009) 73–80.

[28] Australian Government. Attorney-General's Department. National Health Security Act 2007. Available from: http://www.comlaw.gov.au/ComLaw/Legislation/ActCompilation1.nsf/0/BE00CA9C14F9F7D7CA2576510010DE11/$file/NatHealthSecurity2007.pdf.

[29] Australian Government. Department of Health and Aging. Recommendations of the report on the regulation and control of biological agents. Available from: http://www.health.gov.au/internet/main/publishing.nsf/content/0DA6B6F12E71AE2ACA2573CC00819A45/$File/bioreport-recommendations.pdf.

[30] L. Toms, G.A. Lum, National Health Security Act 2007: Registration of laboratories that hold Tier 1 and Tier 2 agents, Microbiol. Aust. 29 (2) (2008) 72–74.

[31] S.M. Darby, M.L. Miller, R.O. Allen, Forensic determination of ricin and the alkaloid marker ricinine from castor bean extracts, J. Forensic Sci. 46 (5) (2001) 1033–1042.

[32] R.C. Johnson, Y. Zhou, R. Jain, S.W. Lemire, S. Fox, P. Sabourin, et al., Quantification of L-abrine in human and rat urine: A biomarker for the toxin abrin, J. Anal. Toxicol. 33 (2009) 77–84.

[33] D.M. Wolk, L.B. Blyn, T.A. Hall, R. Sampath, R. Ranken, C. Ivy, et al., Pathogen profiling: Rapid molecular characterization of *Staphylococcus aureus* by PCR/electrospray ionization-mass spectrometry and correlation with phenotype, J. Clin. Microbiol. 47 (10) (2009) 3129–3137.

[34] M.N. Van Ert, S.A. Hofstadler, Y. Jiang, J.D. Busch, D.M. Wagner, J.J. Drader, et al., Mass spectrometry provides accurate characterization of two genetic marker types in *Bacillus anthracis*, BioTechniques 37 (4) (2004) 642–651.

[35] L.B. Blyn, T.A. Hall, B. Libby, R. Ranken, R. Sampath, K. Rudnick, et al., Rapid detection and molecular serotyping of adenovirus by use of PCR followed by electrospray ionization mass spectrometry, J. Clin. Microbiol. 46 (2) (2008) 644–651.

[36] R. Sampath, T.A. Hall, C. Massire, F. Li, L.B. Blyn, M.W. Eshoo, et al., Rapid identification of emerging infectious agents using PCR and electroionization mass spectrometry, Ann. N.Y. Acad. Sci. 1102 (2007) 109–120.

[37] T.A. Hall, K.A. Sannes-Lowery, L.D. McCurdy, C. Fisher, T. Anderson, A. Henthorne, et al., Base composition profiling of human mitochondrial DNA using polymerase chain reaction and direct automated electrospray ionization mass spectrometry, Anal. Chem. 81 (18) (2009) 7515–7526.

# CHAPTER 38

# Validation of Microbial Forensics in Scientific, Legal, and Policy Contexts

**Randall S. Murch[a] and Elizabeth L. Bahr[b]**

[a] *Center for Technology, Security, and Policy, Virginia Polytechnic Institute and State University, Alexandria, Virginia*
[b] *Office of General Counsel, Department of the Navy, Arlington, Virginia*

## INTRODUCTION

The use of effective, robust, and properly validated methods for the collection, preservation, transport, analysis, interpretation, and communication of probative evidence is a linchpin of reliability and confidence-building measures that contribute to the acceptance, use, and understanding of science by investigators, judges, attorneys, juries, the media, and lay public. Stakeholders expect that forensic methods, protocols, and techniques have been validated properly. All science proposed and admitted to court is subject to discovery and scrutiny under U.S. case law and prescribed legal procedures (1–3). In recent years, as courts and the media have become more aware of the value, power, risks, and uncertainties of forensic science, whether or not methods have been validated properly is receiving increasing attention. Further, traditional forensic science and performers are increasingly being held to higher expectations of performance (4). Microbial forensic laboratories are not exempt from these requirements and expectations simply because their focus is microbes and their by-products related to specialized criminal or national security events (5,6). Issues of confidence and reliability in the science that influences national security decisions are just now being taken up by policy makers.

Although it is not precisely defined, validation fundamentally establishes and defines the properties and performance characteristics of forensic methods, including their limits. Goals for forensic methods include robust collection and preservation, relevant exploitation of sample, high discrimination power, informative comparison of sample's known and questioned origin, utility across known and encountered sample types, accuracy, reliability, a defined and acceptable error rate, speed and responsiveness, repeatability, transferability, and independently established or confirmed validity and that results

Microbial Forensics. DOI: 10.1016/B978-0-12-382006-8.00038-4

are probative (relevant), interpretable, understandable, and defensible (7). Establishing the validity of forensic methods and their use and interpretation contributes to acceptance, admissibility, confidence, value, and weight of physical evidence in the jurisprudence process (8).

Properly designed, validated, applied, and communicated forensic science is increasingly being recognized within the intelligence, counterterrorism, and national security policy communities as crucial because of the important contribution forensic science can make to their planning, actions, and decisions. Although science for these purposes will not necessarily "go to court," the stakes are potentially so high, and the uncertainty from intelligence and unpredictability of political processes potentially so substantial, that some have opined that at least the science should be as defined, relevant, and defensibly applied and interpreted as possible for decision makers all the way up to the president. It is also entirely conceivable that demands for discovery could occur in these domains. A country that is suspected or accused of the proliferation or use of biological weapons could make public demands to produce any and all evidence of these allegations for independent scrutiny on the world stage with a U.S. accusation or prior to imminent punitive action. At the present time, no framework exists for testing the acceptability or use of forensic science for national or international security policy decisions.

## AN ACCEPTED APPROACH TO VALIDATION EXISTS

Validation itself has the following objectives (7):

- Assess the ability of procedures to obtain reliable results under defined conditions
- Rigorously define the conditions required to achieve results
- Determine limitations of the procedures
- Identify aspects of the analysis that must be monitored and controlled
- Form the basis for interpretation guidelines to convey significance of the findings

Validation is addressed in quality assurance guidelines that have been published for a number of forensic disciplines (9–13), including microbial forensics (14).

There are three principal categories of validation guidelines for methods and procedures that should be applied as appropriate (7):

- *Developmental validation* addresses specificity, sensitivity, reproducibility, bias, precision, false positives and false negatives, use of appropriate controls, and use of any reference databases.
- *Preliminary validation* acquires limited test data to enable evaluation of a method that will be used to support the investigation of an event of interest or crime, that is, for lead generation or corroboration of

information; results used beyond this should at least be subjected to external peer experts to assess the utility of the method and define the limits of interpretation and conclusions that can be drawn.

- *Internal validation* is performed when a method is developed elsewhere and transferred to an operational laboratory.

For each of these categories, as appropriate, the following also applies (7):

- Validation should be tested using known, fully characterized samples; reproducibility and precision should be monitored and documented, as well as the range of performance using controls.
- Before introduction of a new method, users should successfully complete a qualifying test to demonstrate proficiency.
- Any substantive modifications to previously validated procedures should be documented and subjected to validation testing commensurate with the modifications.

Validation should demonstrate accuracy, reliability, the basis for transferability and repeatability, performance attributes and limits (including defining uncertainty), and robustness with sample types to be encountered. Peer review, preferably through independent scrutiny by qualified experts, strengthens the validation process. In science, peer review normally occurs through publication in peer-reviewed journals, presentation at professional meetings, or by testing by other scientists. It can be risky for developers or even users to declare a method fully validated without independent assessment, simply because it has met their own conceptions and expectations.

Even during fast-moving events when time constraints and exigencies exist, validation should be undertaken (15), embracing the elements of preliminary validation (7,8), with limitations identified and the imposed constraints duly noted. Full validation can be undertaken at the earliest available time if appropriate or necessary.

It is likely to be very difficult to delineate all possible criteria and processes that could be used to test any particular method against all possible sample types. However, particular focus should be given to two primary criteria: reliability and reproducibility against reasonably representative samples. Method validation should be performed with reference and mock forensic samples to produce data sufficient in quantity and quality to demonstrate that those criteria chosen have been met or exceeded, to performers, reviewers, and those who will rigorously scrutinize the science after the fact (7). A validation plan should be developed for each validation event and thoroughly documented for archival purposes as well as for subsequent inspection or review (7).

Effective and appropriate validation of methods, techniques, protocols, and standard operating procedures should be viewed as an interdependent system, from collection and preservation through interpretation and communication.

The validation of one aspect should be taken as seriously as validation for any other aspect to ensure that no shortfalls, gaps, or vulnerabilities exist across the system, as the system is interdependent for its robustness. If a weakness exists in one area, the entire continuum is potentially at risk.

## VALIDATION AS PART OF THE QUEST FOR QUALITY IN FORENSIC SCIENCE

Since the mid-1980s, forensic science and its stakeholders have increasingly recognized the need to implement and be held accountable to a quality environment to improve the accuracy, reliability, credibility, and confidence (16). By now, the majority of U.S. forensic laboratories have achieved laboratory accreditation as has its one declared microbial forensic laboratory, the National Bioforensics Analysis Center (17). One erroneous assumption that has been stated publicly by some is that "laboratory accreditation confers validity on the methods are being utilized" (paraphrased). The accreditation process identifies whether a laboratory has an infrastructure and follows it but does not examine analytic methods for their validity. Validation of methods is a separate, included requirement of good laboratory practice and quality assurance for forensic or any other diagnostic and calibration laboratories.

ISO/IEC Standard 17025, General Requirements for the Competence of Testing and Calibration Laboratories (18), is the benchmark body of standards against which such laboratories are measured and accreditation is awarded, including forensic laboratories. An entire section of ISO 17025 is devoted to technical requirements; that is, establishment of "the correctness and reliability of tests and calibrations performed in the laboratory," including new test and calibration methods (18). Validation is defined under ISO 17025 as "the confirmation by examination and the provision of objective evidence that the particular requirements for a specific intended use are fulfilled." Laboratories operating under 17025 are expected to "validate nonstandard methods, laboratory-designed and developed methods, standard methods used outside of their intended scope and amplifications and modifications of standard methods to confirm that the methods are fit for the intended use." ISO goes on to state "the validation should be as extensive as is necessary to meet the needs of the given application or field of application" (18).

The primarily U.S.-based microbial forensic community agrees in principle with ISO regarding how the performance of a new or modified method should be determined (7,18):

- Calibration using reference standards or reference materials
- Comparison of results achieved with other methods
- Interlaboratory comparisons

- Systematic assessment of factors influencing the result
- Assessment of the uncertainty of results based on scientific understanding of the theoretical principles of the method and practical experience

ISO makes a crucial point with respect to uncertainty and relevance; "the range and accuracy of the values obtained from validated methods (e.g., the uncertainty of the results, detection limit, selectivity of the method, linearity, limit of repeatability and/or reproducibility, robustness against external influences and/or cross-sensitivity against interference from the matrix of the sample/test object), as assessed for the intended use, *shall be relevant to the customer's needs*" (18).

In addressing the crucial component of uncertainty, ISO 17025 also further states that testing laboratories will:

- Have and shall apply procedures for estimating uncertainty of measurement
- At least attempt to identify all components of uncertainty and make a reasonable estimation
- Ensure that the form of reporting does not give a wrong impression of the uncertainty (or, conversely, certainty) (18)

Among the factors that ISO further details for the degree of rigor to be applied to establish uncertainty are *"the requirements of the customer"* (18), taken to mean what the customer requires or will accept. In the context of this chapter, this would mean both U.S. legal and national security policy communities, which may differ, and perhaps even the respective international communities, which could well differ still.

## LEGAL VALIDATION OF NOVEL SCIENTIFIC TECHNIQUES IN U.S. COURTS

Members of the U.S. government have asserted that "unless the world community acts decisively and with great urgency, it is more likely than not that a weapon of mass destruction will be used in a terrorist attack somewhere in the world by the end of 2013" (19). In addition, these same government officials have concluded that "terrorists are more likely to be able to obtain and use a biological weapon than a nuclear weapon" (19). It is a fair assumption that some time in the near future, the U.S. criminal justice system will be faced with the prosecution of a case involving casualties due to use of a biological weapon. More likely than not, the prosecution of such an event will rely heavily on the use of microbial forensics to attribute the source of an attack to a particular suspect (20). Therefore, ensuring that microbial forensic techniques can be validated properly to have sufficient scientific rigor such that they will be accepted by U.S. courts, judges, and juries is an essential component of the broader legitimacy of this new science.

## Current Law on Validation of Scientific Evidence

The legal standard for validation of novel scientific techniques varies depending on whether a case is being tried in a court that follows the federal standard for scientific evidence or whether a court follows a state-specific legal standard. Additionally, the legal standard for scientific validation could vary depending on whether the evidence is being offered in a criminal or civil case (21). Although it is most likely that forensic microbiology evidence used in the prosecution of a suspect for the alleged use of a biological weapon will occur in a federal court under federal terrorism statutes, it is worth exploring the current law in state courts' evidence standards on validation, as well, in the event that microbial forensics is used in a state court proceeding.

Supreme Court precedent and the Federal Rules of Evidence (FRE, 3) dictate the legal tests for the admissibility of expert and scientific evidence in a federal case. Beginning in 1923, the legal test for admissibility of expert scientific testimony involving novel techniques was the "general acceptance" standard established by the Supreme Court in Frye v. United States (1). In this case, the court ruled that (i) expert testimony deduced from a well-recognized scientific principle or discovery will often be admitted, but (ii) the thing from which the deduction is made must be sufficiently established to have gained "general acceptance in the particular field to which it belongs."

Fifty years later, Congress promulgated the FRE in 1975, which today remain the authority on the admission of evidence in federal courts. Under current federal rules, if an expert scientific witness testifies as to the validity of a novel scientific technique, it must first be proven to the judge that (i) the expert witness can, in fact, be qualified as an expert and (ii) any such testimony by the expert scientific evidence is relevant to the case, as specified by FRE 104(a) and 104(b).

Once qualified as an expert, a judge then determines under FRE 702 whether "the testimony is based on sufficient facts or data, the testimony is a product of reliable principles or methods, and the witness has applied the principles and methods reliably to the facts of the case." Next, the judge determines whether "the facts or data underlying the expert testimony are of a type reasonably relied upon by experts in the particular field in forming opinions or inferences upon the subject, as required by FRE 703." In addition, as addressed earlier, the judge assesses the expert's testimony to ensure that there is "a foundational process showing that a scientific process or system produces an accurate result," as required by FRE 901. Finally, FRE 403 states that even if a judge finds an expert's testimony to be reliable, the judge may exclude it from evidence if its likely prejudicial effect outweighs its probative value.

After the FRE were adopted, there was some confusion in U.S. courts as to whether the new federal rules or *Frye* governed the admissibility of scientific evidence. In 1993, the Supreme Court clarified this confusion in Daubert v. Merrell Dow Pharmaceuticals (2). The Court recognized that, given the often rapid advances being made in science, new discoveries and theories might be perfectly sound but still be new enough that they had not yet gained "general acceptance," as mandated by the *Frye* standard. The *Daubert* court held that FRE Rule 702 controlled the admission of expert testimony in federal courts and that, when applying Rule 702, a "trial judge must ensure that any and all scientific testimony or evidence admitted is not only *relevant*, but *reliable*" (2).

The merits of scientific validation play a significant role in the *Daubert* test. The *Daubert* court "directed federal judges to take a scientific approach to the admissibility of scientific evidence" (22) and insisted that in order for scientific evidence to be legally reliable, it must be found to be scientifically reliable. In other words, "[f]or scientific testimony to be sufficiently reliable, it must be derived by the scientific method and must be supported by appropriate validation" (23). *Daubert* recognized that reliability and validity differ as scientific measures. Whereas validity describes how well the scientific method reasons to its conclusion, reliability describes the ability of the scientific method to produce consistent results when replicated" (22). Therefore, the robust validation—per scientific standards—of any novel scientific technique will be the prerequisite showing for the eventual acceptance and validation of that science by the state and federal courts that follow the *Daubert* test (24).

If a scientific technique has been shown to meet the reliability threshold, a judge then determines whether the scientific evidence is also relevant—the second part of the *Daubert* test. The relevancy prong requires that judges examine "the proffered connection between the scientific research or test result to be presented, and particular disputed factual issues in the case." Therefore, the evidentiary reliability of future forensic microbiology evidence submitted to U.S. courts following the *Daubert* test will turn on whether it has been shown to be validated scientifically by showing that the science supporting the evidence is both (i) relevant—assisting the trier of fact in understanding or determining the pertinent facts—and (ii) reliable—its methodology is based on scientific knowledge (22).

However, not all state courts have adopted the *Daubert* test. The Supreme Court's decision in *Daubert* was based on the language of FRE 702 and therefore was not grounded in a constitutional right mandating adoption by the states. Currently, 25 states have affirmatively adopted *Daubert* or a similar test for use in their courts or had previously abandoned *Frye* and had developed a similar test (25). Fifteen states and the District of Columbia continue to adhere to the "general acceptance test" of *Frye* (25). Additionally, 6 states have not completely rejected the *Frye* standard, but also apply the *Daubert* factors (25). Finally, 4 states have developed their own tests for the admission of novel scientific evidence (25).

Whether forensic microbiology evidence is found to be legally admissible by a court would first depend on whether the court in question has adopted the *Frye* standard, the *Daubert* standard, or its own unique admissibility standard. However, any forensic microbiology evidence—at a minimum—must be shown to be either generally accepted by the relevant scientific community or validated based on reliable scientific techniques and relevant to the case at hand (26).

## Case Precedent

Although the admissibility of forensic microbiology evidence has yet to be tested in a U.S. court, other contemporary cases involving the validation of scientific evidence could give clues as to how courts might handle the submission of such evidence in a criminal prosecution for the use or threatened use of biological weapons. Given the potential importance of microbial evidence in instances with an international connection, scientific evidence in relevant cases from other countries and how such is treated under international law deserve attention in a future study.

In United States v. Mettetal (27), for example, the defendant was convicted by a jury for the illegal possession of ricin in violation of 18 U.S.C. § 175 (28). In trial, a U.S. magistrate ordered the Department of Homeland Security's National Bioforensic Analysis Center (NBFAC) to test seized contraband in order to confirm that it was an illegal toxin under the circumstances described in § 175 (28). Three separate laboratories (29) performed tests on the seized contraband; all three laboratories concluded that the samples tested positive for ricin.

The defendant challenged the admission of the laboratories' analysis of the ricin, asserting that the laboratories' testing methods were unreliable, and that in order to be legally validated, dose–response evidence from live animals was necessary to prove that a substance is a "toxin" for the purposes of 18 U.S.C. §§ 175 and 178 (28). However, the court ruled that

> the Government need only prove by a preponderance of the evidence that the [tested sample]... satisfies the statutory definition of toxin under the circumstances described in § 175(b). Thus, methods of testing that reliably show the amount and purity of a sample and reliably identify it as a substance known to possess the qualities of a "toxin" are sufficient under the plain language of the statute.... (29, page 3)

The court held that the laboratories' testing methods were adequately validated to be considered legally validated and that the law only required the prosecution to show that the evidence was simply a toxin as defined by 18 U.S.C. § 178 (28)—not that the substance seized from the defendant was, in fact, ricin. The defendant's conviction was ultimately overturned on appeal for reasons unrelated to the validity of the laboratories' analysis of the ricin evidence (29).

Cases that involve the validation of microbial genomic evidence also provide insight into how courts handle the validation of relevant scientific techniques (30). Louisiana—a state that has adopted the Daubert standard—has had two notable attempted murder cases for exposure or potential exposure to human immunodeficiency virus (HIV) for which the validation of DNA evidence was at issue. In 1995, in State v. Caine (31), the defendant was charged with attempted second degree murder when, during a botched convenience store robbery, he told the clerk "I'll give you AIDS" and then stabbed the clerk in the arm with a needle attached to a syringe filled with clear liquid. The court heard extensive expert testimony from pathologists who ran blood tests on both the defendant and the clerk. Experts confirmed the defendant was HIV positive through these tests, and the court held that the experts had adequately validated the genome-based HIV analysis of the defendant's blood. The liquid in the syringe that was used to stab the clerk was not tested for or confirmed to contain HIV so the defendant argued that the state failed to prove that the syringe was a dangerous weapon. Regardless, the court found that the defendant had the specific intent to kill the clerk when he stabbed her with a needle that was possibly contaminated with the HIV virus; thus the state proved beyond a reasonable doubt that the defendant was guilty of attempted second degree murder.

In State v. Schmidt (32), genome-based phylogenetic analysis of blood samples was used to determine whether a doctor infected his mistress with the HIV-infected blood of one of his other patients. After hearing the testimony of multiple expert witnesses, the court held that the phylogenetic analysis techniques used to analyze the samples were sufficiently validated to allow this particular type of genome-based forensic analysis into evidence.

In these two cases, the combination of rigorous scientific validation of submitted genome-based evidence and adequate expert testimony regarding the results of this evidence analysis was sufficient for the court to hold the evidence reliable and relevant, as required by *Daubert* and the rules of evidence adopted by the Louisiana legislature.

## Basis for Challenges

Even if a particular scientific method or certain forensic evidence is verifiable by the scientific community, both judge and jury in each case have the discretion to conclude that a novel scientific technique has not been sufficiently validated. Thus, the ultimate reliance of a novel scientific technique and its results is only as strong as the credence in each court for each case, even if similar evidence has been heard in courts elsewhere.

Opposing counsel could directly or indirectly attack the credibility of any forensic microbiology evidence that a prosecution attempts to submit in many ways. For example, opposing counsel could question the professional qualifications of the expert witness who is to testify in support of the technique, thus disqualifying the witness. Additionally, counsel could present available evidence that casts doubt on the reliability of the technique, even if that court has previously made reliability determinations based solely on the consensus of scientists. Here, the testimony of opposing experts or advice to counsel for cross-examination of prosecution experts can be most useful. If the defense can reduce or eliminate the value, weight, and/or credibility of the scientific evidence or the expert presenting the information, than the jury or judge could find that the prosecution could no longer meet its burden of proving culpability "beyond a reasonable doubt."

## OBSERVATIONS OF THE NATIONAL ACADEMY OF SCIENCES (2009)

In 2009, the National Academy of Sciences published a report entitled "Strengthening Forensic Science in the United States: A Path Forward" (4) in response to a 2006 congressional request that asked for a comprehensive assessment of many aspects of the nation's forensic science enterprise. The committee issued 13 recommendations that covered the most salient gaps, shortfalls, and opportunities for systemic improvement as tasked. Microbial forensic performers and stakeholders should take particular note of that report's most relevant recommendations in the context of validation of the science:

- From Recommendation 3: Research is needed to address issues of accuracy, reliability, and validity (establishing the scientific bases demonstrating the validity of forensic methods; establishing quantifiable measures of the reliability and accuracy of forensic analyses reflective of realistic case scenarios; establishing limits of reliability and accuracy as conditions of evidence may vary; developing quantifiable measures of uncertainty in conclusions)
- From Recommendation 5: Encouraging research on the effects of observer bias and human error (determining the extent that results of forensic analyses are influenced by background information or theories and observations of customers and stakeholders; developing standard operating procedures that minimize observer bias and human error)
- From Recommendation 6: Advancing best practices (developing tools for advancing measurement, validation, reliability, information sharing,

and proficiency testing in forensic science; establishing protocols for best practice forensic examinations, methods, and practices)

- From Recommendation 8: Routine quality assurance and control procedures (establishing routine quality assurance and quality control procedures to ensure the accuracy of analyses and work of forensic practitioners to identify mistakes, fraud, and bias; ensure the continued validity and reliability of standard operating procedures and protocols; establish corrective action procedures and protocols when improvement is required or needed)

In the wake of the NAS report cited earlier, the Federal Bureau of Investigation engaged the NAS to conduct a thorough review of the science that was developed and applied to physical evidence from the investigation of anthrax spores sent through the U.S. mail in October 2001. This is currently ongoing; a report will likely not be out until sometime in late 2010. Validation of the methods developed and used in this investigation is a major focus of this NAS study committee.

## VALIDATION: SCIENCE AT INTERSECTIONS OF LEGAL AND POLICY ENVIRONMENTS: SNAPSHOT ILLUSTRATIONS

Consider the implications of these notional scenarios and whether microbial forensics has been or will be properly and defensibly validated.

It is now October 2011, 10 years after the anthrax-letter mailings allegedly perpetrated by Bruce Ivins (33), for which the Department of Justice closed its investigation in February 2010. By this time, the NAS report on the science developed and applied to the anthrax-letter cases has been published for a year, which included a number of concerns regarding the validation of certain key methods. A series of copycat anthrax-laden package attacks occurs across the United States, resulting in 15 dead and 36 others who become infected but are treated successfully. The perpetrator makes some tactical mistakes and is identified through effective investigation, concerned citizen tips, and traditional forensic evidence. She is arrested and indicted in June 2012 and is awaiting trial. As part of its case, the defense intends to attack a number of issues associated with the forensic evidence analyzed both in pretrial admissibility hearings and in trial if necessary, foremost among them is validation of certain analytic methods and results, interpretation, and conclusions related to the microbial evidence. It is known in the relevant community that robustness of the validation of a number of those methods developed and used is in question. The observations and recommendations in the NAS report are still valid: shortcomings in the validation of some of the methods, interpretation of results, and conclusions that were drawn. Even with time and resources available, the government has not followed up on these recommendations. Meanwhile, the defense believes that it has also found significant problems with how the traditional forensic evidence was handled and analyzed. The fate of the trial and a successful conviction hinge on the successful admission of much of the physical evidence that it believes links the accused to the events and outcomes at issue. The defense team is poised.

The U.S. embassy in Islamabad, Pakistan, has just experienced an unusual outbreak of a highly pathogenic strain of what has been identified preliminarily as *Shigella* spp. The Centers for Disease Control and Prevention later identifies the causal agent as *Shigella dysenteriae* type 1 in all samples; this strain is found in the developing world and can cause deadly epidemics (34). There is no vaccine available for *Shigella*. A credible claim of responsibility is issued by Al Qaeda. As this unfolds and response options are weighed, the President and National Security Council are demanding that U.S. government agencies provide rock-solid scientific evidence, investigation, and intelligence, upon which any policy decision and subsequent military or diplomatic action or criminal prosecution would be based. Suppose forensic methods for *Shigella* are not well developed nor have they yet been rigorously validated. Concern exists with the National Security Council and Attorney General that U.S. microbial forensic capabilities will not be particularly supportive, reliable, or defensible for any follow-on action taken, including addressing international and domestic political and public scrutiny.

The notorious foot-and-mouth disease (FMD) (35) outbreaks occur simultaneously in Texas, Oklahoma, Colorado, and Idaho, which has never before been detected in the United States. The U.S. Departments of Agriculture and Homeland Security quickly confirm that the causative agent is FMD virus. The strain is identified as one that is endemic in Afghanistan. Bioterrorism or biological warfare is suspected. Extensive investigation and intelligence gathering implicate North Korea and its supreme leader, Kim Jong-il. The President believes that he has sufficient evidence and intelligence to mount a response. Kim, sensing that retaliation by the United States is imminent, "calls the United States out," demanding a public accounting of its evidence and intelligence against his country. He calls for international scrutiny of the forensic evidence the United States claims it has. The challenges and uncertainties, highly definitive and validated forensic analysis, and source attribution regarding FMD are well known and have become the focus of an international scientific–legal debate fueled by intense international media coverage. Kim is confident he can debunk or neutralize the value and weight of any scientific evidence in possession of the United States. After conferring with its scientific, legal, and policy experts, U.S. leadership is concerned. Kim is winning the political and public perception battle.

At this time, we do not know whether current science could have survived the anticipated withering adversarial scrutiny during U.S. v. Ivins, let alone the other contexts.

## CHALLENGES AHEAD FOR MICROBIAL FORENSICS

Looking ahead, we propose four major challenges for the microbial forensic community and its stakeholders:

- Produce a competent and rapidly adaptive kit of current microbial forensic methods for all priority bioterrorism select agents that have been fully validated and would survive intense, critical, expert-enabled scrutiny in U.S. courts, relevant scientific community, and public eye
- U.S. policy leadership issues expectations for its microbial forensic capabilities to include validated methods and outputs that will effectively inform both legal and national security policy decisions
- Development frameworks that tightly couple microbial forensic capabilities with what is required to make effective, defensible policy decisions with regard to attribution and how validation supports decision making and risk management
- Conduct deep integrated analyses of requirements and expectations of international, legal, policy, diplomatic, and scientific communities for forensic science performance, reliability, and acceptance as applied to problems and events related to illicit biological weapons development, possession, transfer, and use.

## DISCLAIMER

The contents of this chapter are solely the responsibility of the authors and, with regard to the second author, do not necessarily represent the official views of the Department of the Navy or any part of the U.S. government.

## REFERENCES

[1] Frye v. United States, 293F, 1013, 1014 D.C. Cir. (1923).
[2] Daubert v. Merrell Dow Pharmaceuticals, 509 U.S. 509 (1993).
[3] Federal Rules of Evidence, 28 U.S. Code Annotate.
[4] The National Academies, Strengthening Forensic Science in the United States: A Path Forward, National Academies Press, Washington, DC, 2009.
[5] B. Budowle, S.E. Schutzer, A. Einseln, L.C. Kelley, A.C. Walsh, J.A.L. Smith, et al., Building microbial forensics as a response to bioterrorism, Science 301 (2003) 1852–1853.
[6] B. Budowle, S.E. Schutzer, M.S. Asher, R.M. Atlas, J.P. Burans, R. Chakraborty, et al., Toward a system of microbial forensics: From sample collection to interpretation of evidence, Appl. Environ. Microbiol. 71 (2005) 2209–2213.

[7] B. Budowle, S.E. Schutzer, S.A. Morse, K.F. Martinez, R. Chakraborty, B.L. Marrone, et al., Criteria for validation of methods in microbial forensics, Appl. Environ. Microbiol. 74 (2008) 5599–5607.

[8] R. Harmon, Admissibility standards for scientific evidence, in: R.G. Breeze, B. Budowle, S.E. Schutzer (Eds.), Microbial Forensics, Elsevier, Boston, 2005, pp. 382–392.

[9] DNA Advisory Board. DNA Advisory Board Quality Assurance Standards, 1998. Available from: www.cstl.nist.gov/strbase/dabqas.

[10] Federal Bureau of Investigation. Quality Assurance Standards for Forensic DNA Testing Laboratories. Revised 2009. Available from: www.cstl.nist.gov/strbase/QAS/Final-FBI-Director-Forensic-Standards.

[11] Society of Forensic Toxicology and American Academy of Forensic Science. Forensic Toxicology Laboratory Guidelines, 2006. Available from: www.soft-tox.org/doc/Guidelines.

[12] Scientific Working Group for the Analysis of Seized Drugs (SWGDRUG). SWGDRUG Recommendations, 2008. Available from: www.swgdrug.org.

[13] M.A. LeBeau, Quality assurance guidelines for laboratories performing analysis of chemical terrorism, Forensic Sci. Commun. 6 (2) (2004). Available from: www.fbi.gov/hq/lab/fsc.

[14] Scientific Working Group on Microbial Genetics and Forensics, Quality assurance guidelines for laboratories performing microbial forensic work, Forensic Sci. Commun. 5 (4) (2003). Available from: www.fbi.gov/hq/lab/fsc.

[15] S.E. Schutzer, P. Keim, K. Czerwinski, B. Budowle, Use of forensic methods under exigent circumstances without full validation, Sci. Transl. Med. 1 (8) (2009) 1–3.

[16] American Society of Crime Laboratory Directors-Laboratory Accreditation Board. Laboratory Accreditation Program and 2008 Manual. Available from: www.ascld-lab.org.

[17] American Association of Laboratory Accreditation. Accreditation Programs, 2009. Available from: www.a2la.org.

[18] International Standards Organization. General Requirements for the Competence of Testing and Calibration Laboratories, 2005; ISO/IEC 17025:2005(E).

[19] B. Graham, World at Risk: Report of the Commission on the Prevention of WMD Proliferation and Terrorism. Vintage Books, New York, 2008, xv (hereinafter "WMD Report").

[20] The White House. Homeland Security Presidential Directive 33: Biodefense for the 21st Century, 2004.

[21] E. Bahr, R. Katz, Assessing the impact Melendez-Diaz will have on the prosecution of biological weapons incidents, Biosecur. Bioterrorism Biodefense Strategy Practice Sci. 7 (4) (2009) 365–370.

[22] Confronting the New Challenges of Scientific Evidence, 108 Harv. L. Rev. 1481, 1485 (1995).

[23] Perry v. Novartis Pharmaceuticals Corp., 564 F. Supp. 2d 452 (E.D. Pa. 2008).

[24] Kumho Tire Co. v. Carmichael, 526 U.S. 137 (1999).

[25] A.B. Lustre Post-Daubert Standards for Admissibility of Scientific and Other Expert Evidence in State Courts, 90 A.L.R.5th 453, *2.

[26] D.A. Klein, Reliability of Scientific Technique and Its Acceptance within Scientific Community as Affecting Admissibility, 105 A.L.R. Fed. 299, *2a.

[27] United States v. Mettetal, 2006 U.S. Dist. LEXIS 25913 (W.D. Va. May 3, 2006).

[28] 18 U.S.C. §§ 175–178.

[29] See United States v. Mettetal, 2000 WL 530330 (4th Cir., 2002).

[30] See Andrews v. State. 533 So. 2d 841 (Fla. App. 1988).

[31] State v. Caine, 652 So.2d 611 (1995 La. App.).

[32] State v. Schmidt. 699 So. 2d 448 (1997 La. App.).

[33] Wikipedia. Bruce Edwards Ivins, 2009. Available from: http://en.wikipedia.org/wiki/Bruce_Edwards_Ivins.

[34] Centers for Disease Control and Prevention, Division of Foodborne, Bacterial and Mycotic Diseases. Shigellosis, 2009. Available from: www.cdc.gov/nczved/dfbmd/disease_listing%20/shigellosis.

[35] Center for Food Security and Public Health. Foot and Mouth Disease. Iowa State University, College of Veterinary Medicine.

# Microbial Forensics Curricula and Training

# CHAPTER 39

# Microbial Forensics: Educating the Workforce and the Community

**Steven E. Schutzer,[a] Bruce Budowle,[b] and Paul S. Keim[c]**

[a]*Department of Medicine, University of Medicine and Dentistry of New Jersey, New Jersey Medical School, Newark, New Jersey*
[b]*Institute of Investigative Genetics, Department of Forensic and Investigative Genetics, University of North Texas Health Science Center, Fort Worth, Texas*
[c]*Center for Microbial Genetics and Genomics, Northern Arizona University and Pathogen Genomics Division, The Translational Genomics Research Institute, Flagstaff, Arizona*

As the field of microbial forensics evolves, substantial developments in technology and analytical capabilities have occurred. An equally important aspect of preparedness involving microbial forensics that needs a similar commitment is education and training. The scientific bases, applications, interpretations, and lessons learned by those who have been intimately involved in the early years of microbial forensics need to be documented and transferred to the next generation of scientists and decision makers so that we can protect society from potential harm resulting from bioterrorism and biocrime. Thus, the burgeoning field of microbial forensics should be accompanied by a parallel development of educational infrastructure and resources targeted at the next generation of practitioners, as well as diverse elements for the policy, research, and law enforcement communities. A microbial forensics education program can be broad, providing information encompassing all aspects of the field from science to policy, or more focused depending on its purpose and target audience. On one end of the target audience spectrum is the student at an academic center who desires to enter into the discipline of microbial forensics and would like to have options for a career choice. This student may become a forensic scientist analyzing crime scene evidence for a law enforcement or intelligence agency. Alternatively, the student may become an investigator who employs traditional law enforcement approaches merged with those of epidemiology for attribution purposes or crime investigation. An individual may become a law enforcement official whose responsibility is to understand the scope of an investigation and what tools are available to generate investigative leads. Policy makers must have a general understanding of microbial forensics results and better appreciation of their implications in order to effect sound and defensible policy decisions. Finally, an important

667

group that informs the public and government is the news media. They are frequently the primary interface between the scientist and the public, making their observations, insights, or inaccuracies of great importance and impact. Educational efforts will better prepare such individuals to be informed and responsible and must be varied in depth and scope to match the target audience of various entities involved in microbial forensics.

There can be many formats and venues for microbial forensics education. Full academic-style programs should be developed at universities to comprehensively educate individuals in this applied science. Microbial forensics will necessarily cover a broad range of topics (microbiology, epidemiology, evolution, statistics, infectious diseases, etc.) and no one can be an expert in all aspects. However, all interested parties need to have some requisite knowledge in the various aspects of the discipline. A full academic program likely is not practical for working professionals; more abbreviated educational/training activities, similar to that of a continuing education course format, could be very effective at integrating professionals into the microbial forensics discipline. Shorter courses or symposia will be useful for expanding the knowledge base of trained professionals. Microbiologists, epidemiologists, public health, and law enforcement officials are highly trained in relevant aspects of microbial forensics, but may need additional training to integrate effectively their expertise with the demands of this new discipline. To broadly educate as well as to specifically educate those involved in the widely varied aspects of microbial forensics represents an educational challenge that must be met to develop the experts and expertise that we desperately need to combat bioterrorism and biocrime.

Topics in Table 39.1 cover the spectrum of educational opportunities in microbial forensics and could form the template for a comprehensive education and training program. Clearly some areas are more relevant to scientists, others to crime scene investigators, and others to decision makers. Many of the subjects naturally overlap. We briefly identify some general areas and discuss why these should be considered as part of the core curriculum for scientists. Most of these topics are addressed in greater detail in other chapters of this book.

## MICROBIAL FORENSICS CURRICULA AND TRAINING

Microbial forensics is defined as a scientific discipline dedicated to analyzing evidence from a bioterrorism act, biocrime, or inadvertent release of a microorganism/toxin for attribution purposes (1). It is the same as other forensic disciplines except for its focus on a particular type of crime (1,2). Based on

**Table 39.1** Overview and Origins of Microbial Forensics

Basic epidemiology
Molecular epidemiology
Microbial forensics curricula and training
  Basic and advanced
Microbes and their products
  a. Viruses
  b. Bacteria
  c. Fungi
  d. Eukaryotic parasites
  e. Toxins
The host target—how does a person or animal become ill?
  Immunology
The plant as a target—how does a plant or crop get damaged?
The host response as a forensic indicator
  a. Immune system
  b. Pharmacokinetics
  c. Antibiotics
Processes and technology
  a. Sample collection
  b. Forensic handling
  c. Preservation
  d. Extraction
  e. Advanced microscopy
  f. Proteomics
  g. Genomics
  h. Bioinformatics
  i. Statistical analysis and confidence estimations
  j. Indicators of engineering
  k. Synthetic biology
  l. Population genetics
  m. High-throughput sequencing
  n. Nonbiologic tools
  o. Sensitive signature detection and characterization
  p. Evolving, nascent technology
Quality assurance and quality control
Investigative genetics (i.e., forensic genetics)
  a. Interpretation
  b. Forensic science in general
Crime scene investigation
  a. Identify crime scene
  b. Evidence collection
  c. Sampling strategies
  d. Sample storage and transportation
  e. Trial preparation including moot court
Case histories
  a. Civilian
    i. Food safety and public health
      1. Food borne—*Shigella*, *Salmonella* (spinach)
      2. Anthrax
      3. Ricin

*(Continued)*

**Table 39.1** Overview and Origins of Microbial Forensics (*Continued*)

ii. Agriculture
  1. Foot-and-mouth UK 2007
  2. Mad cow disease US 2003 (attribution by host genetics)
iii. Environmental science
  Poultry industry water contamination Arkansas
iv. Emerging infections
  1. H1N1
  2. Severe acute respiratory syndrome
  3. Monkey pox
  4. HIV
b. Criminal
  i. U.S. anthrax 2001 with focus on technology and investigation HIV
  ii. Ricin
c. Biodefense
  Terrorism and biocrimes
Legal issues
 a. United States
 b. International
Select Agent rules
Operational and intelligence issues
National-level capabilities and resources
 a. Country capabilities
  i. What and how should any country be prepared?
  ii. What strategies make sense?
  iii. Planning, implementing, and measuring effectiveness
  iv. Exercises
  v. Where can additional support be sought?
  vi. Epidemiologic investigation as a basic country skill
Public information (media and public)
 Dissemination of accurate information in timely manner
Entertainment industry
 Depiction of accurate information

past history and with current technology capabilities, the potential use of biological weapons is greater than at any other time in history. Only a few semiexpert individuals are needed with access to dual-use equipment (e.g., equipment used in the pharmaceutical or food industries) to produce bioweapons inexpensively. These bioweapons will contain signatures that might be exploited to help identify the perpetrators. One may consider attribution solely to be the "DNA fingerprinting" of a pathogenic agent, but unique genomic identification of a microorganism may not always be possible because of the clonal nature of many microorganisms and, on a case-by-case basis, lack of population and phylogenetic data. Microbial forensics employs the same general practices as other forensic disciplines. Recognizing a crime scene, preserving a crime scene, chain of custody practices, evidence collection and handling, evidence shipping, analysis of evidence, and interpretation of results are carried out in the same general manner as other forensic evidence. A common exception is that evidence will be handled as a biohazard (even more so than, for example, HIV-infected blood). It is anticipated that the majority of microbial forensic evidence will fall into a category with shared characteristics, with some data being very informative and some being less informative. An understanding of the field is essential to determining what type of evidence is collected, how it is analyzed, what the significance of a result is, and what is supportive in identifying a perpetrator and for prosecution.

To support a career in microbial forensics, a university microbial forensics curriculum will necessarily cover a broad range of disciplines, which may include microbiology, chemistry, statistics, epidemiology of infectious diseases, evolution, genetics, genomics, and forensics. These courses could be taught individually or merged into a few dedicated microbial forensic courses. From a practical standpoint, many microbial forensic training programs will be based in other majors or minors in epidemiology, genetics, molecular biology, or microbiology. A major in epidemiology or microbiology could easily become a training platform for microbial forensics with the addition of select courses to include fundamentals in forensics. Alternatively, a forensics science program with additional training in basic sciences such as microbiology and epidemiology could serve to educate microbial forensic scientists. It will be important to emphasize integration of the material toward a specific microbial forensic profession. Concurrent enrollment in microbial forensics seminars, capstone courses, and internships will be needed to provide students the contextual importance of the basic material toward their chosen discipline that will often be taught more generically or under an unrelated discipline.

The depth of the curriculum will vary depending on the level and occupation of the student. High school students may have abbreviated versions that can pique their appetites to learn more. College students will need comprehensive training to prepare them for graduate school or for entering the workforce. Legal experts will require an overview to understand the limitations of the field and how to support or refute scientific findings.

# BASIC EPIDEMIOLOGY

Epidemiology is a cornerstone of public health and is critical to microbial forensics. The goal of epidemiology is to recognize infectious disease outbreaks and to attribute the outbreak to a source in order to prevent additional cases (see Chapter 15). In many aspects, microbial forensics employs the same tools as those used in epidemiology. A training program in microbial forensics will parallel many parts of current programs in epidemiology. Models can be obtained from epidemiology curricula and experience from natural outbreaks will help guide how microbial forensic scientists will perform investigations of biocrimes. Tracing the course of a disease will assist in identifying the index case, cause, and/or time of the outbreak. With many disease outbreaks as well as cases of unusual infections (e.g., monkey pox), the recurring question will be: Is this a natural event or an intentional attack? Epidemiological factors will help distinguish between natural or intentional events and enable more effective responses in either event. A biocrime may only be recognized through surveillance linking multiple unusual disease occurrences in contiguous or noncontiguous geographic areas. Often microbial forensic investigations will be based on initial public health findings and proceed further to address attribution as it applies to identify the perpetrator(s) of a biocrime or bioterrorist act.

## MOLECULAR EPIDEMIOLOGY

Molecular epidemiology focuses on the contribution of potential genetic and environmental risk factors, identified at the molecular level, to the etiology, distribution, and prevention of disease within families and across populations. Molecular epidemiology can be expanded to include the investigation of microbes at their molecular level (3). The field also provides a good example where application of newer technologies may help overcome many of the same problems encountered with traditional epidemiology with respect to study design and interpretation (4). Molecular tools can be employed to characterize and potentially individualize samples and isolates to address forensically relevant questions. This subdivision of epidemiology has special importance in microbial forensics because it is desirable to determine the source of a particular microbe used in a crime. Highly discriminating assays can precisely identify strains and isolates, resulting in a more focused and effective investigation. These types of data could associate a sample with a single geographic area, even possibly a particular laboratory or flask, or with the specific conditions and nutrients used to culture the microorganism. Some of these aspects are discussed in the chapters on anthrax.

## MICROBES AND THEIR PRODUCTS AS BIOLOGICAL WEAPONS

Agents that can be used in biocrimes span the microbial world of viruses, bacteria, fungi, eukaryotic parasites, and toxins. It is important to have a basic understanding of each type of microorganism to appreciate the parameters that make a particular microbe a serious threat as a weapon. These parameters will include accessibility, stability, transmissibility, associated history with weapons programs, and the capacity to produce disease with transient or sustained consequences, including death. Different technologies are needed to culture bacteria and viruses, as they differ greatly in growth requirements. Indeed, some microbes are difficult or impossible to culture. Such information may help an investigator understand what microbes would be considered and how they might have been used in a particular circumstance. A basic understanding of different microbial classes and their products would include human, animal, and plant pathogens.

# HOST FACTORS INCLUDING IMMUNE RESPONSES

It is important to understand how the host responds to microbes and how this can provide unique signatures, including those for a particular microbe or for timing the exposure to a pathogen. For forensic purposes, an immediate goal is to distinguish a potential victim from a perpetrator and to distinguish between a natural or intentional event. A basic understanding of the immune system, how antibodies are generated, and when different classes of antibodies appear may assist in criminal investigations.

# PROCESSES AND TECHNOLOGY

Sophisticated instruments (technology) that reside in the laboratory are only part of the process for obtaining reliable and meaningful information. The process begins with sample acquisition and proceeds with packaging, storage, and analysis and ultimately ends up with interpreting the results. All aspects are important and must be integrated effectively to have high confidence in results.

## Crime Scenes and Chain of Custody

After recognizing that a bioterrorist act or biocrime has occurred, defining the crime scene is the first important part of an investigation. Once the crime scene has been identified and delimited, a plan is needed to properly collect and maintain integrity of the evidence that may be subsequently analyzed. Practices are needed that minimize contamination of the evidence. Microbial contamination may be somewhat different from other types of contamination because the contaminating organisms can replicate confounding results. The nuances of a microbial forensic investigation add a layer on top of traditional crime scene investigations, particularly because of the hazardous nature of the evidence. The need for proper documentation may seem obvious but it is a very important part of maintaining the integrity of the evidence. Crime scenes are chaotic and missteps can occur. To minimize missteps in handling documentation procedures should be established so the crime scene can be reconstructed at a later date for investigators or in a court of law. It is likely that biocrimes and acts of bioterrorism will add another dimension of complexity because (i) there is less experience in crime scene collection due to (fortunately) fewer cases, (ii) addressing the safety of victims will not be trivial, (iii) investigators will be wearing cumbersome but absolutely necessary protective gear, and (iv) the best approaches for collection and preservation of evidence may have to be determined at the scene given the limited extant information available. Thus, preparing for a crime scene investigation and defining the processes of chain of custody should be essential parts of any curriculum.

The first responder community needs to be aware of the safety issues and the methods of collection because they may become involved in performing evidence collection. Laboratorians must understand these processes because better decisions can be made as to what evidence is pertinent for analysis. Lawyers and judges will want to understand the basics of chain of custody to be assured that acceptable handling methods have been exercised to maintain the integrity of the evidence. Those who will have contact with the crime scene, as well as those in the laboratory who require downstream interoperability of collected evidence, will have to learn basic do's and don'ts of crime scene investigation (5) to effect a better systems-based process. Education about crime scene investigation will help ensure use of validated microbial identification practices that will collect the most pertinent evidence and will best preserve the integrity of the evidence for analysis in a forensic laboratory.

## Sample Collection

One must understand the tools available to collect the sample as well as the limitations posed with a collection process or tool. While most approaches focus on collection tools, it is very important to consider sampling strategies to obtain the most relevant data. This involves strategic planning, logistics, and statistics. Conditions that are proper for collection and/or preservation of one microbe may be deleterious for another and, for that matter, to traditional forensic materials such as human DNA, fingerprints, and trace materials. For example, food-borne pathogens are particularly vexing; conditions that are intended to preserve the material may promote growth of natural bacteria in a food product and this overgrowth may destroy or obscure the initial bioweapon. Tools for collection need to be validated for efficient collection and for determining that they do not react with the target of interest. Tools developed for powder collection may be inefficient or ineffective for collecting plant material. Sample collection is not trivial and requires substantial consideration.

## Preservation of Forensic Evidence

The same issues about evidence collection will need to be applied to preservation processes. It is imperative to prevent further degradation of the evidentiary target once collected. Conditions for preservation apply for packaging and shipping, for maintaining of the evidence in the laboratory, and for postanalysis storage.

## Extraction

Extraction efficiency, particularly of interest to the scientist, pertains to obtaining the highest quality and quantity yield as possible of the target of interest. Yield is related to the target and removal from the collection matrix. Targets can include cells, nucleic acids, proteins, nutrients, growth materials, and elements.

## Advanced Microscopy

Various forms of microscopy may be used to visualize the evidence. These may range from basic to electron microscopy to atomic force microscopy and are available for characterization of a microbe. These approaches are rapid and can be used to identify candidate threats as well as to dismiss hoaxes.

## Proteomics

Defining chemical and physical properties of a biological agent can provide information on how and when the agent was produced and can be used to determine if two microbial samples were produced by the same process. Proteomics is a comprehensive study of the protein composition of biological systems at a moment in time or at different stages of the microbe. Many proteins are conserved and can be used for general identification, while other proteins may be expressed based on environmental stimuli, growth state, or growth conditions. Protein profiling can provide information beyond genomic analysis about the conditions of the bioweapon prior to host exposure. Proteomics is a complement to genomics described next.

## Genomics

One of the fastest growing areas with implications for microbial forensics is genomics. More rapid and in-depth sequencing of microbes is possible than was a decade ago; methods such as those used in the anthrax-letter attack seem almost antiquated today. Genomic analyses will continue to be essential in identifying species, strains, isolates, and individual samples to assist in a microbial forensic investigation. The rapid expansion of sequencing capabilities, to where sequencing some microbes within a day at very deep coverage, has raised the importance of genetic identification. It will likely be a mainstay of microbial forensics in another attack with any microbe. The cost of whole genome sequencing has decreased at least 100-fold in just a few years. This technology will be one of the methods of choice to examine the genetic structure of a particular pathogen and to identify those signatures of forensic relevance. Likewise, proteomic analysis can comprehensively determine which proteins are present in a sample and will grow in importance to forensics. The legal profession has to have a basic understanding of the capabilities and limitations to be successful in the courtroom, just as has been necessary for human DNA forensics. Several chapters in this book expand on this technology in detail.

## Statistical Analysis, Interpretation, and Confidence

Central to interpretation is, when possible, a statistical analysis of the findings. These should be performed to provide significance of the result or to convey the strength of the evidential results. A variety of statistical approaches exist and it is imperative to understand which ones apply to particular analyses and interpretations. Interpretation could be as simplistic as positive or negative to very complex evaluations using limits of detection and to complex algorithms for identifying and characterizing protein moieties. A host of answers and additional questions can arise from data interpretation.

Both scientists and legal analysts need to understand (or at least appreciate) the results and their significance. Moreover, the degree of confidence that can be placed on a result must be understood so that the weight of a comparative analysis is not overstated. Basic statistics, probability, and population genetics are essential requirements of any curriculum involving the analysis of forensic biological evidence.

## Bioinformatics

The term "bioinformatics" was developed as a result of the Human Genome Project. Because of the immense amount of data generated, it became necessary to apply more sophisticated computational techniques beyond what the average bench biologist had available. Bioinformatics requires a combination of data handling and analysis skills (including standard statistics) that connect routine biology with high-powered computation. As scientific investigations and data generation expand using high resolution, deep sequencing of genomes of microbes, and large-scale proteomics, computational analyses will be more critical than ever. This subject can be of value in a simplistic form for the biologist or a more complex form for the computationally inclined scientist. All scientists and individuals with interest in microbial forensic sciences will need to have basic training in statistics and bioinformatic tools.

## Indicators of Engineering

With rapid developments in molecular biology to benefit humankind also comes a great potential for manipulating a microorganism for nefarious purposes. Microbes could be engineered to be more potent, and difficult-to-obtain microbes may be synthesized *de novo* in a laboratory. There is a need to detect not only the microbe but to determine if it was genetically engineered or perhaps is a novel chimera. Synthetic biology is a frontier arena, in some respects, and manipulations or synthesis signatures may be detected through sequencing and bioinformatics. The skills and materials needed to manipulate a microbial genome may provide clues about the perpetrator and degree of sophistication used to develop the biothreat agent. This capability should be of interest to law enforcement and the intelligence community for supporting investigative leads.

## Population Genetics

Population genetics is essential for understanding the rarity of a genetic (and sometimes protein) profile derived from an evidence sample. Molecular epidemiology is increasingly applying the principles of evolutionary and population genetics to pathogens. It is important to understand what constitutes a sample population as opposed to a sample collection, the mode of inheritance related to a genetic marker, what significance or weight to apply to a genetic marker, what the mutation rate of a marker is, and how to combine the weight of multiple markers. Training of the student in this discipline will require basic genetics courses and more advanced courses in phylogenetic analyses. Such educational material will be found in population genetics and systematic and evolutionary biology programs. The population genetics of pathogens and its importance for microbial forensics are covered elsewhere in this book.

## Nonbiological Tools

This topic is broad and can encompass tools that characterize a microbe morphologically or chemically. These will range from microscopy to basic chemistry to analytical chemistry applications. The Amerithrax investigation demonstrated the importance of nonbiological measurements on samples of biological agents. A variety of mass spectral, spectroscopic, and other instrumental methods were used in an attempt to answer questions related to how, when, and what materials were used to produce the anthrax powders. Such information can be used to compare evidence directly to a reference sample or, indirectly, to infer something about the processes used to culture, stabilize, and/or disseminate the biothreat agent.

## FORENSIC SCIENCE

Forensic science is the application of science to answer questions of interest to a legal system as well as for military or state decisions (1,6,7). While science may not offer definitive solutions to the problems of society, it does serve a special investigative role, particularly in the criminal justice system. The areas of science that have been traditionally exploited are diverse, but typically include the major disciplines of biology, chemistry, physics, and geology. Within each discipline are many scientific subcategories that may be used in a forensic science investigation. For example, within the discipline of biology are the subdisciplines of medicine, pathology, molecular biology, immunology, odontology, serology, psychology, and entomology. The specific discipline(s) applied depends on the circumstances of the crime. Mathematics, especially statistics, is used to place weight or significance on observations or data retrieved from crime scene evidence. The ultimate question addressed by forensic science is usually "who committed the crime?" (i.e., attribution), and crime scene evidence can play a role in answering that question. Evidence can be any material, physical or electronic, that can associate or exclude individuals, victim, and/or suspect with a crime. It typically comprises materials specific to the crime and to control samples for background information. Types of evidence may be fingerprints, blood, semen, saliva, hair, fibers, documents, photos, computer files, videos, firearms, glass, metals, plastics, paint, powders, explosives, tool marks, and soil. The student needs to be cognizant of the types of evidence, how these different forms of evidence interplay, and how they can be used to help reconstruct the crime and/or identify the perpetrator.

## CASE HISTORIES

A case history is a detailed account of a person or event. Studies of case histories are instructive because they provide analysis of information in the relevant context, including real complexities. The study of each incident can be tailored to the particular group learning about them. The Amerithrax case is likely to be studied for years by many different groups ranging from scientists to law enforcement to lawyers. In addition to this case, many other cases are described in chapters of this book and the previous edition (6), as well as in specific publications (8). Among these threats are food-borne illnesses from bacteria, such as *Shigella* and *Salmonella*, and toxins such as ricin. In addition, there have been events directed at agriculture, including foot-and-mouth outbreaks in the United Kingdom and mad cow disease in the United States in 2003. Environmental contamination is also an area of interest, such as water contamination by the poultry industry in Arkansas. Perhaps the most common area where issues of natural versus intentional events arise is related to emerging infections. This question has arisen with the outbreaks of influenza H1N1, severe acute respiratory syndrome, monkey pox, and specific cases of HIV infection.

## LEGAL ISSUES

Legal issues are of obvious importance to the legal community but are also important to the scientific community. There will be times when the evidence will be used in a court of law to prosecute an individual who has been arrested for a biocrime. There are standards for admissibility of scientific evidence in a legal setting. The scientist may be asked to provide expert testimony. These need to be known and appreciated so that the burden of admissibility of evidence can be achieved. The government will use forensic scientists and their results, other experts, and the scientific literature to support its position. The defense will defend its client vigorously to attempt to achieve an acquittal. Because of the adversary system in the United States and other English-based law countries there will be challenges to the credibility of the science and the practitioners (7,9). Studying the science behind headlines can be a very instructive and creative way to interest students. A current controversy in forensics, which can be used instructively, involves the use of low copy number of DNA (10,11). News headlines have revolved around this topic in relationship to its use in court trials.

The standards and court proceedings, however, will vary for each country. For example, in the United States, possession of unauthorized material can be considered a crime by itself.

## OPERATIONAL AND INTELLIGENCE ISSUES

Evidence derived from a microbial forensic investigation may not necessarily end up in court. For example, such evidence can be used for intelligence purposes. Information can be gathered to determine risk or probability of an individual, a group, or a state that may plan to use (or has used) a bioweapon in an attack. The primary goal is to intercede and thwart the attack before it can happen. Alternatively, if an attack has occurred, a head of state may require some evidence to determine whether to retaliate and to whom retaliation should be directed. Use of microbial evidence is far reaching and has consequences. Training individuals in understanding the strengths and limitations of scientific evidence is essential so that proper decisions and responses can be made. Understanding how information is gathered, analyzed, and acted upon is likely to be of interest to any level of student.

## NATIONAL-LEVEL CAPABILITIES AND RESOURCES

Policy and decision makers need to learn about and support advances in microbial forensic strategies and capabilities, such as was described in the "National Science and Technology Council, National Research and Development Strategy for Microbial Forensics, Office of Science and Technology Policy, 2009." The following should be addressed: (i) What and how should a country be prepared? (ii) What strategies make sense? (iii) Planning, implementating, and measuring effectiveness. (iv) Training and evaluation exercises. (v) Where can additional support be sought? (vi) Leveraging of epidemiological tools.

# CONCLUSION

Disciplines related to microbial forensics are evolving rapidly. This evolution includes technology, analytical capabilities, and, equally as important, education and training. This book is one form of education that should accompany advances in the field of microbial forensics. Other forms of education should include didactic lectures, practical demonstrations, and discussions at specialty societies. The target audience may include college students, bench scientists, law enforcement agents, medical care and first responder personnel, lawyers, and judges. Those who fulfill teaching roles, whether by profession or indirectly as reporters and even entertainment writers, can become informed so that their writings are founded in facts.

# REFERENCES

[1] B. Budowle, S.E. Schutzer, A. Einseln, L.C. Kelley, A.C. Walsh, J.A. Smith, et al., Public health: Building microbial forensics as a response to bioterrorism, Science 301 (2003) 1852–1853.

[2] B. Budowle, S.E. Schutzer, A. Einseln, L.C. Kelley, A.C. Walsh, J.A. Smith, et al., Quality assurance guidelines for laboratories performing microbial forensic work, Science 301 (2003) SOM 1–SOM 17.

[3] L.W. Riley, Molecular epidemiology of infectious diseases: Principles and practices, ASM Press, Washington, DC, 2004.

[4] P. Boffetta, Molecular epidemiology: A tool for understanding mechanisms of disease, Eur. J. Surg. Suppl. (2002) 62–69.

[5] United States, Federal Bureau of Investigation, Laboratory FBI. Handbook of forensic services, 1999. Available from: http://www.fbi.gov/hq/lab/handbook/intro.htm.

[6] R.G. Breeze, B. Budowle, S.E. Schutzer (Eds.), Microbial Forensics, Academic Press, San Diego, 2005.

[7] R. Harmon, Admissibility standards for scientific evidence, in: R.G. Breeze, B. Budowle, S.E. Schutzer (Eds.), Microbial Forensics, Academic Press, San Diego, 2005, pp. 381–392.

[8] S.E. Schutzer, B. Budowle, R.M. Atlas, Biocrimes, microbial forensics, and the physician, PLoS Med. 2 (2005) e337.

[9] E.W. Kirsch, Daubert v. Merrell Dow Pharmaceuticals: Active judicial scrutiny of scientific evidence, Food Drug Law J. 50 (1995) 213–234.

[10] B. Budowle, A.J. Eisenberg, D.A. van, Validity of low copy number typing and applications to forensic science, Croat. Med. J. 50 (2009) 207–217.

[11] J. Gilder, R. Koppl, I. Kornfield, D. Krane, L. Mueller, W. Thompson, Comments on the review of low copy number testing, Int. J. Legal Med. 123 (2009) 535–536.

# CHAPTER 40

# So You Really Want to Be an Expert Witness? A Primer for the Occasional Expert Witness

**Rockne P. Harmon**

*Forensic/Cold Case Consultant, Alameda, California*

Most forensic science disciplines are based on crimes and evidence that occur routinely. DNA typing, toxicology, firearms, and fingerprint identification are among the most common disciplines applied toward solving a variety of crimes. As such, these disciplines are often associated with generally defined approaches and criteria to ensure that results from an analysis can be interpreted properly and that admissible and probative evidence will be produced should a crime be solved, and the case proceed to a jury trial. Microbial forensics, however, has developed a variety of approaches in an attempt to anticipate the nature of an assault and the mechanism used in the incident. Laboratories involved in microbial forensics are not the usual entities that produce forensic evidence for criminal prosecutions. Toward that end, admissibility of microbial forensics evidence has been discussed previously (1).

To continue describing various aspects of the criminal justice system, this chapter discusses what might apply to expert witnesses who have produced microbial forensic evidence for courtroom presentation. The discussion addresses the respective roles of attorneys and their obligations, discovery demands and their roles in the process, and challenges to expert witnesses and their work.

## ATTORNEYS' ROLES

The notion that the legal system necessarily leads to justice or proper outcomes, if it ever existed, was shattered most recently by the trial and the outcome of the O.J. Simpson case in 1995. The outcome of this case can better be appreciated once one understands the respective duties and responsibilities of attorneys once lawsuits are filed. First, a brief description of differences between the civil and the criminal system is in order.

681

Microbial Forensics. DOI: 10.1016/B978-0-12-382006-8.00040-2

Civil law is designed to resolve disputes between parties, often involving monetary awards or orders that tend to resolve the disputes. Some of those issues afford the right to a jury trial, some do not. The burden of proof in a civil action is by a preponderance of the evidence. This means that the civil plaintiff will prevail if the judge or jury decides that his/her evidence slightly outweighs the evidence that does not support his/her case. Civil jury verdicts need not be unanimous.

Criminal law is designed to hold persons accountable for transgressions against society. While the defendant is an individual, the plaintiff in criminal law is the prosecutor, who represents the community at large. Punishments, ranging from fines to execution are possible outcomes. In the United States, there is a presumption of innocence. A criminal defendant is presumed innocent and cannot be convicted unless and until the prosecution has overcome the presumption by proof in court. The burden of proving the prosecution's case is "beyond a reasonable doubt." In almost all jurisdictions, verdicts in criminal cases must be unanimous.

In a criminal case, the verdict by a jury of "not guilty" is not necessarily a finding of innocence, but it is rather a finding that the prosecution's case was not proven "beyond a reasonable doubt." Once a not guilty verdict is returned, the defendant may never be prosecuted for the same crime again under any circumstances.

There are rules of evidence that govern how evidence may or may not be presented. For the most part, rules of evidence for civil and criminal cases are identical. In other words, despite the significant differences between civil and criminal systems, there is no real difference in the rules governing admissibility and presentation of evidence in cases.

While serving the legal system, the roles and obligations of the prosecutor and defense attorney are different. Indeed these differences play out in the adversary system of law we enjoy in the United States. These differences also make it difficult to reach agreement in disputes of science.

## The Prosecutor

The prosecutor has an obligation to the community to see that those who commit crimes are held accountable. The prosecutor has significant discretion in deciding whether to charge crimes, and he/she is precluded from charging a person without having adequate evidence to sustain a conviction (2).

## The Defense Attorney

The defense attorney, in contrast to the prosecutor, has an individual obligation to the client to prevent him/her from being convicted using any and all means available. The fact that the client may be guilty and deserving of significant punishment has absolutely no bearing on this obligation. The recent decision to prosecute Muslim war criminals in federal court is indicative of this point. Despite public statements from the President and the Attorney General that Khalid Sheik Mohammed is guilty and will be convicted and given the death penalty, one prominent defense attorney declared, "But if I'm privileged enough to be asked," he said, "I'll step to the front and gladly represent one of these human beings with the same zest and zeal I would any other human being who is facing the death penalty" (3). Subsequent topical discussion in areas such as discovery will demonstrate how effective criminal defense attorneys have used and will use their "zest and zeal" to prevent their clients from being convicted. Of course competent criminal defense attorneys cannot admit that they are simply approaching their job with zest and zeal; they must convince the jury that they actually believe in the client's innocence.

## DISCOVERY

Discovery in a criminal case involves the disclosure of information necessary to afford the parties due process and a fair trial. The respective burden of the parties differs materially. Because the prosecution has the burden of proving the case beyond a reasonable doubt, it must disclose any and all information deemed necessary to afford the criminal defendant a fair trial. Compliance with this requirement is straightforward in typical criminal prosecution. All materials produced during the investigation must be disclosed. In addition, any known materials related to the credibility of witnesses should also be disclosed. When the prosecution includes scientific evidence, the matter to be disclosed may become more difficult to define. Certainly all laboratory reports and notes supporting testing must be provided. If requested, laboratory protocols, examiner proficiency test results, curriculum vitae, and any published and peer review articles may also be included. In some cases, defendants have requested and been provided with a polymerase chain reaction contamination log for periods before and after laboratory work in the case was performed.

The prosecutor has another well-established responsibility in the discovery process, over and above what has already been discussed. Pursuant to a decision by the U.S. Supreme Court, the prosecutor must disclose any information in his constructive possession that *is* exculpatory (4). This responsibility is known as the *Brady* requirement. There are two important components to this responsibility: those of constructive possession and that the information *is*, in fact, exculpatory. Courts consider any material in the hands of any government entity to be within its constructive possession and that the prosecution has an affirmative obligation to seek out the information. Constructive possession may extend to expert witnesses who are not government employees, but who will testify for the prosecution. It is the second point, however, that often creates confusion. The court's pronouncement in *Brady*, "We now hold that the suppression by the prosecution of evidence favorable to an accused upon request violates due process where the evidence is material either to guilt or to punishment, irrespective of the good faith or bad faith of the prosecution," is clear enough. In the *Brady* prosecution, evidence that was not disclosed was a confession by Brady's companion Boblit that Boblit had actually strangled the victim, not Brady as the prosecution contended. Brady and Boblit were each convicted and sentenced to death in separate trials. During Brady's trial, which occurred first, his counsel was not given or made aware of Boblit's confession. Clearly that information should have been provided. Clearly that information is exculpatory. This emphasis is critical because the usual contention by a defense attorney is that any information that was not disclosed would have or could be exculpatory. This argument is well beyond the protection afforded by *Brady*, but it is always made when any information is not disclosed. A recent case in Maine is discussed later that illustrates this point.

A spectrum of materials might have to be disclosed, starting with the obvious, eventually to more esoteric information. One must remember, in this context, the previous discussion about attorneys' respective roles. A defense attorney has no obligation to see that the truth emerges. In fact, he/she is obliged to obscure the truth if it serves his/her client's interest. In the context of discovery, then, effective criminal defense attorneys have used the discovery process, which is designed to afford a fair trial, to cause sanctions to be imposed if discovery material is not provided. Often the sought-after materials are of little use to the defense attorney. An episode in the human DNA typing history illustrates this point.

Most current forensic DNA typing is done using manufactured kits. One of the major manufacturers is Applied Biosystems (now called LifeTechnologies). Defense attorneys soon realized that the primer sequences for the kits were deemed proprietary by the company and that the company was reluctant to publicly disclose its proprietary information. Once the word got out on this issue, prosecutors were barraged with discovery requests for the primer sequences. Judges seemed stymied. On the one hand, there were defense experts claiming they needed the information to ensure that the tests themselves were reliable. On the other hand, there was a company asserting that this was proprietary and should not be disclosed. There were also government experts that supported that the sequences did not have to be disclosed to carry out studies to test the validity of the use of the primers.

A Colorado case, People v. Schreck (5), demonstrated how the discovery process can be used or abused with "zest and zeal" to represent a defendant and preclude the introduction of powerful DNA evidence, even though it is the technology and hence the same evidence used to exonerate persons.

> The defense strategy in the DNA battle was largely patterned after the *Bokin* case. The strategy begins with excessive and persistent discovery demands and is then followed up with a subpoena to the corporate manufacturer to produce primer sequences, all documentation of developmental validation studies and identifying information related to population databases. The second step is to convince the court that it is the individual commercial kit that must be subjected to *Frye* scrutiny. It was in the middle of these "discovery hearings" that I was appointed as Special Prosecutor on the case.

> Clearly, manufacturers are unwilling to provide such information and regard them as trade secrets. After numerous hearings that took months to complete, the trial court ordered the manufacturer to turn over its validation studies but not its primer sequences. The manufacturer sought relief both from the Colorado Supreme Court and in the California court where the manufacturer was domiciled. The Colorado Supreme Court rejected the appeal. The California court quashed the subpoena from Colorado. In order to prevent preclusion of the evidence based upon a confrontation clause claim or discovery violation, the prosecution persuaded the FBI to allow the defendant's expert to review the FBI's validation study (6).

Another case in Minnesota provided additional insight into this clash of valid positions in the legal process. The court characterized it this way, "What this issue really comes down to is this: Can the State meet its burden of showing the admissibility of the tests run on the PE Biosystem's equipment without the defense subpoenaed material? And even if the State can, is the unavailability of this material, even if the unavailability is not the State's doing, of such a nature that defendants cannot get a fair trial under the due process clause without it?" (7) This judge answered the question thusly, "That although the exact structure of these primers might be of interest, the defense has not demonstrated any particularized need for this information, nor that there is any indication that knowing this information would enable the defense experts to come to any conclusions about the reliability of these kits that could not be reached in other ways." The question to be addressed in this context is purely a legal one. Judges in some states rendered conflicting opinions that are impossible to reconcile. Such is the frailty of the legal system! One thing is certain, any competent defense attorney, using his "zest and zeal," must endeavor to convince a judge to rule in his client's favor.

Once the primer sequences were distributed pursuant to the protective orders, demands for that information promptly ceased (8).[1] No scientists who received this information ever demonstrated through writing or in court that any use was ever made of the protected, proprietary information to undermine either its validity or the reliability of the kits. In the context of the adversary system, the quest to find potentially discoverable materials that will not be disclosed and that lead to nowhere is never ending. A consortium of well-known allies published a letter in *Science* arguing for the release of the seven million DNA profiles contained in the national offender database (9). They suggested that release of this information to them would allow for analyses that "can only strengthen the quality of forensic DNA analysis." The group consisted of defense attorneys who have failed to secure the information in criminal cases and frequent defense experts in those same cases. One of those experts, Dr. Mueller, is discussed later in this chapter. One can easily conclude from these examples that when the "zest and zeal" approach to the adversary system merges with science, the integrity of science is compromised.

## EXPERTS' ROLES

An expert witness is defined as one who, through education, training, skill, or experience, is qualified to provide testimony to assist the trier of fact in evaluating and understanding the subject of their testimony. Expert testimony is not limited to matters of scientific or technical evidence.

### Prosecution Experts

Once an expert is retained by the prosecution, that fact and any opinions of that expert must be disclosed to the defense. If the expert supports the prosecution's case, it is likely that he/she will testify and all of the areas for criminal discovery discussed previously are likely to be invoked. If the expert does not support the prosecution case, then the provisions of *Brady* are triggered and disclosure will be made to the defense for that reason.

### Defense Experts

Once an expert is hired by a criminal defendant, that fact may never be made known to the prosecution. Many states do not have reciprocal discovery requirements wherein the defense has the same obligation that the prosecution has. In those states that do have reciprocal discovery obligations, such as California, that obligation only accrues when the defense has decided that

---

[1]The protective order precluded dissemination of information to any parties outside the defense team. The court-ordered disclosure of trade secrets was so noteworthy to the scientific community that it was discussed in one prominent journal.

the expert will testify. This obligation is almost always undermined when the defense defers deciding that the expert will testify until just before the expert testifies. The purpose of disclosure is further undermined by not having the expert write a report or notes. Under this scenario the defense usually belatedly announces who its expert is and produces no documentation of the subject of the expert's testimony. The sanctions that may be brought to bear on the prosecution for this practice are seldom applied against the defense because it is thought that these sanctions would undermine defendant's right to a fair trial.

It is not uncommon, in criminal cases, for the defense to hire more than one expert for assistance. For example, in the forensic DNA area, defense attorneys will occasionally have evidence in the case retested. Results of the retest almost always corroborate the prosecution's results and, more often than not, remain hidden from disclosure by the work product privilege or some other legal artifice. The defense may then retain another expert, who does not and has not retested the evidence, to provide testimony critical of the prosecution DNA results even though the results have been confirmed by the defense retest. After all, this is just another part of the "zest and zeal" approach to criminal defense work. It is to be expected from a criminal defense attorney and an obligation, even if he is convinced of his client's innocence. It is perfectly permissible, within the parameters of the legal system, for the defense attorney to try to convince the jury that the prosecution's results are wrong, even if the attorney knows and believes that it is not true.

## CONSEQUENCES OF DISCOVERY VIOLATION

Discovery violations may become apparent either before trial or afterward. If they become known before trial, they are several ways to deal with them. After disclosure is made, the court might order an appropriate sanction to be imposed. Belated disclosure may result in the proffered evidence being excluded from presentation to the jury. Another option is continuing (i.e., postponing) the proceedings to afford the aggrieved party an opportunity to examine and review new information.

If the "violation" is uncovered after trial, it may form the basis for having the conviction reversed on appeal or having a motion for a new trial granted. A case in Maine demonstrated this latter remedy. A motion for a new trial was granted when the defense learned after trial that certain information concerning DNA testing was not provided to them in discovery before the trial. "Central to Hunter's decision was the lab's admission that skin samples submitted for DNA in Claridge's case had been mixed up with two others and that crime lab personnel failed to disclose the mix-up in advance of the trial" (10).

A careful reading of this story disclosed that this evidence was not the evidence that led to the conviction. Although Claridge was convicted in February 2009 based mainly on a DNA match in the semen taken from the victim's clothing, the judge said, "Given the significant weight of scientific evidence, there is the possibility that a fully informed jury might have heard that there was a failure to follow a standard of care [by the lab]." Somehow the judge equated evidence that, if the laboratory had made one mistake, they might have made another, with the powerful exculpatory evidence in *Brady*. Decisions such as the Maine case are frequent and are often thought to undermine the public's confidence in the legal system because it seems to exalt form over substance. The guilt of the defendant seems to be ignored in the process.

## GRANT SOLICITATIONS/APPLICATIONS

Many scientists, forensic and otherwise, rely on research funding to further their scientific pursuits. Usually grants are rewarded in response to solicitations that outline the criteria that need to be addressed or what is sought to be accomplished by the research. Sometimes the solicitations include existing deficiencies or gaps in the scientific knowledge that are to be addressed. Seldom will there be a grant funded for research that is unnecessary. It is conceivable that the existence of a solicitation in general, or an application by a prosecution witness, may fall into the realm of discoverable information. Such was the case in some highly charged fingerprint identification challenges several years ago. During the peak of the fingerprint admissibility litigation, a National Institute of Justice solicitation was discovered, titled "Solicitation, Nat'l Inst. of Justice, Forensic Friction Ridge (Fingerprint) Examination Validation Studies (March 2000)." One of the challenged issues involving fingerprint identification testimony concerned whether it had been demonstrated scientifically that fingerprint identity could be established to the level of individualization. When it was learned that the solicitation declared that "the theoretical basis for [fingerprint] individuality has had limited study and needs additional work to demonstrate the statistical basis for identifications," those challenging fingerprint identification testimony thought they had found a pot of gold. Unfortunately for them, the solicitation had negligible impact in challenging fingerprint testimony. However, it seems clear that the document contained discoverable information that should have been disclosed, had the prosecutor been made aware of its existence. While this may seem to be an onerous burden to place on the government, it can probably be dealt with easily by a careful review of the expert's curriculum vitae, which normally lists such endeavors. One can also imagine that had any of the experts for either party submitted proposals for this solicitation, they would have provided grounds for cross examination by the opposing attorney.

# PEER REVIEW ARTICLES/COMMENTS

The role of the peer review process is well understood in the legal setting—it is one of the critical criteria in deciding the admissibility of scientific evidence under any admissibility standard. Attorneys have also begun to recognize that the peer review process itself may contain important relevant information that should be disclosed, if requested, in the discovery process. The following excerpt from cross examination in a recent murder prosecution demonstrates the point. The proffered expert, Dr. Lawrence Mueller, had submitted an article that received scathing reviewer comments and he never revised and resubmitted the article:

Q. Over the years you've been rejected for publications specifically in the area of DNA forensic statistical analysis. Am I right about that?
A. That's happened twice, yes.
Q. One article that you submitted in the area of DNA forensic statistical analysis was entitled—I'm going to take a deep breath before I say this—"Methods of multilocus genotype frequency estimation hypervariable DNA in the application to forensic science." Is that true, sir?
A. Yes, that's the title of the paper.
Q. You endeavored to have that publication published in the journal *Genetics*. Am I correct about that, sir?
A. Back in 1990, that's correct.
Q. And that article dealt with DNA in a forensic setting and the calculation of DNA profiles and the like. Am I right?
A. In a very general sense, yes.
Q. Now, as we discussed earlier, in order to have these publications submitted to journals like *Genetics*, as you well know, you've got to submit them to referees who in turn make comments about your articles and your submissions; correct?
A. Well, the editor sends them to referees, I don't, but that's part of the process, that's correct.
Q. Part of the process is when the editor sends it out to these referees, the referees have to pass on it and recommend that it be published or not published, correct?
A. Right, and give reasoning behind their recommendation, that's correct.
Q. The reasoning that's given by some of these referees who reviewed your article in your effort to be published in *Genetics* made comments in regard to your submission, right?
A. Right, they wrote reviews, that's correct.
Q. Some of the comments included that your submission was, quote, "naive and unintelligible"; am I right?
A. That's one referee's comment.

Q. Yes. Another referee said that you, quote, "made extreme assumptions." Am I right?

Ms. Barlow: I'm sorry, I'm going to object again, your honor. Unless Mr. Merin is going to bring these people in, I don't see the relevance.

The Court: Overruled.

The Witness: I don't know if that was a different referee. I believe that could have been the same referee.

Mr. Merin: Q. Another comment that was made in relation to a ratio that you used in a statistic was, quote, "wildly sensitive"; am I right?

A. Again, a comment I think from the same referee, correct.

Q. And in summarizing this submission on your—of your effort, it was noted that it was "inefficient at best and largely uninformative"; am I right?

A. That was another comment. Again, I don't know, it may have been from the same person, correct.

Q. Well, some of these comments came from Dr. Bruce Weir; am I right?

A. I don't know about any of those comments, but there were comments by Dr. Weir; that's correct.

Q. Now, you got your article back and you then went to the editor of *Genetics*, and you didn't revise or resubmit your article, did you?

A. I didn't, no (11).

This experience, coupled with the near complete absence of any other human DNA peer review articles and combined with the fact that Mueller had previously billed about $750,000 for his services in other cases, afforded the jury some insight into his credibility and the validity of his opinions. Peer review is normally viewed in the scientific community as a constructive process that leads to the publication of sound science. It should be clear from the cross examination of defense expert Mueller that reviewer comments can provide insight into the credentials of the expert. While it might seem that this is an intrusive inquiry into the sanctuary of the peer review process, under the proper circumstances and showing peer review comments like these are not privileged information and are likely to be ordered to be provided. In the case of an expert such as Mueller, they served a legitimate purpose in unmasking his credibility.

Some say that watching the legislative process is like watching sausage being made. Watching an entire trial can be compared to watching animals being butchered and then made into sausage. Except for an occasional televised trial, few people ever watch a trial from start to finish. The topics discussed herein are meant to prepare the scientist who might occasionally become involved in a criminal prosecution in the field of microbial forensics. While they may seem to paint a grim picture about how the criminal legal process functions, they are based on real experiences and provide lessons to be learned by the occasional expert witness. Despite this grim picture, and despite the O.J. Simpson debacle, the true value of scientific evidence provided by qualified experts is usually recognized by courts and juries.

# REFERENCES

[1] R. Harmon, Admissibility standards for scientific evidence, in: R.G. Breeze, B. Budowle, S.E. Schutzer (Eds.), Microbial Forensics, Academic Press, San Diego, 2005.

[2] Uniform Crime Charging Standards (California District Attorneys' Association). See also Corrigan, "Commentary: On Prosecutorial Ethics," 13 Hastings Const. L.Q. 537 (1985–1986).

[3] For Sept. 11 Cases, Calling on a Shortlist of Defense Lawyers. New York Times.

[4] Brady v. Maryland (1963) 83 S.Ct. 1194.

[5] People v. Michael Shreck, 89 CR 2475, District Court, Boulder County.

[6] Tomsic, DNA Amissibility in Colorado: People v. Shreck, Proceedings, The 11th International Symposium on Human Identification.

[7] State v. Dishmon, Hennepin County, Minnesota, number 99047345, court order dated March 3, 2000.

[8] P. Smaglik, Legal protests prompt DNA primer release, Nature 406 (6794) (2000) 336.

[9] D.E. Krane, V. Bahn, D. Balding, B. Barlow, H. Cash, B.L. Desportes, et al., Time for DNA disclosure, Science 326 (5960) (2009) 1631–1632.

[10] DNA mix-up results in mistrial. Bangor Daily News, December 18, 2009.

[11] People v. John Puckett, San Francisco Superior Court SCN 201396, February 4, 2008.

# CHAPTER 41

# Microbial Forensics, What Next?

**Paul S. Keim,[a] Stephen A. Morse,[b] Steven E. Schutzer,[c] Roger G. Breeze,[d] and Bruce Budowle[e]**

[a] Center for Microbial Genetics and Genomics, Northern Arizona University and Pathogen Genomics, The Translational Genomics Research Institute, Flagstaff, Arizona
[b] National Center for Emerging and Zoonotic Infectious Diseases, Centers for Disease Control and Prevention, Atlanta, Georgia
[c] Department of Medicine, University of Medicine and Dentistry of New Jersey, New Jersey Medical School, Newark, New Jersey
[d] Centaur Science Group, Washington, DC
[e] Institute of Investigative Genetics, Department of Forensic and Investigative Genetics, University of North Texas Health Science Center, Fort Worth, Texas

Microbial forensics as a discipline has been affected dramatically by the anthrax-letter attacks and the intense effort associated with the Amerithrax investigation. Prior to October 2001, only a few forensic investigations of microbial agent crimes, or events, had been carried out. These included the Dallas Texas *Shigella* poisoning, the Aum Shinrikyo Kameido "anthrax" attack, and the Sverdlovsk anthrax accident. In these three cases, the subsequent investigation not only involved traditional epidemiology, but the use of molecular genetic typing also played an important role. The technical analyses were primitive by today's standards, and the involvement of law enforcement forensic laboratories was limited or nonexistent. In contrast, the Amerithrax case involved highly sophisticated technologies and the development of novel scientific analytical approaches and was driven by federal law enforcement efforts. In addition to the massive scale of Amerithrax, engagement of the law enforcement community led to new standards for microbial analyses that have begun to effect a change in how epidemiologists and public health officials approach normal disease outbreaks. Doubtlessly, future criminal investigations will quickly result in law enforcement-driven forensic analysis and will capitalize on the very latest in technological innovations. Consequently, analysis standards will be set high in order to support the prosecution of perpetrators within the judicial system. Microbial forensics is no longer just a "side activity" for epidemiologists, but rather a discipline all its own that will need specialists trained in multiple disciplines.

693

Microbial Forensics. DOI: 10.1016/B978-0-12-382006-8.00041-4

This book covered a broad range of topics relevant to microbial forensics and in a manner focused on this topic. As such, we believe that it is the definitive guide on the topic, to date, and will be useful to a broad range of readers interested in microbial forensics. An important difference between a microbial forensics investigation and an epidemiological investigation is that results of the former need to withstand the scrutiny of the adversarial legal system. Controlling the crime scene, maintaining chain of custody, validating methods, proficiency testing, and defendable interpretation of results will need to be performed to very high standards. Likewise, the actual methodologies for microbial forensics are evolving rapidly and new analyses are being devised and applied to biocrime investigations. The explosion in genomics is evident to all, but vast improvements are also being made in the physical and material sciences for understanding the exact evidentiary composition from isotopes to elements. Coupling high-resolution microscopy to elemental analysis was a key to understanding the anthrax-letter spores. A complementary suite of methodologies will ultimately prepare scientists with better investigative strategies. Preparing to investigate future events has largely been focused on a small list of pathogens and toxins, primarily developed from Cold War-era agents. A comparison of these research targets reveals unique characteristics that will require unique investigative approaches. The biology of each pathogen and toxin will necessitate agent-specific expertise and analysis. Inevitably, an unanticipated pathogen will be used in a future attack, requiring microbial forensics to adapt previous approaches to a novel event. A better understanding of the agents that may be used will lead to a quicker response for agents that were not predicted to be used. Finally, microbial forensic investigations must be approached with an eye toward the final legal stage. Investigators must be using methods that will meet judiciary standards for scientific evidence (e.g., the *Daubert* admissibility standard) and with a level of rigor that can be defended successfully against critical objections. Other excellent books exist on various aspects of forensics, pathogen tracking, and molecular epidemiology (1–3), but this book is unique in trying to combine the essential components and timely topics into a coherent body of work.

This leads to the question of whether microbial forensics will become a stand-alone science or whether it will remain a subspecialty of others. It is currently organized as either a subspecialty of epidemiology or of traditional forensics, and its growth into its own discipline will proceed for many years. Even with time, the necessity to be a separate discipline will be dependent on the number of future biocrime events and the resources available for such specific investigations. The specialized need to conduct forensic examinations in a biosafety containment environment does separate it from molecular epidemiology and traditional forensics. The National Bioforensic Analysis Center is the U.S. government's response to this need for an infrastructure with both capabilities. However, if microbial forensics is to be practiced outside this single focused environment, it will likely be closely associated with either traditional epidemiology or forensics. Biocrimes may remain rare events; thus local disease control and law enforcement efforts will need to be prepared continuously but microbial forensics will need to be practiced only occasionally. Maintaining widespread capacity focused solely on microbial forensics would appear to be unneeded and expensive. Rather, we would envision some capacity, training, and preparedness in all regions and locations and across both public health and law enforcement agencies.

Challenges still exist for efficient and effective microbial forensic investigations, including the development of large strain repositories for references to evidentiary material. While genetic methods have become less expensive, faster, and with better discrimination power, strain repositories remain problematic. New federal regulations, agency-specific restrictions, and increasing international paranoia have placed additional barriers for strain acquisition and sharing. Complicating the situation is the need for comprehensive strain collections for each pathogen investigated, which multiplies the size of the challenge. In addition, genomic technologies are changing so fast that the standardization of methods and even data formats remains a hurdle. Error rates and differential quality across data sets must be estimated, captured, and understood to minimize false investigative leads, which would expend resources unnecessarily, degrade public confidence in microbial forensic analysis, and, ultimately, impact the ability to prosecute perpetrators successfully. Evolutionary models, coupled with repository databases, need to be applied to evidence to generate confidence estimations associated with any results. Regardless of analysis results, interpretation of results and their communication to peers, lawyers, judges, and juries will have to be effective. The microbial forensics community must strive for a common language to describe investigative outcomes effectively and accurately. While these challenges must be met, prospects are good if the science is rigorous and the community is open to critical review.

Crimes of all types cost our society valuable resources in terms of money, productivity, and peace of mind. Indeed, in extreme cases the cost is measured in lives lost. Ineffective and inaccurate forensics compounds these societal problems with injustice while failing to curtail crime. Scientific rigor needs to be applied to microbial forensics to maximize its effectiveness and benefits. This will result in fewer biocrimes by removing criminals and providing a deterrent to potential perpetrators. Performing microbial forensics accurately with valid interpretation will not always result in identification of a perpetrator, but inappropriate use of microbial forensics that attributes a crime to an individual erroneously will come with a much greater cost.

# REFERENCES

[1] J.W. Santo Domingo, M.J. Sadowsky (Eds.), Microbial Source Tracking, ASM Press, Washington, DC, 2007.

[2] L.W. Riley, Molecular Epidemiology of Infectious Diseases: Principles and Practices, ASM Press, Washington, DC, 2004.

[3] P. Emanuel, J.W. Roos, K. Niyogi, Sampling for Biological Agents in the Environment, ASM Press, Washington, DC, 2008.

# Index

**FIGURE 1.2**

Blood agar plate of sampling from the Kameido site (4).

**FIGURE 1.3**

Multiple-locus, variable-number tandem repeat analysis of a Kameido isolate and the Sterne strain of *Bacillus anthracis* (4).

**FIGURE 7.2**

Contour map for a primary focus of Asian soybean rust based on 2-unit interval pixel intensity values extracted from an IKONOS satellite image obtained August 27, 2006, over Quincy, Florida (35). Image consists of 22 × 22 pixels, each providing 1-m² resolution.

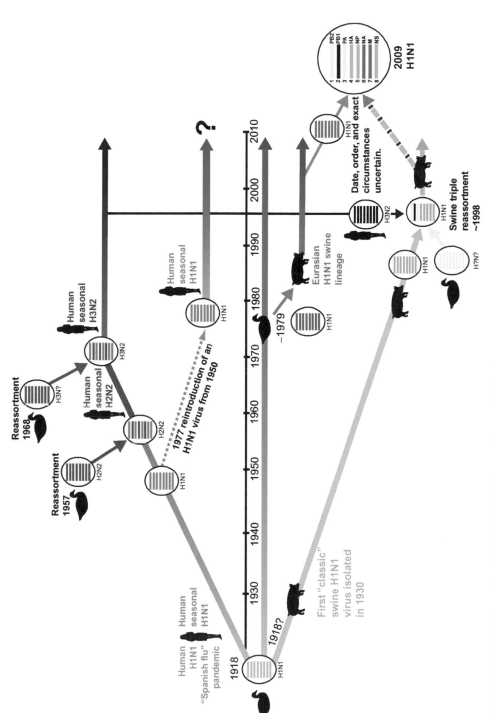

**FIGURE 8.2 Evolution of the 2009 H1N1 influenza A pandemic virus.**

The human 1918 "Spanish flu" and "classic" swine influenza A H1N1 viruses probably evolved from a single avian-adapted ancestor (left side of figure). Since 1918, genetic variation has accumulated in both human and swine influenza A lineages as a result of reassortment (explicitly shown in the figure, see bottom right for key to gene segments) and gradually, via point mutation (suggested by gradual color transitions on the lines that represent individual lineages). The 2009 H1N1 pandemic virus appears to have been derived through the reassortment of several viruses currently known to circulate in swine.

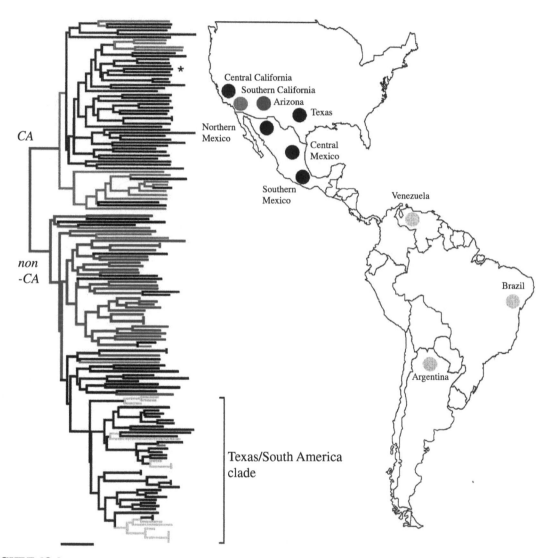

**FIGURE 18.1**

Microsatellite genotyping of *Coccidioides*: Neighbor-joining tree of pair-wise allele-sharing genetic distances calculated with the program MICROSAT (14). (Used with permission from PNAS.)

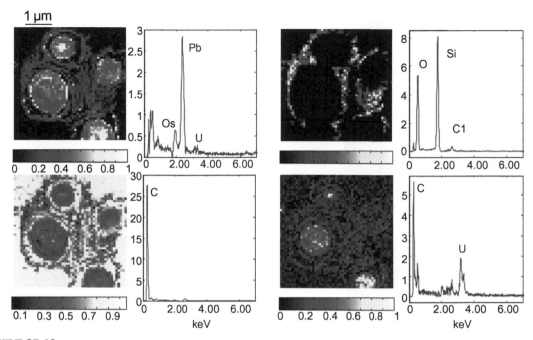

**FIGURE 25.10**

Results of multivariate statistical analysis of the spectral image that shows four chemically significant component image-spectrum pairs. Images describe where the corresponding spectral signatures are found in the microstructure. No operator input was required to calculate this result.

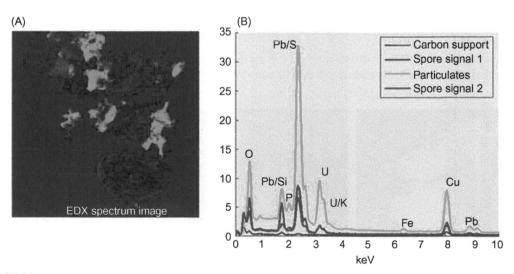

**FIGURE 25.11**

An EDX spectrum image showing features (A) and spectra (B) from microtomed TEM sections of bacterial spores stained with uranium acetate and lead citrate. Note the X-ray peak overlaps for Si, S, and K due to the staining agents.

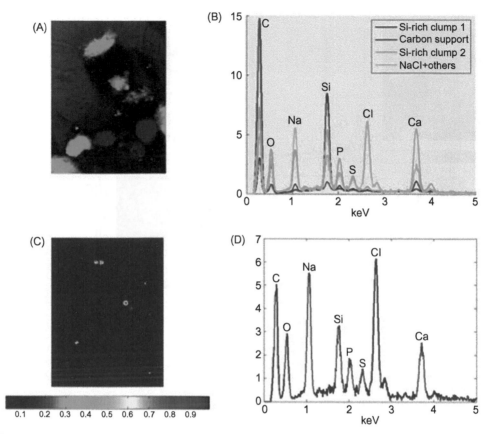

**FIGURE 25.15**

X-ray spectral image components from area in Figure 25.14. (A and B) A composite image and set of spectra from the whole area. (C and D) Features and spectrum from sodium chloride-rich particles within the field of view.

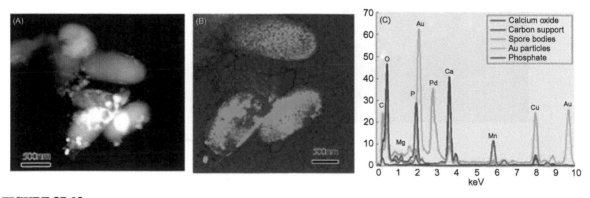

**FIGURE 25.16**

Annular dark-field (ADF) image of spores dispersed on a TEM grid. Component image overlay (B) and component spectra (C) from this region. Note that the copper signal in all spectral components is from the support grid.

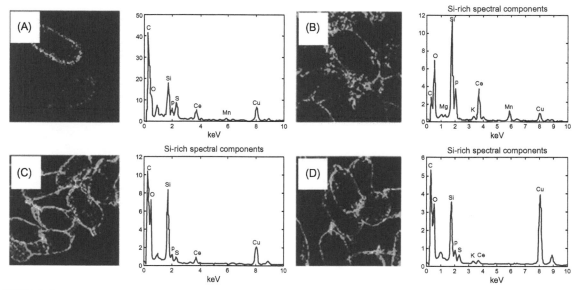

**FIGURE 25.17**

Spectral image components showing distribution of silicon-rich coating on outer portion of *Bacillus thuringiensis* spores for four different processing conditions. Note the difference in coating compositions.

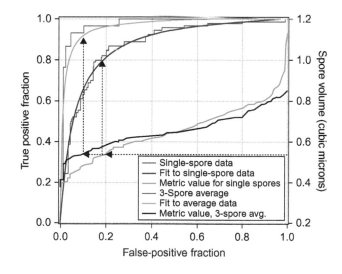

**FIGURE 30.4**

Receiver-operating characteristic for the size-based test for agar plate culture. Dashed arrows indicate the approximate position of the spore volume value ($0.53\,\mu m^3$) that corresponds to a likelihood ratio value of 1.

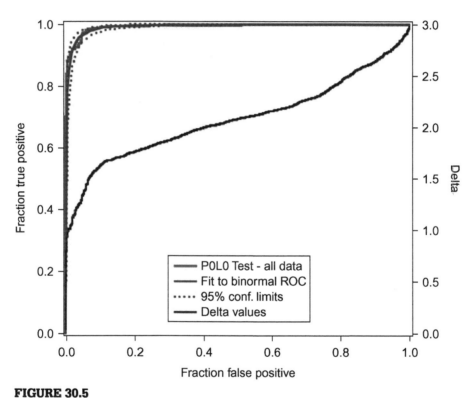

**FIGURE 30.5**

Receiver-operating characteristic curve for testing if two samples were made by the same process in the same laboratory.